5G NR and Enhancements

5G NR and Enhancements
From R15 to R16

Edited by

JIA SHEN

Standard Research Department, OPPO Research Institute, Shenzhen, China

ZHONGDA DU

Standard Research Department, OPPO Research Institute, Shenzhen, China

ZHI ZHANG

Standard Research Department, OPPO Research Institute, Shenzhen, China

NING YANG

Standard Research Department, OPPO Research Institute, Shenzhen, China

HAI TANG

Vice president of OPPO Research Institute, Shenzhen, China

ELSEVIER

Elsevier
Radarweg 29, PO Box 211, 1000 AE Amsterdam, Netherlands
The Boulevard, Langford Lane, Kidlington, Oxford OX5 1GB, United Kingdom
50 Hampshire Street, 5th Floor, Cambridge, MA 02139, United States

Notices
Knowledge and best practice in this field are constantly changing. As new research and experience broaden our understanding, changes in research methods, professional practices, or medical treatment may become necessary.

Practitioners and researchers must always rely on their own experience and knowledge in evaluating and using any information, methods, compounds, or experiments described herein. In using such information or methods they should be mindful of their own safety and the safety of others, including parties for whom they have a professional responsibility.

To the fullest extent of the law, neither the Publisher nor the authors, contributors, or editors, assume any liability for any injury and/or damage to persons or property as a matter of products liability, negligence or otherwise, or from any use or operation of any methods, products, instructions, or ideas contained in the material herein.

British Library Cataloguing-in-Publication Data
A catalogue record for this book is available from the British Library

Library of Congress Cataloging-in-Publication Data
A catalog record for this book is available from the Library of Congress

ISBN: 978-0-323-91060-6

For Information on all Elsevier publications
visit our website at https://www.elsevier.com/books-and-journals

Publisher: Glyn Jones
Editorial Project Manager: Naomi Robertson
Production Project Manager: Kamesh Ramajogi
Cover Designer: Greg Harris

Typeset by MPS Limited, Chennai, India

Working together
to grow libraries in
developing countries

www.elsevier.com • www.bookaid.org

Contents

5. 5G flexible scheduling   167

Yanan Lin, Jia Shen and Zhenshan Zhao

Weijie Xu, Chuanfeng He, Wenqiang Tian, Rongyi Hu and Li Guo

9. 5G radio-frequency design   507

Jinqiang Xing, Zhi Zhang, Qifei Liu, Wenhao Zhan,
Shuai Shao and Kevin Lin

16. Ultra reliability and low latency communication in high layers 741

Zhe Fu, Yang Liu and Qianxi Lu

17. 5G V2X   769

Zhenshan Zhao, Shichang Zhang, Qianxi Lu, Yi Ding and Kevin Lin

18. 5G NR in the unlicensed spectrum   823

Hao Lin, Zuomin Wu, Chuanfeng He, Cong Shi and Kevin Lin

List of contributors

Wenhong Chen
Standard Research Department, OPPO Research Institute, Shenzhen, China

Shengjiang Cui
Standard Research Department, OPPO Research Institute, Shenzhen, China

Yi Ding
Standard Research Department, OPPO Research Institute, Shenzhen, China

Zhongda Du
Standard Research Department, OPPO Research Institute, Shenzhen, China

Yun Fang
Standard Research Department, OPPO Research Institute, Shenzhen, China

Zhe Fu
Standard Research Department, OPPO Research Institute, Shenzhen, China

Li Guo
Standard Research Department, OPPO Research Institute, Shenzhen, China

Yali Guo
Standard Research Department, OPPO Research Institute, Shenzhen, China

Chuanfeng He
Standard Research Department, OPPO Research Institute, Shenzhen, China

Rongyi Hu
Standard Research Department, OPPO Research Institute, Shenzhen, China

Yi Hu
Standard Research Department, OPPO Research Institute, Shenzhen, China

Yingpei Huang
Standard Research Department, OPPO Research Institute, Shenzhen, China

Haitao Li
Standard Research Department, OPPO Research Institute, Shenzhen, China

Xue Lin
Standard Research Department, OPPO Research Institute, Shenzhen, China

Bin Liang
Standard Research Department, OPPO Research Institute, Shenzhen, China

Hao Lin
Standard Research Department, OPPO Research Institute, Shenzhen, China

Kevin Lin
Standard Research Department, OPPO Research Institute, Shenzhen, China

Yanan Lin
Standard Research Department, OPPO Research Institute, Shenzhen, China

Jianhua Liu
Standard Research Department, OPPO Research Institute, Shenzhen, China

Qifei Liu
Standard Research Department, OPPO Research Institute, Shenzhen, China

Wendong Liu
Standard Research Department, OPPO Research Institute, Shenzhen, China

Yang Liu
Standard Research Department, OPPO Research Institute, Shenzhen, China

Qianxi Lu
Standard Research Department, OPPO Research Institute, Shenzhen, China

Shuai Shao
Standard Research Department, OPPO Research Institute, Shenzhen, China

Jia Shen
Standard Research Department, OPPO Research Institute, Shenzhen, China

Cong Shi
Standard Research Department, OPPO Research Institute, Shenzhen, China

Yongsheng Shi
Standard Research Department, OPPO Research Institute, Shenzhen, China

Zhihua Shi
Standard Research Department, OPPO Research Institute, Shenzhen, China

Tricci So
Standard Research Department, OPPO Research Institute, Shenzhen, China

Jinxi Su
Standard Research Department, OPPO Research Institute, Shenzhen, China

Jiejiao Tian
Standard Research Department, OPPO Research Institute, Shenzhen, China

Wenqiang Tian
Standard Research Department, OPPO Research Institute, Shenzhen, China

Shukun Wang
Standard Research Department, OPPO Research Institute, Shenzhen, China

Zuomin Wu
Standard Research Department, OPPO Research Institute, Shenzhen, China

Han Xiao
Standard Research Department, OPPO Research Institute, Shenzhen, China

Jinqiang Xing
Standard Research Department, OPPO Research Institute, Shenzhen, China

Jing Xu
Standard Research Department, OPPO Research Institute, Shenzhen, China

Weijie Xu
Standard Research Department, OPPO Research Institute, Shenzhen, China

Yang Xu
Standard Research Department, OPPO Research Institute, Shenzhen, China

Haorui Yang
Standard Research Department, OPPO Research Institute, Shenzhen, China

Ning Yang
Standard Research Department, OPPO Research Institute, Shenzhen, China

Xin You
Standard Research Department, OPPO Research Institute, Shenzhen, China

Wenhao Zhan
Standard Research Department, OPPO Research Institute, Shenzhen, China

Shichang Zhang
Standard Research Department, OPPO Research Institute, Shenzhen, China

Zhi Zhang
Standard Research Department, OPPO Research Institute, Shenzhen, China

Nande Zhao
Standard Research Department, OPPO Research Institute, Shenzhen, China

Zhenshan Zhao
Standard Research Department, OPPO Research Institute, Shenzhen, China

Zhisong Zuo
Standard Research Department, OPPO Research Institute, Shenzhen, China

Preface

From 1G to 4G, mobile communication systems have gone through four stages: analog, digital, data, and broadband. These systems have brought "unprecedented" experiences to billions of users all over the world. In particular, 4G technology has opened the era of mobile internet and profoundly changed people's way of life. While users have enjoyed the rich mobile internet applications enabled by 4G such as social media networking, e-meetings, mobile shopping/payments, and mobile gaming, the mobile communication industry has been shifting its focus from "2C" to "2B," trying to promote the vertical industries' digitalization and automation with 5G new radio (NR) technology. Therefore compared with 4G, 5G has gained more attention in a wider scope due to its emphasis on increasing support for a "mobile Internet of Things."

What is the core of 5G technology? What innovations has 5G introduced? What's the difference between 5G and 4G? What kind of technical capability can 5G achieve? We believe these issues are of great interest to readers. From our perspective, 5G is not a magic and omnipotent technology. It inherits the system design concepts of 3G and 4G to a large extent, introduces a series of necessary innovative technologies, and makes a series of special optimizations for various vertical applications. Most of these innovations and optimizations are not "big concepts" that can be explained in a few words or sentences but are composed of many detailed engineering improvements. The purpose of this book is to analyze the innovation and optimization points of 5G and explain them to readers.

Some may think that 5G copies the core technology of 4G and is a "broadband 4G." It is true that, theoretically, 5G follows the core technology of 4G long-term evolution (LTE), namely, OFDMA (orthogonal frequency division multiple access) + MIMO (multiple input multiple output). However, compared with the "simplified version" OFDMA in LTE, the 5G system design achieves greater flexibility in both time and frequency domains. It can give full play to the technical potential of the OFDMA system and effectively support rich 5G application scenarios such as eMBB (enhanced mobile broadband), URLLC (ultrareliable and low-delay communication), and mMTC (a large scale Internet of Things). At the same time, the 5G system is much more sophisticated and complex than the 4G system. Based on LTE design, many modifications, enhancements, and extensions have been made. Therefore this book is based on the LTE standard, with the assumption that readers already have basic knowledge of LTE, and focuses on introducing the new and enhanced system design adopted in 5G NR, and interpreting the "incremental changes" of 5G NR relative to 4G LTE.

Unlike most 5G books, this book uses a method of "analyzing the 5G standardization process." The book features not only the 5G NR standard but also the 5G system design and standard drafting process. The experience of writing our 4G book showed us recognition of this writing method by the majority of readers. The authors of this book are the 3rd Generation Partnership Project (3GPP) delegates from the OPPO company who participated in 5G NR standardization and promoted the formation of most 5G technology design details. Many of the technical schemes they proposed were accepted and have become part of 5G standards. The standardization delegates will introduced the process of technology selection and system design of 5G NR standards. The standardization delegates will introduced the process of technology selection and system design of 5G NR standards. 5G is a complex system, in which the technical scheme selection in each part is not isolated. The optimal scheme for a single point is not necessarily the one that contributes the most to the performance of the whole system. The goal of system design is to select the technology combination achieving the overall optimization of the system. In most chapters, this book reviews the various technical options in 5G standardization, introduces the advantages and disadvantages of various options, and tries to interpret the reasons and considerations the 3GPP used to choose the final solution, which included not only the performance factors, but also the complexity of the device implementation, the simplicity of signaling design, and the impacts on the existing standards. If the target is only to interpret the final version of the technical specifications, it is unnecessary to make such a thorough explanation. But the authors hope to help readers "not only know what it is but also *why* it is," through the process of reasoning and selection, and to explore the principles, methods, and means of wireless communication system design.

From another point of view, the technical solutions chosen for 5G today are only the best choices made at the specific moment, for specific service requirements, and considering current product R&D capabilities. In the future, when service requirements change and equipment capabilities become stronger, the rejected "suboptimal" options today may become "optimal" and become our new choices. The 3GPP standard is only a "tool document" to guide product development, and does not have the function of interpreting technical principles and design ideas. If only the final results of standardization were shown to readers, they may mistakenly think that these designs are "the only reasonable choice," as if "the comparison of advantages and disadvantages, the difficult technical trade-offs" have never occurred, giving readers only a "one-sided 5G." Readers would find it difficult to understand in many cases why a certain design choice was made. Was there no other choice? What's the advantage of this design? On the contrary, if readers are allowed to take a critical and objective view of 5G standards through these technology selection processes, fully learn from the experience and lessons of 5G standardization when they design the next-generation system (such as 6G), and have the opportunity to conceive better designs, the authors' review, analysis, and summary in this book will be meaningful and helpful. Because of this feature, I believe this book can not only be used as a reference book for 5G R&D engineers, but also as a reference book for college students majoring in wireless communications to learn 5G.

This book is divided into 20 chapters. In addition to the overview in Chapter 1, Chapters 2—14 are the introduction to 5G core features, which are mainly defined in 3GPP Release 15. The core of Release 15 5G NR is for the eMBB use cases and provides a scalable technical basis for Internet of Things services. From Chapter 15 to Chapter 19, we introduce the technical features of "5G enhancement" defined in Release 16, including URLLC, NR V2X, unlicensed spectrum communication, terminal power saving, etc., which are indispensable parts of a complete 5G technology. In Chapter 20, we also briefly introduce the further 5G enhancement in Release 17, and our preliminary view on the trend of 5G-Advanced and 6G.

The authors listed at the beginning of each chapter are all our colleagues in the standard research department of OPPO. They are 5G standard experts in various fields, and many of them have participated in the standardization of 4G LTE. We thank them for their contributions to 5G international standardization. When readers use 5G mobile phones, a part of the hardware or software design (though it may be only a small part) is based on their innovation and effort. At the same time, we would like to thank Zheng Qin and Ying Ge of the OPPO external cooperation team for their contributions to the publication of this book. Finally, I would like to thank Tsinghua University Press and Elsevier for their support and efficient work in getting this book to readers as soon as possible. Questions or comments about this book should be sent to the authors via sj@oppo.com.

Authors

CHAPTER 1

Overview

Jinxi Su, Jia Shen, Wendong Liu and Li Guo

1.1 Introduction

Mobile communication basically follows the law that a new generation of technology emerges every decade. Since the successful deployment of the first generation of the analog cellular mobile phone system in 1979, mobile communication has evolved through four generations and has entered the fifth generation. Each generation of mobile communication system has its own specific application requirements, and continues to adopt innovative system designs and technical solutions to promote the rapid improvement of the overall performance of mobile communication.

First generation mobile communication technology (1G) was deployed in the 1980s and that was the first time cellular networking was adopted. It provides analog voice services to users. However, its service capacity and system capacity were very limited, and the price for the service was expensive.

About 10 years later, second-generation mobile communication technology (2G) was born. 2G system adopted narrowband digital mobile communication technology for the first time, which not only provided high-quality mobile calls, but also supported short message services as well as low-speed data services at the same time. The cost of mobile communication was greatly reduced, which enabled large-scale commercial use of the 2G systems all over the world. At the end of the 1990s, driven by the tide of the Internet, third-generation mobile communication (3G) came into being. 3G systems finally produced three different communication standards: the European-led WCDMA technology scheme, the American-led CDMA2000 technology scheme, and the TD-SCDMA technology scheme independently proposed by China. The data transmission capacity of 3G can reach dozens of Mbps, which enhances the system capacity of voice services and can also support mobile multimedia services. However, the pace of the development of mobile communication technology did not slow down. With the explosive growth of mobile Internet and intelligent terminals, the transmission capacity of 3G system is

5G NR and Enhancements
DOI: https://doi.org/10.1016/B978-0-323-91060-6.00001-5

increasingly unable to satisfy service requirements. Fourth-generation mobile communication (4G) technology appeared around 2010. The key air interface technologies adopted in 4G systems include orthogonal frequency division multiplexing [Orthogonal Frequency Division Multiplexing (OFDM)] and multiantenna, multiinput and multioutput [Multiple Input Multiple Output (MIMO)]. The transmission rate in 4G systems can reach from 100 Mbps to 1 Gbps, and it can support a variety of mobile broadband data services, which well meet the needs of the current development of mobile Internet.

In summary, after nearly 40 years of rapid development, mobile communication has entered into every corner of social life and has profoundly changed the manner of communication between people and even the normal life. However, new requirements for communication keep emerging, and the communication technologies keep evolving with new innovations. In 2020, the world ushered in the deployment of large-scale of fifth-generation mobile communication (5G) commercial networks. A 5G system can support large bandwidth, low latency, wide connectivity, and the Internet of everything. In the following chapters of this book, we will provide detailed description and discussion on what a 5G system exactly looks like, what problems have been solved, what businesses and requirements are supported, and what technology enhancements and evolution have been made.

At the beginning of the commercial use of the fourth-generation mobile communication technology, the global mainstream telecom enterprises and research institutions have begun to actively invest in the research on the fifth-generation mobile communication (5G) technology. There are many forces driving the emergence of 5G technology, including the emergence of new application scenarios, technological innovation, standards competition, business drive, industrial upgrading, and other factors.

Of course, in addition to the driving force of national strategy and industrial competition, the evolution and development of 5G technology is also the inevitable result of the continuous optimization and enhancement of the technology itself and the evolution toward higher and stronger technical indicators and system performance targets. 5G technology adopts the new radio (NR) design [1−12] based on the fundamental air interface technology framework of LTE OFDM + MIMO. 5G technology also includes numerous technical enhancements and improvements in comparison with LTE from the perspective of system scheme design, including the support of higher frequency band and larger carrier bandwidth, flexible frame structure, diversified numerology, optimized reference signal design, new coding, symbol-level resource scheduling, MIMO enhancement, delay reduction, coverage enhancement, mobility enhancement, terminal energy saving, signaling design optimization, new network architecture, quality of service (QoS) guarantee enhancement, network slicing, vehicle to everything (V2X), industry Internet of things (IIoT), new radio-unlicensed (NR−U) spectrum design, good support for a variety of vertical industries, etc. These advanced technical solutions enable 5G to fully meet the 5G vision requirements of ubiquitous connection and intelligent interconnection between people, between things as well as between people and things in future product development and commercial deployment.

1.2 Enhanced evolution of new radio over LTE

Mobile communication has profoundly changed the life of people, and is infiltrating into every aspect of human society. Fourth-generation mobile communication (4G) is a very successful mobile communication system [13]. It well meets the needs of the development of the mobile Internet and brings great convenience to communication between people. The world as a whole and most industries can enjoy the rewards of the developments of the mobile communication industry. However, the LTE technology adopted by 4G still has some technical shortcomings and some unresolved problems are also seen during the commercial network deployment of LTE. Any technological evolution and industrial upgrade are driven by the strong driving force of business and application requirements to achieve rapid maturity and development. As the two major driving forces of 5G development, mobile Internet and mobile Internet of things provide broad prospects for the development of mobile communications in the future. 5G defines three major application scenarios [14,15]: enhanced mobile broadband (eMBB), ultrareliable and low latency communications (URLLC), and massive machine type communications (mMTC). Among them, eMBB is mainly for the application of mobile Internet, while URLLC and mMTC are mainly for the application of mobile Internet of things. The mobile Internet will build an omnidirectional information ecosystem centered on end users. In recent years, ultrahigh definition video, virtual reality (VR), augmented reality (AR), distance education, telecommuting, telemedicine, wireless home entertainment, and other human-centered needs are becoming more and more popular. These booming new service requirements will inevitably put forward higher requirements for the transmission bandwidth and transmission data rate of mobile communication. On the other hand, vertical industries such as mobile Internet of things, industrial Internet, Internet of vehicles, smart grids, and smart cities are also being rapidly informatized and digitized. In addition to smartphones, mobile terminals such as wearable devices, cameras, unmanned aerial vehicles, robots, vehicle-borne ships, and other terminal modules, as well as customized terminals for the industry, etc., also come with more diverse forms. It is difficult for 4G technology to meet the vision of 5G and the emergence of a variety of new business requirements and new application scenarios, and evolution from the 4G to 5G technology and development of 5G technology is an inevitable trend. The major shortcomings of LTE technology and the corresponding enhancements and optimizations made in 5G NR will be introduced in the following sections.

1.2.1 New radio supports a higher band range

The frequency range supported by LTE is mainly in the low-frequency band, and the highest frequency band supported in LTE is TDD Band42 and Band43 in the range of 3400–3800 MHz. The actual commercial deployment networks of the global LTE are basically deployed in a frequency range below 3 GHz. For mobile communication, the spectrum is the most precious and scarce resource. The available bandwidth range in low-frequency band is limited, and that will be occupied by the existing mobile communication system for a long time. With the vigorous development of follow-up mobile communication Internet services, the requirements of wireless communication and transmission data rate are getting higher and higher, and 4G networks have traffic congestion in high-capacity areas. Therefore there is an urgent need to explore more frequency bands to support the development of mobile communications in the future.

In the current global radio spectrum usage, there is still a wide range of unused frequency bands above 6 GHz. Therefore 5G supports the millimeter wave band in the frequency range 2 (FR2) (24.25–52.6 GHz) to better resolve the difficulties of insufficient wireless frequency band. On the other hand, in order to solve the challenges of poor millimeter wave propagation characteristics, large propagation loss, and signal blocking, the NR protocol introduced a series of technical schemes, such as beam scanning, beam management, beam failure recovery, digital + analog hybrid beamforming, and so on, to ensure the quality of millimeter wave transmission. Supporting a wide millimeter wave band is a huge enhancement of 5G NR in comparison with LTE, with great potential for future 5G deployment and business applications.

1.2.2 New radio supports wide bandwidth

In the LTE standard, the maximum single-carrier bandwidth is 20 MHz. If the system bandwidth exceeds that, multicarrier aggregation (CA) has to be implemented. CA brings extra complexity to both protocol and product implementation due to the addition and activation of auxiliary carriers in the empty port, as well as the joint scheduling between multicarriers. Furthermore, a certain guard period (GP) is inserted between carriers in multiCA, which reduces the effective spectrum efficiency. In addition, the effective transmission bandwidth of LTE carrier signal is only about 90% of the carrier bandwidth, and the spectrum efficiency is also lost due to that. With the development of the semiconductor industry and technology in the past decade, the processing capabilities of semiconductor chips and key digital signal processing devices have been greatly enhanced. Together with the application of new semiconductor materials and devices such as radio-frequency power amplifiers (PAs) and filters, they can make it possible for 5G equipment to handle larger carrier bandwidth. At present, the maximum carrier bandwidth of 5G NR is 100 MHz in the frequency bands below 6 GHz, and the maximum carrier bandwidth of millimeter wave band is 400 MHz, which is an order of magnitude larger than that of LTE. That lays a better foundation for the NR system to support large bandwidth and ultrahigh throughput.

Compared with LTE, NR also greatly improves the effective spectrum efficiency of the system bandwidth by applying a digital filter, which increases the effective bandwidth of the carrier from 90% of the LTE to 98%, thus equivalently increasing the system capacity.

1.2.3 New radio supports more flexible frame structure

LTE supports two types of frame structures, FDD and TDD, which are frame structure type 1 and frame structure type 2, respectively. For the TDD frame structure, the uplink and downlink service capacity is determined by the uplink and downlink time slot ratio, which can be configured and adjusted. Seven kinds of fixed uplink and downlink time slot ratios were specified and the frame structure of TDD in a LTE cell was determined in the process of cell establishment. Although the dynamic TDD frame structure was also designed in the subsequent evolution version of LTE, it is not applicable to traditional UE, and the overall scheme is still not flexible enough and therefore has not been applied in practical LTE commercial networks.

From the beginning, NR design considered the flexibility of the frame structure. First, it no longer distinguishes between the FDD and TDD frame structure. The effective FDD is implemented by configuring OFDM symbols as uplink or downlink in the time slots. Secondly, the uplink and downlink configuration periods of the TDD band can be configured flexibly. For example, various cycle lengths such as 0.5 Ms/ 0.625 Ms/1 Ms/1.25 Ms/2 Ms/2.5 Ms/5 Ms/10 Ms can be configured by signaling. In addition, each symbol in a time slot can be configured not only as an uplink or a downlink symbol, but also as a flexible symbol that can be used as a downlink or an uplink symbol in real time though dynamic indication in the physical layer control channel, so as to flexibly support the diversity of services. It can be seen that 5G NR provides great flexibility for TDD frame structure as well as downlink and uplink resource allocation.

1.2.4 New radio supports flexible numerology

The subcarrier spacing (SCS) of the OFDM waveform defined in the LTE standard is fixed at 15 kHz. Based on the basic principle of the OFDM system, the time-domain length of the OFDM symbol is inversely proportional to the SCS, so the air interface parameter of LTE is fixed and has no flexibility. The services supported by LTE are mainly traditional mobile Internet services, and the possibility of supporting other types of services will be limited by the fixed underlying parameters.

In order to better meet the needs of diversified services, NR supports multiple SCSs. The SCS is extended by the integer power of 2 based on 15 KHz, which includes the values of 15 kHz/30 kHz/60 kHz/120 kHz/ 240 kHz. With the increase of SCS, the corresponding OFDM symbol length is also shortened proportionally. Due to the use of flexible SCS, it can adapt to different service requirements. For example, the low-latency services of URLLC require a larger subcarrier interval to shorten the symbol length for transmission, in order to reduce the transmission air interface delay. On the other hand, the mMTC services of the Internet of things with its large number of connections need to reduce the subcarrier interval and increase the symbol transmission time and power spectral density to extend the coverage distance.

The carrier bandwidth of the millimeter wave band supported by NR is often larger, and the Doppler frequency offset is also relatively large, so larger SCS is suitable for the carrier in the high-frequency band to resist Doppler shift. Similarly, for high-speed scenarios, a larger subcarrier interval is also suitable.

It can be seen that NR has laid a good technical foundation for the flexible deployment and coexistence of multiple various services in the followup 5G through supporting flexible numerology and a unified new air interface framework for both high- and low-frequency bands.

1.2.5 Low-latency enhancements of air interface by new radio

The basic unit of time interval for data scheduling and transmission defined in the LTE protocol is 1 Ms subframe. This is the main reason the air interface data transmission cannot break through the time unit limit of 1 Ms. In addition, due to the design of the timing relationship of at least $N + 4$ for HARQ retransmission in LTE, it is difficult for the air interface delay of LTE to meet the service requirements of low delay. Although LTE introduced the technical scheme of shortening the transmission time interval (TTI) in the subsequent evolution of the protocol, due to the practical factors such as the progress of the whole industry of LTE, the development cost, and the weak deployment demand, the probability of practical application and deployment of shortening TTI technology in an LTE commercial network is very low.

In order to solve the issues of air interface delay, the 5G NR system was designed and optimized in several technical dimensions from the beginning. First, NR supports flexible SCS. For low-delay traffic, large SCS can be used to directly shorten the length of OFDM symbols, and thus the length of a time slot is reduced.

Secondly, NR supports symbol-level resource allocation and scheduling. The time-domain resource allocation granularity of the downlink data channel can support 2/4/7 symbol length, while the uplink can support resource scheduling of any symbol length (1−14). By using symbol-level scheduling, when the data packet arrives at the physical layer, it can be transmitted at any symbol position of the current time slot instead of waiting for the next frame boundary or the next time-slot boundary, which can fully reduce the waiting time of the packet at the air interface.

In addition to increasing the subcarrier interval and symbol-level scheduling mechanism to reduce the air interface delay, NR also reduces the feedback delay of hybrid automatic repeat request (HARQ) by means of a self-contained time slot. In a self-contained time slot all three different direction attributes of symbol, downlink symbol, guard interval, and uplink symbol, are contained in one time slot. The same time slot includes physical downlink shared channel (PDSCH) transmission, GP, and downlink acknowledgement/negative acknowledgement (ACK/NACK) feedback transmission, so that the UE can receive and decode the downlink data and quickly complete the corresponding ACK/NACK feedback in the same time slot. As a result, the feedback delay of HARQ is greatly reduced. Of course, the implementation of self-contained time slots also requires high UE processing capacity, which is very suitable for URLLC scenarios.

1.2.6 Enhancement of reference signals in new radio

The design of reference signal is the most important technical aspect in the design of mobile communication systems, because the wireless channel estimation at the receiver is obtained through the reference signal. The design of the reference signal will directly affect the signal demodulation performance of the receiver. In 4G systems, the cell-specific reference signal (CRS) defined by the LTE protocol can be used for maintaining downlink synchronization and frequency tracking of all users in the cell. The CRS is also used as demodulated reference signals of LTE users in various transmission modes such as space-frequency block code (SFBC) and space division multiplexing (SDM). That is, the channel estimation obtained based on CRS is used for the demodulation and reception of PDSCH data in the downlink traffic channel. The CRS occupies the whole carrier bandwidth in the frequency domain, and the base station sends it steadily after the cell is established. The CRS is transmitted no matter whether there are users and data transmission in the cell or not and it is a type of always-on signal. Due to the transmission of full bandwidth, such an always-on reference signal CRS not only costs large overhead of downlink resources, but also brings cochannel interference in the overlapping areas of the network. Another consequence caused by constant reference signal transmission is that the base station equipment cannot achieve effective energy savings by using technical means such as radio-frequency shutdown when there is no service transmission in the cell.

In view of those issues existing with the CRS, the public reference signal of LTE, 5G NR made fundamental improvements in pilot design to avoid cell-specific public signals as much as possible. For example, in NR system, only synchronization signals are retained as cell-specific public signals, but all the other reference signals in NR are UE-specific. In this way, the system overhead of constant resource occupation of cell-specific public signals can be reduced, and the spectrum efficiency can be improved. For example, when the base station sends data to a UE, the UE-specific demodulation reference signal (DMRS) is only sent within the bandwidth of the scheduled data. In addition, considering that the 5G base station system generally uses the beamforming technology of massive MIMO for data transmission, the same precoding method is applied on both the data symbols and demodulated pilots, and pilot signals are sent only when there is data transmission. The beamformed transmission will also effectively reduce the interference in the system.

Furthermore, NR adopts the design of front-loaded DMRS combined with additional DMRS. The front-loaded DMRS is beneficial for the receiver to estimate channel quickly and reduce the demodulation and decoding delay. The purpose of introducing additional DMRS is to meet the requirement of time-domain DMRS density in high-speed scenarios. For the users moving at different speeds, the base station can configure the number of additional pilots in the time slot to match the user's mobile speed and provide a guarantee for accurate channel estimation at the user side.

1.2.7 Multiple input multiple output capability enhancement by new radio

The air interface technology of LTE is OFDM + MIMO, and the support for MIMO has been constantly evolving and enhanced. The full-dimensional MIMO (FD−MIMO) introduced in the later version of LTE can achieve spatial narrow beamforming in both horizontal and vertical dimensions, which can better differentiate the users spatially. However, as the most important technical mechanism to improve the air interface spectrum efficiency and system capacity of wireless communication, MIMO technology has always been an important direction for pursuing the ultimate performance.

With the maturity of the key components of Massive MIMO as well as the requirements of engineering application and commercial deployment, from the beginning of 5G requirement scenario definition and system design, Massive MIMO is treated as an important technical component of NR and the mainstream product form of large-scale deployment of 5G commercial network. Therefore 5G NR made a lot of optimizations and enhancements to MIMO technology during the process of standardization.

First of all, NR enhances the DMRS. By the methods of frequency division and code division, DMRS can support up to 12 orthogonal ports, which can better meet the performance requirement of multiuser MIMO than LTE. Secondly, NR introduces a new type 2 codebook with higher performance than that in LTE. The CSI−RS-based type2 codebook can feedback the best matching degree of spatial channels. With the high-precision codebook in UE feedback, the base station can implement higher the spatial beam directivity and shaping accuracy, and thus the performance of multiuser multistream SDM can be greatly improved.

One of the major advantages of NR over LTE is the support of millimeter wave band. Millimeter wave has the characteristics of high-frequency band, short wavelength, large space propagation loss, weak diffraction ability, and large penetration loss. So millimeter wave communication must use extremely narrow beam alignment transmission to ensure the quality of the communication link. In order to defeat those challenges, NR adopts the technology of hybrid digital and analog beamforming. NR supports the narrow beam sweeping mechanism for broadcast channels and public channels for coverage enhancement. For the control channel and traffic channel, NR introduces the beam management mechanism, including multibeam scanning, beam tracking, beam recovery, and other technical means and processes, in order to align the beams of both sides of the communication and adaptively track the movement of users. On the basis of multibeam operation, NR further supports the design of multipanel to improve the reliability and capacity of transmission.

It can be seen that a series of enhancement schemes was introduced by 5G for MIMO technology. Combined with the improved capability of large-scale antenna equipment itself, the massive MIMO will inevitably release huge technical advantages and economic benefits in 5G mobile communication system.

1.2.8 Enhancement of terminal power saving by new radio

LTE has limited consideration on the design of terminal power-saving technology. The major terminal power-saving scheme in LTE is discontinuous reception (DRX) technology. Due to the much wider working bandwidth, larger number of antennas, and much higher data rate of the 5G system, the power consumption of the RF module and baseband signal processing chip in the terminal will increase significantly, and the user experience will be seriously affected by the hot heat or short standby time during the working process of the mobile phone.

Aiming at the terminal power consumption problem, 5G designs a variety of technical schemes. From the perspective of power saving in the time domain, 5G includes a new wakeup signal for users in connected-state when DRX is configured. According to the demand for traffic transmission, the network can determine whether to wake up the UE before the arrival of the DRX activation cycle for data reception monitoring, which can prevent users from entering the DRX activation state for additional service monitoring without data transmission, thus spending unnecessary power consumption for PDCCH detection. In addition, 5G also introduces a cross-slot scheduling mechanism, which can reduce the unnecessary reception and processing of PDSCH by UE before decoding PDCCH in the case of discontinuous and sporadic traffic transmission, and reduce the activation time of RF circuit in the time domain.

From the perspective of power saving in the frequency domain, 5G introduces the function of bandwidth part (BWP). As mentioned earlier, the carrier bandwidth of NR is much larger than that of LTE, and many core bands can support typical 100 MHz carrier bandwidth. The advantage of large bandwidth is that high transmission rate can be supported. However, if the business model is a small amount of data transmission or the service is discontinuous, it is very uneconomical for UE to work in large bandwidth mode. The core of BWP function is to define a bandwidth that is smaller than the carrier bandwidth of the cell and also the bandwidth of the terminal. When the amount of data transmitted by the air interface is relatively low, the terminal works in a smaller bandwidth with the dynamic configuration of the network, so that the RF front-end device, the RF transceiver, and the baseband signal processing module of the terminal can all work with a smaller processing bandwidth and a lower processing clock. Thus the UE can work in a state with lower power consumption.

Another technical mechanism for power saving in the frequency domain is the Scell sleep mechanism introduced for the multi-RAT dual connectivity (MR−DC) and NR CA scenarios. The Scell in the active-state can enter the state of dormant Scell when there is no data transmission. UE only needs to measure channel state information (CSI) without monitoring physical downlink control channel (PDCCH) on dormant Scell, and then quickly switches to the normal state for scheduling information monitoring when there is data transmission. That can reduce the power consumption of UE without deactivating Scell.

From the perspective of power saving in the frequency domain and antenna domain, 5G introduces the self-adaptive function of MIMO layers. The network side combines the demand for terminal data transmission and the configuration of BWP to reduce the number of layers of spatial transmission, so that UE can downgrade the MIMO processing capacity and throughput rate, which would reduce the terminal power consumption equivalently.

In addition to the abovementioned terminal power-saving technologies, 5G also supports mechanisms that can relax the requirements for UE radio resource management (RRM) measurements to reduce power consumption. For example, when UE is still or moves at a low speed, the measurement requirements can be relaxed appropriately by increasing the RRM measurement cycle without affecting the mobility performance of UE to reduce UE power consumption. When UE is in IDLE or INACTIVE state, or when UE is not at the edge of the cell, appropriate RRM measurement relaxation can be carried out to reduce UE power consumption.

1.2.9 Mobility enhancement by new radio

The mobility management in LTE is mainly based on the measurement report of UE. The source base station triggers the handover request and sends it to the target base station. After receiving the confirmation reply from the target base station, the source base station initiates the handover process and sends the configuration information of the target base station to the terminal. After receiving the configuration message, the terminal initiates a random access process to the target base station, and when the random access process is successful, the handover process is completed. It can be seen that in the process of cell handover in LTE system, UE needs to complete random access in the target cell before it can carry out service transmission, and there will inevitably be a short duration of service interruption.

In order to meet the requirement of 0 Ms interruption and improve the robustness of handover, 5G NR makes two main enhancements to mobility: a handover mechanism based on dual active protocol stack (DAPS) and a conditional handover mechanism.

The handover mechanism of the DAPS is similar to the LTE handover process, and the terminal determines the type of handover to be performed based on the received handover command. If the handover type is based on the DAPS, the terminal will maintain the data transmission and reception with the source cell until the terminal successfully completes the random access process to the target cell before releasing connection with the source cell. Only after the terminal is successfully connected to the target base station will the terminal release the connection of the source cell and stop the data transmission and reception with the source cell based on the explicit signaling from the network. It can be seen that the terminal will maintain connection and data transmission with the source cell and the target cell simultaneously during the handover process. Through the mechanism of DAPS, NR can meet the performance metric of 0 Ms service interruption latency during the handover process, which greatly improves the service awareness of users in mobility.

The goal of conditional handover is mainly to improve the reliability and robustness of the handover process. It can solve the problem of handover failure caused by too long of a handover preparation time or the sharp decline of channel quality of the source cell in the handover process. The core idea of conditional handover is to preconfigure the contents of handover commands to UE, in advance. When certain conditions are met, UE can independently execute the configuration of handover commands and directly initiate handover to a target cell that meets the preconfigured conditions. Because UE no longer triggers measurement reporting when the switching conditions are met, and UE has obtained the configuration in the switching command in advance, the problem that measurement reporting and switching commands cannot be received correctly is solved. Especially for high-speed cases or cases where the signal experiences fast fading in the switching frequency band, conditional switching can greatly improve the success rate of handover.

1.2.10 Enhancement of quality of service guarantee by new radio

In the LTE system, QoS control is carried out through the concept of evolved packed system (EPS) bearer, which is the smallest granularity of

QoS processing. A single UE supports up to 8 radio bearers at the air interface. The differentiated QoS guarantee corresponding to maximum 8 EPS bearers cannot satisfy more refined QoS control requirements. The operation of the radio bearer and the QoS parameters setting at the base station completely follows the instructions of the core network. For the bearer management request from the core network, the base station can only choose to accept or reject, but cannot establish the radio bearer or adjust the parameters by itself. The standardized QoS class identifier (QCI) defined by LTE has only a limited number of values, so it is impossible to provide accurate QoS guarantee for business requirements that are different from the preconfigured QCI or standardized QCI in the current operator network. With the vigorous development of various new services on the Internet, and the emergence of a variety of new services, such as private network, industrial Internet, vehicle networking, machine communication, and so on, the types of services that the 5G network needs to support and the demand for QoS guarantee of services greatly exceed the QoS control capability that can be provided in the 4G network.

In order to provide a more differentiated QoS guarantee for a variety of 5G services, the 5G network has made more refined adjustments to the QoS model and types. The concept of bearer is removed at the core network side and is replaced with QoS flow. Each PDU session can have up to 64 QoS flows, which greatly improves the QoS differentiation and carries out finer QoS management. The base station determines the mapping relationship between the QoS flow and the radio bearer, and is responsible for the establishment, modification, deletion, and QoS parameter setting of the radio bearer, so as to use the wireless resources more flexibly. Dynamic 5QI configuration, Delay-Critical resource types, Reflective QoS, QoS Notification Control, Alternative QoS profile, and other features are also added in the 5G network, which can provide a better differentiated QoS guarantee for a wide variety of services.

1.2.11 Enhancement of core network architecture evolution by new radio

In the LTE network, the adopted network architecture has no separation of the control plane and the user plane, and the session management of the terminal and the mobility management of the terminal are handled by the same network entity, which leads to the inflexibility and nonevolution of the network evolution.

In the 5G era, the goal of 5G mobile communication is to achieve the Internet of everything and support rich services of mobile Internet and Internet of things. 4G network architecture mainly meets the requirements of voice service and traditional mobile broadband (MBB) services, and is not able to efficiently support a variety of services.

In order to better and more efficiently meet the above requirements, and to support operators to achieve rapid service innovation, rapid on-boarding and on-demand deployment, etc., 3GPP adopts a network architecture with completed separation of control plane and user plane. This design is beneficial to the independent expansion, optimization, and technological evolution of different network elements. The user plane can be deployed with a centralized or distributed manner. In a distributed deployment, the user plane can be sunk into a network entity that is closer to the user so as to improve the response speed to user requests. However, the control plane can be managed centrally and deployed in a unified cluster to improve maintainability and reliability.

At the same time, the mobile network needs an open network architecture. The network capability can be expanded through changing the open network architecture and the service can invoke the network capability through the open interface. On the basis of such a design, 3GPP adopts 5G service-based architecture (SBA). Based on the reconstruction of the 5G core network, the network entity is redefined in the way of network function (NF). Each NF provides independent functionalities and services and the NFs can invoke the functionalities and services of each other. Thus it transforms from the traditional rigid network (fixed function, fixed connection between network elements, fixed signaling interaction, etc.) to a service-based flexible network. The SBA solves the problem of tight coupling of point-to-point architecture, realizes the flexible evolution of the network, and meets the requirements of various services.

1.3 New radio's choice of new technology

As discussed in previous sections, NR has made a lot of enhancements and optimizations compared to LTE technology in the process of standardization. In order to meet the basic goal of large bandwidth, low latency, and high data rate of future mobile communication networks, and to support the diversified services of vertical industry more flexibly, the goal of NR from the beginning of standard research and technical scheme design was to adopt many brand-new key technologies, such as new architecture, new air interface, new numerology, new waveform, new coding, new multiple access, and so on. At the formal standardization discussion stage, a large number of program research reports and technical recommendations were submitted by many companies for each key technology. After many rounds of discussion and evaluation, we reached the final conclusion of standardization based on comprehensive consideration of various factors and certain trade-offs. In the determined final NR standardization, some new technologies, such as new numerology and new coding, were finally formulated as a standardization scheme, but other key technologies that were fully discussed during the standardization process were not finally standardized in the completed versions of R15 and R16, such as new waveform and new multiple access technologies. The following sections presents a high-level discussion and summary of NR's choice of new technologies in the process of standardization.

1.3.1 New radio's choice on new numerology

The motivation for NR to design flexible numerology is that NR needs to better support diverse business requirements. The SCS of the OFDM waveform defined in the LTE standard is a fixed value, 15 kHz. This single value of SCS cannot meet the system requirements of 5G. The three typical services of 5G, eMBB, URLLC, and mMTC, have different requirements for transmission rate, air interface delay, and coverage capability, so different services would require different numerologies (SCS, cyclic prefix length, etc.). Compared with the traditional eMBB service, the low-latency service of URLLC needs a larger SCS to shorten the symbol length for transmission in order to reduce the transmission air interface delay. However, the mMTC services of the Internet of things with large connections often need to reduce the SCS and increase the coverage distance by increasing the symbol transmission time and power spectral density. And NR needs to ensure that the services with different numerologies can coexist well in the air interface and do not interfere with each other.

According to the basic principle of the OFDM system, the SCS of the OFDM waveform is inversely proportional to the length of the OFDM symbols. Since the OFDM symbol length can be changed correspondingly by changing the SCS, the time length of a time slot in the air interface can be directly determined. Considering that NR should better support different air interface transmission delays and large carrier bandwidths, NR finally supports multiple SCSs, which are based on 15 kHz and expand with the integral power of 2, including the values of 15 kHz/ 30 kHz/60 kHz/120 kHz/240 kHz/480 kHz. With the increase of SCS, the corresponding OFDM symbol length decreases proportionally. The purpose of this design is to achieve boundary alignment between OFDM symbols with different subcarrier intervals, so as to facilitate resource scheduling and interference control in frequency division multiplexing of services with different SCSs. Of course, at the beginning of the NR discussion, other SCSs such as 17.5 kHz were also considered. But after evaluation, it was determined that the SCS based on 15 kHz could support the compatible coexistence scenario of LTE and NR and the scenario of spectrum sharing better than other SCSs. Thus the scheme of other SCS numerology was not adopted.

Flexible and variable SCS can adapt to different business requirements. For example, using a larger SCS can shorten the symbol length, thus reducing the air interface transmission delay. On the other hand, the FFT size and SCS of the OFDM modulator jointly determine the channel bandwidth. For a given frequency band, phase noise and Doppler shift determine the minimum SCS. The carrier bandwidth of the high-frequency band is often larger, and the Doppler frequency offset is also relatively large. So, a larger SCS is suitable for the high-frequency band carrier, which not only can meet the limitation of the number of FFT points, but also is able to resist the Doppler frequency shift well. Similarly, in high-speed scenarios, it is also suitable to use a larger SCS to resist the influence of Doppler offset.

Based on the above analysis, we can conclude that with multiple SCSs, the NR has good scalability and the NR with flexible numerology can well meet the needs of various scenarios, such as different service delay, different coverage distance, different carrier bandwidth, different frequency range, different mobile speed, and so on. NR has laid a good technical foundation for the flexible deployment and coexistence of 5G multiservices by supporting flexible numerology and a unified new air interface framework on both high and low frequencies.

1.3.2 New radio's choice on new waveform

NR's requirements for new waveforms have the same starting point as the flexible numerology discussed earlier, which is that NR needs to support diverse business requirements. Those different services transmitted with different numerologies [SCS, symbol length, cyclic prefix (CP) length, etc.] at the air interface need to coexist well and not interfere with each other. Therefore the design goal of the new waveform is not only to support higher spectral efficiency, good intercarrier resistance to frequency offset and time synchronize deviation, lower out-of-band radiation interference, and excellent peak-to-average power ratio (PAPR), but also to meet the requirements of asynchronous and nonorthogonal transmission among users.

As is well known, the CP−OFDM waveform used in the downlink transmission of LTE has some inherent advantages, such as good resistance to inter-symbol interference and frequency selective fading, simple frequency-domain equalization receiver, easy to be combined with MIMO technology, and support of flexible resource allocation. But CP−OFDM waveform also has inherent disadvantages, such as high PAPR, spectral efficiency overhead due to CP, being sensitive to time synchronize and frequency deviation, large out-of-band radiation, and performance degradation caused by intercarrier interference. Based on the motivation of NR to support a variety of new services, the goal of air interface new waveform design is to flexibly select and configure appropriate waveform parameters according to business scenarios and business types. For example, the system bandwidth is divided into several subbands to carry different types of services and different waveform parameters can be selected for different subbands. There is only a very low protection band or no need for protection bands between the subbands, and the system can implement a digital filter on each subband to eliminate the related interference between the subbands to realize the waveform decoupling of different subbands, and satisfy the flexible coexistence of different services.

During the standardization discussion on new NR waveforms, a variety of optimized or brand-new waveform schemes based on CP−OFDM waveforms were proposed [16−26]. As shown in Table 1.1, more than a dozen proposals for new waveforms were submitted, which can be divided into three categories: time-domain windowing processing, time-domain filtering processing, and no windowing or filtering processing.

The candidate new waveforms of multicarrier time-domain windowing are as follows:

1. FB−OFDM: Filter-Bank OFDM
2. FBMC−OQAM: Filter-Bank MultiCarrier offset-QAM

Table 1.1 New radio candidate new waveform

	Time-domain windowing	Time-domain filtering	Without windowing/filtering
Multicarrier	FBMC–OQAMFB–OFDMFC–OFDMGFDMW–OFDMOTFS	F–OFDMFCP–OFDMUF–OFDMOTFS	CP–OFDMOTFS
Single-carrier	DFT spreadingTDW MC candidates	DFT spreadingTDF MC candidates	DFT–s–OFDMZT–s–OFDMUW DFT–s–OFDMGI DFT–s–OFDM

3. GFDM: Generalized Frequency Division Multiplexing

4. FC–OFDM: Flexibly Configured OFDM

5. OTFS: Orthogonal Time Frequency Space.

The candidate new waveforms of multicarrier time-domain filtering are as follows:

1. F–OFDM: Filtered–OFDM

2. UF–OFDM: Universal-Filtered OFDM

3. FCP–OFDM: Flexible CP–OFDM

4. OTFS: Orthogonal Time Frequency Space.

In addition to time-domain windowing and time-domain filtering, the following new waveforms are available for single-carrier waveforms:

1. DFT–S–OFDM: DFT-spread OFDM

2. ZT–S–OFDM: Zero-Tail spread DFT-OFDM

3. UW DFT–S–OFDM: Unique Word DFT–S–OFDM

4. GI DFT–S–OFDM: Guard Interval DFT–s–OFDM.

3GPP evaluates and discusses a variety of candidate new waveform schemes, among which several candidate waveforms were mainly discussed, including F–OFDM, FBMC, UF–OFDM, and so on. The new waveforms do have some advantages in the orthogonality between subbands or subcarriers, spectral efficiency, out-of-band radiation performance, and resistance to time-frequency synchronize errors, but they also have some problems, such as poor performance gain, incompatibility with CP–OFDM waveforms, high implementation complexity of combining with MIMO, underutilization of fragment spectrum, and so on. The final conclusion of the standard is that no new waveform is to be adopted, but only specific performance requirements such as effective carrier bandwidth, adjacent channel leakage, and out-of-band radiation of NR are specified in the standard. In order to meet these technical requirements, the technical schemes that may be used in NR waveform processing, such as time-domain windowing, time-domain filtering, and so on, are left to the product implementation of each manufacturer. Finally, NR maintains the CP–OFDM waveform in the downlink, and supports the CP–OFDM waveform in the uplink in addition to the single carrier DFT–s–OFDM waveform of LTE. The major reason is the fact that the equalization and detection of CP–OFDM waveform is relatively simple, and is more suitable for MIMO transmission. The uplink and downlink use the same modulation waveform, which is also conducive to support a unified interference measurement and interference elimination between the downlink and uplink of the TDD system.

1.3.3 New radio's choice on new coding

Because the spatial propagation channel of wireless communication experiences large-scale fading and small-scale fading, and there may exist intrafrequency or adjacent-frequency interference within and between systems, forward-error correction codes are usually used in wireless communication systems to provide reliability for data transmission. As one of the most important key technologies in wireless communication systems, channel coding has been continuously studied and explored in communications research. Early mobile communication systems such as GSM and IS-95 CDMA generally adopted convolutional coding and Viterbi decoding. In order to support high-speed multimedia services and mobile Internet services, 3G and 4G adopted Turbo coding scheme in data channel, while convolutional coding and tail-bit convolutional code (TBCC) are still used in control channel. 5G needs to satisfy the business requirements of large bandwidth, high speed, low delay, and high reliability, due to which the industry fully expects the new coding of 5G.

In the process of standardization, the discussion of 5G data channel coding mainly focused on the choice between Turbo codes and low-density parity-check codes (LDPC). As the system throughput of eMBB service to be carried by 5G is much higher than that of 4G, the downlink needs to support peak throughput of 20 Gbps and the uplink needs to support peak throughput of 10 Gbps. Therefore despite the mature application of Turbo codes in 4G and the optimization of parallel processing in interleaves, Turbo codes still cannot meet the business requirements of large bandwidth and high throughput in the future in terms of large block decoding performance and ultrahigh throughput decoding delay. Although LDPC coding has not been used in previous generations of mobile communication systems of 3GPP, this coding scheme has been studied for decades, and has been widely used in digital video broadcasting (DVB), wireless local area networks (WLANs), and other communication fields. LPDC has the advantages of low decoding complexity, being very suitable for parallel decoding, good decoding performance of large block and high bit rate, and excellent performance approaching Shannon limit. Thus LDPC is naturally suitable for 5G services that require large bandwidth and high throughput. From the perspective of actual production and industrialization, LDPC has obvious advantages in decoding delay, chip efficiency-area ratio, chip power consumption, device cost, and so on. After several rounds of discussion, 3GPP finally decided that LDPC is the coding scheme of the data channel of NR.

The main characteristic of the control channel coding is that the reliability is higher and the length of the encoded data block is smaller than that of the data channel. Because LPDC has no advantage in short code performance, the discussion on coding for control channel of NR mainly focused on the trade-off between 4G tail-biting convolutional code (TBCC) and Polar code. As a new coding scheme that was proposed in 2008, Polar code has the obvious advantage in short code. Polar code can obtain arbitrarily low code rate, arbitrary coding length, excellent performance of medium and low bit rate, and no error flat layer in theoretical analysis. After sufficient evaluation, it was determined that the performance of Polar code in control channel transmission is better than that of TBCC code. Finally, Polar code was adopted as the coding scheme of NR control channel.

It can be concluded that NR uses a new channel coding scheme to replace the original 4G channel coding scheme. On the one hand, due to the driving force of the new requirements of 5G new services, it is necessary to adopt new technologies to support higher performance requirements. On the other hand, because the channel coding is a relatively independent functional module in the systematic scheme and framework of the whole wireless communication, replacing the channel coding scheme will not affect other functional modules. In summary, 5G adopts a new channel coding scheme that can provide strong underlying technical support for 5G to provide new services and create a powerful air interface capability.

1.3.4 New radio's choice on new multiple access

In addition to new waveforms, new codes, and so on, the technology of nonorthogonal multiple access was also extensively studied and discussed in the industry as early as when defining key technical performance indicators and selecting key technologies in NR was started. In order to improve the spectrum efficiency of the air interface and the access capacity of multiple users, the wireless communication system adopted the multiuser multiplexing technology of time division, code division, frequency division, and space division from 2G to 4G. With the arrival of the era of 5G Internet of everything, the mMTC scenario with large-scale Internet of things needs to provide ultrahigh user capacity per unit coverage area. The nonorthogonal multiple access technology can provide up to several times of user capacity compared with orthogonal multiple access technology, which is very suitable for scenarios with a large number of connections. Many companies proposed their own nonorthogonal multiple access technology, but during the process of standardization, nonorthogonal multiple access technology was only discussed in a study item in R16 work, and the final standardization work has not been completed. It is also not included in the scope of the project in the upcoming R17 work. Nonorthogonal multiple access technology will be discussed in more detail in the chapter on the URLLC physical layer in this book.

In summary, the NR standard adopts a lot of enhancements, optimizations, and upgrades compared to LTE, and NR is a brand-new design that is not compatible with LTE. The designs of NR air interface focus on the overall optimization of the system design, such as the increase of bandwidth, the increase of the number of MIMO layers, the diversification of numerology, flexible frame structure, flexible resource allocation, flexible scheduling, and so on. From the perspective of wireless communication technologies and signal processing, NR still follows the general technical framework of OFDM + MIMO, and the key technologies adopted in NR do not have essential breakthroughs and changes. Of course, that does not deny the innovation of NR technology. The NR technology is a practical and reasonable choice made by the whole industry based on existing technology and demand.

The ultimate goal of mobile communication technology is to deploy and apply large-scale commercial applications in the industry, providing better information services for the entire society, each individual customer, and multiple industries. Every link in the mobile communication industry, including operators, network equipment manufacturers, terminal equipment manufacturers, chip manufacturers, and so on, also needs to be able to obtain certain commercial value and benefits with the development and upgrading of the industry. Through the selection of key technologies in the process of NR standardization, it can be seen that the practicability of technology is more important from the perspectives of productization, engineering, and commercialization. Factors such as the complexity of equipment development, development cost, difficulty, and development cycle are all important factors influencing the choice of new technology. In addition, the introduction of new technology must also fully consider systemicity. Technical upgrades or enhancements in certain direction should be well compatible with the existing systems, so as to avoid a major impact on the existing technology framework and not have major impact on the current status of the industry. Of course, innovative new technologies are always expected. Constantly exploring and researching new technologies, continuously improving system performance, and innovatively solving problems are the goals all communications practitioners and the entire industry are constantly pursuing.

1.4 Maturity of 5G technology, devices, and equipment

As the process of 4G industrialization and commercialization has just entered its initial stage, research on 5G technology in the industry was in full swing. Based on the research results of many companies, scientific research institutions, and universities around the world on 5G technology in the past decade, 5G technology and standards have matured rapidly. As described in section 1.1 of this book, compared with 4G, 5G has made a lot of enhancements and evolutions in standardization, supporting a variety of different technical characteristics such as the millimeter wave band range, flexible frame structure, flexible numerology, pilot optimization design, flexible scheduling, new coding, MIMO enhancement, delay reduction, mobility enhancement, terminal power saving, signaling optimization, new network architecture, unlicensed spectrum (NR−U), vehicle networking (NR−V2X), and Industry Internet of Things (IIoT). The technical solutions ensure that the advanced 5G technology and future 5G commercial deployment can meet the needs of new scenarios and new business services. The entire 5G industry can quickly carry out global commercial deployment after the standardization of the first version (R15), which benefits from the technical accumulation of the entire mobile communication industry in the industrialization of devices, chips, and equipment.

1.4.1 The development and maturity of digital devices and chips have well supported the research and development needs of 5G equipment

Compared with the 4G mobile communication system, 5G needs to meet more diverse scenarios and extreme performance indicators, so it also puts forward higher requirements for the processing capacity of the equipment. 5G needs to support the user experience data rate of 1 Gbps, the user peak rate of several Gbps, the peak throughput of dozens of Gbps, and 10 times higher spectral efficiency than 4G systems. 5G also needs to support millisecond level end-to-end latency, high reliability with 99.999% data transmission accuracy, and massive connected terminals of Internet of things. The satisfaction of all these 5G ultrahigh performance technical indicators requires 5G commercial equipment to have a powerful computing and processing platform. Taking the most mainstream massive MIMO macrobase station deployed in a 5G commercial network as an example, the processing function of the baseband unit (BBU) in a eMBB scenario needs to meet the following technical specifications: single-carrier 100 MHz bandwidth, 64T64R digital channel, uplink 8-stream and downlink 16-stream MIMO processing capacity, system peak rate of 20 Gbps, air interface delay of 4 Ms, and a single cell supporting simultaneous online connection of thousands of users. These performance indicators require a significant increase in the processing capacity of the 5G base station platform, including such as baseband processing application specific integrated circuit chip, system on chip, multicore CPU, multicore DSP, large capacity field programmable array, high-speed transmission switching chip, etc. Similarly, the peak rate of a single 5G user is increased to several Gbps, and thus the processing performance of the communication chips in the terminal devices shall also be greatly improved compared with the 4G terminals. It can be seen that 5G has put forward higher requirements for communication devices and semiconductor chips.

With the continuous evolution of mobile communication technology, the semiconductor industry has been developing rapidly. Especially in recent years, under the huge driving force of the demand for communications, semiconductor materials and integrated circuit (IC) processes are also rapidly completing technological innovation and upgrading. The digital IC process has been upgraded from 14 nm a few years ago to mainstream 10 and 7 nm processes. In the next year or two, the leading chip design companies and semiconductor manufacturers in the world are expected to move toward more advanced 5 and 3 nm processes. The technology level of the advanced semiconductor chip is mature and developed, meeting the communication and computing power requirements of 5G network equipment and terminal equipment, and making it suitable for 5G commercial use.

1.4.2 5G active large-scale antenna equipment can meet the engineering and commercial requirements

In addition to digital devices and chips that can already meet the communication needs of 5G equipment, the equipment engineering problem of 5G Massive MIMO technology, the most representative of 5G, has also been effectively solved. As we all know, the remote radio unit (RRU) devices in LTE networks are generally four antennas (FDD) or eight antennas (TDD). There do not exist technical bottlenecks in the implementation and engineering of devices such as RRU volume, weight, power consumption, and so on. However, 5G massive MIMO requires beam scanning capability and higher spatial resolution in both horizontal and vertical dimensions to support SDM transmission capability of up to dozens of streams. Therefore the size and number of antenna arrays need to be greatly improved. The typical configuration of 5G massive MIMO is 64 digital channels, 192 antenna elements, and 200 MHz operating bandwidth. Compared with 4G RRU equipment, the operating bandwidth is increased by 5−10 times and the number of RF channels is increased eight fold. Due to the large number of RF channels, the traditional engineering application method of separating RRU and antenna array used in mobile communication before 4G is no longer applicable. 5G needs the integration of the radio-frequency unit and the passive antenna, and the equipment form has evolved into an active antenna array active antenna unit (AAU) integrating radio frequency and antenna. In addition, the requirement for commercial network deployment is that the 5G base station can share the site with the 4G network. Because the working frequency band of 5G is higher than that of 4G and the working bandwidth is wider than 4G, this requires 5G AAU equipment to support greater transmission power to meet the same coverage distance as 4G.

Since the later stage of the deployment of LTE network, in order to meet the increasing demand of user capacity, system manufacturers have developed engineering prototype of large-scale active antenna array for LTE TDD band equipment, and have been continuously improving and optimizing them. In recent years, with the improvement of radio-frequency device process and the application of new materials, the working bandwidth and efficiency supported by PA have been continuously improved, and the sampling rate, interface bandwidth, and integration of RF transceiver have been continuously improved. The capabilities and performance indicators of key devices and core chips have made huge breakthroughs, gradually meeting the needs of commercialization and engineering. However, the equipment power consumption and equipment engineering of 5G AAU still face great challenges. On the one hand, because the radio-frequency channel number and working bandwidth of 5G large-scale antenna equipment are several times that of 4G equipment, the power consumption of 5G large-scale antenna equipment is also several times that of 4G equipment. On the other hand, from the perspective of power conservation, environmental protection, and lower basic operating costs, operators not only have low power consumption requirements for equipment, but also have strict restrictions on equipment engineering parameters such as volume, weight, windward area, and so on. Therefore under certain engineering constraints, the power requirements, power consumption requirements, and heat dissipation requirements that 5G AAU equipment must meet are the difficult problems that network equipment manufacturers have been trying to solve in recent years. For this reason, equipment manufacturers optimize the design of RF circuit to improve the efficiency and linearity of the PA by selecting highly integrated transceivers, optimizing the medium RF algorithm, and reducing the peak-to-average ratio (PAPR) and error vector amplitude (EVM) so as to improve the efficiency of the whole machine and reduce the power consumption of the equipment. Secondly, for the heat dissipation problem of equipment, on the one hand, advanced heat dissipation scheme, reasonable structure design, and device miniaturization are adopted from the perspective of equipment development. On the other hand, combined with the working scene and business load, adaptive software technical schemes such as closing carrier wave, channel, time slot, and symbols are adopted to save power and reduce consumption, and ultimately meet the requirements of commercialization and engineering of the equipment.

1.4.3 Millimeter wave technology—devices and equipment are becoming more and more mature

One of the significant enhancements of 5G compared to 4G is that it supports the millimeter wave band in the FR2 (24.25−52.6 GHz) range. Because the spectrum available in the high-frequency band is very wide, it provides greater potential and flexibility for the future deployment and business applications of 5G.

Although the spectrum of the millimeter wave band is wide and the available bandwidth is large, because the millimeter wave is in a high-frequency band, it has the following problems when applied to mobile communication systems compared to the low-frequency band:

1. The propagation loss is large, and the coverage ability is weak.
2. The penetration loss is high and the signal is easily blocked. It is suitable for Line-of-Sight (LOS) transmission.
3. The operating frequency band of the millimeter wave is high and the efficiency of PA is low.

4. The operating bandwidth of the millimeter wave is large, which requires high sampling rate and high working clock of RF devices such as ADC/DAC.
5. The phase noise of the millimeter wave devices is large, and the EVM performance is worse than that of low-frequency devices.
6. Millimeter wave devices are expensive, and the industrial chain is immature.

On the problems of the millimeter wave, the industry has been studying and breaking through for many years. In order to solve the problems of poor millimeter wave propagation characteristics, large propagation loss, and susceptibility to obstruction, the 5G NR protocol includes the mechanism of multibeam, including beam scanning, beam management, beam alignment, beam recovery, and so on to ensure the transmission quality of wireless link from the perspective of air interface communication mechanism. Secondly, in terms of the equipment form, because the wavelength of the millimeter wave frequency band is shorter than that of the sub-6GHz wave, the size of the antenna array is smaller, and the outlet power of the millimeter wave PA is low. Thus the millimeter wave antenna array is designed to have phase modulation at the front end of the radio frequency. The phased array antenna panel synthesizes a high-gain narrow beam with a group of antenna elements, and adjusts the direction of the analog beam to the receiver through a phase shifter to improve the signal quality of the air interface transmission link. In addition, in order to avoid the power loss caused by the connection of discrete devices, the millimeter wave equipment is developed in the form of integration of RF front-end and antenna unit (Antenna in Package) to maximize the efficiency. In order to meet the functional requirements of NR MIMO transmission while reducing the requirement of baseband processing capability, millimeter wave devices adopt the scheme of digital + analog hybrid beamforming; that is, a digital channel signal is extended to a group of antenna units at the front end of the radio frequency and after that, it is transmitted through analog beamforming. In the process of communication, based on the beam management mechanism of NR protocol, the base station and the terminal constantly adjust and align the beam direction to ensure that the beams of both sides can accurately point to each other.

It can be seen that the use of the millimeter wave in mobile communication brings a huge challenge to both the technical solutions and the performance requirements of the equipment. However, after years of unremitting efforts of industry experts, the technical solutions of millimeter wave communication have been introduced and standardized in 5G, and great breakthroughs have been made in the development of millimeter wave devices and equipment. In the past 2—3 years, some equipment manufacturers, operators, and chip manufacturers have demonstrated some demonstration and test results based on millimeter wave prototypes. Due to the shortage of the 5G commercial spectrum, North America, South Korea, Japan, and other countries started the deployment of millimeter wave precommercial networks in 2020. The IMT—2020 (5G) working group of China organized and completed the first phase of 5G millimeter wave technology trial in 2019, and will continue to carry out related work in 2020 to further promote the maturity of millimeter wave technology and equipment, and provide technical solution support and test data reference for the future deployment of 5G millimeter wave commercial networks.

1.5 R16 enhancement technology

The previous sections gave an overall introduction to the evolution and enhancement of 5G key technologies and the current status of 5G equipment and products. This section gives an overview of the progress of 5G standardization. 3GPP has a clear timeline for the standardization of NR. As the first basic version of NR specification, the R15 protocol version was completed and released in June 2018, supporting eMBB services and basic URLLC services. At the time of writing, the standardization of 3GPP R16 protocol version has come to an end. The R16 version has enhanced the eMBB service, and also fully supports the URLLC service. The new functional features supported by the projects completed by the R16 protocol version mainly include the following aspects.

1.5.1 Multiple input multiple output enhancement

The MIMO enhancement of R16 is an enhancement and evolution on the basis of the MIMO of R15. The main enhancements include the following.

1.5.1.1 eType II codebook

In order to solve the problem of the too high feedback cost of the R15 Type II codebook, R16 further introduces the eType II codebook. Unlike the Type II codebook, which decomposes the channel on the broadband or subband into the amplitude and phase of multiple beams, the eType II codebook performs equivalent time-domain transformations on the subband channels and feedbacks the multipath delay and weighting coefficient of each beam. The overhead of feedback signaling is greatly reduced. At the same time, the eType II codebook also supports more refined channel quantization and higher spatial rank, which can further improve the codebook-based transmission performance. The performance improvement is more significant in the MU—MIMO scenario.

1.5.1.2 Multitransmission and reception points enhancement

In order to further improve the throughput and transmission reliability of cell edge UE, R16 introduces MIMO enhancement based on multiple transmission and reception points (TRPs). The typical target scenario is eMBB for noncoherent-joint transmission based on the single downlink control information (DCI) and multiple DCIs. The typical target scenario is URLLC for multi-TRP diversity transmission based on a single DCI. Among them, the NC—JT transmission based on a single DCI can support two TRPs to transmit data on the same time-frequency resource at the same time without increasing the DCI overhead, thereby improving the transmission rate of the edge UE in the scenarios with ideal backhaul. The NC—JT transmission based on multiple DCIs supports two TRPs to schedule and transmit data to the same UE independently, which improves the throughput while ensuring the scheduling flexibility, and can be used for various backhaul assumptions. Multi-TRP-based diversity transmission supports two TRPs to transmit the same data through space division, frequency division or time division, which improves the transmission reliability of edge UE and better meets the requirements of URLLC services.

1.5.1.3 Multibeam transmission enhancement

The beam management and beam failure recovery mechanism based on analog beamforming introduced by R15 enables the high-speed transmission in the millimeter wave band. R16 specified further optimizations and enhancements on the basis of these mechanisms, such as activating the beam information of a group of uplink signals or downlink signals (such as signals on multiple resources or multiple carriers) simultaneously, introducing default uplink beams, reducing the signaling overhead of configuring or indicating beam information, and providing diversified beam measurement and reporting information for the network by introducing a beam measurement mechanism based on L1−SINR. By extending the beam the failure recovery mechanism is extended to the Scell, which improves the reliability of the analog beam transmission on the Scell.

1.5.1.4 Uplink full-power Tx

Based on the uplink transmission power control mechanism of Rel-15, if the number of ports transmitted for the codebook-based physical uplink shared channel (PUSCH) transmission is greater than 1 and less than the number of transmit antennas of the terminal, the PUSCH cannot be transmitted at full power. In order to avoid the performance loss, R16 introduces the enhancement of full-power transmission; that is, UE with different PA architectures can report particular UE capability, so that the network side can schedule full-power PUSCH transmission. Specifically, R16 introduces three full-power transmission modes: full-power transmission supported by UE with single PA (there is no need to use power scaling mode); full-power transmission supported by fully correlated precoding vectors (i.e., full port transmission); and UE reporting precoding vectors that support fullpower transmission. Whether full-power transmission is actually used and which method is used depends on the UE capability report and the configuration provided by the network side.

1.5.2 Ultrareliable and low latency communications enhancement-physical layer

The R15 protocol has limited support for URLLC functions. Based on the flexible framework of NR, it has enhanced the processing capacity for URLLC, and introduced a new modulation and coding scheme (MCS) and channel quality indicator (CQI) corresponding table, and also introduced scheduling-free transmission and downlink preemption technology. R16 established an enhancement project for URLLC, which further broke through the latency and reliability bottlenecks. The URLLC enhancements in R16 mainly include the following directions:

1. Downlink control channel enhancement, including compressed DCI format and downlink control channel monitoring enhancement.
2. Uplink control information enhancement, including support for multi-HARQ−ACK transmission in one time slot, construction of two HARQ−ACK codebooks, and preemption mechanism of uplink control information with different priority services within the users.
3. Uplink data channel enhancement to support back-to-back and repeated transmission across slot boundaries.

4. Enhanced scheduling-free transmission technology to support multiple scheduling-free [configured grant (CG)] transmissions.
5. Enhanced semicontinuous transmission technology to support multiple semicontinuous transmission configurations and short-term, semicontinuous transmission.
6. Priority transmission between different users and enhanced uplink power control.

1.5.3 Ultrareliable and low latency communications enhancement high layer

While the physical layer enhancement project was established, in order to better support applications of vertical industries, such as the IIoT, smart grid, etc., R16 also established a high layer enhancement project for URLLC with the design goal of supporting 1 μs time synchronize accuracy, 0.5 Ms air interface delay, and 99.999% reliable transmission. The main technical enhancements of the project include the following aspects.

1.5.3.1 Supporting time-sensitive communication

In order to support the transmission of services such as industrial automation, the following aspects were studied: Ethernet frame header compression, scheduling enhancement, and high-precision time synchronization. Specifically, the compression of the Ethernet frame header is to support the transmission of Ethernet frame in the air interface to improve the transmission efficiency of the air interface. The scheduling enhancement is to ensure the delay of TSC service transmission. High-precision time synchronization is to ensure the precise delay requirements of TSC service transmission.

1.5.3.2 Data replication and multiconnection enhancement

The R15 protocol version already supports the mechanism of data replication and transmission over air interface links. R16 supports the data replication transmission function of up to four RLC entities, which further improves the reliability of business transmission.

1.5.3.3 Intrauser priority/reuse enhancement

The resource conflict scenarios supported in R15 are the dynamic grant (DG) and resource CG conflict scenarios, and the DG priority is higher than the CG transmission. In the R16 protocol version, it is necessary to support the scenario where URLLC services and eMBB services coexist, and URLLC service transmission can use either DG resources or CG resources. In order to ensure the transmission delay of URLLC services, R16 needs to enhance the conflict resolution mechanism in R15.

1.5.4 UE power-saving enhancement

The main functions of R15 protocol version for terminal power saving are DRX technology and BWP function, which reduce terminal processing from the perspective of time domain and frequency domain, respectively. R16 has also enhanced terminal power saving in the following aspects:

1. A wake-up signal (WUS) is introduced and the network determines whether to wake up UE before the arrival of the DRX activation cycle to monitor and receive data. The UE wake-up mechanism is implemented through the newly added control information format (DCI format 2−6).
2. The mechanism of cross-slot scheduling is enhanced, which can avoid unnecessary reception and processing of PDSCH channel data by the UE in the case of discontinuous transmission of service data.
3. MIMO layer number self-adaption function and the network side combined with the configuration of BWP can inform UE to reduce the number of layers of spatial transmission, thereby reducing the requirements of UE processing capacity.
4. Support RRM measurement mechanism relaxation.
5. Support terminal to report preferred power-saving configuration.

1.5.5 Two-step RACH

In order to shorten the initial access delay, the two-step RACH process was discussed in the early stage of R15 standardization, but R15 only completed the traditional 4-step RACH process standardization. Since 2-step RACH has obvious gain in shortening the delay of random access and reducing the LBT operation of NR−U, 2-step RACH is formally standardized in R16. The random access process is changed from the ordinary four-step process with msg1 to msg4 to the optimized two-step process of msgA and msgB.

1.5.6 Uplink band switching transmission

From the perspective of actual 5G spectrum allocation and availability of spectrum for the operators, the TDD midband (such as 3.5 GHz/ 4.9 GHz) is the most mainstream 5G commercial spectrum in the world.

The TDD frequency band is relatively high, with large bandwidth, but insufficient coverage. The FDD band is low and the coverage is good, but the bandwidth is small. UE can support two-antenna transmission capability in the TDD frequency band, but only single antenna transmission capability in the FDD frequency band. R16 introduces the uplink switching transmission technology; that is, the terminal works on TDD carrier and FDD carrier in time division multiplexing mode in the uplink, which facilitates the advantages of large uplink bandwidth of TDD frequency band and good uplink coverage of FDD frequency band to improve uplink performance. Uplink band switching transmission can improve the uplink throughput and uplink coverage performance, while also achieving lower air interface delay.

The premise of the uplink intercarrier switching transmission scheme is that UE has only two radio-frequency transmission channels, in which the FDD transmission channel shares a set of RF transmitters with one of the TDD dual transmission channels. Since it is scheduled by the base station, the terminal can dynamically switch between the following two working modes:

1. FDD carrier 1 channel transmission + TDD carrier 1 channel transmission (1T + 1T).
2. FDD carrier 0 channel transmission + TDD carrier 2 channels transmission (0T + 2T).

The uplink inteR−carrier switching transmission mechanism can work in three modes: EN−DC, uplink CA, and SUL. When the UE is close to the cell center, the base station can schedule the user to work in the TDD carrier (0T + 2T), so that the user can obtain the high uplink MIMO rate in the TDD carrier; the base station can also schedule the cell-center user in the FDD + TDD CA state (1T + 1T), so that the user can obtain the high rate in the uplink CA mode. When the UE is at the edge of the cell, the base station can schedule users to work on the FDD low-frequency carrier to improve the coverage performance of edge users.

1.5.7 Mobility enhancement

The R16 protocol version mainly introduces the following two new functions for mobility enhancements.

1.5.7.1 Dual active protocol stack enhancement

For the handover process of the capability terminal supporting DAPS, the terminal does not first disconnect the air interface link with the source cell, but after successfully accessing the target cell, the terminal will release the connection to the source cell according to the explicit signaling from the network side and stop sending and receiving data with the source cell. It can be seen that the terminal can maintain connection with the source cell and the target cell and perform data transmission at the same time during the handover process, thereby meeting the 0 Ms service interruption delay in the handover process.

1.5.7.2 Conditional handover

The core idea of conditional handover is that the network side preconfigures the candidate target cell and handover command information to the UE in advance. When certain conditions are met, the UE can independently execute the configuration in the handover command and directly initiate handover access to the target cell that meets the conditions. Because the UE no longer triggers the measurement reporting when the handover conditions are met, and the UE has obtained the configuration in the handover command in advance, the potential issues that measurement reporting or the handover command are not received correctly in the handover process can be avoided, thereby improving the handover success rate.

1.5.8 Multi-RAT dual connectivity enhancement

In the project of R16 MR−DC enhancement, in order to improve the service performance in MR−DC mode, the function of quickly establishing SCell/SCG is supported. That is, the UE is allowed to perform measurements in idle state or inactive state, and immediately report the measurement results to the network side after entering the RRC connected state, so that the network side can quickly configure and establish SCell/SCG.

R16 introduces the SCell dormancy feature. When SCell/SCG is activated but there is no data transmission, a dedicated dormant BWP is configured through RRC; that is, the UE does not monitor PDCCH on the BWP and only performs CSI measurement and reporting, which is helpful for the UE to save power. When there is data transmission, it can quickly switch to the active state through dynamic indication to quickly restore the service.

In order to reduce the service interruption caused by radio link failure, fast MCG recovery function is introduced in R16. When a radio link failure occurs in the MCG, an indication is sent to the network through the SCG link, which triggers the network to quickly restore the MCG link.

In terms of network architecture, R16 has also specified enhancements to support asynchronous NR−DC and asynchronous CA, providing flexible options for the deployment of 5G networks.

1.5.9 New radio−vehicle to everything

3GPP started the standardization of device-to-device (D2D) communication technology in R12, mainly for public safety scenarios. D2D technology is used to support terminal-to-terminal direct communication based on the sidelink. Compared with the traditional cellular communication system, the data communicated by the terminal on the sidelink does not need to be forwarded through the network equipment, so it has higher spectrum efficiency and lower transmission delay. In R14, D2D technology is applied to the LTE-based V2X, namely LTE−V2X. LTE−V2X can realize assisted driving, which provides drivers with information on other vehicles or warning information, and assists drivers to judge road conditions and vehicle safety. The communication performance requirement of LTE−V2X is not high; for example, the required communication delay is 100 Ms. As the demand for autonomous driving increases, LTE−V2X cannot meet the high communication performance requirements of autonomous driving. R16 officially launched the project of NR technology-based NR−V2X. The communication delay of NR−V2X needs to reach 3−5 Ms, and the reliability of data transmission should reach 99.999% to meet the needs of autonomous driving.

R16 NR−V2X defines the frame structure, physical channels, physical layer procedures, and whole protocol stack of the sidelink. NR−V2X supports sidelink resource allocation mechanisms, including network allocating sidelink transmission resources and the UE autonomously selecting the sidelink transmission resources. NR−V2X supports three communication modes: unicast, groupcast, and broadcast. The following techniques have been enhanced and optimized in NR−V2X, such as sensing, scheduling, HARQ-based retransmission and CSI reporting, etc. The NR−V2X provides good standard support for the flexible and reliable deployment of vehicle networking and services.

1.5.10 New radio-unlicensed

The NR technology of R15 protocol is applied to licensed spectrum, which can realize the characteristics of seamless coverage, high spectrum efficiency, and high reliability of cellular network. The unlicensed spectrum is a kind of shared spectrum, and multiple different communication systems can share the resources on the unlicensed spectrum for wireless communication when meeting certain requirements. The NR technology of the R16 protocol can also be applied to the unlicensed spectrum, which is called NR-based access to the unlicensed spectrum (NR−U).

NR−U technology supports two networking modes: licensed spectrum-assisted access and unlicensed spectrum independent access. In the former mode, users need to use the licensed spectrum to access the network, and the carrier on the unlicensed spectrum can be used as a secondary carrier, which provides transmission for big data services for the user as a supplementary band of the licensed spectrum. While the latter mode can be independently networked through the unlicensed spectrum, so that users can access the network directly through unlicensed spectrum.

In addition to the ten new features enhanced by the R16 version listed above, R16 also has completed projects such as integrated access and backhaul, NR positioning, UE wireless capability reporting optimization, network slicing enhancement, remote interference management, cross-link interference measurement, self-organization network, and so on. At the same time, R16 has also carried out research projects such as nonterrestrial network, nonorthogonal multiple access, and so on. In short, the new functions and features of the R16 protocol version standardization provide operators and industry customers with powerful functional options and technology support for NF deployment, network performance improvement, network upgrade and evolution, and expansion of new business in 5G mobile communication networks.

1.6 Summary

As the overview of the book, this chapter focused on the main enhancement and evolution of 5G NR technology and standards compared to LTE. At the same time, it also summarized and analyzed the choices of new technologies in the standardization process of NR. Then the maturity of 5G key devices and equipment was introduced, which is an important factor in promoting the process of 5G standardization. Finally, this chapter summarized the main functional features of the R16 version that has been just standardized by 3GPP, so that readers can have a basic understanding of the technical content of R16 in the subsequent chapters.

References

[1] Liu, X, Sun, S, Du, Z, Shen, Z, Xu, X, Song, X. 5G wireless system design and international standard. Posts & Telecom Press; 2019.

[2] 3GPP TS 38.300 V15.6.0. NR: NR and NG-RAN overall description: stage 2 (release 15); 2019—06.

[3] 3GPP TS 21.915 V15.0.0. NR: summary of Rel-15 work items (release 15); 2019—06.

[4] 3GPP TS 38.211 V15.6.0. NR: physical channels and modulation (release 15); 2019—06.

[5] 3GPP TS 38.212 V15.6.0. NR: multiplexing and channel coding (release 15); 2019—06.

[6] 3GPP TS 38.213 V15.6.0. NR: physical layer procedures for control (release 15); 2019—06.

[7] 3GPP TS 38.214 V15.6.0. NR: physical layer procedures for data (release 15); 2019—06.

[8] 3GPP TS 38.101—1 V15.8.2. NR: UE radio transmission and reception; part 1: range 1 standalone (release 15); 2019—12.

[9] 3GPP TS 38.321 V16.0.0. NR: Medium Access Control (MAC) protocol specification (release 16); 2020—03.

[10] 3GPP TS 38.323 V16.0.0. NR: Packet Data Convergence Protocol (PDCP) specification (release 16); 2020—03.

[11] 3GPP TS 38.331 V16.0.0. NR: Radio Resource Control (RRC) protocol specification (release 16); 2020—03.

[12] 3GPP TS 38.300 V16.1.0. NR: NR and NG-RAN overall description; stage2 (release 16); 2020—03.

[13] Shen, J, Suo, S, Quan, H, Zhao, X, Hu, H, Jiang, Y. 3GPP long term evolution: principle and system design. Posts & Telecom Press; 2008.

[14] 3GPP TS 22.261 V15.6.0. NR: Service requirements for the 5G system; stage1 (release 15); 2019—06.

[15] 3GPP TS 38.913 V15.0.0. Study on scenarios and requirements for next generation access technologies (release 15); 2018—06.

[16] R1—165666. Way forward on categorization of IFFT-based waveform candidates, Orange, 3GPP RAN1 #85, Nanjing, China, May 23—27, 2016.

[17] R1—162200. Waveform evaluation proposals. Qualcomm Incorporated, 3GPP RAN1#84bis, Busan, Korea, 11th—15th April 2016.

[18] R1—162225. Discussion on new waveform for new radio. ZTE, 3GPP RAN1#84bis, Busan, Korea, 11th—15th April 2016.

[19] R1—162199. Waveform candidates. Qualcomm Incorporated, 3GPP RAN1#84bis, Busan, Korea, 11th—15th April 2016.

[20] R1—162152. OFDM based flexible waveform for 5G. Huawei: HiSilicon, 3GPP RAN1#84bis, Busan, Korea, 11th—15th April 2016.

[21] R1—162516. Flexible CP-OFDM with variable ZP. LG Electronics, 3GPP RAN1#84bis, Busan, Korea, 11th—15th April 2016.

[22] R1—162750, Link-level performance evaluation on waveforms for new RAT. Spreadtrum Communications, 3GPP RAN1#84bis, Busan, Korea, 11th—15th April 2016.

[23] R1—162890, 5G waveforms for the multi-service air interface below 6GHz. Nokia, Alcatel-Lucent Shanghai Bell, 3GPP RAN1#84bis, Busan, Korea, 11th—15th April 2016.

[24] R1—162925, Design considerations on waveform in UL for new radio systems. InterDigital Communications, 3GPP RAN1#84bis, Busan, Korea, 11th—15th April 2016.

[25] R1−164176, Discussion on waveform for high frequency bands. Intel Corporation, 3GPP RAN1 #85, Nanjing, China, 23−27 May 2016.

[26] R1−162930, OTFS modulation waveform and reference signals for New RAT. Cohere Technologies, AT&T, CMCC, Deutsche Telekom, Telefonica, Telstra, 3GPP RAN1#84bis, Busan, Korea, 11th−15th April 2016.

CHAPTER 2

Requirements and scenarios of 5G system

Wenqiang Tian and Kevin Lin

The evolution of mobile communication has greatly promoted developments of society in recent decades. From analog communication to digital communication, from 3G to 4G, people's dependence on mobile communication is increasing day by day. Around 2010, given the foreseeable explosive data growth and personalized communication devices used by many vertical industries, various government agencies, international standardization organizations, corporate companies, colleges, and universities have embarked on analyzing different use cases and requirements and conducting advanced research into future generations of wireless communication systems [1−8], and the standardization of 5G mobile communication systems starts accordingly.

2.1 Current needs and requirements in the 5G era

Each new generation of mobile communication system is closely related to its previous generation. The era of "cell phone" fulfills people's desire for wireless communication to establish instant connection with each other. The second-generation mobile communication system represented by global system for mobile communications (GSM) allowed people worldwide to enjoy the convenience offered by wireless services. Rapid developments in 3G and 4G fulfill the needs of wireless multimedia services and pave the way for mobile Internet. But what are the new requirements and driving forces in the 5G era? Firstly, there is no doubt that a broadband communication system should be brought into the scope of 5G, so that people can enjoy better communication experiences. In addition, will there be any new features introduced in 5G? Requirements from various vertical industries are expected to open doors for wireless communication systems in the future that will be more in-depth and extensive.

2.1.1 Requirements of high data rate

High data transfer rate is a core feature for a communication system. We usually compare a wireless communication system to an information highway.

5G NR and Enhancements
DOI: https://doi.org/10.1016/B978-0-323-91060-6.00002-7

Data rate is the capacity of this information highway. In recent years, with the emergence of new requirements, on the one hand, a surge in applications have made the "road" of 4G congested. On the other hand, the limited "road capacity" designed 10 years ago has become a bottleneck of many emerging applications. Researchers in the communications industry are trying to optimize 4G systems by means of using better resource allocation schemes and utilizing more advanced scheduling methods to avoid "traffic jams." However, this kind of optimization is not a fundamental solution to the problem. The demand for a new generation of wireless communication system is gradually appearing on the agenda [3−5].

2.1.1.1 Enhanced multimedia service

Multimedia service generally refers to an integration of various media information, including text, sound, image, video, and other forms. In fact, the evolution of a communication system from one generation to another is to constantly meet people's requirements for different kinds of multimedia services. When a mobile communication system was able to provide voice interaction, then new requirements appeared for images. When the image transmission was realized, new demands came for videos. When the basic video transmission needs were supported, novel multimedia needs started emerging (e.g., 4K video, 8K video). People's ultimate pursuit of sensory experience is always a step ahead, and new demand for "clearer and richer" multimedia services will never stop.

2.1.1.2 Immersive interactive multimedia services

If we regard the enhanced multimedia service as an accumulation of "quantity" or buildup of data rate, then the immersive interactive multimedia services covering augmented reality (AR), virtual reality (VR), and holographic communication would be considered as a "qualitative change." In immersive multimedia services (e.g., AR and VR games), multichannel ultra high definition (UHD) video data streams collected from multiple angles should be timely provided to terminals in order to ensure a good user experience. The required system capacity and communication resources for the above processes are far exceeding the current communication systems.

2.1.1.3 Hotspot services

While providing enhanced data rate services for a single user, it is necessary to consider multiuser requirements in hotspots as well. In scenarios with a large number of users all requiring services like video sharing, real-time, high-definition live broadcast, and AR/VR gaming at the same time such high-density and high-throughput communication services are difficult for existing communication systems to support. Therefore how to ensure a large number of users are able to obtain their desired communication experience at the same time in hotspots such as shopping malls is a problem that needs to be analyzed and solved.

2.1.2 Requirements from vertical industries

How to build a good communication system to satisfy the needs of users is the main target of a wireless communication technology. Additionally, in recent years people have gradually started thinking about making wireless communication technologies more widely used in their daily lives [3—5].

2.1.2.1 Low-latency communication

First, real-time services should be considered in the design of 5G communication systems. For example, a new system must not only "run fast," but to also "start fast." On the one hand, people expect to experience the ultimate feeling of "instant connection," and hence the communication latency needs to be shortened as much as possible. For example, for cloud offices, cloud games, VR, and AR applications, instant transmissions and responses are the key factors to ensure the success of these services. If every mouse click from a user over a cloud office was accompanied by a long delay, the user experience would be unacceptable. On the other hand, for many industry applications such as intelligent transportation, a large communication latency could even cause some serious safety risks.

2.1.2.2 Reliable communication

In addition to the above requirement of a "short latency," the "high reliability" requirement is important as well. For example, in an unmanned driving situation, vehicles need to accurately obtain dynamic driving status and warning information from surrounding vehicles, road side units, and pedestrians in order to adjust and make necessary maneuvers for safe driving on the road. If this kind of information cannot be obtained reliably, delivered on time, or provided accurately serious consequences could result.

2.1.2.3 Internet of Things communication

In order for every potential object to enjoy the convenience of "wireless connection," the scale of Internet of Things (IoT) connections will grow rapidly in the future and far beyond the scale of current communications between human users. These connected "objects" can range from simple devices, such as sensors and switches, to high-performance equipment, such as smart phones, vehicles, industrial machine tools, etc., and more complex systems, such as smart cities, environmental monitoring systems, forest fire prevention systems, etc. Under different scenarios and applications, these devices will require different levels of IoT communication systems, which are then used to fulfill requirements from various industries.

2.1.2.4 High-speed communication

Mobility is one of the main operating characteristics in wireless communication systems. In 4G, movement from users and vehicles is considered in the design of long term evolution (LTE) systems. In 5G, high-speed mobile scenarios include high-speed trains. At present, daily flows of high-speed trains during peak hours can carry more than 10 million people. How to provide better user experience for stationary, moving, and high-speed users has become a new problem. As such, it is necessary to take this into consideration and solve this problem for in next-generation mobile communication systems.

2.1.2.5 High-precision positioning communication

In addition to the above scenarios, demands for high-precision positioning are emerging as well. Positioning capability is important for many applications, such as location-based services, unmanned aerial vehicle (UAV) control, automatic factories, public safety, vehicle driving, and so on, all of which require accurate positioning information to support their corresponding use cases. Previous generations of communication systems cannot fully support this positioning feature, especially since it is difficult to meet some high-precision positioning requirements.

2.2 Typical scenarios

Based on different types of 5G operating scenarios, applications, and service requirements, the international telecommunication union (ITU) divides the typical application scenarios of 5G into three categories by extracting their common features. They are described as enhanced mobile broadband (eMBB),

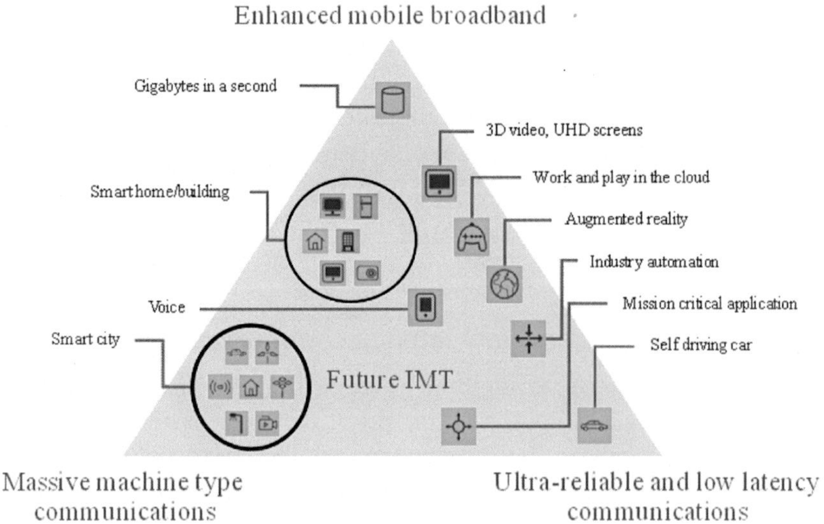

Figure 2.1 Usage scenarios of IMT for 2020 and beyond [3].

ultrareliable and low latency communications (URLLC), and massive machine type communications (mMTC). The basic services and scenarios are shown in Fig. 2.1 [3].

2.2.1 Enhanced mobile broadband

The eMBB feature is used to satisfy user needs for multimedia services. As human society develops, these needs are increasing and evolving. Accordingly, it is necessary to further build an eMBB communication system to fulfill the latest requirements of various applications and to support personalized services for different users. Specifically, current eMBB services can be divided into two categories: wide-area coverage and hotspot communication.

For hotspot communication, typical requirements are high data throughput and user capacity in crowded areas. However, expectation of user mobility in this scenario is lower than that in wide-area coverage scenarios. For example, at sporting events, people gather in the same location for a long time, thus forming a local hot spot area. When people want to share game information and video with others, they need communication services with high data rate in the hot spot area.

For mobile communication in wide-area coverage, seamless service coverage and the support for medium- to high-speed mobility should be considered as users do not expect their services to be discontinuous. When compared with hotspot communication, the data rate requirement in a wide area coverage scenario is slightly more relaxed.

The goal for eMBB service is to achieve better performance than the previous generation of mobile communication system, and further subdivide the wide coverage and hotspot high data rate requirements, in order to create a basic framework and define capabilities of next-generation wireless communication systems.

2.2.2 Ultrareliable and low latency communications

The URLLC feature is an important part of the three typical scenarios of the 5G system. It is mainly used for special deployment and applications, and these applications have stringent requirements for throughput, latency, and reliability. Typical examples include wireless control in industry, autonomous vehicle driving, transportation security, etc. It can be seen that any packet error and delays in these applications will have serious consequences. Therefore when compared with normal broadband services, URLLC is associated with extreme requirements in terms of latency and reliability, and at a high cost. Considering these factors, it is necessary to distinguish URLLC from the ordinary eMBB in the 5G system.

2.2.3 Massive machine type communications

mMTC is another key operating scenario in the 5G era. Its landmark application is large-scale IoT services (e.g., intelligent water network, environmental monitoring system, etc.). The primary feature of this kind of communication system is the number of terminals is extremely large within a network. In addition to the large-scale connected devices, they often only need to transmit a small amount of data, but they are delay sensitive at the same time. The low cost and lower power consumption requirements should also be taken into account, so the needs for the actual deployment of mMTCs can be fulfilled.

eMBB, URLLC, and mMTC are the three expected typical operational scenarios for the 5G era. Fig. 2.2 illustrates the differences and connections of the three scenarios from the perspective of corresponding feature indicators. It can be seen that eMBB service is the basic communication scenario in 5G. It has high requirements on various 5G key performance, but it does not require large-scale connections, extremely low delay, and high reliable communication. URLLC and mMTC services

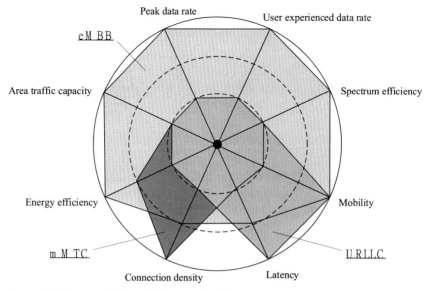

Figure 2.2 Key capabilities of 5G scenarios [3].

correspond to two kinds of special communication requirement scenarios with different characteristics to supplement the deficiencies of eMBB.

2.3 Key indicators of 5G systems

ITU formulated eight basic indicators of the 5G communication system [3,7,8]: peak rate, user experience rate, delay, mobility, connection density, traffic density, spectrum efficiency, and energy efficiency, as shown in Table 2.1.

Peak data rate is the capability with the highest priority for a communication system. The peak data rate of a 5G communication system defined by ITU represents the highest data rate that the terminal is expected to achieve under ideal conditions. The basic assumption is that all the configurable wireless resources are allocated to a user for both downlink reception and uplink transmission, but transmission errors, retransmissions, and other factors are not taken into account. For 5G communication systems, the peak data rate will be greatly improved compared with the previous communication systems. 20 Gbps downlink and 10 Gbps uplink are the peak data rate indicators the design of a 5G system needs to achieve. However, this does not mean that every user is able to enjoy such a high data rate. When considering a multiuser scenario with

Table 2.1 Key indicators of 5G system.

Indicator	Description
Peak data rate	Maximum achievable data rate under ideal conditions per user/device (in Gbit/s)
User experienced data rate	Achievable data rate that is available data rate across the coverage area to a mobile user/device (in Mbit/s or Gbit/s)
Latency	The contribution by the radio network to the time from when the source sends a packet to when the destination receives it (in Ms)
Mobility	Maximum speed at which a defined quality of service (QoS) and seamless transfer between radio nodes that may belong to different layers and/or radio access technologies (RAT) can be achieved (in km/h)
Connection density	Total number of connected and/or accessible devices per unit area (per km^2)
Area traffic capacity	Total traffic throughput served per geographic area (in $Mbit/s/m^2$)
Spectrum efficiency	Average data throughput per unit of spectrum resource and per cell (bit/s/Hz)
Energy efficiency	Energy efficiency has two aspects:on the network side, energy efficiency refers to the quantity of information bits transmitted to/ received from users, per unit of energy consumption of the radio access network (in bit/J);on the device side, energy efficiency refers to quantity of information bits per unit of energy consumption of the communication module (in bit/J).

limited available wireless resources, the peak data rate of 5G system is not the perceived data rate that every user is able to reach.

The user-experienced data rate is the 5% tile point on the cumulative distribution function (CDF) of the user throughput. User throughput (during active time) is defined as the number of correctly received bits over a certain period of time. The user-experienced data rate focuses on the available data rate of multiple users, and considers the impact of limited coverage and mobility. The requirement for user experience data rate in dense urban are 100 Mbit/s for downlink and 50 Mbit/s for uplink.

User plane latency is the work of the radio network from the time when the source sends a user packet to the time when the target destination correctly receives it (in Ms). In 5G systems, the importance of low latency is more prominent due to the requirements from URLLC. In the

4G era, we consider the minimum delay requirement to be 10 Ms. In 5G, when considering a new latency requirement, ITU further compresses the end-to-end delay of the 5G system to the level of 1 Ms only, which brings great challenges to the design of a 5G communication system. Faced with this problem, in the following chapters of this book, with joint efforts of communication researchers all over the world, many technical features are introduced into 5G in order to achieve the low-latency target as much as possible.

Mobility is another key functional and operational index of a communication system. Presently, the most challenging mobility requirement comes from high-speed scenarios such as high-speed trains. For a 5G communication system, the speed of 500 km/h is regarded as the key operational index for mobility. That is to say, when users travel between mountains and rivers on a high-speed train, a 5G communication system needs to ensure high-quality communication for users in the carriage. How to achieve this and how to ensure this quality of service for users in high-speed mobile scenarios should be considered in 5G.

Connection density is the total number of devices fulfilling a specific quality of service per unit area. This requirement is defined for the purpose of evaluation in the mMTC usage scenario. Furthermore, the above-mentioned specific service requirements can be simplified with a successful transmission of a 32Byte packet within 10s. It can be seen that the index has relatively low requirements for both speed and delay, which mainly considers the actual data patterns in IoT scenarios. For the definition of connection size, the current expectation from ITU is "million level connections." However, how to define the unit area in the definition also needs to be clear. To solve this problem, ITU gives the corresponding description, which needs to support the abovementioned million connections within the scope of 1 km^2.

Area traffic capacity refers to the total throughput that can be supported in a unit area, which can be used to evaluate the regional service support capability of a communication network. Specifically, the traffic density is evaluated by the network throughput per square meter, which is directly related to the characteristics of the network such as deployment of base stations, bandwidth configuration, spectrum efficiency, and so on. The area traffic capacity proposed by ITU is the order of magnitude to support a 10 Mbit/s per square meter area. Then the question is, do we need such a relatively strong traffic density index for each application scenario? The answer is No. For example, it is unnecessary to achieve such high traffic density index in an IoT scenario. Generally speaking, only when we consider service requirements of some hotspot scenarios do we need to focus on this area traffic capacity and use corresponding technical solutions to satisfy the requirement of high throughput in hotspot areas.

Spectrum efficiency is the expected data throughput per unit spectrum resource. From the evolution of time division, frequency division, and code division to the continuous iteration of MIMO technology, different organizations are trying to improve the throughput level through various approaches within a limited bandwidth. For a 5G communication system, how to make a better use of its spectrum resources and achieve greater improvements of spectrum efficiency compared with the previous communication system is a new challenge. The design and implementation of the scheme characterized by large-scale MIMO will be an effective method to improve the spectrum efficiency in this new stage. In addition, effective utilization of unlicensed spectrum will be another way to make better use of the spectrum resources. The research on non-orthogonal multiple-access (NOMA) is another attempt to improve the spectrum efficiency as well. This will also be introduced in the following chapters.

Green communication would be one of the directions for future communications development. It is of long-term significance for the healthy development of the communications industry and the main rationale behind the use of social resources to effectively improve communication systems and consider the impact of energy consumption. For example, from the network side, the power consumption of the serving base stations and other equipment has become one of the important costs of network operation. From the terminal side, in the battery-based power supply mode, high energy consumption will lead to the shortening of battery life and other problems, which will inevitably lead to poor user experience. In addition, the proportion of energy consumed by the communication system in the entire society is constantly increasing, which is also a problem that cannot be ignored. Thus the design and development of a 5G network should pay close attention to the energy consumption from the beginning, and take energy efficiency as one of its basic requirements. In the following chapters, we can see that many technical solutions for reducing the energy consumption and improving the efficiency have been introduced into 5G, in order to build a green communication system.

As shown in Fig. 2.3, the next-generation mobile communication network developed for the above indicators will be significantly improved in

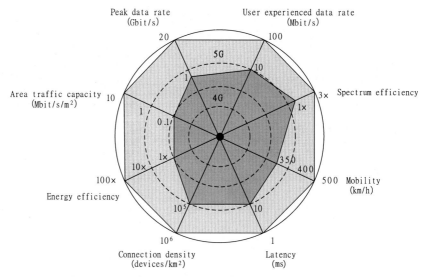

Figure 2.3 Enhancement of key capabilities of 5G [3].

all aspects compared with existing communication networks. The 5G network will be a flexible network to support different user needs, and also a green network with efficient support for connections between "people to people, people to things, and things to things."

2.4 Summary

This chapter analyzed the 5G system from the perspective of use cases and scenarios to show the requirements and challenges of a 5G system. The requirement for high data rate and new applications brought by vertical industry are the core driving forces for the construction of a 5G system. Typical 5G scenarios are represented by eMBB, URLLC, and mMTC. The clear definition of peak data rate, user perceived data rate, latency, mobility, connection density, traffic capacity, spectrum efficiency, energy efficiency, and other indicators discussed here provided basic direction and target guidance for 5G system design.

References

[1] IMT-2020(5G) Promotion Group. White paper on 5G vision and requirements; 2014.
[2] IMT-2020(5G) Promotion Group. White paper on 5G concept; 2015.

[3] Recommendation ITU-R M.2083−0. IMT vision-framework and overall objectives of the future development of IMT for 2020 and beyond; 2009.

[4] Report ITU-R M.2320−0. Future technology trends of terrestrial IMT systems; 2011.

[5] Report ITU-R M.2370−0. IMT traffic estimates for the years 2020 to 2030; 2007.

[6] Report ITU-R M.2376−0. Technical feasibility of IMT in bands above 6 GHz; 2007.

[7] 3GPP TR 37.910 V16.1.0. Study on self evaluation towards IMT-2020 submission (release 16); 2009.

[8] 3GPP TR 38.913 V15.0.0. Study on scenarios and requirements for next generation access technologies (release 15); 2006.

CHAPTER 3

5G system architecture

Jianhua Liu, Ning Yang and Tricci So

3.1 5G system architecture

3.1.1 5G system architecture requirements

In the 4G LTE system, one of the major achievements was to support native IP connectivity between the UE and the network, which enabled Mobile Broadband (MBB) services. While the 4G LTE system is capable of providing good voice services and legacy MBB services, it is no longer sufficient to support the emerging Enhanced MBB (eMBB) services, Ultra Reliable Low Latency Communications (URLLC), and Massive Machine Type Communications (mMTC) services, which are the main network service drivers for 5G systems. eMBB services are data-driven services and require higher capacity, consistent connectivity, and support for higher mobility. URLLC applications impose strict requirements on ultralow latency and high reliability to support mission critical communications (e.g., remote surgery, autonomous vehicles, or the tactile internet). On the other hand, mMTC requires support for a very large number of devices in a small area, which may only send data sporadically, such as the Internet of Things (IoT) use case.

Motivated by the above emerging service requirements, the 3rd Generation Partnership Project (3GPP) recognized that the new generation of MBB communication system must be evolved to address the wide range of conflicting 5G system requirements, and the resulting diverse service profiles of system resources and performance concurrently and efficiently. Modular and programmable network functions (NF) with flexible reconfigurability are essential to enable the new 5G MBB system to support new service variants for different deployment scenarios. 5G network solutions must achieve cost efficiency without sacrificing system performance, so that there is no need to overengineer the 5G system.

One strategy to implement the 5G system to address the above objectives was to leverage technological improvements enabled by new hardware (e.g., COT-based server pool) and new software platforms and infrastructure techniques such as NF virtualization (NFV) and software defined networking (SDN). This strategy facilitated the development of a

5G NR and Enhancements
DOI: https://doi.org/10.1016/B978-0-323-91060-6.00003-9

programmable MBB virtual public or private network that can be tailored for specific business needs.

In addition to the above the infrastructure techniques, another important consideration was to enable the 5G system functional components to be adaptive to fast-changing and industry-specific microservices and performance requirements in order to relieve the tight coupling and dependencies between NFs. As a result, the 5G-Core network (CN) architecture adopted service-based architecture (SBA) for its control plane (CP) between NFs. The SBA decouples interNFs and enables NF self-improvement and self-maintenance. Therefore it improves service agility supported by NFs.

Combining all the above considerations, the following are the key design principles and concepts for 5G system architecture, which are also captured in 3GPP TS 23.501 [1]:

- Separate the User Plane (UP) functions from the CP functions, allowing independent scalability, evolution, and flexible deployments [e.g., centralized location or distributed (remote) location].
- Modularize the function design (e.g., to enable flexible and efficient network slicing).
- Wherever applicable, define procedures (i.e., the set of interactions between NFs) as services, so that their reuse is possible.
- Enable each NF and its NF Services (NFS) to interact with other NF and its NFS directly or indirectly via a Service Communication Proxy if required. The architecture does not preclude the use of another intermediate function to help route CP messages (e.g., similar to Diameter Routing Agent concept in 3G and 4G systems).
- Minimize dependencies between the Access Network (AN) and the CN. The architecture is defined with a converged CN with a common AN−CN interface, which integrates different access types such as 3GPP access and non-3GPP access.
- Support IP packet, non-IP packet, and Ethernet packet transmission.
- Support a unified authentication framework.
- Support "stateless" NFs, where the "compute" resource is decoupled from the "storage" resource.
- Support capability exposure for opening the network capabilities.
- Support system architecture enhancement for vertical industries.
- Support concurrent access to local and centralized services. To support low-latency services and access to local data networks (DN), the UP Function (UPF) can be deployed close to the AN.

- Support roaming with both Home routed traffic as well as Local breakout traffic in the visited PLMN.
- Support different RAT types including 3GPP access type (e.g., E-UTRA) and non-3GPP access type (e.g. Wi-Fi, fixed network access, satellite access, etc.); Besides, GERAN and UTRAN access is not supported;
- Support decoupling AN and CN to allow evolution.
- Support simultaneous connection through multiple radio access technologies at the same time.

By carrying out the design principles and concepts as described above, the 3GPP 5G system supports a modular, scalable, and optimal "network function services" design practice to align with the network service composition. The separation of CP and UPs as well as supporting Service-Based Interface (SBI) for the CP enable operators to scale their network solutions and deployment to enhance user experience and system performance.

The UP entities can be deployed in either a centralized or distributed way based on operator requirements. With distributed deployment, the UP can be localized and situated much closer to the user to significantly reduce traffic latency and signaling processing. With centralized deployment, the cost for maintaining the system could be reduced and the reliability could be improved.

In SBA, the 5G-core NF has both functional behavior and interface. An NF can be implemented either as a network element on a dedicated hardware, as a software instance running on a dedicated hardware, or as a virtual function instantiated on an appropriate platform (e.g., a cloud infrastructure). At the same time, the system architecture becomes more open and scalable to support expansion of network capabilities and is much easier to access by various services via different Application Programming Interfaces (APIs). Each NF is implemented independently and is accessible by other NFs via a dedicated NFS security requirement, thus realizing the change from a traditional system (e.g., fixed function in network elements, fixed connection between network elements, fixed signaling interaction, etc.) to a service-based flexible system. The SBA solves the problem of tight coupling of point-to-point architecture, realizes the flexible network evolution, and satisfies a variety of service requirements.

The new form of network virtualization approach referred to as network slicing has also been adopted by the 5G system architecture.

Leveraging the virtualization of the NF (i.e., the NFV and SDN architectural design concept), the network resources are organized and partitioned into different logically isolated networks (i.e., referred to as slices) that are tailored for the network services with different system characteristics and requirements over the same physical infrastructure. The network slicing enables new business models and use cases across all verticals, and creates new revenue opportunities for MBB service providers. It provides service flexibility and agility to deliver services faster with high security, isolation, and applicable characteristics to meet the contracted Service-Level Agreement (SLA). Network slicing enables operators to maximize return on investment (ROI) via efficient usage and management of the network resources (i.e., lower the CAPEX) and provide differentiated services at scale.

3.1.2 5G system architecture and functional entities

3GPP 5G system architecture representation maintains the legacy reference point approach and also adopts the SBA approach, and both representations are captured in 3GPP TS 23.501 [1]:

- A service-based representation. The NF of the CP allows other authorized NFs to access their services. This representation also includes point-to-point reference points where necessary.
- A reference point representation. The two NFs are described by point-to-point reference points, and the two NFs interact through the reference points.

Fig. 3.1 shows a nonroaming reference architecture, and SBIs are used within the Control Plan [2].

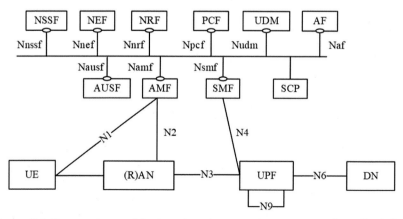

Figure 3.1 Nonroaming architecture based on service representation. *Cited from Figure 4.2.3—1 in TS 22.261, Service requirements for the 5G system: Stage 1.*

Further details on each SBI description can be found in 3GPP TS 23.501 [1].

The 5G SBA offers the following key architecture benefits.

3.1.2.1 Loose coupling and service-oriented network element functions

The traditional CN element consists of a group of functions closely coupled with each other, which limits adaptability for new service requirements. In SBA, the functionality of an NF is self-contained and reusable to provide NFS. Hence, each NFS can be maintained and evolved independently.

3.1.2.2 Open and secure network interface

The interface between different NF entities is implemented based on mature standardized protocols, such as the HTTP-based protocol, which facilitates the rapid development of programmable network interfaces for each NF. At the same time, the HTTP-based open and secure network interface enables secure exposure of the network capabilities to service applications.

3.1.2.3 Unified network function management

The NF Repository Function entity (NRF) is introduced into the 5G CN to support the unified NF registration management and service discovery mechanism for NFS.

3.1.2.4 Enable continuous integration/continuous deployment time to market microservices

As the NFs in 5G Core are programmable and self-contained, the 5G system can support Continuous Integration (CI) and Continuous Deployment to reduce time to market for microservices (Fig. 3.2).

5G system architecture contains the following NFs. For further details on the functionalities of each reference point and the roaming support architecture, 3GPP TS 23.501 [1] can be referenced:

- *Application Function (AF)*: A generic term is used to refer to applications such as IMS or third-party services (e.g., web service, video, games, etc.) that influence traffic routing and interact with NFs (e.g. policy framework for policy control) directly, if the NF is resided in a trusted domain. Otherwise, if the AF does not reside in a trusted domain, the

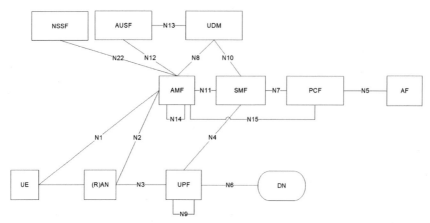

Figure 3.2 Nonroaming architecture based on reference point representation. *Cited from Figure 4.2.3−2 in TS 22.261, Service requirements for the 5G system: Stage 1.*

AF can only interact with other NFs via the Network Exposure Function (NEF).

- *Access and Mobility Management Function (AMF)*: The AMF terminates NAS signaling, NAS ciphering, and integrity protection for a UE. It handles all UE tasks related to connection management, mobility management (MM), registration management, access authentication and authorization, and security context management.
- *Authentication Server Function (AUSF)*: The AUSF is resided in the home network and is used to interact with AMFs to perform authentication with the UE and to request the keying materials from Unified Data Management Function (UDM) and forward them to AMF to support authentication.
- *Binding Support Function (BSF)*: The BSF is used for binding an AF request to a specific Policy Control Function (PCF) instance.
- *Data Network (DN)*: DN is generic term to refer to remote server that handles operator services, third-party services, video and game services, etc.
- *Non-3GPP Interworking Function (N3IWF)*: The N3IWF terminates the IKEv2/IPsec protocols with the UE over NWu to support authentication and authorization for the user of the untrusted non−3GPP access to interwork with the 5G CN.
- *Network Exposure Function (NEF)*: NEF is responsible for securely exposing the events and capabilities of 5G NFS in the form of an API. The

NEF enables external nontrusted applications (i.e., AFs) that need access to the internal data of 5G core NFS. The NEF also interacts with the 5G Core policy framework to support policy control within the 5G system.

- *Network Slice Selection Function (NSSF)*: The NSSF is responsible for selecting the Network Slice Instance(s) [NSI(s)] that serve the UE, determining the Allowed NSSAI and Configured NSSAI mapped to the Subscribed S-NSSAI(s). It also determines the corresponding serving AMF set that serves the Allowed NSSAI by querying the NRF.
- *NF Repository Function (NRF):* The NRF is a repository to collect and maintain the information of the NF profile of available NF instances and their supported services. The NRF is used to support NF registration, management, status update and detection, and discovery of all other NFS. When each NF is instantiated, it registers with the NRF to indicate that it is available to provide services. The registration information includes the type, address, service list of NFs, etc.
- *Network Data Analytics Function (NWDAF)*: The NWDAF represents the operator-managed network analytics logical function. The NWDAF supports data collection from NFs and AFs as well as from OAM. It provides service registration and metadata exposure to NFs/AFs. It supports analytics information provisioning to NFs and AFs.
- *Policy Control Function (PCF)*: The PCF supports a unified policy framework to provide access subscription information for policy decisions taken by Unified Data Repository (UDR), and provides operator network control policies to other NFs and UE (e.g., QoS, charging control, etc.).
- *Security Edge Protection Proxy (SEPP)*: The SEPP is a nontransparent proxy and supports message filtering and policing on the inter-PLMN interface for CP communications in order to enable NF topology hiding between the PLMNs.
- *Service Communication Proxy (SCP)*: The SCP is part of the SBA to assist operators to efficiently secure and manage their 5G network by providing routing control, resiliency, and observability to the CN.
- *Session Management Function (SMF)*: The SMF terminates NAS signaling related to session management (SM). It handles session establishment, modification and release, tunnel maintenance between UPF and AN (Access Node), UE IP address assignment and management, DHCP functions, selection and control of UPF (including the support of traffic steering configuration for the UPF for proper traffic routing), charging data collection and charging interface support, etc.

- *Trusted Non-3GPP Gateway Function (TNGF)*: The TNGF enables the UE to connect to the 5G CN over Trusted WiFi access.
- *Unified Data Management Function (UDM):* The UDM is one of two functions to support a converged repository of subscription-related information management. It is the front end of UDR to handle user subscription data generation and storage, authentication data management, support data retrieval with other NFs and support interaction with external third-party servers.
- *Unified Data Repository (UDR)*: A converged repository of subscription-related information to be made available via the UDM to serve NFs (e.g., AMF, SMF, AUSF, PCF, etc.) that control UE operation within the 5G system. It is used to store Subscription Data, Policy Data, Application data, and Structured data that can be exposed to an NF. Note that Application data is placed into the UDR from external NF via the support NEF.
- *Unstructured Data Storage Function (UDSF)*: A repository for storage and retrieval of unstructured data used by any NF within the 5G Core.
- *User Plane Function (UPF):* UPF handles UP packet routing and forwarding, UP QoS processing, user traffic statistics and reporting, and external data network interconnection to the Data Network (DN). It is an anchor point for intra- and inter-RAT mobility.

3.1.3 5G end-to-end architecture and protocol stack based on 3rd Generation Partnership Project access

The 5G system has made some modifications to its CP and UPs due to architectural improvement (e.g., MM and SM decoupling) and traffic handling (e.g., QoS flow management).

This section presents the end-to-end protocol stack for the 5G system with respect to the point-to-point reference points, when applicable. For further details, refer to 3GPP TS 23.501 [1] (Fig. 3.3).

3.1.3.1 End-to-end protocol stack of 5G control plane based on 3rd Generation Partnership Project access

The CP is used to carry the signaling between the UE and the network. The CP signaling is transmitted through the Uu, N2, and N11 interfaces. As mentioned earlier, the MM and SM CP functionalities have been reorganized and decoupled into two functionally independent NFs, AMF and SMF, respectively. 5G-AN uses N2 (CP) to interface with 5G-Core

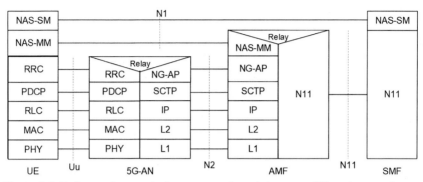

Figure 3.3 End-to-end control plane protocol stack between UE and network over 3rd Generation Partnership Project access. *Cited from figure 8.2.2.3−1 in TS 22.261, Service requirements for the 5G system: Stage 1.*

where N1 is a transparent interface between 5G-UE with 5G-Core (i.e., carrying NAS signaling).

The N1 interface is used by the UE for transmitting nonradio signaling between the UE and AMF/SMF. This includes information related to connection, mobility, and session-related messages to the AMF, which handles cases related to connection and mobility messages, but forwards SM info to SMF.

The N2 interface supports CP signaling between the RAN and 5G core including the scenarios related to UE context management and PDU session/resource management procedures. The N2 uses the Stream Control Transmission Protocol (SCTP) to transport the Next Generation Application Protocol (NG-AP) to support CP communication between 5G-AN and 5G-Core.

Between the SM function (i.e., SMF) and UPF over the N4 interface, it is based on Packet Forwarding Control Protocol (PFCP) to allow the SMF to control UPF for the UP data. Note that the initial design of the PFCP was defined to support the Control and UPs Separation feature in the EPC.

Since 5G-Core architecture has adopted SBA for its CP among the NFs, the CP protocol stacks (e.g., HTTP-based) between NFs has also been updated (Fig. 3.4).

3.1.3.2 End-to-end protocol stack of 5G User Plane based on 3rd Generation Partnership Project access

The UP is used to carry the application data of the UE, and the UP data is transmitted through the Uu, N3, N9, and N6 interfaces. Among them, the N6

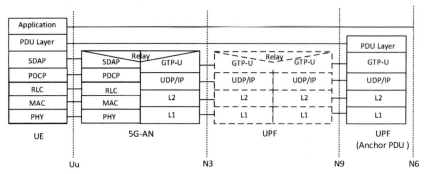

Figure 3.4 End-to-end User Plane protocol stack between UE and core network.

interface is the interface between the 5G network and external Data Network (DN), which imports and exports of application data to/from the DN. In a 5G system, there are different types of PDUs to support application data. Currently supported PDU types are IPv4, IPv6, IPv4v6, Ethernet, unstructure, etc. The protocol stacks of different PDU types correspond to their respective protocol stacks. The GTP-U protocol has been enhanced for the 5G system to transport UP traffic over the backhaul (i.e., over N9 and N3 interfaces).

A UPF handles UP paths on the N3, N9, and N6 interfaces, and the setup is controlled by the SMF via the N4 interface using PFCP. A UP path will be manipulated based on application requirements for the PDU session corresponding to the path. An SMF is also able to receive information regarding routing path with the API from the AF via the NEF, PCF, and other SMFs.

Between the UE and NG-AN, the UP has been enhanced with the Service Data Adaptation Protocol (SDAP) for the air interface to support more advanced QoS flow handling.

3.1.4 5G end-to-end architecture and protocol stack based on non-3rd Generation Partnership Project access

The 5G system promises to be access agnostic and to support a converged CN solution where different access technologies share a common anchor point in the core. In fact, as defined in 3GPP TS 23.501 [1], a 5G AN could comprise a NG-RAN and/or non-3GPP AN connecting to a 5G CN.

The 5G system supports three types of non-3GPP ANs:
- Untrusted non-3GPP networks
- Trusted non-3GPP networks
- Wireline ANs

3GPP works with the Broadband Forum (BBF) to support for Wireline ANs in a 5G system, and BBF has taken the lead on the development. For further details on wireline access type support, TS 23.316 [3] and related BBF specifications should be referenced.

While the UE is able to access 5G-Core through access technologies other than 3GPP access (e.g., WLAN networks), 5G operators are able to control and manage the UE to access to their 5G networks based on operator's policy, UE preference, and service context in order to improve overall network capacity and reliability over multiple accesses (via traffic steering, switching, and aggregation, etc.).

Examples of untrusted non-3GPP access include public hotspots, corporate Wi-Fi, etc., that are not controlled by the 5G mobile operator. However, by enabling convergence to the 5G-Core, which provides IP-based service, the non-3GPP access can supplement the 3GPP access to provide 5G services to UE.

In the case of trusted non-3GPP access, it is deployed and managed by either a 5G mobile operator or by a third party who is trusted by the 5G mobile operator. How trust is established between the trusted non-3GPP access and the 5G mobile operator is not considered in the 3GPP scope.

Fig. 3.5A and B present the system architecture of the UE connecting to 5GC through untrusted and trusted non-3GPP accesses, respectively.

The untrusted network is connected to the CN through the non-3GPP Interworking Function (N3IWF). The trusted network is connected to the CN through the Trusted Non-3GPP AN (TNAN).

When the UE decides to access the 5G CN via untrusted non-3GPP:
- The UE first selects and connects to a non-3GPP network.
- The UE then selects a PLMN and the N3IWF in this PLMN. PLMN selection and non-3GPP AN selection are independent procedures.

When the UE decides to access to the 5G CN with trusted non-3GPP:
- The UE first selects a PLMN.
- The UE selects a TNAN that can be reliably connected to PLMN; TNAN selection is influenced by PLMN selection. The UE can be connected to different PLMNs through 3GPP and non-3GPP and the UE can also be registered to two different AMFs simultaneously via the 3GPP AN and the non-3GPP AN in the case of the home-routed scenario.

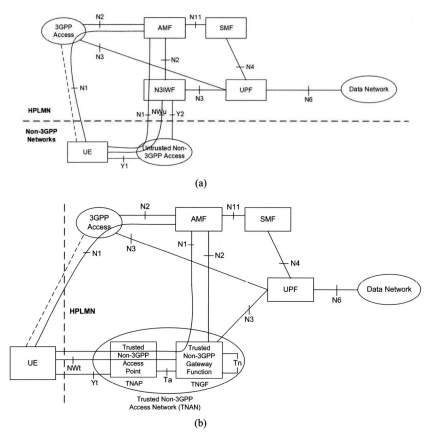

Figure 3.5 (A) System architecture with untrusted non-3GPP access to 5G-Core in nonroaming scenario. (B) System architecture with trusted non-3GPP access to 5G-Core in nonroaming scenario. *(A) Cited from Figure 4.2.8.2.1−1 in TS 23.501, System architecture for the 5G System (5GS): stage 2. (B) cited from Figure 4.2.8.2.1−2 in TS 23.501, System architecture for the 5G System (5GS): stage 2.*

3.1.5 5G system identifiers

In the 3GPP mobile network system, the 5G system identifiers play an important role in identifying the UE in order to support essential network operations to provide services for the UE. Proper design of the UE identifier also ensures the protection of the UE's privacy and confidentiality as well as to safeguard the 5G system security. Prior to 5G, privacy and confidentiality of the UE were guaranteed only when the UE was authenticated. In 5G, this loophole has been fixed with the introduction of the

Subscriber Concealed Identifier (SUCI) and the enhancement of the UE authentication and authorization procedures.

The 5G system defines the subscriber identifier independently of the UE identifier. The format of the UE identifier has been revised to support various access types (e.g., fixed line, WiFi, satellite, etc.) because the 5G system provides converged network solutions.

The following is the list of 5G system identifiers:

- Subscription Permanent Identifier (SUPI)
- Subscriber Concealed Identifier (SUCI)
- 5G Globally Unique Temporary Identity (5G-GUTI)
- Permanent Equipment Identifier (PEI)
- AMF Name
- Data Network Name (DNN)
- Internal-Group Identifier
- Generic Public Subscription Identifier
- AMF UE NGAP ID
- UE Radio Capability ID
- 5G NR Radio Network Temporary Identifier (RNTI)

Each of the above identifiers are briefly described as follows. For further details, TS 23.501 can be referenced.

1. Subscription Permanent Identifier (SUPI):

a. It is a global unique identifier for subscriber provisioned in UDM/UDR

b. It is used only within 3GPP system

c. International Mobile Subscriber Identify (IMSI) can still be used as SUPI, so that interworking with Evolved Packet System (EPS) is feasible.

i. IMSI: = <Mobile Country Code (MCC)> <Mobile Network Code (MNC)>

d. <Mobile Subscriber Identification Number (MSIN)>

e. The use of generic NAI as defined by IETF RFC 7542 provides wider range of SUPI.

ii. NAI: = <user id> @ <domain>

f. SUPI must include the home network identification in order to support roaming

    1. Subscriber Concealed Identifier (SUCI):

g. One-time use subscription identifier that has the following information

    2. SUCI: = <SUPI type> <Home Network Identifier> <Routing Indicator> <Protection Scheme Id> <Home Network Public Key Identifier> <Scheme Output>

a. *SUPI Type* — It identifies the type of the SUPI concealed in the SUCI.

b. *Home Network Identifier:* It identifies the home network of the subscriber. When the SUPI Type is an IMSI, the Home Network Identifier is composed of MCC and MNC. When the SUPI type is a Network Access Identifier, the Home Network Identifier is the domain name (e.g., user@techno.com).

c. *Routing Indicator:* It is assigned by the home network operator and provisioned within the USIM and is used with the Home Network Identifier to route the Authentication traffic to UDM that contains the subscriber's information.

d. *Protection Scheme Identifier:* It is used to identify the profile for the protection scheme on SUPI.

e. *Home Network Public Key Identifier:* It represents a public key provisioned by the HPLMN and is used to identify the key used for SUPI protection. In the case of null-scheme being used, this data field shall be set to the value as 0.

f. *Protection Scheme Output*: It contains the concealed SUPI.

g. Designed for privacy and confidential protection for UE by concealing SUPI.

h. Subscription-specific part is encrypted with home network public key.

i. Encrypted by UE using a ECIES-based protection scheme with the public key of the home network that was securely provisioned to the USIM during the USIM registration.

j. Only the MSIN part of the SUPI is concealed, MCC + MNC part is not concealed.

1. 5G Globally Unique Temporary Identity (5G−GUTI):

   a. A temporary subscription identifier that can be used to identify the last serving AMF which allocation the 5G−GUTI to retrieve the UE's security context

   b. It comprises a Globally Unique AMF Id (GUAMI) and a 5G Temporary Mobile Subscriber Id (5G−TMSI), where GUAMI identifies the assigned AMF and 5G−TMSI identifies the UE uniquely within the AMF.

1. 5G-GUTI: = <GUAMI> <5G-TMSI>

2. GUAMI: = <MCC > <MNC> <AMF Identifier>

3. AMF Identifier: = <AMF Region Id > <AMF Set Id> <AMF Pointer>

4. where AMF Region ID identifies the region, AMF Set ID uniquely identifies the AMF Set within the AMF Region and AMF Pointer identifies one or more AMFs within the AMF Set.

  a. The AMF may decide to assign a new 5G−GUTI to the UE anytime.

  b. In order to enable more efficient radio resources (especially in Paging and Service Request), a new identifier called 5G−S−TMSI is introduced as a shortened form of 5G−GUTI to be used over the air interface. 5G-S-TMSI: = <AMF Set Id> <AMF Pointer> <5G−TMSI >

  c. Common 5G-GUTI can be assigned to both 3GPP and non-3GPP access

1. Permanent Equipment Identifier (PEI):

  a. Defined for 3GPP UE to access the 5G system.

  b. Different formats for different UE types and use cases. For example:

1. IMEI: = <Type Approval Code> <Serial Number >

2. IMEISV: =   <Type Approval Code > < Serial Number > < Software Version No (SVN) >

  a. UE must support a PEI with IMEI format when using 3GPP access technology.

  b. IMEI is the only format supported in Rel-15 for PEI.

1. AMF Name:

  a. It is used to identify an AMF.

  b. It is a globally unique FQDN.

  c. At a given time, GUAMI with distinct AMF Pointer value is associated to one AMF name only.

1. Data Network Name (DNN):

  a. A DNN is equivalent to an Access Point Name (APN).

  b. The DNN may be used to:

iii. Select a AMF and UPF(s) for a PDU Session

iv. Select N6 interface(s) for a PDU Session

 v. Determine policies to apply to the PDU Session

  1. Internal-Group Identifier:

    a. Identify the group(s) that the UE belongs and apply to non-roaming case at this point.

    b. It is part of the subscription data for the UE in UDR.

    c. UE can belong only to a limited number of groups.

    d. It is provided by UDM to the SMF as part of the SM Subscription data and by the SMF to PCF (when PCC applies to a PDU session).

    e. The SMF may use this information to apply local policies and to store this information in the Charging Data Record (CDR).

    f. The PCF may use this information to enforce AF requests.

1. Generic Public Subscription Identifier (GPSI):

    a. A public identifier used both inside and outside of the 3GPP system.

    b. It is either an MSISDN or an External Identifier (e.g., NAI)

1. MSISDN: = <Country Code> <National Dest. Code> <Subscriber Number>

2. External Id: = <user id> @ <domain>

    a. It is used to address a 3GPP subscription in different external data networks.

    b. 5G-Core maintains the mapping between GPSI and SUPI, however, they are not necessarily one-to-one relationship.

1. AMF UE NGAP ID:

    a. An identifier is used to identify the UE in AMF on N2 reference point.

    b. It is allocated by AMF and is sent to 5G-AN.

    c. It is unique per AMF set.

    d. It may be updated by AMF without changing AMF.

1. UE Radio Capability ID:

    a. It is used to uniquely identify a set of UE radio capabilities (i.e., UE Radio Capability information).

    b. Two types − assigned by the serving PLMN (PLMN-assigned) or by the UE manufacturer (UE manufacturer-assigned)

1. gNB Id

    a. It identifies a gNB within Public Land Mobile Network (PLMN).

    b. NR Cell Id (NCI) contain gNB Id field.

1. Global gNB Id

    a. It identifies a gNB at the global scale. It is constructed from PLMN and gNB Id.

    b. It uses same MCC and MNC as used in NR Cell Global Identifier (NCGI). NCGI is used to identify NR cells globally. The NCGI is constructed from the PLMN identity the cell belongs to and the NR Cell Identity (NCI) of the cell.

3.2 The 5G RAN architecture and deployment options

When considering the deployment of the 5G system, operators have proposed a variety of deployment options according to their own network migration requirements and considerations in order to ensure the smooth migration to adopt 5G technologies to improve their system performance and service deployment. Part of the considerations is to reuse existing 4G LTE systems as much as possible in order to maximize existing investment. As a result, many different flavors of the RAN architecture deployment options have been proposed due to different considerations of the lifetime of the 4G LTE system and the migration paths.

The discussion on the 5G RAN architecture deployment options began as early as the RAN3#91bis meeting in April 2016, at which it was decided that TR 38.801 would be responsible for capturing the relevant configuration options [4].

During the discussion in the RAN2#94 meeting in May 2016, a way forward was preliminarily agreed upon [5] that focused on the combination of 4G AN/5G AN and 4G CN/5G CN. It was preliminarily decided that three alternatives based on a combination of nonstandalone Long Term Evolution (LTE) and NR with interconnection to either EPC or 5GC (as shown in Fig. 3.6) and two alternatives based on standalone LTE and NR interconnection to 5GC (as shown in Fig. 3.7) would continue to be studied.

Further discussions in the subsequent meetings on the possible options that were captured in TR38.804 [6] and different combinations of the 4G/5G AN with the 4G/5G CN deployment options are listed Table 3.1.

- Option 1: the architecture is traditional LTE and its enhancements, as shown in Fig. 3.8.
- Option 2: 5G standalone. The AN is 5G NR, and the CN is 5GC. This architecture should be the ultimate mode of deployment for operators to deploy 5G (i.e., the network architecture presented when the whole 4G network evolves into the 5G network in the future, as shown in Fig. 3.9).
- Option 3: EN−DC mode. The AN is mainly 4G LTE, supplemented by 5G NR, and the CN is 4G EPC. This architecture is proposed mainly because operators hope to reuse the existing 4G AN and CN investment as much as possible, and at the same time to improve the network performance by deploying 5G radio technology. It should be

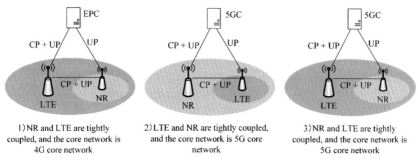

Figure 3.6 Three ways of 4G-5G coordination.

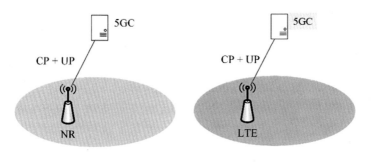

Figure 3.7 Two ways of 5G standalone.

Table 3.1 Combination of 4G/5G access network with 4G/5G core network deployment options.

	4G core network	5G core network
4G access network LTE	Option 1	Option 5
4G access network LTE (main) + 5G access network NR (secondary)	Option 3	Option 7
5G access network NR	Option 6	Option 2
5G access network NR (main) + 4G access network LTE (secondary)	Option 8	Option 4

noted that even though the secondary node has the function of CP in this architecture, when comparing with the master node, the secondary node can only support part of the CP function, and the main CP function and its corresponding signaling are still carried by the master node, as shown in Fig. 3.10.

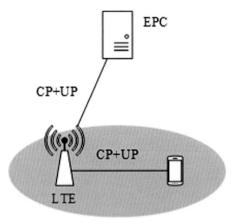

Figure 3.8 4G/5G network architecture deployment option 1 configuration.

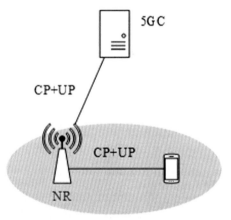

Figure 3.9 4G/5G network architecture deployment option 2.

- Option 4: NE–DC mode. The AN is mainly based on 5G NR, supplemented by 4G LTE, and the CN is 5GC. This architecture is mainly based on the scenario that the operator plans for the large-scale deployment of the 5G system compared to smaller-scale deployment of the 4G LTE system, and there is a small number of LTE UEs in the operator's network. The 4G LTE network resources are sufficient to serve the LTE UE, as shown in Fig. 3.11.
- Option 5: eLTE mode. The AN is upgraded based on 4G LTE, and the CN is 5GC. The motivation for this network architecture deployment option is that some operators only consider upgrading the CN

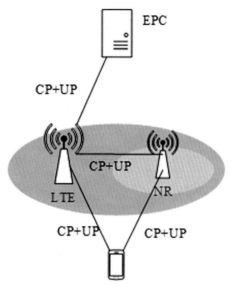

Figure 3.10 4G/5G network architecture deployment option 3.

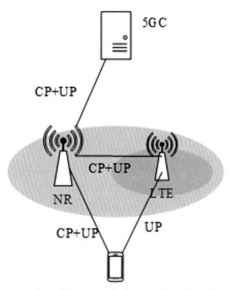

Figure 3.11 4G/5G network architecture deployment option 4.

to support new 5G features, such as the new QoS architecture, but still reuse the capabilities of 4G LTE radio ANs, as shown in Fig. 3.17 (Fig. 3.12).

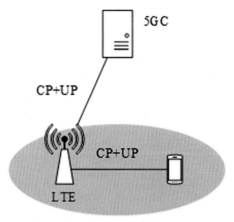

Figure 3.12 4G/5G network architecture deployment option 5.

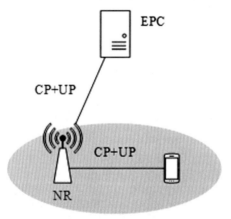

Figure 3.13 4G/5G network architecture deployment option 6.

- Option 6: 5G NR AN is deployed with 4G EPC. This network architecture deployment option was lack of support from operators, so it is excluded at an earlier stage, as shown in Fig. 3.13.
- Option 7: NG EN−DC. The AN is based on 4G LTE, supplemented by 5G NR, and the CN is 5GC. This network architecture deployment option is for the scenario where the EPC is eventually upgraded to 5GC on the basis of option 3, as shown in Fig. 3.14.
- Option 8: the AN is based on 5G NR, supplemented by 4G LTE, and the CN is EPC. There is no operator's proposal for this option, so it is excluded in the early stage, as shown in Fig. 3.15;

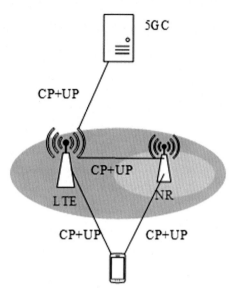

Figure 3.14 4G/5G network architecture deployment option 7.

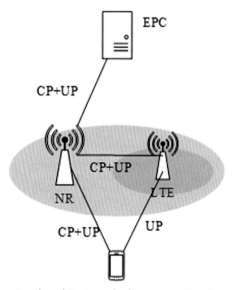

Figure 3.15 4G/5G network architecture deployment option 8.

Operators generally agreed that options 6 and 8 would not be considered as part of the system migration path and hence these options were not considered at all. Option 3 is based on existing 4G LTE deployment and supplemented by 5G NR deployment, and the investment in existing 4G network is protected, therefore this option is widely supported by operators; other options (option 2/4/5/7) were considered for further study for standardization [7].

As time went on and due to time constraints for 5G deployment concerns from operators, option 3 EN-DC with the most endorsements was first completed and options 4/7 were eventually completed in December 2018.

3.2.1 Description of EN-DC and SA arechitecture

The rest of this section will focus on the presentation of option 3 EN-DC and option 2 SA architecture deployment options that operators most supported.

EN−DC stands for E−UTRAN New Radio—Dual Connectivity. This architecture allows UE to connect to an LTE eNB that acts as a master node and a 5G gNB that acts as a secondary node, and it also allows UE to access both LTE and 5G radio access simultaneously on the same spectrum band.

In 4G LTE there are two main types of bearers, namely the Signaling Radio Bearers (SRB) and the Dedicated Radio Bearer (DRB). The DRB is used to carry the user data. And in EN−DC, cell(s) of the LTE network are part of the Master Cell Group (MCG). SRBs are used for reconfiguration including Secondary Cell Group (SCG) addition. SRB is used for the transmission of RRC and NAS messages.

Deployment Option 3, which is based on EN−DC architecture, attracted the most interest from operators as the initial 5G deployment configuration. It is divided into three categories: options 3/3a/3x as shown in Fig. 3.17. The main difference is that the aggregation nodes of the user data are different [8]. From a CP point of view, the protocol stack for option 3/3a/3x is the same, as shown in Fig. 3.20. In this figure, the CP of option 3 is based on 4G LTE base stations eNB (evolved NodeB) and supplemented by 5G NR base stations gNB. (As the synonym for the 5G base station "g" has no special meaning but is the next in alphabetic order. The 4G base station is called eNodeB. Because fNodeB is not good for writing and reading, 5G base station is called gNodeB, or gNB for short.) In this process, the master node is responsible for controlling the CP and UP state transition and connection management, and also for controlling the secondary node. The secondary node can also trigger its own control but mostly for the UP.

From the UE perspective, it establishes the signaling connection of SRB0, SRB1, and SRB2 with the master node eNB; at the same time, it also establishes the signaling connection of SRB3 with the secondary node gNB. Therefore the secondary node gNB may not have an independent access function to support the process of system information transmission, paging message transmission, connection establishment procedure, reestablishment procedure, etc., but the UE can support gNB configuration, measurement configuration, and measurement reporting through SRB3 (Fig. 3.16).

The EN−DC option is divided into option 3, option 3a, and option 3x according to the location of the SCG termination, as shown in Fig. 3.17. Specifically, option 3 is the master node to control the termination of the UP data flow (i.e., the MCG Split Bearer); Option 3a is the CN control termination of the UP data flow (i.e., the MCG bearer +

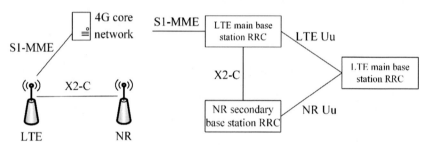

Figure 3.16 Option 3/3a/3x control plane connection and control plane protocol stack.

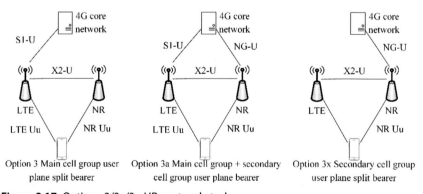

Option 3 Main cell group user plane split bearer

Option 3a Main cell group + secondary cell group user plane bearer

Option 3x Secondary cell group user plane split bearer

Figure 3.17 Options 3/3a/3x UP protocol stack.

SCG bearer); and Option 3x is the secondary node to control termination of the UP data flow (i.e., the SCG Split Bearer).

The Split Bearer is a bearer for which traffic is routed via the LTE and/or NR node. The Split Bearer can be either SCG split or MCG split. The MCG Split refers to the traffic that is split at MN, whereas the SCG Split refers to the traffic split at the SN node.

As for option 2, Standalone follows the same logic similar to 4G LTE. As shown in Fig. 3.18, the overall radio architecture is not much different from 4G LTE. When compared with the initial stage of 4G LTE, carrier aggregation (CA), dual connection (DC), and supplementary uplink mechanism (SUL) can be supported.

The protocol stack of the CP and user interface is shown in Fig. 3.19. The detailed function description of option 2 is shown in Chapter 11, Control plan design.

In the 3GPP protocol, the terms for standardization of these system architecture deployment options are adopted, as shown in Table 3.2.

In the case of Option 3 Dual-Connectivity architectures, the protocol stack of the UP and the CP is relatively complex. A big difference between 4G and 5G systems is the difference in QoS architecture. When MCG is connected to EPC, even if the radio access technology adopts the radio bearer established by NR, it adopts the QoS architecture of the

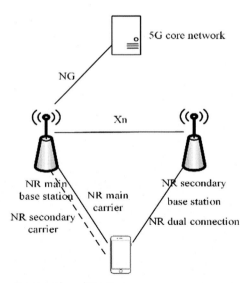

Figure 3.18 Option 2 network architecture.

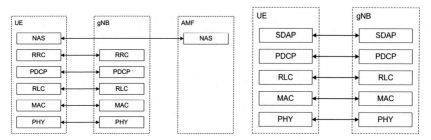

Figure 3.19 Option 2 control plane and user plane protocol stack.

Table 3.2 System architecture terminology.

System architecture	Standardized name	Standardized version
Option 3/3a/3X	EN−DC	Rel15, early drop
Option 5	E−UTRA connected to EPC	Rel15, late drop
Option 7	NGEN−DC	Rel15, late drop
Option 2	SA (Stand-alone)	Rel15
Option 2 + DC	NR−DC	Rel15, late drop
Option 4	NE−DC	Rel15, late drop

LTE system. When the MCG is connected to the 5GC, the established radio bearer needs to adopt the QoS architecture of the NR system. For a detailed description of the QoS of the NR system, refer to Chapter 13, QoS control.

The radio bearer of EN−DC is also different from other DC architectures (i.e., NG EN−DC, NE−DC, and NR−DC). From UE and Base Station point of views, the composition of the radio bearer protocol stack is also different. In the Dual-Connection architecture, the network node is composed of two nodes, namely MN (master node) and SN (secondary node). The PDCP and SDAP protocol stacks and the corresponding protocol stacks below the PDCP (i.e., RLC, MAC, and PHY) may be located on different network nodes. There is no problem in distinguishing different nodes within the UE, but there is a problem of the role of the cell group (i.e., it is necessary to distinguish between the MCG and the SCG. The difference between MCG and SCG comes from the different protocol layers of radio bearer aggregation (i.e., MAC and PHY).

For EN−DC, from the network side, the composition of radio bearer is shown in Fig. 3.20 [3].

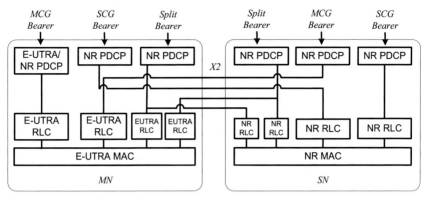

Figure 3.20 EN−DC network side protocol stack.

When a radio bearer RLC, MAC, and PHY protocol stack is resided in MN, such radio bearer is referred as MCG Bearer. Likewise, if the protocol stack is resided in the SN, it is referred to as the SCG bearer. The purpose of the introduction of the SCG bearer is to make full use of the radio access traffic of the 5G NR. When the PDCP protocol stack of the SCG bearer is on the SN, a direct UP data GTP−U channel is established between the SN and the CN, so that the SN can share the processing load of the PDCP protocol. When the PDCP stack is on the MN, the benefit of handling load balancing is lost, and the delay on the user side is increased. When the MCG bearer's PDCP protocol stack is in SN, the radio bearer has a similar problem. However, this type of radio bearer can reduce the loss of user performance in some abnormal cases. For example, from a UE point of view, if there is a radio link failure (RLF) on the SN, then the network can choose to release the split bearer under the PDCP protocol stack on the SN branch, but it can keep the PDCP protocol stack on the SN and the protocol stack on the MN, so that the radio bearer can continue to operate. Otherwise, the network either chooses to release all radio bearers on the SN or transfer these radio bearers to the MN. All of these actions lead to the re-establishment of the user-side protocol stack (including PDCP). The operation of releasing the protocol stack below the PDCP will only lead to the reconstruction of the RLC related to the radio bearer and the reset operation of the HARQ process corresponding to the MAC layer, and the loss of the user side is much smaller. Based on this, the technical feature of fast recovery of MCG is introduced in Rel-16, which is used to deal with the RLF that occurs on MN (when SN is still working normally). By the same token, PDCP's

SCG bearer on MN can also be used as a complementary solution to reduce user-side losses.

For the EN—DC architecture, the separated PDCP adopts the PDCP protocol as in the NR system. The reason for this is to reduce the complexity of the UE. If the separated PDCP adopts LTE PDCP and NR PDCP protocol stacks according to the node location, then there will actually be two kinds of detached bearers from the UE perspective. If only the NR PDCP protocol stack is used, then there is only one separate bearer in the PDCP protocol layer from the UE point of view. This difference can be seen in Fig. 3.21.

Single NR PDCP protocol stack support works well with the architecture of separation of CU and DU on the network side as long as NR PDCP is implemented in CU, as shown in Fig. 3.22. For other DC architectures (e.g., Multi-RAT Dual Connectivity (Mr—DC)), the protocol stacks on the network side and UE side are shown in Figs. 3.22 and 3.23:

In EN—DC architecture, there is SDAP protocol stack on the PDCP protocol layer. This is to support the QoS architecture of the NR system. From a UE point of view, SDAP has only one entity, the PDCP protocol stack is unified to support only the NR PDCP protocol stack. The protocol stack of RLC and MAC does not clearly distinguish between the protocol stack of LTE and NR because both protocol stacks are possible on either MN or SN. In the Stage 3 protocol specification, these protocol stacks will have a clear distinction. Among them, the protocol stack of LTE needs to refer to the protocol specification of TS 36 series, while the protocol stack of NR needs to refer to the protocol specification of TS 38 series. A detailed introduction of the UP in the TS 38 series protocol

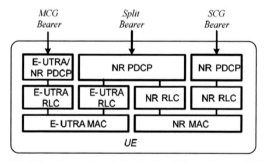

Figure 3.21 EN—DC UE side protocol stack. *Cited from Figure 4.2.2—1 in TR38.801. Study on new radio access technology: radio access architecture and interfaces TS 37.340, multi-connectivity, stage 2.*

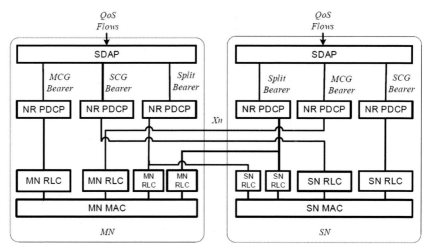

Figure 3.22 Mr–DC network side protocol stack. *Cited from Figure 4.2.2–4 in TR38.801. Study on new radio access technology: radio access architecture and interfaces TS 37.340, multi-connectivity, stage 2.*

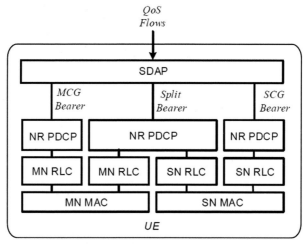

Figure 3.23 Mr–DC UE side protocol stack. *Cited from Figure 4.2.2–2 in TR38.801. Study on new radio access technology: radio access architecture and interfaces TS 37.340, multi-connectivity, stage 2.*

specification can found in Chapter 10, User plane protocol design, and a detailed introduction of the CP can be found in Chapter 11, Control plan design.

3.3 Summary

This chapter described the high-level system architecture of the 5G system. It presented the network service requirements and technical rationales that lead to the system design decisions. It describes the overview of the SBA-based 5G-Core architecture, a significant revolution from the legacy point-to-point architecture, which opens the door for 5G systems to support variety of emerging network services with high efficiency and flexibility.

This chapter also provided some history considerations from the 3GPP standard community on how to evolve the existing 4G LTE AN to 5G—AN in order to address operator requirements for cost-effective migration (e.g., maximum reuse of 4G radio access investment), coexistence, and evolution to high capacity and highly efficient 5G systems. It described the important aspects of the 5G system identifiers that have been significantly improved from prior 4G systems in order to ensure more secure and flexible UE operation within and beyond the 5G network thirdly.

This chapter was based on a nonroaming scenario. It also does not describe the interworking aspects between the 5G-Core and 4G-Core. For further details on these two aspects, TS 23.501 [1] can be referenced.

References

[1] TS 23.501. System architecture for the 5G system (5GS): stage 2.
[2] TS 22.261. Service requirements for the 5G system: stage 1.
[3] TR38.801. Study on new radio access technology: radio access architecture and interfaces TS 37.340, multi-connectivity, stage 2.
[4] R3—160842. 5G architecture scenarios. Ericsson.
[5] R2—163969. Text proposal to TR 38.804 on NR deployment scenarios. NTT DOCOMO, Inc.
[6] RP-161266. Architecture configuration options for NR. Deutsche Telekom.
[7] RP-170741. Way Forward on the overall 5G-NR eMBB workplan. Alcatel-Lucent Shanghai-Bell, Alibaba, Apple, AT&T, British Telecom, Broadcom, CATT, China Telecom, China Unicom, Cisco, CMCC, Convida Wireless, Deutsche Telekom, DOCOMO, Ericsson, Etisalat, Fujitsu, Huawei, Intel, Interdigital, KDDI, KT, LG Electronics, LGU +, MediaTek, NEC, Nokia, Ooredoo, OPPO, Qualcomm, Samsung, Sierra Wireless, SK Telecom, Sony, Sprint, Swisscom, TCL, Telecom Italia, Telefonica, TeliaSonera, Telstra, Tmobile USA, Verizon, vivo, Vodafone, Xiaomi, ZTE.
[8] TS 23.316. Wireless and wireline convergence access support for the 5G system (5GS).

Further reading

R3—160947. Skeleton TR for new radio access technology: radio access architecture and interfaces. NTT DOCOMO Inc.

R3—161008. Next generation RAT functionalities. Ericsson.

R3—160823. Multi-RAT RAN and CN. Qualcomm.

R3—160829. Overall radio protocol and NW architecture for NR. NTT DOCOMO, Inc.

R3—161010. CN/RAN interface deployment scenarios. Ericsson.

R2—162365. Dual connectivity between LTE and the New RAT, Nokia. Alcatel-Lucent Shanghai Bell.

R2—162707. NR architectural discussion for tight integration. Intel Corporation.

R2—162536. Discussion on DC between NR and LTE. CMCC discussion.

R2—164306. Summary of email discussion [93bis#23] [NR] deployment scenarios. NTT DOCOMO, Inc.

R2—164502. RAN2 status on NR study—rapporteur input to SA2/RAN3 joint session. NTT DOCOMO, Inc.

RP-161269. Tasks from joint RAN-SA session on 5G architecture options. RAN chair, SA chair.

CHAPTER 4

Bandwidth part

Jia Shen and Nande Zhao

Bandwidth part (BWP) is one of the most important concepts innovated in 5G NR, which has a profound impact on almost all NR aspects such as resource allocation, uplink and downlink control channels, initial access, Multiple Input Multiple Output (MIMO), Media Access Control (MAC)/ Radio Resource Control (RRC) layer protocols, etc. One can hardly well understand NR specifications without solid knowledge of BWP.

In the early stage of NR standardization, many companies put forward different initial points of views for "subband of operation," and finally agreed to name it BWP. With comprehensive studies and discussions in 3rd Generation Partnership Project (3GPP), the BWP concept was continuously modified and improved, and finally became a powerful system tool. Its potential capability may grow beyond designers' expectations in the future.

The NR specifications of R15 and R16 are fairly said to just define the basic capabilities of BWP. The industry will gradually explore the advantages as well as issues of BWP in the first wave of 5G NR commercial deployments. In the subsequent 5G enhancement releases (such as R17 and R18), the BWP concept may further evolve and play a greater role in 5G technology.

A lot of design problems of BWP were gradually solved in the course of standardization, including how to configure a BWP; the relationship between BWP configuration and numerology; whether the BWP configuration should be different between Time Division Duplex (TDD) and Frequency Division Duplex (FDD) systems; activation/deactivation mechanism for BWP; how to use BWP in the initial access process before RRC connection being established; and the relationship between BWP and carrier aggregation (CA). This chapter explains the corresponding solutions one by one.

4.1 Basic concept of bandwidth part

The core concept of BWP is to define an access bandwidth that is smaller than the system bandwidth of cell operation and an User Equipment (UE)

5G NR and Enhancements
DOI: https://doi.org/10.1016/B978-0-323-91060-6.00004-0

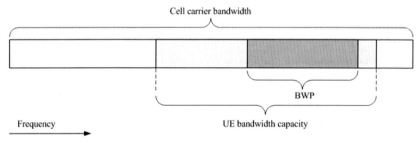

Figure 4.1 Basic concept of BWP. *BWP*, Bandwidth part.

bandwidth capability. All the receiving and transmitting operations of the UE can be carried out within this small bandwidth, so as to achieve more flexible and more efficient operation with lower power consumption in the large-bandwidth system. The maximum single-carrier system bandwidth of Long Term Evolution (LTE) is 20 MHz, which is also the single-carrier bandwidth capability of the UE. So LTE does not have such a situation that the UE bandwidth capability is smaller than the system bandwidth. In contrast, in a 5G NR system, the maximum carrier bandwidth is significantly increased (e.g., to 400 MHz), while the UE bandwidth capability, though improved (e.g., to 100 MHz), cannot catch up with the system bandwidth operated on network side. In addition, the UE does not always need to run with a bandwidth equal to its maximum bandwidth capability. In order to save power consumption and realize the more efficient frequency-domain operation, the UE may work in a smaller bandwidth, which is BWP (as shown in Fig. 4.1).

Before BWP was clearly defined in the specification, similar concepts were proposed from different perspectives such as resource allocation, UE power saving, forward compatibility [1], etc. Because BWP is treated as a flexible "NR specification tool," the purposes for introducing the BWP concept will not be explicitly written in the 3GPP standards.

4.1.1 Motivation from resource allocations with multiple subcarrier spacings

The concept of BWP was firstly proposed in 3GPP discussion for resource allocation. A major innovation in 5G NR is the ability to transmit data with multiple Orthogonal Frequency Division Multiplexing (OFDM) subcarrier spacings (SCS). The problem is how to efficiently allocate the frequency-domain resources using different SCSs, given different SCSs result in different Physical Resource Block (PRB) sizes and therefore

Table 4.1 PRB size and PRB number in 20 MHz with different subcarrier spacings.

Subcarrier spacing (kHz)	PRB size (kHz)	Number of PRBs within 20 MHz bandwidth
15	180	100
30	360	50
60	720	25

PRB, Physical resource block.

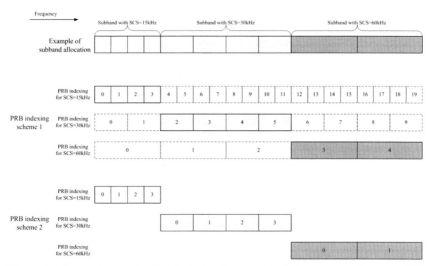

Figure 4.2 Two schemes for PRB indexing for multiple SCS resource allocation. *PRB*, Physical resource block; *SCS*, subcarrier spacings.

different frequency-domain allocation granularities. In the 3GPP meetings from end of 2016 to the beginning of 2017, some companies [2,3] proposed resource allocation schemes with multiple SCSs, among which the key idea is to adopt a two-step allocation scheme, with a coarse-granularity allocation followed by a fine-granularity allocation.

Take an example as shown in Table 4.1. In a bandwidth of 20 MHz (the usable bandwidth is 18 MHz), both the PRB sizes and the numbers of available PRBs vary with different SCSs, which also drives different resource allocation granularities and different PRB indexing for different SCSs. Hence, the LTE resource allocation scheme cannot be directly applied in NR.

In October 2016, two PRB indexing schemes for multiple SCS resource allocation were proposed in the RAN1#86-bis meeting:

- The first scheme is to use different SCSs to index all the PRBs within the whole system bandwidth. As shown in Fig. 4.2, the three SCSs are used in three different subbands. But the starting PRB index (PRB#0) for each SCS is always assigned to the PRB at the beginning of the system bandwidth and the very last PRB index corresponds to the PRB at the end of the system bandwidth. This method can realize "one-step" indication of PRB, and dynamically schedule PRBs of various SCSs in any frequency-domain locations over the whole system bandwidth instead of a subband. (Of course, there should be a frequency-domain gap of 1−2 PRBs wide between the resources of different SCSs to avoid the inter-SCS interference.) However, this scheme needs to define for each SCS a set of PRB indexing in the whole system bandwidth, which results in excessive Downlink Control Information (DCI) overhead in frequency-domain resource allocation indication. In order to reduce the DCI overhead, some complicated resource indication methods need to be introduced.

- The second scheme is to index the PRB of various SCS independently on a per-subband basis. As shown in Fig. 4.2, the PRB index counts from the PRB (PRB#0) at the beginning of every subband to the very last PRB at the end of the subband. This method needs to firstly indicate the size and location of a subband for a SCS to determine the PRB indexing in the subband, and then to use the corresponding PRB indexing to identify the resources in the whole frequency domain. Obviously, the advantage of this method is that the legacy LTE resource allocation method can be directly reused within each subband. However, it requires a "two-step" indication logic.

With discussions and compromises, it was finally decided in the RAN1#88 meeting to adopt a "two-step" resource allocation method similar to scheme 2 [4,5] mentioned above. However, when it comes to naming the "subband" in the first-step allocation, in order to avoid the prejudice impacts from the existing concept to the future design, instead of using the well-known terms such as subband, a new terminology— "bandwidth part," abbreviated as BWP—was proposed in the discussion. Moreover, due to different preferences on this concept, the BWP concept was not agreed upon in the early-stage discussion to associate it with SCS. It was only emphasized that this concept can be used to describe the scenario where the UE bandwidth capability is smaller than the system bandwidth. This was the first time for the "bandwidth part" concept to appear in the discussion of 5G NR with its meaning still somehow unclear. With the development of subsequent study, this concept kept getting modified and expanded, and gradually became clear and complete.

4.1.2 Motivations from UE capability and power saving

As mentioned in the previous section, in the 3GPP RAN1 discussion for the "bandwidth part" concept, the reason for some companies to associate BWP with UE bandwidth capability was the preference to use this concept to describe the case where the UE bandwidth capability is smaller than the system bandwidth as well as the "bandwidth adaptation" operation for UE power saving.

First of all, as the carrier bandwidth of 5G NR becomes larger (single-carrier bandwidth can reach 400 MHz), it is unreasonable to require all classes of UEs to support such a large radio frequency (RF) bandwidth; it is necessary to support UEs to operate in a smaller bandwidth (such as 100 MHz). On the other hand, although the peak data rate of 5G increases, a single UE may still run a relatively low data rate in most of time. Even if the UE has 100 MHz bandwidth capability, it may operate with a smaller bandwidth when the high data rate is not demanded, so as to save the UE power.

The RAN1 meetings from October 2016 to April 2017 progressively reached the consensus to support the bandwidth adaptation in 5G NR [6−11], which is the function for a UE to monitor downlink control channel and receive downlink data channel in two different RF bandwidths. As shown in Fig. 4.3, the UE can monitor the downlink control channel in a narrower RF bandwidth W_1 and receive downlink data channel in a wider RF bandwidth W_2.

This functionality does not exist in the LTE system, because the UE always monitors Physical Downlink Control CHannel (PDCCH) within

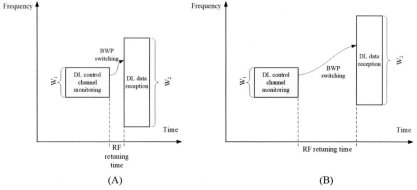

Figure 4.3 Downlink bandwidth adaptation principle (A) Same center frequency (B) Different center frequencies.

the entire system bandwidth (such as 20 MHz). However, in order to achieve higher data rate, the 5G NR bandwidth for data transmission is greatly increased (such as 100 MHz), while the bandwidth for PDCCH monitoring does not need to proportionally increase (such as 20 MHz or even smaller). Therefore it is beneficial to adjust the RF bandwidth of the UE according to different motivations at different times, which is the original intention of bandwidth adaptation. In addition, the central frequencies of W_1 and W_2 do not necessarily align to each other. The switching from W1 to W2 may only show a bandwidth expansion while the center frequency remains unchanged (as shown in Fig. 4.3A), or it can only switch to a new center frequency while keeping the bandwidth unchanged so as to realize load balancing across the system bandwidth and make full use of frequency resources.

The bandwidth adaptation in uplink is similar to that in downlink. When there is no high-speed data traffic in transmission, gNB can limit the frequency-domain scheduling range of Physical Uplink Control CHannel (PUCCH) to a smaller bandwidth W_3 to save UE power consumption, as long as W_3 is not too small to maintain sufficient frequency diversity. When high-speed data traffic needs to be transmitted, the frequency resource of Physical Uplink Shared CHannel (PUSCH) is scheduled in a larger bandwidth W_4, as shown in Fig. 4.4.

3GPP studied the UE power-saving effect brought by bandwidth adaptation [12−14]. Generally speaking, the capacity of downlink control information is much smaller than that of downlink data, while the UE needs to continuously monitor the downlink control channel for a long time, but only receives downlink data occasionally. Therefore if the UE

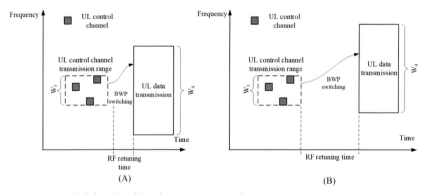

Figure 4.4 Uplink bandwidth adaptation principle.

always works at a fixed RF bandwidth (say W_2 to meet the data reception bandwidth), the RF bandwidth of the UE exceeds the needed size most of the time, which may lead to unnecessary UE power consumption [The power consumption of analog-digital conversion (ADC) and Digital-Analog Conversion (DAC) is proportional to the working bandwidth (the direct impact is sampling rate), and the power consumption of baseband operation is proportional to the number of processing time-frequency resources] [12,13]. An estimation for the LTE system shows that 60% of the UE power consumption comes from PDCCH demodulation, synchronization tracking, and small data rate traffic. Although there are different analysis results on the actual power consumption that can be saved, a common view is that bandwidth adaptation can significantly reduce UE power consumption in wideband operations.

It should be noted that bandwidth adaptation cannot be completed instantaneously because it involves the parameter adjustment of RF devices, which takes a certain amount of time. The RF retuning time includes the time to receive bandwidth adaptation command, the time used to retune the frequency point and RF bandwidth, and the time used to reset the device components such as ADC, DAC, and AGC (automatic gain control) [13]. The study on BWP switching delay was undertaken in 3GPP RAN4 (the working group responsible for RF performance and requirements), and was carried out in parallel with the BWP study in RAN1 (the working group responsible for physical layer design). Table 4.2 shows RAN4's preliminary study results in

Table 4.2 RF retuning time under different conditions.

| | | RF retuning time (μs) | Corresponding number of symbols (Normal CP) | |
			Subcarrier spacing = 30 kHz	Subcarrier spacing = 60 kHz
Intraband operation	Center frequency unchanged, only the bandwidth changed	20	−1 symbols	−2 symbols
	Center frequency changed	50−200	−2−6 symbols	−3−12 symbols
Interband operation		900	−26 symbols	−51 symbols

early 2017, which contains the RF retuning time. The study on bandwidth adaptation and BWP in 3GPP RAN1 was based on these results. The complete study results from RAN4 on RF retuning time were officially rolled out in early 2018 and are described in detail in Section 4.3.4.

As seen from Table 4.2, RF retuning time, which may be several or more than ten OFDM symbols long, is a nonnegligible transition period (BWP switching delay including baseband parameter reconfiguration is even larger). During RF retuning time, the UE cannot perform normal transmitting/receiving operations. This brings a great impact on the subsequent BWP design, because the inability of transmission/reception during BWP switching must be considered in the timeline designs of many channels and signals.

4.1.3 Basic bandwidth part concept

Although the BWP concept was separately generated based on two different technical motivations mentioned in Sections 4.1.1 and 4.1.2, the 3GPP discussion finally converged to a view that these two motivations can be fulfilled with one single BWP concept. At the 3GPP RAN1#88bis and RAN1#89 meetings, a series of agreements established the basic concept and key features of BWP [10,15−18]:

- A UE can be semistatically configured by gNB with one or more BWP(s), for uplink and downlink, respectively.
- The BWP bandwidth is equal to or smaller than the RF bandwidth capability supported by a UE, but larger than the bandwidth of Synchronization Signal/PBCH block (SS/PBCH block), including SS and PBCH; see Chapter 6: NR Initial Access for details).
- Each BWP contains at least one CORESET (Control-Resource Set; see Section 5.4 for details).
- BWP may or may not contain SS/PBCH block.
- BWP consists of a certain number of PRBs, and is associated with a specific numerology [including a SCS and a cyclic prefix (CP)]. Configuration parameters of a BWP include bandwidth (e.g., number of PRBs); frequency location (e.g., center frequency); and numerology (including SCS and CP).
- A UE has only one downlink active BWP and only one uplink active BWP at a time.
- A UE only operates within the active BWP and does not receive/transmit signals in the frequency range outside the active BWP.

- Physical Downlink Shared CHannel (PDSCH) and PDCCH are transmitted in the downlink active BWP, and PUSCH and PUCCH are transmitted in the uplink active BWP.

4.1.4 Use cases of bandwidth part

The use cases of the BWP concept were relatively limited in the early stage, which mainly covered the control channel and data channel. However, the concept was progressively expanded in the subsequent standardization study, and finally became universal by covering almost all aspects of the 5G NR physical layer.

First of all, the basic concept of BWP not only comes from not only "the perspective of resource allocation" but also "the perspective of UE power saving." The BWP scheme designed from "the perspective of resource allocation" only takes into account the use of BWP for the data channels on a per-SCS basis, with no intention to apply BWP to the control channel. The BWP scheme designed from "the perspective of UE power saving" only intends to use BWP for the control channel, without taking into account the relationship with the SCS and the use of BWP for the data channel (the data channel is still scheduled across the UE bandwidth capability). The finally agreed upon BWP concept applies not only to the data channel and control channel, but also to reference signals (RS) and even in the initial access process procedure. BWP becomes one of the most basic concepts of 5G NR specification. Its importance grows beyond the expectations of those who came up with the two original aspects of the concept.

As to whether BWP should be used for data channel scheduling, there was at least one consensus that the UE bandwidth capability may be smaller than the system bandwidth, in which case it is necessary to indicate the location of the UE operation bandwidth inside the system bandwidth or relative to a reference point (this approach is introduced in Section 4.2). However, as long as the BWP is used for data channel resource allocation and PRB indexing and Method 2 shown in Fig. 4.2 is adopted, it is necessary to associate BWP with a SCS so as to realize the advantage of method 2 (i.e., the resource indication method of LTE can be reused within the BWP).

The 3GPP discussion on whether to introduce the BWP concept for the control channel was once controversial. Theoretically, the UE power saving relating to control channel transmission and reception can be realized by other physical layer concepts. For example, the PDCCH resource set (CORESET) can be directly configured to define the frequency range for PDCCH monitoring; and the PUCCH resource set can be directly configured to define the frequency range for PUCCH transmission. As long as the frequency ranges of the two resource sets are configured to be much smaller than the frequency range of data channel reception/transmission, the goal of UE power saving can also be achieved. However, to introduce a separate but universal BWP concept is obviously a more intuitive, clearer, and more systematic method. It is also more efficient and more scalable to unify, based on BWP, the resource indications of various channels and signals such as CORESET and PUCCH resource set. Another advantage of using BWP to define the operation bandwidth during PDCCH monitoring is that a BWP slightly wider than CORESET can be configured, so that the UE can receive a small amount of downlink data while monitoring PDCCH (as shown in Fig. 4.5). The 5G NR system emphasizes low-latency data transmission and aims to transmit the data at any time. This design can effectively support simultaneous reception of PDCCH and PDSCH at any time (the multiplexing of PDCCH and PDSCH is described in detail in Chapter 5: 5G flexible scheduling).

In addition, in the basic concept of BWP, the BWP was only used by UE in RRC connected mode, given BWP was configured through UE-specific RRC signaling. However, in the later 3GPP study, the use of BWP was extended to the initial access procedure. The determination of BWP in initial access follows a different method, discussed in detail in Section 4.4.

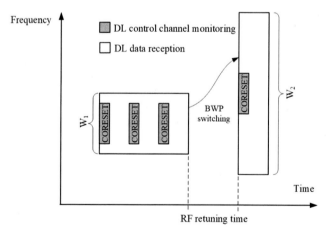

Figure 4.5 BWP to support simultaneous PDCCH monitoring and PDSCH reception. *BWP*, Bandwidth part.

4.1.5 What if bandwidth part contains synchronization signal/physical broadcast channel block?

There were different views on whether each BWP should contain SS/PBCH block (SS/PBCH Block, SSB). The advantage for each BWP to contain SSB is that the UE can carry out SSB-based mobility measurement and PBCH-based system information (SI) update without switching BWP; but the disadvantage is to bring in many restrictions on the flexibility of BWP configuration. As shown in Fig. 4.6A, if each BWP must contain an SSB, all BWPs can only be configured surrounding the SSB, making it difficult to fully utilize the frequency range far apart from the SSB. Only with BWP configuration being allowed having no SSB, as shown in Fig. 4.6B, all frequency resources can be fully utilized and flexible resource allocation can be realized.

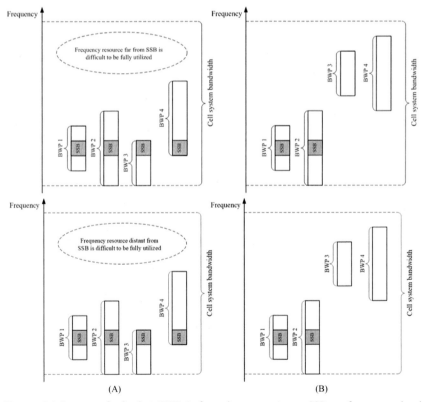

Figure 4.6 Impact of whether BWP is forced to contain an SSB on frequency load balancing (A) Each BWP must contain an SSB (B) BWP may not contain an SSB. *BWP,* Bandwidth part.

4.1.6 Number of simultaneously active bandwidth parts

Another controversial discussion in 3GPP was in regards to whether a UE can simultaneously have multiple active BWPs. The original intention of activating multiple BWPs is to support UE to transmit/receive traffics with different SCSs simultaneously (e.g., eMBB traffic with SCS = 30 kHz and URLLC traffic with SCS = 60 kHz). Because each BWP is associated with only one SCS, it is necessary to activate multiple BWPs simultaneously.

As shown in Fig. 4.7, if only one downlink BWP and one uplink BWP can be active in a carrier at a time (i.e., single active BWP), the UE has to switch between two SCSs in order to use one or another; and it can only switch to another SCS after the RF retuning time (Theoretically, if only numerology is changed without changing bandwidth or center frequency, there is no need for RF retuning, but the current 5G NR specification assumes that RF retuning time should be reserved for all kinds of BWP switching.). Therefore the limitation of one active BWP makes it impossible for the UE to support two SCSs in one symbol simultaneously, or to quickly switch between two SCSs.

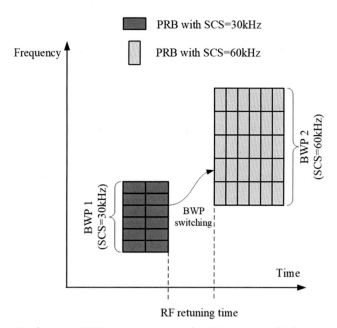

Figure 4.7 Single active BWP cannot support the UE to use multiple subcarrier spacings simultaneously. *BWP*, Bandwidth part.

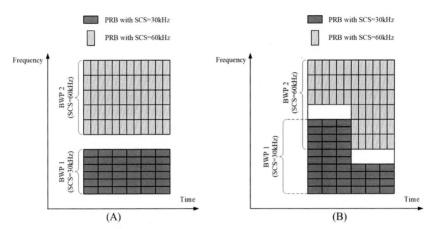

Figure 4.8 Multiple active BWPs support the UE to use multiple subcarrier spacings simultaneously (A) Non-overlapping multiple active BWPs. (B) Overlapping multiple active BWPs. *BWP*, Bandwidth part.

If multiple active BWPs are supported, as shown in Fig. 4.8, the UE can be allocated frequency resources with multiple SCSs simultaneously, and can switch between two SCSs quickly (It can switch to another SCS without going through RF retuning time). The simple case is that two simultaneous active BWPs do not overlap each other in frequency domain (as shown in Fig. 4.8A). But it is also possible for the two active BWPs to either completely or partially overlap with each other in frequency domain (as shown in Fig. 4.8B). In the overlapping part, the gNB can schedule resources for the UE based on either of the two SCSs. In case the gNB uses both BWPs shown in Fig. 4.8B for the same UE, a certain frequency guard band should be reserved between two BWPs with different SCSs to prevent interference.

In the end, whether to support multiple active BWPs mainly depends on whether to support a 5G UE running with multiple SCSs simultaneously. Within 3GPP discussion, most companies shared a view that it was a highly strict requirement for R15 NR UE to run two SCSs simultaneously. Such requirement forces the UE to run two different Fast Fourier transform sizes simultaneously, which may increase the complexity of the UE baseband processing. In addition, activating multiple BWPs simultaneously would complicate the power-saving operation of the UE. A UE needs to calculate the envelope of multiple BWPs to determine the working bandwidth of RF devices. If the multiple active BWPs are scattered in the frequency domain, the power-saving target may not be achieved. Therefore 3GPP finally determined that only single active BWP is supported in R15 NR [19] as well as in R16 NR. Whether to support multiple active BWPs in a future release depends on future study.

4.1.7 Relation between bandwidth part and carrier aggregation

CA is a bandwidth-extension technology supported since the LTE-Advanced era, which can aggregate multiple component carriers (CC) for simultaneous reception or transmission. According to the range of aggregated carriers, CA can be divided into intraband CA and interband CA. One use case of intraband CA is the "wide carrier" operation, in which the cell carrier bandwidth is greater than the single-carrier bandwidth capability of a UE. For example, if the base station supports the 300 MHz carrier while the UE only supports the 100 MHz carrier, the UE can use intraband CA to implement a wideband operation beyond 100 MHz. The CCs can be adjacent or nonadjacent to each other. The eNB can activates/deactivates a CC. As shown in Fig. 4.9, among the three carriers, Carrier 1 and Carrier 2 are activated, while Carrier 3 is not. Then the UE performs CA of two CCs (200 MHz).

In the study of NR CA and BWP, it was suggested to merge the two designs due to the similarity of the configuration parameters of carrier and BWP, for example, both configurations include bandwidth and frequency location. Because BWP is a more flexible concept than CC, the BWP

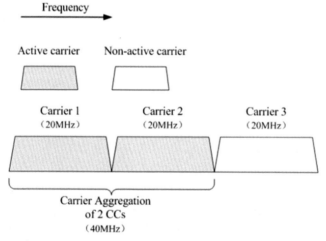

Figure 4.9 Wide carrier operation based on CA. *CA*, Carrier aggregation.

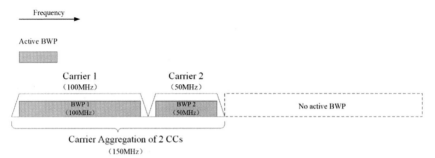

Figure 4.10 Use active multiple BWPs to unify CA and intracarrier BWP operation. *BWP*, Bandwidth part; *CA*, carrier aggregation.

operation can replace CC operation and operation of multiple active BWPs can replace CA operation. Then multiple BWPs can be used to unify the configuration and activation of both CA (to achieve large bandwidth operation) and BWP (to achieve UE power saving and resource allocation of multiple SCSs). As shown in Fig. 4.10, activating two BWPs, which are 100 and 50 MHz, respectively, is equivalent to activating two carriers. In this case, the UE performs CA (100 MHz + 50 MHz = 150 MHz) of two CCs, and there is no active carrier in the frequency range where there is no active BWP. This is equivalent to activating CC by activating BWP.

One reason to use BWP instead of CA process is that the BWP can be activated by DCI, which is faster than MAC−CE (Media Access Control—Control Element) that is used to activate the CC. However, this method requires the support of multiple active BWPs. As mentioned above, it was decided R15 NR does not support multiple active BWPs for the time being.

3GPP finally adopted a "carrier + BWP" two-layer configuration and activation for 5G NR. That is to say, BWP is a concept within a carrier. The configuration and activation/deactivation of the carrier and the configuration and activation/deactivation of the BWP are separately designed. The traditional method to activate a carrier is still inherited. One BWP can be activated within each active carrier (i.e., the BWP in a carrier can be activated only after the carrier itself is activated). If a carrier is deactivated, the active BWP in the carrier is also deactivated. As shown in Fig. 4.11, among the three carriers, Carrier 1 and Carrier 2 are activated [i.e., the UE performs CA of two CCs (200 MHz)]. A BWP of 100 MHz is activated in CC 1 and a BWP of 50 MHz is activated in CC 2. Carrier

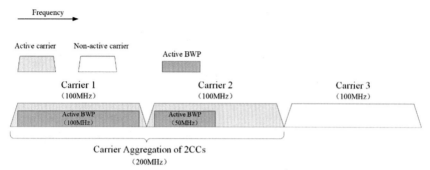

Figure 4.11 "Carrier + BWP" two-layer configuration and activation. *BWP*, Bandwidth part.

3 is not activated, so no BWP can be activated in carrier 3. When a CC is deactivated, all BWPs in the CC are deactivated simultaneously.

The next question is: Can BWP deactivation be used to trigger CC deactivation? Some suggested that although CC activation and BWP activation are independent mechanisms, BWP deactivation can be associated with CC deactivation. For example, if the only active BWP in a CC is deactivated, the CC should also be deactivated at the same time. This may help to further save UE power consumption. However, even after the BWP for scheduling data is deactivated, it does not mean that any transmission/reception operation is no longer needed in the CC. Other transmission/reception such as PDCCH monitoring and SPS (SemiPersistent Scheduling) data may still exist. As we will describe in Section 4.3, in fact, because there is only one single BWP being active, there is no need to design a specific BWP deactivation mechanism. That is to say, as long as a CC is active, there is always an active BWP. There is no case that "there is no active BWP in the active CC."

The NR CA design still basically follows the LTE CA design philosophy (e.g., still using MAC CE for CC activation, rather than supporting DCI-based CC activation). As mentioned above, R15 and R16 NR allow only one BWP being active in a carrier. But in case of N-CC CA, N BWPs can actually be active simultaneously in different CCs.

So far, this section introduces how the BWP concept was established. Many details in the BWP design will be introduced in the rest of this chapter. The frequency-domain characteristics of BWP are defined in the RRC configuration of BWP; and the time-domain characteristics of BWP are defined in the BWP activation/deactivation procedure. These are introduced in Sections 4.2 and 4.3, respectively. Section 4.4 and

Section 4.5 will introduce BWP in the initial access and the impact of BWP on other physical layer designs, respectively. In the 3GPP specification, the core physical process of BWP is defined in Section 13.1 of TS 38.213 [20], while the other BWP-related contents are distributed in different specifications of physical layer and high layer protocols.

4.2 Bandwidth part configuration

As described in Section 4.1.3, when the basic concept of BWP was formed, it had been clear that the basic configuration parameters of BWP should include bandwidth (e.g., number of PRBs); frequency location (e.g., center frequency); and numerology (including SCS and CP). An important function of BWP is to provide the PRB indexing for fine granularity in resource allocation (i.e., the second step of the "coarse + fine" method described in Section 4.1.1). In the LTE system, the PRB is defined within the carrier. While in the 5G NR system, almost all physical layer procedure within a single carrier are described based on BWP, so the PRB is also defined in the BWP. As long as the frequency-domain size and location of BWP are known, various frequency operations can be described in a similar way to LTE based on PRB indexing within the BWP. So the key issues in the NR BWP specification include: how to determine the bandwidth and frequency location of the BWP? and how are PRBs in the BWP mapped to the absolute frequency locations?

4.2.1 Introduction of common RB

First of all, because the bandwidth and frequency location of BWP can be flexibly configured by UE-specific RRC signaling, the configuration of these two parameter needs to be based on a Resource Block (RB) grid known by the UE. That is, the RB grid for indicating where the BWP starts and ends in frequency-domain needs to be informed to the UE when the UE set ups an RRC connection. Given the BWP configuration is UE-specific and already flexible enough, it is fairly reasonable to design this RB grid as a common grid to all UEs.

At the RAN1#89 and RAN1#AH NR2 meetings, it was determined to introduce the basic concept of a common RB (CRB) [16,21−23], with which UEs with different carrier bandwidth and UEs with CA can share a unified RB indexing. Therefore this CRB concept equivalently provides an absolute frequency scale that can cover one or more carrier frequency ranges. One of CRB use cases is to configure BWP, while the

PRB indexing in the BWP is UE-specific to fit the use for resource allocation within the BWP [23]. Consequently, there needs a mapping relationship between UE-specific PRB indexing in the BWP (n_{PRB}) and CRB indexing (n_{CRB}), which can be simply expressed as follows (as described in Section 4.4.4.4 of TS 38.211 [24]):

$$n_{\mathrm{CRB}} = n_{\mathrm{PRB}} + N_{\mathrm{BWP},i}^{\mathrm{start}} \tag{4.1}$$

Of course, BWP configuration is not the only purpose of introducing CRB indexing. Another important consideration lies in the sequence generation procedure for many types of RS, which may need to be based on a universal starting point that does not change with BWP size and location.

4.2.2 Granularity of common RB

As mentioned above, in order to determine the basic characteristics of CRB, it is necessary to firstly determine the frequency-domain granularity (e.g., the unit for frequency allocation) and the SCS of the CRB indexing.

Regarding the granularity of CRB, two options were proposed: One is to directly use RB (i.e., the minimum frequency allocation unit) to define CRB; the other is to use a subband (consisting of several RBs) to define CRB. The first option allows the most flexible BWP configuration, but brings a larger RRC signaling overhead. In contrast, the second option results in smaller RRC signaling overhead. The BWP is only a "coarse allocation" that does not require a fine granularity. If the whole carrier is divided into several subbands with each subband containing a predefined number of RBs, the BWP configuration signaling only needs to indicate which subbands the BWP contains [25]. However, RRC signaling is transmitted in PDSCH, and is not sensitive to signaling overhead. Therefore the 3GPP discussion finally decided to use one RB as the granularity of CRB.

Different BWPs may adopt different SCSs. There are two options to define SCS for CRB:

- Option 1: BWPs with different SCSs adopt different CRB granularities (i.e., each kind of SCS has its own CRB indexing). As shown in Fig. 4.12, three CRB indexing methods are respectively used for the three SCSs (i.e., 15, 30, and 60 kHz) and defined in unit of RB with SCS = 15, SCS = 30, and SCS = 60 kHz. In the example in Fig. 4.12,

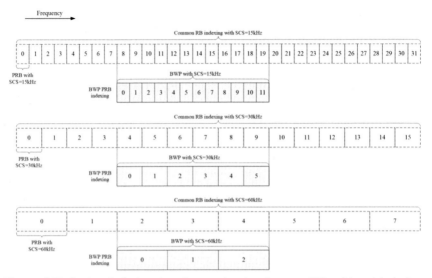

Figure 4.12 Option 1: Each subcarrier spacing has its own CRB grid and indexing. *CRB*, Common RB.

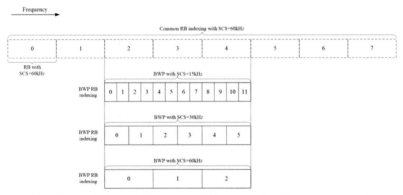

Figure 4.13 Option 2: A unified common RB indexing for all subcarrier spacings.

the BWP of SCS = 15 kHz starts from RB#8 in the SCS = 15 kHz CRB indexing, the BWP of SCS = 30 kHz starts from RB#4 in the SCS = 30 kHz CRB indexing, and the BWP of SCS = 60 kHz starts from RB#2 in the SCS = 60 kHz CRB indexing.

- Option 2: A unified reference SCS (reference SCS) is used to define the CRB in a carrier. All the BWPs in this carrier, regardless of their carrier spacings, adopt the unified CRB indexing. One example is given in Fig. 4.13, where the CRB is defined based on a reference

SCS equal to 60 kHz. As the example shows, a large SCS used to define the CRB can reduce the RRC signaling overhead for BWP configuration.

Both two options can work. In comparison, the first option is simpler and more intuitive. As mentioned above, RRC signaling is not sensitive to overhead. Thus it was decided to adopt the first option (i.e., each SCS has its own CRB indexing).

4.2.3 Reference point—point A

CRB is an "absolute frequency grid" that can be used for any carrier and BWP. The starting point [23] of the CRB indexing is named CRB 0. Once the granularity of CRB is determined, the next task is to determine the position of CRB 0, for which the corresponding position index can be used to indicate the location and size of the carrier and BWP. Assume the CRB 0 is located at "Point A," intuitively, there are two methods to define the Point A [22,26,27].

- Method 1: Define Point A based on carrier location

Because a BWP is part of a carrier, an intuitive method is to define Point A based on location of the carrier (such as the center frequency or carrier boundary). Then the BWP location can be directly indicated from CRB 0 by $N_{BWP,i}^{start}$ (as shown in Fig. 4.14). The first reference point design assumes the traditional system design [i.e., the synchronization signal is always located in the center of the carrier (such as LTE system)]. Thus the UE already knows the location and size of the carrier after the cell search procedure. From the perspective of gNB and UE, the carrier concept is completely the same. In this way, if the gNB wants to configure a BWP for the UE in a carrier, it can directly indicate the BWP location based on

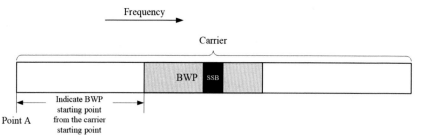

Figure 4.14 BWP starting point indication method based on traditional system design. *BWP*, Bandwidth part.

the carrier location, such as indicating the offset from the carrier starting point to the BWP starting point.

• Method 2: Define Point A based on the initial access "anchor point"

Through the initial access procedure, the UE also has knowledge of more frequency-domain "anchor points," such as location of SSB (center frequency or boundary), location of [remaining minimum SI (RMSI); i.e., SI block type1 (SIB1)]. These "anchor points" can actually provide a more flexible definition of Point A.

One design objectives of NR system is to support a more flexible carrier concept. For example, SSB may not necessarily be located in the center of the carrier; a carrier may contain multiple SSBs (this design can be used in the "wideband" carrier); or a carrier may not contain any SSB at all (this design can be used to implement a more flexible CA system). Moreover, the carrier can be defined from the UE perspective (i.e., the carrier seen from the UE perspective can be theoretically different from the carrier seen from the gNB perspective).

Fig. 4.15A shows the scenario in which a carrier contains multiple SSBs. The UE accesses the network from SSB 1 and operates in a "virtual carrier" that is different from the "physical carrier" seen on the gNB side. The "virtual carrier" may only contain SSB 1 but not other SSBs in the physical carrier (such as SSB 2).

Fig. 4.15B shows the scenario in which a carrier does not contain any SSB. The UE accesses from SSB 1 in Carrier 1, but works on Carrier 2 that does not contain any SSB. In this case, instead of indicating the BWP location relative to the Carrier 2 location, it is better to directly indicate it from SSB 1.

In order to support the above two flexible deployment scenarios, Point A can be defined based on the "anchor point" mentioned above.

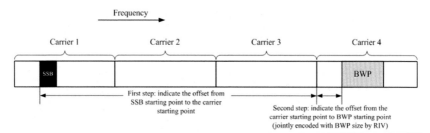

Figure 4.15 BWP starting point indication scenarios (A) Scenario 1: a carrier contains multiple SSBs (B) Scenario 2: some carriers do not contain SSB. *BWP*, Bandwidth part.

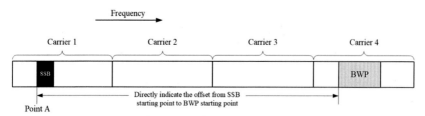

Figure 4.16 The method of directly indicating the BWP location relative to SSB. *BWP,* Bandwidth part.

For example, when the UE accesses from an SSB, it takes the SSB as the "anchor point" (i.e., Point A) to derive the frequency resource locations of other Physical layer (PHY) channels. In this way, the UE does not need to know the range of the physical carrier on the gNB side and the locations of other SSBs. As shown in Fig. 4.16, if Point A is defined by the starting point of SSB 1, and the offset of the BWP starting point relative to Point A is indicated, it is not necessary to provide a UE with the information of the actual carrier for the UE to learn the location and size of the BWP.

When comparing Method 1 and Method 2, it is obvious that the second method is more in line with the original intention of NR design principle. But this method also has some problems. For example, in a NR system with CA, the BWP may be configured at any CC. If a BWP is located in a CC far apart from the CC containing the SSB used by the UE for initial access (as shown in Fig. 4.16), the offset between the BWP starting point and the SSB starting point may contain a large number of CRBs. There would be no problem if the starting point and size of BWP are indicated separately. However, if the starting point and size of BWP need to be indicated with joint encoding [e.g., by resource indication value (RIV)], the large value of the BWP starting RB index may result in a large RIV value and a large signaling overhead, as well as a complicated RIV decoding.

In order to solve this problem, a combination of the above two methods can be used to indicate the BWP location. As shown in Fig. 4.17, the first step is to indicate the offset from the SSB starting point to the starting point of the carrier in which the target BWP locates; and the second step is to indicate the offset from the carrier starting point to the BWP starting point (the second offset is jointly encoded with the BWP size in RIV).

In the example shown in Fig. 4.17, the SSB starting point is directly used as Point A. One method to introduce further flexibility is to allow a

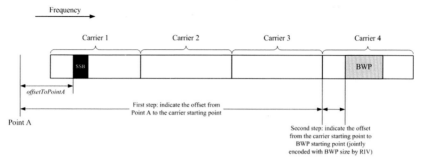

Figure 4.17 "Two-step method" to indicate the BWP location. *BWP*, Bandwidth part.

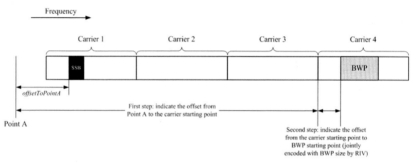

Figure 4.18 BWP indication method with *offsetToPointA*.

certain offset between Point A and SSB (i.e., the relative offset from the SSB starting point to Point A can also be flexibly configured). In this way, the method in Fig. 4.17 can be modified as shown in Fig. 4.18. The UE uses the high-layer signaling parameter *offsetToPointA* to determine the offset from the first subcarrier of the first RB of SSB to Point A. There have been different schemes regarding what kind of RRC signaling is used to indicate *offsetToPointA*. For the sake of flexibility, UE-specific RRC signaling can be used to indicate *offsetToPointA* so that different UEs accessing from the same SSB can even have different Point A and CRB indexing. However, at least as of R16, the necessity of this flexibility is not clear and strong, and it remains sufficient for Point A to be used as a common starting point of CRB for all UEs accessing from the same SSB. Therefore *offsetToPointA* can be carried in RMSI signaling (SIB1) to avoid the waste of overhead due to separate UE-specific configurations.

Fig. 4.18 shows the basic method of indicating the Point A position based on the SSB starting point, but there are still some details to be determined when it comes to the definition of "the SSB starting point." With the consideration of signaling overhead, *offsetToPointA* needs to be indicated in RB units. However, due to restriction of RF design, the frequency grid used in cell search may be different from that of CRB, leaving the subcarrier and RB of SSB possibly not aligned with the subcarrier grid and RB grid of CRB. Therefore the SSB starting point cannot be directly taken as the reference point of *offsetToPointA*. Instead, a certain RB in the CRB indexing should be used as the SSB starting point. Finally, it was determined in 3GPP to use the method shown in Fig. 4.19 to determine the reference point for *offsetToPointA* indication: First, use the SCS indicated by *subCarrierSpacingCommon* to define a CRB grid; then, find the first CRB (called N_{CRB}^{SSB}) overlapped with SSB, and the offset relative to the SSB is given by the high-layer parameter k_{SSB}. The center of the first subcarrier of N_{CRB}^{SSB} is used as the reference point to indicate *offsetToPointA*. The SCS of *offsetToPointA* is 15 kHz for FR1 (Frequency Range 1) and 60 kHz for FR2 (Frequency Range 2).

The method of determining the position of Point A and CRB 0 based on SSB, as shown in Fig. 4.19, can be applied to the TDD system and FDD downlink, but cannot directly applied to FDD uplink. FDD uplink and FDD downlink are in different frequency bands with at least tens of MHz apart, and FDD uplink carrier does not contain SSB. Therefore it is difficult for FDD uplink to determine the position of Point A and CRB 0 from the receiving SSB. In this case, a traditional method inherited from the 2G era [i.e., the indication of absolute RF channel number (ARFCN) [28]] can be used to indicate Point A. As shown in Fig. 4.20, without relying on SSB, the UE can determine the position of Point A for FDD uplink based on ARFCN according to the high-layer signaling parameter *absoluteFrequencyPointA*.

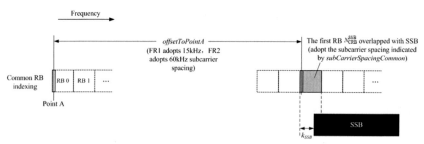

Figure 4.19 The method of determining the reference point indicating *offsetToPointA*.

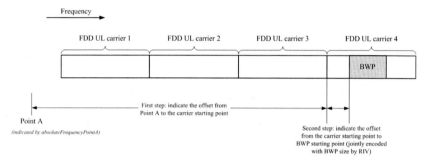

Figure 4.20 BWP indication method for FDD uplink. *BWP*, Bandwidth part.

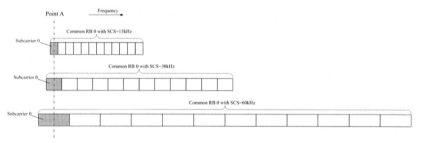

Figure 4.21 Subcarrier 0 of common RB 0 at Point A.

4.2.4 The starting point of common RB—RB 0

As described in Section 4.2.3, the position of Point A can be determined based on SSB or ARFCN, and Point A is where CRB 0 is located. However, Point A is a point in frequency domain, but CRB 0 has a certain frequency width. There is still a detail to be decided about how to determine the position of CRB 0 based on Point A.

As mentioned in Section 4.2.2, each SCS μ has its own CRB indexing. Meanwhile, the positions of CRB 0 with different μ are all determined by Point A (as specified in Section 4.4.4.3 of TS 38.211 [24]) (i.e., as shown in Fig. 4.21), the center frequency of the first subcarrier of CRB 0 (i.e., subcarrier 0) is located at Point A for any μ. In other words, the boundaries of CRB 0 with different μ are actually not completely aligned.

4.2.5 Indication method of carrier starting point

To summarize the methods described in Sections 4.2.1—4.2.4: Point A is determined by SSB or ARFCN; CRB 0 is determined by Point A; the carrier starting point is indicated from CRB 0; and finally the location and size of BWP are indicated from the carrier starting point. With CRB 0 determined, the offset $N_{grid}^{start, \mu}$ of carrier starting point relative to CRB 0 can be indicated by *offsetToCarrier* in the high-layer signaling *SCS-SpecificCarrier*, which is carried by SIB information in the initial access procedure or RRC signaling in the handover procedure. The value range of *offsetToCarrier* is 0−2199 (see TS 38.331 [29]). The RB indexing in 5G NR contains up to 275 RBs, so *offsetToCarrier* can indicate the frequency location of at least eight adjacent CCs to support CA operation.

4.2.6 Bandwidth part indication method

5G NR adopts two frequency resource allocation methods similar to LTE: contiguous resource allocation and noncontiguous resource allocation (respectively named Type 1 and Type 0 resource allocations in NR specification; see Section 5.2). In principle, both contiguous resource allocation (as shown in Fig. 4.22A) and noncontiguous resource allocation (as shown in Fig. 4.22B) can be used for BWP frequency allocation. However, since BWP frequency allocation is only a "coarse allocation," Type 0 resource allocation type can still be used to allocate noncontiguous PRBs when "fine allocation" is carried out within the BWP. It is not necessary to configure a BWP containing noncontiguous RBs. Therefore it is finally determined that the BWP is composed of contiguous CRBs [10,15]. Type 1 resource allocation type is used to configure BWP (i.e., the RIV with "starting position + length" joint encoding is used).

Consequently, the starting point of BWP in CRB unit relative to the carrier starting point can be expressed as:

$$N_{BWP,i}^{start, \mu} = N_{grid,i}^{start, \mu} + RB_{BWP,i}^{start, \mu} \tag{4.2}$$

where $N_{grid, i}^{start, \mu}$ is the CRB number where the carrier starting point is located, and $RB_{BWP,i}^{start}$ is the CRB number of the starting point derived from the RIV value according to formula (4.2).

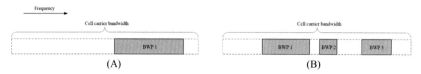

(A)   (B)

Figure 4.22 Contiguous BWPs and noncontiguous BWPs (A) Contiguous BWPs (B) Noncontiguous BWPs. *BWP*, Bandwidth part.

This RIV value is indicated by the high-layer signaling *locationAndBandwidth* in BWP configuration. Then the starting point and size of BWP can be deduced from RIV by using the following formula (see Section 5.1.2.2.2 of TS 38.214 [30]):

$$\text{If} \quad (L_{\text{RBs}} - 1) \le \left\lfloor N_{\text{BWP}}^{\text{size}}/2 \right\rfloor, \quad \text{RIV} = N_{\text{BWP}}^{\text{size}}(L_{\text{RBs}} - 1) + RB_{\text{start}}$$
$$\text{else}, \quad \text{RIV} = N_{\text{BWP}}^{\text{size}}\left(N_{\text{BWP}}^{\text{size}} - L_{\text{RBs}} + 1\right) + \left(N_{\text{BWP}}^{\text{size}} - 1 - RB_{\text{start}}\right)$$

$$(4.3)$$

It should be noted that this formula was written to indicate the PRB allocation within a BWP, and was borrowed to define the RIV indication of the starting point and size of BWP itself. In other words, when Eq. (4.3) is used for interpret *locationAndBandwidth* in BWP configuration, RB_{start} and L_{RBs} in Eq. (4.3) equivalently represent the starting position (i.e., $N_{\text{BWP},i}^{\text{start}}$ in Section 4.2.7) and size (i.e., $N_{\text{BWP},i}^{\text{size}}$ in Section 4.2.7) of BWP, respectively, and $N_{\text{BWP}}^{\text{size}}$ in Eq. (4.3) should be the upper limit of the BWP size. Because the maximum number of RBs that can be indicated in R15 NR is 275, $N_{\text{BWP}}^{\text{size}}$ is fixed to be 275 (see Chapter 12: 5G network slicing of TS 38.213 [20]) when Eq. (4.3) is used for BWP configuration. Mathematically, when $N_{\text{BWP}}^{\text{size}} = 275$, $RB_{\text{start}} = 274$, and $L_{\text{RBs}} = 138$, RIV in Eq. (4.3) reaches the maximum value of 37,949. So the value range of *locationAndBandwidth* is 0−37949.

The frequency resource of each BWP is configured independently. As shown in Fig. 4.8, different BWPs can contain overlapping frequency resources. In this respect, the specification does not set any restriction.

4.2.7 Summary of the basic bandwidth part configuration method

Fig. 4.23 shows the methods for the BWP frequency configuration and the mapping from PRB to CRB. The whole procedure can be summarized as follows:

1. Determine the first CRB $N_{\text{CRB}}^{\text{SSB}}$ overlapping with SSB.
2. Determine downlink Point A from the first subcarrier of $N_{\text{CRB}}^{\text{SSB}}$ or uplink Point A from ARFCN.
3. Determine the CRB 0 position and CRB indexing of each SCS from Point A.
4. Determine the carrier starting position $N_{\text{grid}}^{\text{start}, \mu}$ based on the CRB 0 position and *offsetToCarrier*.

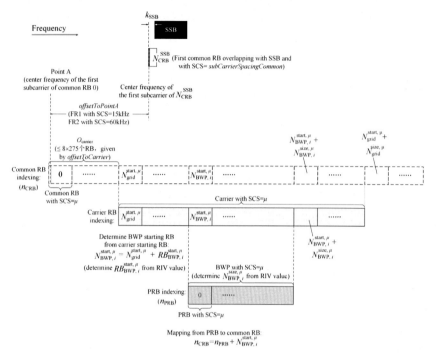

Figure 4.23 PRB indexing determination process in the BWP (taking determining Point A based on SSB as an example).

5. Determine the offset of the BWP starting point relative to $N_{\text{grid}}^{\text{start}, \mu}$ and the BWP bandwidth according to the RIV value indicated by *locationAndBandwidth*, so as to determine the BWP starting point $N_{\text{BWP},i}^{\text{start}}$ and size $N_{\text{BWP},i}^{\text{size}}$ calculated in CRB. The RIV indication assumes the maximum number of CRBs in a BWP is 275.

6. The mapping relationship between PRB indexing and CRB indexing in the BWP is $n_{\text{CRB}} = n_{\text{PRB}} + N_{\text{BWP},i}^{\text{start}}$.

From the above procedure, it can be seen that the UE only needs to use the carrier starting point $N_{\text{grid}}^{\text{start}, \mu}$ to determine the BWP frequency range and the mapping from PRB to CRB. It does not need to know the carrier size $N_{\text{grid}}^{\text{size}, \mu}$. (For the FR requirement of UE transmission, the UE still needs to know the system carrier location, but this information is not needed from the perspective of configuring BWP and CRB, as well as the resource allocation of PHY channels.) Therefore the above procedure only ensures that the BWP starting point is within the carrier range, but cannot ensure that the BWP ending point is also within the carrier range. To ensure that the BWP frequency range configured by gNB for UE is limited to the carrier range, the following formula is introduced in Section 4.4.5 of TS 38.211 [24]:

$$N_{\text{grid},x}^{\text{start},\mu} \leq N_{\text{BWP},i}^{\text{start},\mu} < N_{\text{grid},x}^{\text{start},\mu} + N_{\text{grid},x}^{\text{size},\mu}$$
$$N_{\text{grid},x}^{\text{start},\mu} < N_{\text{BWP},i}^{\text{size},\mu} + N_{\text{BWP},i}^{\text{start},\mu} \leq N_{\text{grid},x}^{\text{start},\mu} + N_{\text{grid},x}^{\text{size},\mu}$$

$$(4.4)$$

To sum up, a basic BWP configuration includes frequency location; size (the corresponding high-layer signaling parameter is *locationAndBandwidth*); and numerology of the BWP. Because the SCS in the numerology is characterized by the parameter μ, and selection from two CPs needs to be indicated by another parameter when $\mu = 2$ (i.e., SCS = 60 kHz), the basic configuration of BWP includes three parameters: *locationAndBandwidth*, *subcarrierSpacing*, and *cyclicPrefix* (see TS 38.331 [29]). Given the original intention to introduce BWP is to facilitate UE power saving and resource allocation, the BWP is configured by UE-specific RRC signaling in normal cases. However, as a special BWP, the initial BWP has its own determination method, as described in Section 4.4.

It can be seen that the three parameters of each BWP are configured independently, providing great flexibility for BWP configuration. Completely different from the traditional subband concept, two BWPs can partially or even completely overlap in frequency domain. Two overlapping BWPs can adopt different SCSs, so that the frequency resources can be flexibly shared among different traffic types.

4.2.8 Number of configurable bandwidth parts

The number of configurable BWPs should be proper, neither too large nor too small. From the perspective of UE power saving, it is sufficient to configure two BWPs for downlink and uplink, respectively (as shown in Fig. 4.3 and Fig. 4.4): A larger BWP is used for downlink or uplink data transmission (e.g., equal to the UE downlink or uplink bandwidth capability respectively); and a smaller BWP is used to save UE power consumption while transmitting PDCCH/PUCCH. However, from the perspective of resource allocation, two BWPs may not be enough. Take downlink BWP in Fig. 4.24 as an example. The frequency resources of different SCSs may be located in different frequency regions of the carrier. If downlink data reception with two to three SCSs needs to be supported, at least two to three DownLink (DL) BWPs need to be configured. With

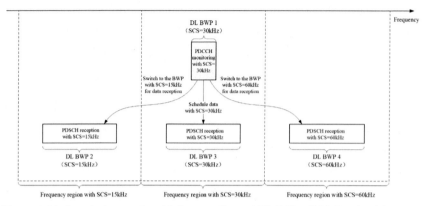

Figure 4.24 A typical scenario with four configured DL BWPs. *BWP*, Bandwidth part.

a so-called "default DL BWP" (described in Section 4.3.6) which is BWP1 in Fig. 4.24 for PDCCH monitoring, a total of three to four DL BWPs needs to be configured. Given three or four BWPs eventually need a 2-bit indicator in DCI (described in Section 4.3.5), it is reasonable to configure four BWPs.

Some proposed to configure more BWPs, such as eight BWPs. When DCI is used to indicate BWP switching, eight BWPs need a 3-bit BWP indicator (BWPI). It was observed in 3GPP discussion that at least for R15 NR system, it is sufficient to configure up to four BWPs for uplink and downlink, respectively. If more BWPs need to be configured in future, such capability can be expanded in later releases.

It should be noted that the limitation of up to four configured BWPs refers to the BWPs configured by UE-specific RRC signaling (also known as UE-dedicated BWP). Besides the UE-dedicated BWPs, a UE additionally has an initial DL BWP and an initial UL BWP (the initial BWP will be described in Section 4.4). Hence, a UE can have five BWPs configured by high-layer signaling in the uplink and downlink, respectively, including the initial DL BWP and initial UL BWP which are not configured by UE-specific RRC-dedicated signaling. This is why the maximum number of BWPs defined in TS 38.331 [29] (maxNrofBWPs) equals to 4, but the value range of the BWP number *BWP-Id* is [0, ..., 4] (i.e., five BWP numbers are supported. *BWP-Id* = 0 corresponds to the initial DL BWP or initial UL BWP, and *BWP-Id* = 1−4 corresponds to four UE-dedicated BWPs configured by RRC signaling).

The specific BWP indication method is described in Section 4.3.3.

4.2.9 Bandwidth part configuration in the TDD system

Even before the BWP concept was established, a consensus was reached in the RAN1#87 meeting to decouple downlink and uplink bandwidth adaptations. However, it was found in the follow-up BWP study that the independent BWP switch on downlink and uplink without restriction may cause problems in TDD system. With regard to the configuration in the TDD system, there are three possible schemes:

- Scheme 1: Uplink BWP and downlink BWP are configured separately and activated independently (i.e., same as FDD).
- Scheme 2: Uplink and downlink share the same BWP configuration.
- Scheme 3: Uplink BWP and downlink BWP are configured and activated in pairs.

In RAN1#AH NR2 meeting in June and the RAN1#90bis meeting in October 2017, it was determined that the UL BWP and DL BWP in the FDD system were switched independently. When BWP switching is performed in the TDD system, the center frequencies of DL BWP and UL BWP should keep the same [22,31] although the bandwidth can change upon the switching. Therefore if the uplink BWP switching and downlink BWP switching in the TDD system are independent, the uplink center frequency and downlink center frequency may become misaligned. As shown in Fig. 4.25, at the first moment, the downlink active BWP and uplink active BWP of the TDD UE are DL BWP 1 and UL BWP 1, respectively, and their center frequencies are the same. At the second moment, the downlink active BWP is switched to DL BWP 2 while the uplink active BWP remains at UL BWP 1. The center

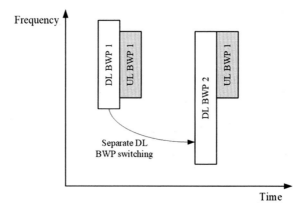

Figure 4.25 The problematic BWP switching in TDD system. *BWP*, Bandwidth part.

frequency of DL BWP becomes different from that of UL BWP. Therefore the independent switching as in scheme 1 is problematic.

On the other hand, scheme 2 [32] is too rigid. It is also unreasonable to mandate 5G NR TDD system to use BWPs of the same size in uplink and downlink (as shown in Fig. 4.26), given 5G NR may have a larger bandwidth (e.g., 100 MHz) demand on downlink than on uplink.

Therefore it was finally decided to adopt scheme 3 (i.e., downlink BWP and uplink BWP in TDD are configured and switched in pairs [33]). Uplink BWP and downlink BWP with the same BWP index are paired. The center frequencies of the paired uplink BWP and downlink BWP must be identical. As shown in Fig. 4.27, at the first moment, the

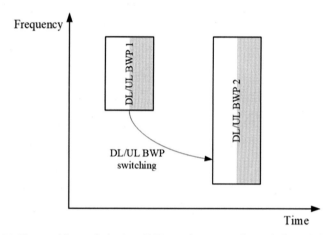

Figure 4.26 The problem of sharing BWP configuration for uplink and downlink in TDD system. *BWP*, Bandwidth part.

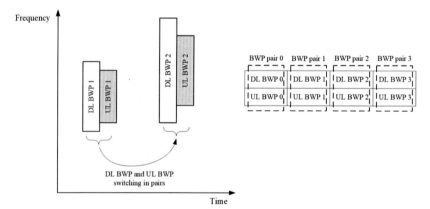

Figure 4.27 TDD DL/UL BWPs configured and switched in pairs. *BWP*, Bandwidth part.

uplink BWP 1 and downlink BWP 1 are active and their center frequencies are the same. At the second moment, the active UL BWP and active DL BWP are switched to uplink BWP 2 and downlink BWP 2 simultaneously, where the center frequencies are still the same between downlink BWP and uplink BWP after the switching. Scheme 3 not only avoids the misaligned center frequencies between the active DL BWP and active UL BWP, but also allows independent sizes of downlink BWP and uplink BWP [34].

4.3 Bandwidth part switching

4.3.1 Dynamic switching versus semistatic switching

Because the R15/R16 NR specification only supports single active BWP, one active BWP must be deactivated before a new BWP can become active. Therefore the BWP activation/deactivation can also be called "BWP switching."

How to activate a BWP from one or more configured BWPs has been discussed as an aspect for bandwidth adaptation before the BWP concept was established. Dynamic activation and semistatic activation were both studied [35,36]. Semistatic activation can be realized by RRC reconfiguration or a method similar to DRX. RRC-based BWP activation can achieve the basic BWP function (i.e., "to allow the UE bandwidth capability smaller than the system bandwidth"). Because the UE bandwidth capability is not time-varying, if BWP is set up only to restrict the UE operating bandwidth within its capability and meanwhile to achieve cell load balancing, it is good enough to semistatically adjust the UE bandwidth frequency location in the system bandwidth; dynamic adjustment is not necessary [37]. Semistatic BWP switching can avoid DCI overhead and simplify the network/UE operation.

However, with the other two motivations to introduce BWP—multiple SCS resource allocation and power saving—it is necessary to introduce a dynamic BWP switching mechanism. First of all, the dynamic change of the BWP size can give UE as many opportunities as possible to fall back to a smaller operating bandwidth and to achieve the power-saving effect. In addition, from the perspective of resource allocation, a more real-time BWP switching is also desired. Of course, because every change of active BWP pays retuning time, BWP switching should not be performed too frequently. But this depends on the specific implementation of gNB and should not be restricted in the specification.

Therefore the NR specification is determined to support both semi-static and dynamic BWP activation. In the R15 specification, semistatic BWP activation based on RRC signaling is a mandatory UE feature, and dynamic BWP activation is an optional UE feature. Semistatic BWP activation based on RRC signaling can be used in specific cases such as BWP activation in the initial access procedure and BWP activation after CA secondary cell (SCell) activation (see Section 4.4 for details).

4.3.2 Introduction of bandwidth part activation method based on DCI

There are two candidate schemes to dynamically activate a BWP—MAC CE activation and DCI activation [18].

MAC CE is used to activate SCell in the LTE CA. The advantage of using MAC CE is the higher reliability (because Hybrid Automatic Repeat-reQuest (HARQ) feedback can be used for MAC CE, but not for DCI) and the avoidance of DCI signaling overhead (MAC CE, as a high-layer signaling, can bear a larger overhead than DCI). The disadvantage of MAC CE is that it takes a period of time (e.g., several slots) to take effect, which impacts the real-time performance. In contrast, as a physical layer signaling, DCI can take effect in several symbol durations [38], which is more conducive to fast BWP switching. Although it is not necessary or even not preferable sometime to switch BWP too frequently, fast BWP switching can still be used for many purposes in the NR system. Therefore compared with SCell, BWP needs a faster activation mechanism.

The disadvantage of DCI is the existence of miss detection and false alarm. Especially if the UE missed the DCI indicating downlink BWP switching, the UE still monitors PDCCH on the original DL BWP when gNB already starts to use the new DL BWP for DCI transmission for the UE. Then the UE can never receive DCI under normal circumstances (e.g., the CORESET and search space of two BWPs do not overlap). In order to solve the problem that gNB and UE have different understandings of downlink active BWP caused by DCI miss detection, the timer-based BWP switching is also adopted in NR specification. The timer-based BWP switching enables the UE to fall back to the default DL BWP in case of DCI miss detection (see Section 4.3.3 for details), and to some extent makes up for the reliability disadvantage of DCI compared to MAC CE. Finally, NR will anyway support the reliable BWP activation based on RRC signaling. The combination of RRC signaling and DCI can balance the tradeoff between reliability and low latency. Therefore NR finally decided to adopt the dynamic BWP activation method based on DCI [39], instead of MAC CE.

As mentioned in Section 4.2.9, DL BWP and UL BWP are switched independently in the FDD system (i.e., UL BWP can remain unchanged while DL BWP is switched, and vice versa). However, in the TDD system, DL BWP and UL BWP must be switched in pair. DL BWP and UL BWP with the same BWPI are always active simultaneously [33]. The BWP indication design in DCI is shown in Section 4.3.5.

In the technical discussion of NR CA, it was also proposed to enhance the CC activation mechanism and support CC activation/deactivation based on PHY signaling (i.e., DCI), but it was not accepted in R15 NR specification. Therefore DCI activation has become an obvious advantage of the BWP mechanism over CA in R15, which makes BWP activation/deactivation faster and more efficient than CC activation/deactivation. This is also one of the reasons why BWP has become an obviously new system tool that is different from CA.

Finally, for the design of DCI activation of BWP, it is also clear that only the BWP "activation" needs to be defined. There is no need to define BWP "deactivation." The key to this question is: is there always a BWP being active? If the answer is yes, activating a new BWP naturally deactivates the previous BWP, without a need for a specific BWP "deactivation" mechanism. However, if "the system activates a BWP for a UE only when the system resource is in use by the UE, and the UE can have no active BWP at all when it does not use any system resource," both BWP activation mechanism and BWP deactivation mechanism are needed. Along with study on BWP, the use cases of BWP concept were gradually extended, not only for data scheduling, but also for the transmission/reception of the control channels and RS. Obviously, BWP has become a universal concept and "necessary at any time." So it is necessary for a UE to always have an active BWP. Thus it is not necessary to design a BWP "deactivation" mechanism for either dynamic BWP switching or semistatic BWP switching.

4.3.3 DCI design for triggering bandwidth part switching—DCI format

The first question to be answered for the DCI-based BWP activation/switching is which DCI format is used to transmit the BWP switching command. There were different proposals for the answer. One scheme is to use the scheduling DCI to trigger BWP switching [40]. Another is to add a new DCI format dedicated to BWP switching purpose.

In R15 5G NR, in addition to the scheduling DCIs (format 0_0, 0_1, 1_0, 1_1, see Section 5.4 for details), several other DCI formats were introduced to transmit certain specific physical layer configuration parameters, including the slot format (see Section 5.6) in DCI format 2_0, URLLC preemption indication (see Chapter 15: 5G URLLC: PHY layer) in DCI format 2_1, and power control command in DCI format 2_2 and format 2_3. The advantages of using a separate DCI format (instead of a scheduling DCI format) to indicate BWP switching include [41]:

1. BWP switching can also be triggered when scheduling data is not needed [e.g., it can be switched to another BWP for channel state information (CSI) measurement].
2. It has smaller payload size than scheduling DCI, which is helpful to improve PDCCH reliability.
3. One DCI can be used to switch DL BWP and UL BWP at the same time, not like the scheduling DCI, which has different DCI formats applicable to uplink and downlink separately.
4. If "multiple active BWPs" is to be supported in the future, a separate DCI is more suitable than the scheduling DCI.

On the other hand, a new DCI format for BWP switching also brings some problems: It introduces extra PDCCH transmission and UE complexity in detecting new PDCCH format, and consumes the limited budgets of blind detections. If the methods such as sequence-based DCI design and fixed DCI resource location are adopted to reduce the complexity of PDCCH detection, this new DCI format will also bring a series of shortcomings and problems [42]. Besides, the group-common DCIs, including existing DCI formats 2_0, 2_1, 2_2, and 2_3, are targeting a group of multiple UEs and meanwhile not related to data scheduling, and hence are not suitable to be reused for BWP switching.

BWP switching operation is admittedly more related to data scheduling. Switching from a smaller BWP to a larger BWP is usually due to the need to schedule a big amount of data. To indicate BWP switching in a scheduling DCI can activate a new BWP and schedule data in the new BWP (i.e., cross-BWP scheduling) at the same time, so as to reduce the DCI overhead and latency. Switching from a larger BWP to a smaller BWP is not necessarily caused by data scheduling and have anything to do with scheduling DCI, but can be realized by timer-based BWP fall-back mechanism (see Section 4.3.6). To sum up, it is appropriate for the first 5G NR release to use the scheduling DCI to indicate BWP switching. Of course, theoretically, the DCI triggering BWP switching does not necessarily contain the scheduling information of PDSCH or PUSCH. If the resource allocation field does not contain any PRB indication, the DCI is only used to trigger BWP switching.

What is cross–BWP scheduling? As described in Section 4.1, a DL BWP can be applied to both PDCCH reception and PDSCH reception. Each BWP contains at least one CORESET so that the PDSCH in a BWP can be scheduled by a PDCCH in the same BWP without requiring BWP switching, and the PDCCH and the scheduled PDSCH have the same SCS. However, there may also be a different situation: BWP switching occurs after a DCI and before the scheduled PDSCH (a typical case is that the DCI contains a BWPI that triggers BWP switching), so that the scheduling PDCCH and the scheduled PDSCH belongs to different BWPs. This is called cross–BWP scheduling. Cross–BWP scheduling brings up a series of new issues. So there were different opinions in 3GPP on whether to support cross–BWP scheduling. However, if cross–BWP scheduling is not allowed, the flexibility of time-domain scheduling during BWP switching would be limited, and the scheduling latency cannot be guaranteed. Therefore 3GPP decided to support cross–BWP scheduling, but restricted it to be performed after RF retuning time [16,18].

As shown in Fig. 4.28A, if cross–BWP scheduling is allowed, the DCI triggering BWP switching in BWP 1 can directly schedule PDSCH in BWP 2 after BWP switching, so that the UE can receive DL data in the scheduled PDSCH immediately after switching to BWP 2. Therefore

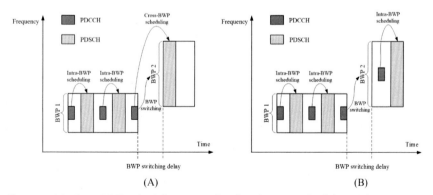

Figure 4.28 Cross-BWP scheduling to realize low-latency scheduling (A) Cross-BWP scheduling is allowed (B) Cross-BWP scheduling is not allowed. *BWP*, Bandwidth part.

using the scheduling DCI to indicate BWP switching is definitely the most efficient way to do "cross-BWP scheduling."

If cross-BWP scheduling is not allowed, using a DCI format dedicated to BWP switching is also a suitable design. But it is difficult to achieve fast scheduling. As shown in Fig. 4.28B, the last DCI in BWP 1 only triggers BWP switching, but is not used to schedule PDSCH. After BWP switching is completed, the UE detects the first DCI that schedules PDSCH in BWP 2 to obtain the scheduling information of PDSCH, and receives data in the scheduled PDSCH at a later time.

It should be noted that some companies in 3GPP believed the advantages of fast cross-BWP scheduling are disputable. One of the concerns is that cross-BWP scheduling cannot obtain CSI of the new BWP in advance, likely resulting in inefficient scheduling. In many scenarios, efficient resource scheduling in the frequency domain and spatial domain is based on real-time CSI and channel quality indication (CQI) information. But according to the R15 NR design, the UE can only activate one DL BWP and CSI measurement only occurs in the active DL BWP, so the CSI in the to-be-activated BWP cannot be obtained in advance. That is to say, even if cross-BWP scheduling is carried out, gNB can only perform relatively inefficient "conservative" scheduling because it does not know CSI and CQI in the target BWP [43]. Of course, even if there is a loss of efficiency, cross-BWP scheduling is still an important means to achieve fast scheduling.

All the above analysis aims at PDSCH scheduling. The situation of PUSCH scheduling is similar (i.e., if the BWPI in the DCI that schedules PUSCH indicates a UL BWP different from the currently active UL BWP, the scheduling becomes cross-BWP scheduling). The corresponding processing method is the same as that in downlink, and not repeated here.

The R15 NR specification was decided to firstly support using the nonfallback scheduling DCI to trigger BWP switching [i.e., the nonfallback DCI format for uplink scheduling (format 0_1) can trigger uplink BWP switching, and the nonfallback DCI format for downlink scheduling (format 1_1) can trigger downlink BWP switching]. The fallback DCI formats (format 0_0 for uplink and format 1_0 for downlink) only provide basic DCI functions excluding BWP switching. Other group-common DCI formats (such as formats 2_0, 2_1, 2_2, 2_3) do not support BWP switching function either. Whether or not to support the DCI format dedicated for BWP switching purpose can also be further studied in future NR releases.

4.3.4 DCI design for triggering bandwidth part switching— "explicitly trigger" versus "implicitly trigger"

To use "scheduling DCI" for BWP switching, there are still two methods for consideration: "explicitly triggering" and "implicitly triggering." Explicitly triggering uses an explicit field in DCI to indicate BWPI [40] of the to-be-activated BWP. Implicitly triggering triggers BWP switching through the existence of the DCI itself or other content in the DCI, but with no BWPI in DCI. As one possible method, UE automatically switches to a larger BWP upon receiving a scheduling DCI that schedules some data [44]. Another method is that a UE switches to a larger BWP if it finds that the frequency range of the data scheduled by the DCI exceeds the size of current BWP [45].

The first "implicitly triggering" method has some disadvantages: First, it can make the switch only between a large BWP and a small BWP. As shown in Fig. 4.29, when the UE does not detect the scheduling DCI, it stays in the small BWP; and when it detects the scheduling DCI, it is switched to a larger BWP. This method cannot meet the corresponding NR specification target described in Section 4.2.7, which is to support switching between up to four BWPs in downlink and uplink, respectively. Second, for a very small amount of data transmission, the UE does not have to be switched to a wideband BWP, and it can stay in the

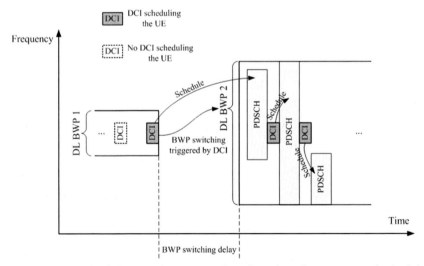

Figure 4.29 "Implicitly triggered" BWP switching based on the existence of scheduling DCI. *BWP*, Bandwidth part.

narrowband BWP, so as to maintain the power-saving effect. If "implicitly triggering" is used, the UE would be unreasonably switched to a wideband BWP regardless of the amount of scheduled data.

The second "implicitly triggering" method can avoid the above problem of "unreasonable switching to a wideband BWP." But it has a problem similar to the first multi-SCS PRB indexing method described in Section 4.1.1 (i.e., the PRB indexing defined in the whole system bandwidth is needed), resulting in excessive DCI overhead on frequency resource allocation, which is not as efficient and concise as the "intra-BWP" resource indication method. Because PRB indexing is only defined within the BWP (as described in Section 4.2.1), this method is not applicable either.

The "explicitly triggering" method can achieve flexible switching among four BWPs by including a 2-bit BWPI in DCI. As shown in the example in Fig. 4.30, in addition to the smaller BWP (DL BWP 1) used to monitor PDCCH in power-saving mode, different larger BWPs (DL BWP 2, DL BWP 3) used to receive PDSCH can be configured in different frequency ranges. Then the UE can be switched to DL BWP 2 or DL BWP 3 through the BWPI.

As described in Section 4.1.7, BWP operation and CA are different concepts in NR, so BWPI and CIF (carrier indicator field) are two separate fields in DCI [19].

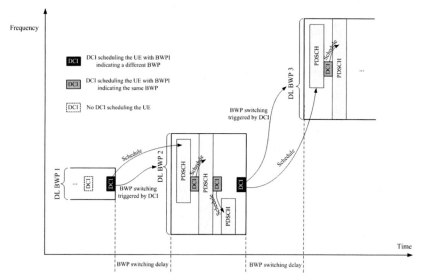

Figure 4.30 "Explicitly triggered" BWP switching based on BWPI. *BWP*, Bandwidth part; *BWPI*, BWP indicator.

4.3.5 DCI design for triggering bandwidth part switching—bandwidth part indicator

There are two methods to define BWPI in DCI: One is that BWPI is contained in DCI only when BWP switching is needed; the other is that BWPI is always contained in DCI. The first method can slightly reduce the DCI overhead in most cases, but it causes the dynamic change of DCI payload and increases the complexity of PDCCH blind detection. Therefore it is finally decided to adopt the second method (i.e., as long as BWP is RRC configured in uplink or downlink, there is always a BWPI field in format 0_1 or format 1_1). If a UE is not configured with any BWP [i.e., the only available BWP is the initial BWP, there is no need for DCI-based BWP switching (and timer-based DL BWP switching)]. In this case the BWPI bitwidth is 0. As long as the UE is configured with some BWPs in the uplink or downlink, every time PUSCH or PDSCH is scheduled with format 0_1 or format 1_1, gNB must use BWPI to indicate which BWP the scheduling is aimed at, even if no BWP switching is performed. For intra-BWP scheduling, the BWP ID of the currently active BWP needs to be filled in BWPI. Only when the BWP ID in BWPI field is different from the currently active BWP, BWP switching is triggered. Of course, as mentioned above, only DCI format 0_1 and 1_1 contain BWPI field. The fallback DCI formats (i.e., format 0_0 and 1_0) do not contain the BWPI field and do not support triggering BWP switching.

As shown in the example in Fig. 4.30, if the UE detects the DCI scheduling PDSCH in DL BWP 1 (DCI format 1_1) and the BWPI points to DL BWP 2, the UE is switched to DL BWP2 and can meanwhile be scheduled to receive PDSCH in DL BWP 2. After switching to DL BWP 2, if the detected downlink scheduling DCI is DCI format 1_0 or the detected DCI is DCI format 1_1 with a BWPI pointing to DL BWP 2, the UE still receives PDSCH in DL BWP 2. When the UE detects a DCI format 1_1 with BWPI field pointing to another BWP (e.g., DL BWP3), the UE is switched to DL BWP3.

The mapping relationship between BWPI and BWP ID is initially very simple, given 2-bit BWPI is just enough to indicate up to four configured BWPs. However, when the initial BWP is taken into account but the BWPI bitwidth is not expanded, the problem becomes a bit complicated which needs to be treated in two cases.

As described in Section 4.2.8, a UE can have five candidate BWPs in uplink and downlink, respectively, including four UE-dedicated BWPs configured by RRC and one initial BWP. As mentioned above, the BWPI bitwidth in DCI is 2, which is determined from the number of UE-dedicated BWPs. The initial design only assumed the switching among the four BWPs, not including the switching from/to the initial BWP. However, as described in Section 4.4, it is necessary to have switching between the UE-dedicated BWP and initial BWP in some scenarios. Because the 2-bit BWPI cannot indicate five target BWPs and meanwhile extending BWPI to 3 bits is not preferred, it was finally decided to support switching to the initial BWP only when the number of configured UE-dedicated BWPs is no more than three.

When the UE is configured with no more than three UE-dedicated BWPs, the mapping rule between BWPI in DCI and *BWP-Id* is shown in Table 4.3. BWPI = 00 can be used in DCI to indicate a switching to the initial BWP. If only one UE-dedicated BWP is configured, BWPI field has only 1 bit, with value 0 corresponding to *BWP-ID* = 0.

When the UE is configured with four UE-dedicated BWPs, the mapping between BWPI and *BWP-Id* is given in Table 4.4, and is different from that in Table 4.3. The UE needs to determine which of the two mapping rules to adopt according to the number of UE-dedicated BWPs configured in RRC signaling.

Compared to LTE, an important enhancement of NR is that DCI can be detected at any position in a slot (although this is an optional UE

Table 4.3 Mapping between BWPI and BWP-Id (three UE-dedicated BWPs are configured).

	BWP-Id	BWPI
Initial BWP	0	00
UE-dedicated BWP 1	1	01
UE-dedicated BWP 2	2	10
UE-dedicated BWP 3	3	11

Table 4.4 Mapping between BWPI and BWP-Id (four UE-dedicated BWPs are configured).

	BWP-Id	BWPI
UE-dedicated BWP 1	1	00
UE-dedicated BWP 2	2	01
UE-dedicated BWP 3	3	10
UE-dedicated BWP 4	4	11

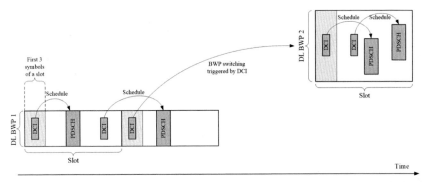

Figure 4.31 DCI indicating BWP switching only appears in the first three symbols of a slot. *BWP, Bandwidth part.*

feature), which greatly improves PDCCH monitoring flexibility and reduces latency (see Section 5.4 for details). But does the DCI indicating BWP switching also needs to be so flexible for its time position? As mentioned in Section 4.3.12, the BWP switching delay is finally specified in the specification in unit of slot. In other words, as long as BWP switching occurs, the UE may have transmission/reception interruption of several slots. Therefore it is unnecessary to allow DCI indicating BWP switching to transmit at any time position in a slot. Finally, it is determined in 3GPP that the DCI indicating BWP switching only appears in the first three symbols of a slot (i.e., the DCI indicating BWP switching follows a slot-based PDCCH detection mechanism similar to LTE, rather than a symbol-based PDCCH detection mechanism). (In the R15 NR specification, it is a mandatory UE feature to monitor PDCCH only at the first three symbols of a slot. Monitoring PDCCH at any symbol of a slot is an optional UE feature). As shown in Fig. 4.31, scheduling DCI (format 0_1, 1_1) in the first three symbols of a slot can be used to schedule data (PDSCH, PUSCH) and trigger BWP switching, but scheduling DCI in other symbols of a slot can only be used to schedule data and cannot be used to trigger BWP switching.

4.3.6 Introduction of timer-based bandwidth part fallback

Scheduling-DCI-based BWP activation is the most flexible BWP activation method, allowing gNB to activate a BWP at any time. However, when BWP switching is indicated at the time with no data scheduling, quite some DCI fields are basically useless except the field of BWPI, resulting in a waste of DCI overhead.

There are two typical scenarios for using "nondata scheduling" DCI to indicate BWP switching: One is DL BWP fallback after data reception is completed; another is DL BWP switching due to a need to receive a periodic channel or signal. During the study of these two scenarios, in addition to DCI activation, timer-based activation method and time-pattern-based activation method were also considered [22,39]. This section describes timer-based downlink default BWP fallback. The time-pattern-based BWP activation method (although not adopted in NR specification) is introduced in Section 4.3.10.

As shown in Figs. 4.3 and 4.4, when gNB schedules downlink or uplink data to the UE, it can switch from a smaller BWP to a larger BWP through DCI triggering. After data transmission is completed, the UE needs to fall back to a smaller BWP to save power consumption. To use DCI to trigger DL BWP fallback certainly generates additional DCI overhead. So to use a timer to trigger DL BWP fallback is attractive [44].

The timer-based DL BWP fallback design can borrow ideas from the discontinuous reception (DRX) mechanism that is already mature in LTE system, where the fallback from active state to DRX state is triggered by a DRX inactivity timer. As a similar mechanism in a DL BWP fallback procedure, when the UE receives the DCI scheduling data, one inactivity timer is restarted to postpone the fallback operation.

In fact, the downlink BWP operation can be considered as a "frequency-domain DRX." Both of the two mechanisms are used for UE power saving, and use a timer to control the fallback from "normal operation state" to "power-saving state." For DRX, the working state is the time duration to monitor PDCCH while the power-saving state is the time duration not to monitor PDCCH; for DL BWP fallback, the working state is the frequency range to receive wideband PDSCH while the power-saving state is the frequency range to only monitor PDCCH or to receive narrow-band PDSCH. For eMBB services, the default state of a UE is monitoring PDCCH without being scheduled PDSCH. So operation within a small BWP should be regarded as the "default state," and this smaller DL BWP should be used as the "default DL BWP."

Both DL BWP fallback operation and DRX operation need the fall back to be prompt when needed to save power consumption, and meanwhile avoid the fallback being performed too frequently. Relatively speaking, because DL BWP switching delay due to retuning time prevents the UE from receiving downlink data for several slots, the negative impact of frequent fallback can be more serious. If the UE falls back to the default DL BWP after one PDSCH reception, and the next downlink traffic just arrives, the PDSCH cannot be scheduled immediately. Therefore it is appropriate to use a timer to control the fallback timing to avoid unnecessarily frequent fallback (i.e., to stay at a larger and nondefault DL BWP before the BWP inactivity timer expires and to fall back to the default DL BWP when the BWP inactivity timer expires).

As shown in the example in Fig. 4.32, when the UE is in a nondefault DL BWP (such as DL BWP 2), the BWP inactivity timer runs according to the status of PDCCH monitoring and PDSCH scheduling. If the UE has not received any DCI scheduling PDSCH for a period of time and the BWP inactivity timer expires, the BWP fallback action is triggered, and the UE falls back to the default DL BWP (DL BWP 1).

It should be noted that the configuration of the default DL BWP is not an independent BWP configuration. The default DL BWP is selected among the configured DL BWPs through *defaultDownlinkBWP-Id*. For example, RRC signaling configures UE with four DL BWPs with BWP-Id being 1, 2, 3, and 4, respectively (as described in Section 4.2.7, BWP-Id = 0 in the initial DL BWP), then *defaultDownlinkBWP-Id* can indicate a BWP-Id from 1, 2, 3, and 4.

Another advantage of timer-based BWP fallback is to provide a fallback function in case of DCI miss detection. DCI is always subject to a probability of miss detection due to a lack of reception confirmation like HARQ-ACK. When a DCI miss detection happens, the UE cannot switch to the correct BWP while the gNB does. In this case, the gNB

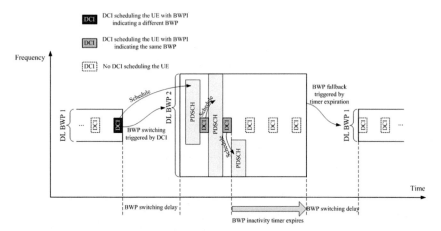

Figure 4.32 Timer-based downlink BWP fallback. *BWP*, Bandwidth part.

and UE may stay in different DL BWPs, resulting in the loss of contact between UE and gNB. As shown in Fig. 4.33, if the DCI indicating UE to switch from DL BWP 2 to DL BWP 3 is not detected, the UE continues to stay in DL BWP 2 while the gNB switches to DL BWP 3 as scheduled. In this case, the DCIs sent by the gNB in DL BWP3 cannot be received by the UE, resulting in a loss of contact.

The above problem of loss of contact is well-handled by timer-based DL BWP fallback mechanism. As shown in Fig. 4.34, after the DCI

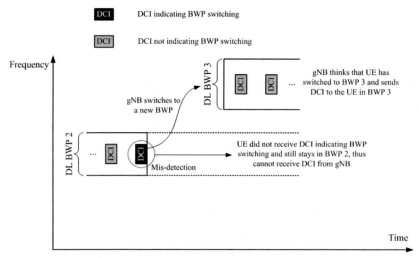

Figure 4.33 DCI miss detection may cause the loss of contact between gNB and UE.

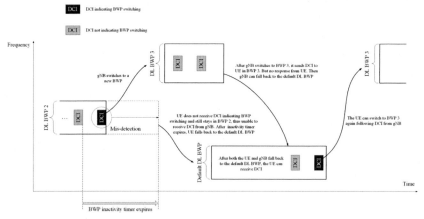

Figure 4.34 Timer-based BWP fallback helps to recover from DCI miss detection. *BWP*, Bandwidth part.

indicating BWP switching is mis-detected, the gNB switches to DL BWP 3 while the UE stays in DL BWP 2. The UE cannot receive the DCI sent by gNB. So the inactivity timer keeps running until expiration. Then the UE automatically falls back to the default DL BWP. Meanwhile the gNB can also return to the default DL BWP when it cannot receive any response from the UE in DL BWP 3. The contact between the gNB and the UE can be resumed in the default DL BWP, in which the gNB can send another DCI to switch the UE to DL BWP 3.

4.3.7 Whether to reuse discontinuous reception timer to implement bandwidth part fallback?

There were two choices in design of BWP fallback timer: Reuse DRX timer, or design a new timer.

4.3.7.1 Review of discontinuous reception timer

The traffic in mobile communications usually arrives in a few bursts followed by a time period with no traffic arrival. It is not conducive to UE power saving for UE to still continue control channel monitoring during the traffic-silent period. Therefore 3GPP introduced the DRX operation long time ago, in which the UE is configured to temporarily stop monitoring the control channel when there is no data transmission.

To be more specific, in 4G LTE and 5G NR, DRX configures a certain "On duration" in a DRX cycle. Then the UE only monitors PDCCH within the "On duration." The time interval in the DRX cycle other than On duration is "DRX opportunity." The UE can perform DRX operation (e.g., a microsleep) in the DRX opportunity, as shown in Fig. 4.35.

However, if the base station still needs to schedule data to the UE at the end of the On duration, the UE needs to continue PDCCH monitoring for a while after the end of the On duration. Generally speaking, it is

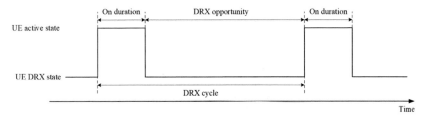

Figure 4.35 Basic working principle of DRX.

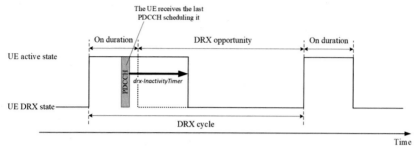

Figure 4.36 Operation principle of DRX inactivity timer.

implemented by a timer *drx-InactivityTimer*. When the UE is scheduled with some data, *drx-InactivityTimer* is started (or restarted) and runs until either it is restarted again upon another scheduling or it expires upon which the UE stops monitoring PDCCH (the actual DRX design also includes another configurable timer—Short DRX Timer—which is not explained here for simplicity) and enters DRX state. As shown in Fig. 4.36, if the UE receives a scheduling DCI during On duration, it starts *drx-InactivityTimer* and does not enter DRX state until the timer expires. If the UE receives a new scheduling DCI before *drx-InactivityTimer* expires, the *drx-InactivityTimer* is restarted to further keep the UE in active state. Therefore when the UE enters DRX state depends on the time when the UE receives the last scheduling DCI and the length of *drx-InactivityTimer*.

4.3.7.2 Whether to reuse discontinuous reception timer for bandwidth part fallback timer?

As mentioned above, DL BWP switching can be regarded as a "frequency-domain DRX." Both mechanisms are used for UE power saving, so it sounds natural and feasible to reuse *drx-InactivityTimer* to trigger DL BWP fallback. Indeed, some companies suggested in 3GPP not to define a dedicated timer for DL BWP fallback operation, but reusing *drx-InactivityTimer* directly [32,43,46]. This is equivalent to unifying framework for power-saving operation in both time-domain and frequency-domain. As shown in Fig. 4.37, when *drx-InactivityTimer* expires, the On duration ends and the UE also falls back from the wideband DL BWP to the default DL BWP.

Even though the reuse of DRX timer can avoid designing a new timer, it may not be reasonable to strictly associate "time-domain power

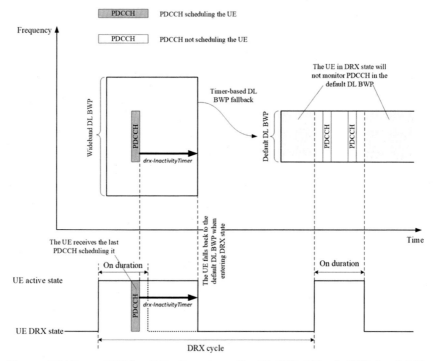

Figure 4.37 Reuse DRX inactivity timer to realize DL BWP fallback. *BWP*, Bandwidth part.

saving" with "frequency-domain power saving," because the UE behavior in DRX state is different from that in default DL BWP. The UE in DRX state completely stops PDCCH monitoring, while the UE falling back to the default DL BWP can still monitor PDCCH in the default DL BWP. The frequency-domain power-saving effect is achieved by reducing the BWP size.

From this perspective, the DL BWP fallback operation is more flexible than the DRX operation. If the same timer is applied to two operations, it does not make much sense for the UE to fall back to the default DL BWP. As shown in Fig. 4.37, when *drx-InactivityTimer* expires, from the perspective of BWP operation, the UE can only fall back to the default DL BWP for power saving while continuing to monitor PDCCH. But if the UE enters DRX state at the same time, it cannot actually continue to monitor any PDCCH.

If DL BWP fallback and DRX adopt different timers, the power-saving gain of DL BWP switching can be fully obtained (i.e., BWP switching operation is nested in the DRX operation). The UE first determines whether to monitor PDCCH according to the DRX configuration and DRX timer status, and if the UE is not in DRX state, determines which DL BWP is used to monitor PDCCH according to the BWP configuration and BWP timer status.

Finally, as mentioned in Section 4.3.6, another function of the DL BWP fallback mechanism is to ensure that the UE can resume contact with gNB in case of DCI miss detection. Unlike the DRX mechanism which is only for UE power saving, the design principle of BWP timer needs to cover more motivations and functionalities.

Therefore 3GPP finally determined to define a new timer, which is called *bwp-InactivityTimer* and different from *drx-InactivityTimer*, for the DL BWP fallback operation. A typical configuration of the two timers is to have a long *drx-InactivityTimer* and a relatively short *bwp-Inactivitytimer*. As shown in Fig. 4.38, after receiving the last DCI scheduling the UE, the UE starts *drx-InactivityTimer* and *bwp-InactivityTimer*. When the *bwp-InactivityTimer*

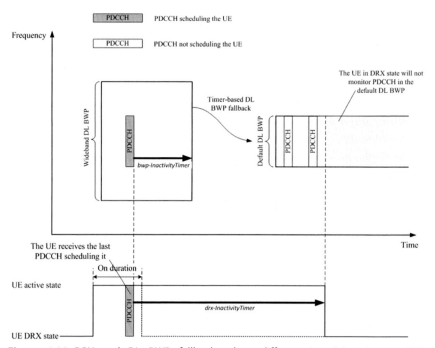

Figure 4.38 DRX and DL BWP fallback adopt different inactivity timers. *BWP,* Bandwidth part.

expires first, the UE falls back to the default DL BWP to continue to monitor PDCCH. If the UE does not receive another scheduling DCI and *drx-InactivityTimer* expires, the UE enters DRX state and completely stops the PDCCH monitoring.

4.3.8 Bandwidth part inactivity timer design

Similar to other timers in the specification, the design of *bwp-InactivityTimer* includes timer configuration as well as its conditions for start, restart, stop, and so on.

4.3.8.1 Configuration of bwp-InactivityTimer

The first problem in the configuration is the timer unit. The timer running in MAC layer, such as *drx-InactivityTimer*, is generally in units of millisecond. As introduced in Section 4.3.12, BWP switching delay is measured in "slot," so it is not necessary to configure *bwp-InactivityTimer* in a too small granularity (like symbol level granularity). Eventually, *bwp-InactivityTimer* is still configured in units of millisecond. The minimum configurable timer value is 2 Ms, and the maximum configurable value is 2560 Ms which is consistent with the maximum value of *drx-InactivityTimer*. If gNB intends to achieve better UE power-saving effect through more frequent DL BWP fallback, a shorter *bwp-InactivityTimer* can be configured; if gNB tries to avoid frequent DL BWP switching to achieve relatively simple BWP operation, a longer *bwp-InactivityTimer* can be configured.

It should be noted that if multiple DL BWPs are configured, the same *bwp-InactivityTimer* is applicable to each DL BWP. Different DL BWPs cannot be configured with different *bwp-InactivityTimer*. This is a simplified design. If all DL BWPs have the same SCS, it is certainly good enough to use the same *bwp-InactivityTimer*. If DL BWPs have different SCSs, in order to allow configuring the active timer durations on different DL BWPs to be the duration of the same number of slots, different DL BWPs should be configured with separate *bwp-InactivityTimer*. However, as mentioned, 3GPP chose to simplify the design by adopting a BWP-Common configuration for *bwp-InactivityTimer*.

In addition, *bwp-InactivityTimer* is not applicable to default DL BWP, because the timer itself is used to control the timing for falling back to the default DL BWP. If the UE has been in the default DL BWP, it makes no sense for UE to determine the fallback timing.

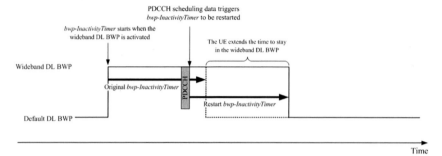

Figure 4.39 The start and restart of *bwp-InactivityTimer*.

4.3.8.2 Condition to start/restart bwp-InactivityTimer

Obviously, the starting condition of *bwp-InactivityTimer* should be the activation of DL BWP. As shown in Fig. 4.39, when a DL BWP (except the default DL BWP) is activated, *bwp-InactivityTimer* starts immediately.

The restart of *bwp-InactivityTimer* is mainly triggered by a reception of data scheduling. As described in Sections 4.3.4 and 4.3.5, the main function of *bwp-InactivityTimer* is to trigger fallback to the default DL BWP when the UE has not been scheduled with data for the given time. Similar to the DRX operation, when the UE receives a DCI scheduling its data, it should expect that there may be subsequent data scheduling soon in the same BWP. Therefore regardless how long *bwp-InactivityTimer* has been running, it should be restarted. As shown in Fig. 4.39, every time a new scheduling DCI is received, *bwp-InactivityTimer* is restarted. This actually extends the time for the UE to stay in the wideband DL BWP. If the UE has not received a new scheduling DCI when *bwp-InactivityTimer* expires, the UE falls back to the default DL BWP.

4.3.8.3 Condition to stop bwp-InactivityTimer

As mentioned above, a successful detection of one scheduling DCI may imply additional DCIs scheduling DL data will arrive soon. So the *bwp-InactivityTimer* should be restarted when a scheduling DCI is received. However, there exist some other PHY procedures, during which the DL BWP switching should not occur. Hence *bwp-InactivityTimer* needs to be stopped temporarily, and resumes running after the completions of the procedures. In the study of NR BWP, two PHY procedures that may require *bwp-InactivityTimer* temporarily stop: PDSCH reception procedure and random access procedure.

If a UE configured with a short *bwp-InactivityTimer* is scheduled to receive a long-duration PDSCH (such as multislot PDSCH) whose ending instance is far from PDCCH, its *bwp-InactivityTimer* may expire before the UE finishes the PDSCH reception (as shown in Fig. 4.40). In this case, according to the *bwp-InactivityTimer* operation rule defined earlier, the UE should have stopped the PDSCH reception in the active DL BWP and fallen back to the default DL BWP. This behavior is certainly unreasonable and undesirable.

One remedy for the problem shown in Fig. 4.40 is to let UE temporarily stop or pause *bwp-InactivityTimer* when the UE starts receiving PDSCH, as shown in Fig. 4.41, and resume the *bwp-InactivityTimer* running after the PDSCH is completely received.

However, after study and discussion in 3GPP, it was generally considered that the chance for a configured *bwp-InactivityTimer* to expire before the end of the PDSCH is quite minor. gNB can even avoid such issue by flexible scheduling. Therefore R15 NR did not take this scenario into account for the condition to stop *bwp-InactivityTimer*.

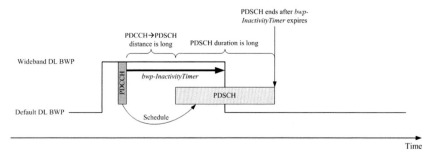

Figure 4.40 *bwp-InactivityTimer* expires before end of PDSCH reception.

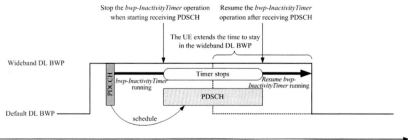

Figure 4.41 Stop *bwp-InactivityTimer* during PDSCH reception.

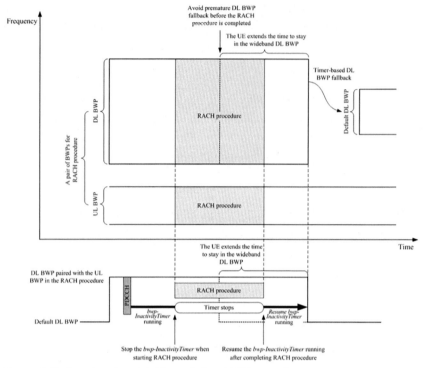

Figure 4.42 Stop *bwp-InactivityTimer* during RACH procedure.

It was the random access (RACH) procedure that the condition of stopping *bwp-InactivityTimer* mainly deals with. As shown in Fig. 4.42, when a UE starts a RACH procedure, it temporarily stops the *bwp-InactivityTimer*, and resumes the *bwp-InactivityTimer* after completing the RACH procedure. The reason for avoiding DL BWP switching in the RACH procedure is that the RACH procedure needs to be completed on paired DL BWP and UL BWP (as described in Section 4.3.11) but the DL BWP fallback may leaves the UE with a pair of mismatched DL BWP and UL BWP before RACH procedure completes. Therefore before the RACH procedure completes, the UE should avoid falling back to the default DL BWP (if the current active DL BWP is not the default DL BWP).

4.3.9 Timer-based uplink bandwidth part switching

Sections 4.3.6—4.3.8 mainly describe timer-based downlink BWP switching. It was suggested in 3GPP to also adopt timer-based BWP switching in uplink (i.e., to control UL BWP activation/deactivation through inactivity timer). For example, the timer can be stopped or paused to extend the active state of UL BWP when HARQ-ACK needs to be transmitted in uplink [47]. However, differently from downlink, it is not necessary to introduce default BWP in the uplink. In downlink, even if no data is scheduled, it is also necessary to monitor PDCCH in the DL default BWP. In contrast, there is no periodic operation such as PDCCH monitoring in uplink. As long as no uplink data is scheduled, the UE can enter power-saving state without monitoring any UL BWP. Hence there is no need to define a UL default BWP. Consequently, NR does not support the timer-based UL BWP switching, simply because the target BWP for the switching (the default UL BWP) does not exist.

As mentioned in Section 4.2.9, in the FDD system, DL BWP and UL BWP are switched independently (i.e., the UL BWP can remain unchanged when the DL BWP is switched). So it is clear the UL BWP switching has nothing to do with timers in FDD system. However, in the TDD system, DL BWP and UL BWP must be switched in pair. When timer-based DL BWP switching occurs, the corresponding UL BWP must be switched as well to maintain the configured pairing [33]. Therefore the UL BWP switching can be a consequence of the expiration of *bwp-InactivityTimer* in NR TDD system.

4.3.10 Time-pattern-based bandwidth part switching
4.3.10.1 The principle of time-pattern-based bandwidth part switching
In the early stage of BWP study, time-pattern-based BWP switching was also a candidate scheme for BWP switching [22,39,42,48]. The principle of time-pattern-based BWP switching is that the UE can be switched between two or more BWPs according to a predetermined time pattern, instead of performing BWP switching according to explicit switching commands. As shown in Fig. 4.43, with a time pattern defined over a certain period, UE switches from the current active BWP (such as BWP 1) to another BWP (such as BWP 2) at a specific time point given in the pattern to complete a specific operation (such as reception of SSB/SI), and then switches back to BWP 1 at another specific time point given in the pattern after completing the operation. Like timer-based BWP switching, time-pattern-based BWP switching has some similarities with the legacy DRX and SPS operations, and can share some common design principles and ideas.

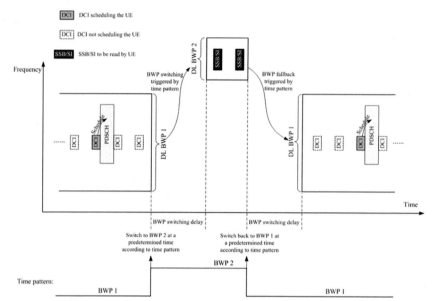

Figure 4.43 Time-pattern-based BWP switching. *BWP*, Bandwidth part.

4.3.10.2 The competition between time-pattern-based bandwidth part switching and timer-based bandwidth part switching

Time-pattern-based BWP switching is suitable for the scenario where the UE has an predetermined operation in BWP 2, which occurs at a predetermined time and lasts for a predetermined period of time, such as SI update (as shown in Fig. 4.43), cross-BWP channel sounding, cross-band mobility measurement, and so on. Admittedly, the BWP switching can always exhibit a certain switching pattern in time domain. The question is whether it is necessary to explicitly standardize the dedicated time pattern configuration method and the corresponding BWP switching procedure. One view is that BWP switching based on "DCI + timer" can equivalently achieve the similar effect of time-pattern-based BWP switching so that a dedicated BWP switching method based on time pattern is redundant. However, the counter-arguments to this view listed some issues in the switching method based on "DCI + timer," including:

- The DCI-based BWP switching brings additional DCI overhead, when there is no data to be scheduled and the DCI runs with meaningless fields except BWPI. In addition, as described in Section 4.3.5, when four UE-dedicated BWPs are configured, it is impossible to rely on DCI to switch to the initial BWP.

- The function of timer-based BWP switching is only about "falling back from a DL BWP of PDSCH reception to a default DL BWP of PDCCH monitoring." It cannot be used for switching to a DL BWP other than the default. Moreover, the timer design mainly considers the uncertainty of DCI arrival (i.e., gNB cannot accurately predict when to schedule the UE via DCI). Thus the UE needs to stay in the larger DL BWP for a period of time after every DCI reception to prepare for receiving further scheduling from gNB. In this case the timer can be used to avoid frequent BWP switching. However, the ending times of some PHY procedures such as SI update and cross-BWP channel sounding are anticipated and deterministic. Hence the UE can switch to the target BWP at a predetermined time, and return to the original BWP immediately after the operation is completed. There is no need to continue to stay in the BWP until a timer expires. Using the timer delays the return to original BWP, thus impacting the system efficiency. Additionally, the granularity of the timer is too coarse (1 or 0.5 Ms) to realize fast BWP switching.

4.3.10.3 The reason why time-pattern-based bandwidth part switching was not adopted

The first BWP switching scenario suitable for using of time pattern is the reception of SI update. When the active BWP is mainly used for UE to transmit and receive UE-specific uplink and downlink data or control channel, this BWP may not contain some common control channels and signals. For example, a DL BWP may or may not contain SSB or RMSI, as described in Section 4.1.3. Suppose that only the initial DL BWP contains SSB and RMSI, while the current DL active BWP does not. When the UE needs to update cell SI (MIB, SIB1, etc.), it needs to switch back to the initial DL BWP to access SI, and then return to the previous DL active BWP after completing SI update. Because the time positions of SSB and RMSI are either relatively fixed or known to the UE via higher-layer configurations, the UE can predict the time window in which the UE needs to switch to the initial DL BWP. Compared with DCI-based and timer-based BWP switching, time-pattern based BWP switching brings some but slight advantages. DCI-based BWP switching requires gNB to send DCI twice. Timer-based method allows the UE switching to the initial DL BWP to access SI, but only if the UE is not configured with default DL BWP. The shortcomings in both methods are avoided in time-pattern based BWP switching. However, the SI update occurs infrequently in practical deployment, which just drives occasional DCI-based BWP switching, and therefore leaves little impact on the overall DCI overhead.

Another periodic UE procedure that can possibly involve with time-pattern based BWP switching is RRM measurement, which is a relatively frequent UE behavior. In the BWP concept at the very beginning, when a DL BWP is activated, all downlink signals should be received in the active DL BWP. But this restriction was not strictly followed in the later study. Then logically, if the UE needs to perform RRM measurement based on a certain channel or signal [such as SSB or CSI−RS (channel state information reference signal)], which is not contained in the current active BWP, the UE needs to switch to the BWP containing the signal. After study and discussion in 3GPP, it was decided to treat RRM measurement as an exception which can be carried out outside the active DL BWP. That is to say, the RRM measurement bandwidth can be configured separately, which is however not treated as another BWP. This decision is also reasonable because the interfrequency measurement definitely cannot be performed within the current active DL BWP anyway.

Finally, cross-BWP channel sounding is also a suitable scenario for using time-pattern-based BWP switching. If two BWPs configured for the UE do not overlap in frequency domain, during the time when the UE operates in BWP 1, it can periodically measure and estimate the CSI in BWP 2 so as to quickly enter an efficient operation state after switching to BWP 2. Because CSI estimation can only be carried out in the active BWP, it is necessary for the UE to periodically switch to BWP 2 for the CSI estimation. Given the time point and duration of cross-BWP CSI estimation are preknown, it is appropriate to switch between two BWPs according to a preconfigured time pattern. However, this motivation did not become strong enough because cross-BWP channel sounding is treated as a feature that does not have to be implemented in R15.

To sum up, time-pattern-based BWP switching and timer-based BWP switching share some interchangeability. The time pattern has the advantage of high efficiency and low overhead in dealing with periodic BWP switching, while the timer has the advantage of being more flexible and can avoid unnecessary BWP switching delay [42]. At the end, R15 NR specification only defines DCI-based BWP switching and timer-based BWP switching, but not time-pattern-based BWP switching.

4.3.11 Automatic bandwidth part switching

In addition to RRC-based BWP switching (as described in Section 4.3.1), DCI-based BWP switching (as described in Sections 4.3.2—4.3.5) and timer-based DL BWP switching (as described in Sections 4.3.4—4.3.6), automatic BWP switching may also be triggered in some cases.

4.3.11.1 Paired switching of DL bandwidth part and UL bandwidth part in TDD

As described in Section 4.2.9, an uplink BWP and a downlink BWP in TDD system are configured and activated in pair. In order to activate "a pair of BWPs" with one DCI, one way is to define a DCI or BWPI that can indicate "a pair of BWPs" and is specifically used for BWP activation in NR TDD.

As mentioned in Section 4.3.3, if a "neither uplink nor downlink" DCI format dedicated for BWP switching is introduced, it can naturally activate a pair of DL BWP and UL BWP at the same time. However, due to the additional complexity of adding the new DCI format, this design was not adopted.

Another method of paired BWP activation is to define a new BWPI that specifically activates "BWP pair" in DCI, which still causes BWP switching signaling in TDD to be different from the one in FDD. Therefore it was finally decided to reuse the BWPI design of FDD, which is "DL scheduling DCI triggers DL BWP" and "UL scheduling DCI triggers UL BWP," with a minor modification that the DL BWP switching also triggers the switching between paired UL BWPs, and the UL BWP switching also triggers the switching between paired DL BWPs [1,6−19,21−64]. As shown in Fig. 4.44, if either DCI format 1_1 or DCI

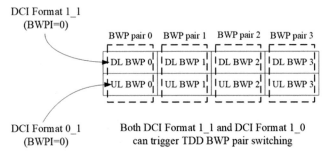

Figure 4.44 Both DL scheduling DCI and UL scheduling DCI can trigger switching of a pair of BWPs simultaneously. *BWP*, Bandwidth part.

format 0_1 show BWPI = 0, DL BWP 0, and UL BWP 0 are activated simultaneously.

Finally, it should be noted that the paired BWP activation is a general rule in TDD, not only applicable to both RRC-based and DCI-based BWP switching, but also to the timer-based BWP switching. When the timer expires and the downlink active BWP of the UE falls back to the default DL BWP, the uplink active BWP also switches to the UL BWP paired with the default DL BWP (i.e., the UL BWP with the same BWPI as the default DL BWP).

4.3.11.2 DL BWP switching caused by random access

As shown in Section 4.3.8, in order to avoid DL BWP switching during random access (RACH), *bwp-InactivityTimer* is stopped during RACH because the UL BWP and DL BWP for RACH need to be activated in pairs. Why restrict paired activation?

DL BWP and UL BWP in FDD can be generally configured and activated independently. However, during the contention-based random access procedure when gNB has not yet fully identified the UE, some issues may occur. As shown in Fig. 4.45, assume there are three BWPs in the uplink and downlink, respectively, and a UE sends a random access preamble (i.e., Msg.1) in UL BWP 2 to initiate random access. According to the random access procedure, gNB cannot determine which UE sends the Msg.1 when gNB sends the random access response (RAR, i.e., Msg.2), which means the gNB does not know either which DL BWP is the active DL BWP on UE side. So the gNB needs to send Msg.2 in all possible DL BWPs, resulting in a great waste of resources.

To avoid this problem, 3GPP finally decided that the RACH procedure must be based on paired DL BWP and UL BWP. Same as BWP

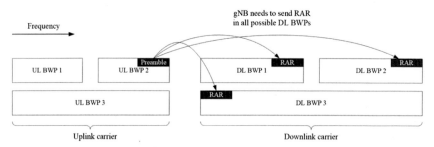

Figure 4.45 If DL BWP is configured independently in FDD, RAR needs to be sent in multiple DL BWPs. *BWP*, Bandwidth part.

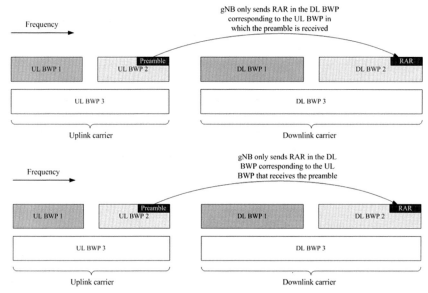

Figure 4.46 Pairing UL BWP and DL BWP in RACH procedure. *BWP*, Bandwidth part.

pairing in TDD, the BWPI-based pairing method is adopted (i.e., the DL BWP and UL BWP with the same BWPI are paired for the RACH procedure). As shown in Fig. 4.46, if the UE sends preamble in UL BWP 1, it must receive RAR in DL BWP 1. In this way, gNB only needs to send RAR in DL BWP 1 and does not need to send RAR in other DL BWPs.

Based on the above design, when the UE sends preamble in the active UL BWP and starts the RACH procedure, if the current active DL BWP and active UL BWP have different BWPI, the current active DL BWP automatically switches to the DL BWP with the same BWPI as the active UL BWP.

4.3.12 Bandwidth part switching delay

As described in Section 4.1.2, the UE cannot transmit or receive signals during BWP switching delay. So the gNB should not schedule uplink or downlink data during this delay interval. If the UE finds that the starting time of the scheduled PDSCH or PUSCH falls in the BWP switching delay, the UE regards this as an error case and may not receive PDSCH or transmit PUSCH as scheduled.

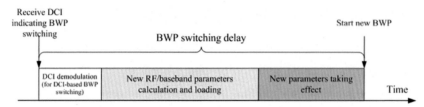

Figure 4.47 The composition of BWP switching delay. *BWP*, Bandwidth part.

As for the cross-BWP scheduling, its details are given in Section 5.2.3. This section only relates to BWP switching delay. Given BWP switching delay mainly depends on the UE implementation, 3GPP RAN4 is in charge of the related study. BWP switching delay can be composed of three parts (as shown in Fig. 4.47):

- The first part is the time when the UE demodulates the DCI containing BWP switching command. If BWP switching is triggered by other methods (such as timer-based BWP switching), this part of delay may vanish.
- The second part is the time for the UE to calculate and load new BWP parameters.
- The third part is the time for new BWP parameter to take effect.

Four BWP switching scenarios are considered, as shown in Fig. 4.48, including:

- Scenario 1: The BWP switching changes center frequency but not bandwidth (regardless of whether the SCS changes or not).
- Scenario 2: The BWP switching changes bandwidth but not center frequency (regardless of whether the SCS changes or not).
- Scenario 3: The BWP switching changes both center frequency and bandwidth (regardless of whether the SCS changes or not).
- Scenario 4: The BWP switching changes SCS but not center frequency and bandwidth.

The RAN4 study results for the above four scenarios are shown in Table 4.5. Type 1 delay and Type 2 delay correspond to two UE capabilities (i.e., the UE with high capability needs to meet the requirement corresponding to the Type 1 delays in the table), and the UE with basic capability needs to meet the requirement corresponding to the Type 2 delays in the table. BWP switching delay requirements corresponding to the two UE capabilities are quite different. However, the requirements are the same for both FR1 (Frequency Range 1) and FR2 (Frequency Range 2). Take Type 1 UE capability as an example. Table 4.5 shows

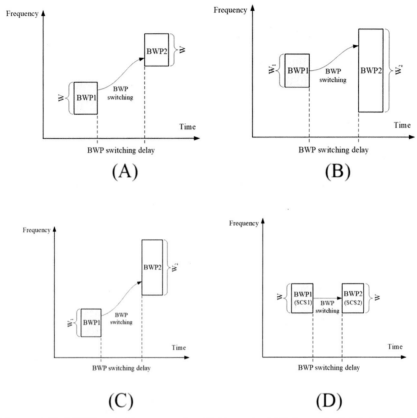

Figure 4.48 4 BWP switching scenarios (A) Scenario 1 (B) Scenario 2 (C) Scenario 3 (D) Scenario 4. *BWP*, Bandwidth part.

Table 4.5 BWP switching delay under various scenarios.

Frequency range (FR)	BWP switching scenario	Type 1 delay (μs)	Type 2 delay (μs)	Notes
FR1	Scenario 1	600	2000	Affect baseband and RF
	Scenario 2	600	2000	Affect baseband and RF
	Scenario 3	600	2000	Affect baseband and RF
	Scenario 4	400	950	Only affect baseband
FR2	Scenario 1	600	2000	Affect baseband and RF
	Scenario 2	600	2000	Affect baseband and RF
	Scenario 3	600	2000	Affect baseband and RF
	Scenario 4	400	950	Only affect baseband

Table 4.6 BWP switching delay in units of slots.

Subcarrier spacing	Slot length (Ms)	BWP switching delay (slots)	
		Type 1 delay	Type 2 delay
15 kHz	1	1 slot	3 slots
30 kHz	0.5	2 slots	5 slots
60 kHz	0.25	3 slots	9 slots
120 kHz	0.125	6 slots	17 slots

that a portion of around 200 μs of BWP switching delay is relating to UE RF retuning, and the parameter reconfiguration and validation of baseband module takes no more than 400 μs. To be more specific, in scenarios 1, 2, and 3 the BWP switching changes either bandwidth or center frequency, which makes the switching delay to include both RF retuning time and baseband reconfiguration time. This switching delay needs to be upper-bounded by 600 μs. Scenario 4 only changes the SCS, which means the BWP switching only includes baseband reconfiguration time but not RF retuning time, thus only results in no more than 400 μs delay. It was a bit beyond intuitive expectation that the longest baseband reconfiguration delay in BWP switching can be 400 μs, even significantly longer than that of RF reconfiguration and therefore suggesting the baseband reconfiguration delay to be the dominant part of total BWP switching delay. Of course, because these values only reflect the minimum performance requirements for all devices, the RAN4 study was based on the worst-case analysis. The BWP switching delay of the actual products may be smaller.

For 30 kHz SCS, Type 1 delay upper-bound of 600 μs is converted into 16.8 OFDM symbol durations. Because the UE requirement in the RAN4 specification is calculated in unit of slot, this Type 1 delay upper-bound is two slots. Table 4.6 shows the BWP switching delays, in unit of slot, corresponding to different SCSs in the final RAN4 specification [56]. If SCSs before and after BWP switching are different, the delay in unit of slot in Table 4.6 assumes the larger one between the two SCSs.

4.4 Bandwidth part in initial access

4.4.1 Introduction of initial DL bandwidth part

As mentioned in Section 4.1, the original motivation of introducing BWP is to support power saving and multiSCS resource scheduling in the

RRC connection mode. There is also a need for UE to have a specific BWP operation for the initial access procedure before RRC connection is set up. With RRC connection already established, the UE can obtain the BWP configuration from the RRC configuration, and then activate a BWP following the DCI indication. However, before RRC connection is set up, the UE cannot obtain BWP configurations from RRC signaling. Meanwhile, because CORESET is configured in the BWP, before DL BWP is activated via UE-specific DCI, the UE cannot determine the configuration of UE-specific search space, thus cannot monitor the DCI indicating BWP activation. This becomes a "chicken and egg" problem. In order to solve this problem, an approach to automatically determine the operating bandwidth in the initial access procedure must be designed.

Assume there is such a working bandwidth, called "initial BWP," which is automatically determined for initial accesses. Its BWP operation procedure can be shown in Fig. 4.49 (take downlink as an example). The UE automatically determines the initial DL BWP for the initial access procedure, where the determination is based on SSB detection (e.g., the initial DL BWP may be located in the frequency-domain location next to detected SSB; see Section 4.4.3 for details). After the initial access procedure is completed and RRC is set up, multiple UE-specific BWPs can be configured through RRC signaling, and DL BWPs for wideband operation can be activated by DCI. Because BWP is designed to achieve more flexible resource scheduling and to define the UE operating bandwidth, the gNB can also use BWP to achieve load balancing in a large 5G system bandwidth. As a flexible gNB choice in frequency-domain load balancing,

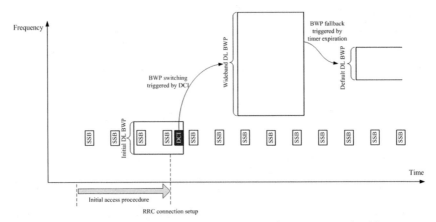

Figure 4.49 BWP operation in the initial access procedure. *BWP*, Bandwidth part.

the wideband DL BWP may or may not contain SSB, given the NR specification does not specify whether BWP configuration should contain SSB or not. After data transmission is completed and the timer expires, the active DL BWP falls back to the default DL BWP with a smaller bandwidth, which again may or may not include SSB.

It can be seen that before RRC connection is set up, the UE needs to determine an initial operating bandwidth in both downlink and uplink. In the early stage of BWP study, BWP was mainly considered with the background use cases for data channel and control channel, rather than the initial access procedure. Therefore when it came up to describe a bandwidth concept in initial access, the question to be answered is: Should a new concept of "initial BWP" be added? Or can an existing bandwidth concept be used? After cell search, the UE at least knows the PBCH or SS/PBCH block (SSB) bandwidth. If the UE only needs to complete downlink operations of the initial access in the PBCH or SSB bandwidth, there is no need to define a new bandwidth concept [60].

For the further details relating to this initial operating bandwidth, three possible schemes were proposed in RAN1#89 [16]:

- Scheme 1: Both CORESET#0 (full name in the specification is "CORESET for Type0-PDCCH CSS set") and RMSI are confined within SSB bandwidth (as shown in Fig. 4.50A).
- Scheme 2: CORESET#0 is confined within SSB bandwidth, but RMSI is not (as shown in Fig. 4.50B).
- Scheme 3: Both CORESET#0 and RMSI are not confined within SSB bandwidth (as shown in Fig. 4.50C).

It was found in 3GPP study that the downlink bandwidth of the initial access operation should not be limited to the bandwidth and frequency location of SSB, and it is necessary to define a separate BWP for the initial access procedure (i.e., a dedicated "initial DL BWP" concept is needed).

First of all, the initial DL BWP needs to contain PDCCH CORESET (referred to as CORESET#0) during initial access procedure. It was supposed in the BWP study that CORESET#0 may be used to schedule many types of information delivery, including RMSI (i.e., SIB1); OSI [i.e., SIBs starting from system information block type2 (SIB2)]; and configuration information for PDSCH transmissions corresponding to Msg.2 and Msg.4 in the initial access procedure, etc.). Although high scheduling flexibility and large PDCCH capacity are both generally targeted in NR, the number of control channel element (CCE) contained in the SSB bandwidth, which is determined as large as 20 PRBs, is even less than the

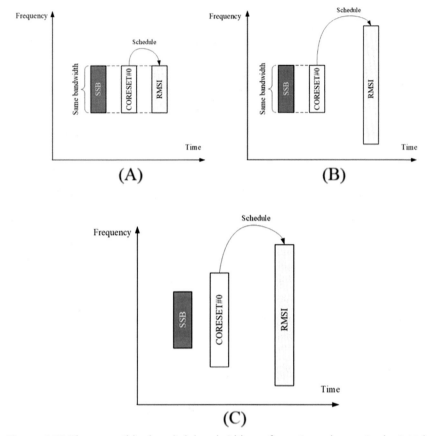

Figure 4.50 Three possible downlink bandwidth configuration schemes in the initial access process (A) Scheme 1 (B) Scheme 2 (C) Scheme 3.

number of CCE in the LTE PDCCH common search space (CSS) (in the case of three OFDM symbols in time domain). This is obviously insufficient for NR system. Therefore CORESET#0 needs to be carried in a larger bandwidth than SSB.

At the same time, the frequency-domain location of CORESET#0 and SS/PBCH block may also be different. For example, CORESET#0 needs to perform frequency division multiplexing (FDM) with SS/PBCH block to achieve transmission of CORESET#0 and SS/PBCH block in the same time resource (e.g., due to the need of beam sweeping in the millimeter wave system, cell search and system information broadcasting need to be completed in a time that is as short as possible), as shown in Fig. 4.51. In this case, the frequency-domain location of the initial DL

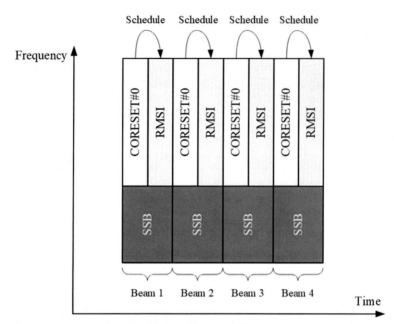

Figure 4.51 FDM between CORESET#0, RMSI and SSB. *FDM,* Frequency division multiplexing; *RMSI,* remaining minimum system information.

BWP should be flexible [e.g., indicated by MIB (master information block)] carried in PBCH. Of course, it is also possible to choose not to define a dedicated initial DL BWP and to directly use the CORESET#0 bandwidth instead, provided that the frequency-domain scheduling range of RMSI does not exceed that of CORESET#0. But a separate definition of initial DL BWP can ensure a clearer signaling structure.

Similarly, the bandwidth required to transmit RMSI is also likely to exceed that of SSB. In the high-frequency multibeam NR system, RMSI may also need to perform FDM with SSB, as shown in Fig. 4.51. From the perspective of forward compatibility, RMSI may need to be expanded in the future 5G enhancements, so it is unwise to limit RMSI within a very narrow bandwidth.

With the above reasons, at least an initial DL BWP needs to be defined to transmit CORESET#0 or RMSI.

4.4.2 Introduction of initial UL bandwidth part

In the initial access procedure, the first uplink bandwidth determined by a UE is the bandwidth (e.g., the bandwidth where Msg. 1 is located) used

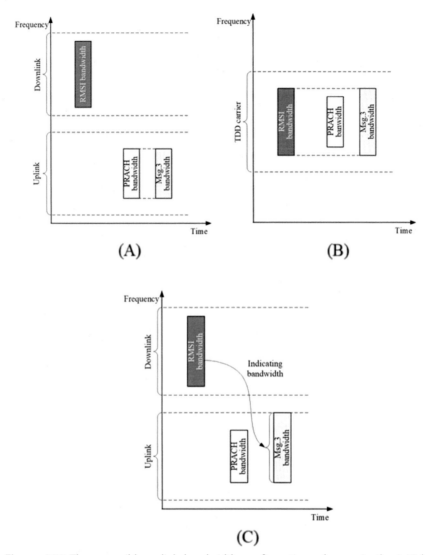

Figure 4.52 Three possible uplink bandwidth configuration schemes in the initial access procedure (A) Scheme 1 (B) Scheme 2 (C) Scheme 3.

to transmit PRACH (physical random access channel). One question in the 3GPP study is about the bandwidth in which the second uplink transmission (i.e., Msg.3) is scheduled. Similar to downlink, this bandwidth configuration can also be considered in several schemes:

- Scheme 1: Msg.3 is scheduled within PRACH bandwidth (as shown in Fig. 4.52A).

- Scheme 2: Msg.3 is scheduled within a known downlink bandwidth (such as RMSI bandwidth) (as shown in Fig. 4.52B). This scheme is for TDD spectrum only.
- Scheme 3: Msg.3 is scheduled within the known bandwidth, which is flexibly configured by RMSI (as shown in Fig. 4.52C).

Scheme 1 is the simplest. It does not need to define a separate uplink initial BWP and reuses the frequency range of PRACH instead [42]. However, the PRACH bandwidth may have different configurations (2−24 PRBs). In some configurations, the bandwidth is very limited, which results in undesirable scheduling restriction for Msg. 3.

Scheme 2 may be a feasible scheme in the TDD system, but cannot be directly used in the FDD system. Even for TDD system, to align uplink bandwidth and downlink bandwidth in initial access is not conducive to scheduling flexibility and load balancing in frequency domain.

Relatively speaking, scheme 3 is the most flexible. A separate uplink initial BWP is also helpful to form a clearer signaling structure.

Therefore it was determined in RAN1 #90 that the initial active BWP concept would be introduced in both downlink and uplink [58,59], which is mainly used for BWP operations before RRC connection is set up. The initial UL BWP can be configured in SIB1. After decoding SIB1, the UE can complete the remaining initial access operation in the initial UL BWP.

The initial UL BWP must be configured with RACH resources. If the currently active UL BWP is not configured with RACH resources, the UE needs to switch to the initial UL BWP to initiate random access. When this happens, as shown in Section 4.3.11, the UE may also need to switch its active DL BWP to the initial DL BWP simultaneously.

4.4.3 Initial DL bandwidth part configuration

As described in Section 4.4.1, the initial DL BWP bandwidth should be appropriately larger than the SSB bandwidth to accommodate CORESET#0 and the initial downlink information transmitted in PDSCH, such as RMSI, OSI, random access messages (Msg.2 and Msg.4), paging information, etc. Specifically, there are two schemes: the first is to define one single initial DL BWP to hold CORESET#0 and downlink information such as RMSI, as shown in Fig. 4.53A; the second is to define two initial DL BWPs, respectively, for CORESET # 0 and down-link information such as RMSI, as shown in Fig. 4.53B.

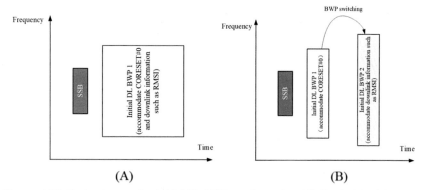

Figure 4.53 Two schemes for initial DL BWP configuration (A) Single initial DL BWP (B) Two initial DL BWPs. *BWP*, Bandwidth part.

From the perspective of the need in initial access, the initial DL BWP should support an access procedure that is "flexible," "simple," "fast," and "universal." To be more specific:

- The location of initial BWP should be close to SSB but with certain flexibility (i.e., TDM or FDM with SSB).
- The signaling to inform the structure of initial BWP should be simple, so that the UE can obtain it quickly and reliably in the downlink initial access procedure.
- The UE can independently determine the initial DL BWP before receiving the first scheduling command.
- The initial DL BWP should be applicable to UEs with any capability (i.e., it should be limited within the RF bandwidth of the UE with the minimum capability).

Based on above requirements, a scheme was proposed and then became prominent in RAN1#89 and RAN1#AH NR2 to indicate initial DL BWP based on the frequency range of CORESET#0. Because CORESET#0 bandwidth can be identified through the initial access procedure (e.g., based on the SSB bandwidth or PBCH indication), directly taking the frequency range of CORESET#0 as the initial DL BWP is the simplest and the most straightforward solution. This scheme can be applied to the UE with any RF capability and can avoid extra BWP switching. Therefore the preliminary study in 3GPP reached a consensus that NR adopts this scheme (i.e., the first scheme) to determine the initial DL BWP.

However, this decision was modified somehow in the follow-up study and actually turned to the second scheme, in which the initial DL BWP can be reconfigured once to contain a larger bandwidth than CORESET#0. This modification was related to the change of the functionality of the initial DL BWP, which will be described later. Here we first focus on the first scheme (i.e., to determine the initial DL BWP bandwidth based on the frequency range of CORESET#0).

In cell search procedure, the only physical channel a UE can detect is SSB. So the UE can determine CORESET#0 only based on the information conveyed by SSB. On one hand, the payload of PBCH is very limited, and is only used to convey the most important system information [called master information block (MIB)]. Therefore the information field that can be used in MIB to indicate CORESET#0 must be limited to a few bits. It is neither realistic nor necessary to adopt the commonly used configuration method like the indication of "starting position + length" as described in Section 4.2.6. On the other hand, after completing cell search, the time-frequency position of the SSB has been determined as an "anchor point." Hence it is natural to predefine several possible time-frequency "patterns" surrounding SSB for CORESET#0, and then to use a few bits to indicate one of the patterns, as described in Section 6.2.2.

Based on this method, the UE can successfully determine its own initial DL BWP according to the frequency range of CORESET#0 in the initial access procedure. However, in the subsequent standardization work, 3GPP RAN2 found that for some use cases (such as P/Scell addition and cell handover), the bandwidth of CORESET#0 is still too small (up to 96 RBs) for initial DL BWP, and suggested to provide an opportunity for the gNB to reconfigure a larger initial DL BWP [65]. After study, RAN1 decided in RAN1#94 to support this improved design by "decoupling" the bandwidth of initial DL BWP and the bandwidth of CORESET#0. The gNB can configure another initial DL BWP to the UE after cell search, which can be different from CORESET#0 bandwidth. Before this second initial DL BWP is configured, the UE keeps using CORESET#0 bandwidth as the initial DL BWP [66].

Obviously, the advantage of a separately configured initial DL BWP is that the configured initial DL BWP can be much wider than CORESET#0 bandwidth. In addition to better supporting operations such as P/Scell addition and cell handover, this method provides the possibility for UEs to work in a simple "single BWP" mode. As mentioned in Section 4.1, BWP is one of the most important new concepts introduced by 5G NR in frequency domain, which can effectively support multiSCS resource allocation and UE power-saving operation. However, for the NR UEs pursuing simplicity and low cost, it can also fall back to the simple operation mode as in LTE, with "fixed bandwidth and single subcarrier spacing." The reconfiguration of initial DL BWP enables such simple operation mode. If the CORESET#0 of small-bandwidth is used as the initial DL BWP, in order to achieve a larger NR operating bandwidth, it is necessary to configure at least one larger downlink UE-dedicated BWP to the UE, resulting in at least two DL BWPs which means the UE may need to support the dynamic switching between them. In contrast, if the initial DL BWP can be reconfigured to a larger bandwidth, all UE operations preferring large bandwidth can be realized in the initial DL BWP. If the gNB does not configure any UE-dedicated BWP for the UE, the UE can still operate in the initial DL BWP in a simple "single BWP" mode, without operating any BWP switching.

Therefore the NR specification finally supports two methods to determine the initial DL BWP:

- First, if the initial DL BWP is not configured in high-layer signaling, the initial DL BWP bandwidth is based on the frequency-domain range of CORESET#0, and the numerology of CORESET#0 (SCS and CP) is used as the numerology of the initial DL BWP.
- Second, if the initial DL BWP is configured in the high-layer signaling, the initial DL BWP is determined according to the high layer configuration.

Obviously, the second method is more flexible. It can be used to determine the initial DL BWP in the PCell as well as the initial DL BWP in the SCell. It can support not only dynamic switching between different BWP configurations (if UE is additionally configured with UE-dedicated BWPs and default DL BWP), but also the simple "single BWP" mode (if UE is not configured with any UE-dedicated BWP). In comparison, the first method can only be used to determine the initial DL BWP in PCell and is normally not applicable to the "single BWP" mode. The two initial DL BWP determination methods support different deployment requirements from operators and different product implementations in different types of UEs.

Fig 4.54A shows the first method, where, if a wider initial DL BWP is not configured by the gNB, the UE keeps using the initial DL BWP determined based on CORESET#0; if a UE-dedicated DL BWP is additionally configured, the UE can be dynamically switched between the UE-dedicated DL BWP and initial DL BWP (which is taken as the default DL BWP when the default DL BWP is not configured). This

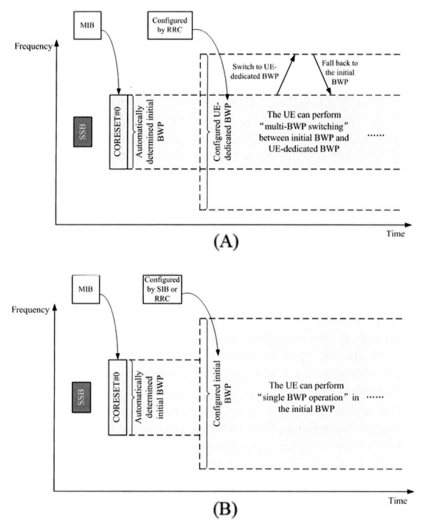

Figure 4.54 Two methods to determine the initial DL BWP (A) Initial DL BWP based on CORESET#0 (B) Initial DL BWP based on high-layer configuration. *BWP,* Bandwidth part.

method is suitable for the UE supporting "multiBWP switching" function. Fig 4.54B shows the second method, where the gNB configures a wider initial DL BWP to the UE through SIB1 without further configuring any UE-dedicated DL BWP, leaving the UE to operate in the initial DL BWP. This method is suitable for the UE only supporting the "single BWP operation" function.

4.4.4 Relationship between the initial DL bandwidth part and default DL bandwidth part

After the initial access procedure, a UE always has a downlink initial BWP. From this perspective, the initial DL BWP can take role of a default DL BWP. In the BWP study, it was once proposed to directly use initial DL BWP to completely replace the concept of default DL BWP. However, the default DL BWP and initial DL BWP provide different functionalities. As mentioned in Section 4.4.1, if the default DL BWP is limited to the frequency location of the initial DL BWP, as shown in Fig. 4.55A, all the UEs in power-saving state could be gathered into a narrow initial DL BWP, which is not conducive to load balancing in frequency domain. Frequency-domain load balancing can be achieved if

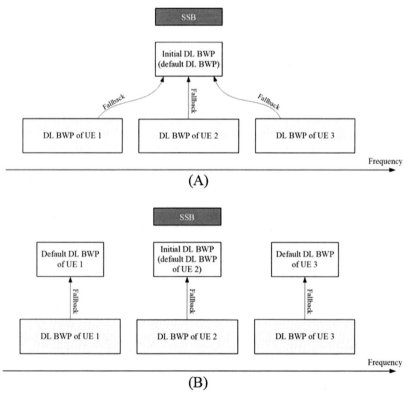

Figure 4.55 The impact of whether the DL default BWP can be configured separately (A) Initial DL BWP taking role of default DL BWP (B) Default DL BWP separately configured from initial DL BWP. *BWP*, Bandwidth part.

different UEs can be configured with different default DL BWPs, as shown in Fig. 4.55B.

In addition, the initial DL BWP only uses one specific SCS, while the UE may be more suitable to work upon another SCS. If all UEs need to fall back to the initial DL BWP, some of them have to endure the unsuitable SCS. If a UE can be configured with default DL BWP with a SCS that is different from the one used in initial DL BWP, the UE can be prevented from losing its suitable SCS when falling back to the default DL BWP.

Therefore NR specification finally supports the separately configured default DL BWP. If gNB does not configure default DL BWP to a UE after RRC connection is set up, the UE treats the initial DL BWP as the default DL BWP [58,59]. This mechanism enables gNB to flexibly choose the configuration mode of the default DL BWP. For example, in a carrier with a relatively small bandwidth, if gNB decides to live with the situation in which the default DL BWPs of all UEs are centralized, it can use the initial DL BWP as the DL default BWP for all UEs.

4.4.5 Initial bandwidth part in carrier aggregation

As mentioned in Section 4.1.7, NR adopts the "carrier + BWP" two-layer configuration and activation method (i.e., the separate designs of CC configuration/activation/deactivation and BWP configuration/activation/deactivation). Therefore when a CC is activated, it is also necessary to determine the active BWP in the CC. In the case of intracarrier scheduling, it is necessary to wait for the UE to receive the first DCI in the CC to determine the active BWP. However, because CORESET is configured in the BWP, the UE cannot determine the CORESET and search space configuration for PDCCH monitoring before knowing the activate DL BWP. Therefore this is another "chicken and egg" problem, which is similar to how to determine the "initial DL BWP" in the initial access procedure in a single carrier.

Therefore an "initial BWP" similar to that in the initial access procedure in the PCell is also needed during the activation procedure of a SCell, so that the UE has an available operating bandwidth before a DCI activates the UE-specific BWP. In order to distinguish the initial BWP in the PCell, the "initial BWP" in the Scell is called "the first active BWP." As shown in Fig. 4.56, in NR CA, RRC can configure not only the BWP in the PCell, but also the first active BWP in the SCells. When one SCell is activated by MAC CE in the PCell, the corresponding first active

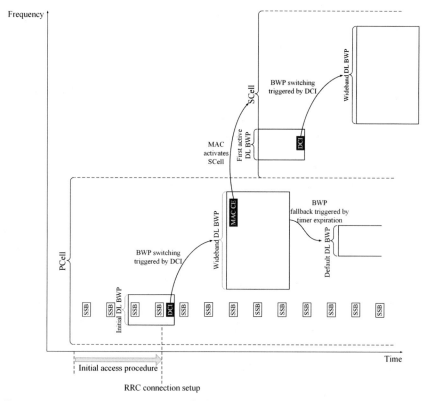

Figure 4.56 Start BWP operation of the SCell through first active BWP. *BWP, Bandwidth part.*

DL BWP is activated simultaneously. The UE can immediately monitor PDCCH in the CORESET configured in the first active DL BWP. After receiving DCI indicating BWP switching, the UE can be switch to the BWP with a larger bandwidth, with the subsequent BWP switching steps being the same as in the PCell.

4.5 Impact of bandwidth part on other physical layer designs

4.5.1 Impact of bandwidth part switching delay

As mentioned in Sections 4.1.2 and 4.3.12, gNB needs to ensure the cross-BWP scheduling does not schedule the uplink or downlink data transmission during BWP switching delay interval. Take downlink scheduling as an example, as shown in Fig. 4.57A. If the PDSCH scheduled by DCI triggering

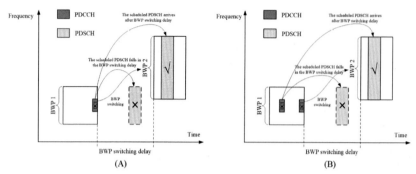

Figure 4.57 Error case of cross-BWP switching (A) DCI triggering BWP switching schedules PDSCH (B) Other DCIs schedule PDSCH. *BWP*, Bandwidth part

BWP switching is transmitted after BWP switching delay, the UE can receive it normally; however, if the PDSCH scheduled by DCI falls within the BWP switching delay interval, the UE may regard this as a scheduling error case and does not receive PDSCH as scheduled.

If the DCI scheduling PDSCH is another DCI before the DCI triggering BWP switching, as shown in Fig. 4.57B, same rule applies—if the scheduled PDSCH falls in the BWP switching delay interval, the UE may regard this as an error case.

It should be noted that, BWP switching delay is the BWP switching interruption duration defined as a UE performance requirement (see Section 4.3.12) (i.e., as long as the UE can resume normal operation in the new BWP before this required interruption duration ends, the UE meets the requirement). However, the interruption duration of BWP switching in actual operation depends on gNB scheduling (i.e., the slot offset field in the DCI triggering BWP switching determines the interruption duration of the BWP switching).

4.5.2 Bandwidth part-dedicated and bandwidth part-common parameter configuration

In 3GPP NR specifications, most of PHY parameters are configured in framework of BWP. Each of these PHY parameters is either "BWP-dedicated" or "BWP-common." BWP-dedicated parameter can be configured "per BWP." If four BWPs are configured for a UE, a BWP-dedicated parameter can be configured as one "copy" for each of four BWPs. The values of the parameter copies can be configured to be different in different BWP, so that the parameter value taking in effect can change along with

BWP switching. A parameter with its value always staying unchanged after BWP switching is BWP-common parameter, also known as cell-specific parameter. In the signaling structure, some BWP-common parameters need to be configured for each BWP as well, but BWP-common parameters can only be configured with the same value across different BWPs.

In the standardization of BWP, an important study topic is: which PHY parameters should be BWP-dedicated parameters and which should be BWP-common? A typical BWP-dedicated PHY parameter is CORESET. It was determined in the RAN1#AH NR2 meeting that each DL BWP should be configured with at least one CORESET that contains UE-specific search space to support normal PDCCH monitoring in the DL BWP [22,31]. On the other hand, CORESET containing CSS only needs to be configured in one DL BWP, which is similar to configuring CORESET containing CSS in at least one CC in CA.

In the NR specification, most of the PHY parameters in RRC connection state are finally defined as BWP-dedicated parameters. BWP-common parameters mainly include those parameters used in the initial access procedure. The UE without RRC connection cannot be configured with BWP-dedicated parameters, but can have BWP-common parameter configurations via SIB.

4.6 Summary

BWP is one of the most important new concepts introduced in 5G NR, and can more effectively realize UE power saving and multinumerology resource allocation in large bandwidth. The key designs include BWP configuration method, BWP switching method, BWP operation in initial access procedure, and so on. Although the complete BWP functions may need to be gradually deployed in 5G era, BWP has almost become the basis of the whole NR frequency-domain operation. Various PHY procedures, such as the initial access, control channel monitoring, resource allocation, etc., were described in a way relating to the BWP concept. Therefore comprehensive knowledge of BWP is indispensable for in-depth understanding of the existing 5G specification and future 5G enhancements.

References

[1] R1−1711795. On bandwidth parts and "RF" requirements, Ericsson, 3GPP TSG RAN WG1#AH NR2, Qingdao, China, 27th−30th June 2017.

[2] R1−1609429. Discussion on resource block for NR, Huawei, HiSilicon, 3GPP RAN1#86bis, Lisbon, Portugal, 10−14 October 2016.

[3] R1−1600570. DL resource allocation and indication for NR, OPPO, 3GPP RAN1 NR Ad Hoc#1701, Spokane, USA, 16th−20th January 2017.

[4] Report of 3GPP TSG RAN WG1#88. Athens, Greece, 13th−17th February 2017.

[5] R1−1703781. Resource allocation for data transmission, Huawei, HiSilicon, OPPO, InterDigital, Panasonic, ETRI, 3GPP RAN1#88, Athens, Greece 13th−17th February 2017.

[6] Report of 3GPP TSG RAN WG1#86bis. Lisbon, Portugal, 10th−14th October 2016.

[7] R1−1611041. Way forward on bandwidth adaptation in NR, MediaTek, Acer, CHTTL, III, Panasonic, Ericsson, Nokia, ASB, Samsung, LG, Intel, 3GPP RAN1#86bis, Lisbon, Portugal, 10th−14th October 2016.

[8] Report of 3GPP TSG RAN WG1#87. Reno, USA, 14th−18th November 2016.

[9] R1−1613218. WF on UE bandwidth adaptation in NR, MediaTek, Acer, AT&T, CHTTL, Ericsson, III, InterDigital, ITRI, NTT DoCoMo, Qualcomm, Samsung, Verizon, 3GPP RAN1#87, Reno, USA, 14th−18th November 2016.

[10] Report of 3GPP TSG RAN WG1#88bis. Spokane, USA, 3rd−7th April 2017.

[11] R1−1706427. Way forward on UE-specific RF bandwidth adaptation in NR, MediaTek, AT&T, ITRI, 3GPP RAN1#88bis, Spokane, USA, 3rd−7th April 2017.

[12] R1−1611655. Mechanisms of bandwidth adaptation for control and data reception in single carrier and multi-carrier cases, Huawei, HiSilicon, 3GPP RAN1#87, Reno, USA, 14−18 November 2016.

[13] R1−1700158. UE-specific RF bandwidth adaptation for single component carrier operation, MediaTek Inc, 3GPP RAN1 NR Ad Hoc#1701, Spokane, USA, 16th−20th January 2017.

[14] R1−1612439. Bandwidth adaptation for UE power savings, Samsung, 3GPP RAN1#87, Reno, USA, 14−18 November 2016.

[15] R1−1706745. Way forward on bandwidth part in NR, MediaTek, Huawei, HiSilicon, Ericsson, Nokia, 3GPP RAN1#88bis, Spokane, USA, 3rd−7th April 2017.

[16] Report of 3GPP TSG RAN WG1#89. Hangzhou, China, 15th−19th May 2017.

[17] R1−1709519. WF on bandwidth part configuration, OPPO, Ericsson, Huawei, HiSilicon, MediaTek, Intel, DOCOMO, LGE, ETRI, CATR, NEC, ZTE, CATT, Samsung, 3GPP RAN1#89, Hangzhou, China, 15th−19th May 2017.

[18] R1−1709802. Way forward on bandwidth part for efficient wideband operation in NR, Ericsson, 3GPP RAN1#89, Hangzhou, China, 15th−19th May 2017.

[19] Report of 3GPP TSG RAN WG1 #AH_NR3 v1.0.0. Nagoya, Japan, 18th−21st September 2017.

[20] 3GPP TS 38.213 V15.3.0 (2018−09). NR: physical layer procedures for control (release 15).

[21] R1−1709625. WF on PRB grid structure for wider bandwidth operation, LG electronics, OPPO, ASUSTEK, Ericsson, Intel, DOCOMO, Huawei, HiSilicon, 3GPP RAN1#89, Hangzhou, China, 15th−19th May 2017.

[22] Report of 3GPP TSG RAN WG1#AH NR2. Qingdao, China, 27th−30th June 2017.

[23] R1−1711855. Way forward on PRB indexing, Intel, Sharp, Ericsson, MediaTek, NTT DOCOMO, Panasonic, Nokia, ASB, NEC, KT, ETRI, 3GPP TSG RAN WG1#AH NR2, Qingdao, China, 27th−30th June 2017.

[24] 3GPP TS 38.211 V15.3.0 (2018−09). NR: physical channels and modulation (release 15).

[25] R1−1706900. On bandwidth part and bandwidth adaptation, Huawei, HiSilicon, 3GPP TSG RAN WG1 Meeting #89, Hangzhou, China, 15−19 May 2017.

[26] R1—1711812. WF on configuration of a BWP in wider bandwidth operation, LG Electronics, MediaTek, 3GPP TSG RAN WG1#AH NR2, Qingdao, China, 27th—30th June 2017.

[27] R1—1710352. Remaining details on wider bandwidth operation, LG Electronics, 3GPP TSG RAN WG1#AH NR2, Qingdao, China, 27th—30th June 2017.

[28] R1—1716019. On bandwidth part operation, Samsung, 3GPP TSG-RAN WG1 Meeting NRAH#3, Nagoya, Japan, 18th—21st September 2017.

[29] 3GPP TS 38.331 V15.3.0 (2018—06). NR: Radio Resource Control (RRC) protocol specification (release 15).

[30] 3GPP TS 38.214 V15.3.0 (2018—09). NR: physical layer procedures for data (release 15).

[31] R1—1711802. Way Forward on Further Details for Bandwidth Part, MediaTek, Huawei, HiSilicon, 3GPP TSG RAN WG1#AH NR2, Qingdao, China, 27th—30th June 2017.

[32] R1—1715755. On remaining aspects of NR CA/DC and BWPs, Nokia, Nokia Shanghai Bell, 3GPP TSG-RAN WG1 Meeting NRAH#3, Nagoya, Japan, 18th—21st September 2017.

[33] Report of 3GPP TSG RAN WG1 #90bis v1.0.0. Prague, Czech Rep, 9th—13th October 2017.

[34] R1—1700371. Scheduling and bandwidth configuration in wide channel bandwidth, Intel Corporation, 3GPP RAN1 NR Ad Hoc#1701, Spokane, USA, 16th—20th January 2017.

[35] R1—1704091. Reply LS on UE RF bandwidth adaptation in NR, RAN4, MediaTek, 3GPP RAN1#88, Athens, Greece, 13—17 February 2017.

[36] R1—1700497. Further discussion on bandwidth adaptation, LG Electronics, 3GPP RAN1 NR Ad Hoc#1701, Spokane, USA, 16th—20th January 2017.

[37] R1—1716601. On CA related aspects and BWP related aspects, Ericsson, 3GPP TSG-RAN WG1 Meeting NRAH#3, Nagoya, Japan, 18th—21st September 2017.

[38] R1—1700011. Mechanisms of bandwidth adaptation, Huawei, HiSilicon, 3GPP RAN1 NR Ad Hoc#1701, Spokane, USA, 16th—20th January 2017.

[39] R1—1711853. Activation/deactivation of bandwidth part, Ericsson, 3GPP TSG RAN WG1#AH NR2, Qingdao, China, 27th—30th June 2017.

[40] R1—1710164. Bandwidth part configuration and frequency resource allocation, OPPO, 3GPP TSG RAN WG1#AH NR2, Qingdao, China, 27th—30th June 2017.

[41] R1—1716327. Remaining aspects for carrier aggregation and bandwidth parts, Intel Corporation, 3GPP TSG-RAN WG1 Meeting NRAH#3, Nagoya, Japan, 18th—21st September 2017.

[42] R1—1713654. Wider bandwidth operations, Samsung, 3GPP TSG RAN WG1#90, Prague, Czech Rep, 21st—25th August 2017.

[43] R1—1716258. Remaining details of BWP, InterDigital, Inc., 3GPP TSG-RAN WG1 Meeting NRAH#3, Nagoya, Japan, 18th—21st September 2017.

[44] R1—1709054. On bandwidth parts, Ericsson, 3GPP RAN1#89, Hangzhou, China, 15th—19th May 2017.

[45] R1—1700709. Bandwidth adaptation in NR, InterDigital Communications, 3GPP RAN1 NR Ad Hoc#1701, Spokane, USA, 16th—20th January 2017.

[46] R1—1716019. On bandwidth part operation, Samsung, 3GPP TSG-RAN WG1 Meeting NRAH#3, Nagoya, Japan, 18th—21st September 2017.

[47] R1—1716109. Remaining issues on bandwidth parts for NR, NTT DOCOMO, INC., 3GPP TSG-RAN WG1 Meeting NRAH#3, Nagoya, Japan, 18th—21st September 2017.

[48] R1—1715892. Discussion on carrier aggregation and bandwidth parts, LG Electronics, 3GPP TSG-RAN WG1 Meeting NRAH#3, Nagoya, Japan, 18th—21st September 2017.

[49] R1—1700362. On the bandwidth adaptation for NR, Intel Corporation, 3GPP RAN1 NR Ad Hoc#1701, Spokane, USA, 16th—20th January 2017.

[50] R1—1707719. On bandwidth part configuration, OPPO, 3GPP TSG RAN WG1#89, Hangzhou, P.R. China, 15th—19th May 2017.

[51] R1—1701491. WF on resource allocation, Huawei, HiSilicon, OPPO, Nokia, Panasonic, NTT DoCoMo, InterDigital, Fujitsu, 3GPP RAN1 NR Ad Hoc#1701, Spokane, USA, 16th—20th January 2017.

[52] R1—1704625. Resource indication for UL control channel, OPPO, 3GPP RAN1#88bis, Spokane, USA, 3rd—7th April 2017.

[53] R1—1711788. Way forward on further details of bandwidth part operation, Intel, AT&T, Huawei, HiSilicon, 3GPP TSG RAN WG1#AH NR2, Qingdao, China, 27th—30th June 2017.

[54] R1—1711795. On bandwidth parts and "RF" requirements Ericsson, 3GPP TSG RAN WG1#AH NR2, Qingdao, China, 27th—30th June 2017.

[55] R1—1803602. LS on BWP switching delay, RAN4, Intel, 3GPP TSG RAN WG1#93, Sanya, China, 16th—20th April 2018.

[56] 3GPP TS 38.133 V15.3.0 (2018—09). NR: requirements for support of radio resource management (release 15).

[57] R1—1710416. Design considerations for NR operation with wide bandwidths, AT&T, 3GPP TSG RAN WG1#AH NR2, Qingdao, China, 27th—30th June 2017.

[58] Report of 3GPP TSG RAN WG1 #90 v1.0.0. Prague, Czech Rep, 21st—25th August 2017.

[59] R1—1715307. Way forward on bandwidth part operation MediaTek, Intel, Panasonic, LGE, Nokia, Ericsson, InterDigital, 3GPP TSG RAN WG1#90, Prague, Czech Rep, 21st—25th August 2017.

[60] R1—1712728. Remaining details of bandwidth parts, AT&T, 3GPP TSG RAN WG1#90, Prague, Czech Rep, 21st—25th August 2017.

[61] R1—1712669. Resource allocation for wideband operation, ZTE, 3GPP TSG RAN WG1#90, Prague, Czech Rep, 21st—25th August 2017.

[62] R1—1709972. Overview of wider bandwidth operations, Huawei, HiSilicon, 3GPP TSG RAN WG1#AH NR2, Qingdao, China, 27th—30th June 2017.

[63] R1—1713978. Further details on bandwidth part operation, MediaTek Inc., 3GPP TSG RAN WG1#90, Prague, Czech Rep, 21st—25th August 2017.

[64] R1—1714094. On the remaining wider-band aspects of NR, Nokia, Nokia Shanghai Bell, 3GPP TSG RAN WG1#90, Prague, Czech Rep, 21st—25th August 2017.

[65] R1—1807731. LS on bandwidth configuration for initial BWP, RAN2, 3GPP TSG RAN WG1 #93, Busan, Korea, 21st—25th May 2018.

[66] R1—1810002. LS on bandwidth configuration for initial BWP, RAN1, 3GPP TSG RAN WG1 #94, Gothenburg, Sweden, 20th—24th August 2018.

CHAPTER 5

5G flexible scheduling

Yanan Lin, Jia Shen and Zhenshan Zhao

Similar to Long Term Evolution (LTE), 5G New Radio (NR) is a wireless communication system based on Orthogonal Frequency Division Multiple Access (OFDMA). The time−frequency resource allocation and scheduling are the core components of system design. As mentioned in Chapter 1, Overview, due to its relatively coarse resource allocation granularity and relatively simple scheduling method, LTE does not fully explore the potential of OFDMA system, and in particular has difficulty in supporting low-latency transmission. For the design of 5G NR system, one primary goal is to introduce more scheduling flexibility, to fully realize the potential of multidimensional resource allocation and multiplexing of OFDM, and to meet the requirements of 5G application scenarios.

5.1 Principle of flexible scheduling

5.1.1 Limitation of LTE system scheduling design

The minimum resource allocation granularity in the LTE system is Physical Resource Block (PRB): 12 subcarriers × 1 slot, where one slot contains seven symbols of normal Cyclic Prefix (NCP) or six symbols of extended Cyclic Prefix (ECP). The LTE resource scheduling method is shown in Fig. 5.1. The basic unit of time-domain resource allocation (TDRA) is a subframe, including two slots (14 symbols using NCP or 12 symbols using ECP). The time−frequency resource scheduled for Physical Downlink Shared Channel (PDSCH), Physical Uplink Shared Channel (PUSCH), and Physical Uplink Control Channel (PUCCH) contains at least 12 subcarriers × 1 subframe (i.e., two PRBs). A couple of earliest symbols of each downlink subframe (up to three symbols when the carrier bandwidth is not 1.4 MHz) form a "control region," which is used to transmit Physical Downlink Control Channel (PDCCH), and the remaining symbols in the downlink subframe are used to transmit PDSCH.

The LTE scheduling method has the following limitations:

1. The time-domain granularity of LTE data channel (PDSCH and PUSCH) is one subframe. It is difficult to carry out fast data transmission. Even if the

scheduling bandwidth of one 5G LTE carrier has been increased to more than 100 MHz, the transmission duration cannot be shortened.

2. LTE PDSCH and PUSCH must occupy all time-domain resources of one subframe within a certain bandwidth, and Time Division Multiplexing (TDM) between two PDSCHs or two PUSCHs in a subframe cannot be realized.

3. The frequency-domain scheduling method of LTE PDSCH and PUSCH is suitable for carrier bandwidth of 20 MHz and below. But it is difficult to meet the requirements of 100 MHz bandwidth operation.

4. LTE PDCCH can only be transmitted at the beginning of a subframe, which limits the time-domain density and location flexibility of PDCCH. Downlink Control Information (DCI) cannot be transmitted as on demand.

5. LTE PDCCH must occupy the whole system bandwidth (up to 20 MHz), and PDSCH cannot use symbols in the control region even if fast transmission is required. Even if there are spared resources in the control region, the simultaneous transmission of PDSCH and PDCCH by Frequency Division Multiplexing (FDM) is not supported. With a large carrier bandwidth in 5G (such as 100 MHz), this limitation can cause a huge spectrum of waste.

Considering the minimum frequency-domain and time-domain resource units of the OFDM signal are "subcarrier" and "symbol" respectively [1], the granularity of LTE resource allocation is obviously too coarse. This is not caused by the principle of OFDM, but an artificial choice for simplified design. It was compatible in 4G with the product

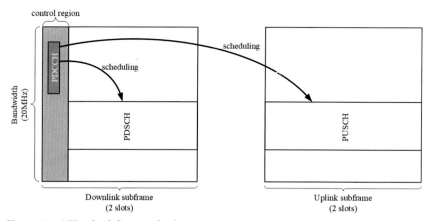

Figure 5.1 LTE scheduling method.

capabilities of software and hardware as well as business requirements, but it is not flexible enough in 5G to satisfy the design requirements for large bandwidth, low latency, and multiservice multiplexing.

5.1.2 Scheduling flexibility in the frequency domain

The 5G NR system introduces more flexible resource scheduling in the frequency domain to better support large bandwidth operations. Compared with LTE, the time−frequency resources allocated to a user in 5G NR can be simply characterized as "widening in the frequency domain and shortening in the time domain." The typical NR resource allocation is shown in Fig. 5.2, exhibiting the capability of carrying out large bandwidth transmission to make full use of the increased bandwidth and meanwhile shortening the time-domain transmission duration to improve the data rate.

The main problems to be solved from the specification perspective for "widening in the frequency domain" are the increase of complexity caused by large bandwidth control and the increase of control signaling overhead. In the 4G LTE system, a carrier bandwidth is 20 MHz and the bandwidth beyond 20 MHz is realized by carrier aggregation; the bandwidth size for a resource allocation of data channel and control channel is mostly in a relatively moderate range. Based on the fact that both PDCCH detection and PDSCH/PUSCH resource scheduling are within the range of 20 MHz, the terminal complexity and the control signaling overhead can be tolerable. However, when it comes to the 5G NR system, where a single-carrier bandwidth is extended to 100 MHz or even larger, to clone the LTE design within such a large bandwidth will greatly increase the complexity of PDCCH detection and the signaling overhead of PDSCH/PUSCH scheduling.

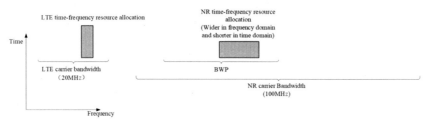

Figure 5.2 Comparison of typical resource allocations for LTE and 5G NR data channels.

Another purpose of 5G NR flexible frequency-domain scheduling is to support forward compatibility. Certain parts of time—frequency resources in the carrier can be flexibly reserved for future business and technologies. The specific design of resource reservation is described in Section 5.7.2.

To support frequency-domain resource allocation with large bandwidth, high scheduling flexibility, and forward-compatible resource reservation without significantly increasing complexity and signaling overhead, the 5G NR mainly adopts the following three techniques.

5.1.2.1 Resource allocation based on bandwidth part

Different from the LTE system where the resource allocation range is the whole system bandwidth, 5G NR has carrier bandwidth much larger than the LTE system bandwidth, which means that a user terminal may not always need to transmit in such a large full bandwidth. Thus a small bandwidth part (BWP) can be configured to limit the size of frequency-domain resource allocation range to reduce scheduling complexity and signaling overhead. In addition, frequency-domain resources within the carrier but outside the configured BWP cannot be allocated by scheduling signaling, which is equivalent to resource reservation. The relevant content of BWP has been described in Chapter 4, Bandwidth Part.

5.1.2.2 Increase granularity of frequency-domain resource allocation

The granularity size is determined mainly based on the signaling overhead and the complexity of resource allocation. A large granularity can be used to alleviate a significant increase in the number of resource particles caused by the increase of bandwidth. On one hand, the larger Subcarrier Spacing (SCS) adopted in 5G NR can increase the absolute frequency-domain size of one PRB. For example, 5G NR PRB is composed of 12 subcarriers as in LTE, but with 30 kHz SCS, the size of a PRB in the frequency domain is 360 kHz, which is double the size in LTE. On the other hand, 5G NR can use a larger Resource Block Group (RBG) to achieve a scheduling with more PRBs, which will be described in Section 5.2.3.

It should be noted here that the concept of PRB in 5G NR specification is different from LTE. In LTE specification, PRB is a time—frequency, 2D concept. One PRB refers to a rectangle resource block over 12 subcarriers × 7 symbols (assuming normal Cyclic Prefix (CP)) and includes 84 Resource Elements (REs), where 1 RE equals to 1 subcarrier × 1 symbol. To describe the number of time—frequency resources in LTE, the number of PRBs is sufficient. But in NR specification, due to the introduction of the symbol-based flexible time-domain resource indication method (see Section 5.2.5), it is not appropriate to define a single resource allocation unit over both frequency domain and time domain. A PRB in NR specification, which contains 12 subcarriers, is completely a frequency-domain concept and does not have any time-domain meaning. Time—frequency resource scheduling cannot be described only by PRB, but needs to use "the number of PRBs + the number of slot/symbols." This is a big conceptual difference between 5G NR and LTE, and therefore demands attention.

5.1.2.3 Adopt more dynamic resource indication signaling
Due to the needs of large bandwidth scheduling and fine scheduling granularity, 5G NR adopts a more flexible signaling structure than LTE. As shown in Table 5.1, many fixed parameters in LTE are changed to semistatic configurations in NR, and some semistatic configurations in LTE become dynamic indications in NR.

5.1.3 Scheduling flexibility in the time domain
In the time domain, an important concept introduced in the Study Item stage of 5G NR is minislot, which is a scheduling unit that is significantly shorter than a slot.

In LTE system, time-domain resource scheduling is based on subframe (i.e., two slots). Although this helps to save downlink control signaling overhead, it greatly limits the flexibility of scheduling. Why does 5G NR pursue higher time-domain scheduling flexibility than LTE? The following scenarios are considered as the main motivations and the use cases in which minislot is beneficial.

5.1.3.1 Low-latency transmission
As described in Chapter 1, Overview, 5G NR is very different from LTE in that in addition to Enhanced Mobile Broadband (eMBB) service, 5G NR supports Ultra Reliable and Low-Latency (URLLC) service. To achieve low-latency transmission, a feasible method is to use smaller TDRA granularity. PDCCH adopts smaller TDRA granularity to help achieve faster downlink signaling transmission and resource scheduling; data channels (PDSCH and PUSCH) adopt smaller TDRA granularity to help achieve faster uplink or downlink data transmission; PUCCH adopts smaller TDRA granularity to help achieve faster uplink signaling transmission

Table 5.1 LTE and 5G NR resource allocation signaling structure comparison.

	4G LTE	**5G NR**
PDSCH/ PUSCH frequency scheduling	• The scheduling range is fixed as the carrier bandwidth and the granularity of resource allocation is a function of the carrier bandwidth. • The scheduling type is semistatically configured. • The specific resource allocation is directly indicated by DCI.	• The scheduling range is fixed as a configured BWP, and the granularity of resource allocation is a function of the BWP size, with the function being configurable. • The scheduling type is semistatically configured or dynamically switched. • The specific resource allocation is directly indicated by DCI.
PDSCH/ PUSCH time scheduling	• Fixed length and subframe-level scheduling. • The resource allocation is directly indicated by DCI.	• Flexible length and location, and subframe + symbol-level scheduling. • The candidate resource allocations are semistatically configured as in a table and dynamically selected by DCI.
PDCCH resource configuration	• The monitoring range in the frequency domain is fixed as carrier bandwidth. • The time-domain control region is located at the beginning of a subframe, and the length of 1−3 symbols is dynamically variable.	• The monitoring range in the frequency domain is semistatically configured. • Time-domain monitoring range can be located at any position in a slot, and the length and positions can be semistatically configured.
PUCCH resource scheduling	• The length is fixed and the time-domain position relative to PDSCH is fixed.	• Flexible length and location, and subframe + symbol-level scheduling. • The candidate resource allocations are semistatically configured as in a table and dynamically selected by DCI.

(Continued)

Table 5.1 (Continued)

	4G LTE	5G NR
Reserved resource	• Multicast Broadcast Single Frequency Network (MBSFN) subframe reservation can be configured. • Frequency-domain granularity is the whole carrier and time-domain granularity is the whole subframe.	• The candidate resources are semistatically configured in a table and dynamically selected by DCI. • Frequency-domain granularity is a PRB and time-domain granularity is a symbol.

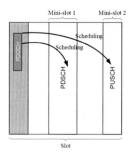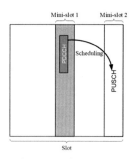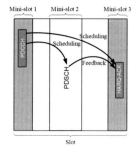

Figure 5.3 Fast scheduling based on minislot.

and Hybrid Automatic Repeat Request-Aknowledgement (HARQ−ACK) feedback.

Compared with the subframe-level resource allocation shown in Fig. 5.1, the resource allocation using minislot as the granularity can reach much shorter processing latency of each physical process. As shown in Fig. 5.3, the time-domain resource of a slot can be further partitioned into minislots.

• As shown in the left figure, PDCCH allocated in the beginning of the slot can quickly schedule PDSCH in the same slot (with Minislot 1 as the resource unit) and PUSCH at the end of the same slot (with Minislot 2 as the resource unit).
• As shown in the middle figure, Minislot 1 that contains PDCCH can be allocated at any suitable location of a slot. This allows the PDCCH to be sent at any time when an urgent scheduling data channel

transmission is needed, even in the middle or at the end of the slot. Minislot 2 containing the scheduled PDSCH can use the remaining time-domain resources at the end of the slot.

- As shown in the right figure, after PDSCH transmission in Minislot 2, if there are sufficient time-domain resources within the remaining slot, Minislot 3 can be allocated at the end of the slot for a PUCCH transmission carrying the corresponding HARQ−ACK information, to achieve fast HARQ−ACK feedback within a slot.

The resource allocations for PDSCH/PUSCH (*left* figure), PDCCH (*middle* figure), and PUCCH (*right figure*) are described in Section 5.2, Section 5.4, and Section 5.5, respectively. In addition, to achieve schemes in the left and right figures in Time Division Duplexing (TDD) system, both downlink transmission and uplink transmission in a slot need to be supported, which is a so-called self-contained slot format (TD−LTE special subframe can be regarded as a special case) and is introduced in Section 5.6.

5.1.3.2 Multibeam transmission

When 5G NR system is deployed in millimeter wave spectrum (i.e., FR2), the equipment generally uses analog beam-forming technology to aggregate power in space to overcome the coverage defect of high frequency band, which means the whole cell can only be beamed in one direction at a certain time. To support multiuser access, beam sweeping in multiple directions needs to be supported. As shown in Fig. 5.4, if beam scanning is carried out based on slot, each beam needs to occupy at least one slot. When the number of users is large, the transmission gap of each user can be quite long, making the latency in resource scheduling and signaling feedback too large to be acceptable. If minislot-based beam scanning is adopted, the time granularity occupied by each beam is reduced to minislot and multiple-beam transmission can be completed even within one slot, which can greatly improve the transmission efficiency of each beam and maintain the latency of resource scheduling and signaling feedback within the acceptable range.

5.1.3.3 Flexible multiplexing between channels

As mentioned above, minislot can support self-contained slot structure. PDSCH can be multiplexed with PUCCH or PUSCH in the same slot, which is an example of flexible multiplexing between channels. Minislot structure can be also used in multiplexing of channels with the same transmission direction in a slot, including PDCCH and PDSCH, PDSCH, and

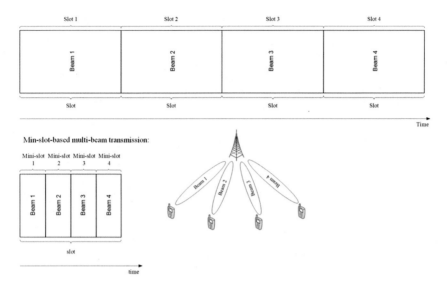

Figure 5.4 Comparison between beam sweeping based on slot and beam sweeping based on minislot.

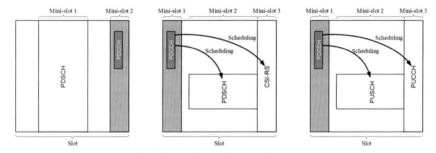

Figure 5.5 Multiplexing of multiple channels in a slot.

Channel State Information Reference Signal (CSI—RS), and PUSCH and PUCCH.

The minislot structure allows more flexibly multiplexing of PDCCH and PDSCH. Within a slot, the terminal can even receive PDCCH after receiving PDSCH (as shown in Fig. 5.5 *left*), which can facilitate the transmission of downlink control signaling at any time and make more efficient use of fragment resources.

CSI—RS can be used to detect the change of downlink Channel State Information (CSI) for more efficient PDSCH scheduling. With the multiplexing of PDSCH and CSI—RS in a slot, the terminal can receive CSI—RS (as shown in the middle of Fig. 5.5) in the remaining symbols after receiving PDSCH, so that the base station can get the report of the latest downlink CSI in time. In LTE, PDSCH and CSI—RS can be multiplexed in a subframe by rate matching. However, minislot-based multiplexing is a more flexible way, for example, to support the two channels to use different beams.

PUCCH is used to transmit Uplink Control Information (UCI) including HARQ—ACK feedback of PDSCH, Scheduling Request (SR) and CSI report. If PUSCH and PUCCH cannot be transmitted simultaneously in one slot, the terminal needs to multiplex UCI into PUSCH. With the use of minislot to multiplex PUSCH and PUCCH in a slot, the terminal can transmit PUCCH in the remaining symbols of the slot after transmitting PUSCH (as shown in Fig. 5.5 right), so that the terminal can feed back HARQ—ACK or initiate SR in time. This PUCCH, which consumes only a few symbols, is called "Short PUCCH."

5.1.3.4 Effectively support unlicensed spectrum operation

Starting from 4G, 3rd Generation Partnership (3GPP) has been committed to expanding from licensed spectrum to unlicensed spectrum. Although the first release of 5G NR specification (R15) did not yet contain a component for NR unlicensed (NR—U) spectrum operation, it has been widely considered that the basic design of 5G NR needs to provide better support for the introduction of NR—U features in the future release. Transmission in the unlicensed spectrum is subject to the transmission rules of the unlicensed spectrum, such as the Listen-Before-Talk (LBT) principle. The transmission window obtained after LBT detection may be very short, and the transmission in the window needs to be completed as soon as possible. Theoretically speaking, the minislot structure is more advantageous to the LBT transmission scheme and the effective NR—U transmission than the slot structure. The NR—U operation is defined in the 3GPP R16 specification, as described in Chapter 18, 5G NR in Unlicensed Spectrum.

5.2 5G resource allocation

As described in Section 5.1, the flexible scheduling design of 5G NR is reflected in the resource allocation of various data channels and control channels. This section mainly introduces the resource allocation for data channels, and the resource allocation for control channels is introduced in

Section 5.4 and Section 5.5. In terms of frequency-domain resource allocation, 5G NR mainly improves the flexibility in the selection of allocation types and the determination of allocation granularity, which are described throughout Sections 5.2.1—5.2.3. Meanwhile, the improvement of time-domain scheduling flexibility is more significant than in the frequency domain. For the flexible scheduling operating on symbol level, the resource allocation method and signaling structure are redesigned, as described in Sections 5.2.4—5.2.7.

5.2.1 Optimization of resource allocation types in the frequency domain

The frequency-domain resource allocation method for 5G NR data channel is optimized on top of LTE. The LTE system uses OFDMA technology in the downlink and Single Carrier—Frequency Division Multiple Access (SC—FDMA) technology based on Discrete Fourier Transform—Spread Orthogonal FDM (DFT—S—OFDM) in the uplink. DFT—S—OFDM is a kind of variation of OFDM. To maintain the single-carrier characteristic of SC—FDMA, LTE PUSCH uses continuous resource allocation in the frequency domain (i.e., occupying contiguous PRBs; see Ref. [1] for details). PDSCH can adopt either continuous or noncontinuous (i.e., noncontiguous PRBs are occupied) frequency-domain resource allocation. There are three types of downlink frequency-domain resource allocation in the LTE specification:

1. Type 0 resource allocation: RBG indicated by bitmap is used to realize continuous or noncontinuous resource allocation in the frequency domain.
2. Type 1 resource allocation: It is an extension of Type 0 resource allocation, where the allocated PRBs are indicated by a bitmap that is defined over a dispersed subset of all PRBs in the frequency domain. This type of resource allocation can only realize noncontinuous resource allocation, but guarantee the frequency selectivity of the allocated resources.
3. Type 2 resource allocation: A set of contiguous PRBs in the frequency domain are indicated by two pieces of information: starting PRB index and the number of contiguous PRBs. This type of resource allocation can only realize continuous resource allocation.

The above three types of resource allocation in LTE have some functionality overlapping and therefore specification redundancy. With study in NR, Type 0 resource allocation and Type 2 resource allocation are supported, and Type 2 resource allocation in LTE is renamed to Type 1 resource allocation in NR. Therefore the mechanisms of Type 0 and Type 1 resource allocations in 5G NR basically share the same principle as Type 0 and Type 2 resource allocations in LTE. The main changes made in 5G NR include:

1. OFDMA is introduced for 5G NR uplink transmission. Therefore the same resource allocation method can be used by PDSCH and PUSCH (in the early release of LTE, PUSCH does not support noncontiguous resource allocation).
2. The frequency-domain resource allocation range is changed from system bandwidth to BWP, as described in Chapter 4, Bandwidth Part.
3. The dynamic switching between Type 0 resource allocation and Type 1 resource allocation is supported, as described later in this chapter.
4. The method for determining the size of Type 0 RBG is adjusted, as described in Section 5.2.2.

Both Type 0 and Type 1 (i.e., LTE Type 2) resource allocation have their own advantages and disadvantages:

- As shown in Fig. 5.6, Type 0 resource allocation uses a bitmap to indicate the selected RBGs, where for each bit in the bitmap "1" represents the corresponding RBG assigned to the terminal and "0" represents the opposite. It can realize the flexible frequency-domain resource allocation in BWP, and support noncontinuous resource allocation. However, the disadvantages are as follows: (1) the bitmap has a larger number of bits than the Type 1 Resource Indicator Value

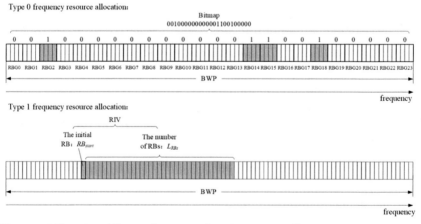

Figure 5.6 Type 0 and Type 1 frequency-domain resource allocation.

(RIV), resulting in the Downlink Control Information (DCI) over-head greater than that of Type 1 resource allocation; (2) the granularity of resource allocation is coarse because an RBG contains 2−16 RBs, and resources cannot be selected on a per-RB basis. The Type 0 resource allocation is normally considered more suitable in using frag-mented frequency-domain resources. For example, once the base sta-tion scheduler allocates the continuous and noncontinuous resources to some terminals, and the remaining resources in the BWP are divided into small fragments, Type 0 resource allocation can be used to fill in the fragments for other terminals and then make full usage of all spectrum resources.

- As shown in Fig. 5.6, Type 1 uses an RIV to jointly encode the initial RB (RB_{start}) and RB number (L_{RBs}) (calculating RIV based on RB_{start} and L_{RBs} is the same as LTE, except that BWP is used instead of system band-width). The advantage of Type 1 is that it can indicate RB-level resource with a small number of indication bits, and the base station scheduler algo-rithm is simple. The disadvantage is that only contiguous frequency-domain resource can be allocated. When the amount of allocated resources is small, the frequency diversity is limited, and the transmission can be eas-ily affected by frequency selective fading. When the amount of allocated resources is large, Type 1 is a simple and efficient way to allocate resources, especially for high data rate service.

Accordingly, Type 0 and Type 1 resource allocations fit into different scenarios with different amounts of data and different service types, which however may change rapidly for a terminal. In LTE specification, the resource allocation type is semistatically configured by Radio Resource Control (RRC), so only one resource allocation type can be used for a certain period. 5G NR supports a method to dynamically indicate the resource allocation type by using 1 bit in DCI, and then to dynamically switch between the two types of resource allocations. As shown in Fig. 5.7, the base station can configure the frequency-domain resource allocation to a terminal as follows:

- Type 0: when Type 0 is configured, the Frequency-domain Resource Assignment (FDRA) field in DCI contains a bitmap for Type 0 resource allocation. The bitmap contains N_{RBG} bits, where N_{RBG} is the number of RBGs included in the current activated BWP.
- Type 1: when Type 1 is configured, the FDRA field in DCI contains an RIV value for Type 1 resource allocation. According to the calculation method of RIV, the FDRA field contains $\left\lceil log_2 \left(N_{RB}^{BWP} \left(N_{RB}^{BWP} + 1 \right)/2 \right) \right\rceil$

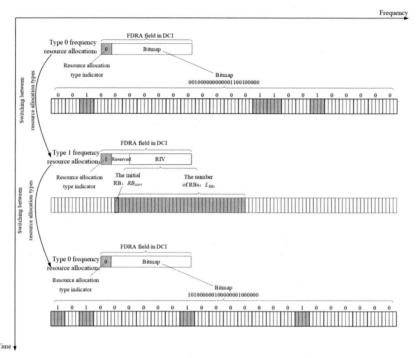

Figure 5.7 Dynamic switching of frequency-domain resource allocation types based on DCI indicator.

bits, where N_{RB}^{BWP} is the number of RBs included in the current activated BWP.

- Dynamic switching between Type 0 and Type 1: when both Type 0 and Type 1 are configured, the first bit in the FDRA field is an indicator of the resource allocation type. When the value of this bit is 0, the remaining bits of the FDRA field contain a bitmap for Type 0 resource allocation. When the value of this bit is 1, the remaining bits of the FDRA field contain an RIV value for Type 1 resource allocation, as shown in Fig. 5.7. It should be noted that, because Type 0 bitmap and Type 1 RIV may require different amounts of bits, the size of the FDRA filed equals to the larger number of bits that are needed in two types (i.e., $\max\left(\left\lceil \log_2\left(N_{RB}^{BWP}\left(N_{RB}^{BWP} + 1\right)/2\right)\right\rceil, N_{RBG}\right) + 1$), respectively. One example is shown in Fig. 5.7. If $N_{RBG} > \left\lceil \log_2\left(N_{RB}^{BWP}\left(N_{RB}^{BWP} + 1\right)/2\right)\right\rceil$, the RIV occupies the last $\left\lceil \log_2\left(N_{RB}^{BWP}\left(N_{RB}^{BWP} + 1\right)/2\right)\right\rceil$ bits of the FDRA field when Type 1 allocation is applied.

5.2.2 Granularity of resource allocation in the frequency domain

As mentioned in the previous section, Type 0 and Type 1 frequency-domain resource allocations of PDSCH and PUSCH in 5G NR follow Type 0 and Type 2 frequency-domain resource allocation of PDSCH in LTE. Before the application scope of BWP was clarified in the early study of NR resource allocation, the resource allocation scheme still assumes to "allocate resources within the entire carrier and determine the RBG size based on the carrier bandwidth" [2]. When the applicability of BWP became clear, all signal transmissions were restricted in one BWP, and consequently "using BWP as the frequency-domain resource allocation range and determining the RBG size based on the BWP size" became a logical choice.

As shown in Fig. 5.6, the number of bits of the FDRA field in the DCI for Type 0 resource allocation depends on the number of RBGs contained in the activated BWP. To reduce the overhead of the FDRA field in DCI, the RBG size (i.e., the number of RBs included in one RBG is P) should be increased as BWP grows. In LTE specification, the relationship between RBG size and system bandwidth is defined in Table 5.2.

As seen from the above table, the RBG size in the LTE varies slightly, and the maximum RBG contains four RBs and the bitmap contains at most 28 bits. The maximum scheduling bandwidth of 5G NR is 275 RBs, which is the same as the maximum NR carrier bandwidth and significantly larger than that of LTE (110 PRBs). It was considered too complicated to support every RBG size that is incremented from 1. Meanwhile, some RBG sizes such as 3 and 6 have been proposed [2] in the NR study stage, mainly for better compatibility with the size of the PDCCH resource allocation unit [i.e., Control Channel Element (CCE)] to reduce the resource fragmentation between PDSCH and PDCCH

Table 5.2 Relationship between RBG and system bandwidth in LTE specification.

System bandwidth (the number of RBs)	RBG size (the number of RBs)
≤ 10	1
$11-26$	2
$27-63$	3
$64-110$	4

when PDSCH and PDSCCH are multiplexed. However, the later study turns to using a more flexible rate-matching method (described in Section 5.7) to solve the multiplexing between PDSCH and PDCCH, and 3 and 6 are not accepted as candidate values of RBG size. It was finally decided to support RBG size equal to 2^n RBs, where $n = 1, 2, 3, 4$.

The next question is: How to keep the RBG size changing with BWP size? In response to the question, several methods were proposed [3]:

- Option 1: The RBG size is directly configured by RRC or even directly indicated by DCI [4], and then the number of RBGs is determined according to the RBG size. The relationship between RBG size and BWP size is hidden in base station implementation.
- Option 2: Similar to RBG size determination in LTE, the RBG size in NR is determined based on the BWP size according to a mapping table, and then the number of RBGs is determined according to the RBG size.
- Option 3: A target number of RBGs is first set, and then the RBG size is determined according to the target number of RBGs and the BWP size [5].

Option 1 can achieve flexible RBG configuration. The number of RBGs (or equivalently the number of bits in the Type 0 bitmap) is controlled within a predetermined range, and the number of RBGs can be further reduced to achieve additional DCI overhead saving. DCI overhead reduction was one of the design targets of 5G NR, which aims to reduce the number of DCI bits (although the final overhead reduction is not significant comparing to LTE) when achieving high scheduling flexibility. DCI payload reduction also helps to improve the reliability of DCI transmission. Under the same channel condition, the smaller the DCI payload size, the lower the channel coding rate to consequently achieve a lower DCI transmission Block Error Rate (BLER). The enhancement to improve the reliability of PDCCH transmission by compressing the DCI size is supported in the R16 URLLC, described in Section 15.1. In R15 NR, a compressing method based on RRC configuration was proposed to apply to the FDRA field in DCI. Simply speaking, a large RBG size can be configured to reduce the number of RBGs so as to compress the size of the Type 0 bitmap. As shown in Fig. 5.8, for a given BWP size (96 PRBs), the base station configures the RBG size to 4 at the first moment and the FDRA field in the DCI contains 24 bits. At the second moment, the base station reconfigures the RBG size to 16 by RRC signaling, and reduces the FDRA field in the DCI to 6 bits. The DCI

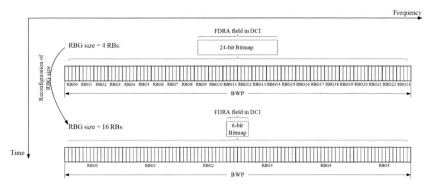

Figure 5.8 Dynamic switching of frequency-domain resource allocation type based on DCI indication.

overhead is smaller and the PDCCH transmission reliability becomes higher. The disadvantage of Option 1 is that the BWP size and RBG size are configured by two RRC configurations, without standardized binding relationship between them. The base station needs to ensure that the two configurations are compatible with each other (i.e., the result of BWP size divided by the RBG size does not exceed the predetermined maximum FDRA field length).

Option 2 follows the LTE mechanism, only with the system bandwidth replaced by the BWP size. It is simple and reliable, but the disadvantage is a lack of configuration flexibility.

Option 3 reverses the determination procedure in Option 1 and Option 2 by determining the number of RBGs first that is then used to determine the RBG size. Option 3 does not provide configuration flexibility either, but the advantage is that the size of the FDRA field can be fixed. Because LTE has quite some different DCI sizes, the PDCCH blind detection complexity in LTE is relatively high. Therefore it was proposed in the early SI stage of NR that the number of different DCI payload sizes should be as few as possible to reduce the complexity of blind PDCCH detection. In Option 3, once the number of bits of a Type 0 bitmap (i.e., the upper limit of the number of RBGs in the BWP) is set, the RBG size can be determined from {2, 4, 8, 16} according to the BWP size and the upper limit of the number of RBGs. The unused bits in the bitmap are processed as zero padding. The FDRA overhead due to those unused padding bits can be relatively large in some cases, which is the disadvantage of Option 3.

Eventually, 3GPP Radio Access Network Layer 1 (RAN1) decided to make a small enhancement on top of Option 2, by defining two sets of mapping relationship between BWP size and RBG size with the selection being configured by RRC. As shown in Table 5.3, the first set of mapping relationships uses RBG size = 2, 4, 8, and 16 RBs for different BWP sizes; the second set of mapping relationships uses RBG size = 4, 8, and 16 for different BWP sizes. The base station configures one of these two mapping relationships by RRC signaling. Compared with mapping relationship 1, mapping relationship 2 adopts a larger RBG size, which helps to reduce the size of the FDRA field. Furthermore, mapping relationship 2 only includes three RBG sizes, and the RBG size of 16 RBs is adopted for the larger BWP sizes (>72 RBs). This is beneficial to reduce the complexity of the base station scheduler. In the LTE system, the RBG size is a function of system bandwidth and therefore is the same for all terminals in a cell, and the base station scheduler performs multiuser frequency-domain resource scheduling based on one RBG size. However, in 5G NR system, because RBG size is determined by BWP size, which may be different among terminals, the RBG size may not be unique in a cell. The base station performs multiuser frequency-domain based on multiple RBG sizes, and the design complexity of the scheduler is significantly increased. Mapping relationship 2 can cut this complexity increase by limiting the total number of different RBG sizes.

Once the RBG size P is determined based on the BWP size according to the configured mapping relationship in Table 5.3, the size of RBG bitmap in Type 0 is calculated by the following formula, which is BWP size dependent. This formula does not use zero padding to align the number of RBGs as in Option 3 above. This FDRA field-size problem relating to BWP switching is described with more detail in Section 5.2.3.

Table 5.3 Mapping relationships between RBG and system bandwidth in the 5G NR Specification.

BWP size (the number of RBs)	RBG size P (the number of RBs)	
	Mapping relationship 1	Mapping relationship 2
1−36	2	4
37−72	4	8
73−144	8	16
145−275	16	16

$$N_{\mathrm{RBG}} = \left\lceil \left(N_{\mathrm{BWP},i}^{\mathrm{size}} + \left(N_{\mathrm{BWP},i}^{\mathrm{start}} \mathrm{mod} P \right) \right) / P \right\rceil \tag{5.1}$$

In 5G NR, another discussion point was whether Type 1 resource allocation should also support larger resource allocation granularity. Type 1 resource allocation is indicated by two pieces of information <starting position, length>. Both the "starting position" and the "length" in LTE Type 2 resource allocation are defined in RB unit. Because the 5G NR carrier has a significantly larger bandwidth than LTE, the "starting position" and the "length" can have larger variation ranges than in LTE. However, due to the specific way to calculate RIV, RIV size is not quite sensitive to the range variations of the "starting position" and the "length," which is different from the inverse proportion between Type 0 bitmap size and RBG size. In Type 1 resource allocation, the number of bits saved by adopting larger resource granularity is limited. Therefore the usage of larger resource granularity for Type 1 resource allocation is not accepted in R15. In the R16 URLLC, due to the stringent PDCCH reliability requirements, DCI payload size needs to be further compressed, and larger resource granularity is supported in Type 1 resource allocation.

5.2.3 Frequency-domain resource indication during BWP switching

This section deals with the problem caused by the simultaneous occurrence of "BWP switching" and "cross-BWP scheduling." It is noted in Section 4.3 that the base station can trigger a "BWP switching" by a DCI and simultaneously allocate resources for the data channel, as shown in Fig. 5.9. Section 4.3 mainly focuses on why the BWP switching is needed and how the BWP switching is triggered, but does not give the details on the difference between the resource allocation in this special DCI and the one in normal DCI. The issue comes from the fact that the DCI used to trigger the BWP switching is not the DCI dedicated to triggering purpose but the DCI used to schedule data channel in the "new BWP"; in other words, this DCI is transmitted in the "current BWP" before the BWP switching but is used to allocate resources in the "new BWP" after the BWP switching. The size of the FDRA field of the DCI is determined according to the size of the "current BWP" from which the DCI is transmitted, as shown in Table 5.3. When the size of the FDRA field calculated based on the size of the "new BWP" is different from the one calculated based on the size of "current BWP," the number of bits in the

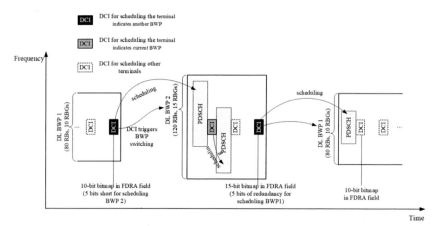

Figure 5.9 Insufficient or redundant FDRA field bits caused by BWP switching.

FDRA field can be either inadequate or excessive. This problem happens within both Type 0 FDRA field and Type 1 FDRA field. In an example shown in Fig. 5.9 for Type 0 FDRA, before BWP switching, the active downlink BWP of the terminal is BWP 1, which contains 80 RBs. Assume the mapping relationship 1 in Table 5.3 is configured. Accordingly, the RBG size is 8 and the Type 0 FDRA field containing a 10-bit bitmap is used to determine the DCI payload size. If the BWP indicator (BWPI) in the DCI indicates a BWP switching to BWP 2 that contains 120 RBs, the resource allocation should follow the RBG size still equal to 8 RBs but a 15-bit Type 0 bitmap. Therefore a 15-bit bitmap cannot directly fit into a 10-bit FDRA field. Conversely, when the DCI indicates the terminal to switch from BWP 2 back to BWP 1, it needs to put the 10-bit bitmap into a 15-bit FDRA field.

To solve this problem caused by the "cross-BWP scheduling," the following two solutions have been proposed [6]:

1. Scheme 1: The number of bits in the FDRA field is always equal to the maximum FDRA field size across various BWP sizes, so that the FDRA field size does not change upon any BWP switching. As shown in Fig. 5.10, two BWPs of 80 RBs and 120 RBs are configured, and the corresponding FDRA field requires 10 bits and 15 bits, respectively. According to Scheme 1, the size of the FDRA field is always 15 bits and independent from which BWP is active. When the FDRA field is used to allocate resources in BWP 1, there are 5 redundant bits. When it is used to allocate resources in BWP 2, the size of the

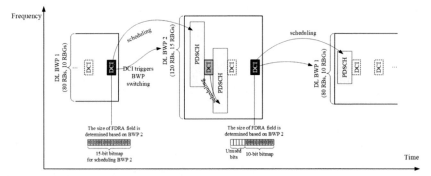

Figure 5.10 Scheme 1 of frequency-domain resource indication during the BWP switching.

FDRA field fits exactly what is needed. As for the bitmap size, when the DCI in BWP 1 allocates the frequency-domain resources in BWP 2, the actual Type 0 bitmap is determined by the size of the scheduled BWP 2 (i.e., 15 bits), rather than the size of BWP 1 that contains DCI. Conversely, when the DCI in BWP 2 allocates the frequency-domain resources in BWP 1, the Type 0 bitmap is determined by the size of the scheduled BWP 1 (i.e., 10 bits).

2. Scheme 2: The size of the FDRA field changes with the size of the BWP in which the DCI is transmitted. When the size of the FDRA field is inadequate, the FDRA field is transmitted as is but on the terminal side zero bits are inserted at the Most Significant Digit (MSD) of the FDRA field. When the FDRA field size is more than needed, the excessive MSD bits of the FDRA field are truncated on the terminal side. As shown in Fig. 5.11, when the active BWP is BWP 1, the size of the FDRA field is 10 bits. If the cross-BWP scheduling is used to allocate the resources in BWP 2, 5 zero bits are inserted by the terminal at the MSD of the original 10 bits to form a 15-bit Type 0 bitmap. When the active BWP is BWP 2, the size of the FDRA field is 15 bits. If the cross-BWP scheduling is used to allocate the resources in BWP 1, the original 15 bits need to be truncated from the MSD on the terminal side to form a 10-bit Type 0 bitmap.

The advantage of Scheme 1 is that it can obtain enough scheduling flexibility by avoiding the size change of the FDRA field. The disadvantage is that the FDRA field overhead is relatively large in some cases. The advantage of Scheme 2 is that the FDRA field overhead is small. The disadvantage is that zero padding is performed on terminal side when the

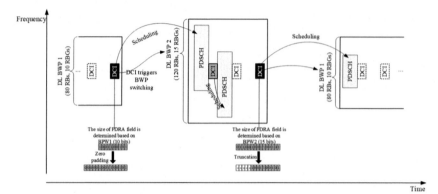

Figure 5.11 Scheme 2 of frequency-domain resource indication during the BWP switching.

number of bits is inadequate, and some scheduling flexibility is lost. As shown in Fig. 5.11, the zero-padding operation can only allow the terminal converting the FDRA field to a bitmap with correct bitmap size, but the frequency-domain resources allocation cannot go beyond what 10-bit bitmap can support. This means there are five RBGs in the BWP that never get chance to be scheduled.

It was finally decided in 3GPP RAN1 to adopt Scheme 2. Although this section uses the Type 0 resource allocation for the description, the mentioned problem is solved for Type 1 resource allocation in the same way.

5.2.4 Determination of frequency-hopping resources in BWP

The normal frequency-domain resource allocation in BWP is described in Sections 5.2.2 and 5.2.3. For the uplink channel, due to the small transmission bandwidth and insufficient frequency diversity, frequency hopping is often used to obtain additional frequency-diversity gain. Uplink frequency hopping is widely used in OFDM system, including LTE system. 3GPP RAN1 agreed that the uplink frequency-hopping technology should be supported in 5G NR. Meanwhile, due to the introduction of BWP, the transmission of any uplink signal needs to be limited within one uplink BWP. How to keep the frequency-hopping resources within a BWP while fully obtain frequency-hopping gain should be considered.

Take LTE PUCCH frequency hopping as an example. As shown in Fig. 5.12, the first hop and the second hop of LTE PUCCH are mirror-symmetrical with respect to the center of the system bandwidth. The distance between first hop and lower edge of the system bandwidth is the same as the distance between the second hop and the upper edge of the system bandwidth. PUCCH is distributed on both sides of the system bandwidth, and the central part of the system bandwidth is reserved for the data channel (such as PUSCH). In a system that uses a fixed system bandwidth such as LTE, this method can make the PUCCH hop with the possibly largest offset, thereby to maximize the frequency diversity gain.

But this hopping design also has some problems. The first is that the PUCCH frequency-hopping offset is uneven for different terminals. As shown in Fig. 5.12, some terminals have a larger frequency-hopping offset than the others. The PUCCH that is allocated closer to the edge of the system bandwidth owns the stronger frequency-domain diversity effect and therefore the better transmission performance. The PUCCH that is closer to the center of the system bandwidth has the weaker frequency-domain diversity effect and therefore the worse transmission performance. This problem is not so prominent in LTE system which usually uses 20 and 10 MHz system bandwidth. However, in the 5G NR system, when the bandwidth of the active uplink BWP is small, the hopping offset of the PUCCH not allocated close to BWP edge is further reduced, and the PUCCH transmission performance is impacted.

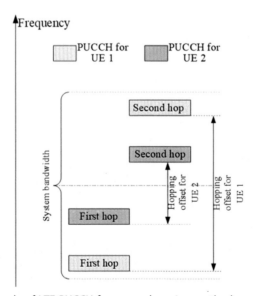

Figure 5.12 Example of LTE PUCCH frequency-hopping method.

Therefore it is necessary to consider flexible configuration of frequency-hopping offset by RRC signaling to adapt to different BWP sizes. Three options are proposed:

1. Option 1: RRC configures the absolute value of the frequency-hopping offset;
2. Option 2: RRC directly configures the frequency-domain position of each hop;
3. Option 3: Specification defines the frequency-hopping offset based on the BWP size.

Interestingly, these three options were all finally adopted by the 5G NR specification, and they are used for different channels and in different scenarios. Option 1 is used for PUSCH frequency-hopping resource indication after RRC connection is established. Option 2 is used for PUCCH frequency-hopping resource indication after RRC connection is established. Option 3 is used for PUSCH and PUCCH before RRC connection is established (i.e., during initial access).

Option 1 is to configure the number of RBs, RB_{offset}, between the first hop and the second hop. When DCI indicates the frequency-domain position, RB_{start}, of the first hop, the frequency-domain position of the second hop can be calculated as $RB_{start} + RB_{offset}$, as shown in Fig. 5.13.

In Fig. 5.13, if the first hop position, RB_{start}, is close to the lower boundary of the UL BWP, the second hop position, $RB_{start} + RB_{offset}$,

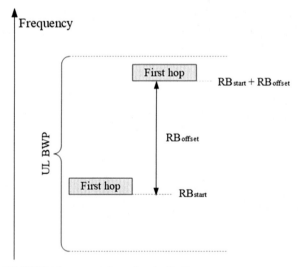

Figure 5.13 NR PUSCH frequency-hopping indication.

does not exceed the higher boundary of the UL BWP. However, if RB_{start} is close to the higher boundary of the UL BWP, the position of the second hop may exceed the higher boundary of the UL BWP, and a frequency-domain resource indication error happens. 5G NR specification takes the second hop position modulo BWP size, such as the formulation in (5.2), to solve this problem. As shown in Fig. 5.14, if the second hop position, $RB_{start} + RB_{offset}$, exceeds the UL BWP boundary, after the modulus, the second hop position is wrapped around back into the UL BWP.

$$RB_{start} = \begin{cases} RB_{start} & i = 0 \\ (RB_{start} + RB_{offset}) \mathrm{mod} N_{BWP}^{size} & i = 1 \end{cases} \tag{5.2}$$

Nevertheless, Formula (5.2) still has some problems: It only ensures that the starting RB of the second hop of PUSCH falls within the UL BWP range, but may still misbehave in a case where the starting RB of the allocation does not wrap-around in modulo but the ending RB of the allocation exceeds the BWP boundary. Assume the frequency-domain resource of each hop contains L_{RBs} contiguous RBs, as shown in Fig. 5.15, the starting position of the second hop is within the UL BWP, so that the modulus operation in Formula (5.2) does not move the second hop to the lower side of the BWP. Then a part of RBs in the second hop

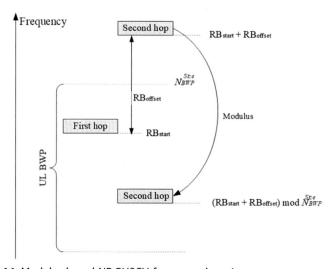

Figure 5.14 Modulus-based NR PUSCH frequency hopping.

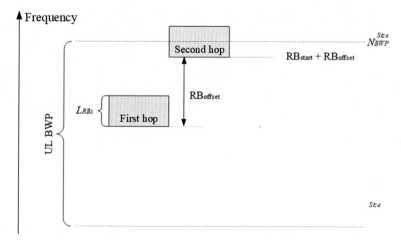

Figure 5.15 The second hop part exceeds the UL BWP range.

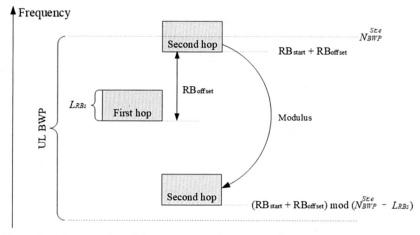

Figure 5.16 Improved modulus operation of NR PUSCH frequency hopping.

can exceed the UL BWP range and frequency-hopping operation falls into an exception.

A correction can be a modification of Formula (5.2) into Formula (5.3), by taking the number of RBs of the allocation into account when doing modulo calculation, as shown in Fig. 5.16 [7]. However, this method was a late proposal, which is concerned to have impact on R15 NR product development, so it was not accepted to R15 NR. Consequently, scheduling restrictions have to be imposed to ensure any part of the second hop after Formula (5.2) being applied is not outside the UL BWP:

$$RB_{start} = \begin{cases} RB_{start} & i = 0 \\ (RB_{start} + RB_{offset}) \bmod (N_{BWP}^{size} - L_{RBs}) & i = 1 \end{cases} \quad (5.3)$$

In general, the advantage of Option 1 is that DCI can directly indicate the frequency-domain position of the first hop and therefore the scheduling flexibility is the highest among the three options. However, the semi-static configuration of the frequency-hopping offset may lead to scheduling restrictions as explained above.

The advantage of Option 2 is that the frequency-domain position of each hop is configured separately per-UL BWPs, so that the frequency-domain resources of any single hop can be prevented from falling outside the UL BWP. The disadvantage of Option 2 is that it cannot directly indicate the frequency-domain position of the first hop by DCI. Only one of several configured candidate positions can be selected for the first hop. The scheduling flexibility is worse than that of Option1. Thus Option 2 is not suitable for PUSCH resource allocation. But it meets the requirement of the resource allocation for PUCCH. Therefore this option is finally used for NR PUCCH resource allocation, as described in Section 5.5.4.

Compared with Option 1 and Option 2, the advantage of Option 3 is that the frequency-hopping offset is bound to the BWP size and does not depend on the RRC configuration. As shown in Fig. 5.17, the frequency-hopping offset can be defined as 1/2 or 1/4 of the BWP size. As described in Chapter 4, Bandwidth Part, the terminal knows the active UL BWP at any time, so it can calculate the frequency-hopping offset without waiting for any terminal-specific configurations. After the establishment of RRC connection, although this option can save some RRC signaling overhead, it is less flexible than Option 1 and Option 2. However, before the RRC connection is established (e.g., during the initial access process), the frequency-hopping offset or the frequency-domain position of each hop cannot be configured by RRC signaling, which makes Option 3 attractive. Therefore in the random-access process, the NR PUSCH frequency-hopping offset in the third step (Msg3) is defined as $\lfloor N_{BWP}^{size}/2 \rfloor$, $\lfloor N_{BWP}^{size}/4 \rfloor$, or $-\lfloor N_{BWP}^{size}/4 \rfloor$, where the real-time selection is indicated by Random Access Response (RAR).

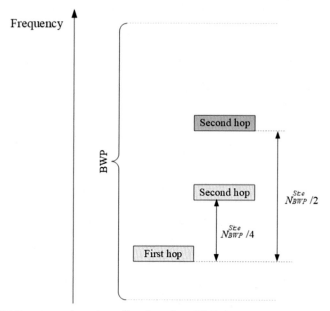

Figure 5.17 Frequency-hopping offset based on BWP size.

5.2.5 Introduction to symbol-level scheduling

As mentioned in Section 5.1.3, a new time–domain scheduling granularity, minislot, which is shorter than a slot, was proposed in 5G NR Study Item. How to implement flexible time-domain scheduling based on minislot needs to be investigated.

One scheme is to follow the design philosophy of LTE short Transmission Time Interval (sTTI), and to divide a slot into several small minislots. As shown in Fig. 5.18, a slot is divided into four minislots which contain 4, 3, 3, and 4 symbols, respectively, and then resources are allocated based on minislot. PDSCH 1 contains Minislot 0 in the first slot. PDSCH 2 contains Minislots 2 and 3 in the first slot. PDSCH 3 contains Minislots 1, 2, and 3 in the second slot. This scheme reduces the Transmission Time Interval (TTI) by using a smaller minislot length, but it does not achieve the arbitrary flexibility of the minislot length and time–domain position. Data channel cannot be transmitted at any time. This is considered an incomplete innovation.

The other scheme is to design a channel structure that "floats" in the time domain (i.e., a channel starts at any symbol position and ends at any

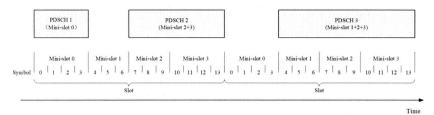

Figure 5.18 Minislot format based on LTE sTTI.

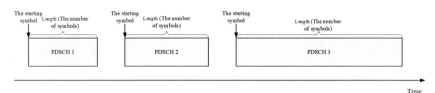

Figure 5.19 "Floating" minislot format.

symbol position). As shown in Fig. 5.19, the channel looks like a variable-length ship, which can float freely on the time axis.

This "floating" channel is an idealized concept. How to realize this "floating" depends on whether a realistic system design can be found, including whether equipment software and hardware can implement such flexible operation and whether the signaling can effectively indicate such flexible resource allocation.

Firstly, the "floating" channel requires finer resource allocation granularity, motivating to directly use OFDM symbol as the resource allocation unit. 3GPP first introduced the design of "symbol-level scheduling" [8], such as indicating the "starting symbol + length (i.e., the number of symbols)" of the channel (as shown in Fig. 5.19) or indicating the "starting symbol + ending symbol." Theoretically, "symbol-level scheduling" can allocate a time-domain resource with any length at any position to a channel. It not only goes beyond the LTE "slot-level scheduling," but also surpasses the concept of minislot. Therefore there is no need to use the concept and terminology associated with minislot in R15 NR specifications. In R16 URLLC, to transmit multiple PUCCHs for HARQ−ACK feedback in one slot and to simplify the handling of collision between multiple PUCCHs, subslot is introduced, as described in Chapter 15, 5G URLLC: PHY Layer.

5.2.6 Reference time for indication of starting symbol

As shown in Fig. 5.19, the first problem needs to be solved for "symbol-level scheduling" is how to indicate its starting symbol of the "floating" channel. Just like a ship floating in a river, it needs an "anchor point" to determine its relative position. There are two options for this reference position: the slot boundary or the position of PDCCH scheduling this channel.

- Option 1: The starting symbol is indicated relative to the slot boundary.

    As shown in Fig. 5.20, the starting boundary (Symbol 0) of the slot in which the starting of the channel is located is used as the reference time to indicate the starting symbol of the channel. In the example shown in the figure, the starting symbol of PDSCH 1 is Symbol 0, the starting symbol of PDSCH 2 is Symbol 7, and the starting symbol of PDSCH 3 is Symbol 3. The index of the starting symbol of the PDSCH is not relative to the position of the PDCCH scheduling the PDSCH.

- Option 2: The starting symbol is indicated relative to PDCCH.

    As shown in Fig. 5.21, the starting point of the PDCCH scheduling the data channel is used as the reference time to indicate the offset between the starting point of the data channel and the starting point of the PDCCH. In the example shown in the figure, PDSCH 1 and corresponding PDCCH start at the same time, so the starting symbol is Symbol 0. PDSCH 2 starts at two symbols after corresponding PDCCH, so the starting symbol is Symbol 2. PDSCH 3 starts at seven symbols after corresponding PDCCH, so the starting symbol is Symbol 7. The index of the

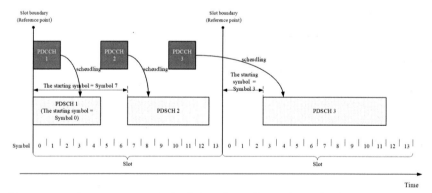

Figure 5.20 The starting symbol of the channel is indicated relative to the slot boundary.

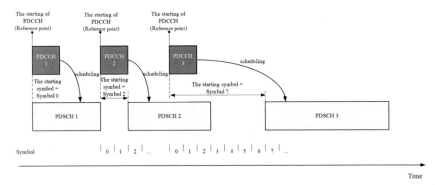

Figure 5.21 The starting symbol of the channel is indicated relative to PDCCH.

starting symbol of the PDSCH is not relative to the position of the PDSCH in the slot, and the slot boundary does not need to be relevant in Fig. 5.21.

Option 1 allows simple design logic and implementation, but the disadvantage is that it still relies on the slot as resource allocation unit or container, and therefore the channel is not completely "floating" in the time domain. Option 2 adopts "symbol-level scheduling" to achieve unrestricted "floating channel," especially for low-latency transmission. When the PDCCH and the scheduled data channel are close to each other, "symbol-level scheduling" has higher efficiency. Option 2 can completely get rid of the slot concept for resource allocation. It should be noted that, as the basic resource allocation granularity in LTE, subframe has been transformed into an ms-level timing tool in the NR specification without providing the function of resource allocation. In the NR, the OFDM symbol period and slot length vary with the OFDM SCS. A larger SCS leads to a finer granularity of TDRA. However, the length of the subframe does not vary with the SCS and is always 1 ms, because it is not used for resource allocation in the NR. If Option 2 is adopted, the slot would likely quit from resource allocation as well, and leave OFDM symbol as the granularity of TDRA.

Option 2 can indicate the position of the data channel more efficiently when the PDCCH and the scheduled data channel (such as PDSCH, PUSCH) are closer to each other in time. When the data channel is far from the PDCCH, the signaling overhead can be very large. For example, if the offset between the PDCCH and the data channel is dozens of symbols, a lot of bits are needed to express the offset in binary, and the DCI overhead is too large. Option 1 adopts two-level indication (i.e., "slot + symbol"), which is equivalent to a 14-ary indication. Even if there are dozens or even hundreds of symbols apart, resource indication can still be implemented with lower DCI overhead. But in fact, the difference on NR resource indication overhead between the two options may not be that big anyway, because the time-domain scheduling of the NR already adopts another kind of two-level resource indication (i.e., "RRC configuration + DCI indication"), which means the TDRA field in DCI only needs to tell from a limited number of candidates of the time-domain resources that are configured by RRC. The only difference is the overhead of the RRC configuration signaling, which is usually not a sensitive issue.

R15 NR specification finally adopts Option 1. The main reason is the difficulty of product implementation with Option 2. With the current software and hardware capabilities of 5G equipment, both base stations and terminals need to rely on the time-domain grid, such as slot, for the timing reference. To completely eliminate the concept of slot and to perform flexible "symbol-level" operation is too difficult to implement in the near future.

The slot-level indication information tells the slot-level offset from the PDCCH to the data channel. Specifically, the slot-level offset from PDCCH to PDSCH is defined as K_0, and the slot-level offset from PDCCH to PUSCH is defined as K_2. Take the PDSCH scheduling in Fig. 5.22 as an

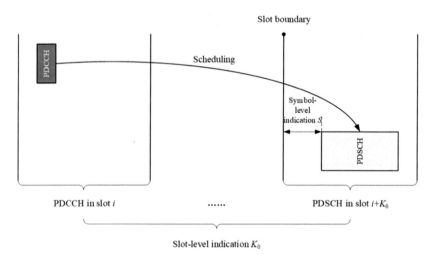

Figure 5.22 "Slot + symbol" two-level time-domain resource indication.

example. The slot-level indicator, K_0, is used to indicate the offset between the slot containing the PDCCH and the slot containing the scheduled PDSCH; the symbol-level indicator, S, is used to indicate a few symbols from the boundary of the slot where the scheduled PDSCH is located to the starting symbol of the PDSCH. In this way, the terminal can determine the time-domain position of the PDSCH by the pair of parameters (K_0, S).

Although R15 NR specification adopts Option 1 for data channel time-domain resource indication, Option 2 still has promising advantages in low-latency scheduling, especially within the two-level indication signaling structure of "RRC configuration + DCI indication." Related content is described in Section 5.2.9.

5.2.7 Reference SCS for indication of K_0 or K_2

Section 5.2.3 describes how to deal with the FDRA field in case of two BWPs involving BWP switching with different sizes. If the two involving BWPs have different SCSs, another problem needs to be solved regarding which SCS should be used to calculate slot offsets K_0 (for PDSCH) and K_2 (for PUSCH).

For PUSCH scheduling, PDCCH and the scheduled PUSCH are transmitted in DL BWP and UL BWP, respectively, so they are always in different BWPs. For PDSCH scheduling, PDCCH and the scheduled PDSCH are in different DL BWPs in the case of cross-BWP scheduling. Because SCS is attributed to BWP, one commonality between downlink scheduling and uplink scheduling is that the SCS of scheduling PDCCH and SCS of the scheduled data channel can be different. Take PDSCH scheduling in Fig. 5.23 as an example. Assume the SCS of the BWP holding PDCCH transmission is 60 kHz, and the SCS of the BWP holding PDSCH transmission is 30 kHz. The PDCCH slot length is half of the PDSCH slot length. If K_0 is calculated based on the SCS of the PDCCH, $K_0 = 4$; if K_0 is calculated based on the SCS of the PDSCH, $K_0 = 2$. 3GPP finally decided to use the SCS of the scheduled data channel to calculate K_0 and K_2, because the symbol-level scheduling parameters (such as the starting symbol, S, and the number of symbols, L) must be calculated based on the SCS of the data channel.

It should be noted that various time-domain offsets are calculated based on the SCS of the channel that is transmitted at the end of the offset. This commonality applies not only to K_0 and K_2, but also to the time

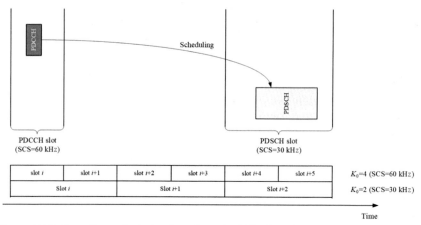

Figure 5.23 Slot-offset calculation based on the SCS of PDCCH or PDSCH.

offset K_1, which is the delay from PDSCH to the corresponding HARQ−ACK. When the SCS of the DL BWP holding a PDSCH transmission and the SCS of the UL BWP holding the PUCCH that carries HARQ−ACK for the PDSCH are different, K_1 is calculated based on the SCS of the UL BWP in which the PUCCH is transmitted.

5.2.8 Resource mapping type: type A and type B

In Section 5.2.6, we discussed the "floating" channel structure that supports flexibility scheduling. Due to the difficulty in product implementation, the reference time for a scheduling is still based on slot. The channel can float within the slot, but cannot float freely across slot boundary. If the starting time of a channel is close to the end of a slot and the length of the channel is long enough, part of the channel may enter into the next slot. The TDRA across slot boundary is not difficult to support from the perspective of scheduling signaling, but it brings additional complexity to product implementation. The various channel processing also needs to be implemented in a timeline based on slot. Therefore it was finally determined that a transmission cannot cross the slot boundary in R15, as shown in Fig. 5.24. In the R16 URLLC topic, in order to reduce time latency and improve scheduling flexibility, a transmission crossing the slot boundary is supported in uplink data transmission.

Even if the channel is only allowed to "float" within a slot, it is still a tough requirement for product implementation. It is unreasonable to mandate this requirement for all 5G terminals. Although low-latency

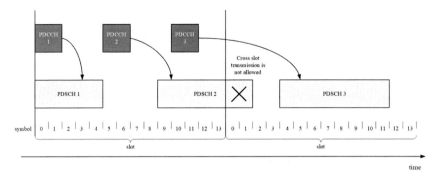

Figure 5.24 Cross-slot transmission not allowed.

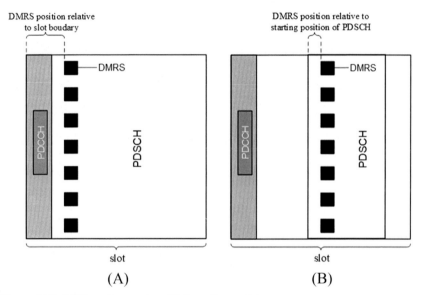

Figure 5.25 (A) Type A mapping. (B) Type B mapping.

services (such as the URLLC service) require transmit PDSCH and PUSCH at any time in a slot, a large number of eMBB services do not have such strict latency requirements. Therefore it was finally decided to define two scheduling modes: slot-based scheduling and nonslot-based scheduling [9].

Slot-based scheduling actually maintains a time–domain structure similar to LTE, including the PDCCH symbols locating at the starting part of the slot and the PDSCH/PUSCH occupying the remaining symbols of the slot (as shown in Fig. 5.25A). Nonslot-based scheduling can start from any symbol in a slot, and a PDSCH or PUSCH can be based on minislot or symbol-based channel structure (as shown in Fig. 5.25B). Though a set of resource allocation methods of nonslot-based scheduling can be designed, which are completely different from those of slot-based scheduling, to make nonslot-based scheduling get rid of the limitation of LTE design and to obtain more flexibility, it was decided to adopt a unified design as much as possible for the two scheduling methods to simplify the design and implementation. The only difference between these two scheduling methods lies in the indication of DMRS position in the time domain.

As shown in Fig. 5.25, assume that DMRS is located in a certain symbol in the PDSCH (the DMRS pattern in the figure is only for illustrative purposes, and details on the DMRS design of the NR system is given in 3GPP TS38.211). For slot-based scheduling, PDSCH occupies all available symbols in a slot and DMRS is placed in a relatively fixed position in the slot with the position of the first DMRS symbol being defined relative to the slot boundary (as shown in Fig. 5.25A). This DMRS mapping is called Type A mapping. For nonslot-based scheduling, the PDSCH may start from one of a certain symbols in the middle of the slot and the position of DMRS floats along with the PDSCH. Therefore the position of the first DMRS symbol cannot be defined relative to the slot boundary; instead, the position of the first DMRS symbol can only be defined relative to the starting time of PDSCH (as shown in Fig. 5.25B). This DMRS mapping is called Type B mapping. Therefore the 5G NR specification supports both DMRS mapping Type A and mapping Type B, which is equivalent to supporting slot-based scheduling and nonslot-based scheduling. The terminals dedicated to eMBB services can just support DMRS mapping Type A and slot-based scheduling. The terminals also supporting low-latency services, such as URLLC service, need to additionally support DMRS mapping Type B and nonslot-based scheduling.

5.2.9 Time-domain resource allocation

Throughout Sections 5.2.5—5.2.8, we introduced the issues and solutions in 5G NR TDRA. The key issue to be solved is how to provide a unified solution for signaling.

As mentioned in Section 5.1, a major enhancement of 5G NR resource allocation compared to LTE is to use DCI dynamic indication as much as possible to support flexible scheduling. However, to only use

DCI to indicate so many resource allocation parameters can generate unacceptable DCI signaling overhead. A compromise solution is to use the "RRC configuration + DCI scheduling" method to balance signaling overhead and scheduling flexibility. RRC signaling is used to configure a candidate list of resources (or resource set) that can be scheduled, and DCI is used to select a resource from the candidate resource list or resource set. For example, if RRC configures a list of 16 resources, DCI only needs 4 bits to indicate the resource. Each resource in the list may require tens of bits to configure, but RRC signaling is carried by PDSCH, which can bear a large signaling overhead.

The next question is which resource parameters are configured by RRC with candidate values in the resource list and which resource parameters are directly indicated by DCI. Two types of channels are considered separately: data channels (PDSCH and PUSCH) and PUCCH. PUCCH resource allocation will be introduced in Section 5.5.5. For the resource allocation of PDSCH and PUSCH, according to Sections 5.2.5−5.2.8, the PDSCH and PUSCH resource parameters that need to be indicated to the terminal by gNB (next generation Node BaseStation) include:

- Slot-level indication information: K_0 (the slot-level offset from PDCCH to PDSCH), K_2 (the slot-level offset from PDCCH to PUSCH);
- Symbol-level indication information: the starting symbol and length (number of symbols) of PDSCH or PUSCH;
- Mapping type: Type A versus Type B.
  The potential signaling structures include (Table 5.4).
- Option 1 is to configure candidates for all three parameters in RRC. This option can minimize the DCI overhead and configure all three resource parameters in combinations (4 bits can indicate 16 combinations). The disadvantage is that all three parameters, although in dynamic signaling, can only select the values from the semistatic configuration. If a parameter combination desired to be used is not configured in the candidate resource list, the candidate resource list needs to be reconfigured through RRC signaling, which results in large latency.
- Option 2 is to configure candidates for slot-level indication information (K_0, K_2) and symbol-level indication information (starting symbol S and L) by RRC, and meanwhile to use 1 bit in DCI to indicate the PDSCH/PUSCH mapping type. This option can indicate the mapping type more flexibly than Option 1. For each combination of K_0/K_2, S, and L, Type A or Type B can be independently selected. The

Table 5.4 Potential signaling structures for time-domain resource allocation.

	Mapping type (Type A or Type B)	Slot-level indication (K_0, K_2)	Symbol-level indication (starting symbol + length)
Option 1	Configured in RRC and then indicated by DCI		
Option 2	Directly indicated by DCI	Configured in RRC and then indicated by DCI	
Option 3	Directly indicated by DCI		Configured in RRC and then indicated by DCI

disadvantage is that the mapping type still consumes 1 bit in DCI. If the TDRA field has a total budget of 4 bits, the remaining 3 bits can only allow up to eight combinations of K_0/K_2, S, L.

- Option 3 further uses DCI to directly indicate K_0 and K_2. Once K_0 and K_2 are completely dynamically indicated, the number of combinations over S and L that can be configured is further reduced comparing to that in Option 2.

The final decision in 3GPP RAN1 was to adopt Option 1 as the TDRA method for PDSCH and PUSCH. It is fully up to gNB to configure the candidate time-domain resources. If the gNB can "smartly" gather the optimal resources in the candidate list, the best resource allocation effect can be achieved. In comparison, a method similar to Option 3 was adopted for the PUCCH TDRA. The slot-level offset from PDSCH to HARQ–ACK (K_1) is directly indicated in DCI, while symbol-level information is configured by RRC into the PUCCH resource set. It can be seen that there is no absolute pros and cons among the above three options. The selection from these options can consider the tradeoff between the complexity and flexibility. The minimum values of K_1 and K_2 are also dependent on the terminal's processing capabilities, and the relevant content is given in Section 15.3.

Table 5.5 is an example of a typical PUSCH time-domain resource list (taken from 3GPP specification TS 38.214 [10]). It can be seen that the list contains 16 candidate PUSCH time-domain resources, and each resource is associated with a combination of <mapping type, K_2, S, and L>.

The 16 candidate resources given in Table 5.5 are shown in Fig. 5.26, where it can be seen that these 16 resources include resources that have different positions and lengths and that are distributed in different slots after the

Table 5.5 Example of PUSCH time-domain resource table.

Resource index	PUSCH mapping type	Slot-level indication K_2	Symbol-level indication	
			Starting symbol S	Length L
1	Type A	0	0	14
2	Type A	0	0	12
3	Type A	0	0	10
4	Type B	0	2	10
5	Type B	0	4	10
6	Type B	0	4	8
7	Type B	0	4	6
8	Type A	1	0	14
9	Type A	1	0	12
10	Type A	1	0	10
11	Type A	2	0	14
12	Type A	2	0	12
13	Type A	2	0	10
14	Type B	0	8	6
15	Type A	3	0	14
16	Type A	3	0	10

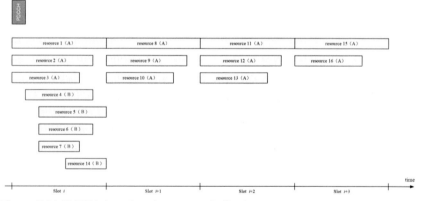

Figure 5.26 PUSCH time-domain resource indication.

PDCCH. It can also be found that all the resources corresponding to mapping Type B belong to the slot (Slot i with $K_2 = 0$) where the PDCCH is located. This is because the purpose of Type B mapping is for low–latency use case. The resources from Slot $i + 1$, Slot $i + 2$, and Slot $i + 3$ may not be able to achieve low latency, so it is meaningless to configure resources with Type B mapping in these slots. On the other hand, the specification does not preclude the configuration of Type B mapping in a slot with $K_0 \neq 0$ or $K_2 \neq 0$, which can be motivated by other considerations such as flexible channel multiplexing, instead of low latency.

In addition, it should be noted that in the example shown in Fig. 5.26, Resource index 1, 2, and 3 cannot be used in TDD system, because PUSCH cannot be transmitted at the same time as PDCCH reception. It is also difficult to use Resource index 4 in the TDD system. The PUSCH resources with $K_2 = 0$ and $S = 0$ can only be used in the FDD system.

With a configured PUSCH resource list containing 16 candidate resources, a 4-bit TDRA field in DCI can dynamically indicate any resource from the 16 candidates, and the terminal uses the resource indicated by the DCI to send PUSCH. The resource list configuration and indication method for PDSCH TDRA is similar to those of PUSCH. Therefore this section will not repeat the details for PDSCH.

It should be clarified that the candidate resources listed in Table 5.5 are predefined "Default PUSCH TDRA." These default resource lists are often used in the initial access process or some resource scheduling that does not require high flexibility. For example, during the initial access process, RRC connection has not been established and the resource list cannot be configured through RRC signaling; therefore only the default resource list can be used. The default lists for PDSCH and PUSCH are directly defined in the specification (see Sections 5.1.2.1.1 and 6.1.2.1.1 of Ref. [10]), requiring no RRC signaling overhead. However, to obtain higher scheduling flexibility, the default lists are not sufficient and the resource list still needs to be configured on demand using RRC signaling. When it comes to RRC signaling, the separate configurations of the starting symbol S and the length L can result in extra signaling overhead. Although the RRC signaling overhead is not as sensitive as DCI overhead, it is still beneficial to keep the signaling overhead as small as possible.

The overhead compression method is to jointly encode the starting symbol S and the length L (i.e., to use a Start and Length Indicator Value (SLIV) to represent a pair of S and L). The algorithm is formulated as follows:

if $(L - 1) \leq 7$ then

$$SLIV = 14\,(L - 1) + S$$

else

$$SLIV = 14\,(14 - L + 1) + (14 - 1 - S)$$

where $0 < L \leq 14 - S$

Table 5.6 "Starting symbol + length" information expressed in SLIV.

S (start symbol)	Length L (symbols number)													
	1	2	3	4	5	6	7	8	9	10	11	12	13	14
SLIV formula	$SLIV = 14\ (L-1)+S$							$SLIV = 14\ (14-L+1)+(14-1-S)$						
0	0	14	28	42	56	70	84	98	97	83	69	55	41	27
1	1	15	29	43	57	71	85	99	96	82	68	54	40	26
2	2	16	30	44	58	72	86	100	95	81	67	53	39	25
3	3	17	45	45	59	73	87	101	94	80	66	52	38	24
4	4	18	32	46	60	74	88	102	93	79	65	51	37	23
5	5	19	33	47	61	75	89	103	92	78	64	50	36	22
6	6	20	34	48	62	76	90	104	91	77	63	49	35	21
7	7	21	35	49	63	77	91	105	90	76	62	48	34	20
8	8	22	36	50	64	78	92	106	89	75	61	47	33	19
9	9	23	37	51	65	79	93	107	88	74	60	46	32	18
10	10	24	38	52	66	80	94	108	87	73	59	45	31	17
11	11	25	39	53	67	81	95	109	86	72	58	44	30	16
12	12	26	40	54	68	82	96	110	85	71	57	43	29	15
13	13	27	41	55	69	83	97	111	84	70	56	42	28	14

Joint coding of SLIV uses a generation method similar to RIV (Type 1 frequency-domain resource indication method), which can express various combinations of S and L with as few bits as possible, as shown in Table 5.6. Theoretically, S has 14 possible values from 0 to 13, and L has 14 possible values from 1 to 14. According to the generation formula $SLIV = 14(L-1) + S$, 128 values of a 7-bit indicator cannot cover all $14 \times 14 = 196$ combinations over S and L. However, as described in Section 5.2.8, the R15 NR specification does not allow PDSCH and PUSCH crossing the slot boundary, so $S+L \leq 14$, which only corresponds to the gray entries in Table 5.6. There are 105 gray entries in total, which means 7 bits are enough to represent SLIV. Further, the entry values in Table 5.6 (except for $L=1$ column) are central-symmetric (i.e., the table entry at $<\text{row} = S, \text{column} = L>$ has the same number as in table entry $<\text{row} = 13\text{-}S, \text{column} = 16\text{-}L>$ for $2 \leq L \leq 7$, and all the gray entry values for $L \geq 9$ are central-symmetrically mapped to nongray entries for $L \leq 7$). This is why the SLIV uses different calculation formulas for $L \leq 8$ and $L \geq 9$.

It should be noted that, according to different requirements for scheduling flexibility and channel multiplexing, different channels have different S and L value ranges. As shown in Table 5.7, the PDSCH and PDSCH of Type A mapping start from the beginning portion of the slot. Given the beginning of a downlink slot may contain PDCCH, Type A PDSCH can have its starting symbol position S from $\{0, 1, 2, 3\}$, while Type A PUSCH can only start from Symbol 0. If PDCCH and PUSCH are to be multiplexed in the same slot, Type B PUSCH can be applied. In addition, Type B PUSCH supports various lengths in $\{1, \ldots, 14\}$, while Type B PDSCH only supports three lengths in $\{2, 4, 7\}$, which suggests that PUSCH has a more flexible minislot scheduling capability than PDSCH. The scheduling flexibility of minislot PDSCH is expected to be further enhanced in later releases of the NR.

Table 5.7 S and L value ranges of different channels (normal CP).

Channel	Mapping type	S	L
PDSCH	Type A	$\{0, 1, 2, 3\}$	$\{3, \ldots, 14\}$
	Type B	$\{0, \ldots, 12\}$	$\{2, 4, 7\}$
PUSCH	Type A	$\{0\}$	$\{4, \ldots, 14\}$
	Type B	$\{0, \ldots, 13\}$	$\{1, \ldots, 14\}$

As seen from the example shown in Table 5.5, 16 candidates in a resource list cannot cover all combinations of K_0/K_2, S, L, and one parameter can enumerate only a few candidate values. As shown in Fig. 5.26, the starting symbol S only covers four values including 0, 2, 4, and 8, instead of any possible symbol. This brings bottlenecks to the achievable "ultra-low latency."

The candidate resource list must be configured by RRC to the terminal in advance and cannot be changed dynamically once configured. On the other hand, only a few possible starting positions can be roughly estimated before being configured to the terminal. Take PDSCH resource scheduling as an example. Assume the four candidate starting symbol positions are $S = 0$, 4, 8, and 12, then the PDSCH transmission can only start at every fourth symbol in a slot. It becomes impossible to schedule a PDSCH between two successive transmission opportunities. Although the value range of S for Type B PDSCH is theoretically $\{0, \ldots, 12\}$, the PDSCH starting position that can be actually applied is limited to a few starting positions configured in the PDSCH resource list. As shown in Fig. 5.27A, even if a DCI is detected in Symbol 5, the corresponding PDSCH cannot be scheduled in Symbols 5, 6, and 7; Symbol 8 is the earliest symbol that the PDSCH can be scheduled.

To solve this problem and quickly schedule PDSCH at any time, one can turn to the mechanism using the PDCCH as a reference time to indicate PDSCH, as discussed in Section 5.2.6. As shown in Fig. 5.27B, if the

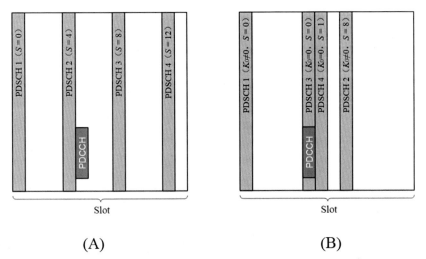

(A)   (B)

Figure 5.27 (A) Slot boundary as reference point. (B) PDCCH as reference point when $K_0 = 0$.

PDSCH resource is indicated relative to the PDCCH, no matter which symbol the PDCCH is in, the PDSCH can be transmitted immediately after the PDCCH. In other words, the nature of PDSCH to "float" with the PDCCH can achieve the most real-time PDSCH transmission. As mentioned in Section 5.2.6, it is not efficient to indicate PDSCH relative to PDCCH if the PDSCH is far from the PDCCH. It can be a work-around solution that PDCCH is used as the PDSCH scheduling reference time only when $K_0 = 0$ [11], and slot boundary is used as the PDSCH scheduling reference time when $K_0 \neq 0$. A PDSCH resource list can be configured to contain both PDSCH resources that takes $K_0 \neq 0$ and the slot boundary as the reference time (such as PDSCH 1, PDSCH 2 in Fig. 5.27B) and PDSCH resources that takes $K_0 = 0$ and the scheduling PDCCH as the reference time (such as PDSCH 3 and PDSCH 4 in Fig. 5.27B). If low-latency PDSCH scheduling is required, PDSCH 3 and PDSCH 4 can be used as the PDSCH resources that float with PDCCH.

Because the R15 NR focuses on eMBB service, it did not specifically optimize the low-latency performance. Therefore this scheme of determining the PDSCH resource configuration reference time based on the value of K_0 was not accepted, but it was adopted in the R16 URLLC enhancement to reduce the DCI signaling overhead, as described in Chapter 15, 5G URLLC: PHY Layer.

5.2.10 Multislot transmission

Multislot scheduling is a coverage enhancement technology adopted in 5G NR. With multislot scheduling, one DCI can schedule one PUSCH or one PUCCH to be transmitted in multiple slots, so that gNB can combine the PUSCH or PUCCH signals in multiple slots to equivalently obtain a better Signal to Interference plus Noise Ratio (SINR). Multislot scheduling has been used in previous OFDM wireless communication systems such as LTE, and includes two types: slot aggregation and slot repetition. With slot aggregation, the data transmitted in multiple slots is jointly encoded to obtain decoding gain at the receiver. Slot repetition does not jointly encode data in multiple slots, but simply repeats transmission of data in multiple slots to achieve energy accumulation, which is equivalent to repetition encoding. R15 NR uses slot repetition for PUSCH, PUCCH, and PDSCH. PUSCH and PUCCH are repeatedly transmitted in N consecutive slots and the value of N is semistatically configured by RRC. The reason for N to be semistatic configured by RRC signaling instead of dynamically indicated by DCI is that the number of repetition slots is mainly determined by the slow channel fading experienced by the terminal, which cannot change much during N slots, so semistatic configuration is sufficient and can save DCI overhead.

In a system with slot being the scheduling unit, multislot scheduling only needs to indicate the number of slots. However, in the 5G NR system, due to the use of symbol-level scheduling, flexible TDD DL/UL assignment ratio and various slot formats, symbol-level resource allocation for PUSCH and PUCCH in each slot becomes a more complicated task.

The first question is whether the relative position of the symbol-level resource in a slot is allowed to be different over N slots. The first option is to allow different symbol-level resource positions in different slots. As shown in Fig. 5.28, assume $N = 4$ slots are scheduled for a terminal to transmit a PUSCH, different symbol-level resource positions can be assigned to four PUSCH repetitions in four slots. Although this option can achieve maximum scheduling flexibility, it needs to include four symbol-level indications in the TDRA field, which in principle increases the DCI overhead by three times. The second option is to use the same symbol-level resource positions in each slot for each repetition, as shown in Fig. 5.29. The advantage of this option is that no additional TDRA field overhead is required, but it has a great impact on scheduling flexibility.

In the end, 5G NR decided to adopt the second option (i.e., "all repetitions using the same symbol-level allocation per slot"), which is to sacrifice scheduling flexibility to achieve low-overhead scheduling. This method imposes big restriction on base station scheduling, and makes it

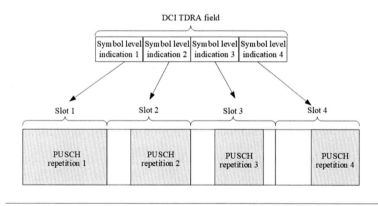

Figure 5.28 TDRA indicates different symbol-level resource positions for different slots.

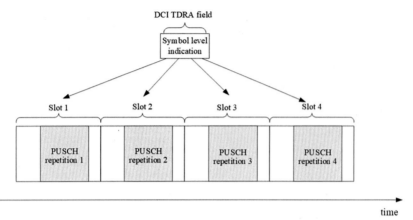

Figure 5.29 TDRA indicate the same symbol-level resource position for different slots.

difficult to resolve the overlapping between a transmission repetition and a conflicting symbol assigned by TDD Downlink (DL)/Uplink (UL) configuration.

The 5G NR system supports flexible TDD operation. Instead of following a fixed TDD frame structure over time like LTE, 5G NR can rely on both semistatic configuration and dynamic indication to claim which slots and symbols are used for uplink and which slots and symbols are used for downlink. The flexible TDD design of 5G NR is described in Section 5.6. This section only discusses how to resolve the confliction between multislot scheduling and uplink or downlink assignment. Due to the flexible TDD slot format, it is difficult to ensure that the allocated symbols for a scheduled transmission within all of the N slots are all uplink resources or all downlink resources. If there is overlapping between uplink transmission and downlink resource, the overlapped symbols originally allocated for uplink transmission cannot be used. In this case, there are two possible options:

1. **Option 1:** Abandon the repetition within the slot in which the conflict occurs, that is, use as many slots as possible, but subject to the resource conflict, to perform transmission within the originally allocated N slots.
2. **Option 2:** Postpone the transmission of the slots in which the conflict occurs to next slot, that is, transmission is only performed in the N nonconflicting slots.

These two options are both adopted in the R15 5G NR. Option 1 is adopted for PUSCH multislot transmission, and Option 2 is adopted for PUCCH multislot transmission.

An example of Option 1 is shown in Fig. 5.30. The base station schedules the terminal to transmit a PUSCH in 4 consecutive slots. Assume the TDD UL/DL assignment is as given in the figure (the actual NR TDD system still has flexible symbols between downlink and uplink. This figure is only for illustration purpose. How to judge whether a symbol can be used for uplink transmission is described in Section 5.6). There is no conflict in slots 1, 2, and 4 for the symbol-level resources allocated to PUSCH. But the first part of Slot 3 is downlink and cannot be used for PUSCH transmission. According to Option 1, the terminal discards the PUSCH repetition 3, and only transmits the other three repetitions in slots 1, 2, and 4. This scheme is actually a kind of "best effort" multislot transmission, which is suitable for the PUSCH data channel.

An example of Option 2 is shown in Fig. 5.31. The base station schedules the terminal to send a PUCCH in four consecutive slots. There is no conflict in slots 1, 2, and 4 for the symbol-level resources allocated to PUCCH. But a resource conflict occurs in Slot 3, which cannot satisfy the transmission of PUCCH repetition 3. According to Option 2, PUCCH repetition 3 and all remaining repetitions are postponed to the subsequent slots that have no resource collisions. As shown in the figure, PUCCH repetition 3 and 4 are transmitted in Slots 4 and 5, respectively. This option can ensure that the number of PUCCH repetitions is obeyed as configured, which is suitable for PUCCH control channels.

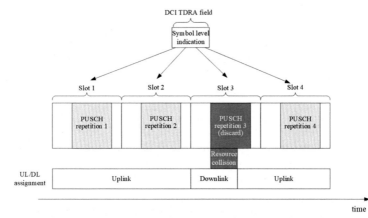

Figure 5.30 PUSCH repetition discarded in case of conflict with downlink resource.

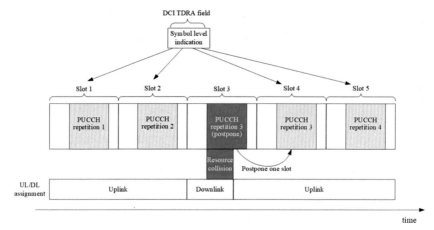

Figure 5.31 PUCCH repetition postponed in the case of conflict with downlink resource.

5.3 Code Block Group

5.3.1 Introduction of Code Block Group transmission

The bandwidth of 5G NR system can be significantly larger than that of LTE. When a large bandwidth is scheduled for data transmission, the size of a Transport Block (TB) can be very large. In practice, to keep the implementation complexity of the FEC encoder and decoder low, the length of payload bits fed into the encoder and decoder is limited. When the number of bits contained in a TB exceeds a threshold, the TB is segmented into multiple small code blocks (CBs), with the size of each CB not exceeding the threshold. In the LTE system, the segmentation threshold is 6144 bits for processing chain with Turbo coding. In the NR system, the segmentation threshold in the processing chain of LDPC code is 8448 bits or 3840 bits. As the size of a TB increases, the number of CBs contained in the TB also increases.

The LTE system adopts TB-level ACK/NACK feedback (i.e., if any CB in a TB fails in decoding, the whole TB is retransmitted). Based on the statistics of 10% BLER, it is found that the error reception caused by the decoding failure of one single CB in the TB accounts for 45% of all error receptions. Therefore in the TB-level retransmission, a large number of correctly decoded CBs are retransmitted. For NR systems, due to the significant increase in the number of CBs in a TB, there can be quite a few unnecessary retransmissions, which severely affects the system efficiency [12−14].

At the early phase of NR system design, in order to improve the retransmission efficiency of a large TB, an accurate ACK/NACK feedback mechanism (i.e., multibit ACK/NACK feedback for a TB) was adopted. At the RAN1 Ad-Hoc meeting in January 2017, three methods were proposed to achieve multibit ACK/NACK feedback for a TB [15,16]:

1. Method 1: Code block group (CBG)-based transmission. A TB is divided into multiple CBGs, and the number of CBGs is determined based on the maximum number of ACK/NACK bits to be reported. Method 1 is beneficial in reducing the impact from a succeeding URLLC transmission to an ongoing eMBB transmission. As shown in Fig. 5.32, when a URLLC transmission overlaps with an eMBB transmission, only the conflicted CBGs need to be retransmitted.

2. Method 2: Decoder State Information (DSI) feedback. DSI information is collected during the decoding process and fed back to the transmitter. According to the DSI information, the transmitter can determine the coding bits that need to be retransmitted based on an optimization function.

3. Method 3: Outer Erasure code [17]. With outer erasure codes (such as Reed–Solomon code and Raptor code), the receiver does not need to feed back the exact CBs that fail in decoding, but only reports the number of CBs that need to be retransmitted. In addition, the feedback of outer erasure codes can also be based on CBG, therefore the feedback overhead of method 3 is less than method 1. However, one of the disadvantages of the outer erasure code is that it makes HARQ soft combining difficult.

For method 2 and method 3, the actual gains are unclear, and both are complicated for specification and implementation. Therefore it was agreed upon in the RAN1 #88bis meeting to support CBG-based multibit ACK/NACK feedback.

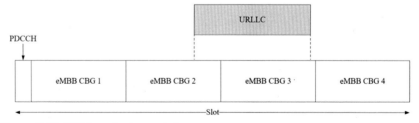

Figure 5.32 CBG-based retransmission.

5.3.2 CBG construction

The following design principles were first established in RAN1 for CBG-based transmission in the RAN1 #88bis meeting [18–20]:

- Only allow CBG-based (re)-transmission for the same TB of a HARQ process.
- CBG can include all CBs of a TB regardless of the size of the TB — In such a case, UE reports single HARQ–ACK/NACK bit for the TB.
- CBG can include one CB.
- CBG granularity is configurable.

The next problem to be solved is how to divide a TB into multiple CBGs. During the RAN1 discussion, the following three options of CB grouping were proposed:

1. Option 1: With configured number of CBGs, the number of CBs in a CBG changes according to TBS (TB size).
2. Option 2: With configured number of CBs per-CBG, the number of CBGs changes according to TBS.
3. Option 3: The number of CBGs or the number CBs per-CBG is determined in specification according to TBS.

Because the number of CBGs is directly related to the number of ACK/NACK bits, Option 1 provides a best control on the overhead of uplink signaling. For Option 2, as the value range of TBS is large, the number of feedback information bits may vary greatly. In addition, when ACK/NACK bits are multiplexed, Option 2 may cause ambiguity on the ACK/NACK codebook between gNB and UE. Method 3 is complex, and the benefit of adjusting both the number of CBGs and the number of CBs per-CBG according to TBS is unclear. Consequently, it was agreed in the RAN1 #89 meeting to adopt Option 1 to determine the number of CBGs in a TB via configuration, and to make the number of CBs contained in each CBG as uniform as possible over all CBGs in the TB.

After further discussions in the RAN1 #90 and RAN1 #90-bis meetings, the complete method of CB grouping was approved. In the approved method, the maximum number of CBGs per-TB is configured by RRC signaling. For single-TB transmission, the maximum number of CBGs included in a TB can be 2, 4, 6, or 8. For multiple-TB transmission, the maximum number of CBGs included in a TB can be 2 or 4, and the maximum number of CBGs per-TB remains the same between TBs.

For a TB, the number of CBGs is equal to the minimum of the number of CBs contained in the TB and the maximum number of CBGs configured by RRC signaling. Then $K_1 = \lceil C/M \rceil$ CBs are contained in each of the first $M_1 = \mod(C, M)$ CBGs, and $K_2 = \lfloor C/M \rfloor$ CBs are contained in each of the last $M - M_1$ CBGs, where C is the number of CBs contained in the TB and M is the number of CBGs. Take Fig. 5.33 as an example. The maximum number of CBGs included in a TB is configured as 4. A TB is segmented into 10 CBs based on the TBS. Each of CBG 1 and CBG 2 contains three CBs (i.e., CB 1−3 and CB 4−6, respectively). Each of CBG 3 and CBG 4 contains 2 CBs (i.e., CB 7− 8 and CB 9−10, respectively).

5.3.3 CBG retransmission

The main reason for supporting CBG-based transmission is to improve the retransmission efficiency. After the CB construction, the design should focus on how to achieve efficient retransmission. In the RAN1 discussion, two approaches for determining the retransmitted CBGs were proposed:

1. Approach 1: The retransmitted CBGs are determined according to CBG-based ACK/NACK feedback. Specifically, the UE always expects the base station to retransmit the CBGs corresponding to NACK in the latest ACK/NACK feedback.
2. Approach 2: The retransmitted CBGs are explicitly indicated in a DCI.

The advantage of Approach 1 is that it does not increase DCI overhead. However, the scheduling flexibility of Approach 1 is significantly limited, because the base station has to retransmit all CBGs corresponding to NACK. Furthermore, if the base station does not correctly receive UCI feedback, the terminal and the base station will have different understandings of the retransmitted CBGs, which leads to decoding failure and the loss of retransmission efficiency. Therefore it was determined in RAN1 that the retransmitted CBGs are indicated in the scheduling DCI, and the following two implementation methods for DCI indication were proposed [21]:

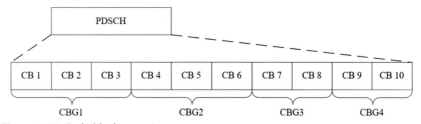

Figure 5.33 Code block grouping.

1. Method 1: The CBG indication field is always contained in the DCI, regardless whether it schedules an initial transmission or a retransmission. In other words, once the CBG-based transmission is configured, the CBG indication field based on bitmap is always included in the DCI for determining the retransmitted CBGs (Fig. 5.34).

2. Method 2: Reusing some existing field in the DCI as CBG indication. The assumption of this method is that for a TB, all CBGs should be included in the initial transmission, and CBG-based transmission is only used for retransmission. Therefore there is no need to indicate CBGs in the DCI scheduling initial transmission. For the retransmission, because TBS (TB size) has been indicated to the UE during the initial transmission, only the modulation order from the MCS information field is needed. Therefore redundant information exists in MCS information field in the DCI scheduling a retransmission. Reusing MCS information fields to indicate the retransmitted CBG can effectively reduce the DCI overhead. The specific implementation steps of Method 2 include:

   a. Firstly, the UE will determine whether a TB-based transmission (initial transmission) or a CBG-based transmission (retransmission) is scheduled based on an explicit indication (e.g., 1-bit CBG flag) or implicit indication (e.g., according to NDI) in the DCI.

   b. If a CBG-based transmission is scheduled, some bits in the MCS information field of the DCI are used to indicate the retransmitted CBGs, and the remaining bits are used to indicate the modulation order, as shown in Fig. 5.35.

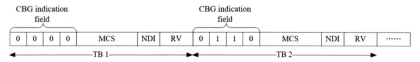

Figure 5.34 CBG indication field in a DCI.

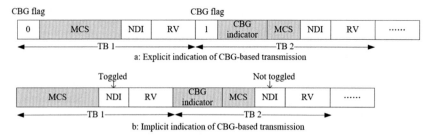

Figure 5.35 Reusing existing field in the DCI as CBG indication.

Method 2 can reduce the DCI overhead. However, with the Method 2, the correct decoding of a CBG-based retransmission depends on the correct receiving of the DCI scheduling the initial transmission, since TBS is determined from the DCI scheduling the initial transmission. If only a CBG-based retransmission but not the initial transmission is received by the UE, the retransmission cannot be decoded. Furthermore, Method 1 can keep the implementation complexity low, and the CBG-based retransmission can be self-decoded. It was then agreed on in RAN1 #91 that using MCS/TBS information field to indicate the retransmitted CBGs was not supported.

5.3.4 DL control signaling for CBG-based transmission

For CBG-based transmission, one more question needs to be discussed: What indication fields need to be included in the DCI to support efficient CBG-based transmission?

First of all, in order to support the CBG-based retransmission, a Code Block Group Transmission Information (CBGTI) should be included in the DCI to tell the UE which CBGs are retransmitted. It was agreed that the CBGTI is indicated by a bitmap (i.e., the Method 1 described in Section 5.3.3).

Second, when the decoding failure of an eMBB CBG is caused by a collision with a URLLC transmission, the retransmitted CBG cannot be soft-combined with the previously received signals, because the previously received signal is corrupted with URLLC transmission. To prevent the UE from combining the corrupted signal with the retransmission signal, CBG Flushing out Information (CBGFI) needs to be added to the scheduling DCI as well. CBGFI is a 1-bit indicator. If the CBGFI is set to "0," it indicates that the retransmission cannot be combined with the earlier transmission; otherwise, it indicates that the previously retransmitted CBGs can be used for HARQ soft combining.

Once CBG-based transmission is configured to a UE, the CBGTI field is always included in the DCI. However, whether the CBGFI field is included in the DCI is configured by high layer signaling.

5.3.5 UL control signaling for CBG-based transmission

In the NR system, HARQ timing can be dynamically indicated. Specifically, when a PDSCH or a DCI indicating SPS release is received in slot n, the corresponding ACK/NACK information is transmitted in the slot $n + k$, where the value of k is indicated by the DCI.

If the corresponding HARQ timing of multiple PDSCHs or the DCIs indicating the release of SPS resources configured in multiple downlink slots point to the same slot for HARQ—ACK transmission, the corresponding ACK/NACK information will be multiplexed through one PUCCH. NR R15 supports two ways to transmit ACK/NACK feedback.

The first way is using the Type-1 HARQ—ACK codebook. With the Type-1 codebook, the number of ACK/NACK information bits carried in an uplink slot is determined based on the number of a set of occasions for candidate PDSCH receptions, where the occasions for candidate PDSCH receptions are determined based on high layer parameters (e.g., TDD UL-DL configuration, HARQ timing set K_1, and PDSCH TDRA table). In other words, the size of the Type-1 HARQ—ACK codebook is semistatically determined. As shown in Fig. 5.36, the first serving cell supports CBG-based transmission, and a PDSCH contains up to four CBGs. The second serving cell does not support CBG-based transmission, and a PDSCH contains up to two TBs. Then, each slot of the first serving cell, as an occasion for candidate PDSCH, corresponds to 4 ACK/NACK bits in the HARQ—ACK codebook, and each slot of the second serving cell corresponds to 2 ACK/NACK bits in the HARQ—ACK codebook.

The second way is using the Type 2 HARQ—ACK codebook. With the Type 2 codebook, the number of ACK/NACK information bits carried in an uplink slot is determined by the number of scheduled PDSCHs. The following methods were proposed for determination of the Type 2 HARQ—ACK codebook with CBG-based ACK/NACK information [22]:

- Option 1: The serving cells are divided into two groups, where the first group includes the cells configured with CBG-based transmission and the second group includes the cells not configured with CBG-

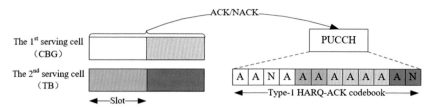

Figure 5.36 Type-1 HARQ—ACK codebook with CBG-based ACK/NACK.

based transmission. Correspondingly, the HARQ—ACK codebook contains two subcodebooks. The first HARQ—ACK subcodebook contains the CBG-based ACK/NACK feedback information of the first cell group, where the number of ACK/NACK bits equals to the maximum number of CBGs. In other words, the number of the first HARQ—ACK subcodebook is semistatically configured. The second HARQ—ACK subcodebook contains the TB-based ACK/NACK feedback information of the second cell group, and the number of ACK/NACK bits is determined based on the DAI that corresponds to the accumulative number of PDSCHs. CBG-based DAI is not used. The UCI overhead of Option 1 is high, because the first HARQ—ACK subcodebook is semistatically configured.

- Option 2: The HARQ—ACK codebook is generated based on CBG-based DAI, where DAI denotes the accumulative number of CBGs. The advantage of Option 2 is that the ACK/NACK feedback overhead is kept minimal. However, additional bits are required to indicate DAI in a DCI, even for the DCI scheduling the PDSCH on the cell that is not configured for CBG-based transmission.
- Option 3: Two HARQ—ACK subcodebooks are generated and joined into one HARQ—ACK codebook. The first HARQ—ACK subcodebook contains the TB-based ACK/NACK for all serving cells, including:
  - ACK/NACK corresponding to the DCI indicating a SPS PDSCH release.
  - ACK/NACK corresponding to the SPS PDSCH.
  - ACK/NACK corresponding to the dynamic PDSCH on the cell that is not configured with CBG-based transmission.
  - ACK/NACK corresponding to the PDSCH, scheduled by a DCI format 1_0 or 1_2, on the cell that is configured with CBG-based transmission.
- If two-codeword transmission is configured on any serving cell, the number of ACK/NACK bits corresponding to each PDCCH or dynamical PDSCH in the first subcodebook equals to 2; otherwise, each PDCCH or dynamical PDSCH corresponds to 1 ACK/NACK bit.
- The second HARQ—ACK subcodebook includes CBG-based ACK/NACK feedback information, corresponding to the PDSCHs scheduled by DCI format 1_1, on the serving cell that is configured with CBG-based transmission. If multiple serving cells are configured with

CBG-based transmission, the number of ACK/NACK bits corresponding to each PDSCH in the second subcodebook equals to the maximum number of CBGs configured for the multiple serving cells. For the first subcodebook and the second subcodebook, DAI counts independently.

- Fig. 5.37 shows an example, where serving cell 1 is configured with CBG-based transmission and a PDSCH supports up to four CBGs, and serving cell 2 is configured with CBG-based transmission and a PDSCH supports up to six CBGs. Then the number of ACK/NACK bits corresponding to each dynamical PDSCH in the second subcodebook is 6. Serving cell 3 is not configured with CBG-based transmission and a PDSCH supports up to two TBs, then the number of ACK/NACK bits corresponding to each dynamical PDSCH in the first subcodebook is 2. In the first subcodebook, the ACK/NACK information corresponding to the SPS PDSCH is appended after the ACK/NACK information corresponding to the dynamic PDSCHs. In the figure, $b_{i,j}$ represents the jth ACK/NACK bit corresponding to the PDSCH i.

- Option 4: CBG-based ACK/NACK feedback information is generated for all serving cells. The number of ACK/NACK bits corresponding to a TB equals to the maximum number of CBGs configured for all serving cells. This option has the highest feedback overhead.

Among the above options, Option 1 and Option 4 cannot satisfy the design principle (i.e., UCI overhead reduction) of the Type 2 HARQ—ACK codebook, since a lot of redundant information exists in the UCI feedback. Option 2 has to pay additional DCI overhead on the DAI field. Option 3 introduces no additional DCI overhead, and bears only a little redundant information in the UCI. Therefore in the 3GPP RAN1#91 meeting, Option 3 was adopted for Type 2 HARQ—ACK codebook with CBG-based ACK/NACK information.

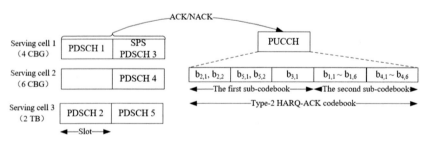

Figure 5.37 Type 2 HARQ—ACK codebook with CBG-based ACK/NACK feedback.

5.4 Design of NR PDCCH

5.4.1 Considerations of NR PDCCH design

Section 5.2 introduces the resource allocation design in 5G NR, which is mainly suitable for PDSCH, PUSCH, PUCCH, and other channels that need to be dynamically scheduled by the base station. Although the signaling mechanism of "RRC configuration + DCI indication" is widely used in 5G NR, this mechanism does not apply to PDCCH, simply because DCI is carried in the PDCCH and the terminal cannot precisely know in advance the time—frequency resources for the DCI transmission. The terminal can only search for DCI in a resource region, which is known as blind detection.

The PDCCH blind detection mechanism has been well designed in LTE. The goal of 5G NR PDCCH design is to enhance the LTE mechanism. The main considerations for enhancements include [7].

5.4.1.1 Changing from cell-specific PDCCH resources to UE-specific PDCCH resources

In LTE, each terminal search for DCI in a PDCCH resource region. The frequency-domain span of the region is equal to the system bandwidth, and the time-domain span is the first 1—3 OFDM symbols in every downlink subframe, where the exact number of symbols in the time-domain span is indicated by Physical Control Format Indication Channel (PCFICH) that is broadcasted in the cell. Therefore all terminals in the cell would use the same control region to search for DCI. This design, if applied to 5G NR, may lead to several shortcomings. First, the 5G NR data channel design becomes more UE-specific in many aspects, such as resource allocation, beam-forming, and reference signal transmission. If PDCCH still maintains a cell-specific structure (as shown on the left in Fig. 5.38), there can be significant differences between PDCCH and PDSCH in terms of link performance, resource allocation, etc. The base station may have to schedule PDCCH and PDSCH separately, resulting in an increased scheduling complexity. In addition, even when only a small number of DCIs need to be sent, the LTE base station has to use the entire system bandwidth for PDCCH transmission, and the terminal also needs to monitor the PDCCH in the entire system bandwidth, which is not beneficial to both resource utilization and power saving in both base station and terminal. If PDCCH is not bound to cell-specific transmission, both PDCCH and PDSCH can be transmitted in a UE-specific

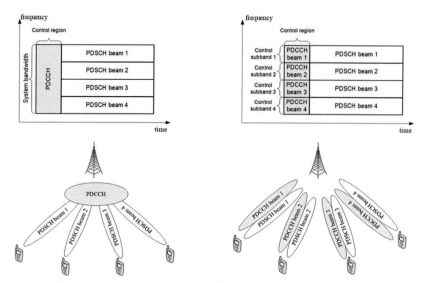

Figure 5.38 Cell-specific and UE-specific PDCCH.

manner (as shown on the right in Fig. 5.38), which can support PDCCH beam-forming, customize PDCCH reference signal pattern, simplify the base station scheduling, and reduce the power consumption in base station and terminals. In terms of resource allocation, the most significant improvement is to change the PDCCH-monitoring range in the frequency domain from system bandwidth to a "control subband." This concept becomes the base of Control Resource Set (CORESET), which is introduced in Section 5.4.2.

5.4.1.2 PDCCH "floating" in the time domain

As mentioned in Section 5.2, the minislot and "floating" channel structure is introduced into 5G NR to achieve low latency. In the entire physical procedure of resource scheduling, besides the fast transmission of PDSCH, PUSCH, and PUCCH, the fast transmission of PDCCH is also very important. Only if a terminal can receive the DCI scheduling signaling at any time, it becomes possible to realize the fast scheduling of PDSCH, PUSCH, and PUCCH. If the PDCCH can only be transmitted in a first few symbols of the subframe as in LTE, it is impossible to meet the requirements of URLLC and low-latency eMBB services. Moreover, the LTE terminal needs to monitor the PDCCH in each downlink subframe, which causes the terminal to consume unnecessary power in the continuous PDCCH monitoring. For services that are not sensitive to latency, if the terminal can be configured to monitor the PDCCH once every several slots, the terminal can save power through microsleep. This also motivates the NR PDCCH to be able to "float" in the time domain to support the on-demand transmission at any time. Such flexibility is finally reflected in the design of PDCCH search-space set, which is introduced in Section 5.4.3.

• More flexible multiplexing between PDCCH and PDSCH:

In LTE, because the PDCCH control region occupies the entire system bandwidth in the frequency domain, FDM between PDCCH and PDSCH cannot be supported. The number of symbols used for PDCCH can be dynamically changed by PCFICH based on the number of terminals and the overall load of PDCCH, subject to the Time Domain Multiplexing (TDM) between PDCCH and PDSCH in a subframe, as shown in Fig. 5.39A. Because one single LTE carrier is up to 20 MHz, it is fairly reasonable to adopt TDM instead of FDM between PDCCH and PDSCH. However, with the NR carrier bandwidth being more than 100 MHz, PDCCH and PDSCH cannot be efficiently multiplexed in NR with only TDM, since a large amount of frequency-domain resources that share the same symbols with PDCCH but not used for PDCCH transmission are all wasted. Therefore FDM between PDCCH and PDSCH should be supported in 5G NR. If a PDCCH is restricted to its control subband, its scheduled PDSCH can be transmitted outside the

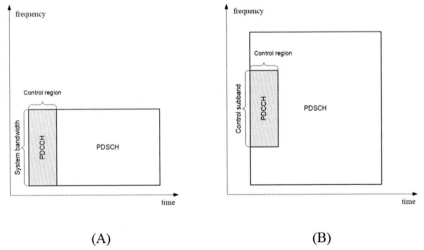

Figure 5.39 (A) LTE only supports TDM. (B) NR supports both TDM and FDM.

control subband, as shown in Fig. 5.39B. The multiplexing between the scheduling PDCCH and the scheduled PDSCH can be booked by the configured CORESET information, which is described in Section 5.4.2. The multiplexing between the PDSCH for one terminal and the PDCCH for another terminal is more complicated due to the necessity for one terminal to know the PDCCH resource allocated for another terminal. Such multiplexing is described in Section 5.7.

5.4.1.3 Reduced complexity of DCI detection

The fundamental principle of the DCI blind detection is to try the decoding based on each possible DCI size in the PDCCH search space until the desired DCIs are found. As the LTE specification continues to evolve, more and more DCI formats and new DCI sizes are defined, resulting in increasing complexity of DCI blind detection. Therefore the 5G NR specification intends to control the number of DCI formats/sizes. At the same time, a common control channel is used for common DCI transmission to reduce the detection complexity and save the terminal power. The details are given in Section 5.4.4.

Some other enhancements were proposed and studied in the design of NR PDCCH, such as enhanced search space design, and two-stage DCI, but they were not accepted in the 5G NR R15 and therefore will not be the focus here. This chapter mainly focuses on the time—frequency resource configuration and DCI design of NR PDCCH.

5.4.2 Control Resource Set

CORESET (Control Resource Set) is one of the main innovations introduced for NR PDCCH. As mentioned in Section 5.4.1, in order to support high-efficient UE-specific PDCCH transmission and FDM between PDCCH and PDSCH, a NR PDCCH transmission should be confined to a control subband instead of the entire system bandwidth; meanwhile, the PDCCH resource should be able to "float" in the time domain based on minislot. These two design targets drive an enhancement direction of having a more flexible "time—frequency region" for PDCCH. This time—frequency region is ultimately defined as CORESET. The design of CORESET mainly involves the following aspects:

- The external structure aspects include: the relationship between CORESET and DL BWP, the relationship between CORESET and search space, the frequency-domain range of CORESET, the time-domain position of CORESET, whether to support discontinuous frequency-domain resource allocation, and whether to dynamically adjust the time-domain length of CORESET.

- The internal structure aspects include: internal structure levels of CORESET, frequency-domain granularity, and the resource mapping order.

5.4.2.1 External structure of CORESET

As mentioned in Section 5.4.1, the NR CORESET follows the control region concept in LTE and serves the same functionality as the LTE control region in providing a time—frequency resource region for terminal to search for PDCCH. On the other hand, NR CORESET is designed as a more flexible control region in the time—frequency domain than LTE control region. The main differences between CORESET and LTE control region are the following:

- CORESET does not need to occupy the entire system bandwidth. Its frequency-domain occupancy can be as small as one subband.
- CORESET has more flexible position in the time domain. It can be located not only at the beginning of a slot, but also at other positions in the slot.
- The CORESET configuration is UE-specific, instead of cell-specific.

For the frequency-domain properties of CORESET, the first question that needs an answer is: What is the relationship between CORESET and DL BWP? Or specifically, is it a good design to combine the functions of DL BWP and CORESET into one? In fact, CORESET and BWP are two concepts that were formed independently in the 5G NR design phase, and the CORESET concept was created even earlier than BWP. When the BWP concept was proposed, it was indeed suggested to combine the two concepts into one. The reason comes from the similarities between the two concepts: for example, both are kinds of subbands smaller than the carrier bandwidth (as described in Chapter 4: Bandwidth Part, BWP was initially called a subband before the concept of "bandwidth part" was clearly defined); both are UE-specifically configured. However, as the NR study went in-depth, the two concepts were found to still bear different key essences: First, DL BWP is defined as the frequency-domain range accommodating all downlink channels and signals (including control channels, data channels, reference signals, etc.), while CORESET is defined to just describe the resource range for PDCCH detection. Secondly, given the NR system has much higher downlink traffic volume than LTE, the typical size of DL BWP is likely larger than 20 MHz; in comparison, the capacity of NR PDCCH is expected to be only slightly higher than that of LTE, so the typical size of CORESET is less than 20 MHz, such as 10, 5, or even 1.4 MHz (such as for NR-IoT systems like NB-IoT). Thus combining CORESET and DL BWP into one concept may force the terminal to search for PDCCH in the entire DL BWP, which is unnecessary in most of cases and conflicts with the original intention of introducing the concept of CORESET. At last, BWP is a one-dimensional concept in the frequency domain, which cannot attribute a 2D time—frequency resource region that the CORESET targets to identify for the terminal to search for PDCCH. Consequently, DL BWP and CORESET remains as two separate concepts in NR.

Because the typical bandwidth of CORESET is not large, if CORESET can only be allocated with contiguous frequency-domain resources, the frequency diversity may be insufficient, which affects the transmission performance of PDCCH. Therefore it was finally decided that CORESET can occupy either contiguous or noncontiguous frequency-domain resources. A bitmap in RRC signaling is used to indicate the PRBs occupied by CORESET in DL BWP, where each bit in the bitmap indicates a RB group (RBG) containing six RBs. The bitmap can select any RBGs in the DL BWP. The granularity of this bitmap is different from the RBG size mentioned in Section 5.2 for PDSCH/PUSCH Type 0 frequency-domain resource allocation. The RBG size used for PDSCH/PUSCH has several choices in {2, 4, 8, 16}. The reason for the CORESET RBG to contain six RBs is related to the granularity of the internal resources constructing a candidate PDCCH, which is described later for CORESET internal structure. Differently from CORESET, as mentioned in Section 4.2, BWP can only be composed of contiguous RBs. This is simply because BWP mainly identifies an operating frequency range of the terminal, and it is sufficient to adopt contiguous frequency-domain range.

The next question is what time-domain characteristics a CORESET should have. CORESET in NR is considered a more flexible control region in the time—frequency domain. The region length in time (1—3 symbols) is somehow inherited from the LTE control region. But the difference is that the length of NR CORESET is semistatically configured by RRC instead of dynamically indicated by a physical channel. It seems, at first glance, that the configuration of NR CORESET length in time is not as dynamic as that of LTE control region. But the decision was made due to the fact that the frequency-domain size of CORESET can already be configured semistatically, which is much more flexible than LTE control region that is fixed to carrier bandwidth and is sufficient to adapt to different PDCCH capacities. It is unnecessary to use a dedicated physical channel (similar to LTE PCFICH) to dynamically indicate the length of CORESET. The 5G NR specification does not define a physical channel with similar function of LTE PCFICH.

In attempts to achieve one of design targets of CORESET (i.e., the capability of "floating" based on minislot), two options were considered:

1. Option 1: The RRC signaling for a CORESET can configure its frequency-domain position, time-domain length, and time-domain position.

2. Option 2: The frequency-domain position and time-domain length of a CORESET are configured by RRC, but the time-domain position is derived from the configuration of the search space.

The advantage of Option 1 is that a CORESET configuration parameter can be used to completely describe all the time−frequency-domain characteristics of the PDCCH control region, including the position of the starting symbol. The search space does not have to own any physical meaning or measure, but is just a logical concept. The advantage of Option 2 is that the CORESET concept and LTE control region have good inheritance; that is, it only has frequency-domain characteristics (corresponding to the system bandwidth in LTE) and time-domain length (corresponding to the number of control region symbols indicated by PCFICH in LTE), but the CORESET itself does not provide information for the starting symbol position. Both options can work. The difference is only in concept definition and signaling design. The NR specification finally chose Option 2. Meanwhile, in order to distinguish from the logical concept of search space, the time-domain positions of the PDCCH is called search-space set. With Option 2, while the specification makes the CORESET have good inheritance from LTE control region, it divides a complete description of a time−frequency-domain region into two concepts: CORESET and search-space set. Neither of these two concepts can independently give a full description for a time−frequency control region; they must be combined together to define the time−frequency characteristics of CORESET. This may impact the readability of the specification, and make it a bit difficult for the first-time reader of the specification to understand it well. This is a disadvantage of Option 2.

Fig. 5.40 is used here to illustrate the concept of CORESET (assuming that the CORESET of terminal 1 and terminal 2 are configured with the same DL BWP): The "frequency-domain resource" parameter (i.e., *frequencyDomainResources*) in the RRC parameter CORESET uses a

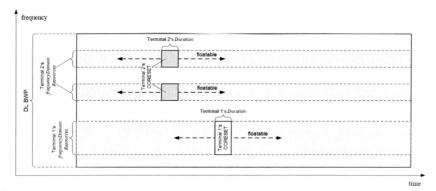

Figure 5.40 Example of external structure of CORESET.

bitmap to describe the frequency-domain positions of CORESET, which occupy part of the RBs in the DL BWP. The frequency-domain positions can be either contiguous (such as the CORESET of terminal 1 in the figure) or noncontiguous (such as the CORESET of terminal 2 in the figure). The "length" field (namely *duration*) in parameter CORESET describes the time-domain length of the CORESET. Assume *duration* = 2 in the figure. The parameter CORESET itself does not carry the information for the time-domain positions; instead, the time position of CORESET can "float" in the time domain with a length of 2 OFDM symbol. Its location in the time domain can be determined by the time-domain information in the search space, which is introduced in Section 5.4.3.

The CORESET configured for the two terminals can overlap in the frequency domain and the time domain, which is similar to BWP. Although CORESET is a UE-specific configuration, it is unnecessary for the base station to configure a different CORESET for each terminal. In fact, the base station may only plan a few typical CORESET configurations for a given DL BWP, with different frequency-domain sizes and positions. For a certain terminal, only one of these CORESET configurations is assigned. As shown in Fig. 5.41 (assuming that the CORESETs of terminal 1 and terminal 2 are configured within the same DL BWP), the base station plans the CORESET configurations of 5 and 10 MHz in the 20 MHz DL BWP, and configures terminal 1 and terminal 2 with a 5 MHz CORESET and a 10 MHz CORESET, respectively. The CORESET resources for the two terminals overlap in the frequency domain by 5 MHz.

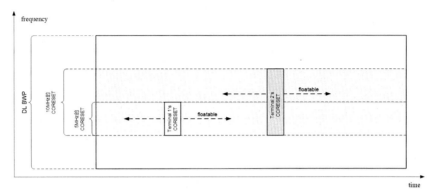

Figure 5.41 CORESET of different terminals can overlap in the frequency domain.

5.4.2.2 Internal structure of CORESET

With the given external structure information of CORESET, the terminal knows in which time—frequency region to detect PDCCH. CORESET is composed of PDCCH candidates. To detect PDCCH in CORESET, terminals must also know the internal structure of CORESET, which is also the structure of PDCCH. In general, the time—frequency domain structure of the NR PDCCH basically follows the design of the LTE PDCCH, and the NR PDCCH still adopts the two-level resource allocation granularities including RE Group (REG) and CCE. REG is the basic frequency-domain granularity and is composed of several REs. One CCE is composed of several REGs, and one PDCCH candidate occupies several CCEs. The REG-to-CCE mapping of NR PDCCH is similar to that of LTE PDCCH, except that some optimizations have been made on the sizes of REG and CCE.

To determine the two-level resource granularities of the NR PDCCH (i.e., the number of REs per-REG and the number of REGs per-CCE), the following two factors are mainly considered:

1. As 5G NR needs to achieve higher scheduling flexibility than LTE, the size of NR DCI is inevitably larger than LTE DCI.
2. Part of the resource allocation method for LTE PDCCH is either too complicated or unnecessary, and therefore needs modification when applying to NR PDCCH.

In the LTE specification, an REG is composed of four consecutive REs on the same symbol, except for the REs occupied by DMRS. A PRB can accommodate two or three REGs in a symbol (depending on whether these 12 REs contain DMRS). A CCE contains nine REGs, having a total of 36 REs. As mentioned in Ref. [1], in order to achieve the necessary coverage performance in a limited-length control region (up to three symbols), the REGs in each CCE are arranged to fill all OFDM symbols in the control region.

As mentioned in Ref. [1], the reason for LTE to adopt a small REG size of four REs is to efficiently support resource allocations for PCFICH and Physical HARQ indication channel (PHICH), which carry a very small amount of data (only several bits). However, PCFICH and PHICH are not used in the design of R15 NR. Therefore if only the size of PDCCHs (at least dozens of bits) is considered in determination of the REG size, the size of four REs is unnecessarily small. On the other hand, because PRB (12 subcarriers) is used as the resource allocation unit for PDSCH, a REG size smaller than PRB may cause difficulty in FDM multiplexing between PDCCH and PDSCH and leave unusable resource fragments. Therefore it was finally determined that one REG in the NR PDCCH is equal to 1 PRB in the frequency domain and still 1 symbol in the time domain.

The number of REGs contained in a CCE depends on the capacity of a typical DCI in such a way that one CCE should be large enough to accommodate a coded and modulated DCI (QPSK modulation is used for PDCCH). Given that 5G NR needs to support more flexible resource scheduling than LTE, the number of DCI fields and the number of bits per-DCI field may be increased, which inevitably increases the overall capacity of DCI. Therefore it was finally determined that a CCE contains six REGs or a total of 72 REs. However, unlike LTE, these 72 REs include DMRS REs, and the actual REs that can be used for PDCCH are the non-DMRS REs.

Similar to LTE, NR PDCCH can improve transmission performance by repeating multiple CCEs. The number of CCEs contained in a NR PDCCH candidate (i.e., aggregation level) includes five levels: 1, 2, 4, 8, and 16. Compared to LTE that includes only four levels of {1, 2, 4, 8}, NR additionally supports the aggregation level of 16 CCEs for a PDCCH candidate to further enhance the PDCCH link performance.

The next question is where the six REGs in a CCE are mapped to (i.e., the mapping from REG to CCE). This mapping mainly involves two aspects:

1. Which mapping order should be adopted: "frequency-first mapping" or "time-first mapping"?

2. In addition to localized mapping, is it necessary to support interleaved mapping?

The selection of the REG-to-CCE mapping order actually involves the CCE-to-PDCCH mapping order, because the two levels of mapping are complementary. At least for PDCCH with a relatively large aggregation level, if time-first REG-to-CCE mapping cannot obtain sufficient frequency-domain diversity, the insufficiency can be compensated through frequency-first CCE-to-PDCCH mapping. Conversely, if frequency-first REG-to-CCE mapping cannot obtain sufficient time-domain diversity, time-first CCE-to-PDCCH mapping can also be used for the compensation. The combination options theoretically start from:

- Option 1: Time-first REG-to-CCE mapping + time-first CCE-to-PDCCH mapping.
- Option 2: Frequency-first REG-to-CCE mapping + frequency-first CCE-to-PDCCH mapping.
- Option 3: Time-first REG-to-CCE mapping + frequency-first CCE-to-PDCCH mapping.
- Option 4: Frequency-first REG-to-CCE mapping + time-first CCE-to-PDCCH mapping.

First of all, because CORESET has no more than three symbols, Option 1 is impossible to implement. Each of the other three options has its own advantages and disadvantages. Take CORESET with a length of 3 symbols (*duration* = 3) as an example, for which Options 3 and 4 are shown in Fig. 5.42. The advantage of frequency-first mapping is that the entire CCE is concentrated in one symbol, and a PDCCH can be received with the shortest latency. It has some advantages for low-latency and power saving in the detection of PDCCH. It occupies six RBs in the frequency domain, corresponding to the maximum frequency-domain diversity allowed for a CCE. Another potential benefit is that, when the number of PDCCHs to be transmitted is relatively small, the last one or two symbols in CORESET can be saved and used to transmit PDSCH. But this benefit is only valid when the CORESET length can be dynamically adjusted. Should NR have the ability to dynamically indicate the CORESET length via a PCFICH-alike channel, once the base station finds that the CORESET of two symbols is enough to accommodate all the PDCCHs to be transmitted in a slot, the terminal can be notified to start receiving PDSCH from the third symbol instead of searching for PDCCH in the third symbol. But once NR was decided to have no functionality similar to what LTE PCFICH provides, and only to support

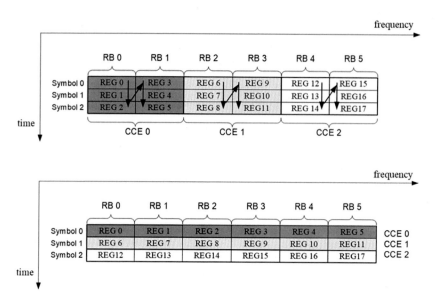

Figure 5.42 Time-first and frequency-first REG-to-CCE mapping.

semistatic configuration of CORESET, this advantage of frequency-first mapping does not exist anymore.

The advantage of time-first mapping is that it can maximize the use of the time-domain length to obtain better coverage performance, and also realize power sharing between different CCEs. When a CCE does not have PDCCH transmission, the power used for transmission of this CCE can be applied to the other CCEs with PDCCH transmissions. The power sharing in the frequency-domain can only be realized in the time-first REG-to-CCE mapping mode.

Another alternative for the specification is to support both frequency-first and time-first REG-to-CCE mapping modes, with a selection through RRC configuration. But to simplify the design, it was decided in the end to only support time-first REG-to-CCE mapping instead of frequency-first REG-to-CCE mapping. It can be seen that this choice basically inherits the design philosophy of LTE PDCCH, which stretches each CCE to all the symbols of CORESET to achieve the sufficient coverage performance. It is one of important requirements of NR PDCCH design. This mapping method is reflected by the time-first arrangement of REG order in CORESET, as shown in the upper figure of Fig. 5.42.

Take CORESET composed of contiguous RBs, shown in Fig. 5.43, as an example, where the external structure of CORESET can have

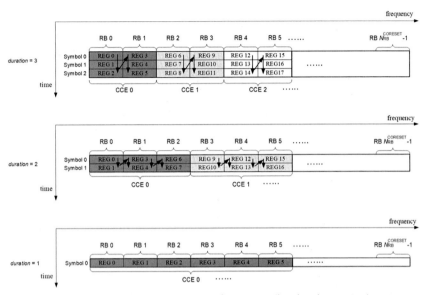

Figure 5.43 Example of internal structure of CORESET (localized mapping).

various lengths (*duration*) from {1, 2, 3}. When *duration* = 1, it is actually a pure frequency-domain REG-to-CCE mapping.

The mappings shown in Fig. 5.43 are all localized REG-to-CCE mappings, where the six REGs in a CCE are contiguous and have consecutive RB indices. In order to solve the problem of insufficient frequency diversity in time-first mapping, NR PDCCH also supports interleaved REG-to-CCE mapping, which distributes the REGs in a CCE to scattered RBs. However, if the six REGs in CCE are completely isolated from each other, a new problem would arise: the PDCCH channel estimation becomes REG-based and the estimation performance can be worse. Because an important difference between NR PDCCH and LTE PDCCH is that NR PDCCH uses UE-specific precoding to achieve beam-forming for each terminal, the terminal can only perform PDCCH channel estimation based on the DMRS within the REGs for the concerned PDCCH. If all the REGs in a CCE are not adjacent to each other, the terminal can only perform channel estimation and interpolation between the DMRS REs within each REG, and cannot perform joint channel estimation over multiple REGs. Therefore in order to ensure the channel estimation performance, even if interleaving is performed, the REGs in the CCE should not be completely isolated, which means the REG continuity must be maintained to a certain extent. Therefore it was finally decided to use REG bundle as the unit for interleaving. A REG bundle contains two, three, or six REGs to ensure two, three, or six REGs to be adjacent, over which the joint channel estimation can be implemented. Fig. 5.44 is a schematic diagram of an interleaved REG-to-CCE mapping, where the REG bundle size is 3, and two REG bundles in each CCE are scattered to different frequency-domain positions.

Interleaved REG-to-CCE mapping uses a simple block interleaver. The interleaving-related parameters, including REG bundle size, interleaver size, and shift index, are all contained in *cce-REG-MappingType* in the CORESET configuration. For simplicity, a CORESET can only use a single REG-to-CCE mapping parameter set. The set of parameters, including interleaving enabling flag, REG bundle size, interleaver size, shift index, etc., must remain the same for all REGs and CCEs within the CORESET. Different REG-to-CCE mapping methods cannot be mixed in a CORESET.

The base station can configure up to three CORESETs for each DL BWP. A total of 12 CORESETs can be configured for four DL BWPs of a terminal. Different CORESETs can use different configurations (frequency-domain resources, duration, REG-to-CCE mapping parameter set, etc.).

5.4.3 Search-space set

As shown in Fig. 5.41, the RRC parameter CORESET does not contain the complete information of the time—frequency region of a CORESET. It only configures the frequency-domain characteristics and the time-domain duration of a CORESET, but the specific time-domain position of the CORESET is not contained in this RRC parameter. The 5G NR

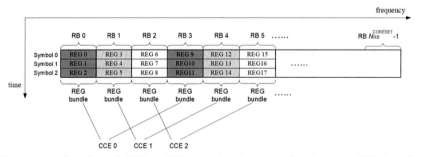

Figure 5.44 Interleaved REG-to-CCE mapping (assume duration = 3, REG bundle size = 3).

specification introduces another time-domain concept, search-space set, to configure the time-domain positions for the terminal to detect the PDCCH. The search space in the LTE specification is a pure logical concept. Although it can be mapped to the CCE through a certain mapping relationship, the search space itself does not determine a concrete time–frequency resource. The LTE specification has no concept of search-space set. This is because all search spaces are confined in the control region starting from the first symbol of the subframe and have no chance to appear in the other time-domain positions in the subframe. The concept of search space is also defined in the NR specification, which is exactly the same as LTE search space. Although there is only one word difference between search-space set and search space, they are completely different concepts. Search-space set is used to determine time-domain starting position of CORESET. The search space of NR may appear in any position within a slot, so a new concept of search-space set is needed to describe the time-domain position of search space.

The resource allocation of the search-space set is very similar to that of symbol-level "floating" PDSCH and PUSCH, and also needs to indicate the "starting position + length" of a channel. Theoretically, the indication of time-domain position of the search-space set can also be directly indicated as "appearing once every X symbols," which is independent from the slot, so that the PDCCH-monitoring position is completely "floating" in the time domain. However, as mentioned in Section 5.2, in the discussion of NR's PDSCH and PUSCH TDRA methods, in order to balance scheduling flexibility and equipment complexity, the concept of slot is still reserved and a two-level indication method of "slot + symbol" is adopted. For the same reason, the TDRA of the search-space set finally adopts the two-level "slot + symbol" configuration method, which is however different from the time-domain resource indication of the data channel in the following aspects:

- PDSCH and PUSCH are "one-shot" scheduling, which require a "reference point" (such as the scheduling PDCCH) to indicate the time-domain position. The search-space set occupies a set of periodic time-domain resources similar to semipersistent scheduling (SPS, semipersistent), so the slot-level time-domain position of the search-space set is not indicated relative to a certain "reference point," but is indicated using a "period + offset" method, which is configured by the parameter *monitoringSlotPeriodicityAndOffset*. The period k_s refers to how many slots there will be in a search-space set, and the offset O_s refers to the slot in which the search–space set appears in k_s slots.

- The symbol–level resource configuration of the search–space set does not need to indicate the symbol–level length of the search–space set, because it has been indicated in the CORESET configuration parameter *duration*. The search–space set only needs to configure the starting symbols of the search–space set. Therefore the search–space set does not use a SLIV joint encoding of "starting point + length," but uses a 14-bit bitmap to directly indicate which symbols in a slot are used as the starting symbols of the search–space set. This parameter is called *monitoringSymbolsWithinSlot*.

- The configuration parameter of the search–space set also introduces another parameter T_s, also called duration but with a different meaning from the configuration parameter duration of CORESET. It is used to indicate the slot-level length of the search–space set in each slot-level period, that is to say, the slot-level PDCCH-monitoring positions for the terminal are "the T_s consecutive slots starting from the O_s slot in every period consisting of k_s slots." It can be seen that this parameter is similar to the "number of slots" in the "multislot PDSCH/PUSCH transmission" introduced in Section 5.2.10. Fig. 5.45 illustrates the relationship between these two *duration* parameters.

- At last, PDSCH and PUSCH are scheduled by DCI, while PDCCH is blindly detected in CORESET, and CORESET is semistatically configured by RRC signaling.

To sum up, the terminal needs to determine the time–frequency search region based on the parameters in both the CORESET configuration and the search–space set configuration, as shown in Table 5.8. For a clear description, the NR specification introduces a terminology of PDCCH-monitoring occasion. A PDCCH-monitoring occasion is equal to a segment of continuous time-domain resources in the search–space set, and its length is equal to the length of CORESET (1−3 symbols). On the contrary, the search–space set is composed of many PDCCH-monitoring occasions that appear periodically.

Fig. 5.45 summarizes how a terminal determines where to monitor the PDCCH:

- The terminal determines the time–frequency position for monitoring PDCCH according to the two sets of RRC configuration parameters, CORESET and search–space set. Each DL BWP can be configured with three CORESETs and ten search–space sets. The association

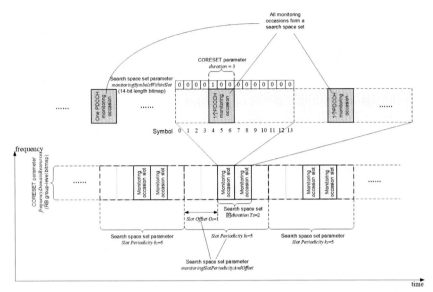

Figure 5.45 Determine the time–frequency region of PDCCH detection through CORESET and search-space set.

between these 10 search-space sets and three CORESETs can be flexibly configured. A terminal can be configured with four DL BWP, resulting in at most 12 CORESETs and 40 search-space sets. It should be noted that the physical layer specification (such as TS 38.213) and the RRC layer specification (TS 38.331) do not use the unified names, respectively, for these two parameters. The CORESET configuration in the physical layer specification is called *ControlResourceSet* in the RRC layer specification, and the search-space set in the physical layer specification is called *SearchSpace* in the RRC layer specification. The naming difference among specifications is a leftover problem from the specification writing in 3GPP RAN1 and RAN2, which has some impacts on the readability of the specifications.

- The terminal monitors the PDCCH in contiguous or noncontiguous RBs, whose positions are defined by the *frequencyDomainResources* parameter in the CORESET configuration. This parameter is a RBG-level bitmap, with a RBG containing six RBs. Fig. 5.45 illustrates contiguous RB allocation for an example.
- The terminal monitors the PDCCH in a series of PDCCH-monitoring occasions in the time domain. A PDCCH-monitoring occasion lasts for one to three symbols, and its length is defined by the *duration* parameter in

Table 5.8 Configuration parameters for CORESET and search-space set.

	Parameter	Indication	Note
CORESET configuration	*frequency DomainResources*	Frequency region for searching PDCCH	RBG-level bitmap to indicate continuous or discontinuous RB
	duration	The length of PDCCH-monitoring occasion	1—3 Symbols, the length of one monitoring occasion
Search-space set configuration	*monitoring SlotPeriodicity AndOffset*	The first slot that search-space set occurs	Slots for monitoring search space configured as periodicity k_s and offset O_s within a period
	duration	Number of consecutive slots that a Search-space set lasts in every occasion	A PDCCH-monitoring occasion repeats in consecutive T_s slots
	monitoring SymbolsWithinSlot	The starting symbol of a monitoring occasion within a slot	14-Bit symbol-level bitmap

the CORESET configuration. In Fig. 5.45, *duration* $= 3$ means a PDCCH-monitoring occasion lasts for three symbols.

- One or more PDCCH-monitoring occasions can appear in a slot, and the corresponding starting symbol positions are defined by the *monitoringSymbolsWithinSlot* parameter in the search-space set configuration, which is a 14-bit symbol-level bitmap. In Fig. 5.45, the bitmap value is 00001000000000, which means only one PDCCH-monitoring occasion appears in the slot and it starts from Symbol 4 of the slot and has three consecutive symbols.

- PDCCH-monitoring occasion can appear in consecutive T_s slots, where T_s is defined by the *duration* parameter in the search-space set configuration. In Fig. 5.45, $T_s = 2$ configures the PDCCH-monitoring occasions in two consecutive slots.
- These consecutive T_s slots recur with a period of k_s slots, starting from the O_s slot in the T_s slot period. In Fig. 5.45, $k_s = 5$ and $O_s = 2$. Then the two consecutive slots for PDCCH-monitoring occasions are Slot 2 and Slot 3 in every five slots.
- In each PDCCH-monitoring occasion, the terminal determines the REG-to-CCE mapping mode (interleaving or noninterleaving, REG bundle size, interleaver size, shift index) and other CORESET structure information based on the *cce-REG-MappingType* parameter set in the CORESET configuration.

The procedure for the NR terminal to search for PDCCH candidates in the PDCCH-monitoring occasion is similar to LTE, which is not described here. In the NR system, the number of PDCCH candidates included in each PDCCH aggregation level can be configured separately and included in the search-space set configuration. The mapping from PDCCH candidate to CCEs of a search-space set is still determined by a hash function with the similar principle as in LTE. The terminal uses the configured Radio Network Temporary Identifier (RNTI) in the attempt of decoding the PDCCH candidate in the CCE, which is determined according to the hash function. The terminal can receive the DCI contents upon a decoding success.

5.4.4 DCI design

DCI of the 5G NR basically follows the structure and design of LTE DCI. Multiple DCI formats are defined, as listed in Table 5.9. Among them, only four UE-specific DCI formats for data scheduling and four group-common DCI formats for basic power control, flexible TDD, and URLLC functions were defined in R15 NR, while R16 enhancements brought in four new UE-specific DCI formats and three new Group-common DCI formats, and also expanded one legacy Group-common DCI format.

In order to reduce the complexity of PDCCH blind detection, the NR specification intends to make the number of DCI formats as few as

Table 5.9 DCI format defined by 5G NR R15 and R16.

		DCI format	Function	Note
R15 NR DCI format	UE-specific DCI	0_0	UL grant (fallback format)	Include only essential fields; the size of each field is fixed as much as possible; depends less on RRC configuration
		0_1	UL grant (normal format)	Include all fields for flexible scheduling; the dependency of field sizes on RRC configuration is not restricted
		1_0	DL assignment (fallback format)	Include only essential fields; the size of each field is fixed as much as possible; depends less on RRC configuration
		1_1	DL assignment (normal format)	Include all fields for flexible scheduling; dependency of field sizes on RRC configuration is not restricted
	Group-common DCI	2_0	Slot format indication	Introduced for flexible TDD
		2_1	DL preemption indication	Introduced for URLLC
		2_2	Power control for PUCCH and PUSCH	
		2_3	Power control for SRS	
R16 NR DCI format	UE-specific DCI	0_2	UL grant (compact format)	Introduced for high reliability of PDCCH in enhanced URLLC technology
		1_2	DL assignment (compact format)	Introduced for high reliability of PDCCH in enhanced URLLC technology
		3_0	NR sidelink scheduling	Introduced for NR V2X
		3_1	LTE sidelink scheduling	Introduced for NR V2X
	Group-common DCI	2_0 enhancement	Indication of available RB set, Channel Occupation Time (COT), or search-space set group switching	Introduced for IAB, NR–U, and search-space set group switching
		2_4	UL cancellation indication	Introduced for URLLC enhancement
		2_5	Soft resource indication	Introduced for IAB
		2_6	Power-saving indication	Introduced for UE power saving

possible. Therefore R15 NR defined only four UE-specific scheduling DCI formats, including two DCI formats for downlink scheduling and two DCI formats for uplink scheduling. Similar to LTE, DCI formats 0_0 and 1_0 are two fallback DCI formats, which are used to provide the most basic scheduling DCI functions in special situations. Compared to the more advanced scheduling DCI formats 0_1 and 1_1, some DCI fields supporting enhanced performance are excluded from the fallback DCI formats, to make the remaining DCI field sizes as fixed as possible and less dependent on RRC configuration. DCI format 2_0 is used to transmit Slot Format Indicator (SFI) to dynamically indicate the slot format, which is described in Section 5.6. DCI format 2_1 is used to transmit downlink preemption indication to supports flexible multiplexing between URLLC and eMBB services, which is described in Section 15.6. DCI format 2_2 is used to transmit Transmit Power Control (TPC) commands for PUSCH and PUCCH, and DCI format 2_3 is used to transmit TPC commands for Sounding Reference Signal (SRS).

The new DCI formats introduced in R16 are mainly designed for various 5G vertical industrial enhancements. In order to improve the transmission reliability of the PDCCH as an URLLC enhancement technique, two new compact DCI formats, DCI formats 0_2 and 1_2, were introduced. This enhancement is described in Section 15.1. With V2X sidelink technology introduced into R16 NR protocol stack, two new sidelink scheduling DCI formats, DCI formats 3_0 and 3_1, were added, which are described in Chapter 17, 5G V2X. The other new R16 features, such as Integrated Access-backhaul (IAB), NR−U(NR unlicensed band, described in Chapter 18, 5G NR in Unlicensed Spectrum), and search-space set group switching, motivated the addition of new DCI format 2_5 as well as the extension of existing DCI format 2_0. DCI format 2_4 is designed to support UL cancellation as an URLLC enhancement, which is described in Section 15.7. DCI format 2_6 is designed to support UE behaviors for power saving, which is described in Chapter 9, 5G RF Design.

From the perspective of DCI structure, the following two design directions were explored for NR.

5.4.4.1 The choice of two-stage DCI

Two-stage DCI is a DCI design philosophy that was extensively studied in the early phase of NR standardization. The details of the traditional PDCCH mechanism, which is single stage, were already provided in this chapter. The single-stage PDCCH has been used in 3GPP for decades. All scheduling information in DCI can be obtained through blind detection of PDCCH candidates in the search space. The disadvantages of single-stage DCI design include high blind detection complexity, full absence of scheduling information without completing the entire DCI decoding, and large detection latency. Therefore it was proposed in the LTE standardization (such as the sTTI project) as well as the NR SI stage to introduce two-stage DCI technique.

The fundamental principle of two-stage DCI is to divide a scheduling DCI into two partial transmissions. The structures of two-stage DCI can be categorized into two types: The first type is shown in the left figure of Fig. 5.46, where the two stages are located in the PDCCH resource region, and are encoded and transmitted independently from each other. A typical design is to allocate the first-stage at the first symbol of the PDCCH and the second-stage at the second and the third symbols. Terminals can decode the first-stage DCI in the shortest time and start preparing for PDSCH reception without waiting for the reception of the whole DCI. After the second-stage DCI is decoded, the PDSCH demodulation can be completed earlier given some preparation steps are already done, which is beneficial for low-latency services. The first-stage DCI usually contains the "fast scheduling information" that the terminal must know as soon as possible to start PDSCH decoding, such as time−frequency resource allocation, MIMO transmission mode, and Modulation and Coding Scheme (MCS). Even before the remaining "slow scheduling

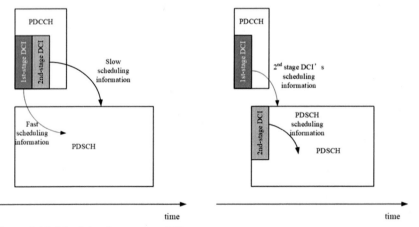

Figure 5.46 Principle of two-stage DCI.

information" [such as Redundancy Version (RV), New Data Indicator (NDI)] is received, the terminal can still use the "fast scheduling information" contained in the first-stage DCI to facilitate the preliminary demodulation processing on the PDSCH. After the second-stage DCI is received, the terminal applies the contained "slow scheduling information" to the full decoding of the PDSCH.

A typical application scenario of this first type of two-stage DCI is HARQ operation [23]. When the initial transmission of PDSCH has been sent to the terminal, the base station needs to wait for the HARQ−ACK information in the PUCCH, before which it does not know whether the transmission is successfully received or not. But during this time period, first-stage DCI can be used to schedule time−frequency resources, MIMO mode, and MCS for the next PDSCH transmission (these parameters usually only depend on channel condition). When HARQ−ACK information is received, if it is NACK, the PDSCH resource scheduled by first-stage DCI is used to retransmit the last downlink data (the second-stage DCI indicates the RV and NDI matching the retransmitted data); if it is ACK, the PDSCH resource scheduled by first-stage DCI is used to transmit new downlink data (indicating the RV and NDI matching the new data through the second-stage DCI). This method can speed up the reception of PDSCH in the same slot as PDCCH, but it cannot speed up the scheduling of PUCCH and PDSCH in subsequent slots, because sufficient reception time has already been reserved for the terminal to process these channels. In addition, one single DCI is divided into two parts and each part has a smaller DCI size than the original, which is beneficial to align the sizes of different DCI formats and to reduce the complexity of blind PDCCH detection.

Another type of two-stage DCI structure is shown in the right figure in Fig. 5.46, where the first-stage DCI contains the resource information of the second-stage DCI so that the terminal can directly find the resources for second-stage DCI without running blind detection on the second-stage DCI. A typical structure is that the second-stage DCI is not located in the PDCCH resource region, but is located in the PDSCH resource region scheduled by the first-stage DCI. The terminal obtains the PDSCH time−frequency resource through the "fast scheduling information" in the first-stage DCI. The second-stage DCI is located in a predetermined position in the PDSCH resource. Terminal can find and decode the second-stage DCI within the PDSCH resource, and obtain "slow scheduling information" from the second-stage DCI. Compared with the first type of two-stage DCI structure, this structure further simplifies the reception of second-stage DCI, saves the terminal's power consumption in searching for second-stage DCI, and reduces the overhead of PDCCH due to the offloading of the second-stage DCI into PDSCH resource region. Moreover, the transmission efficiency of second-stage DCI can be improved in the second type of two-stage DCI structure because PDSCH uses a more flexible transmission format (for example, PDSCH can use various modulation orders, while PDCCH can usually use QPSK only). The second type of two-stage DCI can only be used for PDSCH scheduling and not for PUSCH scheduling, because the terminal needs to obtain all the scheduling information of PUSCH in advance to prepare and encode PUSCH, and it is impossible to place second-stage DCI in PUSCH resource region.

On the other hand, two-stage DCI is also found to have following shortcomings [23]:

- First, the contents of a DCI is divided into two parts that are encoded separately, which reduces the coding efficiency of DCI and causes performance loss of PDCCH detection.
- Second, the overall reliability of two-stage DCI may be reduced, because the terminal has to decode both parts of two-stage DCI correctly to obtain the whole scheduling information. As long as the decoding fails on any part of the two stages, the terminal needs to behave as if the whole DCI is incorrectly received.
- Third, the two-stage DCI has gains in scenarios with high latency requirements for PDCCH decoding (as shown in the example above), while single-stage DCI, as a mature and reliable method, must be supported in the NR specification (e.g., fallback DCI is still suitable for single-stage DCI). In other words, the two-stage DCI can only play a supplementary role in addition to the single-stage DCI, and cannot replace the single-stage DCI. The terminal needs to support two DCI structures at the same time, which increases the complexity of the terminal.
- Finally, two-stage DCI increases the DCI overhead, because the two parts of the two-stage DCI need to add Cyclic Redundancy Check (CRC), respectively.

The final decision in 3GPP was to adopt single-stage DCI, instead of two-stage DCI, as DCI structure in R15 NR specification. However, a two-stage sidelink control information structure, namely two-stage Sidelink control information (SCI), was adopted on the sidelink in R16 NR V2X specification. Details can be found in Chapter 17, 5G V2X.

5.4.4.2 Introduction of group-common DCI

Conventional DCI mainly focuses on the resource allocation of the individual terminal (including downlink, uplink, and sidelink), but there is also some information that is not directly related to terminal resource scheduling but needs to be dynamically indicated to the terminal. To transmit this kind of information in DCI, there are two alternative methods:

1. Option 1: Add a field in the DCI designed for scheduling the data channel and send the information to the terminal along with the scheduling information.
2. Option 2: Design a separate DCI format to specifically transmit this kind of information.

In the NR specification, both above options are used. A typical example of Option 1 is the BWP switching indication. As described in Section 4.3, this switching indication is implemented by inserting a BWPI in the scheduling DCI. DL BWP switching is triggered by BWPI in the downlink scheduling DCI (DCI format 1_1), and UL BWP switching is triggered by the BWPI in uplink scheduling DCI (DCI format 0_1). The advantage of this method is to avoid introducing a new DCI format. The number of PDCCH blind detection that the terminal can complete in a time unit is limited. The more DCI formats that need to be searched for, the shorter detection time budget that can be shared by each DCI format. Reusing scheduling DCI to transmit some physical layer commands can prevent the terminal from detecting too many DCI formats at the same time and save the limited budget of PDCCH blind detections of the terminal. On the other hand, the shortcomings of this method are also obvious. The scheduling DCI is sent only when there is data transmission between base station and terminal. When there is no PDSCH or PUSCH to be scheduled, there is no need to send the scheduling DCI. If a scheduling DCI is sent with the only intention to transmit a physical layer command without any data transmission scheduling, it is necessary to design an "empty scheduling" indication, such as setting the resource scheduling related fields (e.g., TDRA and FDRA) to specific values like zeros. In this case, most fields in DCI are meaningless, which wastes a lot of DCI capacity. Although the physical layer commands, such as BWP switching, are usually UE-specific and only affect the DCI overhead of one terminal, these physical layer commands can still be individually transmitted to many terminals in UE-specific scheduling DCI, in which case the waste of DCI capacity or DCI overhead remains serious.

Due to the issues with Option 1, Option 2 has been always in the study, with the focuses on common DCI for cell-specific or group-common information, and UE-specific DCI of small size to carry dedicated command. A typical example is PCFICH in the LTE specification, which is design to specifically inform the PDCCH control region duration of the entire cell. It is wasteful and unnecessary to send this information with UE-specific DCI. In the NR SI stage, there was a discussion whether it is necessary to retain a PCFICH-alike channel. If it is, a common DCI [similar to Control Format Indicator (CFI) in LTE] can be used to indicate the length of the control region holding UE-specific DCIs, as well as the slot format, reserved resource and configuration information for aperiodic CSI−RS, etc. Therefore the 5G NR specification introduces Group-common DCI, which is used together with UE-specific DCI. The studies showed that the UE-specific DCI is more suitable for reserved resource (introduced in Section 5.7) and Aperiodic CSI−RS, while the group-common DCI is more suitable for common information like SFI. Meanwhile, although downlink preemption indication, uplink cancellation indication, and TPC command are UE-specific information, they are decided to be transmitted in the group-common DCI due to their small payload sizes. The TPC commands applied to PUSCH/PUCCH can be transmitted in either the group-common DCI (DCI format 2_2), or the two UE-specific DCIs (DCI format 0_1 and DCI format 1_1). This is one example of using both UE-specific DCI and group-common DCI to deliver the UE-specific control information of small payload size. The base station can flexibly choose any kind of DCI according to different application scenarios. For example, when there is uplink and downlink data scheduling, it can send TPC through DCI format 0_1 and DCI format 1_1. TPC command can also be sent through DCI format 2_2 when there is no uplink and downlink data scheduling.

Although group-common DCI is received by a group of UEs, it can be actually a "common container" that carries multiple UE-specific control information and is composed of multiple UE-specific information blocks. Take DCI format 2_2 as an example. The terminals configured with TPC−PUSCH−RNTI or TPC−PUCCH−RNTI can open this "public container," and find their own TPC commands from this "public container." As shown in Fig. 5.47, a DCI of format 2_2 carries four blocks of TPC commands for 4 terminals. Each terminal is configured to fetch the TPC command from a corresponding block. When receiving DCI format 2_2, a terminal reads the TPC command in the

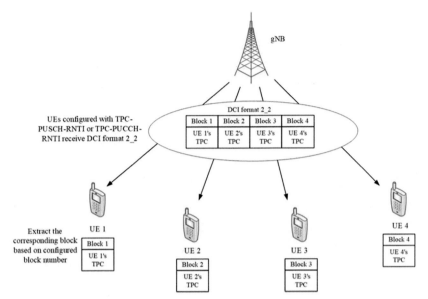

Figure 5.47 Transmission of UE-specific control information through group-common DCI.

corresponding block based on configured block number and ignores the information in other blocks.

5.5 Design of NR PUCCH

5.5.1 Introduction of short-PUCCH and long-PUCCH

NR system adopts flexible resource allocation methods, including: flexible HARQ feedback timing, flexible TDD uplink-to-downlink configuration, flexible physical resource allocation in the time and frequency domain, etc. In addition, NR system supports multiple types of services with different requirements for latency and reliability. To match these flexibilities, the following requirements were taken into account for NR PUCCH design [24,25]:

- High reliability

    For FR1 (Frequency Range 1, up to 6 GHz but later extended to 7.125 GHz), NR PUCCH should have comparable coverage with LTE PUCCH. LTE PUCCH format 1/1a/1b occupies one PRB in the frequency domain and 14 OFDM symbols in the time domain, and uses CAZAC sequences in the code domain, resulting in high reliability. NR PUCCH needs to consume additional time-domain symbols to achieve similar coverage.

For FR2 (Frequency Range 2, between 24.25 and 52.6 GHz), there is no LTE coverage reference for comparison. However, due to the transmission characteristics in FR2 (larger propagation/penetration losses, larger phase-noise, and lower power spectral density, etc.), NR PUCCH in FR2 also needs to rely on additional symbols to provide sufficient coverage.

- High efficiency

   When the terminal performance is not limited by uplink coverage on PUCCH, less time-domain resources will be used to transmit PUCCH to improve the system efficiency, by reducing the amount of physical resources occupied by a PUCCH and using the time-domain resource fragments. In addition, for URLLC service, the duration of a physical channel should not be long in time, or it may not be able to satisfy the latency requirement.

- High flexibility

In LTE system, ACK/NACK multiplexing is used only for TDD system or FDD system with carrier aggregation. From the system perspective, the ACK/NACK information corresponding to multiple PDSCHs is multiplexed into one PUCCH, which leads to higher uplink transmission efficiency. In NR systems, due to the introduction of flexible HARQ feedback timing, ACK/NACK multiplexing can occur for both TDD serving cell and FDD serving cell. Furthermore, the ACK/NACK payloads transmitted in different uplink slots may vary widely. Adjusting the duration of PUCCH according to the actual payload is also beneficial to improve the system efficiency.

In order to achieve high reliability, high flexibility, and high efficiency, it was agreed on in the RAN1#86bis meeting that NR system supports two types of PUCCH, namely long PUCCH and short PUCCH.

5.5.2 Design of short-PUCCH

For the design of short-PUCCH, the PUCCH duration will be determined first. Two options were discussed at the RAN1 #AH_NR meeting in January 2017:

1. Only 1-symbol PUCCH is supported.
2. PUCCH with more than one symbol is additionally supported.

If short PUCCH only supports 1-symbol length, the main problem is that the performance gap of coverage between short-PUCCH and long-PUCCH is large and the system efficiency is low. It was then determined that the duration of NR short-PUCCH can be configured as 1 symbol or 2 symbols. With the duration of short-PUCCH determined, the following approaches for short-PUCCH structure were proposed in the 3GPP RAN1#88 meeting [26−29] (Fig. 5.48):

- Approach 1: UCI and RS are FDM-multiplexed in each OFDM symbol.
- Approach 2: UCI and RS are TDM-multiplexed in different OFDM symbols.
- Approach 3: RS and UCI are FDM-multiplexed in one symbol and only UCI is carried in another symbol with no RS.
- Approach 4: Sequence-based design without RS, only for UCI with small payload size.
- Approach 5: Sequence-based design with CDM-multiplexing between RS and UCI only with small payload size.
- Approach 6: Pre-DFT multiplexing in one or both symbol(s), as shown in Fig. 5.49.

Approach 1 is the most flexible one among the above approaches. It can be used for both 1-symbol PUCCH and 2-symbol PUCCH. By adjusting the frequency-domain resource allocation, different number of UCI bits can be transmitted and various coding rates can be achieved. However, Approach 1 is applicable to CP-OFDM and exhibits high Peak to Average Power Ratio (PARR). Approaches 2 and 3 can be considered as extensions of Approach 1, but fixed to 2-symbol length. Approach 2 can achieve single-carrier property with the same RS overhead (50% as in Approach 1), while Approach 3 can own lower RS overhead. Front-loaded RS is used in Approach 2 and Approach 3 to reduce the

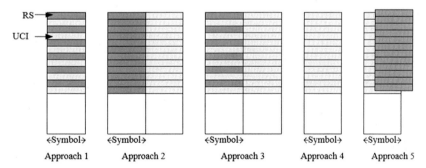

Figure 5.48 Approaches for short-PUCCH structure.

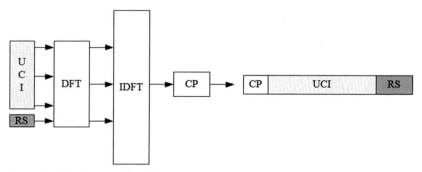

Figure 5.49 Pre-DFT multiplexing.

demodulation latency. But in fact, given the duration of 2-symbol PUCCH is already very short, the latency gain caused by the front-loaded RS is negligible. In addition, the structure of Approach 2 and Approach 3 cannot be applied to 1-symbol PUCCH, which leaves multiple structures being defined for different PUCCH lengths and therefore leads to additional complexity of specification. Approach 4 and Approach 5 use sequences to achieve lower PAPR and multiplexing capability, but the disadvantage is that the UCI payload size is limited. Based on the detection performance [29], either Approach 1 or Approach 4 has the best performance in different individual scenario. Approach 6 has the comparable performance with Approach 1, but needs more complex implementation.

The RAN1 discussion on short-PUCCH structure was then split into two focuses on 1-symbol PUCCH design and 2-symbol PUCCH design. The first conclusion (made in the RAN1#88bis meeting) was that 1-symbol PUCCH with more than 2 UCI bits adopts the structure of Approach 1. Later, it was agreed for 1-symbol PUCCH that the DMRS overhead is 1/3 and the PRB allocation is configurable. In the RAN1 #90 meeting, it was further agreed that the Approach 4 structure with Zadoff-chu (ZC) sequence of 12 cyclic shifts is used for the 1-symbol PUCCH with up to 2 UCI bits. For the design of 2-symbol PUCCH, the following two design principles were considered:

1. Option 1: 2-symbol PUCCH is composed of two 1-symbol PUCCHs conveying the same UCI.
   a. Option 1-1: Same UCI is repeated across the symbols using repetition of a 1-symbol PUCCH.
   b. Option 1-2: UCI is encoded and the encoded UCI bits are distributed across the symbols.

Table 5.10 Mapping of values for one HARQ—ACK information bit to sequences for PUCCH format 0.

ACK/NACK	NACK	ACK
Sequence cyclic shift	$m_{CS} = 0$	$m_{CS} = 6$

Table 5.11 Mapping of values for two HARQ—ACK information bits to sequences for PUCCH format 0.

ACK/NACK	NACK, NACK	NACK, ACK	ACK, ACK	ACK, NACK
Sequence cyclic shift	$m_{CS} = 0$	$m_{CS} = 3$	$m_{CS} = 6$	$m_{CS} = 9$

2. Option 2: 2-symbol PUCCH is composed of two symbols conveying different UCIs. Time-sensitive UCI (e.g., HARQ—ACK) is in the second symbol to get more time for preparation.

Because Option 2 is equivalent to transmission of two 1-symbol PUCCHs independently, it is not necessary to define Option 2 as a single PUCCH format. Finally, Option 1-1 was agreed upon in RAN1 #89 meeting for 2-symbol PUCCH with up to 2 UCI bits (i.e., the ZC sequences conveying the same UCI are transmitted on two symbols). For 2-symbol PUCCH with more than 2 UCI bits, the DMRS structure of 1-symbol PUCCH is used for each symbol. UCI information is encoded and mapped to two symbols for transmission (RAN1#AH3, September 2017).

In the 3GPP specification, short-PUCCH with up to 2 UCI payload bits is named PUCCH format 0, and short-PUCCH with more than 2 UCI bits is named PUCCH format 2. The mappings for ACK/NACK information to sequence cyclic shits for PUCCH format 0 is shown in Tables 5.10 and 5.11. About 1—16 PRBs can be configured for PUCCH format 2. The structure of PUCCH format 2 is shown in Fig. 5.50.

5.5.3 Design of long-PUCCH

As described in Section 5.5.1, long-PUCCH is supported to meet the coverage requirement of uplink control channel. Therefore at the very beginning of the discussion on the design of long-PUCCH it was agreed that the PAPR/CM of long-PUCCH will be low. Furthermore, in order to obtain sufficient energy in the time domain, a long-PUCCH can be transmitted repeatedly over multiple slots. In contrast, a short-PUCCH cannot be transmitted over multiple slots. Additionally, in order to transmit the UCI with different payload sizes and to meet different coverage

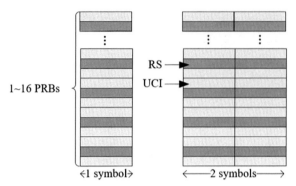

Figure 5.50 The structure of PUCCH format 2.

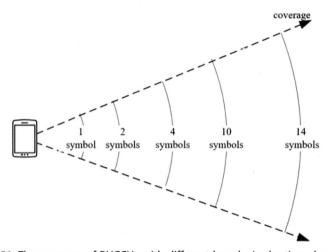

Figure 5.51 The coverage of PUCCHs with different lengths in the time domain.

needs, long-PUCCH supports various lengths in the time domain, which can improve the uplink spectrum efficiency, as shown in Fig. 5.51. In the RAN1 #88bis meeting, it was agreed on that a long-PUCCH may have variable number of symbols, within a range of 4–14 symbols, in a given slot. The long-PUCCH structure that is scalable for different number of symbols can avoid the introduction of multiple PUCCH formats.

In order to realize the design target on low PAPR for long-PUCCH, the following two designs can be straightforward:

1. Reuse the structure of LTE PUCCH format 1a/1b, where ZC sequences are used in the frequency domain, OCC sequences are used in the time domain, and DMRS and UCI are mapped to different symbols.

2. Reuse the structure of LTE PUSCH, where DMRS and UCI are mapped to different symbols, and DFT–S–OFDM is adopted.

The first structure uses sequences to distinguish carried information, so it is only applicable to the UCI with small payload, but it can achieve superior detection performance and multiuser multiplexing. It is considered the optimal structure for the UCI of up to 2 payload bits.

On the other hand, the second structure relies on channel coding and therefore can carry more than 2 UCI bits. As mentioned earlier, the targets of PUCCH design include large coverage and large capacity. However, from the perspective of a UE, these two targets do not need to be reached always at the same time.

- For the power-limited scenario with medium UCI payload size, it is reasonable to reduce the number of resources in the frequency domain and to use more resources in the time domain. It becomes more beneficial in system efficiency improvement to be capable of supporting multiuser multiplexing.
- For the scenario with large UCI payload, it is necessary to use more physical resources (in both time domain and frequency domain) to transmit UCI.

NR system supports three long-PUCCH formats, namely PUCCH formats 1, 3, and 4, which are applicable to the different scenarios (Table 5.12).

Another important issue of channel structure design for long-PUCCH is DMRS pattern. For PUCCH format 1, the following two options were proposed during the RAN1 discussion (Fig. 5.52):

1. Option 1: Similar to LTE PUCCH format 1/1a/1b, DMRS occupies multiple consecutive symbols in the middle of the PUCCH.
2. Option 2: DMRS always occurs in every other symbol in the long-PUCCH.

With the same RS overhead, the performance of Option 1 and Option 2 is almost the same in the low-speed scenario. However, Option 2 outperforms Option 1 when the terminal moves in high speed. Therefore NR PUCCH format 1 adopts the distributed DMRS pattern, in which DMRS occupies the even-indexed symbols (symbol index starts from 0; i.e., a pattern of front-load DMRS). In the frequency hopping of PUCCH format 1, the symbols in two frequency hops are made as even as possible. When the number of symbols in a PUCCH is even, the number of symbols in the first frequency-hop is the same as that in the second frequency-hop; when the number of symbols in a PUCCH is odd, the second frequency-hop has one more symbol than the first frequency-hop.

Table 5.12 NR long-PUCCH formats.

	Length in OFDM symbols	Number of UCI bits	Number of PRBs	Multiuser multiplexing capacity
PUCCH format 1	4–14	1–2	1	ZC sequences of length 12 are used in the frequency domain; OCC sequences are used in the time domain, where the spreading factor is 2–7.
PUCCH format 3		>2	$2^\alpha \cdot 3^\beta \cdot 5^\gamma$, where α, β, γ is a set of nonnegative integers	Not supported.
PUCCH format 4		2	1	OCC sequences are used in frequency domain, where the spreading factor is 2 or 7.

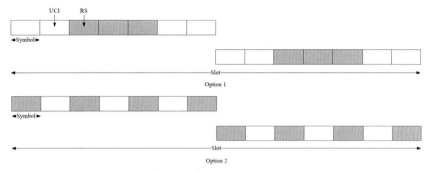

Figure 5.52 DMRS patterns of PUCCH format 1.

In the RAN1 #AH2 meeting, the following two DMRS patterns are proposed for PUCCH formats 3 and 4 that are transmitted with intraslot frequency hopping:

1. Option 1: One DMRS per frequency hop. The location of DMRS is around the middle of the frequency hop.
2. Option 2: One or two DMRS per frequency hop.

In general, the more RS resources lead to higher channel estimation accuracy but less physical resources used to transmit UCI followed by higher UCI coding rate. In order to obtain optimal detection performance, both PUCCH duration and UCI payload size need to be considered upon channel condition. Based on the discussion, it was determined that additional DMRS used by both PUSCH and PUCCH can be configured by high layer signaling. Without additional DMRS, one DMRS symbol is included in each frequency-hop of PUCCH formats 3 and 4. With additional DMRS, if the number of symbols in PUCCH is not more than 9, one DMRS symbol is included in each frequency-hop; otherwise, two DMRS symbols are included in each frequency-hop. Except for 4-symbol PUCCH, the DMRS symbol positions of PUCCH formats 3 and 4 with intraslot frequency hopping are also adopted for the PUCCH formats 3 and 4 without intraslot frequency hopping (Table 5.13).

5.5.4 PUCCH resource allocation

Before carrier aggregation is introduced into the LTE system, the ACK/NACK information corresponding to the PDSCH scheduled by a DCI is transmitted on a PUCCH format 1a/1b resource, which is determined based on the first CCE occupied by the DCI. After the introduction of carrier aggregation, ACK/NACK information corresponding to the

Table 5.13 DMRS symbol position patterns of PUCCH format 3/4.

PUCCH length	DMRS position (symbol index)			
	No additional DMRS		Additional DMRS	
	No hopping	Hopping	No hopping	Hopping
4	1	0, 2	1	0, 2
5	0, 3		0, 3	
6	1, 4		1, 4	
7	1, 4		1, 4	
8	1, 5		1, 5	
9	1, 6		1, 6	
10	2, 7		1, 3, 6, 8	
11	2, 7		1, 3, 6, 9	
12	2, 8		1, 4, 7, 10	
13	2, 9		1, 4, 7, 11	
14	3, 10		1, 5, 8, 12	

PDSCH scheduled by a DCI can be transmitted on a PUCCH format 3/4/5 resource that is determined based on both semistatic configuration and an indication in the DCI. In the early stage of NR design, it was determined to reuse LTE mechanism to indicate the PUCCH resource for ACK/NACK transmission. Specifically, a set of PUCCH resources are configured by high layer signaling firstly, then one PUCCH resource in the set is indicated by DCI.

As mentioned above, NR PUCCH can be flexibly configured. During the 3GPP discussion, the following options were proposed on how to configure the PUCCH resource set [30]:

- Option 1: K PUCCH resource sets are configured, and each PUCCH resource set is used to transmit UCI with a specific range of payload sizes. One PUCCH resource set can apply to multiple PUCCH formats. A UE selects one PUCCH resource set from K PUCCH resource sets based on the UCI payload size, and then picks one PUCCH resource from the selected set based on an indication in DCI (Fig. 5.53).
- Option 2: One or more PUCCH resource sets are configured for each PUCCH format.
  - Option 2-1: Multiple PUCCH resource sets are configured for each PUCCH format.
  - Option 2-2: Multiple PUCCH resource sets are divided into two groups. The first group is used to transmit up to 2 UCI bits, and the

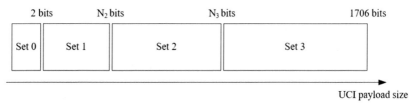

Figure 5.53 Option 1 for PUCCH resource set allocation.

second group is used to transmit more than 2 UCI bits. A MAC-CE indicates two resource sets, one in each group. One resource set from the two MAC-CE indicated resource sets is selected according to the number of UCI bits. Finally, a PUCCH resource is picked from the selected resource set according to an indication in DCI.

- Option 2-3: One PUCCH resource set is configured for one PUCCH format. The determination of PUCCH resource from the resource set reuses the implicit PUCCH resource indication scheme for LTE PUCCH format 1a/1b for short-PUCCH.
- Option 2-4: Two PUCCH resource sets, targeting to be used for short-PUCCHs and long-PUCCHs, respectively, can be configured to a UE. A threshold of UCI payload size (e.g., 100 bits) is used for the UE to select between the two resource sets.

The advantage of Option 1 is that the dynamic switching between long and short PUCCH formats can be supported, since both short-PUCCH and long-PUCCH can be configured in one resource set. Furthermore, the semi-static configuration of long-PUCCH and short-PUCCH allow the resources being configured uneven, which provides additional flexibility for the gNB scheduling. Assume 3-bit PUCCH resource indicator in the DCI. Option 1 can configure up to eight PUCCH resources in one resource set, where the numbers of resources used for long-PUCCHs and short-PUCCH can be flexibly configured by the gNB. For Option 2-1, one explicit bit in the PUCCH resource indicator is used to distinguish whether long-PUCCH or short-PUCCH is used, then the two remaining bits are used to indicate resource within a set, that is, the long-PUCCH set and the short-PUCCH set contain up to four PUCCH resources, respectively. In addition, Option 1 can offer more resource sets for a given range of UCI payload sizes, which can provide more scheduling flexibility. At the end, Option 1 was adopted for NR PUCCH resource allocation. Up to four PUCCH resource sets can be configured by high layer signaling, where resource set 0 is used for up to 2UCI bits, the payload ranges of resource sets 1 and 2 are configured by high layer signaling, and the maximum payload of resource set 3 is 1706, which comes from the restriction of Polar code.

In addition, some companies believe that in the practical system, a large number of UEs need to feedback 1-bit or 2-bit ACK/NACK information at the same time. If 2-bit PUCCH resource indicator in DCI is configured, up to four candidate PUCCH resources can be configured for the UCI of up to 2 bits, which can make the PUCCH resource collision a nonnegligible problem. Therefore it was recommended to consider using an implicit resource indication to extend the number of PUCCH candidates:

- Option 1: The PUCCH resource is determined based on the index of CCE used to transmit the DCI.
- Option 2: The PUCCH resource is determined based on the index of REG used to transmit the DCI.
- Option 3: The PUCCH resource is determined based on the TPC field in the DCI.
- Option 4: The PUCCH resource is determined based on the COREST or search space used to transmit the DCI.
- Option 5: The PUCCH resource is determined based on a 3-bit explicit indicator in the DCI.

In RAN1#92 meeting, it was agreed that a 3-bit explicit indicator in DCI is used to indicate the PUCCH resource. For PUCCH resource set 0, corresponding to the UCI with up to 2 bits, up to 32 PUCCH resources can be configured by the high layer signaling. When the number of configured PUCCH resources is no more than 8, the PUCCH resource is determined based on the 3-bit explicit indicator in the DCI. When the number of configured PUCCH resources is more than 8, the PUCCH resource is determined based on both the 3-bit explicit indicator in the DCI and the index of CCE used to transmit the DCI. The specific method is as follows:

$$
r_{\text{PUCCH}} = \begin{cases} \left\lfloor \dfrac{n_{\text{CCE},p} \cdot \lceil R_{\text{PUCCH}}/8 \rceil}{N_{\text{CCE},p}} \right\rfloor + \Delta_{\text{PRI}} \cdot \left\lceil \dfrac{R_{\text{PUCCH}}}{8} \right\rceil \\ \quad \text{if} \quad \Delta_{\text{PRI}} < R_{\text{PUCCH}} \bmod 8 \\[4pt] \left\lfloor \dfrac{n_{\text{CCE},p} \cdot \lfloor R_{\text{PUCCH}}/8 \rfloor}{N_{\text{CCE},p}} \right\rfloor + \Delta_{\text{PRI}} \cdot \left\lfloor \dfrac{R_{\text{PUCCH}}}{8} \right\rfloor + R_{\text{PUCCH}} \bmod 8 \\ \quad \text{if} \quad \Delta_{\text{PRI}} \geq R_{\text{PUCCH}} \bmod 8 \end{cases}
$$

(5.4)

where r_{PUCCH} is the index of a PUCCH resource, R_{PUCCH} is the total number of PUCCH resources in PUCCH resource set 0, $N_{\text{CCE}, p}$ is the number of CCEs in the COREST, $n_{CCE, p}$ is the index of the first CCE occupied by the DCI, and Δ_{PRI} is the value of the 3-bit indicator in the DCI.

For PUCCH resource sets 1, 2, and 3, corresponding to the UCI with more than 2 bits, up to eight PUCCH resources can be configured by the high layer signaling. The PUCCH resource is determined based on the 3-bit explicit indicator in the DCI, and the implicit indication is not supported.

5.5.5 PUCCH colliding with other UL channels

In LTE system, when multiple uplink channels need to be transmitted simultaneously in a subframe, UCI will be multiplexed in a physical channel to maintain single-carrier transmission. In addition, because an LTE physical channel occupies all available resources in the time domain, the overlapped channels are aligned in the time domain. Therefore when a PUCCH overlaps with other uplink channels, it is only necessary to determine a channel with a large capacity to transmit all of the uplink information from all the overlapped channels. For example, PUCCH format 2a/2b/3/4/5 is used to transmit CSI and ACK/NACK, and PUSCH is used to transmit UCI and uplink data.

In NR system, multiple uplink channels cannot be transmitted simultaneously by a terminal within a carrier. During 3GPP discussion, the conclusion for the overlapped channels with the same starting symbol was firstly agreed, that is, UCI and the other overlapped channel are multiplexed into one physical channel. For the case where the overlapped channels have different starting symbols, a new problem arises: whether the required processing time of each channel can be satisfied when the information from each overlapping channel is multiplexed into one channel. Figs. 5.54 and 5.55 show the examples of the processing time requirements corresponding to two overlapping channels.

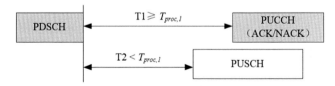

Figure 5.54 PDSCH processing procedure time.

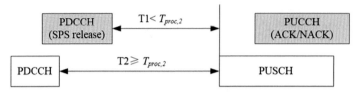

Figure 5.55 PUSCH preparation procedure time.

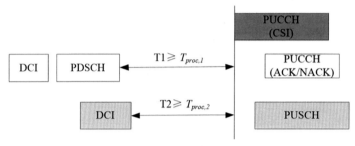

Figure 5.56 Time conditions for NR UCI multiplexing.

1. In Fig. 5.54, the gap between the ending of a PDSCH and the starting of a PUCCH with corresponding ACK/NACK is no less than the PDSCH processing procedure time, $T_{proc,1}$. The value of $T_{proc,1}$ is related to the UE capacity, as specified in Section 5.3 of TS 36.214. Meanwhile, the gap between the ending of the PDSCH and the starting of a PUSCH that overlaps with the concerned PUCCH is less than $T_{proc,1}$. If the ACK/NACK is required to be multiplexed in the PUSCH, the UE may not be able to provide a valid HARQ−ACK corresponding to the PDSCH.

2. In Fig. 5.55, the gap between the ending of a PDCCH with a UL grant and the starting of the corresponding PUSCH is no less than the PUSCH preparation procedure time, $T_{proc,2}$. The value of $T_{proc,2}$ is related to the UE capacity, as specified in Section 6.4 of TS 36.214. Meanwhile, the gap between the ending of a PDCCH with a DCI indicating SPS release and the starting of the concerned PUSCH is less than $T_{proc,2}$. This timeline exceeds the UE capability if the ACK/NACK corresponding to the DCI indicating SPS release is multiplexed in the PUSCH.

In order to solve the above processing time issues and to minimize the impact on the UE, a working assumption was approved for UCI multiplexing in the 3GPP RAN1 #92bis meeting. When multiple uplink channels are overlapped, all UCIs are multiplexed in one physical channel if the following timeline conditions are satisfied, as shown in Fig. 5.56:

1. If the first symbol of the earliest PUCCH(s)/PUSCH(s) among all the overlapping channels starts no earlier than symbol $T_{\text{proc},1} + X$ after the last symbol of PDSCH(s). Beyond the normal preparation time for an UCI transmission, additional processing time is needed for UCI multiplexing. Therefore the value of X is determined as 1.
2. If the first symbol of the earliest PUCCH(s)/PUSCH(s) among all the overlapping channels starts no earlier than symbol $T_{\text{proc},2} + Y$ after the last symbol of PDCCHs scheduling the overlapping channels. Similarly, the value of Y is determined as 1.

If multiple overlapping uplink channels do not satisfy the above timeline conditions, the UE behavior is not defined. In other word, 3GPP assumes it is gNB's responsibility to avoid the uplink channel overlapping if the above timeline conditions cannot be satisfied.

Another discussed issue was how to determine the set of overlapping channels. The determination of the set of overlapping channels directly affects the number of UCI to be multiplexed as well as the multiplexing mechanism itself. One example is shown in Fig. 5.57 with four uplink channels scheduled in one slot, where the first channel does not overlap with the fourth channel, and the third channel overlaps with both the first channel and the fourth channel. In determining the channels overlapping with the first channel, if the fourth channel is not counted as overlapping channel, the second round of uplink channel collision may occur. For example, if the information carried by the first three channels is multiplexed in the third channel, it collides with the fourth channel again.

In order to systematically solve the above problem, 3GPP adopted the following method to determine the multiplexing of overlapped uplink channels in a slot:

- First, determine a set of nonoverlapping PUCCH resources in the slot based on all the given PUCCH resources in the slot (provided as

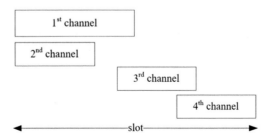

Figure 5.57 Uplink channel collision.

inputs to this method). The determination is done with following steps:

- Step 1: Put all given PUCCH resources in the slot into a set, namely set Q;
- Step 2: Identify a PUCCH resource A, which has the earliest starting symbol among all the PUCCH resources in set Q that overlaps with at least one of other PUCCH resource in set Q. If more than one PUCCH resource meets this condition, the longest PUCCH resource is selected. If multiple PUCCH resources meet this condition and have the same duration, any one of them can be selected as PUCCH resource A;
- Step 3: All PUCCH resources that are in set Q and meanwhile overlapping with PUCCH resource A are also recorded into another set, namely set X;
- Step 4: A PUCCH resource B and its corresponding multiplexed UCI are determined based on the PUCCH resource A and all the PUCCH resources in the set X. Then in set Q, use PUCCH resource B and the corresponding UCI to replace the PUCCH resource A and all the PUCCH resources that are recorded in the set X;
- Step 5: After the replacement in Step 4, if there are at least two PUCCH resources in set Q that overlap with each other, reset set X to null and repeat steps 2–4, until there is no PUCCH resource overlapping in set Q.
- Second, for a PUCCH resource in the set Q, if it overlaps with PUSCH, its corresponding UCI is multiplexed in the PUSCH for transmission; otherwise, the corresponding UCI is transmitted in PUCCH on the PUCCH resource.

5.6 Flexible TDD

5.6.1 Flexible slot

At the beginning of the design of NR system, the targets to having better performance than the LTE system in the following aspects: data rate, spectrum utilization, time latency, connection density, power consumption, etc. In addition, the NR system needs to have good forward compatibility and can support the introduction of other enhanced technologies or new access technologies in the future. Therefore the concept of self-contained slot and flexible slot are introduced in the NR system.

The so-called self-contained slot means that scheduling information, data transmission, and feedback information (if any) corresponding to the data transmission are all transmitted in one slot so as to reduce time latency. The typical self-contained slot format is mainly divided into two types: downlink self-contained slot and uplink self-contained slot, as shown in the figure below.

As shown in Fig. 5.58A, a PDCCH sent by the network schedules a PDSCH in a slot, and the HARQ−ACK information for this PDSCH is fed back through a PUCCH. The PDCCH, PDSCH and PUCCH are in the same slot (i.e., data scheduling, transmission and feedback are completed in one single slot). In Fig. 5.58B, the scheduling PDCCH transmitted by the network and the scheduled PUSCH transmitted by the terminal are in the same slot.

In the downlink self-contained slot, the terminal is required to complete HARQ−ACK feedback in the same slot after receiving the PDSCH. In the LTE system, the terminal receives the PDSCH in the subframe n and sends the HARQ−ACK feedback information in the subframe $n + 4$. Therefore the NR system has higher requirements on the processing capability of the terminal than the LTE system.

The self-contained slot requires that both downlink symbols and uplink symbols are included in the slot. Due to the time-varying needs of the different types of self-contained slot, the numbers of downlink symbols and uplink symbols in the slot are also time-varying. Thus one prerequisite of self-contained slot is the flexible slot format.

The NR system introduces the flexible slot format, which defines a slot to include downlink symbols, flexible symbols, and uplink symbols.

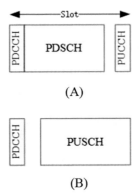

(A)

(B)

Figure 5.58 (A) DL self-contained slot. (B) UL self-contained slot.

Among the three kinds of symbols, the flexible symbol has the following characteristics:

1. The flexible symbol indicates the undecided direction of the symbol, which can be changed to a downlink symbol or an uplink symbol through other signaling.
2. The flexible symbols can also be marked as symbols reserved for future use, as a mean to support forward compatibility.
3. The flexible symbol can be used for the terminal's transceiver switching gap, similar to the guard period symbol in the LTE TDD system, and the terminal completes the transceiver switching within the flexible symbol.

In the NR system, a variety of flexible slot formats are defined, including full downlink slot, full uplink slot, full flexible slot, and slot formats with different numbers of downlink symbols, uplink symbols and flexible symbols. One slot format corresponds to one slot format index and contains one or two "DFU" patterns, where each "DFU" pattern has a number of downlink symbols (including the case of no downlink symbol) at the beginning of the pattern, a number of uplink symbols (including the case of no uplink symbol) at the end of the pattern, and a number of flexible symbols (including the case of no flexible symbol if the slot is full downlink slot or full uplink slot) after the downlink symbols (if any) and prior to the uplink symbols (if any) in the pattern. Fig. 5.59 shows two slot format examples, respectively, with one and two "DFU" patterns. As of R16, NR TDD does not support a slot format with a downlink symbol immediately followed by an uplink symbol. In addition, NR TDD supports a special slot format indication (slot format index is 255), which tells the terminal that the slot format determination for the corresponding slot is postponed to the moment when the terminal receives any downlink/uplink scheduling DCI for the slot.

5.6.2 Semistatic uplink−downlink configuration

The NR system supports multiple methods to configure the slot format, including: the slot format configuration through semistatic uplink−downlink configuration signaling and the slot format configuration through dynamic

Figure 5.59 Slot format examples with one or two "DFU" patterns.

uplink—downlink indication signaling. The semistatic uplink—downlink configuration signaling includes *tdd-UL-DL-ConfigurationCommon* and *tdd-UL-DL-ConfigurationDedicated*, and the dynamic uplink—downlink indication signaling is the group-common DCI format 2-0. This section introduces the semistatic uplink—downlink configuration method, and the next section introduces the dynamic uplink—downlink configuration method.

The network sends *tdd-UL-DL-ConfigurationCommon* signaling, which configures a common slot format applicable to all terminals in the cell. The signaling can configure one or two slot format patterns, with each pattern applied within a period. In each pattern, the configuration of the slot format mainly includes the following parameters: a reference SCS (μ_{ref}), a period (P, the period of the pattern in unit of ms), a number of downlink slots (d_{slots}), a number of downlink symbols (d_{sym}), a number of uplink slots (u_{slots}), and a number of uplink symbols (u_{sym}).

According to the reference SCS and period, the total number of slots S included in the period can be determined. The first d_{slots} of the S slots are the full downlink slots, and the first d_{sym} symbols in the ($d_{slots}+1$)th slot are downlink symbols; the last u_{slots} slots in the S slots are full uplink slots, and the last u_{sym} symbols in the ($u_{slots}+1$)th to last slot are uplink symbols; the remaining symbols in this period of S slots are flexible symbols. Therefore in a pattern period of S slots, the overall frame structure is that the downlink portion (if any) starts from the beginning of the period, the uplink portion (if any) is at the end of the period, and the flexible slots and symbols are in-between. The determined slot format is periodically repeated in the time domain.

Fig. 5.60 shows an example of a slot format configuration. The period of the pattern is $P=5$ ms. For the 15 kHz SCS, the pattern period includes five slots. The other parameters in *tdd-UL-DL-Configuration-Common* give $d_{slots}=1$, $d_{sym}=2$, $u_{slot}=1$, and $u_{sym}=6$. This configuration tells that in a period of 5 ms, the first slot is a full downlink slot, the

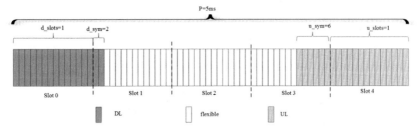

Figure 5.60 TDD uplink and downlink configuration pattern 1.

first two symbols in the second slot are downlink symbols, and the last slot is the full uplink slot, the last six symbols in the second-to-last slot are uplink symbols, and the remaining symbols are flexible symbols. This pattern repeats in the time domain.

The network can configure two different patterns at the same time through *tdd-UL-DL-ConfigurationCommon* signaling, with two patterns having different periods, *P* and *P2*, as well as two independent sets of slot format parameters. When the network is configured with two patterns, the two patterns are concatenated and repeated together periodically in every $(P + P2)$ of time, thereby the slot format can be indicated for all slots. The total period $(P + P2)$ of the two patterns needs to be able to divide 20 ms to ensure the single period of combined pattern does not cross 20 ms boundary.

In addition, the network can further configure a different slot format for a specific terminal through the UE-specific RRC signaling, namely *tdd-UL-DL-ConfigurationDedicated*. This signaling is intended to configure the use of flexible symbols in a group of slots in the period configured by *tdd-UL-DL-ConfigurationCommon*, and it mainly includes the following parameters for each slot:

- Slot index: This parameter is used to identify a slot in the period configured by *tdd-UL-DL-ConfigurationCommon*, for which the next parameter of "Symbol direction" applies.
- Symbol direction: This parameter is used to configure a group of symbols in the identified slot. This parameter can be configured to indicate a full downlink slot, a full uplink slot, or the number of downlink symbols and the number of uplink symbols in a slot.

The *tdd-UL-DL-ConfigurationDedicated* signaling can only change the directions of flexible symbols configured by *tdd-UL-DL-ConfigurationCommon*. The symbol already configured as a downlink symbol (or uplink symbol) by the *tdd-UL-DL-ConfigurationCommon* signaling cannot be changed to an uplink symbol (or downlink symbol) or a flexible symbol through *tdd-UL-DL-ConfigurationDedicated* signaling.

For example, based on the slot format pattern configured by the network through *tdd-UL-DL-ConfigurationCommon* configuration signaling as shown in Fig. 5.60, the network additionally configures two slots, Slot 1 and Slot 2, through *tdd-UL-DL-ConfigurationDedicated* signaling. The new slot format pattern is shown in Fig. 5.61. It still has a 5 ms period.

1. In Slot 1, *tdd-UL-DL-ConfigurationDedicated* signaling converts two flexible symbols to downlink symbols and four flexible symbols to uplink symbols.

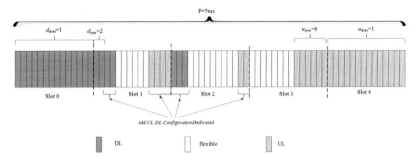

Figure 5.61 TDD uplink and downlink configuration pattern 2.

2. In Slot 2, *tdd-UL-DL-ConfigurationDedicated* signaling converts three flexible symbols to downlink symbols and two flexible symbols to uplink symbols.

5.6.3 Dynamic slot format indicator

In addition to semistatic uplink—downlink configuration signaling, the network can also dynamically indicate the slot format of each slot through SFI. The SFI is carried in DCI format 2_0, with DCI CRC scrambled with SFI—RNTI. The dynamic SFI signaling can only change the direction of a symbol that is left as a flexible symbol by the semistatic uplink—downlink configuration signaling, and cannot change the direction of a symbol that is configured as uplink or downlink by the semistatic configuration signaling.

The dynamic SFI can indicate the slot formats of multiple serving cells at the same time. The network configures the cell index and the associated position in DCI format 2_0 for a SFI indication field. The network also configures multiple *slotFormatCombination*, where each *slotFormatCombination* corresponds to a *slotFormatCombinationId* and contains a set of slot format configurations. Each slot format configuration is used to configure the slot format of a slot.

The SFI indication field includes an SFI index, which corresponds to the *slotFormatCombinationId*. Thus a set of slot formats can be determined according to the index. The slot format indicated by the SFI is applicable to multiple consecutive slots starting from the slot carrying the SFI signaling, and the number of slots indicated by the SFI is greater than or equal to the monitoring period of the PDCCH carrying the SFI. If a slot is indicated by two SFI signals, the slot format indicated by the two SFI signaling should be the same.

When the network configures the slot format of a serving cell, the slot format is interpreted in terms of a configured reference SCS, μ_{SFI}, which is less than or equal to the SCS μ of the serving cell (i.e., $\mu \geq \mu_{SFI}$). Then the slot format of an indicated slot in SFI is applicable to $2^{(\mu - \mu_{SFI})}$ consecutive slots, and each downlink, uplink, or flexible symbol indicated by SFI signaling corresponds to $2^{(\mu - \mu_{SFI})}$ consecutive downlink, uplink, or flexible symbols.

Take the slot indication shown in Fig. 5.62 for an example. The network configures a slot format that includes four downlink symbols, seven flexible symbols, and three uplink symbols, and configures this slot format to associate with a reference SCS of 15 kHz (i.e., $\mu_{SFI} = 0$). This SFI is then used to indicate the slot format of a TDD cell, which runs on a 30 kHz SCS (i.e., $\mu = 1$). Consequently, the slot format indicated by the SFI is applicable to two consecutive slots, and each downlink, flexible or uplink symbol indicated by the SFI, respectively, corresponds to two consecutive downlink, flexible, or uplink symbols.

5.7 PDSCH rate matching

5.7.1 Considerations for introducing rate matching

Rate matching is an existing technology in the LTE system. When there is a difference between the amount of allocated resources and the amount of resources required for data transmission, the data transmission can be

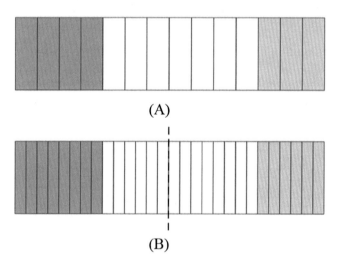

(A)

(B)

Figure 5.62 Dynamic slot indication.

adapted to the allocated resources through an adjustment of the coding rate. This resource amount difference is usually caused by some unusable "resource fragments" within time—frequency resources allocated to the terminal. Generally, the "resource fragments" can be reference signals, synchronization signals, etc., and rate matching is to avoid using these resources. Rate matching is applicable to a wider range of applications in the 5G NR specification, including: PDCCH and PDSCH multiplexing, resource reservation, etc. This section focuses on these application scenarios. In general, rate matching is used in the following two cases:

1. The first case is that when the base station allocates resources for a data channel (such as PDSCH), it cannot accurately determine which resources within the allocated resource region are unusable and should be excluded from being used for data channel. Therefore only a resource region with a roughly estimated resource amount can be allocated to the terminal. When the exact location and size of the unusable resources are later determined, the terminal excludes these resources during resource mapping. Through rate matching, the data transmission is adjusted to remain inside the allocated resources but meanwhile outside of the unusable resources, as shown in Fig. 5.63.

2. The second case is that when the base station allocates resources for PDSCH and PUSCH, it already knows which resources within the allocated resource region are unusable and should have been able to avoid allocating those resources. However, in order to control the

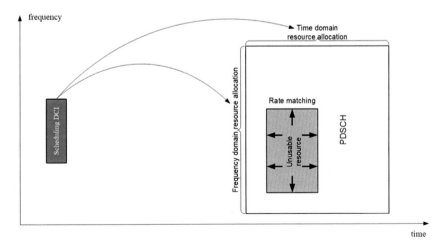

Figure 5.63 The basic principle of rate matching.

overhead of resource allocation signaling, only one rectangle time-—frequency region can be allocated. In the frequency domain, bitmap-based noncontiguous allocation (i.e., Type 0 frequency-domain resource allocation) is supported. In the time domain, only "starting position + length" indication is supported to indicate a contiguous time interval, while the noncontiguous TDRA is not supported. If the resource to be avoided appears in the middle of the time-domain span of the allocated rectangle resource region, it is impossible to "exclude" this resource through resource allocation signaling. During the study of NR resource allocation, there were also suggestions to support noncontiguous TDRA (e.g., based on bitmap signaling), but they were not accepted into specification. Then in this second case, rate matching is a tool for the terminal to bypass the unusable resources.

5.7.1.1 Rate matching in PDCCH/PDSCH multiplexing

As mentioned in Section 5.4.1, one of the big differences between NR PDCCH and LTE PDCCH is that the multiplexing between PDCCH and PDSCH can be not just TDM but also FDM. This can help to fully use of time—frequency resources and reduce resource fragments that cannot be eventually utilized. However, this also causes a more complicated issue in multiplexing between PDCCH and PDSCH. The CORESET can be surrounded by PDSCH. If the rate-matching technique is used to exclude the resources occupied by CORESET from the resources allocated for PDSCH, the terminal decoding the PDSCH needs to know the resources occupied by CORESET. This includes two cases:

1. Case 1: The CORESET embedded in the PDSCH resource is configured for the terminal decoding the PDSCH.
2. Case 2: The CORESET embedded in the PDSCH resource is configured for other terminals, but not the one decoding the PDSCH.

Case 1 is a relatively simple case. As shown in Fig. 5.64A, only the CORESET configured for Terminal 1 exists in the PDSCH resource allocated to Terminal 1, and the time—frequency region of this CORESET is known by Terminal 1 through CORESET configuration. No additional signaling is required for Terminal 1 to learn the exact resource used for PDSCH. What is needed is to define in the specification a rate-matching rule, such as "PDSCH resource that overlaps with the CORESET configured for the terminal is not used for PDSCH transmission." The terminal needs to

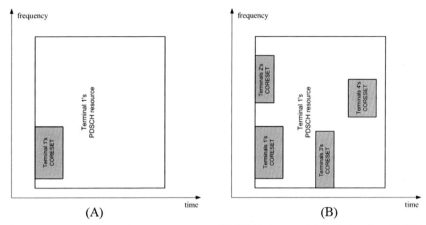

Figure 5.64 (A) Only the same user's CORESET is multiplexed in the PDSCH. (B) Other users' CORESETs are also multiplexed in the PDCCH.

follow this rule when performing rate-matching around the time–frequency region of CORESET.

Case 2 is more complicated. As shown in Fig. 5.64B, there are CORESETs allocated to Terminals 2, 3, and 4 in the PDSCH resource region allocated to Terminal 1. The time–frequency regions of these CORESETs are unknown to Terminal 1, unless to be provided to Terminal 1 through additional signaling.

5.7.1.2 Rate matching in resource reservation

Resource reservation is a new requirement introduced in 5G NR. The specification of a mobile communication system needs to satisfy backward compatibility within each specification release, to ensure that an early-release terminal can connect to the new-release system so that users who have purchased early-release terminals are not mandated to buy a new terminal upon a system upgrade to the new release. Nevertheless, this requirement brings a disadvantage, which is the restriction to the technique evolution in the life cycle of a specification. This is because the technique evolution has to be made invisible to the early-release terminals. In the early stage of 5G NR standardization, the design requirement of forward compatibility was taken into account to allow any technology evolutions in the future. This is mainly due to the fact that 5G systems may have to provide service for various vertical industrial businesses. Many businesses are difficult to predict today, and the 5G terminals manufactured upon early NR release may not be able to identify new technologies adopted in the future. At the time these new technologies are put into practice, the existing 5G terminals must be able to access the system as normal without recognizing these new technologies (i.e., to achieve the effect of forward compatibility). One way to achieve this goal is to reserve resources for the "unknown future service or technology." When the resource allocated to the terminal overlaps a reserved resource, the terminal can skip this resource through rate matching. In this way, even if the early-release terminal does not recognize the signals for new services or technologies transmitted that are transmitted in the reserved resource, its operation is not affected, nor does its operation affect the signals transmitted in the reserved resource (Fig. 5.65).

5.7.2 Rate-matching design

It can be seen in Section 5.7.1 that in order to support PDCCH/PDSCH multiplexing and resource reservation, a new set of signaling needs to be introduced to notify the terminal of the resource region that needs to be rate matched. From the perspective of the signaling type, it should mainly rely on semistatic RRC configuration, supplemented by a small size of DCI indication. A large size of DCI indication for rate-matching purpose will cause significant DCI overhead and is not suitable for rate matching, which is only an auxiliary enhancement.

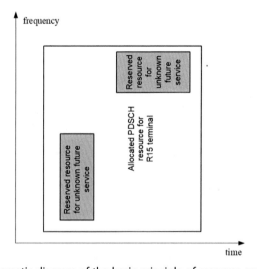

Figure 5.65 Schematic diagram of the basic principle of resource reservation.

In the frequency domain, the reserved resource may appear in any position, and can be contiguous or noncontiguous. Such reserved resources can be indicated by a bitmap similar to Type 0 resource allocation of data channel. For frequency-domain granularity, if only PDCCH/PDSCH multiplexing is considered, it seems good enough to use the CORESET granularity of six RBs as the bitmap granularity. However, if the requirement for resource reservation is also considered, given it is hard to determine the resource granularity of "unknown new services/technologies" that may appear in the forward compatibility scenario, it is better to adopt a universal granularity, which is one RB. If a unified resource indication method is to be used to indicate rate-matching resource for various rate-matching application scenarios, RB-level bitmap should be a safe bet in the frequency domain.

In the time domain, it can be seen in Sections 5.2 and 5.4 that both the NR data channel and control channel adopt a symbol-level "floating" structure that can appear at any time-domain position, and there may be multiple noncontiguous reserved resources. Therefore the "start position + length" indication method is not sufficient. It must be indicated by a symbol-level bitmap like the one in search-space set configuration.

The combination of this pair of bitmaps constructs one or more rectangle resource blocks arranged in a matrix in the time—frequency domain, which can be considered a matrix-shaped rate-matching pattern. None of the resource blocks on the matrix grid entry can be used for PDSCH transmission. However, a single rate-matching pattern may bring up the problem of "overreservation." As shown in Fig. 5.66, there are three resource blocks that need to be reserved within a resource region (as shown in the *left* figure), according to the size and location of these resource blocks, a two-dimensional rate-matching pattern based on the time-domain bitmap (00110110011100) and frequency-domain bitmap (0111011000001100) is shown in the right-hand figure. There are nine resource blocks in the rate-matching pattern, among which six blocks do not contain the reserved resources but are still marked as reserved. This overreservation makes many resources underutilized.

What Fig. 5.66 shows is the rate-matching pattern defined for 1 slot. The NR specification also supports a rate-matching pattern with duration of two slots, for which a symbol-level bitmap with length covering two slots in the time domain. In addition, because it is usually unnecessary to reserve resources in every slot, it is allowed to configure a slot-level bitmap to indicate the slots to which the rate-matching pattern applies. As

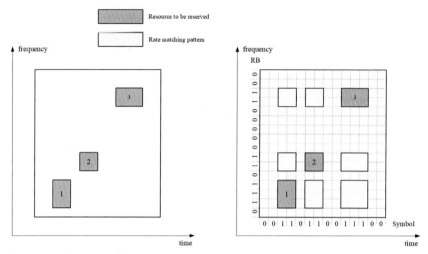

Figure 5.66 Rate-matching pattern causing overreservation.

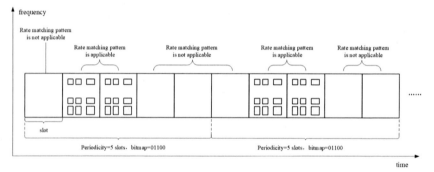

Figure 5.67 Use slot-level bitmap to configure which slots use rate-matching pattern.

shown in Fig. 5.67, a rate-matching pattern with a duration of 1 slot is used; meanwhile, the pattern only applies to the second and third slots in every five slots. Therefore in this example the slot-level bitmap should be configured as 01100, and only 40% of the slots can reserve resources according to the rate-matching pattern. The slot-level bitmap can consist of up to 40 bits, corresponding to the maximum indication period of 40 ms.

In order to solve the problem of "overreservation" to a certain extent, multiple rate-matching patterns can be combined. In the early example in Fig. 5.66, two 2D rate-matching patterns can be configured and combine together. As shown in Fig. 5.68, the rate-matching pattern 1 is configured to cover reserved resources 1 and 2, with time-domain bitmap = 001101100 00000 and frequency-domain bitmap = 0111011000000000; the rate-matching pattern 2 is configured to cover reserved resource 3 only, with time-domain bitmap = 0000000000011100 and frequency-domain bitmap = 0000000000001100. A rate-matching pattern group is then configured to contain a union of these two patterns. It can be seen that the pattern group configures only five resource blocks for reservation. Although there are still

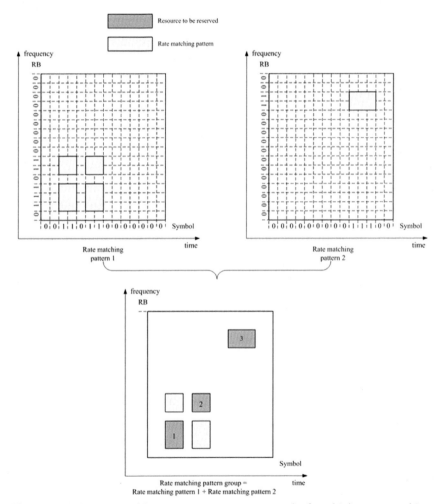

Figure 5.68 A rate-matching pattern group composed of multiple rate-matching patterns.

two resource blocks that are "overreserved," the resource waste becomes much smaller compared to using one pattern. To avoid wasting resources at all, the pattern group should include three patterns that are configured to exactly cover the reserved resources 1, 2, and 3, respectively.

3GPP finally determined that for a terminal, a maximum of four rate-matching patterns can be configured in each BWP and a maximum of four cell-level rate-matching patterns can be configured. In addition, two rate-matching pattern groups can be configured, and each rate-matching pattern group can be composed of several rate-matching patterns.

Although the reserved resources are configured through rate-matching pattern and rate-matching pattern group, whether these reserved resources can transmit the PDSCH or not is still dynamically indicated by the base station through DCI. The DCI 1_1 contains a 2-bit rate-matching indicator. When the base station schedules the PDSCH for the terminal through DCI 1_1, it indicates whether the reserved resources in the two rate-matching pattern groups can be used for this scheduled PDSCH transmission. As shown in Fig. 5.69, the rate-matching indicator in the first scheduling DCI is "10," then the actual transmission resource of the PDSCH scheduled by this DCI needs to exclude the reserved resources indicated by rate-matching pattern group 1. In the second scheduling DCI, the rate-matching indicator is "01," then the actual transmission resource of the PDSCH scheduled by this DCI needs to exclude the reserved resources indicated by rate-matching pattern group 2. The rate-

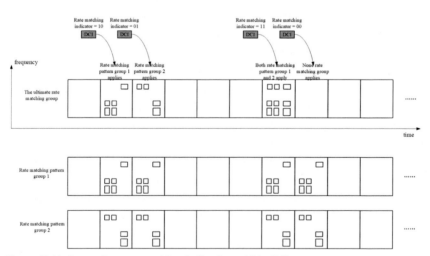

Figure 5.69 Dynamic rate-matching indication within DCI.

matching indicator is "11" in the third scheduling DCI, then the actual transmission resource of PDSCH scheduled by this DCI needs to exclude the reserved resources indicated by the union of rate-matching pattern groups 1 and 2. The rate-matching indicator in the fourth scheduling DCI is "00," then the actual transmission resource of the PDSCH scheduled by this DCI is equal to the full allocation given by TDRA and FDRA, without excluding any reserved resources.

The above discussion shows the RB-symbol-level rate-matching technology. The NR specification also defines the RE-level rate-matching technology, which targets for NR resource avoidance around LTE reference signals. The principle is similar to the RB-symbol-level rate-matching technology, so its details are not described here.

It should be noted that from the perspective of PDCCH/PDSCH multiplexing, only the rate-matching mechanism related to PDSCH resource allocation needs to be defined. From the perspective of resource reservation, the rate-matching mechanism related to PUSCH can also be defined. However, R15 NR specification only defines the rate-matching technology for PDSCH.

5.8 Summary

This chapter introduces various flexible scheduling technologies adopted by 5G NR, including the basic time—frequency resource allocation schemes of data channel, CBG transmission design, PDCCH resource configuration, PUCCH resource scheduling, flexible TDD and PDSCH rate matching. It can be seen that these flexible scheduling technologies make the NR specification more capable than LTE of fully exploring the technical potential of the OFDMA system, and have become the basis for many other designs in the NR system.

References

[1] Shen J, Suo S, Quan H, Zhao X, Hu H, Jiang Y. 3GPP Long Term Evolution (LTE) technology principle and system design. Peoples Posts and Telecommunications Press China; 2008.
[2] R1-1709740. Way forward on RBG size, NTT DOCOMO, 3GPP RAN1#89, Hangzhou, China, May 15–19, 2017.
[3] R1-1711843. Outputs of offline discussion on RBG size/number determination, OPPO, Nokia, LGE, Samsung, vivo, ZTE, CATR, 3GPP RAN1#AH_NR2, Qingdao, China, June 27–30, 2017.

[4] R1-1710323. Discussion on frequency-domain resource allocation, LG Electronics, 3GPP RAN1#AH_NR2, Qingdao, China, June 27−30, 2017.

[5] R1-1710164. Bandwidth part configuration and frequency resource allocation, Guangdong OPPO Mobile Telecommunication, 3GPP RAN1#AH_NR2, Qingdao, China, June 27−30, 2017.

[6] R1-1801065. Outcome of offline discussion on 7.3.3.1 (resource allocation)—part I, Ericsson, RAN1#AdHoc1801, Vancouver, Canada, January 22−26, 2018.

[7] R1-1610736. Summary of offline discussion on downlink control channels, Ericsson, 3GPP TSG RAN WG1 Meeting #86bis, Lisbon, Portugal, October 10−14, 2016.

[8] R1-1706559. Way forward on dynamic indication of transmission length on data channel, ZTE, Microelectronics, Ericsson, Nokia, AT&T, vivo, Panasonic, Convida Wireless, Intel, KT Corp, CATT, RAN1#88bis, Spokane, WA, April 3−7, 2017.

[9] Final Report of 3GPP TSG RAN WG1 #AH_NR2 v1.0.0, Qingdao, China, June 27−30, 2017.

[10] 3GPP TS 38.214 v15.6.0. NR: physical layer procedures for data (release 15).

[11] R1-1800488. Text proposal on DL/UL resource allocation, Guangdong OPPO Mobile Telecom., 3GPP TSG RAN WG1 Meeting AH 1801, Vancouver, Canada, January 2−26, 2018.

[12] R1-1810973. Text proposal for DL/UL data scheduling and HARQ procedure, Guangdong OPPO Mobile Telecom., 3GPP TSG RAN WG1 Meeting #94bis, Chengdu, China, October 8−12, 2018.

[13] R1-1609744. HARQ operation for large transport block sizes, Nokia, Alcatel-Lucent Shanghai Bell, 3GPP RAN1 #86bis, Lisbon, Portugal, October 10−14, 2016.

[14] R1-1700958. TB/CB handling for eMBB, Samsung, 3GPP RAN1 NR Ad Hoc#1701, Spokane, WA, January 16−20, 2017.

[15] R1-1702636. Multi-bit HARQ−ACK feedback, Qualcomm Incorporated, 3GPP RAN1 NR Ad Hoc#1701, Spokane, WA, January 16−20, 2017.

[16] R1-1701874. On HARQ and its enhancements, Ericsson, 3GPP RAN1 #88, Athens, Greece, February 13−17, 2017.

[17] R1-1701020. Enriched feedback for adaptive HARQ, Nokia, Alcatel-Lucent Shanghai Bell, 3GPP RAN1 NR Ad Hoc#1701, Spokane, WA, January 16−20, 2017.

[18] R1-1707725. Discussion on CBG-based transmission, Guangdong OPPO Mobile Telecom, 3GPP RAN1 #89, Hangzhou, P.R. China, May 15−19, 2017.

[19] R1-1706962. Scheduling mechanisms for CBG-based re-transmission, Huawei, HiSilicon, 3GPP RAN1 #89, Hangzhou, China, May 15−19, 2017.

[20] R1-1707661. Consideration on CB group based HARQ operation, 3GPP RAN1 #89, Hangzhou, P.R. China, May 15−19, 2017.

[21] R1-1721638. Offline discussion summary on CBG based retransmission, 3GPP RAN1 #91, Reno, NV, November 27−December 1, 2017.

[22] R1-1721370, Summary on CA aspects, Samsung, 3GPP RAN1#91, Reno, NV, November 27−December 1, 2017.

[23] R1-1612063. Single-part/multi-part PDCCH, Qualcomm, 3GPP RAN1 #87, Reno, NV, November 14−18, 2016.

[24] R1-1610083. Initial views on UL control channel design, NTT DOCOMO, INC, 3GPP RAN1 #86bis, Lisbon, Portugal, October 10−14, 2016.

[25] R1-1611698. Discussion on uplink control channel, Guangdong OPPO Mobile Telecom, 3GPP RAN1 #87, Reno, NV, November 14−18, 2016.

[26] R1-1700618. Summary of [87-32]: UL L1/L2 control channel design for NR, NTT DOCOMO, INC, 3GPP RAN1 AH_NR, Spokane, NV, January 16−20, 2017.

[27] R1-1703318. On the short PUCCH for small UCI payloads, Nokia, Alcatel-Lucent Shanghai Bell, 3GPP RAN1#88, Athens, Greece, February 13—17, 2017.
[28] R1-1706159. Short duration PUCCH structure, CATT, 3GPP RAN1 #88bis, Spokane, NV, April 3—7, 2017.
[29] R1-1705389. Performance evaluations for short PUCCH structures with 2 symbols, Samsung, 3GPP RAN1 #88bis, Spokane, NV, April 3—7, 2017.
[30] R1-1719972. Summary of email discussion [90b-NR—29] on PUCCH resource set, 3GPP RAN1 #91 R1-1719972 Reno, NV, November 27—December 1, 2017.

CHAPTER 6

NR initial access

Weijie Xu, Chuanfeng He, Wenqiang Tian, Rongyi Hu and Li Guo

The cell-search procedure is the first step for a UE (user equipment) to access 5G services, during which the UE searches and determines an appropriate cell and then accesses the cell. The cell-search procedure involves frequency scanning, cell detection, and system information acquisition. In Section 6.1 of this chapter, we go through synchronization raster and channel raster related to the cell-search process, the design of SSB (Synchronization Signal/Physical Broadcast Channel Block), the characteristics of SSB transmission, the positions of actually transmitted SSBs, cell-search procedure, and other aspects. Then in Section 6.2, Type0 PDCCH (physical downlink control channel) CORESET (control resource set) and Type0 PDCCH search spaces related to the transmission of SIB1 (system Information block 1) will be discussed.

After cell search is completed and an appropriate cell is selected, the UE initiates the random access procedure to the selected cell in order to establish the RRC (radio resource control) connection with the network. Section 6.3 of this chapter will go through various unique aspects of NR (new radio) Random access on PRACH (physical random access channel) channel design, PRACH resource allocation, SSB and RO (PRACH occasion) mapping, and power control of PRACH.

Radio Resource Management (RRM) measurement is the basis for a UE to evaluate the signal quality of multiple cells and determine an appropriate cell for initiating random access. After the UE is connected to one cell, it is also necessary for the UE to carry out continuous RRM measurement on the serving cell and neighboring cells to assist the network to make decisions on scheduling and mobility management. Section 6.4 of this chapter introduces various aspects of NR RRM measurement, including reference signals for RRM measurement, NR measurement interval, NR intrafrequency measurement and interfrequency measurement, and the scheduling restrictions caused by RRM measurement.

The Radio Link Monitoring (RLM) procedure is used to continuously monitor and evaluate the wireless link quality of the serving cell after the UE is connected to a cell and enters RRC connected state. Similar to the

5G NR and Enhancements
DOI: https://doi.org/10.1016/B978-0-323-91060-6.00006-4
283

initial access process, as an important mechanism for a UE to monitor and maintain link quality, the RLM procedure is the guarantee of communication between UE and the network. Section 6.5 of this chapter presents various aspects of the NR RLM procedure, including the reference signals for RLM and the RLM procedure.

6.1 Cell search

NR supports larger system bandwidth, flexible subcarrier spacing (SCS), and beam sweeping. These aspects have critical impacts on the design of NR initial access, and also introduce unique technical characteristics to the NR. This section introduces various aspects of the cell-search process, including synchronization raster and channel grid, the design of synchronization signal block (SSB), the transmission characteristics of SSB, the actual transmission locations of SSB, the cell-search process, etc.

6.1.1 Synchronization raster and channel raster

During the cell-search procedure, the UE needs to find a cell and then tries to access that cell. One key question here is which frequency location(s) will the UE try to detect a cell? It is not desired and practical for a UE to blindly search for cells at all the frequencies. Therefore how to effectively guide the cell-search procedure is one of the key issues that needs to be considered in NR design.

6.1.1.1 Synchronization raster and channel raster

First of all, it is important to clarify two terminologies. The first one is "synchronization raster," and the other one is "channel raster." Understanding the meaning of them and the relationship between them is helpful for understanding the whole 5G system cell-search procedure.

Synchronization raster is a list of frequency points that can be used to transmit synchronization signals. For example, point X in the frequency domain can be a synchronization raster. When an operator deploys a cell with carrier frequency close to point X, the synchronization signal of the cell can be configured at point X. Accordingly, when the UE searches for a cell in the frequency band where point X is located, the UE can find the cell through the synchronization signal at point X. Channel raster is the frequency point that can be used to deploy a cell. For example, if a cell bandwidth is 100 MHz, the center frequency point can be configured on a channel raster at frequency point Y, and the range of the cell in frequency domain can be set as [Y − 50 MHz, Y + 50 MHz]. In summary, synchronization raster indicates the frequency-domain location where a synchronization signal can be transmitted, while channel raster indicates the frequency-domain location where a cell-center carrier frequency can be configured.

Synchronization raster is used to help the UE reduce the access delay and power waste caused by the uncertainty of blind cell searching. The larger the granularity of synchronization raster is, the smaller the synchronization raster in the frequency domain. This will decrease the time that is consumed by cell search. However, the granularity of synchronization raster cannot be enlarged without limitation. There should be at least one synchronization raster within the frequency bandwidth of one cell. For example, when a cell bandwidth is 20 MHz, if the granularity of synchronization raster is 40 MHz, it is obvious that some frequency resources would not be usable for cell deployment.

Channel raster indicates the carrier frequency for cell deployment. From the perspective of network deployment, the smaller the granularity of channel raster, the better the deployment flexibility. For example, when an operator deploys a cell in the range of 900−920 MHz, if the operator wants to deploy a cell with a bandwidth of 20 MHz to occupy the whole spectrum, the operator expects that a channel raster could be found at 910 MHz. However, if another operator obtains a frequency band in the range of 900.1−920.1 MHz, this operator would expect a channel raster at 910.1 MHz. Therefore if the granularity of channel raster is too large, it is difficult to meet different network deployment requirements caused by different frequency band allocations. Therefore adopting smaller channel raster granularity is necessary for reducing the restriction of cell deployment.

To summarize, from the perspective of the UE, a larger synchronization raster granularity is better since it is helpful for speeding up the cell-search procedure. While from the perspective of network deployment, a smaller channel raster granularity is better since it can support more flexible deployment of 5G networks.

6.1.1.2 Raster in LTE and changes in 5G NR

In LTE (long term evolution), the center frequency of a synchronization signal is fixed to be the center frequency of a cell. The synchronization raster and the channel raster are bound together. The raster size is 100 kHz. It can ensure the flexibility of network deployment. The LTE bandwidth is only tens of megahertz. Therefore the complexity of cell search with 100 kHz synchronization raster is acceptable.

In contrast, the bandwidth of 5G NR greatly exceeds that of the LTE system. For example, a typical frequency bandwidth of 5G NR system in sub-6 GHz bands is hundreds of megahertz, and a typical frequency bandwidth in high-frequency scenario is several gigahertz. If the initial cell-search interval of 100 kHz is reused by the UE, it will inevitably lead to too long and unacceptable cell-search latency. In 3GPP (3rd generation partnership project) discussions, companies estimated that the initial search time of a cell may be more than ten minutes if no improvement of the raster design is introduced. Therefore optimizing and improving the design of synchronization raster and channel raster of the 5G system is a basic requirement of 5G systems.

In addition, because of the flexibility of the NR system, the location of synchronization signal is no longer constrained in the center of a cell. For a 5G cell, the synchronization signal only needs to be located within the system bandwidth of the cell. Such a change, on the one hand, supports the flexibility of synchronization signal resources. On the other hand, it also gives the possibility of redesigning 5G system synchronization raster. If the 5G synchronization signal is still constrained to be located at the center frequency of a cell, the synchronization raster can be enlarged only at the cost of reducing the network deployment flexibility. But if the synchronization raster and the channel raster are unbounded, the contradiction between the flexibility of network deployment and the complexity of initial search can be solved independently.

6.1.1.3 Synchronization raster and channel raster in 5G NR

For channel raster, 5G NR adopts two basic concepts: "global frequency channel raster" and "NR channel raster." The global frequency channel raster defines a set of RF (radio frequency) reference frequencies, F_{REF}. The RF reference frequency is used to identify the position of RF channels, SSB, and other elements. Usually, this set of F_{REF} will be presented in the form of NR absolute radio-frequency channel number (NR−ARFCN). As shown in Table 6.1, in the frequency domain from 0 to 3 GHz, the basic frequency-domain granularity ΔF_{Global} is 5 kHz, which corresponds to the absolute channel number from 0 to 599999. In the frequency range from 3 to 24.25 GHz, the granularity ΔF_{Global} is expanded to 15 kHz, which corresponds to the ARFCN from 600000 to 2016666. In the frequency domain from 24.25 to 100 GHz, the granularity ΔF_{Global} is expanded to 60 kHz, which corresponds to the ARFCN from 2016667 to 3279165, respectively.

Based on the design of global channel raster, the NR channel raster defines the channel raster for each frequency band. Table 6.2 shows the result of the NR channel raster in some frequency bands. Taking band n38 as an example, 100 kHz is taken as the granularity of channel raster. The first NR channel raster in the uplink of this band is located at

Table 6.1 NR−ARFCN parameters for the global frequency raster.

Frequency range (MHz)	ΔF_{Global} (kHz)	$F_{REF\text{-}Offs}$ (MHz)	$N_{REF\text{-}Offs}$	Range of N_{REF}
0−3000	5	0	0	0−599999
3000−24250	15	3000	600000	600000−2016666
24250−100000	60	24250.08	2016667	2016667−3279165

Table 6.2 Applicable NR−ARFCN per operating band.

NR operating band	ΔF_{Raster} (kHz)	Uplink range of N_{REF} (first— < step size > — last)	Downlink range of N_{REF} (first— < step size > — last)
n1	100	384000— <20> — 396000	422000— <20> — 434000
n3	100	342000— <20> — 357000	361000— <20> — 376000
n8	100	176000— <20> — 183000	185000— <20> — 192000
n34	100	402000— <20> — 405000	402000— <20> — 405000
n38	100	514000— <20> — 524000	514000— <20> — 524000
n39	100	376000— <20> — 384000	376000— <20> — 384000
n40	100	460000— < 20 > — 480000	460000— <20> — 480000
n41	15	499200— <3> — 537999	499200— <3> — 537999
	30	499200— <6> — 537996	499200— <6> — 537996

(*Continued*)

Table 6.2 (Continued)

NR operating band	ΔF_{Raster} (kHz)	Uplink range of N_{REF} (first— <step size> — last)	Downlink range of N_{REF} (first— <step size> — last)
n77	15	620000— <1> — 680000	620000— <1> — 680000
	30	620000— <2> — 680000	620000— <2> — 680000
n78	15	620000— <1> — 653333	620000— <1> — 653333
	30	620000— <2> — 653332	620000— <2> — 653332
n79	15	693334— <1> — 733333	693334— <1> — 733333
	30	693334— <2> — 733332	693334— <2> — 733332
n258	60	2016667— <1> — 2070832	2016667— <1> — 2070832
	120	2016667— <2> — 2070831	2016667— <2> — 2070831

ARFCN 514000, which is the starting point in this band. After that, every two adjacent NR channel rasters are separated by 20 global channel rasters. The last channel raster in this band is located at the frequency point corresponding to the ARFCN 524000.

It can be observed in Table 6.2 that in most of the sub-6G bands, 5G NR reuses 100 kHz granularity for the NR channel raster. For those bands, maintaining a unified design with LTE is conducive to minimizing the impact of refarming bands on network deploying. For the new frequency bands (such as n77, n78, n79, n258) defined for 5G NR, channel raster granularity is determined corresponding to the SCS adopted in each band, such as 15, 30, 60, and 120 kHz, which can provide more flexible network deploying for these new 5G bands.

For synchronous raster, analogous to the relationship between "global channel raster" and "NR channel raster," 5G NR defines "global synchronization raster" as a basic frequency-domain granularity that is used construct the "NR synchronization raster." The frequency-domain position of each global synchronization raster corresponds to a specific global synchronization channel number (GSCN). The relationship between them is given in Table 6.3.

Table 6.3 GSCN parameters for the global frequency raster.

Frequency range	SSB frequency position SS$_{REF}$	GSCN	Range of GSCN
0–3000 MHz	N × 1200 kHz + M × 50 kHz, N = 1:2499, M ∈ {1,3,5}	3 N + (M−3)/2	2–7498
3000–24250 MHz	3000 MHz + N × 1.44 MHz N = 0:14756	7499 + N	7499–22255
24250–100000 MHz	24250.08 MHz + N × 17.28 MHz, N = 0:4383	22256 + N	22256–26639

The method of calculating GSCN of each global synchronization raster varies in different frequency ranges. In Table 6.3, we can observe that the granularity of global synchronization raster is 1.2 MHz in the range of 0–3 GHz, 1.44 MHz in the range of 3–24.25 GHz, and 17.28 MHz in the range of 24.25–100 GHz, respectively. Larger synchronization raster is used in higher frequency bands to reduce the complexity of cell search. In addition, an additional offset of ± 100 kHz is added in the range of 0–3 GHz. Since the synchronization raster and the channel raster are decoupled, the offset between the center frequency of the synchronization signal and the center frequency of the cell is not always an integral multiple of the SCS of the cell. To deal with that, an additional offset is added to fine tune the frequency position. For example, a cell with a SCS of 15 kHz is deployed in band n38 and the cell-center frequency is located at 2603.8 MHz (NR ARFCN = 520760). If the center frequency of the synchronization signal is located at 2604.15 MHz (GSCN = 6510, M = 3, N = 2170), the frequency offset between the channel raster and the synchronization raster would be 350 kHz. As shown in Fig. 6.1, 350 kHz is not an integral multiple of the SCS of 15 kHz. That is not conducive to using a unified FFT (fast Fourier transform) to process the synchronization signal in the cell and the data channel that are multiplexed on the same symbols, which introduces additional signal generation and detection complexity. By introducing the offset of ± 100 kHz in the design of the synchronization raster, the above problems can be solved. In this example, the synchronization signal block can be transmitted at 2604.25 MHz (GSCN = 6511, M = 5, N = 2170). The frequency-domain offset between the channel raster and the synchronization raster is revised to 450 kHz, which is an integral multiple of the SCS of 15 kHz.

The synchronization raster for each NR operating band is illustrated in Table 6.4. The distance between two applicable GSCN entries is given

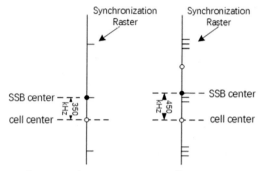

Figure 6.1 An example on synchronization raster offset.

Table 6.4 Applicable synchronization raster entries per operating band.

NR operating band	SSB SCS	Range of GSCN (first—<step size>—last)
n1	15 kHz	5279—<1>—5419
n3	15 kHz	4517—<1>—4693
n8	15 kHz	2318—<1>—2395
n34	15 kHz	5030—<1>—5056
n38	15 kHz	6431—<1>—6544
n39	15 kHz	4706—<1>—4795
n40	15 kHz	5756—<1>—5995
n41	15 kHz	6246—<3>—6717
	30 kHz	6252—<3>—6714
n77	30 kHz	7711—<1>—8329
n78	30 kHz	7711—<1>—8051
n79	30 kHz	8480—<16>—8880
n258	120 kHz	22257—<1>—22443
	240 kHz	22258—<2>—22442

by the <Step size>. In most frequency bands, the position of the NR synchronization raster corresponds to that of the global synchronization raster. However, exceptions do exist in some frequency bands, such as n41 and n79. In order to further reduce the complexity of cell search, the synchronization raster of those bands adopt the integral multiple of the global synchronization raster for the raster granularity.

6.1.2 Design of SSB

SSB plays a fundamental role in the initial access procedure. It supports very important functions, such as carrying cell ID (identifier), time-frequency synchronization, indicating symbol level/time-gap level/frame timing, cell/beam signal strength/signal quality measurement, etc. To support those functions, SSB is comprised of Primary Synchronization Signal (PSS); Secondary Synchronization Signal (SSS); Physical Broadcast Channel (PBCH); and reference signal Demodulation Reference signal (DMRS) of the PBCH. The PSS and SSS carry cell ID (1008 cell IDs) and are used by the UE to complete time-frequency synchronization and obtain symbol-level timing. SSS and the DMRS of the PBCH can be used to measure the signal quality of a cell or a beam. The PBCH indicates slot boundary or frame boundary and also the frame number. It should be noted that at the beginning of the 3GPP standardization discussion, it was called SSB. Since the PBCH channel is also included in SSB, the Synchronization Signal Block/PBCH Block was renamed as the SS/PBCH Block at a later specification stage. In this book they are equivalent, and for the sake of simplicity, we call them SSBs.

Due to the fundamental role of SSB in initial access procedure, the design of signals and structure of SSB were fully discussed during the standardization process.

When the NR system transmits in the manner of beam sweeping, each downlink beam will transmit one SSB. The SSB in each beam needs to contain PSS, SSS, and PBCH. The sequences of both PSS and SSS have a length of 127 and occupy 12 PRBs (including the guard subcarriers). In order to provide sufficient resources so that the PBCH can be transmitted with a sufficiently low code rate, simulation evaluations showed that the performance requirements can be satisfied with two symbols when the bandwidth occupied on each symbol is 24 PRBs. Therefore the bandwidth of the PBCH was determined to be 24 PRBs.

The 3GPP standardization discussion focused on the design of structure of SSB, especially the mapping order of PSS, SSS, and PBCH symbols. The PSS will be first processed by the UE before the SSS is processed during UE receiving and processing SSB. Therefore if PSS is located behind SSS in the time domain, the UE would need to buffer the SSS first but process SSS after finishing the processing of PSS [1]. To avoid that, companies first agreed that PSS should be transmitted before SSS in the time domain. However, regarding the specific mapping order of PSS, SSS, and PBCH in a SSB in the time domain, companies proposed different options. A few common mapping schemes are as follows:
- Option 1: The mapping order is PSS-SSS—PBCH—PBCH, as shown by option 1 in Fig. 6.2.

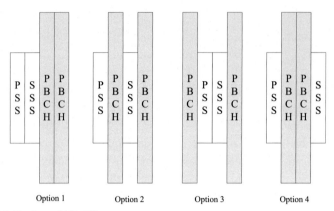

Option 1   Option 2   Option 3   Option 4

Figure 6.2 Design of NR SSB structure.

- Option 2: The mapping order is PSS−PBCH−SSS−PBCH, as shown by option 2 in Fig. 6.2.
- Option 3: The mapping order is PBCH−PSS−SSS−PBCH, as shown by option 3 in Fig. 6.2.
- Option 4: The mapping order is PSS−PBCH−PBCH−SSS, as shown by option 4 in Fig. 6.2.

The main differences of the above schemes are the symbol spacing between PSS and SSS, the spacing between two PBCH symbols, and the relative relationship between PBCH symbols and PSS and SSS symbols. According to Refs. [2,3], the SSS can be used to assist the channel estimation of two PBCH symbols and thus improve the demodulation performance of the PBCH, so the SSS should be placed between two PBCH symbols. Ref. [4] pointed out that appropriately increasing the symbol interval between PSS and SSS is beneficial to improving the accuracy of frequency offset estimation. Option 2 in Fig. 6.2 can meet those requirements, and thus the time-domain mapping order corresponding of option 2 is adopted by the standard.

By then, the design of the SSB structure was completed in 3GPP. However, in the RAN1 group meeting of September 2017, some chipset vendors pointed out that because it was agreed on in RAN4#82bis meeting that the minimum channel bandwidth supported by NR is 5 MHz in FR1 (frequency range 1) and 50 MHz in FR2, the adopted SSB structure would force the UE to perform a huge amount of cell search, and consequently, the latency of cell-search delay would not be acceptable. A detailed analysis of that is given as follows [5]:

Synchronization Raster is determined by the following formula:

$$\text{Synchronization Raster} =$$

$$\text{Minimum Channel Bandwidth} - \text{SSB Bandwidth} + \text{Channel raster} \quad (6.1)$$

Taking FR1 as an example, the minimum channel bandwidth supported by FR1 is 5 MHz and the SSB bandwidth is 24 PRBS (for SCS of 15 kHz, the corresponding bandwidth is 4.32 MHz). Therefore the Synchronization Raster will have to be less than 0.7 MHz according to the above equation. The 5G system generally supports a wide bandwidth. The sum of the bandwidth of typically 5G bands (N1, N3, N7, N8, n26, n28, n41, n66, n77, N78, etc.) in FR1 is close to 1.5 GHz. Therefore based on the existing design, there would be thousands of Synchronization Rasters for those bands in FR1. The typical periodicity of SSB is 20 Ms. The time for the UE to complete a whole frequency scanning in FR1 would be 15 minutes [5]. In high-frequency bands, the UE needs to try multiple receiving panels. Therefore the delay of cell search will be further prolonged, which is obviously unacceptable from the perspective of implementation.

It can be observed from Formula 6.1 that reducing the bandwidth of SSB can increase the bandwidth of the Synchronization Raster and thus reduce the number of searching SSB in the frequency domain.

Therefore a half year after the SSB structure was designed, the 3GPP decided to reverse the previous design and redesign the SSB structure to reduce the complexity of cell search. For that, a few typical schemes of redesigned SSB structure were discussed:

- Option 1: The bandwidth of the PBCH is reduced (e.g., to 18 PRBs without any other changes in SSB structure).
- Option 2: The bandwidth of the PBCH is reduced (e.g., to 18 PRBs, and the number of PBCH symbols is increased).
- Option 3: The bandwidth of the PBCH is reduced and additional transmission of the PBCH is added in the frequency-domain resources on both sides of SSS transmission in SSS symbol.

Option 1 reduces the complexity of cell search by reducing the bandwidth of the PBCH. However, the demodulation performance of the PBCH is impaired due to the decreasing of the total resources of the PBCH. Thus the coverage of the PBCH in negatively impacted.

As shown in Fig. 6.3, Option 2 can also reduce the complexity of cell search through reducing the bandwidth of the PBCH. To compensate for the reduction of PBCH resources caused by reduced bandwidth, Option 2 increases the number of symbols of the PBCH so that demodulation performance of the PBCH is not impaired. On the other hand, increasing the number of symbols of one SSB would result in unavailability of candidate positions of SSB that was designed based on the original SSB structure design. Therefore the pattern of the candidate positions of SSB has to be redesigned too, which will inevitably increase the workload of standardization.

Option 3, shown in Fig. 6.4A, also uses reduced total bandwidth occupied by a SSB to reduce the complexity of initial cell search. On the other hand, to compensate the reduction of PBCH resources caused by reduced PBCH bandwidth, Option 3 uses the remaining frequency-domain resources adjacent to each side of SSS signal for PBCH transmission. Four PRBs on each side of the SSB in frequency domain are used for PBCH transmission, which can compensate the decreasing of PBCH resources caused by the reduced bandwidth of PBCH. Such a design can keep the total resources of PBCH transmission consistent with the

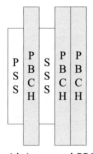

Figure 6.3 SSB redesign scheme with increased PBCH symbols.

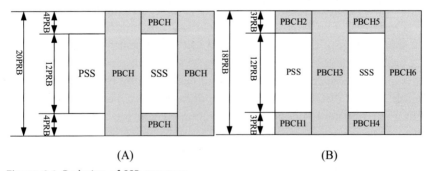

Figure 6.4 Redesign of SSB structure.

previous SSB design and thus ensure that the coverage performance of the PBCH is not impaired. It is worthwhile to note that during the standard discussion, one scheme similar to Option 3 was also proposed, which is shown in Fig. 6.4B. That scheme proposed to reduce the bandwidth of the PBCH and add more frequency resources for PBCH transmission on both sides of PSS and SSS. In addition to the remaining frequency-domain resources in SSB symbol, that scheme also utilized the remaining frequency-domain resources on both sides of PSS. That scheme has the following benefits. On one hand, the total amount of resources of the PBCH can be further increased, and thus the transmission performance of the PBCH can be improved. On the other hand, if we keep the total amount of resources of the PBCH to be consistent with the scheme shown in Fig. 6.4A, the bandwidth of SSB can be reduced, and thus the complexity of cell search can be further reduced. However, this scheme was not adopted because the PSS is usually used for automatic gain control (AGC) adjustment during SSB reception. The additional PBCH transmission on the frequency-domain resources on both sides of the PSS may not be used to enhance the demodulation performance of the PBCH. Furthermore, the gain brought by scheme of Option 3 shown in Fig. 6.4A for reducing the complexity of cell search is sufficient.

Eventually, the design of Option 3 was adopted, which is shown in Fig. 6.4A. Given this design, the bandwidth of SSB is reduced from 24 PRBs to 20 PRBs.

6.1.3 Transmission of SSB

In comparison with LTE, 5G system supports a higher frequency range in order to pursue larger available bandwidth to support higher data rate. But in a higher frequency band, the energy loss of radio transmission is larger and thus the communication distance is shorter. Such a problem would cause negative impact on the coverage distance of a wireless communication system. After several rounds of discussions, the 3GPP finally decided to adopt centralized energy transmission to deal with this issue in high-frequency bands. As illustrated in Figs. 6.5 and 6.6, the issue of limited coverage distance of high-frequency band transmission is resolved through the method of "spatial domain beam forming to improve transmission distance" and "time-domain sweeping to expand the spatial domain coverage."

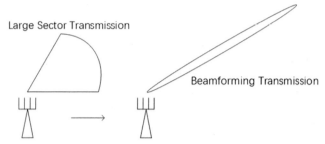

Large Sector Transmission

Beamforming Transmission

Figure 6.5 Beam forming is used to improve the transmission distance.

Scanning different beam directions at different times to ensure long-distance and large-scale data transmission

Figure 6.6 Beam sweeping is used to expand the coverage.

"Spatial domain beam forming to improve transmission distance" means that in a 5G system, beam-forming technology can be used to concentrate transmission power in a specific direction. In comparison with omnidirectional transmission or large sector transmission, the directional transmission with beam forming can increase the communication distance but also reduce the angle of coverage area, as shown in Fig. 6.5. That results in another issue, which is how to extend the coverage area. If a 5G cell can only cover an area with a limited angle, we still meet find in the network deployment and 5G coverage. To deal with that issue, the method of "time-domain sweeping to expand the spatial domain coverage" is adopted. Multiple different directional beams are used to indirectly achieve omnidirectional or large sector coverage through the beam sweeping in the time domain, which is illustrated in Fig. 6.6.

According to the design principle of beam sweeping, the SSBs in NR are transmitted through different beams, which are shown in Fig. 6.7. A group of SSBs forms an SSB burst set, which are sent in the time domain in the form of beam sweeping.

The SSBs in an SSB burst set are transmitted within a half system frame. In each SSB burst set, all the SSBs carry the same system information. One problem here is how to identify different SSBs within an SSB burst set. For example, if a UE detects the fourth SSB within a SSB burst set, how does the UE determine that this detected SSB is in the fourth

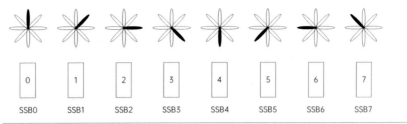

Scanning different beam directions at different times to ensure long-distance and large-scale data transmission

Figure 6.7 SSB is transmitted by beam sweeping.

SSB, not the eighth SSB? If the UE misjudges the timing of the detected SSB, it will inevitably lead to an overall system timing error.

To solve this problem, each SSB is given a unique index within an SSB burst set, which is called the SSB index. When a UE detects an SSB, the UE can determine the location of the SSB within an SSB burst set by identifying its SSB index. In frequency bands below 6 GHz, there are at most eight SSBs within an SSB burst set, and thus at most three bits are needed to indicate the SSB index within an SSB burst set. These three bits are implicitly carried by the DMRS sequence of the PBCH. There are eight different DMRS sequences of the PBCH, corresponding to eight different SSB index, respectively. In the frequency bands above 6 GHz, beams with higher beam-forming gain are utilized to combat larger path loss to support long-distance transmission. Thus the angle of area covered by a single beam is smaller than that in sub-6 GHz bands. More beams will be used to cover the area of one cell. Per the design, up to 64 SSBs can be configured in the frequency bands above 6 GHz. Six bits are needed to indicate the index of up to 64 SSBs. The three LSB (least significant bit) of those six bits are implicitly carried by the DMRS sequence of the PBCH, and the additional three MSB bits are directly indicated in the payload of the PBCH.

In each half system frame, only some slots are available for SSB transmission. As illustrated in Fig. 6.8, if a system can transmit up to four SSBs within a half-frame, the first two slots in each half-frame can be used to transmit SSBs. If a system can transmit up to eight SSBs within a half-frame, the first four slots in each half-frame can be used to transmit SSBs.

In each slot, not all the symbols are available for SSB transmission. The 3GPP specification specifies the symbols that can be used to transmit SSB and the symbols that cannot be used for SSB transmission. In the frequency bands of sub-6 GHz, there are three transmission modes, as shown in Fig. 6.9. The available symbols for SSB transmission are determined

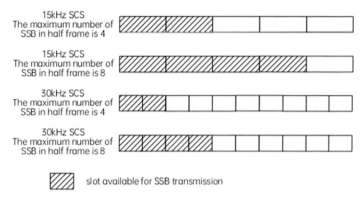

Figure 6.8 Slots for SSB transmission.

Figure 6.9 Symbols for SSB transmission.

based on various factors. In each of those modes, the first two symbols of each slot are reserved for the control channel transmission and thus can not be used for SSB transmission. The last two symbols in each slot are reserved for uplink control channel resources and thus can not be used for SSB transmission as well. Furthermore, a certain symbol interval is reserved between any two consecutive SSBs to support flexible transition of uplink and downlink transmission and to support URLLC services. In addition, considering the NR deployment in the refarming bands, mode B is introduced for the SSB transmission with 30 kHz SCS to avoid the conflict between the transmission of NR SSBs and LTE cell-specific reference signal resources.

According to the specification described above, when a UE determines the SSB index, the UE can derive the specific timing information within one half-frame according to the specified protocol. For example, if a UE determines an SSB index to be 3, the UE can claim that the detected SSB corresponds to the second candidate SSB transmission position in the second slot within a 5 Ms window. That can determine the system timing within a half-frame.

After the timing within a half-frame is resolved, the UE needs to determine the timing beyond a half-frame. In the 5G system, the UE can obtain the SFN through the broadcast channel. Ten bits are used here to indicate an SFN. The six MSB (most significant bit) bits are indicated in the MIB (master information block), while the four LSB bits are carried through the physical layer contents of the PBCH. In addition, the UE also needs to determine if the detected SSB is within the first half-frame or the second half-frame of a system frame. For that, 1-bit half-frame indication is introduced in 5G NR. The UE can obtain the 1-bit half-frame indication in the payload of the PBCH.

To summarize, in a 5G NR system, a total of 14 (or 17) bits are used to indicate the system timing information. There are ten bits for SFN, one bit for determining half-frame, up to three bits to indicate SSB index in sub-6 GHz bands, and six bits to indicate SSB index above 6 GHz bands. After a UE obtains that timing information, it determines the system timing and then obtains the time synchronization to the 5G network.

Furthermore, it worth clarifying some other aspects related to SSB transmission. The first aspect is the mechanism for a UE to determine the QCL (quasi co-located) hypothesis between SSBs. The QCL relationship between signals is used to describe the similarity of their large-scale channel characteristics. If the two signals are QCLed, the large-scale channel characteristics of these two signals can be considered to be similar. Regarding the SSB transmission, the SSB index indicates not only the timing information of each SSB, but also the beam information. All the SSBs with the same SSB index can be considered to be QCLed to each other, and the UE can assume that those SSBs are transmitted with the same transmit beam. The second aspect is the mechanism to determine the transmission periodicity of SSB burst set. The SSB burst set is transmitted periodically. We need to consider two scenarios for the UE determine the time-domain periodicity. The first scenario is during cell-search procedure, when the UE has not obtained the configuration of SSB. According to the 3GPP specification, in this scenario, the UE can assume that the SSB periodicity is 20 Ms for the cell-search procedure, which can reduce the complexity of cell search and scanning. The second scenario is when the UE is configured with other operations such as RRM on SSB. The transmission periodicity of SSB can be flexibly configured by the network. 5G NR currently supports the SSB periodicities of {5Ms, 10Ms, 20Ms, 40Ms, 80Ms, 160Ms}. The base station can configure it to the UE through high-layer signaling.

6.1.4 Position of actually transmitted SSBs and indication methods

As described in Section 6.1.2, candidate positions for SSB burst transmission are specified in the NR. The number of candidate positions for SSB transmission varies in different frequency bands. The system in a higher frequency band needs to support more transmit beams on SSB and thus the number of candidate positions of SSB is also more. However, in actual network deployment, the operators can flexibly determine the number of SSBs that are actually transmitted within a cell according to various factors, which include the radius of one cell, the coverage requirements, the transmission power of the base station, and the number of transmit beams supported by the base station. For example, for a macrocell deployed in a higher frequency band, the base station can use more narrow beams to achieve higher beam-forming gain to defeat the large path propagation loss and then improve cell coverage. On the other hand, in a low-frequency band deployment even a small number of beams can provide sufficient cell coverage.

From the perspective of a UE, it is necessary to know the locations of SSBs that are actually transmitted by the network. One of the reasons for that is the PDSCH cannot be transmitted on the symbols and PRBs occupied by the transmission of SSBs. Therefore we need to specify one mechanism that indicates the locations of the actually transmitted SSBs.

All of the system information, RAR response message, and paging message are carried in the PDSCH channel. To minimize the impact on the reception of this information and messages, it is desired to indicate the location of actual SSB transmission to the UE as early as possible. Thus indicating the locations of the actual SSB transmission in the system message becomes essential. The payload size of the system message is very limited. Therefore signaling overhead is the important factor that will be considered when the method of indicating location of actual SSB transmission is designed.

In FR1, the number of candidate positions of SSB is four or eight. Even if a full bit bitmap is used, the overhead is only four or eight bits, which seems not to be a big issue. Therefore companies agreed that a FR1 system uses full bitmap to indicate the positions of actual SSB transmission. Each bit in the bitmap corresponds to a candidate position. The value of one bit being "1" indicates that a SSB is actually transmitted at the corresponding candidate position, while the value of the bit being "0" indicates that no SSB is transmitted at the corresponding candidate position.

The discussion on indication method design focused on the design for the FR2 system due to the fact that there could be up to 64 candidate SSB positions in the FR2 system. If the indication method with a full bitmap is also used here, 64-bit overhead would be needed in system information, which is obviously not acceptable. Therefore different optimization methods were proposed by companies, and the typical methods are given as follows:

- Method 1: Group indication bitmap and intragroup indication bitmap

In this method, the candidate positions of SSB are first partitioned into groups. The positions of actually transmitted SSB are indicated through two bitmaps: a first bitmap used to indicate group(s) where the SSB are actually transmitted and an intragroup bitmap used to indicate the positions of the SSBs that are actually transmitted within each group. This method requires that the positions of the actually transmitted SSB within each indicated group will be the same. An example is illustrated in Fig. 6.10. In the example, 64 SSB candidate positions are divided into eight groups and one 8-bit bitmap are used to indicate the group(s) among those eight groups where the SSBs are actually transmitted. The value "10101010" of this bitmap indicates that the SSB is transmitted in the first, third, fifth, and seventh groups. Another bitmap with eight bits is used to indicate the SSB positions within one group. For example, the value "11001100" of the second bitmap can indicate that SSB is transmitted at the first, second, fifth, and sixth positions within each group. To summarize, a total of 16 bits are used to indicate the positions of actually transmitted SSBs in this example.

- Method 2: Group indication bitmap and the number of actually transmitted SSBs within each group

In this method, the candidate positions of SSB are also first partitioned into groups and a bitmap is used to indicate the group(s) where the SSBs are actually transmitted, which is the same as Method 1. However, the method of indicating positions of SSB transmission within each group is different from that of Method 1. In this method,

Figure 6.10 Indication of actually transmitted SSBs.

the positions of actual SSB transmissions within each group are continuous. The number of SSBs that are actually continuously transmitted in each group is indicated and the starting position of the transmitted SSBs within each group is predefined.

Consider an example where 64 SSB candidate positions are partitioned into eight groups. A bitmap of eight bits is used to indicate the group(s) where the SSBs are actually transmitted. In each group, up to eight SSBs can be actually transmitted. Therefore three bits are used to indicate the number of SSBs that are actually transmitted in one group. A value "110" of those three bits can indicate that the SSBs are transmitted from the first to sixth positions in each indicated group. In this example, a total of 11 bits are used to indicate the positions of actually transmitted SSBs.

- Method 3: Indicating the number of actually transmitted SSBs, the starting position, and the interval between two adjacent SSBs

  This method needs six bits to indicate the number of SSBs that are actually transmitted, six bits to indicate the start transmission position of SSBs, and six bits to indicate the interval between two adjacent SSBs. Therefore a total of 18 bits are needed for the indication.

- Method 4: Only group indication bitmap

In this method, the candidate positions of SSBs are first partitioned into groups and one bitmap is used to indicate the group(s) where SSBs are actually transmitted. Each group comprises the candidate positions of continuous SSB transmissions. In a group indicated by the corresponding bit in that bitmap, the SSBs are transmitted at all SSB candidate positions within that group. For a system with 64 SSB candidate positions, the SSB positions can be divided into eight groups and only eight bits are needed for the indication.

The different methods described above have different signaling overhead and also provide different indication granularity and flexibility. Methods 2 and 4 require smaller signaling overhead but only allow continuous SSB transmission within each group. That would cause unbalanced distribution of SSB transmission, which is not beneficial for intercell coordination to avoid mutual interference. In contrast, method 1 supports a group-level indication and also supports an intragroup indication that can provide refined controlling on the SSB transmission position within each group. That is helpful for achieving a good balance between the indicating overhead of indication signaling and the flexibility of implementation. Due to those reasons, method 1 was adopted in the 3GPP specification.

The indication information of positions of actually transmitted SSBs can be carried in the SIB1 message. When a UE receives the PDSCH-carrying SIB1 message, the UE is not aware of the positions of actually transmitted SSBs yet. Therefore the UE can assume that there is no SSB transmission at the resource allocation location where the PDSCH-carrying SIB1 message is transmitted. However, after that, when the UE receives other system information (OSI), RAR messages, and paging messages, the UE can assume rate matching of PDSCH according to the information of transmission positions of SSB indicated in SIB1.

In addition, the positions of actually transmitted SSBs can also be indicated through RRC signaling. For that, a full bit bitmap can be used since signaling overhead is not a big issue for RRC signaling.

6.1.5 Cell-search procedure

The structure and time-frequency location of SSB were introduced in Sections 6.1.1−6.1.3. During the initial access procedure, the UE first searches the SSB transmissions according to specified potential time-frequency positions of SSB transmission, and then obtains the time and frequency synchronization, radio frame timing, and physical cell ID through the detected SSB. The UE can also determine the search space configuration information of Type0−PDCCH through the MIB information carried in the PBCH.

To scan and search the SSB, the UE should first determine the SCS used by the SSB transmission. During the initial access procedure, the UE determines the SCS of the SSB transmission according to the frequency band in which the cell search is carried out. In most of the frequency bands in FR1, only one SCS is supported for SSB transmission. As specified in the 3GPP standard, the SCS of SSB transmission in most of those frequency bands in FR1 is one of the SCSs of 15 and 30 kHz. However, in some frequency bands of FR1, both 15 and 30 kHz SCSs are supported due to various requirements of some operators. One example is that in some frequency bands of FR1, both 15 and 30 kHz can be supported in order to support the compatibility between NR and LTE deployment. In all the frequency bands of FR2, both 120 and 240 kHz SCSs are supported. In a frequency band that supports two SCS values for SSB transmission, the UE would have to try to search the SSB with either of those two SCSs. Readers can find the specification on SCS of SSB transmission in different frequency bands in the literature [6,7].

Each SSB index corresponds to a predefined time location within a half-frame. Thus the SSB index can be used to obtain frame synchronization. Section 6.1.3 introduced the location of SSB burst sets in half a frame. A one-to-one mapping between the SSB index and time location of SSB burst set is specified. Each SSB carries its SSB index. Therefore after a UE finishes detecting one SSB, it can determine the symbol position of the detected SSB within a radio frame based on the SSB index and half-frame indication carried therein. Then the UE can determine the frame boundary of the radio frame and complete the frame synchronization.

The UE can combine multiple PBCHs carried in different SSBs and then decode the combined PBCH, which can improve the performance of PBCH reception. At the RAN1#88bis meeting, it was agreed that the UE can combine the PBCHs carried in different SSBs within the same SSB burst set, or combine the PBCHs carried in different SSBs in different SSB burst sets. The SSB burst set is transmitted periodically. Thus the UE will combine the received PBCHs with each certain time period. At the RAN1#88bis meeting, it was agreed that the periodicity of SSB transmission can be one of the {5,10,20,40,80,160} Ms and the value of the periodicity can be configured through high-level signaling. However, during initial cell search, the UE cannot receive high-level signaling indicating the periodicity of the SSB burst set yet. Therefore it is necessary to specify a default periodicity value that can be used by the UE to receive and combine the SSB transmission from different SSB burst sets. As specified for the NR system, a UE can assume the default periodicity of SSB is 20 Ms during initial cell search. In use cases other than the initial cell search, the UE can be provided with a periodicity of SSB burst set that is used for reception of the SSB of the serving cell. If the UE is not configured with a periodicity of SSB burst set in those cases, the UE can assume a periodicity of 5 Ms, as specified in the NR specification.

In the NR system, the SSB can be used for the initial access, and can also be configured in radio resource measurement. In the functionality of initial access, the UE accesses a cell, and the SSB used for initial access is associated with SIB1 information and its frequency-domain position is a synchronization raster. In the functionality of radio resource measurement, the SSB is not associated with SIB1 information even through its frequency-domain position can also be on a synchronization raster. But such a SSB can not be used for accessing a cell. In the discussion of the NR standardization, the SSB used for initial access was referred to as cell definition SSB (Cell-Defining SSB), while the SSB used in radio resource measurement was referred to as noncell definition SSB (Non Cell-Defining SSB). In other words, UE can access a cell only through Cell-Defining SSB. Due to the flexible design in the NR system, Non Cell-Defining SSB can also be transmitted on the synchronization raster. During initial access procedure and cell search, when the UE searches the SSB on the synchronization raster, it is possible that both of these two types of SSBs (Cell-Defining SSB and Non Cell-Defining SSB) are detected by the UE. As explained in previous section, during initial access procedure, the UE obtains the MIB information carried in the PBCH in the detected SSB and then use the control information indicated in MIB to receive SIB1 message. However, if a Non Cell-Defining SSB is detected by the UE during initial access, we would meet a problem because the Non Cell-Defining SSB is not associated with any SIB1 information. The UE would fail to obtain the control information for receiving SIB1 through the MIB information carried in the PBCH of the detected SSB. The UE would have to continue searching for Cell-Defining SSB for accessing a cell. At the RAN1#92 meeting, a network-assisted cell-search scheme was proposed by some companies to resolve that issue. In the proposed scheme, indication information is inserted in the transmission of Non Cell-Defining SSB, which indicates a frequency offset between the GSCN where the Cell-Defining SSB is located and the GSCN where this Non Cell-Defining SSB is located. When a UE detects a Non Cell-Defining SSB, the UE can determine the GSCN of the Cell-Defining SSB according to the indication information carried in the detected Non Cell-Defining SSB. Based on this auxiliary information, the UE can directly search Cell-Defining SSB at the target GSCN location and skip the blind search on Cell-Defining SSB. That can reduce the latency and power consumption of cell search. Specifically, in that network-assisted cell-search scheme, the GSCN offset information is indicated in the information payload carried by the PBCH.

The information carried by the PBCH includes the MIB information and the 8-bit physical layer-related information. Physical layer-related information includes SFN, half-frame indication, SSB index, and so on. The MIB information carried by the PBCH includes 6-bit SFN information, 1-bit SCS configuration information, 4-bit SSB subcarrier offset information, and 8-bit pdcch-ConfigSIB1 information. The 8-bit pdcch-

ConfigSIB1 information field indicates the search space configuration information of the PDCCH that carries the scheduling information of the PDSCH carrying the SIB1 message. The 1-bit SSB subcarrier offset information indicates the subcarrier offset k_{SSE} between the subcarrier 0 in resource block of SSB and the subcarrier 0 in the resource block of CORESET#0. In the FR1, the value of k_{SSE} is 0−23, which is indicated by a 5-bit field (4-bit subcarrier offset information in MIB and 1-bit physical layer-related information). In the FR2, the value of k_{SSE} is 0−11, which is indicated by a 4-bit subcarrier offset information contained in the MIB. If the value of k_{SSE} carried in one SSB is in the range of 0−23 in the FR1 system or in the range of 0−11 in the FR2 system, that means the SSB is associated with SIB1 information. In contrast, if a SSB carries $k_{SSB} > 23$ in the FR1 system or $k_{SSB} > 11$ in the FR2 system, this SSB is not associated with SIB1 information. As discussed in a previous section, in a Non Cell-Defining SSB, the network can indicate the synchronization raster of Cell-Defining SSB. Particularly, the design of indication information in the fields of Subcarrier offset and pdcch-ConfigSIB1 information in MIB are reused to indicate the GSCN of a Cell-Defining SSB. The field of pdcch-ConfigSIB1 information has eight bits. Thus reusing the field can indicate up to 256 values. Reusing the field of Subcarrier offset can further extend to support indicating up to $N \times 265$ values, where each value is indicated with an offset of the GSCN of a Cell-Defining SSB and the GSCN of the current Non Cell-Defining SSB. Tables 6.5 and 6.6 list the mapping between k_{SSE}, pdcch-ConfigSIB1, and N_{GSCN}^{Offset} for the FR1 system and FR2 system, respectively. According to the indicated GSCN offset, the UE can determine the GSCN of Cell-Defining SSB through $N_{GSCN}^{Reference} + N_{GSCN}^{Offset}$. The range of GSCN offset is

Table 6.5 Mapping between k_{SSE}, pdcch-ConfigSIB1 and N_{GSCN}^{Offset} (FR1)

k_{SSE}	pdcch-ConfigSIB1	N_{GSCN}^{Offset}
24	0, 1, ..., 255	1, 2, ..., 256
25	0, 1, ..., 255	257, 258, ..., 512
26	0, 1, ..., 255	513, 514, ..., 768
27	0, 1, ..., 255	−1, −2, ..., −256
28	0, 1, ..., 255	−257, −258, ..., −512
29	0, 1, ..., 255	−513, −514, ..., −768
30	0, 1, ..., 255	Reserved, reserved, ..., reserved

$-768\ldots-1,1\ldots768$ for the FR1 system in Table 6.5, and $-256\ldots-1$, $1\ldots256$ for the FR2 system in Table 6.6. The values of $k_{SSB} = 30$ and $k_{SSB} = 14$ are shown in Tables 6.5 and 6.6, respectively [8].

The procedure of determining Cell-Defining SSB GSCN based on the GSCN offset indicated in Non Cell-Defining SSB is illustrated in Fig. 6.11.

During standardization discussions, another scenario of cell search was also discussed [9], where Cell-Defining SSB may not be transmitted in some frequency carriers (e.g., the carriers that are only used as auxiliary cells) or frequency bands (i.e., the system only transmits SSB that is not on the synchronization raster). In such a scenario, the UE does not actually need to conduct a cell-search procedure within those frequency ranges. Therefore the network can indicate these frequency ranges to the UE and the UE can skip cell-search procedures on these frequency bands.

Table 6.6 Mapping between k_{SSE}, pdcch-ConfigSIB1, and N_{GSCN}^{Offset} (FR2).

k_{SSE}	pdcch-ConfigSIB1	N_{GSCN}^{Offset}
12	0, 1, ..., 255	1, 2, ..., 256
13	0, 1, ..., 255	$-1, -2, \ldots, -256$
14	0, 1, ..., 255	Reserved, reserved, ..., reserved

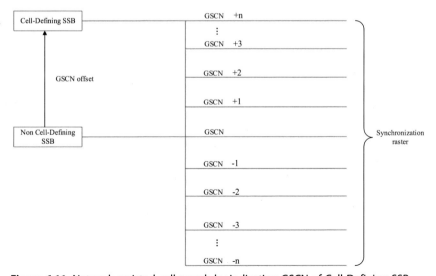

Figure 6.11 Network-assisted cell search by indicating GSCN of Cell-Defining SSB.

That would reduce the latency and power consumption on cell search. Because of those potential benefits, the standard finally adopted that scheme. The fields of subcarrier offset and pdcch-ConfigSIB1 information contained in MIB are reused together to indicate the frequency range with no Cell-Defining SSB. The value of $k_{SSE} = 31$ in the FR1 system and $k_{SSE} = 15$ in the FR2 system indicates that the UE can assume there is no Cell-Defining SSB in the GSCN range $[N_{GSCN}^{Reference} - N_{GSCN}^{Start}, N_{GSCN}^{Reference} + N_{GSCN}^{End}]$, wherein N_{GSCN}^{Start} and N_{GSCN}^{End} are indicated by four MSB bits and four LSB bits of the field of pdcch$-$ConfigSIB1 information, respectively.

The procedure for determining GSCN range with no Cell-Defining SSB through the indication information carried in Non Cell-Defining SSB is illustrated in Fig. 6.12.

6.2 Common control channel during initial access

During the initial access procedure, a UE has not established the RRC connection with the network yet and the UE is not provided with a UE-specific control channel. Thus the UE needs to receive common control information of the cell in the common control channel to complete the initial access procedure. The UE receives common control channel

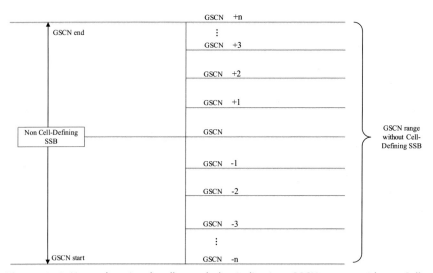

Figure 6.12 Network-assisted cell search by indicating GSCN range without Cell-Defining SSB.

transmission in the common search space (CSS). In the NR system, the following CSSs related with the initial access procedure are specified:

- Type0−PDCCH CSS: Type0−PDCCH indicates the scheduling information for a PDSCH transmission that carries the SIB1 message. The search space of Type0−PDCCH is indicated in the pdcch-ConfigSIB1 information field contained in the MIB information. When the UE enters the RRC connected state, the Type0−PDCCH CSS can also be provided through RRC signaling to a UE and the UE can use it to receive SIB1 message during cell handover or Secondary Cell group (SCG) addition. The CRC of the DCI formats associated with Type0−PDCCH is scrambled by a SI−RNTI.
- Type0A−PDCCH CSS: Type0A−PDCCH indicates the scheduling information for a PDSCH transmission that carries the OtherSystemInformation. The CRC of the DCI formats associated with Type0A−PDCCH is scrambled by a SI−RNTI.
- Type1-PDCCH CSS: Type1-PDCCH indicates the scheduling information for a PDSCH transmission that carries the RAR. The CRC of the DCI formats associated with Type1-PDCCH is scrambled by a RA−RNTI, MsgB−RNTI, or TC-RNTI.
- Type2-PDCCH CSS: Type2-PDCCH indicates the scheduling information for a PDSCH transmission that carries paging messages. The CRC of the DCI formats associated with Type2-PDCCH is scrambled by a P-RNTI.
- In each CSS, UE can detect PDCCH according to the configuration of control resource set at the corresponding monitoring occasions of PDCCH. In the rest of Section 6.2, we will focus on discussing the indication and determination methods for Type0−PDCCH CSS.

The configuration of Type0A−PDCCH CSS, Type1-PDCCH CSS, and Type2-PDCCH CSS can be provided in SIB1 message. For each of them, a dedicated PDCCH search space can be provided or the same search space as Type0−PDCCH CSS is reused. The network can also configure the aforementioned PDCCH search spaces in a noninitial DL BWP through RRC signaling to a UE in RRC connected state if the UE is configured to receive system information, RAR, or paging in that DL BWP.

6.2.1 SSB and CORESET#0 multiplexing pattern

Section 6.1.5 introduces the cell-search procedure. During the initial access procedure, a UE needs to detect SSB to obtain the configuration CORESET of Type0−PDCCH, which is referred to as CORESET#0.

Regarding multiplexing CORESET#0 and SSB in time-frequency resources, the factors to be considered include the minimum bandwidth supported by a UE and the minimum bandwidth of carrier frequency. At the RAN1#90 and #91 meetings, it was agreed that the CORESET#0 does not have to be in the same bandwidth as the corresponding SSB, but the bandwidth of CORESET#0 and the PDSCH-carrying SIB1 will be limited within the minimum bandwidth supported by a UE in a given frequency band. The frequency-domain location and bandwidth of the initial DL BWP is defined by the frequency-domain location and bandwidth of the CORESET#0. The frequency-domain resources of a PDSCH carrying the SIB1 will also be within this bandwidth. The definition and configuration of the BWP can be found in Chapter 4, Bandwidth part. At the RAN1#91 meeting, the following requirements on the configuration of CORESET#0 were agreed upon:

- The bandwidth that accommodates the resources of SSB and CORESET#0 (i.e., initial DL BWP) will be confined within minimum carrier bandwidth.
- The bandwidth that accommodates the resource of the SSB and CORESET#0 (i.e., initial DL BWP) will be confined within the minimum bandwidth of a UE.

An example of frequency-domain location of SSB and initial BWP is shown in Fig. 6.13. As shown in the example, the total bandwidth containing SSB and initial DL BWP is X. The RB offset between SSB and initial DL BWP is Y.

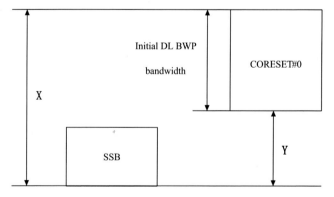

Figure 6.13 Frequency-domain location of SSB and initial DL BWP.

The minimum bandwidth of the UE is defined as the minimum of all the maximum bandwidths that a NR UE must support. The bandwidth of CORESET#0 must not exceed the minimum bandwidth of UE, otherwise a UE may not be able to access the network. According to the literature [10], during the discussion of designing CORESET#0, it was assumed that the minimum bandwidth of the UE is 20 MHz for the FR1 and 100 MHz for the FR2.

The minimum channel bandwidth is defined as the minimum channel bandwidth that can be used in a given frequency band. In the FR1, the minimum channel bandwidth is 5 MHz [6], and in the FR2, the minimum channel bandwidth is 50 MHz [7]. In fact, the channel bandwidth used in a real-field deployment of NR systems is often greater than the minimum channel bandwidth. Thus some bandwidth configuration of CORESET#0 that is larger than the minimum channel bandwidth is also supported.

The design of CORESET#0 takes into account the minimum bandwidth of UE and the minimum channel bandwidth of frequency carriers. At the RAN1#90 meeting, it was decided that the multiplexing of SSB and CORESET#0 will support at least TDM mode. The TDM mode allows us to consider whether the bandwidth of SSB and CORESET#0 meets the above minimum bandwidth restrictions separately. In contrast, if the multiplexing mode of SSB and CORESET#0 is FDM (frequency division multiplex), we would need to consider whether the total bandwidth with multiplexing SSB and CORESET #0 in the frequency domain can meet the above minimum bandwidth restrictions. Considering the minimum bandwidth can be frequency band-specific, it was agreed at the RAN1#91 meeting that FDM multiplexing SSB and CORESET#0 is not supported in the FR1, but in the FR2, both TDM and FDM multiplexing can be supported. Eventually, three multiplexing patterns were specified for SSB and CORESET#0. Multiplexing pattern 1 is a TDM mode, multiplexing pattern 2 is a TDM + FDM mode, and multiplexing pattern 3 is a FDM mode.

- Multiplexing pattern 1: In the time domain, the SSB and the associated CORESET#0 occur at different time instances. In the frequency domain, the bandwidth of the SSB completely or nearly completely overlaps with the bandwidth of the associated CORESET#0.
- Multiplexing pattern 2: In the time domain, SSB and associated CORESET#0 occur at different time instances. In the frequency domain, the bandwidth of SSB does not overlap with that of the

associated CORESET#0, but their frequency-domain locations are as close as possible to each other.

- Multiplexing pattern 3: In the time domain, SSB and associated CORESET#0 occur at the same time instance. In the frequency domain, the bandwidth of SSB does not overlap with that of associated CORESET#0, but their frequency-domain locations are as close as possible to each other.

The schematic diagrams of those three multiplexing patterns are shown in Fig. 6.14.

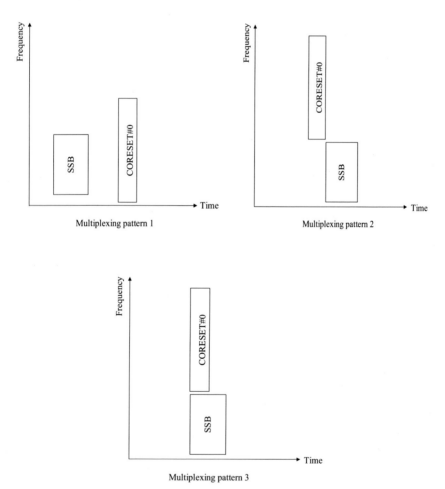

Figure 6.14 Multiplexing patterns of SSB and CORESET#0.

In Section 6.2.3, we will discuss more specific configuration parameters of CORSET#0 for those three multiplexing patterns, including the number of RB, the number of symbols, and the resource location in frequency domain. The method of indicating CORESET#0 in MIB will also be presented there.

6.2.2 CORESET#0

In Section 6.2.1, we discussed the multiplexing patterns of SSB and CORESET#0 specified in NR. In the NR specification 38.213 [8], Tables 13−1 to 13−10 specify the configurations of CORESET#0 for different multiplexing patterns and different combinations of SCS of SSB and CORESET#0. In each of those mapping tables, the frequency location, bandwidth, and symbol number of CORESET#0 are jointly encoded. Such a design not only takes into account the flexible configuration of CORESET#0, but also considers the overhead of CORESET#0 information bits carried in MIB [11,12,13].

Multiple different options were discussed for the determination of SCS of CORESET#0 during standardization discussions. At the RAN1#88bis meeting, the issue of indicating SCS of SIB1-related PDCCH and PDSCH channels was discussed, and the following two alternatives were adopted in the specification:

- Alternative 1: The SCS of SIB1-related channels is indicated in the PBCH.
- Alternative 2: The SCS of SIB1-related channels is the same as that of the PBCH.

At the RAN1#89 meeting, it was agreed upon that the CORESET#0 and the PDSCH-carrying SIB1 will use the same SCS and that SCS is indicated by the PBCH. Furthermore, it was decided that we will try to reduce the switching of SCS during the initial access phase and apply the same SCS on all the channels during the initial access procedure as much as possible. The possible combinations of SCS used by SSB and CORESET#0 were discussed by considering all possible values of SCS of SSBs and CORESET#0 in the FR1 and FR2. Eventually, the following combination of different SCSs of SSB and CORESET#0 were agreed upon at the RAN1#91 meeting:

- {SSB, PDCCH} SCS = {{15,15}, {15,30}, {30,15}, {30,30}, {120,60}, {120,120}, {240, 60}, {240, 120}} kHz°

The configuration of CORESET#0 is indicated by the four MSB bits of pdcch-ConfigSIB1 contained in the MIB. Tables 13−1 to 13−10 in NR specification 38.213 [8] specify the bandwidth, number of symbols, and frequency locations of CORESET#0. Each of those tables specifies the configurations of CORESET#0 for a particular combination of SCS of SSB, SCS of PDCCH, and the minimum channel bandwidth, as is summarized in the Table 6.7.

To obtain the configuration of CORESET#0, a UE will first determine the mapping table according to the SCS of SSB, the SCS of PDCCH, and frequency band with the minimum channel bandwidth. Then the UE determines the RB and symbol number of the corresponding CORESET#0 from the determined table according to the 4-bit CORESET#0 information contained in the MIB. The 4-bit CORESET#0 information indicates one of the rows in the table and each row corresponds to the number of RB and symbols, the multiplexing pattern, and the RB offset between the starting frequency position of the CORESET#0 and the starting frequency position of the SSB. The configuration parameters of the CORESET#0 specified in Tables 13−1 to 13−10 in 38.213 [8] are summarized in Table 6.8, which include the multiplexing pattern, the bandwidth size, the number of symbols, and the RB offset.

A UE needs the RB offset and the subcarrier offset between the RBs of the CORESET#0 and the RBs of the SSB to determine the frequency-domain position of the CORESET#0. As explained in Section 6.1.1, synchronization raster and channel raster are supported in NR for the flexible deployment. Such flexibility could result in the

Table 6.7 CORESET#0 mapping table.

Mapping table	SCS of SSB	SCS of PDCCH	Minimum channel bandwidth
13−1	15 kHz	15 kHz	5 MHz/10 MHz
13−2	15 kHz	30 kHz	5 MHz/10 MHz
13−3	30 kHz	15 kHz	5 MHz/10 MHz
13−4	30 kHz	30 kHz	5 MHz/10 MHz
13−5	30 kHz	15 kHz	40 MHz
13−6	30 kHz	30 kHz	40 MHz
13−7	120 kHz	60 kHz	−
13−8	120 kHz	120 kHz	−
13−9	240 kHz	60 kHz	−
13−10	240 kHz	120 kHz	−

Table 6.8 CORESET#0 configuration.

Mapping table	Multiplexing pattern	CORESET#0 bandwidth (RB)	CORESET#0 symbol number	RB offset
13−1	1	24	2,3	0,2,4
		48	1,2,3	12,16
		96	1,2,3	38
13−2	1	24	2,3	5,6,7,8
		48	1,2,3	18,20
13−3	1	48	1,2,3	2,6
		96	1,2,3	28
13−4	1	24	2,3	0,1,2,3,4
		48	1,2	12,14,16
13−5	1	48	1,2,3	4
		96	1,2,3	0,56
13−6	1	24	2,3	0,4
		48	1,2,3	0,28
13−7	1	48	1,2,3	0,8
		96	1,2	28
	2	48	1	−41, −42, 49
		96	1	−41, −42, 97
13−8	1	24	2	0,4
		48	1,2	14
	3	24	2	−20, −21, 24
		48	2	−20, −21, 48
13−9	1	96	1,2	0,16
13−10	1	48	1,2	0,8
	2	24	1	−41, −42, 25
		48	1	−41, −42, 49

common RB (CRB) of the CORESET#0 not necessarily being aligned with the RB of the SSB. The subcarrier offset k_{SSE} between them is indicated by the information carried in the PBCH. Considering all the possible different combinations of SCSs used by SSB and CORESET#0, the value of k_{SSB} is in the range of 0–23 for the FR1 system, and the value of k_{SSB} is in the range of 0–11 for the FR2 system. The RB offset between the CORESET#0 and the SSB is defined as the offset between the smallest RB index of the CORESET#0 for Type0–PDCCH CSS set and the smallest RB index of the CRB overlapping with the first RB of the corresponding SSB. Taking the multiplexing pattern 1 of SSB and CORESET#0 as an example, the SCSs used by SSB and CORESET#0 are 15 and 30 kHz, respectively. The frequency offset between them is shown in Fig. 6.15, where the RB offset between SSB and CORESET#0 is 2 and k_{SSE} is 23.

6.2.3 Type0–PDCCH search space

The search space of Type0–PDCCH can be referred to as SearchSpace#0. The following aspects were considered for designing SearchSpace#0:

- The number of slots included in a Type0–PDCCH monitoring occasion

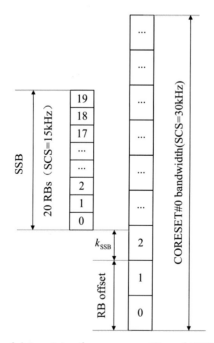

Figure 6.15 Example of determining frequency position of CORESET#0.

- The slots where a monitoring occasion is located
- The symbols where a monitoring occasion is located
- The periodicity of monitoring occasion
- The relationship between monitoring occasion and SSB

When we design any of those above aspects, the following factors will be taken into account: the multiplexing pattern of SSB and CORESET#0, SCSs used by SSB and CORESET#0, and the frequency range. For instance, if multiplexing pattern 1 is implemented for SSB and CORESET#0, the monitoring occasion of Type0−PDCCH should be multiplexed with SSB through TDM mode. Furthermore, the Type0−PDCCH monitoring occasions associated with different SSBs will be multiplexed through TDM as much as possible so that the chance of different monitoring occasions overlapping with each other can be minimized.

Tables 13−11 to 13−15 in specification 38.213 [8] specify the mapping between the PDCCH monitoring occasions of Type0−PDCCH and SearchSpace#0. The four LSB bits of the field of pdcch−ConfigSIB1 information carried in the PBCH indicates the configuration information of SearchSpace#0, which can be used by a UE to determine the PDCCH monitoring occasions of Type0−PDCCH according to the specification defined in those mapping tables. Each of the Tables 13−11 to 13−15 specify the configuration of SearchSpace#0 for a particular set of parameter values of the multiplexing pattern of SSB and CORESET#0, SCS, and frequency range, which is summarized in Table 6.9. A UE will first determine the table according to the values of those parameters and then determines the configuration parameters of the corresponding Type0−PDCCH monitoring occasions according to the 4-bit SearchSpace#0 information carried in the PBCH and the determined table.

Table 6.9 SearchSpace#0 mapping table.

Tables	SSB SCS	PDCCH SCS	Multiplexing pattern	Frequency range
13−11	−	−	1	FR1
13−12	−	−	1	FR2
13−13	120 kHz	60 kHz	2	−
13−14	240 kHz	120 kHz	2	−
13−15	120 kHz	120 kHz	3	−

A UE can determine the Type0−PDCCH monitoring occasions through the following steps. If the multiplexing of SSB and CORESET#0 follows multiplexing pattern 1, the UE monitors PDCCH occasions in the Type0−PDCCH CSS set over two consecutive slots starting from slot n_0.

$$n_0 = (0.2^\mu + \lfloor i \cdot M \rfloor)\mathrm{mod}\ N_{slot}^{frame,\ \mu} \tag{6.2}$$

where the $N_{slot}^{frame,\ \mu}$ is a number of slots per frame for SCS configuration μ. M and O are derived from SearchSpace0 configuration information and one of the mapping Tables 13−11 to 13−12 specified in 38.213 [8]. The value of O can be one of the {0, 2, 5, 7} for the FR1, and one of the {0, 2.5, 5, 7.5} for the FR2. The value of M can be one of {1/2, 1, 2}. Taking Table 13−11 in 38.213 [8] as an example, 4-bit SearchSpace#0 information indicates an index in Table 6.10 that corresponds to a set of parameters of PDCCH-monitoring occasions.

Those two consecutive slots starting from slot n_0 will be in a frame with system frame number (SFN) SFNc and the value of SFNc will satisfy the following conditions:

Table 6.10 SearchSpace#0 mapping table: Multiplexing pattern 1, FR1.

cIndex	O	Number of search space sets per slot	M	First symbol index
0	0	1	1	0
1	0	2	1/2	{0, if i is even}, {$N_{symb}^{CORESET}$, if i is odd}
2	2	1	1	0
3	2	2	1/2	{0, if i is even}, {$N_{symb}^{CORESET}$, if i is odd}
4	5	1	1	0
5	5	2	1/2	{0, if i is even}, {$N_{symb}^{CORESET}$, if i is odd}
6	7	1	1	0
7	7	2	1/2	{0, if i is even}, {$N_{symb}^{CORESET}$, if i is odd}
8	0	1	2	0
9	5	1	2	0
10	0	1	1	1
11	0	1	1	2
12	2	1	1	1
13	2	1	1	2
14	5	1	1	1
15	5	1	1	2

$$\text{SFN}_C \bmod 2 = 0, \text{ if } \left\lfloor (O \cdot 2^\mu + \lfloor i \cdot M \rfloor)/N_{\text{slot}}^{\text{frame, } \mu} \right\rfloor \bmod 2 = 0, \text{ or}$$

$$\text{SFN}_C \bmod 2 = 1, \text{ if } \left\lfloor (O \cdot 2^\mu + \lfloor i \cdot M \rfloor)/N_{\text{slot}}^{\text{frame, } \mu} \right\rfloor \bmod 2 = 1.$$

If the SSB and CORESET#0 are multiplied through multiplexing patterns 2 or 3, a UE monitors the Type0−PDCCH in a slot, and the periodicity of that slot is equal to that of the associated SSB. A SSB with index i is associated with a starting symbol in the slot with index n_c in the frame with number SFN_C, which are used to define the location of PDCCH monitoring occasion. The configuration information of PDCCH monitoring association is specified in SearchSpace#0 mapping Tables 13−13 to 13−15 in 38.213 [8]. Taking Table 13−13 of 38.213 [8] as an example, 4-bit SearchSpace#0 information in the PBCH indicates a table index in Table 6.11 that corresponds to a set of parameters of PDCCH-monitoring occasions.

Table 6.11 SearchSpace#0 mapping table: multiplexing pattern 2 and {SSB, PDCCH} SCS {120, 60} kHz.

Index	PDCCH monitoring occasions (SFN and slot number)	First symbol index ($k = 0, 1, \ldots 15$)
0	$\text{SFN}_C = \text{SFN}_{\text{SSB},i} \; n_C = n_{\text{SSB},i}$	0, 1, 6, 7 for $i = 4k, i = 4k + 1,$ $i = 4k + 2, i = 4k + 3$
1	Reserved	
2	Reserved	
3	Reserved	
4	Reserved	
5	Reserved	
6	Reserved	
7	Reserved	
8	Reserved	
9	Reserved	
10	Reserved	
11	Reserved	
12	Reserved	
13	Reserved	
14	Reserved	
15	Reserved	

6.3 NR random access

Random access is a critical function during the initial access procedure. In addition to supporting the traditional functions such as establishing RRC connection and maintaining uplink synchronization and handover, the NR random access also supports some NR unique functions such as preliminary alignment of uplink/downlink beams and requesting system messages, etc.

In comparison with LTE, the NR system supports flexible slot structure, beam-sweeping operation, variable SCSs, and much more rich deployment scenarios. Therefore the design of NR random access will fully take into account those characteristics of NR systems and also the requirements of various application scenarios.

NR random access will be discussed in this section. In particular, we will discuss the NR PRACH channel design, PRACH resource allocation, SSB and PRACH resource mapping, and PRACH power control.

6.3.1 Design of NR PRACH

6.3.1.1 NR preamble format

The design of the PRACH preamble format was considered first during the design of the PRACH channel. The preamble format for the LTE PRACH channel consists of a cyclic prefix (CP) and preamble, and multiple preamble formats are supported in LTE. Different LTE preamble formats have different numbers of preamble sequences in each PRACH transmission, which can provide different coverage distances. To avoid the interference of PRACH to other signals due to the lack of accurate uplink timing during initial access in PRACH transmission, guard time (GT) is usually reserved after preamble. A general structure of the LTE preamble format is shown in Fig. 6.16.

To support various cell radius sizes in NR deployment, flexible SCS, and flexible deployment scenarios, the 3GPP first agreed on the following design principles for the design of the NR PRACH preamble format:

• NR system will support multiple different preamble formats, including long and short preamble formats.

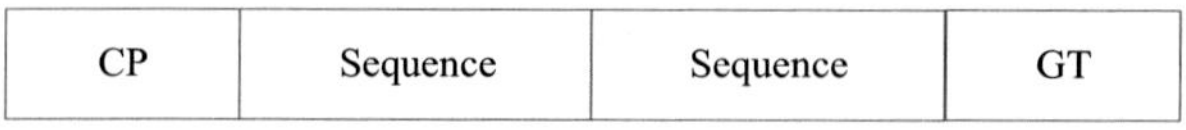

CP	Sequence	Sequence	GT

Figure 6.16 LTE preamble format.

- NR system will support repeated RACH preambles in one PRACH resource, which is used to support beam-sweeping operations and enhance the coverage of NR PRACH.
- NR system will support different SCSs in PRACH preamble in different frequency bands.
- The PRACH preamble could use the same or different SCS as the control channels or data channels in the same cell.

NR preamble formats could contain a single sequence or repeated/multiple sequences. For a NR preamble format with a single sequence, the design of LTE preamble format with CP + sequence + GT format are reused. However, for the NR preamble formats containing multiple or repeated preambles, multiple different design options were discussed in 3GPP, which are illustrated in Fig. 6.17:

- Option 1: In the time-domain structure, a CP is in the first part of the preamble format and GT is inserted at the last part of the preamble format. Between the CP and GT, there are multiple/repeated continuous RACH sequences.
- Option 2: In the time-domain structure, a GT is inserted at the end of the preamble format. A CP is inserted before each PRACH sequence and then multiple identical PRACH sequences are repeated continuously until the GT.
- Option 3: In the time-domain structure, one preamble format consists of multiple and repeated PRACH sequences. For each PRACH sequence, a pre-CP and a post GT are inserted before and after the PRACH sequence, respectively.
- Option 4: A CP is inserted before each RACH sequence and multiple different RACH sequences are continuously mapped in time-domain. A GT is inserted at the end of the preamble format.

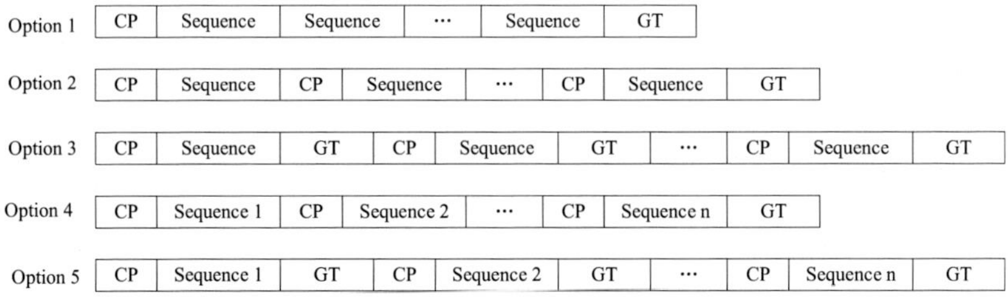

Figure 6.17 Different preamble format design options.

- Option 5: In the time-domain structure, one preamble format consists of multiple different PRACH sequences. For each PRACH sequence, a pre-CP and a post GT are inserted before and after the PRACH sequence, respectively.

Design option 1 follows the design principle of LTE's preamble format. Thus option 1 has good forward compatibility. Design option 1 also has the better resource utilization efficiency than the other options because only one CP and one GT are inserted in the preamble format per the design of option 1. Therefore a preamble format designed according to option 1 can carry more preamble sequences, which is beneficial for supporting the deployment scenarios requiring long-distance coverage and supporting the system that receive preamble in a beam-sweeping manner. Continuous transmission of preamble sequences can result in continuous preamble waveforms. Since option 1 only consists of one CP and one GT in one preamble format, generally the resource overhead of CP and GT is small. Thus with reasonable resource overhead, the system can choose a longer CP and longer GT to provide large coverage distance (such as cell radius of 100 km). In addition, because the same preamble sequence is repeated in one preamble format, one preamble sequence can provide the equivalent function of CP to the adjacent preamble sequence in the time domain.

In Options 3 and 5, the GT provides redundant function of interference protection between two adjacent sequences because the CP can provide sufficient protection between two different adjacent preamble sequences. Thus Options 3 and 5 have the lowest resource utilization efficiency. In Options 2 and 4, the GT overhead is small, but the resource overhead of CP is still large. The main advantage of Options 2 and 4 is the RACH capability can be boosted through applying different OCC (orthogonal cover code) masks by different UEs. As shown in Fig. 6.18, we can apply OCC sequences on the basis of different preamble sequences so that different UE can still use the same sequence, but different OCC masks. However, the scheme of OCC would potentially encounter some challenges that could impair the performance. For instance, due to residual frequency deviation or high-frequency band

User 1	CP	+Sequence	CP	+Sequence	CP	+Sequence	CP	+Sequence	GT
User 2	CP	+Sequence	CP	-Sequence	CP	+Sequence	CP	-Sequence	GT

Figure 6.18 Boosting PRACH capacity with OCC.

phase noise, the channel would be time-varying and thus the orthogonality between different masks applied in time domain would be lost, which would impair the performance of the RACH reception.

Considering all the above factors, we can observe that overall Option 1 has obvious advantages in many aspects. Therefore 3GPP eventually chose Option 1 for the NR RACH design. The NR RACH capacity can be boosted by configuring more RACH resources in the time domain or frequency domain.

Based on the determined basic format of the NR preamble, various preamble formats were designed by adjusting CP length, length of sequence, number of repetitions of sequence, and other parameters. Those different NR preamble formats are used to support different application scenarios, such as large cells, small cells, high-speed scenario, and medium/low speed scenarios.

NR supports four preamble formats with long sequences, which are illustrated in Table 6.12. Among them, format 0 and 1 follow the LTE format. NR format 0 is used for covering typical macrocells while NR format 1 is used for superlarge cell coverage. Format 2 uses more sequence repetition for coverage enhancement, and format 3 is applicable for high-speed mobile scenarios, such as high-speed railways. As shown in Table 6.13, NR supports short preamble formats of series A, B, and C, which are suitable for different application scenarios.

6.3.1.2 NR preamble sequence determination

During the discussion of NR standardization, the sequence selection for NR preamble transmission was also discussed. In addition to the zadoff Chu (ZC) sequences that are adopted in LTE, the m sequence and some variant of ZC sequence (such as mask ZC sequence) were also evaluated during the

Table 6.12 NR preamble formats with long sequence.

Format	SCS (kHZ)	CP (Ts)	Sequence length (Ts)	GT (Ts)	Applied scenario
0	1.25	3168	24576	2975	LTE refarming
1	1.25	21024	2×24576	21904	Macrocell up to 100 km
2	1.25	4688	4×24576	4528	Coverage enhancement
3	5	3168	4×6144	2976	High speed

Table 6.13 NR preamble formats with short sequence.

Preamble format	No. of Sequence	CP (Ts)	Sequence length (Ts)	GT (Ts)	Cell radis supported (meters)	Applied scenario
A						
1	2	288	4096	0	938	Small cell
2	4	576	8192	0	2109	Normal cell
3	6	864	12288	0	3516	Normal cell
B						
0	1	144	2048	0	469	TA is known or very small cell
1	2	192	4096	96	469	Small cell
2	4	360	8192	216	1055	Normal cell
3	6	504	12288	360	1758	Normal cell
4	12	936	24576	792	3867	Normal cell
C						
0	1	1240	2048	0	5300	Normal cell
1	2	1384	4096	0	6000	Normal cell

discussion. The ZC sequence has advantages of low peak-to-average ratio (ZC sequence has the property of constant envelope) and good characteristics of cross-correlation and autocorrelation. Furthermore, the mature design adopted in LTE can be reused including the selection of preamble root sequence and the mapping from logical root sequence to physical root sequence of ZC sequence, which can reduce the effort of standardization and impact on product implementation. Therefore it was decided that the ZC sequence should still be used in the NR preamble sequence.

6.3.1.3 Subcarrier spacing of NR preamble transmission

The first factor that will be considered in designing preamble SCS is to resist Doppler frequency offset. Thus supporting larger SCS is critical for high-speed scenarios. In the FR1, the preamble SCS of 5 kHz can support up to 500 km/h of UE speed. But in the FR2, since the wavelength of carrier frequency is shorter, larger preamble SCS will be adopted.

Secondly, NR supports beam-sweeping operation. In the scenario where the base station does not support beam correspondence, the base station may apply beam sweeping to receive the preamble sequence transmitted by a UE, so as to obtain the best receiving beam. Choosing larger subcarriers results in shorter OFDM symbol length. Thus more preamble sequences can be transmitted within each time unit, which can be used to support the beam-sweeping operation well. For instance, with 15 kHz SCS of preamble transmission, 12 beams can be supported within 1 Ms [14].

Thirdly, different cell coverage distance requires different SCS. A preamble with smaller SCS would occupy narrower channel bandwidth in frequency domain. Thus high-power spectrum density can be applied on PRACH transmission and a larger cell coverage can be supported. A preamble transmission with larger SCS would occupy shorter preamble symbols, which can effectively reduce radio resource overhead and is suitable for supporting smaller cell coverage.

The last factor that will be considered here is that a preamble transmission with smaller SCS can support longer preamble sequence, and thus more orthogonal preamble sequences and larger PRACH capacity would be supported.

In addition, preamble transmission with larger SCS and shorter preamble sequence can be helpful for supporting scenarios with stringent delay requirements. It is also more feasible to multiplex PRACH with a shorter sequence with other channels.

The design of NR PRACH sequence was determined through considering all the above factors. For the FR1, SCSs of 1.25 and 5 kHz for long sequence and 15 and 30 kHz are supported, and for the FR2 the SCSs of 60 kHz/120 kHz are supported.

6.3.1.4 Length of NR preamble sequence

The length of preamble sequence was also intensively discussed in the standardization process of 3GPP. For the macrocell scenario in the FR1, Ref. [15] showed that a preamble using long sequence (sequence length is 839) has better performance than a preamble with short sequence and large SCS. Ref. [16] also pointed out that the NR system in the FR1 should strive to provide similar cell coverage and system capacity as LTE. A long sequence is necessary to provide sufficient RACH preamble capacity. Therefore NR supports a preamble sequence with a length of 839, which is the same as LTE.

For the FR2, a shorter preamble sequence should be adopted to restrict the bandwidth occupied by the PRACH channel. Regarding the PRACH capacity in the FR2 system, the method of configuring multiple PRACH resources that are FDMed in the frequency domain can be utilized to increase the capacity. During the standardization discussion, two choices for the length of short sequences, 139 and 127, were discussed. The major benefit of adopting 127 is that it can support the UE to apply a mask code of M sequence on top of the ZC sequence and thus the PRACH capacity is increased. However, the PRACH capacity in the FR2 system can be resolved by allocating more frequency resources for PRACH channel. Furthermore, one drawback of adopting 127 is the Peak to Average Power Ratio (PAPR) is increased due to adding the mask of the M sequence to the ZC sequence. Therefore the length for short sequence was determined to be 139 [17].

6.3.2 NR PRACH resource configuration

6.3.2.1 Periodicity of PRACH resource

The design of periodicity of PRACH resources has an impact on both the latency of random access and the resource overhead of PRACH. Choosing a shorter PRACH periodicity can reduce the random access latency. In contrast, choosing a longer PRACH periodicity would lead to large random access delay. A unique feature of NR is that the beam-sweeping operation is supported. In order to support random access request transmission from UE distributed in the coverage of each individual beam, the system needs to configure PRACH resources corresponding to each individual beam. Thus in comparison with LTE where multibeam operation is not supported, the PRACH resource overhead in NR is significantly increased.

The PRACH periodicities of {10, 20, 40, 80, 160} milliseconds are specified in the NR specification. The network can choose an appropriate PRACH periodicity value by considering the balance among various factors including latency requirements and system overhead.

6.3.2.2 Time-domain resource configuration for PRACH

In addition to the PRACH periodicity, the distribution of PRACH resources in the time domain within each PRACH period will also be considered in order to determine the time-domain resource configuration of PRACH. A similar mechanism was adopted in NR as in LTE. As shown in Fig. 6.19, for the FR1 system, the subframe numbers of the subframes where PRACH resources are located are provided in PRACH resource configuration information. For the FR2, a reference slot with slot length of 60 kHz SCS is used in the indication of time-domain resource configuration of PRACH. Slot index(s) of reference slots where PRACH resources are located are provided in the configuration. One subframe in the FR1 corresponds to one PRACH time slot of SCS 15 kHz or two PRACH time slots of SCS 30 kHz. One reference time slot in FR2 corresponds to one PRACH time slot with SCS 60 kHz or two PRACH time slots with SCS 120 kHz.

Within each PRACH slot, as shown in Fig. 6.19, the network can configure one or more ROs (PRACH occasions). The so-called PRACH occasion is the time-frequency resources for PRACH preamble transmission. NR supports mixed DL/UL time-slot structure. If downlink control

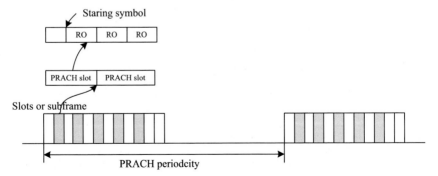

Figure 6.19 Time-domain resource configuration for PRACH.

information is transmitted in the front symbols in a PRACH slot, the corresponding resources occupied by downlink control information transmission can be reserved through configuring a proper starting symbol for the PRACH occasion in the PRACH slot.

6.3.2.3 Frequency-domain resource configuration for PRACH

As specified in the NR specification, we can multiplex 1, 2, 4, or 8 PRACH occasions in the frequency domain. An example for that is shown in Fig. 6.20. Multiplexing multiple ROs in the frequency domain can boost the PRACH capacity. Those PRACH occasions multiplexed in the frequency domain will be continuously distributed in the frequency domain. The network provides the offset of the starting PRB of the first RRACH occasion in the frequency domain relative to the starting PRB of the BWP to the UE.

6.3.2.4 Configuration of PRACH format

As mentioned in Section 6.3.1, NR supports flexible and diverse preamble formats to support various deployment scenarios. The NR network generally configures only one preamble format within one cell, which is similar to LTE. The only difference is that NR can support a mixed preamble format by packing format A and format B together. One example is shown in Fig. 6.21. Within one PRACH slot, preamble format A is located in the front ROs while preamble format B is located in the last RO.

A preamble format A contains a CP part and sequence part but no reserved guard interval GT, while a preamble format B contains a CP

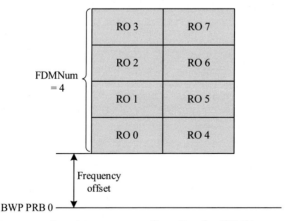

Figure 6.20 Frequency-domain resource configuration for PRACH.

part, sequence part, and reserved GT part. Format A can share an FFT window with other uplink channels. That is beneficial for simplifying the implementation of the base station, but may bring in potential interference to other channels that immediately follow the preamble transmission, as shown in Fig. 6.22. In contrast, format B needs a separate FFT window, but it is beneficial for avoiding interference. Therefore by mixing these two formats, the advantages of both formats can be fully explored: Format A is applied in the front ROs to simplify the implementation of the base station, and format B is applied on the last RO to avoid intersymbol interference on the following channels.

6.3.2.5 Mapping between SSB and PRACH occasion

NR supports multibeam operation. Before the communication between the network and a UE is built, the network needs to know the beam of which the UE is located in the coverage, and then chooses the appropriate beam direction for the subsequent data transmission. The PRACH preamble during the random access procedure is the first message transmitted by a UE to the network. Before the transmission of PRACH msg2, the network needs to know the beam direction of the UE so that the network can use the proper beam to transmit the PRACH msg2. Thus the function of reporting the beam in which the UE is located will naturally be carried in the PRACH preamble transmission. The preamble is just a sequence and it is not able to carry information explicitly.

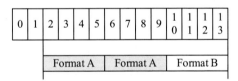

Figure 6.21 Hybrid preamble format.

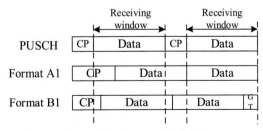

Figure 6.22 Pros and cons of preamble formats A and B.

However, the time-frequency resources occupied by preamble transmission or different preamble sequences can be used to carry beam information implicitly. A mapping between SSB and PRACH occasions is needed in the NR specification to support such implicit beam information reporting.

Before a UE initiates random access procedure, it measures and evaluates the signal quality of one cell and the signal strength of each SSB in the cell. Then, after initiating the random access procedure, the UE sends a preamble on the PRACH occasion corresponding to a SSB with the strongest or relatively stronger signal strength. If the network successfully receives that PRACH preamble transmission, the network can obtain the downlink beam information of the UE based on the PRACH occasion where the preamble is detected, and then use that beam information to transmit subsequent transmissions, such as PRACH msg2 and PRACH msg4.

Many possible proportional mapping relationships between SSB and PRACH occasions were discussed: (1) one-to-one mapping between SSB and PRACH occasion; (2) many-to-one mapping between SSB and PRACH occasion; and (3) one-to-many mapping between SSB and PRACH occasions. Considering NR needs to support diversified deployment scenarios, these three proportional relationships are all adopted in the NR specification. For instance, in a scenario with few users, the system can implement many-to-one mapping (i.e., multiple SSBs corresponding to the same PRACH occasion) to reduce the overhead of PRACH resources. The SSB can be differentiated through the preamble sequence in the same PRACH occasion and different SSBs correspond to different preamble subsets in the same PRACH occasion. In a scenario with a large number of users, the system can implement a method of one-to-many mapping and one SSB can correspond to multiple PRACH occasions, which can provide sufficient PRACH capacity.

There are generally multiple actually transmitted SSBs, multiple configured PRACH occasions, and also the corresponding preamble resources in a system. Both the network and the UE need to know the mapping between each PRACH occasion as well as the corresponding preamble resources and each SSB (i.e., which PRACH occasion and the corresponding preamble resources that each individual SSB corresponds to). Thus the mapping rules between SSBs and PRACH occasions/the corresponding preamble resources will be clearly specified in the NR specification. During the standardization discussion, the following three options were mainly discussed:

- Option 1: the scheme of mapping in the frequency domain first (as shown in Fig. 6.23)

- First in an ascending order of the preamble index within each PRACH occasion; then,
- In an ascending order of the index of the PRACH occasions that are multiplexed in the frequency domain; then,
- In an ascending order of the index of the time-division multiplexed PRACH occasions.

Fig. 6.23 shows a diagram of this method, where SSBs and ROs are one-to-one mapped. Those four SSBs are mapped with the four ROs at the first RO time position in the order of RO index from low frequency to high frequency. Then, those four ROs at the second RO time position are mapped to the SSBs again in the order of RO index from low frequency to high frequency.

- Option 2: The scheme of mapping in the time domain first (as shown in Fig. 6.23)
  - First in an ascending order of the preamble index within each PRACH occasion; then,
  - In an ascending order of the index of PRACH occasions that are multiplexed through time division; then,
  - In an ascending order of the index of PRACH occasions that are multiplexed in the frequency domain.

Fig. 6.24 shows a diagram of this method, where SSB and RO are one-to-one mapped. Those four SSBs are mapped with ROs in the order

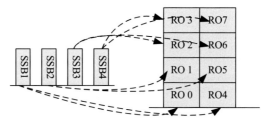

Figure 6.23 The scheme of mapping in frequency domain first.

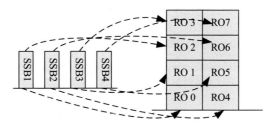

Figure 6.24 Time domain first scheme.

of time domain first and then in the order of RO index from low frequency to high frequency.

- Option 3: Whether mapping in time domain first or frequency domain first is configurable

In this scheme, it is up to network implementation to choose mapping in time domain first or mapping in frequency domain first.

Both options 1 and 2 may have restrictions in real deployment scenarios. For instance, Option 1 may result in the case where multiple PRACH occasions that are mapped to different SSBs occur at the same time position and thus would require the base station to have the capability of receiving PRACH signals with multiple beam directions simultaneously. Option 2 would cause a large time interval between two adjacent PRACH occasions that correspond to the same SSB. That would lead to large PRACH transmission latency. In contrast, Option 3 is a compromised solution. Eventually, Option 1 was adopted in the NR specification. A system implementation method can be utilized to resolve the issue when the base station can only receive with a single beam. The multiple PRACH occasions that are multiplexed in the frequency domain can be mapped to the same SSB, so that the base station can use a single receiving beam to receive them.

6.3.3 Power control of PRACH

An open-loop power control mechanism is applied to PRACH transmission. The UE determines the uplink transmission power for the PRACH transmission based on the expected receiving power provided by the network and the path loss measured from the downlink reference signal. During the random access procedure, if a UE transmits a PRACH preamble but does not receive the Random Access Response (RAR) or does not successfully receive the conflict resolution message, the UE can retransmit the PRACH preamble. If the UE supports multiple transmit beams, it can choose between switching the transmit beam on PRACH retransmission or keeping the transmit beam unchanged. If the UE does not change the transmit beam on PRACH retransmission, the transmission power of the PRACH transmission climbs up on the basis of power of the last PRACH transmission until the random access process is successfully completed. If the UE switches the transmit beam for a PRACH retransmission, whether the transmit power of the PRACH retransmission will be increased or not, is a problem and was extensively discussed during the standardization process. Increasing transmit power along with switching the transmit beam on PRACH transmission may cause more change in interference level to the target beam and thus affect the signal transmission of other users. The following four options were discussed, which are illustrated in Fig. 6.25:

- Option 1: reset the power-ramping counter.
- Option 2: the power-ramping counter remains unchanged.
- Option 3: the power-ramping counter increase.
- Option 4: a separate power-ramping counter is used for each beam [18].

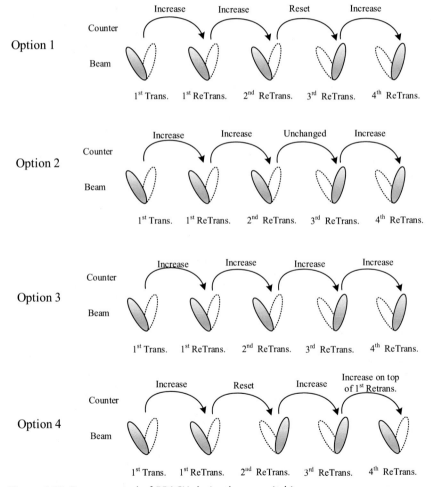

Figure 6.25 Power control of PRACH during beam switching.

Option 1 is the most conservative power control scheme among those four options. It can minimize the interference to the target beam. But due to the reset of the power-ramping counter after switching to the target beam, the UE may need to try many times to achieve the transmission power level with which the PRACH transmission can be successfully received by the network. That would increase the latency of the random access procedure. In Option 3, the power-ramping counter is increased after the transmit beam is switched. That would cause an increase of the interference to the target beam. Option 2 can be considered as a compromised solution between Option 1 and Option 3. In Option 2, the power-ramping counter remains unchanged for the first retransmission after transmit beam switching, and if the first retransmission is attempted but does not go through, the counter is incremented for the following retransmission. The method in Option 2 can take into account both of the interference-level control and the latency of random access. Option 4 maintains an independent separate power-ramping counter for each beam. After switching to a new transmit beam, the power-ramping counter is reset for the first transmission on the target beam, which can minimize the interference on the target beam. On the other hand, one improvement over Option 1 is that if the UE switches the PRACH transmission to a beam that was previously used for PRACH transmission, the power can be climbed on the basis of the power-ramping counter of that beam, instead of resetting the power-ramping counter. Therefore the method in Option 4 is helpful to restrict the latency of random access to a certain extent.

Eventually, Option 2 was adopted in the NR standard because it takes into account both the control of interference during beam switching and the latency of random access. In comparison with Option 4, Option 2 only requires maintaining a single power-ramping counter for PRACH. That can reduce the effort of standardization and complexity of implementation.

6.4 RRM measurement

The function of NR RRM measurement will be discussed in this section. Particularly, we will discuss the RRM measurement reference signal, MG, NR intrafrequency and interfrequency measurement, and other requirements (such as scheduling restrictions).

6.4.1 Reference signals for RRM

Accurate measurements of cell quality and beam quality are the fundamental basis of resource management and mobility management in a wireless mobile communication system. The 5G NR system uses two types of reference signals for RRM measurement: SSB and CSI−RS (channel state information reference signal) resources.

For SSB-based RRM measurement, a base station could configure SSB-based measurement resources to UE through higher layer signaling. Some specific parameters and mapping relationships for SSB-based RRM measurement are illustrated in Fig. 6.26.

The parameter ssbFrequency provides the center frequency of SSB transmission that is supposed to be measured, and the parameter SSB SubcarrierSpacing indicates the SCS of the SSB (e.g., 15 kHz, 30 kHz, etc.).

The parameter SSB Measurement Timing Configuration (SMTC) indicates a time-domain resource for the SSB-based measurement, which is a novel concept that was newly introduced in 5G NR. It is mainly used to indicate a group of measurement windows. The duration, position, period, and other parameters of the window can be adjusted. An example of the SMTC is shown in Fig. 6.27.

It should be noted that SMTC is provided for each frequency layer. The SMTC configured for one frequency layer indicates the available measurement windows for the UE to measure on that frequency layer.

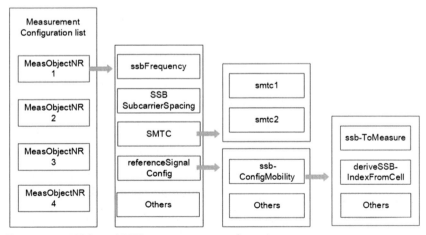

Figure 6.26 SSB-based RRM measurement configuration.

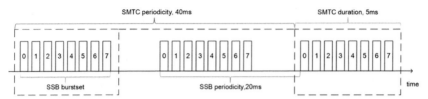

Figure 6.27 An example of SMTC.

However, this restriction was gradually relaxed during recent discussions by 3GPP. In 3GPP Release 15, it is allowed to configure up to two sets of SMTC for measurement in one frequency layer in order to match different SSB periods of different cells. For instance, in addition to a basic SMTC configuration, a second SMTC configuration with more intensive measurement window can be configured for the serving cell and the cells indicated in a specific cell list. In 3GPP Release-16, the maximum number of SMTC configurations for each individual frequency layer is extended for the idle state measurement, which can further meet the flexibility of network operation.

In addition, higher layer signaling can indicate some specific RRM measurement configurations through *ReferenceSignalConfig*. For SSB-based measurement, *ssb-ToMeasure* indicates the location of SSB sent in an SSB burst set through a bitmap. The UE can get to know which SSB candidate positions were sent SSB and which SSB candidate positions were not sent SSB through this bitmap. The parameter *DeriveSSB−IndexFromCell* indicates whether a UE can utilize the timing of its serving cell to derive the index of the SS block transmitted by neighboring cells. The advantage is that the UE can use the known timing information to determine the timing of neighboring cells, so that a large number of cell synchronization works and unnecessary measurements can be avoided.

For the measurement based on CSI−RS, a base station can configure one or more CSI−RS resources for CSI−RS-based RRM measurement. As shown in Fig. 6.28, first, the CSI−RS configuration is given at cell level, and parameters such as cell ID, measured bandwidth, and resource density are configured. Moreover, since each cell can be configured with multiple CSI−RS resources, resource-level configuration will be given as well, and parameters such as specific CSI−RS index, time-domain and frequency-domain resources for each CSI−RS resource, and sequence generation function are configured.

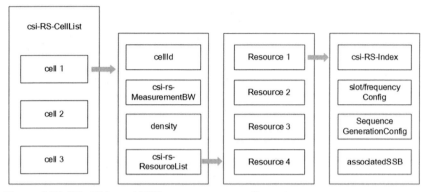

Figure 6.28 CSI—RS-based RRM measurement configuration.

In addition, 5G NR introduced *associatedSSB* to indicate the associated SSB of a CSI—RS resource. When this field is set to true, the UE needs to detect the SSB associated with the CSI—RS to determine the target cell timing. Then determine the resource location of the target CSI—RS according to the timing of the target cell, and finally measure the CSI—RS to get a corresponding measurement results. Otherwise, if the UE fails to detect the SSB associated with the CSI—RS, the CSI—RS does not need to be measured. If this field is not set to true, the UE can directly determine the location of the CSI—RS by using the timing of the cell indicated by *refServCellIndex*, and then directly perform the CSI—RS-based RRM measurement.

6.4.2 Measurement gap in NR

To support mobility management, a UE can be configured by the network to perform intrafrequency measurement, interfrequency measurement, or inter-RAT measurement within a particular time window for measurement, and then report measurement results of RSRP, RSRQ, or SINR. That particular measurement time window is referred to as measurement gap (MG) [19].

6.4.2.1 Configuration of measurement gap

As specified in 3GPP release 15, the operating frequency ranges of the NR include both frequency bands below 6 GHz of FR1 and frequency bands above 6 GHz for the millimeter wave frequency band of the FR2. 3GPP RAN4 specified per-FR MG and per-UE MG, which can be determined according to the UE's capability of frequency ranges. Three

MG configurations are supported in 5G NR: *gapFR1*, *gapFR2*, and *gapUE*. Specifically, UE-capability reporting of supporting independent gap (*independentGapConfig*) was also specified, where a UE can indicate whether the UE supports per-FR1 and/or -FR2 MG. If a UE supports the capability of independent gap, the UE can perform measurement on the FR1 and FR2 independently.

Configuration of MG includes the following higher layer parameters (Table 6.14):

- *MGL* (measurement gap length): the length of MG in Ms.
- *MGRP* (measurement gap repetition period): indicates the periodicity at which the MG is repeated.
- *MGTA* (measurement gap timing advance): indicates the timing advance of the MG. The UE can assume the MG starts MGTA Ms earlier with respect to the subframe timing.
- *gapOffset*: indicates the slot offset of the MG pattern. Its value range is from 0 to MGRP-1.

Based on the above parameters provided by the network, the UE can calculate the SFN and subframe number of the first subframe for each gap. Specifically, the SFN and subframe number are calculated as:

$$\text{SFN mod T} = \text{FLOOR}\left(gapOffset/10\right); \quad \text{subframe} = gapOffset \bmod 10; \quad T$$

$$= \text{MGRP}/10. \tag{6.3}$$

If the parameter MGTA is provided to the UE, the UE is requested to start the measurement MGTA milliseconds before the start of the gap.

6.4.2.2 Measurement gap pattern

Different MG configurations could correspond to different gap patterns, which are indicated through gap pattern IDs. In LTE, the UE only supports per-UE MG, and there are in total four gap patterns. The gap

Table 6.14 Measurement gap configuration.

Gap configuration	Values (Ms)
MGL	1.5, 3, 3.5, 4, 5.5, 6
MGRP	20, 40, 80, 16
MGTA	0, 0.25, 0.5
gapOffset	0−159

Note: the maximum value of *gapoffset* should not exceed MGRP.

pattern ID 0 and 1 are mandatory feature. Then in 3GPP release 14, the gap Pattern 2 and 3 with a shorter gap length 3 ms were added as short MG (indicated as *shortMeasurementGap-r14*). The NR supports more flexible measurement configurations than LTE. In addition to the MG configuration parameters that were already specified in LTE, NR introduced two more parameters, MG length (MGL) and MG repetition period (MGRP), to define a NR MG.

The gap patterns supported in the NR are summarized in Table 6.15. For measurement in the FR1, the NR supports MGL = 3, 4 or 6 ms. Three MGL values dedicated for the FR2 were included: {1.5, 3.5, 5.5} ms. It can be observed that the MGL in the FR2 is 0.5 ms shorter than that in the FR1. The reason for that is the RF switching time. The MG will contain the RF switching time used by the UE (including the latency of switching to target frequency and latency of switching away).

Table 6.15 NR gap pattern.

Gap pattern ID	MGL (ms)	MGRP (ms)
0	6	40
1	6	80
2	3	40
3	3	80
4	6	20
5	6	160
6	4	20
7	4	40
8	4	80
9	4	160
10	3	20
11	3	160
12	5.5	20
13	5.5	40
14	5.5	80
15	5.5	160
16	3.5	20
17	3.5	40
18	3.5	80
19	3.5	160
20	1.5	20
21	1.5	40
22	1.5	80
23	1.5	160

Generally, it is assumed that the RF switching time in the FR1 is 0.5 ms while the RF switching time in the FR2 is 0.25 ms. Thus the length of the MG in the FR2 can be 0.5 ms shorter than that in the FR1.

There are in total 24 gap patterns specified in the NR. As shown in Table 6.15, the gap patterns 0−11 are used for measurement in the FR1 while the gap patterns 12−23 are used for measurement in the eFR2.

In 3GPP release 15, the gap patterns 0 and 1 are a mandatory feature that a UE has to support. The UE is required to report the capability of supporting other gap patterns. Therefore one 22-bit bit map is used for reporting gap patterns 2−23. As an exception, the gap patterns 13 and 14 are mandatory with signaling for the FR2. It should be noted that according to common understanding in 3GPP RAN4, the terminology of "gap patterns in FR1" means "gap patterns GP#2-GP#11," and the terminology of "gap patterns in FR2" means "gap patterns GP#13-GP#23."

In 3GPP release 16, some additional gap patterns were specified as a UE mandatory feature in the RAN4 work item of RRM enhancement. It was agreed that those newly added mandatory gap patterns should be used for NR-only measurement. Here, the NR-only measurement means the target measurement objects (MOs) to be measured within the MG are all NR carriers. All release 16 UEs are mandated to support the additional mandatory gap patterns.

In practical network deployment, the most commonly used gap periodicities are 40 and 80 Ms. Thus the FR1 gap patterns 2 and 3 and FR2 gap patterns 17 and 18 are supported by almost all companies during the 3GPP standardization discussion. However, the discussion on whether to support other gap patterns got heated. Considering the flexibility of RRM measurement and complexity of UE implementation and testability, a compromised solution was reached in the 3GPP RAN4#95 e-meeting among the companies. An additional three mandatory gap patterns were added for the FR1 and FR2, respectively:

- The gap patterns GP#2, GP#3, and GP#11 will be additionally mandatory for NR-only measurement.
- The gap patterns GP#17, GP#18, and GP#19 will be additionally mandatory for NR FR2 measurement.

Furthermore, to resolve the issue of backward compatibility and avoid misalignment on different release versions between the UE and network (i.e., release 15 and release 16), RAN4 also triggered the discussion on UE behavior in different modes. All the companies agreed that the introduction of additional mandatory gap patterns in release 16 does not change the release 15 gap pattern applicability rule. Due to the interaction between LTE and NR network and different UE capabilities of measuring gap configuration, the aspects of UE capability definition, applicability of additional mandatory gap patterns, and related signaling design in 3GPP RAN group 2 needs further clarification. Therefore RAN4 discussed different UE operation modes in NR SA (standalone) and LTE SA/EN−DC/NE−DC network deployment.

- In NR SA and NR−DC deployment, the UE capability of NR-only measurement is introduced to indicate whether those additional gap patterns can only be used in NR-only measurement.
- For the LTE SA, EN−DC, and NE−DC scenarios, some companies pointed out that it is necessary to differentiate the Use that can support the new mandatory gap so that we can avoid compatibility issues with the existing release 15 networks. Eventually, RAN4 reached a compromised solution and decided to introduce additional one bit in UE capability signaling to indicate that the UE supports a full set of "mandatory additional gap patterns defined for NR SA and NR−DC for NR only measurement," which is an optional UE capability [20,21].

6.4.2.3 Applicability of measurement gap

The applicable MG configuration (including per-UE or per-FR gap and gap pattern ID) for a UE is determined according the configuration of serving cell and measurement target cells. For a UE operating in different deployment modes, for example, in EN−DC/NE−DC and NR SA mode, the applicable MG configuration would also be different.

For instance, for a UE operating in NR SA mode, if the serving cell has only FR1 cells or both FR1 and FR2 cells, and the measurement target cell is FR1-only cell, the network can configure per-UE or per-FR MG and the gap pattern can be one of the gap pattern IDs 0−11. If the serving cell has only FR1 cells or both FR1 and FR2 cells, and the measurement target cell has only non-NR RAT cells, the network can configure the per-UE gap or per-FR1 gap with gap pattern ID 0, 1, 2, or 3. But per-FR2 gap is not applicable (indicated as "No gap" in the table) here. If the serving cell has only FR2 cells and the target cell has only non-NR (such as LTE) cells, the non-NR (such as LTE) cells can be measured without MG.

The gapFR1 MG configuration is not suitable for the measurement of FR2 cell only, while the gapFR2 MG configuration is not suitable for the measurement of non-NR RAT or FR1 cell. In addition, since non-NR measurement does not support MG configurations with a periodicity of 160 Ms, the available gap patterns for measuring non-NR and FR1 simultaneously are 0, 1, 2, 3, 4, 6, 7, 8, and 10.

The available gap patterns for a UE with various configurations of serving cell and measurement target cell in various operation modes including EN−DC, NE−DC, and NR SA are summarized in Tables 6.16 and 6.17. For more details on the applicability of gap patterns, refer to Tables 9.1.2−2 and 9.1.2−3 in 3GPP specification TS 38.133.

In addition, if the network configures per-FR or per-UE MG for a UE measurement, that could cause service interruption in the serving cell. Generally, the service interruption does not exceed the configured MGL. under synchronization. For detailed interruption requirements with different values of SCS and MGTA, see the Section 9.1.2 in 3GPP specification TS 38.133.

6.4.2.4 Measurement gap sharing

When a UE performs both intrafrequency and interfrequency measurement within the same MG, the UE has to coordinate the time allocated for each measurement within that gap. Basically, the mechanism of MG sharing in NR follows that in LTE. Similar to the configuration of MG, the configuration of MG sharing (*gapsharing*) can be the configuration of per-UE and per-FR, which are *gapSharingUE*, *gapSharingFR1*, or *gapSharingFR2*.

- In the E−UTRA−NR dual-connectivity scenario, for a UE configured with per-UE MG, the MG sharing will be applied to the following:
  - Intrafrequency measurement: it can be a NR intrafrequency measurement that requires MG, or a measurement with configured SMTC that are fully overlapped with the per-UE MGs, and
  - Interfrequency measurement: it can be a NR interfrequency measurement, E−UTRA interfrequency measurement, UTRA measurement or GSM measurement.
- In NR standalone mode, for a UE without NR−DC operation and configured with per-UE MG, the MG sharing will be applied to:
  - Intrafrequency measurement: it can be a NR intrafrequency measurement that requires MG, or a measurement with configured SMTC that are fully overlapped with the per-UE MGs, and
  - Interfrequency measurement: it can be a NR interfrequency measurement or a E−UTRA measurement that requires a MG. However, the measurement of 2G or 3G is not included here (Fig. 6.29).

Table 6.16 Applicability for gap pattern configurations supported by the E−UTRA−NR dual-connectivity UE or NR−E−UTRA dual-connectivity UE.

Measurement gap pattern configuration	Serving cell	Measurement purpose	Applicable gap pattern Id
Per-UE measurement gap	E−UTRA + FR1, or	non-NR RAT [Note1,2]	0,1,2,3
	E−UTRA + FR2, or	FR1 and/or FR2	0−11
	E−UTRA + FR1 + FR2	non-NR RAT [Note1,2] and FR1 and/or FR2	0, 1, 2, 3, 4, 6, 7, 8,10
Per-FR measurement gap	E−UTRA and, FR1 if configured	non-NR RAT [Note1,2]	0,1,2,3
	FR2 if configured		No gap
	E−UTRA and, FR1 if configured	FR1 only	0−11
	FR2 if configured		No gap
	E−UTRA and, FR1 if configured	FR2 only	No gap
	FR2 if configured		12−23
	E−UTRA and, FR1 if configured	non-NR RAT [Note1,2] and FR1	0, 1, 2, 3, 4, 6, 7, 8,10
	FR2 if configured		No gap
	E−UTRA and, FR1 if configured	FR1 and FR2	0−11
	FR2 if configured		12−23
	E−UTRA and, FR1 if configured	non-NR RAT [Note1,2] and FR2	0, 1, 2, 3, 4, 6, 7, 8,10
	FR2 if configured		12−23
	E−UTRA and, FR1 if configured	non-NR RAT [Note1,2] and FR1 and FR2	0, 1, 2, 3, 4, 6, 7, 8,10
	FR2 if configured		12−23

Table 6.17 Applicability for gap pattern configurations supported by the UE with NR standalone operation (with single carrier, NR CA, and NR−DC configuration).

Measurement gap pattern configuration	Serving cell	Measurement purpose NOTE 2	Applicable gap pattern Id
Per-UE measurement gap	FR1 NOTE5, or	FR1 + FR2 only NOTE3	E−UTRA 0,1,2,3
	E−UTRAN and FR1 and/or FR2 NOTE3	FR1 and/or FR2	0−11
	FR2 NOTE5	E−UTRA only NOTE3	0,1,2,3
		FR1 only	0−11
		FR1 and FR2	0−11
		E−UTRAN and FR1 and/or FR2 NOTE3	0, 1, 2, 3, 4, 6, 7, 8,10
FR2 only			12−23
Per-FR measurement gap	FR1 if configured	E−UTRA only NOTE3	0,1,2,3
	FR2 if configured		No gap
	FR1 if configured	FR1 only	0−11
	FR2 if configured		No gap
	FR1 if configured	FR2 only	No gap
	FR2 if configured		12−23
	FR1 if configured	E−UTRA and FR1 NOTE3	0, 1, 2, 3, 4, 6, 7, 8,10
	FR2 if configured		No gap
	FR1 if configured	FR1 and FR2	0−11
	FR2 if configured		12−23

(Continued)

Table 6.17 (Continued)

Measurement gap pattern configuration	Serving cell	Measurement purpose NOTE 2	Applicable gap pattern Id
FR1 if configured	E−UTRA and FR2 NOTE3	0, 1, 2, 3, 4, 6, 7, 8,10	
FR2 if configured		12−23	
FR1 if configured	E−UTRA and FR1	0, 1, 2, 3, 4, 6, 7, 8,10	
FR2 if configured	and FR2 NOTE3	12−23	

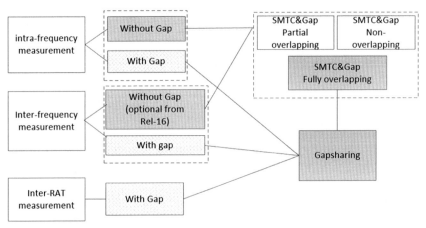

Figure 6.29 Measurement gap sharing.

The MG sharing is configured through a 2-bit RRC parameter *MeasGapSharingScheme*. Note that this parameter will be applied to the calculation of carrier-specific scaling factor, which has an impact on the measurement period of different frequency layers.

The mechanism of MG sharing causes shorter measurement duration for the measurement on each frequency layer within one MG. To compensate for the negative impact of the reduced time duration, the measurement period for each intrafrequency or interfrequency measurement must be scaled if MG sharing is configured for the UE.

The value of RRC parameter *MeasGapSharingScheme* indicates the amount of gap sharing. If the value of *MeasGapSharingScheme* is "01," "10," or "11," one value of X is indicated by this parameter as defined in Table 6.18. The

Table 6.18 Value of parameter X for NR measurement gap sharing.

measGapSharingScheme	Value of X (%)
"00"	Equal splitting
"01"	25
"10"	50
"11"	75

Note: It is left to UE implementation to determine which measurement gap sharing scheme in the table *to be applied*, when *MeasGapSharingScheme* is absent and there is no stored value in the field.

value of X and $100-X$ are the proportions of MG for intrafrequency and interfrequency measurement, respectively. Here, $K_{intra} = 1/X \times 100$ and $K_{inter} = 1/(100 - X) \times 100$, where K_{intra}和K_{inter} are used to scale the measurement period for intrafrequency and interfrequency measurement, respectively. When the value of *MeasGapSharingScheme* is set to "00," equal splitting will be applied.

For example, if the RRC parameter *MeasGapSharingScheme* is set to "01" the value of X is 25 according to the specification in Table 6.18. This indicates 25% of the gap is used for intrafrequency measurement and 75% of the gap is used for others, which leads to $K_{intra} = 4$, $K_{inter} = 4/3$. As a result, the time needed for intrafrequency measurement is four times longer, and the time needed for interfrequency measurement is 4/3 times longer. The mechanism of scaling can ensure that the actual measurement delay of each type of measurement is consistent no matter whether the MG sharing is configured or not.

6.4.3 NR intrafrequency and interfrequency measurement

The NR RRM measurement is based two types of reference symbols: SSB and CSI–RS. In 3GPP release 15, only the standardization of SSB-based RRM measurement was completed. The RAN4 requirements of CSI–RS-based L3 measurement was not defined in release 15 and was shifted to release 16 timeline.

6.4.3.1 SSB-based measurement

1. Definition of intrafrequency and interfrequency measurement.

   The frequency domain resource occupied by NR SSB has a fixed size of 20 RBs. Thus the definition of NR SSB-based intrafrequency and interfrequency measurement is mainly determined by the center frequency and SCS of the SSB. If the center frequency of the SSB of the serving cell is the same as that of the neighboring cell, and the SCSs of these two SSBs are the same, the measurement is called intra-frequency measurement. Otherwise, the SSB-based NR measurement is interfrequency measurement.

2. MG for intrafrequency and interfrequency measurement.

The NR supports intrafrequency measurement both with gap and without gap starting from release 15, which is different from LTE. If the SSBs of the neighboring cell are completely contained within the active BWP of the UE, or the active downlink BWP is the initial BWP, the intrafrequency measurement does not need a MG. Otherwise, a MG is needed.

In contrast, all the interfrequency measurements need MG for a release 15 UE. Some operators proposed to support interfrequency measurement without MG in some scenarios so that we can improve the UE through-put. The reason for that is even though the center frequency of the neighboring cell is different from that of the serving cell, the SSBs of the neighboring cell could be completely contained within the active BWP of UE. In this case, the UE does not need to adjust the RF chain for measuring the SSB of the neighboring cell. The operators expect to involve the enhancement of interfrequency measurement without MG in release 16 to improve the performance of interfrequency measurement. That was finally captured as an important objective in release 16 RRM enhancement WID.

It was acceptable to everyone that the requirement of intrafrequency measurement without MG should be a baseline. However, many issues arised during the discussion, which included new UE behaviors such as how to handle the backward-compatibility issue, the UE-capability signaling, and the relation between UE capability of "*NeedForGap*" and "interfrequency without MG" [22−25]. In addition, the existing release 15 MG-sharing mechanism and scheduling restrictions will also be clarified.

Although the operators proposed to define this function as a mandatory feature, all the companies finally compromised to define the "*interfrequency measurement without MG*" feature as an optional feature with UE capability signaling, which can be explicitly enabled with a release 16 configuration flag.

Regarding the MG-sharing mechanism, if the SMTC is fully over-lapped with configured MG, it is allowed to perform interfrequency measurements not requiring gaps within the MG and interfrequency measurements without requiring gaps can participate in MG sharing. For the case when the SMTC of the interfrequency measurement SSB does not overlap with MG configured by the network, the UE behavior is also clearly defined on how to perform interfrequency measurement outside the MG.

In particular, the controversial issue during the discussion was the UE behavior when the MG and SMTC are partially overlapped. Some companies thought that in that case, whether the UE can perform measurement outside the gap or inside the gap also depends on the capability of carrier aggregation (CA). If a UE has such capability (indicated as *interFrequencyMeas-NoGap-r16*), the UE can perform interfrequency measurement outside the gap, and the measurement requirements can be defined according to the measurement delay scaling requirements outside the gap (e.g., CSSF outside the gap). Otherwise, the requirements of CSSF within the gap will apply. After two meetings of discussion, a consensus was reached for the case where SMTC is partially overlapped with a configured MG:

- If the network configures the flag *interFrequencyConfig-NoGap-r16*, the UE will perform interfrequency measurement outside gap.
- If the network does not configure the flag *interFrequencyConfig-NoGap-r16*, the UE will perform interfrequency measurement within gap.

The relationship between "*NeedForGap*" (from RAN2) and "interfrequency without MG" (from RAN4) was also discussed in 3GPP. RAN4 replied LS on "*NeedForGap*" capability based on the agreements.

As discussed in the previous section, a UE capability indicator for interfrequency measurement without gap *interFrequencyMeas-NoGap-r16* was specified in RAN4. In addition, RAN2 also defined gap configuration parameters that indicate whether the UE needs MG for SSB-based intrafrequency or interfrequency measurement. Those parameters are *needForGapsInfoNR* reported by the UE to the network and *NeedForGapsConfigNR* provided by the network to the UE, which indicate the NR cells of intrafrequency or frequency band list of interfrequency measurement that need MG or not.

As specified in the NR specification, when a UE is provided with higher layer parameter *NeedForGap*:

- If the UE indicates "gap" but indicates the capability of supporting "interfrequency without MG," the UE can perform measurement without gap when the SSB of the target cell is completely contained in the active BWP.
- If the UE indicates "no gap," MG is not required for performing measurement on the neighboring cell regardless of the new RAN4 gapless mechanism. In other words, the signaling of *NeedForGap* in RAN2 is

independent from the discussion of interfrequency measurement without gap in release 16 RRM enhancement work item in RAN4. In this case, the measurement behavior of the UE is simpler if the UE follows the principle of *"network configuration comes first."*

6.4.3.2 CSI–RS-based L3 measurement

Due to the time limitation and that there are no urgent requirements from the industry, the CSI–RS-based mobility measurement was not fully completed in release 15. Although the work groups of RAN1 and RAN2 have completed the specification of basic functions and procedures of CSI–RS-based measurement in the physical layer and high-level protocols, respectively, there are some incomplete RAN4 definitions and requirements of measurement. RAN4 continued discussing relevant requirements for CSI–RS-based L3 measurement at the release 16 stage until the first half of 2020.

1. Definitions of intrafrequency and interfrequency measurement.

The configuration of CSI–RS resources is much more flexible than that of the SSB. The following issues were controversial during the 3GPP discussion:

- If the definition of intrafrequency and interfrequency measurement for SSB-based measurement can be reused for CSI–RS-based measurement;
- If some restrictions on the time-and frequency-domain configuration of CSI–RS resource are needed; and
- The UE capability of the minimum number of frequency layers, cells, and/or resources for CSI–RS-based measurement.

For the definition of CSI–RS-based intrafrequency measurement, the companies had the following diversed views during the 3GPP discussions [26,27]:

- Option 1: The center frequencies of the CSI–RS resources on the serving cell and neighboring cell are the same.
- Option 2: The SCSs of the CSI–RS on the serving cell and neighboring cell are the same, and the CP types of the CSI–RS on serving cell and neighboring cell are the same.
- Option 3: The bandwidth of the CSI–RS on the neighboring cell is within the active BWP of the UE.
- Option 4: The bandwidth of CSI–RS on the neighboring cell is the same to that of CSI–RS resources on the serving cell.
- Option 5: Any combination of options 1–4 described above.

Each of those five options has advantages and disadvantages. Restricting the same bandwidth of the CSI−RS resource of serving cell and neighboring cell can reduce the implementation complexity. But that would impose a limitation on the CSI−RS measurement configuration and impair the advantage of flexibility of CSI−RS resources configuration. The method of assuming CSI−RS resources of the neighboring cell being contained within the active BWP can simplify the requirements for intrafrequency measurement, but can also cause frequent changes of intrafrequency or interfrequency MOs due to BWP switching.

The definitions of CSI−RS intrafrequency measurement and interfrequency measurement were finalized at the RAN4#94-e-bis meeting, and the framework of measurement requirements was also decided. A CSI−RS-based measurement is defined as intrafrequency measurement if the following conditions are satisfied:

- The SCS of CSI−RS of the serving cell and target neighboring cell is the same.
- The CP type of CSI−RS of the serving cell and target neighboring cell is the same.
- The center frequency of CSI−RS resources in the target neighboring cell that is configured for measurement is the same as the center frequency of the CSI−RS resource in the serving cell.

Otherwise, the measurement is an interfrequency measurement. Furthermore, if the CSI−RS resource of the serving cell is not available in all configured MOs, there would be no requirements defined for the MOs that are configured for the CSI−RS-based RRM measurement.

2. Measurement scenarios.

There may exist various relationship between the bandwidth of the CSI−RS resources and the bandwidth of the active BWP. Thus some scenarios may require MG for intrafrequency measurement or interfrequency measurement and some other scenarios may not require MG [28]. Due to the limited timeline of release 16, it was decided to prioritize the following two cases:

- Define the requirements of intrafrequency measurement for the following scenarios:
  - All the CSI−RS resources in the same MO have the same bandwidth.
  - The bandwidth of the CSI−RS of the neighboring cell is within the active BWP of the UE.
  - No requirement is defined when the bandwidth of intra-MO is different from that of the CSI−RS resources configured in the serving cell in release 16.

- Define the requirements of interfrequency measurements for the following scenarios:
  - All the CSI—RS resources in the same MO have the same bandwidth.
  - Interfrequency CSI—RS resources to be measured with MG that are not confined within the active BWP.

An example of intrafrequency MO configuration is illustrated in Fig. 6.30, where different cells can support different CSI—RS bandwidths and all the CSI—RS resources are contained within the active BWP of the UE. In this example, the MG is not needed. In release 16, RAN4 only defined the measurement requirements for Cell 1 of which the CSI—RS resource has the same bandwidth as the CSI—RS resource in the serving cell, but not for Cells 2 and 3.

For the case where both SSB and CSI—RS measurements are configured in the same MO, it has been considered whether a unified

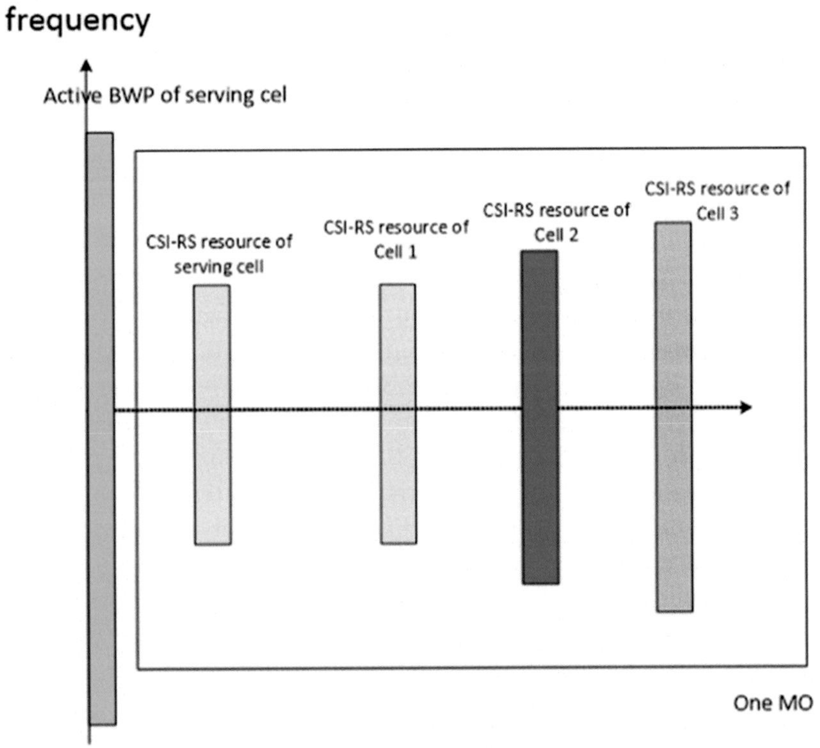

Figure 6.30 An example of MO for CSI—RS-based L3 intrafrequency measurement.

definition for the SSB-based and CSI−RS-based intrafrequency measurement/interfrequency measurement is necessary. Some companies have suggested that a unified definition is needed so that intrafrequency (or interfrequency) measurement of CSI−RS and SSB can be supported to save the UE's RF tuning time by sharing the same UE bandwidth. Another viewpoint was that CSI−RS and SSB measurement are two different types of measurement and their requirements on frequency layers, number of cells, number of resources, and other UE measurement capabilities are separate and independent. For instance, the SSB and CSI−RS for mobility configured in the same MO should be considered as two different layers. Eventually, at the RAN4 96 e-meeting, companies compromised to the second option where the CSI−RS-based measurement and SSB-based measurement have independent UE measurement capabilities [29].

3. Time-domain restriction.

The configuration of periodic CSI−RS resource can be very flexible, which could cause high implementation complexity of UE measurement. To simplify the UE measurement, imposing some restrictions on CSI−RS resource configuration in the time domain is one potential solution. That was supported by almost all companies in RAN4, even though different options for the solution were proposed:

- One option is to introduce a measurement window for CSI−RS-based measurement, which may be called CMTC (CSI−RS Measurement Timing Configuration), similar to the SMTC for SSB-based measurement. The parameters for configuring CMTC is still under discussion. For instance, one possible configuration is a length with 5 Ms and a periodicity of no more than 40 Ms.

- Another option is to reuse the SMTC as CSI−RS measurement windows to avoid additional signaling. If a UE is allowed to perform both CSI−RS-based and SSB-based measurement within the same measurement windows, the implementation complexity at the UE can be reduced but that would limit the flexibility of CSI−RS configuration. This solution was supported by the UE and chipset vendors, but was strongly objected to by network vendors [30].

A compromised conclusion, which is a simpler solution, was reached in RAN4 right before release 16 was completed. The adopted solution is that the CSI−RS resources per frequency layers are always configured within a 5 Ms window at any location.

6.4.3.3 Scaling of measurement delay

The measurement windows (e.g., SMTC or CSI−RS resource period) of SSB-based or CSI−RS-based measurement may not completely overlap with MG. They could be nonoverlapping, fully overlapping, or partial overlapping. Therefore we need scale the measurement period to meet the measurement accuracy requirements. That is the motivation for the mechanism of measurement delay scaling.

Through a mechanism of measurement delay scaling, a UE can receive enough samples of reference signal for evaluation and then report the measurement results to the network within per intrafrequency or interfrequency measurement period. With that, the measurement can meet the measurement accuracy requirements (e.g., measurement accuracy of RSRP, RSRQ, and SINR).

Two types of measurement delay scaling are defined. The first type of scaler is applicable to the scenarios of measurement without gap (e.g., Kp). The second type of scaler is applicable to the case where a UE with capability of CA is configured to monitor multiple MOs (e.g., carrier-specific scaling factor, CSSF). The second type can be applied on the relaxation of cell identification and measurement period of intrafrequency or interfrequency measurements. The configuration scalers could be impacted by the length/period of measurement reference signal and the length/period of the MG.

- Kp

    If any SSB or CSI−RS resource in the SSB or CSI−RS MO fully overlaps with a MG, the SSB or CSI−RS MO will be measured within the gap. Otherwise, the UE will follow:
    - If intrafrequency SMTC does not overlap with MGs or intrafrequency SMTC fully overlaps with MGs, or if intrafrequency CSI−RS resource does not overlap with MGs, Kp = 1.
    - If intrafrequency SMTC partially overlaps with MGs, $Kp = 1/[1 − (SMTC \ period/MGRP)]$, where SMTC period $<$ MGRP. If intrafrequency CSI−RS resource partially overlaps with MGs, $Kp = 1/[1 − (CSI−RS \ resource \ period/MGRP)]$.
    - For the other cases, no requirement is defined in 3GPP and thus is up to UE implementation.
- CSSF (carrier-specific scaling factor)

    If a UE is configured to monitor multiple MOs, the CSSF values are categorized into $CSSF_{outside\_gap,i}$ and $CSSF_{within\_gap,i}$ for the measurements being performed outside MGs and within MGs, respectively. If a UE is expected to perform measurement of MO i only outside the MG, the UE can derive the cell identification and measurement period based on $CSSF_{outside\_gap,I}$ for MO i. If a UE is expected to perform measurement of MO i only within the MG, the UE can derive the cell identification and measurement period based on $CSSF_{within\_gap,i}$ for MO i.

For a UE in SA operation mode, the $CSSF_{outside\_gap,i}$ for intrafrequency SSB-based measurements performed outside MGs are listed in Table 6.19.

The value of $CSSF_{within\_gap,i}$ is affected by the configuration of *measGapSharingScheme* and the number of intrafrequency or interfrequency MOs within the gaps configured to a UE. For example, if *measGapSharingScheme* is set to equal sharing, $CSSF_{within\_gap,i} = \max[ceil(R_i \times M_{tot,i,j})]$, where

- $j = 0 \ldots (160/MGRP)-1$.
- R_i is the maximal ratio between the number of MGs within which MO i is a candidate for measurement and the number of MGs within which the MO i is a candidate but is not used for a long-periodicity measurement of positioning as defined in 3GPP TS 38.133.
- $M_{tot,i,j} = M_{intra,i,j} + M_{inter,i,j}$ is the total number of intrafrequency, interfrequency, and inter-RAT MOs that are candidates to be measured within a gap j, if the MO I is also a candidate within the gap j. Otherwise, $M_{tot,i,j}$ equals 0.
- It should be noted that the positioning measurements were specially considered, and their impact is still under discussion in 3GPP RAN4 (e.g., how to define long periodicity). For more details, refer to Section 9.1.5.2 of the latest version of TS 38.133.

For cases if the parameter *measGapSharingScheme* is not set to equal sharing, a similar calculation method can be applied for $CSSF_{within\_gap,i}$ that can be derived based on K_{intra} (or K_{inter}) additionally.

6.4.4 Scheduling restrictions caused by RRM measurement

As mentioned in the previous section, the scheduling restriction only happens in the scenarios of measurement without MG. The cause for scheduling restrictions could be one of the followings factors: whether a UE supports simultaneous reception of data and measurement reference signals (SSB or CSI−RS) with different SCSs, whether the UE supports Rx beam sweeping, or whether the measurement is in a TDD band.

Table 6.19 CSSF$_{outside\_gap}$ in SA mode.

Scenario	CSSF$_{outside\_gap,i}$ for FR1 PCC	CSSF$_{outside\_gap,i}$ for FR1 SCC	CSSF$_{outside\_gap,i}$ for FR2 PCC	CSSF$_{outside\_gap,i}$ for FR2 SCC where neighbor cell measurement is required	CSSF$_{outside\_gap,i}$ for FR2 SCC where neighbor cell measurement is not required	CSSF$_{outside\_gap,i}$ for interfrequency MO with no measurement gap
FR1 only CA	1	Number of configured FR1 SCell(s) + Y	N/A	N/A	N/A	Number of configured FR1 SCell(s) + Y
FR2 only intraband CA	N/A	N/A	1	N/A	Number of configured FR2 SCell(s) + Y	Number of configured FR2 SCell(s) + Y
FR2 only inter band CA	N/A	N/A	1	2 Note 3,5	2 × [Number of configured SCell(s) + Y−1]	2 × [Number of configured SCell(s) + Y −1]
FR1 + FR2 CA (FR1 PCell) Note 1	1	2 × [Number of configured SCell(s) + Y−1]	N/A	2 Note 3	2 × [Number of configured SCell(s) + Y − 1]	2 × [Number of configured SCell(s) + Y − 1]

Note 1: Only one FR1 operating band and one FR2 operating band are included for FR1 + FR2 interband CA.
Note 2: Selection of FR2 SCC where neighbor cell measurement is required follows clause 9.2.3.2.
Note 3: CSSF$_{outside\_gap,i}$ = 1 if only one SCell is configured and no interfrequency MO without gap.
Note 4: Y is the number of configured interfrequency MOs without MG that are being measured outside of MG.
Note 5: Only two NR FR2 operating bands are included for FR2 interband CA.

In the following discussions on scheduling restrictions, the SSB-based intrafrequency measurement without MG will be used as an example. The same principle can be applied on CSI—RS-based measurement.

In general, a UE cannot transmit or receive data on the OFDM symbols that are occupied by SSBs measured by the UE or on any OFDM symbol within a SMTC window. Furthermore, in some particular scenarios, the scheduling restriction requests the UE not to transmit or receive data on the symbols before and after the OFDM symbols carrying SSB due to the consideration on the issue of synchronization on signal reception.

In 3GPP release 15, simultaneous reception of data and SSB with different SCSs was introduced as an optional UE capability. From the perspective of UE implementation, a UE supporting such capability is able to perform two sets of FFTs simultaneously to process different SCSs. If the UE does not support this capability, the UE is not expected to transmit PUCCH/PUSCH/SRS or receive PDCCH/PDSCH/TRS or CSI—RS for CSI feedback on any OFDM symbols where the UE measures SSB. Furthermore, such a UE is not expected to transmit PUCCH/PUSCH/SRS or receive PDCCH/PDSCH/TRS or CSI—RS for CSI feedback on the OFDM symbol immediately before and after the OFDM symbols carrying SSB to be measured within SMTC window duration (if the higher layer parameter deriveSSB_IndexFromCell is enabled). It should be noted that that rule is applicable to SSB-based intrafrequency measurements without MGs in release 15 and also SSB-based interfrequency measurements without MGs in release 16 [19].

6.5 Radio link monitoring

Radio link monitoring (RLM) is the procedure that a UE in RRC_CONNECTED state monitors the downlink radio link quality of the primary cell. In this section, the RLM reference signals and RLM procedures specified in NR will be discussed.

6.5.1 RLM reference signal

The reference signal used in NR RLM (RLM Reference Signal) is configured through higher layer parameter *RadioLinkMonitoringRS*. There are two types of RLM—RS: CSI—RS and SSB. The configuration of an RLM—RS includes a resource index of CSI—RS, or an index of SSB. Multiple RLM—RSs can be configured for a UE in each BWP. The maximum number of configured RLM—RS is determined by the frequency range: two RLM−RS in the frequency bands below 3 GHz; four RLM−RS in the frequency bands between 3 and 6 GHz; and eight RLM−RS in the frequency bands above 6 GHz [19]. The operation of beam sweeping and the limitation of UE capability were considered when determining the maximum number of configured RLM−RS in different frequency ranges. The measurement results of RLM−RS are used to evaluate the hypothetical BLER of PDCCH. When multiple RLM−RSs are provided, UE can assume all the RLM−RSs have the same antenna port for evaluating the hypothetical BLER of PDCCH.

For SSB-based RLM, the network can provide one or more indexes of SSB as the RLM−RS for a UE. Because of the multibeam transmission in NR, the network has to configure a plurality of SSBs as the RLM−RS for the beams serving the UE during a time period. The UE can measure those SSBs and determine the in-sync/out-of-sync (IS/OOS) status of the link according to the signal quality of those SSBs.

CSI−RS resources are configured UE-specifically. Thus in CSI−RS-based RLM, the network has more flexibility in configuring RLM−RS resources for a certain UE. The network can provide one or more indexes of CSI−RS resources for a UE as the RLM−RS to support the measurement of radio link quality of multibeam transmission. Moreover, the CSI−RS can provide a better match to the PDCCH to be evaluated PDCCH in both spatial and frequency domain than the SSB. However, the CSI−RS resources used for RLM require the following restrictions on configuration. The cdm-Type of CSI−RS resources used for RLM will be "noCDM." The frequency-domain resource density can only be 1 or 3, and the number of antenna ports can only be single antenna port [12].

If a UE is not configured with higher layer parameter *RadioLinkMonitoringRS* and the UE is configured with TCI states for PDCCH reception, the UE will determine the RLM−RS by following some rules specified in the NR specification TS 38.213. Those TCI states configured for PDCCH contain one or more CSI−RS. The UE will determine the RLM−RS as follows, as specified in TS 38.213:

- The UE uses the RS provided in the active TCI state for PDCCH reception as the RLM−RS if the active TCI state for PDCCH reception includes only one RS.
- If the active TCI state for PDCCH reception includes two RS, the UE expects that one RS has QCL-TypeD. The UE uses the RS with QCL-TypeD for RLM.

- The UE is not required to use an aperiodic or semipersistent RS for RLM.
- For $L_{max} = 4$, the UE selects the N_{RLM} RS provided for active TCI states for PDCCH receptions in CORESETs associated with the search space sets in an order from the shortest monitoring periodicity. If more than one CORESET is associated with search space sets having the same monitoring periodicity, the UE determines the order of the CORESET from the highest CORESET index.

If a UE is configured with multiple downlink BWPs in a serving cell, the UE only performs RLM on the active DL BWP using RLM−RS configured for this BWP for RLM measurement. If no RLM−RS is configured on the active BWP, the CSI−RS corresponding to the active TCI state for PDCCH reception is used as the RLM−RS, as explained above.

6.5.2 RLM procedure

During RLM, the UE measures the configured RLM−RS, and then compares the measurement results with the threshold of the IS/OOS to determine the IS/OOS status of the radio link. The UE reports the determined IS/OOS status periodically to the higher layer. If the measurement results of any one of the configured RLM−RS is higher than the IS threshold, the IS status is determined and then reported to the higher layer. If the measurement result of all configured RLM−RS is lower than the OOS threshold, OOS status is determined and reported to the higher layer.

In non-DRX mode operation, the periodicity of reporting IS/OOS status is the maximum of the shortest periodicity of the RLM resources and 10 msec. In DRX mode operation, the periodicity of reporting IS/OOS status is the maximum of the shortest periodicity for RLM resources and the DRX period.

Similar to LTE, the IS/OOS threshold for RLM in NR are determined by BLER of the hypothetical PDCCH. The difference is that the NR supports two groups of BLER of hypothetical PDCCH. The first group is the same as that in LTE, which is that the IS threshold corresponds to 2% BLER of hypothetical PDCCH, and the OOS threshold corresponds to 10% BLER of hypothetical PDCCH [19]. The purpose of introducing one additional group of thresholds is to maintain the radio link connection and also avoid radio link failure caused by poor radio link quality. That is beneficial for maintaining continuity of services such as VoIP. The network can use the higher layer parameter *rlmInSyncOutOfSyncThreshold* to indicate the IS/OOS threshold group that the

UE will use in RLM [31]. The SINR value corresponding to the IS/OOS threshold is not specified in the NR specification. Each vendor can determine the SINR value of IS/OOS by implementation according to the BLER of hypothetical PDCCH and the performance of the UE receiver.

6.6 Summary

This chapter introduced cell search for NR initial access, type-0 PDCCH coreset and type-0 PDCCH search space, NR random access process, and NR RRM measurement and NR RLM measurement. To summarize, the design of NR initial access can well support the features of large bandwidth, multibeam transmission, and flexible deployment scenarios in the NR system.

References

[1] R1−1708161. Discussion on SS block composition and SS burst set composition. Huawei, HiSilicon RAN1#89.
[2] R1−1707337. SS block composition. Intel Corporation, RAN1#89.
[3] R1−1708569. SS block and SS burst set composition consideration. Qualcomm Incorporated, RAN1#89.
[4] R1−1708720. SS block composition and SS burst set composition. Ericsson.
[5] R1−1718526. Remaining details on synchronization signal design. Qualcomm Incorporated, RAN1#90bis.
[6] 3GPP TS 38.101−1. User equipment (UE) radio transmission and reception: part 1: range 1 standalone, V16.2.0 (2019−12).
[7] 3GPP TS 38.101−2. User equipment (UE) radio transmission and reception: part 2: range 2 standalone, V16.2.0 (2019−12).
[8] 3GPP TS 38.213. Physical layer procedures for control, V16.1.0 (2020−03).
[9] R1−1802892. On indication of valid locations of SS/PBCH with RMSI. Nokia, RAN1#92.
[10] R1−1721643. Reply LS on minimum bandwidth. RAN4, 2017, CATT, NTT DOCOMO.
[11] R1−1717799. Remaining details on RMSI. CATT, RAN1#90bis.
[12] 3GPP TS 38.214. Physical layer procedures for data, V16.1.0 (2020−03).
[13] R1−1720169. Summary of offline discussion on remaining minimum system information, CATT, RAN1#91.
[14] R1−1700614. Discussion on 4-step random access procedure for NR. NTT DOCOMO, INC., RAN1Ad-hoc #1 2017.
[15] R1−1704364. PRACH evaluation results and design. ZTE, ZTE Microelectronics, RAN1#88.
[16] R1−1705711. Discussion and evaluation on NR PRACH design. NTT DOCOMO, INC., RAN1#88bis.
[17] R1−1716073. Discussion on remaining details on PRACH formats. NTT DOCOMO, INC., RAN1Ad-hoc #3 2017.
[18] R1−1706613. WF on power ramping counter of RACH Msg.1. Retransmission/ Mitsubishi Electric, RAN1#88bis.

[19] 3GPP TS 38.133. Requirements for support of radio resource management. V16.2.0 (2019−12).

[20] R4−2008992. LS on mandatory of measurement gap patterns. RAN4(ZTE), 3GPP RAN4#95e.

[21] R4−2005846. LS on mandatory of measurement gap patterns. RAN4(ZTE), 3GPP RAN4#94e-bis.

[22] R4−1912739. WF on interfrequency without MG. CMCC, 3GPP RAN4#92bis, Chongqing, China.

[23] R4−1915853. WF on interfrequency without MG. CMCC, 3GPP RAN4#93, Reno, USA.

[24] R4−2002250. WF on interfrequency without MG was agreed in RAN4#94e meeting. CMCC, 3GPP RAN4#94^e.

[25] R4−2005348. WF on R16 NR RRM enhancements—inter-frequency measurement without MG. CMCC, 3GPP RAN4#94e-bis.

[26] R4−2005355. WF on CSI−RS configuration and intra/interfrequency measurements definition for CSI−RS-based L3 measurement. CATT, 3GPP RAN4#94e-bis.

[27] R4−2009037. Email discussion summary for [95e][225] NR_CSIRS_L3meas_RRM_1. Moderator (CATT), 3GPP RAN4#95e.

[28] R4−2009256. WF on CSI−RS configuration and intra/interfrequency measurements definition for CSI−RS-based L3 measurement. CATT, 3GPP RAN4#95e.

[29] R4−2009038. Email discussion summary for [95e][226] NR_CSIRS_L3meas_RRM_2. Moderator (OPPO). 3GPP RAN4#95e.

[30] R4−2009009. WF on CSI−RS-based L3 measurement capability and requirements. OPPO, 3GPP RAN4#95^e.

[31] 3GPP TS 38.331. Radio Resource Control (RRC) protocol specification, 16.0.0 (2020−03).

CHAPTER 7

Channel coding

Wenhong Chen, Yingpei Huang, Shengjiang Cui and Li Guo

Since entering the era of digital communication, channel coding has been the most fundamental technology in the evolution of communication systems. The huge improvement in system performance, reliability, and capacity of each generation of wireless communication system is inseparable from the constant enhancement of channel coding scheme. In the 4G system, in order to meet the performance requirements under different block sizes and code rates, the control channel adopts Reed—Muller (RM) code and Tail-bit Convolutional Code (TBCC), while the data channel adopts the Turbo code. For both data and control information, NR puts forward higher requirements in terms of robustness, performance, complexity, and reliability than the 4G system. Therefore in this chapter new channel coding schemes are discussed and specified for different channel in NR.

7.1 Overview of NR channel coding scheme

Starting from the RAN1#84bis meeting (2016—04), it took the 3GPP five meetings to finally determine the channel coding scheme for the NR data channel and the control channel. After that, starting from the first RAN1 AdHoc meeting in January 2017, the 3GPP finally confirmed the design details of the determined channel coding schemes and completed the standardization of NR channel coding after eight RAN1 meetings (six normal meetings and two additional meetings), thereby the standardization of NR channel coding was completed. So how were the channel coding schemes for different channel type determined in NR? Why were such channel coding schemes selected? This section will reveal the answers one by one.

7.1.1 Overview of candidate channel coding schemes

In this section, we will first introduce a few common channel coding schemes, including RM code, TBCC code, and Turbo code adopted in LTE as well as the newly proposed outer code, Low-Density Parity-Check

5G NR and Enhancements
DOI: https://doi.org/10.1016/B978-0-323-91060-6.00007-6
© 2022 Elsevier Inc.
All rights reserved.

(LDPC) code, and Polar code. These coding schemes are considered as candidate channel coding schemes in NR.

RM code is a type of linear block code that can correct multiple errors and that was proposed by Reed and Muller in 1954. This type of code is simple in structure and rich in structural characteristics, and can be decoded by soft decision or hard decision algorithm. It has been widely used in practical engineering. In LTE, both channel quality indicator (CQI) and hybrid automatic repeat request acknowledgment (HARQ–ACK) apply the RM encoding method [1].

Outer code refers to the coding method comprised of a layer of other code in addition to the main coding scheme. In the wireless mobile communication system, due to the influence of multipath, Doppler shift, obstacles, etc., in the wireless channel, random errors (single scattered errors) or burst errors (large number of errors in a piece) will often occur during the data transmission. In these cases, the outer code can improve the decoding performance. Outer code generally can be categorized as explicit outer code and implicit outer code. Common explicit outer codes are Cyclic Redundancy Check (CRC), Reed–Solomon (RS) code, Bose Ray–Chaudhuri Hocquenghem (BCH) code, etc., which are usually applied outside the main coding scheme (inner code). Implicit outer code refers to the mixing of inner code and outer code, and one of them can be processed first or both can be processed iteratively. In practical applications, the outer code and the inner code may not be in the same protocol layer. For example, the inner code works in the physical layer, while the outer code works in the MAC layer. The inner code and outer code can also be applied on different data. For example, the inner code is applied for each coding block, while the outer code is applied on multiple coding blocks [2].

Convolutional code was first proposed by MIT Professor Peter Elias in 1955 [3]. It has been widely used in the dedicated control channel of CDMA2000 [4], data channel of Wideband Code Division Multiple Access (WCDMA), and also in LTE [1]. The previous conventional convolutional encoder needs to be initialized before it starts to work. At the end of encoding, the registers will be set to zeros by using the tail bits. The tail bits increase the coding overhead, resulting in decoding performance degradation and the loss of coding rate. In order to solve this problem, some researchers proposed TBCC code [5,6]. The basic principle is to directly initialize the encoding registers with the last few bits of the code word during the encoding process, and the encoder does not need to input additional "0" at the end of encoding, thereby improving the code rate.

Turbo code is a concatenated code proposed by Claude Berrou et al. In 1993 [7]. Its basic idea is to concatenate two component convolutional codes in parallel through an interleaver, and then the decoder performs iterative decoding between the two convolutional code decoders. Turbo code usually adopts a deterministic interleaver, so that the process of interleaving and deinterleaving can be derived by an algorithm without storing the entire interleaver table. With excellent performance and implementability in engineering, Turbo codes are widely used in WCDMA [8], 4G LTE [1], and other communication standards. In the choice of interleaver, WCDMA uses a block interleaver, while LTE adopts a quadratic permutation polynomials (QPP) interleaver.

LDPC code is a type of linear block code invented by Gallager in 1963 [9] and is often described by parity check matrix (PCM) or Tanner graph. The PCM of LDPC code is generally a sparse matrix, through which the constraint relationship between systematic bits and parity bits can be clearly described. At the same time, the sparsity of PCM can effectively reduce the complexity of decoding based on message passing algorithm. The Tanner graph divides the check node (the row in the PCM to indicate the parity check equation) and the variable node (the column in the PCM corresponding to the code bit in the code word) into two sets, and then connects them through the constraint relationship of the parity check equation. If a variable node is located in the nonzero position of the row constraint equation in the check matrix corresponding to a check node, the variable node and the check node are connected. The construction of LDPC code is equivalent to the construction of sparse PCM. The LDPC code based on Raptor-like structure can well support multiple code rate, multiple code lengths, and incremental redundancy hybrid automatic repeat request (IR−HARQ), while quasicyclic (QC) structure makes it easy to implement a low-complexity and high-throughput codec. LDPC code has been widely adopted in communication systems such as WiMAX, Wi-Fi, DVB−S2, etc. [10−12], and was also discussed as a candidate scheme in early LTE [13].

Polar code is a channel coding method invented by Professor Erdal Arikan in 2008 [14]. It is currently the only channel coding method that has been proven to be able to reach the Shannon limit in binary erasure channel (BEC) and binary discrete memoryless Channel (B−DMC). Polar code is mainly comprised of channel merging, channel separation and channel polarization. Channel merging and channel polarization are completed during encoding and channel separation is completed during decoding. The encoding of Polar code is a process of selecting an appropriate subchannel to carry data and an appropriate subchannel where frozen bits are placed, and then performing logical operation. The evaluation and ranking of reliability for subchannel will directly affect the selection of information bit set, which in return affects the performance of Polar code. The decoding of Polar code usually adopts the successive cancellation (SC) decoding, which can achieve Shannon capacity when the code length is infinite. When the code length is limited, the decoding performance can be improved by decoding algorithms such as SC list (SCL). For the short code, we can directly use the searching method to the concatenated Polar code with the best code distance. For the long code, adding CRC bit or parity bit can improve the performance of Polar code.

7.1.2 Channel coding scheme for data channel

In the previous section, we introduced several common channel coding schemes, which were widely discussed as candidates at the early stage of NR standardization. At the first 3GPP meeting on NR channel coding (RAN1#84bis meeting), companies proposed the preliminary recommendations from the above candidate solutions based on their own research results, including Polar code [15], TBCC code [16], Outer code [17], LDPC code [16], and Turbo code [18]. Some companies also proposed RM code, TBCC code, and Polar code as candidate channel coding schemes for broadcast and control channels, and Turbo code, LDPC code, and Polar code as candidate channel coding schemes for data channel [19]. At the same time, the contributions from different companies also discussed the evaluation methodology of channel coding, including the requirements and performance metrics of NR channel coding.

Based on the viewpoints of various companies, several companies jointly proposed a technical framework [20], suggesting LDPC code, Polar code, and TBCC code be used as candidate coding technologies for NR data channel, and also proposed corresponding performance evaluation metrics. After long discussion on the proposal, the meeting decided to use LDPC code, Polar code, convolutional code (LTE convolutional code or enhanced convolutional code), and Turbo code (LTE Turbo code or enhanced Turbo code) as the candidate channel coding schemes for NR data channel. Accordingly, performance, implementation complexity, encoding, and decoding delay and flexibility (such as supported code length, code rate, and HARQ) will be considered as the factors of down selection. Although other channel coding schemes had not been excluded for the time being, it was difficult for them to be the focus of 3GPP research. In the subsequent 3GPP meetings, various companies conducted many technical discussions and performance evaluations around these channel coding schemes.

At the RAN1#85 meeting, based on different service types (such as eMBB, URLLC, and mMTC) and different bit rates (high/low bit rates), companies performed a lot of performance evaluation and comparison on the candidate coding schemes from the perspective of complexity, performance, and flexibility.

- Some companies believed that different coding schemes had similar performance in different scenarios, but LTE Turbo code had higher decoding complexity than LDPC code. Thus LDPC code can be superior to LTE Turbo code in terms of delay and high data throughput [21].
- Some other companies believed that small block codes (such as blocks smaller than 1000 bits) should give priority to Polar code and TBCC code for better performance [22].
- Some companies thought that TBCC has certain performance advantages over LDPC and Polar when the block size is small [23].
- Based on LTE Turbo code, some companies proposed an enhanced Turbo code [24–27]. They believed that Turbo code can provide flexibility under various code rates and support encoding of 40–8192 bits.

Because the simulation results from various companies are quite divergent, the output format of the simulation results was unified at this meeting, and each company was required to provide the complexity analysis of the used algorithms. At the same time, some companies also presented specific design schemes for Polar code, LDPC code, Turbo code, outer code, and TBCC code. Among them, the LDPC code, which failed in the competition with Turbo code in 4G, is the most concerned channel coding scheme. And other coding schemes lacked many supporting companies in the early stages of channel coding discussions.

In the following RAN1 meeting, companies provided more comprehensive results, and discussed the advantages and disadvantages of Turbo code, LDPC code, and Polar code in details. The discussion focused on whether the channel coding schemes can meet the performance requirements of NR.

- LDPC code is a well-known channel coding technology that has been deeply studied for decades. Twenty-four companies suggested LDPC code be used in eMBB data channel [28] to provide better performance at high code rate and large data block. Almost all participating companies proposed their own LDPC design solutions, including companies that support other schemes.

- Polar code was an emerging channel coding technology that had only been raised for a few years. More and more companies have been participating in the design of Polar code. The proposal [29] jointly signed by nine companies suggested that Polar code be used as channel coding method for various services of NR, and planned to further optimize the solution in terms of performance, technical maturity, stability, implementation complexity, and power consumption, so as to better compete with LDPC code.

- At the same time, some companies still recommended LTE Turbo code as the channel coding method for NR at low code rate [30], but the enhancement schemes of supporting companies were different and overall lack competitiveness.

- The number of supporting companies for outer code gradually decreased, which almost withdrew from the competition.

Considering that channel coding is a fundamental technology of the NR, the 3GPP planned to first determine the channel coding scheme of eMBB data channel at the RAN1#86bis meeting in order to ensure the other design work of NR could be carried out smoothly. However, Turbo code, LDPC code, and Polar code have their own advantages and characteristics in terms of performance, flexibility, technical maturity, and implementability, and it was difficult decision to make. After a long and in-depth technology discussion, at the RAN1#86bis meeting, a relatively objective summary on these three candidate solutions was made [31] as follows:

- From the aspect of performance, it is hard to draw conclusions about which scheme has better performance than another scheme (in fact, this is because the simulation assumptions of various companies are not completely consistent, and implementation algorithms used are also different).

- In terms of flexibility, LDPC, Polar, and Turbo codes can all provide acceptable flexibility (i.e., they can support a certain range of code rates and block sizes).

- In terms of HARQ support, all of these three methods can support both CC−HARQ and IR−HARQ. But some companies believe that the HARQ of Turbo code has been widely used in 4G LTE, while LDPC code and Polar code lack practical application.

- Each scheme has its own advantages in terms of implementation complexity:
  - LDPC codes are widely implemented in commercial hardware supporting throughput of several Gbps and in some cases have attractive area and energy efficiency with flexibility and features that are more limited than that in NR. However, companies have concerns about it from the following aspects. The area efficiency is low at low code rates and the complexity increases with increasing flexibility. Some companies have worried the applicable flexibility is limited because a flexible switching network has impact on the power, area, and latency. It is difficult for some companies to implement BP and sum-product decoders in NR systems. For LDPC, how to achieve a certain attractive area and energy efficiency may be a challenge when simultaneously meeting the peak throughput and flexibility requirements of NR.
  - Polar code is implementable and can achieve error-free coding in theory, although it has no commercial implementation currently. The main problems of Polar code are: similar to LDPC, the area efficiency will be reduced at small block blocks and low code rates; the implementation complexity of list decoders increases with the increase of list size, especially with large block sizes; for decoding hardware that can achieve acceptable latency, performance, and flexibility, some companies have concerns about the area efficiency and energy efficiency that are achievable with Polar code.
  - Turbo code is widely implemented in commercial hardware, and can support the HARQ and flexibility required for NR, but does not meet the high data rate or low latency required by NR. Proponents consider some specific implementations of Turbo code to meet the flexibility requirements of NR with the most attractive area and energy efficiency particularly at lower code rates. The concerns are summarized as follows: some companies have worried that the latency and area and energy efficiency are not adequate for NR, and the area and energy efficiency will reduce at small code blocks; the decoding complexity increases linearly with the information block size and increases as the constraint length increases; implementation with attractive area and energy efficiency is challenging when meeting the high-throughput requirements of NR; some companies have worried that a Turbo decoder capable of

decoding both LTE and small information block sizes of NR has many problems and is difficult to achieve.

- In terms of delay, proponents believe that their respective schemes can meet the delay requirements of NR. The advantage of Turbo code and LDPC code is that intelligent delay and highly parallel decoders can help to reduce delay. Although Polar code is not highly parallelizable, supporters believe that other designs can be used to reduce the delay of the Polar decoder.

- In other aspects, the design of Turbo code and LDPC code has been relatively stable, and Polar code is the latest technology among the three schemes. Its engineering implementation is still being gradually improved. In order to meet the requirements of NR, a lot of work needs to be done in standard design for all schemes.

At the same time, many companies began to consider some combined solutions that take into account multiple technologies. For example, seven companies supporting Turbo code proposed to adopt LDPC code and Turbo code together for eMBB data channel [32]. Twenty-nine companies still recommended LDPC code as the only encoding scheme for eMBB data channel [33]. At the same time, 28 companies supported Polar code as channel coding scheme for eMBB data [34], which was close to the number of companies supporting LDPC code. Subsequently, some companies proposed the combine scheme of "LDPC + Polar" [35] (i.e., small block transmission adopts Polar code while large block transmission adopts LDPC code). As an attractive alternative, this scheme can get the advantages of both coding technologies. Overall, the discussion has been focused on two solutions of LDPC code and mixing of LDPC and Polar. After a long technical discussion, a preliminary conclusion was finally reached: LDPC code was used at least for data transmission of large block, and the coding scheme for data transmission of small block and the threshold of large and small block will be decided at the RAN1 #87 meeting.

After deciding that the LDPC code should be used for large block of eMBB data, in the RAN1#87 meeting, companies continued research and discussion on the coding scheme of small block transmission of eMBB. The competition mainly existed between two candidate solutions: LDPC code and Polar code.

- Some companies were still worried about the performance of the HARQ mechanism of Polar code (incremental convergence method and subsequent IR—HARQ method). They proposed to only use the LDPC code for NR eMBB. This proposal [36] was supported by 31 companies.

- Many other companies proposed to adopt Polar code as channel coding method for small code blocks of eMBB data channel, which was supported by 56 companies [37]. The division of large and small code blocks and the value of threshold X were discussed many times, but no agreement was reached. In this scheme, LDPC code is used when the code block size is above the threshold value, while Polar code can be used when the code block size is below the threshold value.

In the last two meeting days, after several rounds of long technical discussions, a widely acceptable joint scheme was proposed: LDPC code was used as the only coding technology of the data channel, while Polar code was used as the coding technology of the control channel. In the early morning of the last day of the meeting, this proposal was agreed on and the technology selection finally came to an end. The detailed designs of Polar code and LDPC code adopted in NR will be described in the following sections.

7.1.3 Channel coding scheme for the control channel

In the previous section, we introduced how the channel coding scheme of the data channel was determined. Compared with the data channel, the discussion of the control channel is relatively relaxed.

The channel coding scheme of the control channel was discussed from the RAN1#86 meeting. TBCC code, which had been widely used in the LTE control channel, and Polar code were discussed as the main candidate schemes. LDPC code was almost out of consideration due to performance loss and energy efficiency at small code blocks. Some companies suggested using TBCC code as the coding scheme of the control channel [38], and some proposals also provided enhancement schemes of TBCC code [39] to improve performance. Polar code was recommended by most other companies [40]. The meeting focused on the data channel, but the unified evaluation assumption was only agreed on for the control channel. So the companies were encouraged to continue to compare the performance of TBCC code and Polar code. The schemes evaluated for the control channel included repetition coding, simplex coding, TBCC, Turbo code, LDPC code, RM code, and Polar code. The coding schemes of the control channel were selected mainly based on the demodulation performance, instead of considering multiple factors for the data channel.

In the RAN1 #87 meeting, the channel coding scheme for the control channel and the channel coding for small block data were discussed as key issues of NR, and the main competition was between Polar code and TBCC code.

- Polar code was widely supported, and 58 companies jointly proposed [41] to adopt Polar code as the channel coding scheme for both the uplink and downlink control channel of eMBB.
- Some companies proposed to use TBCC code for the uplink and downlink control channels, but only a few companies supported that proposal [42].
- Some companies proposed a compromised scheme [43] (i.e. Polar code to be used for uplink control channel and TBCC code to be used for the downlink control channel). The proposal was supported by more than 20 companies.
- Some other proposals [44] suggested that TBCC code be used for small code block of the control channel, and LDPC code be used for large code block. However, since the performance of LDPC has no advantage when the code block is small, the use of LDPC in the control channel has always been a marginalized solution.

Different from the discussion on the data channel, the evaluation results of most companies showed that the performance of the Polar code on the control channel was significantly better than that of TBCC code and LDPC code. However, some companies insisted on determining the coding scheme of the control channel and data channel together [i.e., if the Polar code is adopted in the control channel, a unified coding technology (LDPC) should be applied to the data channel regardless of the size of the code block]. After a long period of research and discussion, until the last minute of the meeting, a combined scheme was finally accepted by all companies (i.e., Polar code is adopted for the uplink and downlink control channels, and only LDPC is used for the data channel). From the perspective of technology and standards, introducing two complex coding methods for different code block sizes of a channel does have a significant impact on the complexity of the equipment implementation. Adopting single optimized coding technology for each channel can be a more reasonable choice. That was one of the reasons companies can finally reach an agreement.

7.1.4 Channel coding scheme for other information

Although it was agreed to use Polar code on the control channel in the RAN1#87 meeting, for ultrashort code length (such as control information with less than 12 bits), it is necessary to further discuss which channel coding scheme to use according to the working assumption agreed in the RAN1#87 meeting, such as whether the repetition code or block code in LTE can be reused.

In the RAN1#88bis meeting, companies provided a large number of evaluation results for the encoding method of ultrashort codes. The encoding schemes were mainly applied to control information of 3-X bits, and the discussion focused on three schemes: Polar code [45−47], LTE RM code, and enhanced block code [48,49]. For control information of one or two bits, companies basically had no objection to reusing the scheme of LTE (i.e., one bit uses repetition coding and two bits uses simplex coding). Although companies still had different views on the value of X and whether to optimize the LTE RM code, overall, the majority of companies agreed to reuse the RM code for cases of more than two bits. At the meeting, it was finally agreed to reuse the LTE RM code as the channel coding scheme for control information with three to 11 bits. The details for the corresponding channel coding schemes can be found in LTE specification [50].

In addition, in the first RAN1 AdHoc meeting in 2017, it was clarified that Polar code was only adopted for PDCCH in downlink, and the channel coding scheme for PBCH needed to be further discussed in subsequent meetings. In the RAN1#88bis meeting, companies began to discuss the channel coding scheme of PBCH. Unified evaluation assumptions and two candidate schemes were determined at this meeting:

- Scheme 1: reuse the Polar code of the control channel with the maximum code length of 512 bits.
- Scheme 2: reuse LDPC code of the data channel with the same decoder (i.e., there is no new shift network), but a new basic graph can be considered.

Since the performance requirements and payload sizes of PBCH are similar to that of the PDCCH, the companies supporting Polar code still proposed Polar code for PBCH. Only a few companies that initially supported LDPC scheme still insisted on LDPC code for PBCH, while other companies were neutral or changed position to Polar code to avoid introducing two different channel coding schemes in downlink signaling. At the same time, some companies [51,52] suggested to indicate SSB index to the UE to obtain the combination gain of decoding PBCH, so as to further improve the demodulation performance of Polar code in PBCH. Finally, at the RAN1#89 meeting, it was agreed that the Polar code adopted in the control channel would be reused for the coding of PBCH. Thus the channel coding schemes for all NR channels have been settled.

7.2 Design of polar code

7.2.1 Background

Arikan first proposed the method of channel polarization to construct Polar code [14] in 2008. For Polar code, synthesized channels are constructed from two independent copies of B−DMC: one channel with higher reliability (so-called upgrading channel), while the other with lower reliability (Fig. 7.1) (so-called degrading channel [53]). A length of $N = 2^m$ subchannel can be generated by performing channel polarization recursively.

The recursive channel polarization can be expressed as:

$$G_2 = \begin{bmatrix} 1 & 0 \\ 1 & 1 \end{bmatrix}$$

$$G_N = \begin{bmatrix} G_{N/2} & 0 \\ G_{N/2} & G_{N/2} \end{bmatrix} \tag{7.1}$$

$$x = uG_N$$

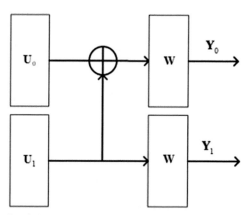

Figure 7.1 Channel polarization.

where u and x are the source input and channel codeword, respectively. The degrading channel is $W^-:U_0 \rightarrow (Y_0, Y_1)$, whereas the upgrading channel is $W^+:U_1 \rightarrow (Y_0, Y_1, U_0)$ given U_0. For W^-, Y_0 is the parity check of U_0 and U_1, so that U_0 suffers interference from U_1; for W^+ with interference cancelation, Y_0 and Y_1 are two independent observations of U_1 by repetition. The channel capacity of those two synthesized channels satisfy the following conditions:

$$I(W^-) \leq I(W) \leq I(W^+)$$
$$I(W^-) + I(W^+) = 2I(W) \qquad (7.2)$$

Without losing the total channel capacity, W^+ is better than the original W while W^- is worse than the original W. For B−DMC W, when the number of copies of independent channel, N, is large, the probability of subchannels with capacity being approximately 1 will approach $I(W)$, thus such noiseless channel can be used to transfer data without error. The remaining $1-I(W)$ subchannels with capacity approaching 0 are pure noise channels, which are unable to carry any data. The exponent of error rate of a channel constructed from a 2×2 polarization kernel is $2^{-N^\beta}, \beta \leq 1/2$ [54], which can be approach 1 [55] when a large transform matrix is used. The capacity of sorted channel is shown in Fig. 7.2.

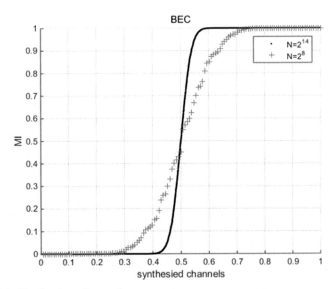

Figure 7.2 Distribution of capacity.

In order to ensure the performance of Polar code, subchannels with good channel quality after channel polarization are selected for data transmission. The subchannels that send data are information bits while the subchannels that do not send data are frozen bits and are generally assumed to be 0. The optimal subchannels for information bits varies with the channel and are usually selected according to bit error rate (BER) [14], or alternatively the reliability of each subchannel may be estimated by density evolution/Gaussian approximation (DE/GA) [56−58].

Arikan proposed a simple decoding algorithm for Polar code—the successive cancelation decoding algorithm—and proved that the Polar code can achieve the Shannon capacity via this algorithm [14]. The decoding route may be expressed as a depth-first traversal of a complete binary tree, where the root is channel likelihood. The 2^N leaf nodes are searched to successively estimate the likelihood of information bits. The decoding tree is shown in Fig. 7.3 where the forward probability on the left branch for parity check is:

$$\alpha_t^i = f(\alpha_{t+1}^i, \alpha_{t+1}^{i+2^t}) = 2\tanh^{-1}\left(\tanh(\alpha_{t+1}^i/2)\tanh(\alpha_{t+1}^{i+2^t}/2)\right) \qquad (7.3)$$

The forward probability on the right branch for repetition is:

$$\alpha_t^{i+2^t} = g(\alpha_{t+1}^i, \alpha_{t+1}^{i+2^t}, \beta_t^i) = \alpha_{t+1}^{i+2^t} + (1 - 2\beta_t^i)\alpha_{t+1}^i \qquad (7.4)$$

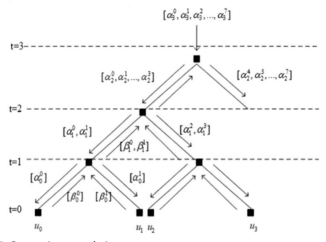

Figure 7.3 Successive cancelation.

Backward messages are calculated by encoding rule, where left and right branches are

$$\beta_t^i = \beta_{t-1}^i \oplus \beta_{t-1}^{i+2^t}$$
$$\beta_t^{i+2^t} = \beta_{t-1}^{i+2^t}$$

(7.5)

where $f(x, y)$ can be simplified by min–sum rule as [59]:

$$f(x, y) = \text{sign}(x)\text{sign}(y)\min(|x|, |y|)$$

(7.6)

To further simplify the complexity of the SC [60,61], Polar code can be decomposed into several types of component codes, such as subtree with all frozen bits (rateless) or all information bit (uncoded), single parity code, or repeated code. On these subtrees, forward recursion can be stopped since reverse messages and decision bits can be calculated directly. Although in theory the Polar code using SC decoding can achieve Shannon capacity, the performance in the case of finite length is still not satisfactory. On the basis of SC decoding, some contributions proposed an SC−List decoding algorithm [62,63]. The decoder searches multiple paths simultaneously and makes a decision from multiple surviving paths through CRC. The decoding performance of SCL can approach the maximum likelihood, which greatly improves the performance of Polar code.

Based on the abovementioned basic principles of Polar code, the NR conducted an in-depth discussion on the standardization of Polar code, and carried out corresponding designs for the structure, sequence, and rate matching of Polar code.

7.2.2 Sequence design

The aforementioned bit-reversal permutation of Polar code [14] only affects the order of constructing the channel, but has no effect on the actual performance. Therefore the NR did not specify the bit-reversal operation. The NR encoder is equivalent to the Arikan encoder from right to left. The Polar code encoder [64] used in the NR is shown in Fig. 7.4.

The optimal positions of frozen bits of the Polar code are different under different channel states, and the positions may vary greatly under different signal-to-noise ratios. Theoretically, the device will calculate DE/GA in real time to determine the reliability of each subchannel for a given channel, but it will increase the computational complexity and delay significantly. It is desired to adopt a channel-independent method (instead of freezing bits that vary with

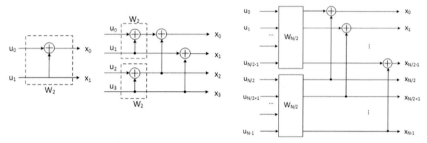

Figure 7.4 NR Polar encoder.

the channel and code length, etc.) to reduce the overhead caused by the UE storing and calculating the channel reliability in practical implementation.

For the construction methods of Polar code, methods based on partial order sequence and fractal construction are discussed. For SC decoding, the reliability of a partial of all subchannels is independent of channel [56,65], and the reliability order of such subchannels constitutes a partial sequence. Partially order sequence is nested, and short sequence can be obtained from the long sequence. DE/GA can be used to determine the remaining sequences related with the channel [66]. The sequence adopted by the NR is basically the partial order sequence, and the stability of its performance has been verified through a large number of simulations. The NR specified the order of a mother code, and the sequence of short codeword was determined by the order of the mother code.

7.2.3 Assistant polar code

The NR mainly considered two types of cascaded Polar code: one is the CA–Polar code with CRC auxiliary check [63] and the other is PC–Polar with parity check [67]. Both of them can improve the performance of SCL decoding.

The CA–Polar code uses CRC to assist in selecting L surviving paths, which will increase the false alarm probability during blind detection. Given a target false alarm rate (FAR), an additional $\log_2 L$ bit CRC is needed to ensure that FAR does not increase along with size of the list. For downlink transmission, a CRC with a length of $21 + 3$ bits can meet the FAR requirements. But when the number of information bits is relatively small, the additional overhead caused by CRC cannot be ignored, and it will lead to significant loss in error rate. For uplink transmission, when the number of information bits is $12 \leq K \leq 19$, 6-bit CRC can meet the FAR requirement of 2^{-3}; when the information bit size $K > 19$,

11-bit CRC can meet the requirement of 2^{-8}. Furthermore, if the initial state of the CRC register is set to all zeroes, the length of the information bit cannot be determined for blind detection. Thus the NR decided that the CRC register of Polar code will be initialized to all 1s [68].

Compared with CA−Polar, PC−Polar introduces parity check bits in some positions in the codeword, and can improve decoding performance by pruning paths that do not satisfy the parity check. The NR uses 3-bit parity check, where one bit is associated with the codeword with the minimum code weight, and two bits are placed on the subchannel with low reliability [69]. Since the parity check bits are located in the information bit, the decoding process can decide in advance to terminate the decoding. The check equation is determined by a circular shift register of length 5 [70]. PC−Polar is only used for PUCCH with small block transmission (information bits less than 19 bits) in NR.

Distributed CRC−Polar is similar to PC−Polar. After a row and column exchange elementary transformation is conducted on the CRC generation matrix, the check matrix forms an upper triangle. Each check bit only relies on the previous information bits, instead of all the formation bits, which forms a nested structure. In the process of decoding, the path that does not satisfy the check equation can be pruned, so that the decoder can terminate the decoding in advance to reduce the delay of blind detection and false alarm probability [71]. An interleaver is added before the polar encoding, so that each check bit of CRC is appended after its information bit. Thus the CRC bits are distributed in the codeword. The interleaver is specified according to the maximum code length. When the code length is less than the predefined code length, the interleaving sequence is obtained from the bottom to the top of the generation matrix [72].

7.2.4 Code length and rate

Large code blocks will increase the complexity, latency, and power consumption of decoding. For the control channel, the payload is generally small, and the code rate is very low when the aggregation level is high. The additional coding gain from optimizing the structure of large code blocks over repetition coding is not significant. Therefore the maximum code length of Polar code in the NR is limited to 512 for downlink transmission and 1024 for uplink transmission, and the minimum designate code rate is 1/8. When the code rate of the mother code is relatively high (e.g., when a Channel State Information (CSI) of several hundred bits is transmitted), the efficiency of repetition coding is low. In this case, multiple low-rate codewords are formed by segmentation, such that more coding gain can be achieved than with repetition coding. In NR, when the number of channel bits is larger than 1088 bits and payload is greater than 360 bits, or when the length of information bits exceeds the maximum mother code length of 1024, the large code block is divided into two shorter code blocks with independent rate-matching and channel interleaving. Finally, the encoder concatenates [73] the code blocks to form the codewords.

7.2.5 Rate matching and interleaving

In a practical system, a flexible code length needs to be configured to match the configured physical resources. The code length of Polar code G2 kernel is a power of 2. When the target code length does not match the mother code length, rate matching is required to adapt the number of channel bits, which includes two functional parts: bit selection and subblock interleaving. The rate matching in NR mainly considers three basic methods: puncturing, shortening, and repetition.

- Puncturing: The length of information bits remains unchanged. By removing some of the parity bits after encoding, the code length can be shortened. For example, the first N channel bits of Polar code can be punctured [74]. The mutual information of the deleted channel is 0 (pure noise channel, likelihood to the decoder is 0). Since the polarization is a capacity preserving transformation, the punctured channel with a capacity of 0 would float at input and a set of bit with likelihood 0 is formed. The receiver cannot detect the original information bits at these positions, resulting in bit errors [75]. The number of punctured channels is equal to the number of subchannels that cannot be used. Therefore these positions need to be frozen before transmission to ensure performance. The optimal puncturing pattern can be obtained by an exhaustive search method [76].

- Shortening: The number of check bits remains unchanged. The code length is shortened by removing some of the information bits after encoding. Several bits in codeword are fixed to 0. Since the receiver knows these positions, the transmitter does not need to send these information bits. The corresponding channel mutual information is 1 and likelihood ratio is infinity. The encoding matrix of the Polar code is a lower triangular matrix. If the last n bits of the input encoder are fixed to 0, the last n bits at output are also 0 [77]. There may be performance loss since reliability of the last subchannels is higher.

- Repetition: When the number of channel bits exceeds the designed code length, repetition is used for rate matching. Repetition includes the repetition of codeword bits and intermediate node bits.

When the code rate is relatively low (e.g., R < 7/16), better performance can be achieved by puncturing. When the code rate is relatively high, a shortening method has better performance. Considering the complexity of implementation, the NR adopts circular buffering to select channel bits [78], as shown in Fig. 7.5.

- Puncturing: select M channel bits from the circular buffer positions N-M to N-1.
- Shortening: select M channel bits from circular buffer positions 0 to M-1.
- Repetition: select M channel bits circularly.

Before the codeword enters the circular buffer, it is shuffled by a subblock interleaver to improve the error correction capability after rate matching [79]. Specifically, the channel bits are divided into subblocks with a length-32 to form four groups. Then those two groups in the middle are interleaved [80] as shown in Fig. 7.6.

For uplink high-order modulation, due to unequal bit channel protection, the performance of the Polar code degrades significantly under the assumption of same channel condition. Some companies proposed to use an interleaver to randomize different channel reliability to improve the

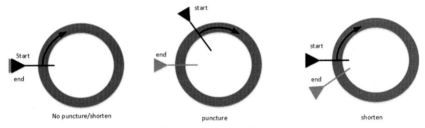

Figure 7.5 NR Polar circular buffer and rate matching.

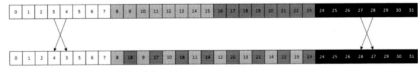

Figure 7.6 NR Polar subblock interleaver.

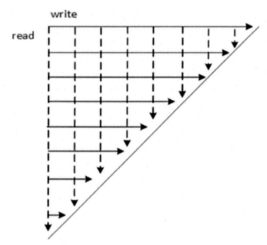

Figure 7.7 NR Polar channel interleaver.

performance of Polar code, which is analogue to Bit Interleaved Code Modulation (BICM). Candidates include a rectangular interleaver with a prime length and an isosceles triangle interleaver. NR adopted isosceles right triangular interleaver [81], as shown in Fig. 7.7, which basically achieves similar performance as random interleaving. Since the channel interleaver does not obviously improve the performance of the downlink frequency selective channel, NR does not use the channel interleaver in the downlink.

7.3 Design of low-density parity-check codes

7.3.1 Basic principles of low-density parity-check codes

The LDPC code is a linear block code proposed by Gallager in 1963 [9]. Its encoder can be described as: after introducing m check bits, a coding sequence c with length n is generated from a binary information sequence u with a length of k, and thus the code rate is k/n. Through following the characteristics of linear code, the codeword c can be expressed by the multiplication of generation matrix **G** and information sequence u:

$$c = G^T \cdot u \qquad (7.7)$$

The generation matrix **G** is made up divided into two parts as follows:

$$G^T = \begin{bmatrix} I_{k \times k} \\ P_{m \times k} \end{bmatrix} \qquad (7.8)$$

where, the matrix $\mathbf{I}$ is an identity matrix corresponding to the systematic bits, and the matrix $\mathbf{P}$ is a matrix corresponding to the check bits. The corresponding PCM $\mathbf{H}$ can be expressed as:

$$H = [P_{m \times k}, I_{m \times m}] \qquad (7.9)$$

The PCM $\mathbf{H}$ and the codeword c satisfy the following equation:

$$\mathbf{H} \cdot c = 0 \qquad (7.10)$$

Generally, a LDPC code is defined through a sparse PCM, which can be mapped to a bipartite or **tanner** graph comprised of check nodes and variable nodes, as shown in Fig. 7.8.

LDPC code needs to support a variety of information block lengths and code rate sizes. A large number of check matrices would be needed to meet the granularity requirements of information block scheduled by 5G NR if the check matrix is designed directly according to the information block and code rate size. That is not feasible for the specification of LDPC code and the implementation of coding and decoding. The proposal of QC–LDPC code has solved this problem. The QC–LDPC code is defined by a base matrix H_b with size $m_b \times n_b$, an expanding factor (lifting size) Z and a permutation matrix $\mathbf{P}$ with size $Z \times Z$. The PCM $\mathbf{H}$ of QC–LDPC can be obtained through replacing each element hb_{ij} in the base matrix H_b with an all–zero subblock matrix with size $Z \times Z$ or the subblock matrix $P^{hb_{ij}}$. The base matrix H_b, the PCM $\mathbf{H}$ and the permutation matrix $\mathbf{P}$ are shown as following.

$$H_b = \begin{bmatrix} hb_{00} & hb_{01} & \cdots & hb_{0(n_b-1)} \\ hb_{10} & hb_{11} & \cdots & hb_{1(n_b-1)} \\ \cdots & \cdots & \cdots & \cdots \\ hb_{(m_b-1)0} & hb_{(m_b-1)1} & \cdots & hb_{(m_b-1)(n_b-1)} \end{bmatrix} \quad H = \begin{bmatrix} P^{hb_{00}} & P^{hb_{01}} & \cdots & P^{hb_{0(nb-1)}} \\ P^{hb_{10}} & P^{hb_{11}} & \cdots & P^{hb_{1(nb-1)}} \\ \cdots & \cdots & \cdots & \cdots \\ P^{hb_{(mb-1)0}} & P^{hb_{(mb-1)1}} & \cdots & P^{hb_{(mb-1)(nb-1)}} \end{bmatrix} \qquad (7.11)$$

$$P = \begin{bmatrix} 0 & 1 & 0 & \cdots & 0 \\ 0 & 0 & 1 & \cdots & 0 \\ \cdots & \cdots & \cdots & \cdots & \cdots \\ 0 & 0 & 0 & \cdots & 1 \\ 1 & 0 & 0 & \cdots & 0 \end{bmatrix}$$

$$H = \begin{bmatrix} 1 & 0 & 0 & 1 & 0 & 1 \\ 1 & 0 & 1 & 0 & 1 & 0 \\ 0 & 1 & 0 & 1 & 1 & 0 \\ 0 & 1 & 1 & 0 & 0 & 1 \end{bmatrix}$$

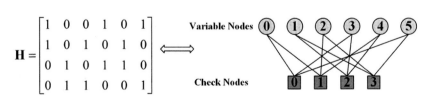

Figure 7.8 Mapping between check matrix and Tanner graph.

In the base matrix $\mathbf{H}_b$, the value of hb_{ij} can be -1, 0, or a positive integer. If $hb_{ij} = -1$ in the base matrix, $\mathrm{P}^{hb_{ij}}$ in $\mathbf{H}$ equals an all-zero matrix with size $z \times z$; if $hb_{ij} = 0$ in the base matrix, $\mathrm{P}^{hb_{ij}}$ in $\mathbf{H}$ equals an identity matrix with size $z \times z$; if hb_{ij} is a positive integer, $\mathrm{P}^{hb_{ij}}$ equals to the hb_{ij} power matrix of the permutation matrix $\mathbf{P}$, that is, the matrix obtained after matrix $\mathbf{P}^0$ is shifted to right by hb_{ij}.

Considering that the QC−LDPC code has been widely used in high-throughput systems such as IEEE802.11n, IEEE802.16e and IEEE802.11ad, it was confirmed that NR adopts the structured design of QC−LDPC at the RAN1#85 meeting [82]. The design of LDPC code adopted in 5G NR can be mainly partitioned into the design of the base matrix $\mathbf{H}_b$, the design of the permutation matrix $\mathbf{P}$ and the determination of the lifting value Z. In addition, the design for LDPC code has to take into account the support of flexible code block length and code rate size, as well as the processing of CRC addition and rate matching in actual transmission. These contents will be introduced in the following sections.

7.3.2 Design of parity check matrix

LDPC code in NR needs to support multiple code rates and different code block sizes. There are three options for the initial design:

- Scheme 1: Multiple base parity check matrices are first designed for some code rates and block sizes, and then all other code rates are supported via reusing the base PCMs. For each of all other code rates, the PCM is generated through repetitions, and puncturing techniques from the base PCM of the nearest code rate [83]. For the code rates higher than 1/2 (such as 2/3, 3/4, 5/6, etc.), PCMs are linked to each other to minimize the implementation effort required for supporting multiple code rates. As shown by the example in Fig. 7.9A, the PCM of 2/3 can be obtained through extending the PCM of code rate 5/6, where the PCMs of rate 3/4 and 2/3 can reuse the PCM of higher code rate. For the lower code rates, the PCMs have a similar relationship to each other. As shown in Fig. 7.9B, the PCM for rate 1/6 is used as the base PCM, and the PCMs of rate 1/3 and 1/2 are generated through similar reusing technique.

- Scheme 2: A uniform base PCM is first designed for low code rate. To support other code rate, a subbase matrix of corresponding rows and columns is extracted from the uniform base matrix [84]. Different expanding factors (lifting size) are used to support different code block

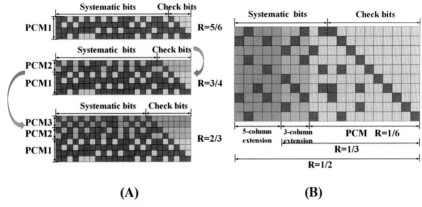

(A)   **(B)**

Figure 7.9 Scheme1 for PCM design: base PCMs and reusing.

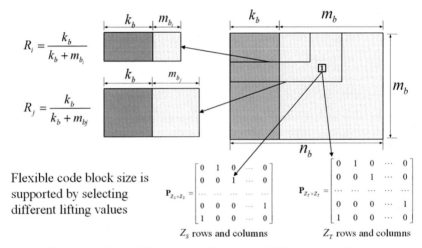

Figure 7.10 Scheme 2 for PCM design: uniform base PCM and extracting.

sizes. An example for supporting code rates R_i and R_j is shown in Fig. 7.10, where an example of different expanding factors (lifting size) of Z_s and Z_t is also illustrated.

- Scheme 3: The LDPC code is first designed for high code rate and then it is expanded to support low code rate. The matrix of LDPC code adopts the structure of "Raptor-Like" [85–94] (as shown in Fig. 7.11). The PCM consists of five submatrices (A, B, C, D, E), which is serially concatenated with higher rate codes and Single Parity Check (SPC) codes. Among those submatrices, **A** and **B** describe the higher code rate matrix. **C** is an all-zero matrix of appropriate size,

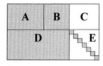

Figure 7.11 Scheme 3 for PCM design: "Raptor-Like" structure.

and **E** is an identify matrix with the same number of rows as **D**. All variable nodes not involved in **A** and **B** (also known as incremental redundant variable nodes) have a degree of 1. **A** and **B** give the highest code rate and a lower code can be supported rate by transmitting other variable nodes from the incremental redundancy part.

Among the above schemes, scheme 3 is the mainstream design idea and is finally adopted in NR. The matrices **A**, **B** and **D** could have different designs. Accordingly, the scheme 3 can be divided into different subschemes:

- According to the structure, the matrix **A** can have two different categories: one type of matrix A contains systematic bits only [85] and the other type of matrix A contains both systematic and parity bits [91].
- The matrix **B** in the base matrix can have two different structures: a lower triangular structure or a dual diagonal structure (i.e., main diagonal and off diagonal). IEEE802.11as adopts a lower triangular structure for the base matrix while IEEE802.16e and IEEE802.11n use the dual diagonal structure. The performance of these two structures is almost the same, and both structure can support linear coding without requiring additional storage of the generator matrix. Some companies [84] suggested that the dual diagonal structure has more limitations than the lower triangular structure in terms of the construction and hardware implementation of Base Graph (BG), so it is recommended to adopt the lower triangular structure for the matrix **B**. However, other companies [85,86,88] believed that using PCM with dual diagonal structure can promote linear time coding and provide more stable decoding performance. Thus it is recommended to use a dual diagonal structure for design.
- The design of matrix **D** focused on the design of row orthogonality and quasirow orthogonality (removing some columns to satisfy row orthogonality). Some companies [83,91,93] believe that adopting orthogonal characteristics can support the high throughput and low delay of the decoder. Other companies [88,94,95] consider that most of LDPC codes with good performance have variable node puncturing

processing, and quasirow orthogonal LDPC codes can also maintain good performance. At the same time, adopting row parallel decoder that is suitable for row orthogonal PCM has the risk of high complexity and high power consumption. So it is suggested to adopt quasirow orthogonal design.

Considering many factors such as the flexibility of code rate and code block, performance and implementation complexity, it was determined at the RAN1#88 meeting that PCM in NR adopts the structure shown in Fig. 7.11 [96]. In the adopted PCM structure, the matrix **A** corresponds to the systematic bits (the part of the coded codeword that is the same as the unencoded information bit). The matrix **B** is a square matrix with dual diagonal structure, corresponding to the parity check bit part. The matrix **C** is an all-zero matrix. The matrix **D** can be divided into row orthogonal part and quasirow orthogonal part and the matrix **E** is an identity matrix. **A** and **B** constitute the high-rate core part of QC−LDPC code, and **D** and **E** together form a single parity check relationship, which can be combined with **C** to achieve low bit-rate extension. Matrix **B** with a dual diagonal structure can have two different types of reference designs according to whether it contains columns with the weight of 1 (as shown in Fig. 7.12):

- If there exists one column with the weight of 1, the nonzero value is in the last row, and the weight of this row is 1 (which means that the weight of the row in matrix **B** is 1). The remaining columns form a square matrix, in which the first column has a weight of 3, and the other columns form a dual diagonal structure, as shown in Fig. 7.12 (1A) and (1B).
- If there is no column with a column weight of 1, the weight of the first column is 3, and the other columns form a dual diagonal structure, as shown in Fig. 7.12 (2).

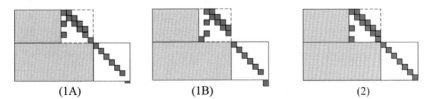

| (1A) | (1B) | (2) |

Figure 7.12 Designs of LDPC code submatrix **B**.

7.3.3 Design of permutation matrix

The circulant weight of a permutation matrix refers to the number of cyclic shift identity matrices superimposed by the permutation matrix. Regarding the circulant weight of permutation matrix $\mathbf{P}$, NR discussed two different design schemes with maximum cyclic weight of 1 and 2, respectively. Some companies [83,89] consider using PCM with a maximum cyclic weight of 2 to improve the performance of LDPC codes with medium and short block length. If the maximum circulant weight of 2 is used, the submatrices could be comprised of super-imposed cyclic permutations of identity matrices or null matrices. A compact base graph can reduce the implementation complexity of the decoder. In the case of a circulant weight of 2, if the lifting size is 5, an example of permutation matrix $\mathbf{P}^{2,4}$ (equivalent to the superposition of $\mathbf{P}^2$ and $\mathbf{P}^4$) is as below:

$$\mathbf{P}^{2,4} = \begin{bmatrix} 0 & 0 & 1 & 0 & 1 \\ 1 & 0 & 0 & 1 & 0 \\ 0 & 1 & 0 & 0 & 1 \\ 1 & 0 & 1 & 0 & 0 \\ 0 & 1 & 0 & 1 & 0 \end{bmatrix} \tag{7.12}$$

In addition, many existing LDPC codes (including 802.11n/ac/ad, etc.) adopt a design with a circulant weight of 1. For a layered LDPC decoding algorithm, if the maximum circulant weight is 2, the same variable node will participate in two check equations in a layer, which will lead to conflicts in Log Likelihood Ratio (LLR) updating during the processing of that layer. These conflicts require special handling such as splitting the LLR memory into two storage area, which can complicate implementation or reduce the degree of parallelism. However, the design with the maximum cycle weight of 1 does not need any special processing at the submatrix level. In addition, studies have shown that LDPC code design with maximum circulant weight of 1 has very good performance and can meet the requirements of eMBB [85,97,98]. Taking into account the impact of circulant weight on performance, implementation complexity and parallel processing, NR finally adopts the LDPC code design with maximum circulant weight of 1 [99].

7.3.4 Design of base graph

This section will focus on the base graph design, including the size, quantity, structure and puncturing method of the base graph. The design of base graph is essential to the design of the PCM.

7.3.4.1 Size of base graph

The total number of columns in the base matrix is proportional to the number of systematic columns (kb) and inversely proportional to code rate (R). A base matrix with more columns will likely have larger average row weight (which is the average number of non -1 in all the rows), resulting in higher decoder complexity and higher decoding latency during parity node updating. If the number of total columns is larger, the number of total rows will be larger under the same code rate, thereby increasing the number of layers in layered decoder and leading to high decoder latency. Although the total number of rows can be reduced by reducing number of the systematic columns (kb), this may destroy the unity of the base matrix. Therefore in a system where QC−LDPC codes are applied, such as IEEE 802.16e/11n/11ac, the maximum number of the number of system columns is limited to 20. Some companies [84,87−89] propose that the base graph of NR LDPC code should be as small as possible, that is, a compact design for the base graph will be adopted to meet the requirements of high throughput and low complexity. Compared with noncompact matrix, compact matrix has the following advantages:

- It is easier and more efficient to implement row parallel decoder for compact matrix because of the smaller number of rows, smaller maximum row weights and less nonnegative elements.
- It can be used to improve the potential maximum parallelism of block parallel decoders.
- The simulation results [100] of several companies show that the compact matrix provide comparable performance than the noncompact matrix.
- Compact matrix has many mature implementation methods with Gbps throughput since compact matrix has been adopted in many wireless wideband communications using LDPC code.
- It can support more Modulation Coding Scheme (MCS) levels and can meet the requirement of peak throughput of 20 Gbp.
- Compact matrix uses less ROM to store the base PCM.
- Compact matrix uses simpler expression and simpler control circuit on cycle shift operations.
- It is shown in Central Processing Unit (CPU) or Digital Signal Processing (DSP) based software simulations that compact matrix obviously needs less time.

Considering the above aspects, NR adopts the compact matrix design, that is, a base graph with a small number of system columns (specifically 22 and 10).

7.3.4.2 Design of base graph

LDPC codes can use a single [84,87,101] base graph or multiple [83,85,102,103] base graphs to support flexible block sizes and code rates. Among them, the method using a single base graph has similar performance as the method using multiple base graphs. It is simple and unified, and it is suitable for parallel decoders and requires less Read-only Memory (ROM) for storage. On the other hand, using multiple base graphs will lead to higher complexity. However, the range of data rate, block size and code rate supported in NR is very large. The base graph can be optimized separately for different block sizes and different coding rates, that is, selecting different base graphs to support different ranges of the data rate, block size and code rate. Thus there is no need to extend the PCM from a very high code rate to a very low code rate.

Considering the impact of BG design on implementation complexity and performance, three candidate solutions based on using single base graph and using multiple base graph were determined at the RAN1#88bis meeting [104]:

- Scheme 1: Using one base graph which cover the range of code rate: $\sim 1/5 \leq R \leq \sim 8/9$.
- Scheme 2: Using two nested base graphs (BG1 and BG2), where the covered range of block size is $K_{min} \leq K \leq K_{max}$, and range of code rate is $\sim 1/5 \leq R \leq \sim 8/9$.
  - BG1 covers info block size K: $K_{min1} \leq K \leq K_{max1}$, where $K_{min1} > K_{min}$, $K_{max1} = K_{max}$, and covers code rate R: $\sim 1/3 \leq R \leq \sim 8/9$. The larger code blocks and higher code rates are supported with high priority. Whether the code rate of $\sim 1/5$ can be supported is further confirmed;
  - BG2 needs to be nested within BG1, and it can cover the info block size K: $K_{min2} \leq K \leq K_{max2}$, $K_{min2} = K_{min}$, $K_{max2} < K_{max}$, where $512 < K_{max2} \leq 2560$. The code rate covered by base graph 2 is: $\sim 1/5 \leq R \leq \sim 2/3$. Smaller code blocks and lower bit rates are supported with high priority. When designing BG2, the maximum number of system columns in the initial design is $K_{bmax} = 16$, and lower value in the range $10 \leq K_{bmax} < 16$ is allowed when it is feasible.

- Scheme 3: Two independent base graphs are used, where the block sizes and code rates covered by BG1 and BG2 are similar to those in Scheme 2, but BG2 does not need to be nested in BG1.

Block Error Rate (BLER) performance is the main criterion for selecting from the above three candidate schemes (since it is already assumed that complexity is not increased significantly by the addition of a second smaller base graph). Companies conducted simulation evaluations for BLER performance and decoding delay, and finally determined to adopt the scheme of using two independent BG based on the performance. Specifically, the size of BG1 is 46×68, and the minimum code rate is 1/3, which is mainly used in scenarios with high throughput requirements, high code rate and large code block. The size of BG2 is 42×52, and the minimum code rate is 1/5, which is mainly used in scenarios with low throughput requirements, low code rate and small code block. BG2 further supports code blocks of different lengths through deleting different numbers of system columns. In particular, when the number of bits in information block is less than or equal to 192, the number of system columns is 6. When the number of bits in information block is greater than 192 and less than or equal to 560, the number of system columns is 8. When the number of bits in information block is greater than 560 and less than or equal to 640, the number of system columns is 9. When the number of bits in information block is greater than 640, the number of system columns is 10.

The BG1 and BG2 matrices used in NR LDPC codes are shown in Figs. 7.13 and 7.14, respectively, where the element "0" indicates that the permutation matrix is an all-zero matrix, and the element "1" indicates that the permutation matrix is a cyclic shift matrix (the specific value of BG can be obtained from the table [105]). The first two columns in BG have large column weight, that is, the number of 1s in these two columns is significantly larger than that of the other columns. The benefit of that is to the flow of messages in the decoding process is enhanced and the efficiency of message transfer between check equations is increased. The matrix at the lower left corner can be divided into two parts: row orthogonal design and quasirow orthogonal design. The lower right corner is a diagonal matrix and is used to support IR–HARQ. In each IR–HARQ retransmission, only more parity bits are sent.

7.3.4.3 The elements of base graph

In the LDPC decoder, each element (non-1, i.e., not being-1) in the base matrix corresponds to a cyclic shift. The value of element being "0"

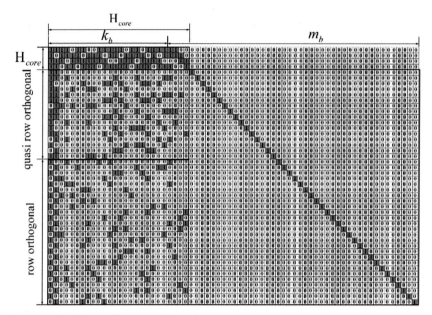

Figure 7.13 BG1 used in NR LDPC code.

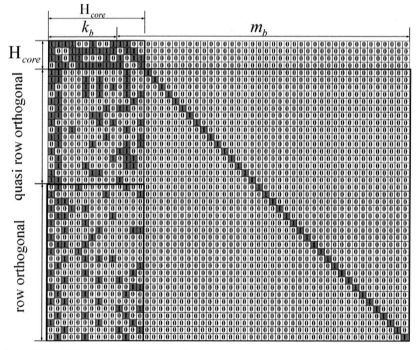

Figure 7.14 BG2 used in NR LDPC code.

means that there is no need for cyclic shift. Therefore if more elements in the base matrix equal to "0", the complexity of LDPC decoder would be lower. For any column in the base matrix, if the first non-1 element of each column is equal to "0", the systematic bits in the original order can be derived when the first two columns are updated. Therefore some companies recommend that for the base matrix, the first non-1 element of each column is fixed to be "0" [84]. Considering the performance of LDPC code and the complexity of implementation, NR finally did not adopt this scheme.

7.3.4.4 Puncturing method of base graph

Puncturing one or a few systematic nodes can improve the threshold of the LPDC code significantly. In order to provide the improved performance, the punctured nodes should be the ones that are well connected to the rest of the graph. Selecting the number of punctured systematic nodes will consider the trade-off between BLER performance (i.e., reduced threshold), and the additional decoder hardware that is needed to handle the longer codeword due to adding punctured systematic nodes. The system bits corresponding to variable nodes with large column weight are not transmitted, and thus the performance of QC−LDPC codes can be further improved [106]. Specifically, there are four candidate schemes for the puncturing of base graphs as below:

- Fixed puncturing is not considered [94].
- Puncturing the first Z systematic bits [89].
- Puncturing the first 2Z systematic bits [84,85,88,90,91,93].
- Puncturing the last 2Z systematic bits [94].

Based on evaluation performance and the considerations on matrix structure design, in order to ensure that the check nodes can achieve the smooth flow of soft information between each other through the full connection with the first few variable nodes, the column weight of the leftmost column of the system bit matrix is designed to be very large. And finally NR determines to puncture the first 2Z system bits at the beginning.

7.3.5 Lifting size

For each basic graph, a specific parity matrix can be obtained by using the corresponding basic matrix and replacing each element with the corresponding $Z \times Z$ cyclic shift matrix according to the lifting value Z. Different block sizes can be supported by changing the lifting size. The lifting value also affects the throughput and latency of encoder/

decoder throughput through the built-in parallelism, which has impact on the performance and implementation of LDPC. Generally, a larger shift size results in lower latency. However, latency can also be further reduced by other techniques such as processing multiple check/variable nodes in parallel or using multiple separate decoders, etc. Thus it is not strictly necessary to support very large shift sizes. NR considers the maximum lifting value to be selected from the candidate set <256, $512, 1024, \sim 320, \sim 384 <$. Among them, a large value of Z (e.g., $Z_{max} = 1024$) may result in a very small PCM (in terms of number of row/column), thereby affecting the performance. Taking into account the throughput, latency, performance, and flexibility, NR finally agreed to use $Z_{max} = 384$.

It is necessary to design multiple lifting sizes in order to support flexible block sizes for LDPC code. In the selection of the granularity of the lifting size, some companies have proposed that the granularity is 1 [92] (i.e., $Z_{min}:1:Z_{max}$). From the perspective of implementation and resource utilization, other companies do not recommend continuous lifting size [97,98,107], but propose that the lifting size can be extracted between the minimum and maximum values with different granularity, such as {1: 1: 8, 8: 1: 16, 16: 2: 32, 32:4:64, 64:8:128, 128:16:256, 256:32:512}. The granularity of the lifting size is 2^L with $L \geq 0$. For a large range of lifting size, the supported lifting size granularity is 2^L with $L \geq 3$. For LDPC codes, if the parallelism is the power of 2, the shift network can be implemented very efficiently with a Banyan network, for example. And, for a layered decoder, the lifting size Z should be equal to an integral multiple of decoder parallelism when the decoder parallelism is less than Z. If Z is equal to a prime integral multiple of 2^j with j being a positive integer, more positive integer factors will be provided for parallelism choice. Finally, NR determined that the lifting value is in the format of $Z = a \times 2^j$, where $a = \{2,3,5,7,9,11,13,15\}$ and $j = \{0,1,2,3,4,5,6,7\}$. The value of 2^j can be designed with a banyan network as much as possible and the introduction of factor a may reduce the number of padding bits.

Optimizing a set of circulant matrices for each lifting size is the most flexible. The spikes in the performance curve can be eliminated by selecting an appropriate offset for each lifting size, and the performance of different code block sizes can also be very smooth. However, to achieve a good trade-off between BLER performance and hardware area efficiency, it is critical to reduce the number of lifting sizes Z that the

Table 7.1 Lifting sizes supported by different offset coefficient tables.

i_{LS}	Lifting size value Z
0	$\{2,4,8,16,32,64,128,256\}$ $(a = 2)$
1	$\{3,6,12,24,48,96,192,384\}$ $(a = 3)$
2	$\{5,10,20,40,80,160,320\}$ $(a = 5)$
3	$\{7,14,28,56,112,224\}$ $(a = 7)$
4	$\{9,18,36,72,144,288\}$ $(a = 9)$
5	$\{11,22,44,88,176,352\}$ $(a = 11)$
6	$\{13,26,52,104,208\}$ $(a = 13)$
7	$\{15,30,60,120,240\}$ $(a = 15)$

decoder must handle. To solve this problem, companies have proposed different solutions:

- Some companies propose clustering or grouping design of lifting sizes with each group consisting of lifting sizes close to each other, using the same offset coefficient design [91].
- Some companies suggest grouping according to the different values of a [94]. In each group, the shift coefficient is designed based on the maximum lifting size $Z_{a,\max}$ in the group so as to obtain the corresponding shift coefficient table of each group. The shift coefficients of other lifting sizes Z in the group are obtained according to $P_{a,Z}^{m,n} = P_a^{m,n} \bmod Z$, where $P_a^{m,n}$ represents the (m, n) shift coefficient in the shift coefficient table of the group.
- Some companies propose to construct a basic offset coefficient table. Different lifting sizes can be derived from a module or div-floor operation [92] from that table.

Considering both the BLER performance and implementation, the NR constructed a total of eight offset coefficient tables ($i_{LS} = \{0,1,2,3,4,5,6,7\}$) according to the different values of a. The value of lifting size supported by each offset coefficient table is shown in (Table 7.1). In each offset coefficient table, the offset coefficient corresponding to the lifting size Z is calculated according to $P_{a,Z}^{m,n} = P_a^{m,n} \bmod Z$.

7.3.6 Code block segmentation and code block CRC attachment

7.3.6.1 Maximum code block size

The maximum code block size $K_{\max}$ at the input of the channel encoder is an important design parameter, which affects the code block segmentation and the implementation complexity of the decoder. The maximum code block size of Turbo codes in LTE is 6144. When the code block size is larger than 6144, the code block is segmented. Similarly, the maximum code block size of NR LDPC codes should be defined for code block segmentation. For the maximum code block size of LDPC codes, companies have proposed the following various design solutions:

- $K_{max} = 12288$ bits [89] (i.e., the number of bits in an IP packet).
- From the perspective of hardware implementation, it is more reasonable to choose $K_{max} = 2^n$, and it is recommended to use $K_{max} = 8192$.
- For LDPC decoders, the larger the maximum code block size K_{max}, the higher the memory consumption. Therefore it is recommended that the maximum code block size of the LDPC code is similar to the LTE turbo code (i.e., K_{max} satisfies $6144 \leq K_{max} \leq 8192$) [108].

The choice of maximum code block size mainly considers the performance and implementation complexity. Some companies [109] find that the performance difference between the code block sizes 8k and 12k is negligible. Since the number of memory and update logic of the LDPC decoder is proportional to the maximum code block size, the implementation complexity of the LDPC decoder for maximum 12K code block size would increase 50% compared to a maximum code block size 8K. Combined with the base graph design, NR finally agreed that the maximum block size of LDPC is $K_{max} = 8448$ (BG1) and $K_{max} = 3840$ (BG2).

7.3.6.2 Location for CRC additional

If the size of a Transport block (TB) is greater than the maximum code block size, the TB will be divided into several segmented Code block (CBs) and the CBs are coded independently. Regarding how to add CRC check codes in TB and CB, there are two candidate schemes for determining the location of adding CRC:

- CRC is added at TB level and Code Block Group (CBG) level [110,111]. At the CB level, a built-in error detection function in the LDPC code is used for error detection and as the decoding stop criteria, which can reduce the overhead of CRC and improve the overall performance. This solution can provide three levels of error detection (as shown in Fig. 7.15): LDPC parity check at CB level, CRC check at CBG level, and CRC check at TB level.
- CRC is added at TB level and CB level [112,113]. For medium and small TB sizes, the CBs may not be grouped. If so, CB level CRC will be considered instead of CBG level CRC. For large TB sizes, if the CBs need to

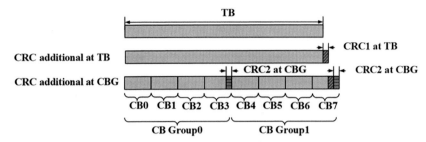

Figure 7.15 CRC addition at TB and CBG level.

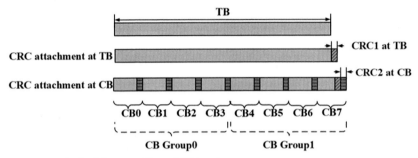

Figure 7.16 CRC addition at TB and CB level.

be grouped, the CB level CRC detection can be used to support CBG level HARQ. If an earlier CB within a CBG has an error and is considered unrecoverable, the remaining CBs in the CBG do not need to be decoded and a Negative Acknowledgement (NACK) information is generated directly, as shown in Fig. 7.16.

CBG-level CRC attachment only works in the scenario that all CBs in the CBG pass the inherent LDPC parity check but at least one CB in the CBG has error. This kind of situation is rare, and TB-level CRC can effectively avoid sending the wrong TB to the high level. On the other hand, although the overhead of CBG-level CRC detection is lower than that of CB−level CRC detection, the overhead of CRC itself is not large, and the overhead is not the determining factor when determining the location of CRC attachment. In the end, NR adopts TB level and CB level CRC addition [113], not the CBG level CRC addition [114].

7.3.6.3 Length for CRC

For LDPC codes, the number of CRC bits to achieve a given undetected error probability varies with the block size and code rate. Considering the actual performance requirements, NR finally determines that when the information block size $A > 3824$, the length of CRC added to a TB is $L = 24$ bits, and the generating polynomial is:

$$g_{CRC24A}(D) = D^{24} + D^{23} + D^{18} + D^{17} + D^{14} + D^{11} + D^{10} + D^7 + D^6 + D^5 + D^4 + D^3 + D + 1$$

$$(7.13)$$

When the information block size $A \leq 3824$, a CRC with a length of $L = 16$ bits is added after the TB, and the generating polynomial is:

$$g_{CRC16}(D) = D^{16} + D^{12} + D^5 + 1 \qquad (7.14)$$

If code segmentation (i.e., the number of blocks $C > 1$) is required, a 24-bit CRC is added after each CB, and the generating polynomial is:

$$g_{CRC24B}(D) = D^{24} + D^{23} + D^6 + D^5 + D + 1 \qquad (7.15)$$

7.3.6.4 Segmentation of code block

The segmented code blocks can be processed independently with lower hardware complexity. Before performing block segmentation, it is necessary to first determine the number of CB according to the TB size $B = A + L$ (where A is the size of the information block, L is the size of the additional TB level CRC) with the added CRC, the maximum block length K_{cb}, and the length of CRC added to one segmented code block CRC_{CB}. The number of CBs is calculated as: $C = \lceil B/(K_{CB} - CRC_{CB}) \rceil$. NR considered three candidate methods for code block segmentation:

- All CBs except the last one have the same number of original bits [115].

Each of the first $C-1$ CBs contains $\lceil B/C \rceil$ original bits, while the last CB contains the remaining $B - (C - 1)\lceil B/C \rceil$ original bits. $C\lceil B/C \rceil - B$ zeros are padded as the filling bits$^+$ to the end of the last CB so that it has the same length as the other CBs. The minimum value of Z that can meet the condition $K_b \cdot Z \geq K = \lceil B/C \rceil + CRC_{CB}$ (where K_b is the number of systematic columns in the base matrix used to encode) is chosen as the lifting size. Some filling bits* are attached at the end of each CB to match the selected lifting size Z. The difference between the filling bits$^+$ and the filling bits* is that the filling bits$^+$ will be sent with the original bit, and the filling bits* could be removed after LDPC encoding. This method can ensure that all the transmitted CBs have the same length, which simplifies the operation of the receiver. The schematic diagram is shown in Fig. 7.17.

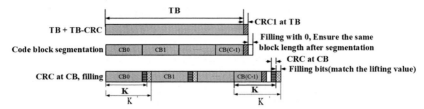

Figure 7.17 Dividing TB equally for all CB except the final CB.

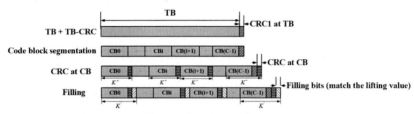

Figure 7.18 Dividing TB roughly equally among all CB's.

- Divide TB roughly equally among all CBs [115−117].

Two block sizes $K^+ = \lceil (B + C \times CRC_{CB})/C \rceil$ and $K^- = \lfloor (B + C \times CRC_{CB})/C \rfloor$ are defined so that each of the first C^+ CBs contains K^+-CRC_{CB} systematic bits and each of the remaining CBs contains K^--CRC_{CB} systematic bits. The minimum value Z that can meet the condition of $K' = K_{b,\max} \cdot Z \geq K^+$ is chosen to be the lifting size. In each of the first C^+ CBs, $K' - K^+$ filling bits are inserted, while $K - K$ filling bits are inserted in each of the remaining CBs. The disadvantage of this method is that the actual length of the transmitted CBs may be different and thus additional operations are required at the receiving side. The schematic diagram is shown in Fig. 7.18.

- Divide TB equally for among all CBs [118].

Each CB contains $CBS = \lceil B/C \rceil$ original bits. Since B is usually not an integral multiple of CBS, we would have to insert zero bits to one or more CB. One method to resolve this issue is to adjust the size of the TB so that padding zeros is not needed for CB segmentation, as shown in Fig. 7.19.

Finally, the third method of equal-length segmentation is adopted in NR. The number of code blocks is $C = \lceil B/(K_{CB} - CRC_{CB}) \rceil$, the total transmission block size after segmentation is $B' = B + C \times CRC_{CB}$, and the size of each code block is $K' = B'/C$. Before LDPC encoding, the value of K_b is determined according to the selected BG and the size of B, and then the minimum lifting size Z (expressed by Z_C) is also determined by satisfying $K_b Z_C \geq K'$ (Table 7.2). After that, the size of the segmented

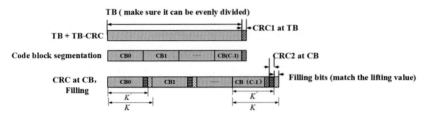

Figure 7.19 Dividing TB equally among all the CBs.

Table 7.2 Number of system columns in different BGs.

BG1		BG2		
All B	$B \leq 192$	$192 < B \leq 560$	$560 < B \leq 640$	$640 < B$
$K_b = 22$	$K_b = 6$	$K_b = 8$	$K_b = 9$	$K_b = 10$

code block is determined according to Z_C. The size $K = 22Z_C$ is set for BG1, and the size $K = 10Z_C$ is set for BG2. If $K \geq K'$, $K - K'$ filling bits are inserted after each original segment code block. The remaining operations after the code block segmentation such as LDPC encoding and rate matching are all performed at the code block level.

7.3.7 Rate matching and hybrid automatic repeat request process

7.3.7.1 Padding techniques

The size of information blocks that can be transmitted in a single transmission depends on the amount of available physical resources. The LDPC code need to flexibility configure block size and code rate to adapt to the configured physical resources. Since the value of lifting size in the specified set is discrete, only relying on the lifting size adjustment often cannot meet all the requirements. Thus other processing such as shortening, puncturing, or repeating is usually needed.

According to the previous section, the length of a segmented code block K' is not necessarily equal to the length of the actual code block K. If $K' < K$, some bits would be padded after the systematic bits whose length is K'. Three solutions are discussed on the design of padding bits.

7.3.7.1.1 Zero padding

Padding zero bits is the most practical method of padding or filling, and is commonly used in many standards [119,120]. In particular, the padding bits are set to be zeros (or bit sequence with known values). In most cases, these bits are punctured after encoding. At the decoder, these punctured padding bits are added back with corresponding LLR values, which are often set to be maximum or minimum LLR value. The method of zero padding can be applied to NR when the number of CBs is small, the overhead of padding bits is significant, and simple operations are required.

7.3.7.1.2 Repeat padding

In LTE, single HARQ ACK/NACK is generated for each TB, and decoding error in a single CB would cause retransmission of the entire TB [119]. In NR system, the similar error protection can be provided by properly using the expected padding bits of LDPC to improve the transmission efficiency. Specifically, the systematic bits in the two adjacent CBs are used as the padding bits to satisfy encoding requirements, and such additional information bits can improve the decoding performance of both code blocks. For example, the padding bits of the one CB are the repetition of part of the systematic bits in the next adjacent CB. At the receiver, if one code block finishes decoding and successfully passes the parity check (or CRC check), the initial LLR of repeated systematic bits can be updated with the latest LLR values. Then, the repeated bits can considered as known bits when decoding next received block. In this way, the padding bits can improve the performance of all CBs. An example of repeat padding is illustrated in Fig. 7.20. After the code segmentation, some of the systematic bits in CB 1 are inserted in CB0 as part of the required information block in encoding CB0 and the similar procedure is conducted on all the CBs until the last one. Specifically, the repeat padding bits added to the last CB is the repetition of some systematic bits taken from the CB 0.

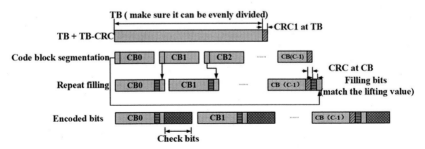

Figure 7.20 Schematic diagram of repeat padding.

After the CB0 is decoded successfully, the LLRs of repetition bits in the CB1 can be updated with the LLR values contained in CB0, and thus more reliable information can be used for decoding CB1. This procedure can be repeated for all other received blocks. If the one CB is not correctly decoded, the decoder can continue to decode the next CB so that the known repetition bits in the correctly decoded CB can be used in the next attempt of decoding the CB that was not correctly decoded. The same principle can be applied to parallel pipeline processing. To support that, inter-CB group padding is performed, instead of interadjacent CB padding. The CBs are first divided into sets. The CBs in the first set are padded with systematic information bits of the CBs in the second set.

7.3.7.1.3 RNTI padding

In this method, the bit sequence that identifies the receiver (such as RNTI) and/or transmitter can be used as the known bits for channel code shortening and used for bit padding [119]. Specifically, the data bits are multiplexed or padded with UR RNTI bits which are known to both transmitter and receiver.

Considering the impacts of different padding bits on performance and implementation, NR finally selects the method of zero-bit padding.

7.3.7.2 Hybrid automatic repeat request transmission

Similar to the LTE Turbo code, NR implements the HARQ and rate matching through a circular buffer. The encoding output is placed in the buffer and a Redundant Version (RV) is defined to indicate the address in the circular buffer which the transmit data starts from. For each transmission, the data are read sequentially from the circular buffer according to the value of RV and then realize rate matching. In order to easily address the RV in the QC−LDPC Code, NR uses the lifting size Z as the unit of the RV.

For HARQ transmission of LDPC, NR mainly considers the following candidate schemes.

7.3.7.2.1 Sequential retransmission

In order to exploit the coding gain of HARQ−IR [110,121], the start index starts from the position where the last transmission ended. In the initial HARQ transmission, an information block of length N is transmitted with all the systematic bits, which is illustrated in Fig. 7.21.

7.3.7.2.2 Retransmission with systematic bits

In this method, as illustrated in Fig. 7.22, the systematic bits are included in every transmission and retransmission [122]:

- In the first transmission, all the systematic bits except the punctured bits and some parity check bits are transmitted. The total length of the first transmission block is N.
- In the second transmission, the systematic bits of length I are transmitted, which include the punctured bits and some other information bits (the length of other information bits can be 0), and some new parity check bits of N-I length are also transmitted.
- In the third transmission, the starting point of the retransmission is moved to the right side by I bits compared to the second transmission. The third transmission includes I systematic bits and N-I new parity check bits.
- In the fourth transmission, the starting point of the retransmission is moved toward the right side by I bits compared to the third transmission. Similarly, the fourth transmission includes I systematic bits and N-I new parity check bits.

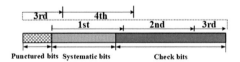

Figure 7.21 Retransmission in sequence.

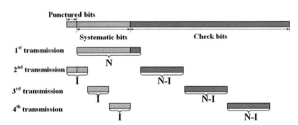

Figure 7.22 Retransmission always with systematic bits.

7.3.7.2.3 Retransmission based on redundant version position

The starting positions of RV values can be uniformly or nonuniformly distributed in the circular buffer. There are two different types of candidate solutions on selecting RV for HARQ retransmission:

- The RV is selected sequentially. Fig. 7.23 illustrates an example of circular buffer with uniform distribution of the RV starting positions, where the first and second transmissions are based on RV0 and RV1, respectively.

The punctured bits can be carried in different HARQ transmissions. According to the index of the HARQ transmission that carries the punctured bits, we can have four different ways of generating HARQ transmission and retransmissions, which are illustrated in Fig. 7.24A−D, respectively. In the method of (A), the punctured bits are transmitted in the first retransmission, and in the method of (B), the punctured bits are in the last retransmission. In the method of (C), the punctured bits are transmitted in both the first and last retransmission while in the method of (D), the punctured bits are not transmitted.

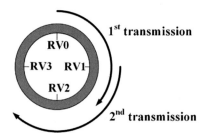

Figure 7.23 RV number is sequential.

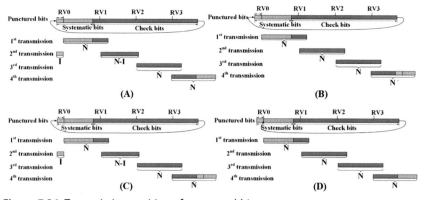

Figure 7.24 Transmission position of punctured bits.

- RV is selected discontinuously. An example is shown in Fig. 7.25 [123,124], where the starting positions of RVs are uniformly distributed in a circular buffer. As shown in the example, the first transmission starts at RV0 while the second transmission starts at RV2.

Considering the impact of retransmission on the performance, complexity of implementation and self-decoding characteristics, NR finally adopts the method that the RV is selected discontinuously (generally follows the order of $0 \to 2 \to 3 \to 1$) and the punctured systematic bits are not be transmitted. Specifically, the adopted solution is:

- The first 2Z punctured bits are not input into the loop buffer, but the padding bits are placed into the circular buffer.
- The number of starting indexes for retransmission is fixed to four and the starting positions for retransmission are fixed in the circular buffer.
- RV#0 and RV#3 are self-decoding (i.e., they contain both systematic bits and parity bits).
- The starting position of each RV is an integral multiple of Z. The starting position of RV in a limited buffer is scaled from the position in a complete buffer, while the value will be an integral multiple of Z.

The RV starting position configuration with nonuniform distribution is shown in Table 7.3, where N_{cb} represents the size of the circular

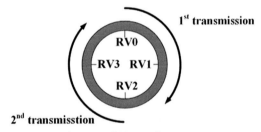

Figure 7.25 Discontinuous selection of RV number.

Table 7.3 RV starting position with nonuniform spacing.

RV_{id}	k_0	
	BG1	**BG2**
0	0	0
1	$\left\lfloor 17N_{cb}/66Z_C \right\rfloor Z_c$	$\left\lfloor 13N_{cb}/50Z_C \right\rfloor Z_c$
2	$\left\lfloor 33N_{cb}/66Z_c \right\rfloor Z_c$	$\left\lfloor 25N_{cb}/50Z_c \right\rfloor Z_c$
3	$\left\lfloor 56N_{cb}/66Z_c \right\rfloor Z_c$	$\left\lfloor 43N_{cb}/50Z_c \right\rfloor Z_c$

buffer. If the receiver uses a limited circular buffer for rate matching, then $N_{cb} = \min(N, N_{ref})$. Otherwise, $N_{cb} = N$. Here the N_{ref} is the size of a limited circular buffer and N is the code block size after encoding (the first 2Z systematic bits are removed). For BG1, $N = 66Z_c$, and for BG2, $N = 50Z_c$.

The starting positions of HARQ transmissions adopted in NR LDPC code are shown in Fig. 7.26.

7.3.7.3 Design of interleaver

In an actual transmission through radio channel, some subcarriers or OFDM symbols may experience deep fading, which would cause severe burst errors in the codeword. A well-designed interleaver can eliminate the consecutive errors and spread them to the entire codeword. Those scattered errors can be easily corrected by the decoder. When using high-order modulation in a transmission, the transmission energy is not evenly distributed over all the bits and thus different bits could have different levels of reliability. Reliability of each bit is different according to its position in the bit labeling. Among the bits corresponding to one symbol, a low-order bit has lower reliability than a high-order bit. Finally, the NR adopts the QAM-Interleaver with the number of rows equal to the modulation order, which uses the method of writing along the rows and then reading out along the columns, as shown in Fig. 7.27.

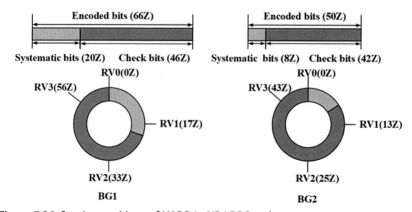

Figure 7.26 Starting positions of HARQ in NR LDPC code.

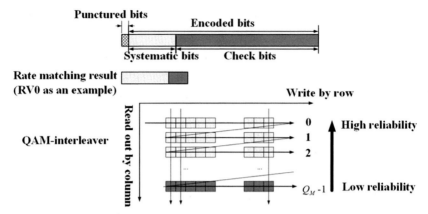

Figure 7.27 QAM-Interleaver in NR.

7.4 Summary

As the basic technology of 5G, the NR channel coding is the result of the hard work of all the participating companies and technical researchers in the 3GPP organization. The requirements of NR for high throughput, flexible code block size range, and small decoding latency are met by introducing LDPC codes in the data channel. At the same time, the Polar code is used to encode the control information in the control channel, which can achieve a low bit-error rate in the case of small code blocks and ensure the transmission reliability of the control information. These two channel coding techniques ensure the basic performance of transmission in NR and have laid a solid foundation for the subsequent design of advanced technologies such as multiantenna technology and beam management.

References

[1] 3GPP. TS36.212 V14.0.0—multiplexing and channel coding (release 14). September 2016.
[2] R1−164280. Consideration on outer code for NR, ZTE Corp., ZTE Microelectronics, 3GPP RAN1#85, Nanjing, China, 23rd−27th May 2016.
[3] Elias P. Coding for noisy channels. Ire Convention Rec 1995;6:33−47.
[4] 3GPP2. C.S0002−0 V1.0 physical layer standard for cdma2000 spread spectrum systems. October 1999.
[5] Solomon G. A connection between block and convolutional codes. Siam J Appl Math 1979;37(2):358−69.
[6] Ma H. On tail—biting convolutional codes. IEEE Trans Commun 2003;34 (2):104−11.

[7] Berrou C, et al. Near shannon limit error—correcting coding and decoding: turbo codes. In: Proc. IEEE Intl. Conf. Communication (ICC93); May 1993, p. 1064—1070.

[8] 3GPP. TS25.212 V5.10.0—multiplexing and channel coding (FDD) (release 5); June 2005.

[9] Gallager RG. Low-density parity-check codes. MIT; 1963.

[10] IEEE. 802.16e.

[11] IEEE. 802.11a.

[12] ETSI. EN 302 307 V1.3.1—Digital Video Broadcasting (DVB) second generation framing structure, channel coding and modulation systems for broadcasting, interactive services, news gathering and other broadband satellite applications (DVB-S2), 2013. 03.

[13] 3GPP. TR 25.814—V710—physical layer aspects for E-UTRA, 2006. 9.

[14] Arikan E. Channel polarization: a method for constructing capacity achieving codes for symmetric binary-input memoryless channels. IEEE Trans Inf Theory 2009;55:3051—73.

[15] R1—162162. High level comparison of candidate FEC schemes for 5G. Huawei, HiSilicon, 3GPP RAN1#84bis, Busan, Korea, 11th—15th April 2016.

[16] R1—162230. Discussion on channel coding for new radio interface. ZTE, 3GPP RAN1#84bis, Busan, Korea, 11th—15th April 2016.

[17] R1—162397. Outer erasure code. Qualcomm, 3GPP RAN1#84bis, Busan, Korea, 11th—15th April 2016.

[18] R1—163232. Performance study of existing turbo codes and LDPC codes. Ericsson, 3GPP RAN1#84bis, Busan, Korea, 11th—15th April 2016.

[19] R1—163130. Considerations on channel coding for NR, CATR, 3GPP RAN1#84bis, Busan, Korea, 11th—15th April 2016.

[20] R1—163662. Way forward on channel coding scheme for 5G new radio. Samsung, Nokia, Qualcomm, ZTE, Intel, Huawei, 3GPP RAN1#84bis, Busan, Korea, 11th—15th April 2016.

[21] R1—165637. Way forward on channel coding scheme for new radio. Samsung, Nokia, Qualcomm, Intel, ZTE, 3GPP RAN1#85, Nanjing, China, 23rd—27th May 2016.

[22] R1—165598. WF on small block length. Huawei, HiSilicon, Interdigital, Mediatek, Qualcomm, 3GPP RAN1#85, Nanjing, China, 23rd—27th May 2016.

[23] R1—165726. Code type for small info block length in NR. Ericsson, Nokia, ASB, 3GPP RAN1#85, Nanjing, China, 23rd—27th May 2016.

[24] R1—164251. Performance evaluation of binary Turbo codes with low complexity decoding algorithm. CATT, 3GPP RAN1#85, Nanjing, China, 23rd—27th May 2016.

[25] R1—164361. Turbo code enhancements. Ericsson, 3GPP RAN1#85, Nanjing, China, 23rd—27th May 2016.

[26] R1—164635. Improved LTE turbo codes for NR. ORANGE, 3GPP RAN1#85, Nanjing, China, 23rd—27th May 2016.

[27] R1—165792. WF on turbo coding. LG, Ericsson, CATT. Orange, 3GPP RAN1#85, Nanjing, China, 23rd—27th May 2016.

[28] R1—167999. WF on channel coding selection. Qualcomm Incorporated, Samsung, Nokia, ASB, ZTE, MediaTek, Intel, Sharp, MTI, Interdigital, Verizon Wireless, KT Corporation, KDDI, IITH, CEWiT, Reliance-jio, Tejas Networks, Beijing Xinwei Telecom Technology, Vivo, Potevio, WILUS, Sony, Xiaomi, 3GPP RAN1#85, Nanjing, China, 23rd—27th May 2016.

[29] R1—168040. WF on channel coding selection. Huawei, HiSilicon, CMCC, CUCC, Deutsche Telekom, Orange, Telecom Italia, Vodafone, China Unicom, Spreadtrum, 3GPP RAN1#86, Gothenburg, Sweden, 22nd—26th August 2016.

[30] R1−168164. WF on turbo code selection. LG Electronics, Ericsson, CATT, NEC, Orange, IMT, 3GPP RAN1#86, Gothenburg, Sweden 22nd−26th August 2016.

[31] R1−168164. Final report of 3GPP TSG RAN WG1#86bis. Lisbon, Portugal, 10th−14th October 2016.

[32] R1−1610604. WF on channel codes for NR eMBB data. AccelerComm, Ericsson, Orange, IMT, LG, NEC, Sony, 3GPP RAN1#86bis, Lisbon, Portugal, 10th−14th October 2016.

[33] R1−1610767. Way forward on eMBB data channel coding. Samsung, Qualcomm Incorporated, Nokia, Alcatel-Lucent Shanghai Bell, Verizon Wireless, KT Corporation, KDDI, ETRI, IITH, IITM, CEWiT, Reliance Jio, Tejas Network, Xilinx, Sony, SK Telecom, Intel Corporation, Sharp, MTI, National Instrument, Motorola Mobility, Lenovo, Cohere Technologies, Acorn Technologies, CableLabs, WILUS Inc, NextNav, ASUSTEK, ITL, 3GPP RAN1#86bis, Lisbon, Portugal, 10th−14th October 2016.

[34] R1−1610850. WF on channel codes. Huawei, HiSilicon, Acer, Bell, CATR, China Unicom, China Telecom, CHTTL, Coolpad, Deutsche Telekom, Etisalat, InterDigital, III, ITRI, MediaTek, Nubia Technology, Nuel, OPPO, Potevio, Spreadtrum, TD Tech, Telus, Vivo, Xiaomi, Xinwei, ZTE, ZTE Microelectronics, 3GPP RAN1#86bis, Lisbon, Portugal, 10th−14th October 2016.

[35] R1−1610607. Way forward on channel coding. ZTE, ZTE Microelectronics, Acer, Bell, CATR, China Unicom, China Telecom, CHTTL, Coolpad, Deutsche Telekom, Etisalat, Huawei, HiSilicon, InterDigital, III, ITRI, MediaTek, Nubia Technology, Neul, OPPO, Potevio, Shanghai Tejet, Spreadtrum, TD Tech, Telus, Vivo, Xiaomi, Xinwei, IITH, IITM, CEWiT, Reliance Jio, Tejas Network, 3GPP RAN1#86bis, Lisbon, Portugal, 10th−14th October 2016.

[36] R1−1613342. WF on channel coding for eMBB data. Samsung, Acorn Technologies, Alcatel-Lucent Shanghai Bell, Ceragon Networks, Cohere Technologies, Ericsson, ETRI, European Space Agency, HCL Technologies limited, IAESI, Intel Corporation, ITL, KDDI, KT Corporation, Mitsubishi Electric, Motorola Solutions, NextNav, NEC, Nokia, Nomor Research, NTT Docomo, Prisma telecom testing, Qualcomm Incorporated, Reliance Jio, Sharp, SK Telecom, Sony, Straight Path Communications, T-Mobile USA, Verizon Wireless, WILUS Inc, 3GPP RAN1#87, Reno, United States, 14th−18th November 2016.

[37] R1−1613342. WF on channel coding. Huawei, HiSilicon, Acer, ADI, Aeroflex, Alibaba, Bell Mobility, Broadcom, CATR, CATT, Coolpad, Coherent Logix, CHTTL, CMCC, China Telecom, China Unicom, Dish Network, ETISALAT, Fiberhome, Hytera, IAESI, III, Infineon, InterDigital, ITRI, Irdeto, Lenovo, Marvell, MediaTek, Motorola Mobility, National Taiwan University, Netas, Neul, Nubia Technology, OOREDOO, OPPO, Potevio, SGS Wireless, Skyworks, Sporton, Spreadtrum, SRTC, Starpoint, STMicroelectronics, TD-Tech, Telekom Research & Development Sdn. Bhd, Telus, Toshiba, Turk Telekom, Union Telephone, Vivo, Xiaomi, Xilinx, Xinwei, ZTE, ZTE Microelectronics, 3GPP RAN1#87, Reno, United States, 14th−18th November 2016.

[38] R1−168170. WF on coding technique for control channel of eMBB. Ericsson, Nokia, ASB, LG, NEC, Orange, IMT, 3GPP RAN1#86, Gothenburg, Sweden, 22nd−26th August 2016.

[39] R1−166926. Further discussion on performance and complexity of enhanced TBCC. Ericsson, 3GPP RAN1#86, Gothenburg, Sweden, 22nd−26th August 2016.

[40] R1−168024. WF on code selection for control channel. Huawei, HiSilicon, CMCC, CUCC, Deutsche Telekom, Vodafone, MTK, Interdigital, Spreadtrum, 3GPP RAN1#86, Gothenburg, Sweden, 22nd−26th August 2016.

[41] R1−1613211. WF on channel coding. Huawei, HiSilicon, Acer, ADI, Aeroflex, Alibaba, Bell Mobility, Broadcom, CATR, CATT, Coolpad, Coherent Logix, CHTTL, CMCC, China Telecom, China Unicom, Dish Network, ETISALAT, Fiberhome, Hytera, IAESI, III, Infineon, InterDigital, ITRI, Irdeto, Lenovo, Marvell, MediaTek, Motorola Mobility, National Taiwan University, Netas, Neul, Nubia Technology, OOREDOO, OPPO, Potevio, SGS Wireless, Skyworks, Sporton, Spreadtrum, SRTC, Starpoint, STMicroelectronics, TD-Tech, Telekom Research & Development Sdn Bhd, Telus, Toshiba, Turk Telekom, Union Telephone, Vivo, Xiaomi, Xinwei, ZTE, ZTE Microelectronics, 3GPP RAN1#87, Reno, United States, 14th−18th November 2016.

[42] R1−1613577. WF on coding technique for control channel for eMBB. LG, AT&T, Ericsson, NEC, Qualcomm, 3GPP RAN1#87, Reno, United States, 14th−18th November 2016.

[43] R1−1613248. WF on NR channel coding. Verizon Wireless, AT&T, CGC, ETRI, Fujitsu, HTC, KDDI, KT, Mitsubishi Electric, NextNav, Nokia, Alcatel-Lucent Shanghai Bell, NTT, NTT DOCOMO, Samsung, Sierra Wireless, T-Mobile USA, 3GPP RAN1#87, Reno, United States, November 14th−18th, 2016.

[44] R1−1613248. Investigation of LDPC codes for control channel of NR. Ericsson, 3GPP RAN1#87, Reno, United States, 14th−18th November 2016.

[45] R1−1706194. On channel coding for very small control block lengths. Huawei, HiSilicon, 3GPP RAN1#88bis, Spokane, United States, 3rd−7th April 2017.

[46] R1−1704386. Consideration on channel coding for very small block length. ZTE, ZTE Microelectronics, 3GPP RAN1#88bis, Spokane, United States, 3rd−7th April 2017.

[47] R1−1705528. Performance evaluation of channel codes for very small block lengths. InterDigital Communications, 3GPP RAN1#88bis, Spokane, United States, 3rd−7th April 2017.

[48] R1−1705427. Channel coding for very short length control information. Samsung, 3GPP RAN1#88bis, Spokane, United States, 3rd−7th April 2017.

[49] R1−1706184. Evaluation of the coding schemes for very small block length. Qualcomm Incorporated, 3GPP RAN1#88bis, Spokane, United States, 3rd−7th April 2017.

[50] 3GPP TS 36.212 V8.8.0 (2009−12). NR: multiplexing and channel coding (release 8).

[51] R1−1709154. Coding techniques for NR-PBCH. Ericsson, 3GPP RAN1#89, Hangzhou, P.R. China, 15th−19th May 2017.

[52] R1−1707846. Channel coding for NR PBCH. MediaTek Inc., 3GPP RAN1#89, Hangzhou, P.R. China, 15th−19th May 2017.

[53] Tal I, Vardy A. How to construct polar codes. IEEE Trans Inf Theory 2013;59 (10):6562−82.

[54] Arikan E, Telatar E. On the rate of channel polarization. In: 2009 IEEE international symposium on information theory. Seoul; 2009, p. 1493−1495.

[55] Korada SB, Şaşoğlu E, Urbanke R. Polar codes: characterization of exponent, bounds, and constructions. IEEE Trans Inf Theory 2010;56(12):6253−64.

[56] Mori R, Tanaka T. "Performance of polar codes with the construction using density evolution. IEEE Commun Lett 2009;13(7):519−21.

[57] Trifonov P. Efficient design and decoding of polar codes. IEEE Trans Commun 2012;60(11):3221−7.

[58] Sae-Young Chung TJ, Richardson, Urbanke RL. "Analysis of sum-product decoding of low-density parity-check codes using a Gaussian approximation. IEEE Trans Inf Theory 2001;47(2):657−70.

[59] Leroux C, Tal I, Vardy A, Gross WJ. Hardware architectures for successive cancellation decoding of polar codes. In: 2011 IEEE international meetingon acoustics, speech and signal processing (ICASSP), Prague; 2011, p. 1665−1668.

[60] Alamdar-Yazdi A, Kschischang FR. A simplified successive-cancellation decoder for polar codes. IEEE Commun Lett 2011;15(12):1378—80.

[61] Sarkis G, Giard P, Vardy A, Thibeault C, Gross WJ. Fast polar decoders: algorithm and implementation. IEEE J Sel Areas Commun 2014;32(5):946—57.

[62] Tal I, Vardy A. List decoding of polar codes. IEEE Trans Inf Theory 2015;61 (5):2213—26.

[63] Niu K, Chen K. CRC-aided decoding of polar codes. IEEE Commun Lett 2012;16 (10):1668—71.

[64] Final report of 3GPP TSG RAN WG1 #AH1_NR, Spokane, United States, 16th—20th January 2017.

[65] Schürch C. A partial order for the synthesized channels of a Polar code. In: 2016 IEEE international symposium on information theory (ISIT). Barcelona; 2016, p. 220—224.

[66] R1—1705084. Theoretical analysis of the sequence generation. Huawei, HiSilicon, 3GPP RAN1#88bis, Spokane, United States, 3rd—7th April 2017.

[67] Wang T, Qu D, Jiang T. Parity-check-concatenated polar codes. IEEE Commun Lett 2016;20(12):2342—5.

[68] R1—1721428. DCI CRC initialization and masking. Qualcomm Incorporated, 3GPP RAN1#91, Reno, United States, 27th November—1st December 2017.

[69] R1—1709996. Parity check bits for polar code. Huawei, HiSilicon, 3GPP RAN1#AH1706, Qingdao, China, 27th—30th June 2017.

[70] R1—1706193. Polar coding design for control channel. Huawei, HiSilicon, 3GPP RAN1#88bis, Spokane, United States, 3rd—7th April 2017.

[71] R1—1708833. Design details of distributed CRC Nokia. Alcatel-Lucent Shanghai Bell, 3GPP RAN1#89, Hangzhou, China, 5th—19th May 2017.

[72] R1—1716771. Distributed CRC for polar code construction. Huawei, HiSilicon, 3GPP RAN1#AH1709, Nagoya, Japan, 18th—21st September 2017.

[73] R1—1718914. Segmentation of polar code for large UCI. ZTE, Sanechips, 3GPP RAN1#90bis, Prague, Czechia, 9th—13th October 2017.

[74] Niu K, Chen K, Lin J. Beyond turbo codes: rate-compatible punctured polar codes. In: 2013 IEEE international meetingon communications (ICC). Budapest; 2013, p. 3423—3427.

[75] Shin D, Lim S, Yang K. Design of length-compatible polar codes based on the reduction of polarizing matrices. IEEE Trans Commun 2013;61(7):2593—9.

[76] Zhang L, Zhang Z, Wang X, Yu Q, Chen Y. On the puncturing patterns for punctured polar codes. In: 2014 IEEE international symposium on information theory. Honolulu, HI; 2014, p. 121—125.

[77] Wang R, Liu R. A novel puncturing scheme for polar codes. IEEE Commun Lett 2014;18(12):2081—4.

[78] R1—1711729. WF on circular buffer of polar code. Ericsson, Qualcomm, MediaTek, LGE, 3GPP RAN1#AH1709, Qingdao, China, 27th—30th June 2017.

[79] R1—1715000. Way forward on rate matching for polar coding. MediaTek Qualcomm, Samsung, ZTE, 3GPP RAN1#90, Prague, Czech Republic, 21st—25th August 2017.

[80] R1—1713705. Polar rate-matching design and performance MediaTek Inc. 3GPP RAN1#90, Prague, Czech Republic, 21st—25th August 2017.

[81] R1—1708649. Interleaver design for polar codes. Qualcomm Incorporated, 3GPP RAN1#89, Hangzhou, China, 15th—19th May 2017.

[82] Final report of 3GPP TSG RAN WG1 #85 v1.0.0, Nanjing, China, 23rd—27th May 2016.

[83] R1—1612280. LDPC design for eMBB. Nokia, 3GPP RAN1#87, Reno, United States, 14th—18th November 2016.

[84] R1−1611112. Consideration on LDPC design for NR. ZTE, 3GPP RAN1#87, Reno, United States, 14th−18th November 2016.

[85] R1−1611321. Design of LDPC codes for NR. Ericsson, 3GPP RAN1#87, Reno, United States, 14th−18th November 2016.

[86] R1−1612586. LDPC design for data channel. Intel, 3GPP RAN1#87, Reno, United States, 14th−18th November 2016.

[87] R1−1613059. High performance and area efficient LDPC code design with compact protomatrix. MediaTek, 3GPP RAN1#87, Reno, United States, 14th−18th November 2016.

[88] R1−1700092. LDPC design for eMBB data. Huawei, 3GPP RAN1#AdHoc1701, Spokane, United States, 16th−20th January 2017.

[89] R1−1700237. LDPC codes design for eMBB. CATT, 3GPP RAN1#AdHoc1701, Spokane, United States, 16th−20th January 2017.

[90] R1−1700518. LDPC codes design for eMBB data channel. LG, 3GPP RAN1#AdHoc1701, Spokane, United States, 16th−20th January 2017.

[91] R1−1700830. LDPC rate compatible design. Qualcomm, 3GPP RAN1#AdHoc1701, Spokane, United States, 16th−20th January 2017.

[92] R1−1700976. Discussion on LDPC code design. Samsung, 3GPP RAN1#AdHoc1701, Spokane, United States, 16th−20th January 2017.

[93] R1−1701028. LDPC design for eMBB data. Nokia, 3GPP RAN1#AdHoc1701, Spokane, United States, 16th−20th January 2017.

[94] R1−1701210. High performance LDPC code features. MediaTek, 3GPP RAN1#AdHoc1701, Spokane, United States, 16th−20th January 2017.

[95] R1−1700111. Implementation and performance of LDPC decoder. Ericsson, 3GPP RAN1#AdHoc1701, Spokane, United States, 16th−20th January 2017.

[96] Final report of 3GPP TSG RAN WG1 #88 v1.0.0; February 2017.

[97] R1−1700108. LDPC code design. Ericsson, 3GPP RAN1#AdHoc1701, Spokane, United States, 16th−20th January 2017.

[98] R1−1700383. LDPC prototype matrix design. Intel, 3GPP RAN1#AdHoc1701, Spokane, United States, 16th−20th January 2017.

[99] RAN1 chairman's notes of 3GPP TSG RAN WG1 # AH_NR Meeting, January 2017.

[100] R1−1701597. Performance evaluation of LDPC codes for eMBB. ZTE, ZTE Microelectronics, 3GPP RAN1#88, Athens, Greece, 13th−17th February 2017.

[101] R1−1704250. LDPC design for eMBB data. Huawei, 3GPP RAN1#88bis, Spokane, United States, 3rd−7th April 2017.

[102] R1−1706157. LDPC codes design for eMBB. LG, 3GPP RAN1#88bis, Spokane, United States, 3rd−7th April 2017.

[103] R1−1705627. LDPC code design for larger lift sizes. Qualcomm, 3GPP RAN1#88bis, Spokane, United States, 3rd−7th April 2017.

[104] Final report of 3GPP TSG RAN WG1 #88b v1.0.0, Spokane, United States, 3rd−7th April 2017.

[105] 3GPP. TS38.212 V15.0.0—multiplexing and channel coding (release 15). December 2017.

[106] Divsalar S, Dolinar CR, Jones, Andrews K. Capacity-approaching protograph codes. IEEE J Sel Areas Commun 2009;27(6):876−88.

[107] R1−1700245. Consideration on flexibility of LDPC codes for NR. ZTE, 3GPP RAN1#AdHoc1701, Spokane, United States, 16th−20th January 2017.

[108] R1−1700247. Compact LDPC design for eMBB. ZTE, 3GPP RAN1#AdHoc1701, Spokane, United States, 16th−20th January 2017.

[109] R1−1700521. Discussion on maximum code block size for eMBB. LG, 3GPP RAN1#AdHoc1701, Spokane, United States, 16th−20th January 2017.

[110] R1−1701030. CRC attachment for eMBB data. Nokia, RAN1#AdHoc1701, Spokane, United States, 16th−20th January 2017.

[111] R1−1702732. eMBB encoding chain. Mediatek, 3GPP RAN1#88, Athens, Greece, 13rd− 17th February 2017.

[112] R1−1703366. On CRC for LDPC design. Huawei, 3GPP RAN1#88, Athens, Greece, 13rd− 17th February 2017.

[113] R1−1704458. eMBB encoding chain. Mediatek, 3GPP RAN1#88bis, Spokane, United States, 3rd−7th April 2017.

[114] Draft report of 3GPP TSG RAN WG1 #AH_NR2 v0.1.0; August 2017.

[115] R1−1714167. Code block segmentation for data channel. InterDigital, 3GPP RAN1#90, Prague, Czech Republic, 21st−25th August 2017.

[116] R1−1712253. Code block segmentation. Huawei, 3GPP RAN1#90, Prague, Czech Republic, 21st−25th August 2017.

[117] R1−1714373. Code block segmentation principles. Nokia, 3GPP RAN1#90, Prague, Czech Republic, 21st−25th August 2017.

[118] R1−1714547. Code block segmentation for LDPC codes. Ericsson, 3GPP RAN1#90, Prague, Czech Republic, 21st−25th August 2017.

[119] R1−1701031. Padding for LDPC codes. Nokia, 3GPP RAN1#AdHoc1701, Spokane, United States, 16th−20th January 2017.

[120] R1−1712254. Padding for LDPC codes. Huawei, 3GPP RAN1#90, Prague, Czech Republic, 21st−25th August 2017.

[121] R1−1701706. LDPC design for eMBB data. Huawei, 3GPP RAN1#88, Athens, Greece, 13rd−17th February 2017.

[122] R1−1700240. IR-HARQ scheme for LDPC codes. CATT, 3GPP RAN1#AdHoc1701, Spokane, United States, 16th−20th January 2017.

[123] R1−1707670. On rate matching with LDPC code for eMBB. LG, 3GPP RAN1#89, Hangzhou, China, 15th−19th May 2017.

[124] R1−1710438. Rate matching for LDPC codes. Huawei, 3GPP RAN1#AH1706, Qingdao, China, 27th−30th June 2017.

CHAPTER 8

Multiple-input multiple-output enhancement and beam management

Zhihua Shi, Wenhong Chen, Yingpei Huang, Jiejiao Tian, Yun Fang, Xin You and Li Guo

Starting from 4G LTE, multiple-input multiple-output (MIMO) transmission has been one of the key technologies that continuously improves the spectrum efficiency by exploiting the freedom in the spatial-domain. In NR, the data can be transmitted in high-frequency band (e.g., millimeter wave), which leads to new challenges in MIMO technology. On the one hand, the bandwidth in high-frequency band is larger and an antenna array can consist of a larger number of antenna elements due to the smaller antenna size in the high-frequency band (i.e., Massive MIMO). However, the path loss is greater, and the coverage distance is consequently smaller. On the other hand, with the increase of the number of antennas, the complexity of MIMO processing may increase exponentially, and it is harder to implement many MIMO transmission schemes (such as massive digital precoding and spatial-multiplexing) in this scenario. In order to address the abovementioned issues, the NR adopted analog beam-forming as a key feature to overcome the limitations on complexity and coverage. Through using narrower beam, larger beam-forming gain can be obtained. Moreover, the complexity of analog beam will be much lower than the traditional MIMO schemes based on fully digital processing.

In order to support new scenarios, and also to improve the performance in frequency bands below 6 GHz, the NR introduced several enhancements to the MIMO technologies: more refined and more flexible channel state information (CSI) feedback, beam management, beam failure recovery (BFR), enhanced reference signal (RS), and multiple transmission and reception point (multi-TRP) cooperative transmission. Among them, the RS design in NR basically reuses the design principles of LTE with some optimization, which will not be introduced in detail

5G NR and Enhancements
DOI: https://doi.org/10.1016/B978-0-323-91060-6.00008-8

here. This chapter will focus on the new features of NR that are not supported in LTE, including CSI feedback enhancement (e.g., R15 Type II codebook), beam management and BFR introduced in R15/R16, enhanced eType II codebook in R16, and multi–TRP transmission.

8.1 CSI feedback for NR MIMO enhancement

CSI feedback is an important part of NR MIMO enhancement. In order to support a large number of antenna ports, more flexible CSI reporting, and more accurate CSI, the NR introduced various enhancements for CSI feedback, such as configuration signaling, triggering scheme, measurement method, codebook design, UE capability reporting, etc. This section will introduce the enhancements and optimizations made in NR on the basis of the LTE CSI feedback mechanism, and the Type II codebook that is newly introduced in R15 will be the focus of the discussion.

8.1.1 CSI feedback enhancement in NR

8.1.1.1 Enhancement on CSI configuration signaling

The overall framework of NR CSI feedback is established on the basis of the CSI feedback mechanism of the LTE system. In the initial framework of CSI configuration [1] agreed on at the beginning of NR discussion, CSI-related resource allocation and reporting configuration were supported through the following four parameter sets:

- N CSI reporting configurations are used to configure the resources and modes of CSI reporting, similar to the CSI process of LTE.
- M channel measurement resource configurations are used to configure the RS for channel measurement.
- J interference measurement resource (IMR) configurations are used to configure RSs resources for interference measurement, similar to IMR in LTE.
- One CSI measurement configuration indicates the association between the above N CSI reporting configurations, M channel measurement resource configurations and J IMR configurations.

At the RAN1#AdHoc1701 meeting, the abovementioned channel measurement resource configuration and IMR configuration are combined into a CSI resource configuration [2]. It was also clarified that the CSI measurement configurations can contain L association indications, each of which is used to associate a CSI reporting configuration with a CSI resource configuration. However, RAN2 does not fully adopt the framework designed by RAN1

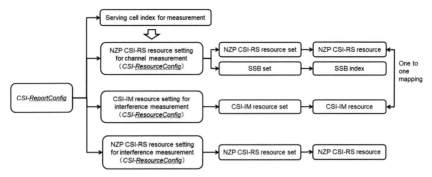

Figure 8.1 CSI resource configuration of NR.

for designing RRC signaling, and instead a few CSI resource configurations for CSI measurement are explicitly included in the CSI reporting configuration, thereby avoiding the additional signaling needed to indicate the relationship between CSI reporting configuration and CSI resource configuration. The UE can obtain all the configuration information needed for CSI measurement and reporting directly from a CSI reporting configuration. Each CSI resource configuration can contain multiple nonzero-power CSI RS (NZP CSI—RS) resource sets for channel measurement (where each set can contain several CSI—RS resources corresponding to the above RS configuration), and multiple CSI—RS resource sets for interference measurement (CSI—IM) (where each set can contain multiple resources for interference measurement, corresponding to the interference measurement configuration mentioned above). For aperiodic CSI reporting, a CSI resource configuration can further include multiple NZP CSI—RS resource sets for interference measurement (as shown in Fig. 8.1).

8.1.1.2 Enhancement on interference measurement

During the discussion of CSI feedback, there were three candidate schemes for interference measurement:

- Similar to LTE, IMR based on zero-power CSI—RS is used to measure the intercell interference.
- The interference measurement is based on the NZP CSI—RS resources [3]. In this scheme, the network adopts the prescheduling method to precode the CSI—RS port(s), so that the UE can measure the interference of multiplexed users based on the precoded CSI—RS port(s). Then the network side can use the measurement results for subsequent data scheduling.

- The interference measurement is based on the demodulation RS (DMRS) port(s) [4]. In this scheme, the UE estimates the possible interference of multiplexed users on the unused DMRS port(s).

Among them, the first scheme follows the LTE method and was agreed on at the early stage of NR standardization. The latter two schemes have similar effects, and they support the same interference type. The advantage of the scheme using NZP CSI–RS is that the interference of each multiplexing port in subsequent scheduling can be estimated accurately, while its disadvantage is that it requires additional resources needed for interference measurement and increases the measurement complexity. The DMRS-based scheme does not require additional resources, and can guarantee high-measurement accuracy in case of continuous PDSCH scheduling. However, this method is only able to measure the channel in the current scheduled bandwidth. In practical systems, it is difficult for the physical resources, the paired users, and the precoding matrix of the two continuous scheduling to remain consistent. After discussions in several meetings, the interference measurement scheme based on NZP CSI–RS was finally agreed on at the RAN1#89 meeting [5], and DMRS-based interference measurement was excluded. In the subsequent standardization process, interference measurement based on NZP CSI–RS resources was further limited to be only used for aperiodic CSI reporting (Fig. 8.2).

8.1.1.3 Enhancement on reporting periodicity

Both periodic and aperiodic CSI reporting are supported in the LTE system. In addition to these two methods, NR supports semipersistent CSI

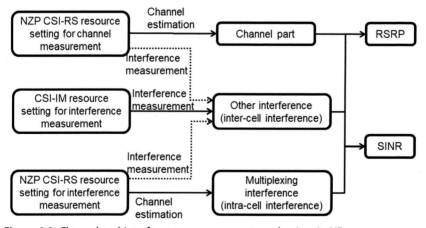

Figure 8.2 Channel and interference measurement mechanism in NR.

reporting, and also further enhances the mechanism of aperiodic CSI reporting. Specifically, NR supports the following three CSI reporting schemes:

- The periodic CSI reporting follows the LTE method, and RRC signaling is used to indicate the configuration of CSI reporting.
- Semipersistent CSI reporting can be implemented via two different ways: the method of PUSCH reporting based on DCI scheduling and the method of PUCCH reporting based on MAC CE activation. The necessity and benefits of the former method was extensively discussed in 3GPP. Finally, both methods were adopted in NR as a compromise. The specific choice depends on the UE capability and network side configuration.
- For aperiodic CSI reporting, two different triggering schemes were discussed: the method of RRC + DCI and the method of RRC + MAC + DCI. The latter allows higher configuration flexibility, but also introduces additional signaling design. Since both schemes are supported many companies, a compromise solution was agreed on in the last two RAN1 meetings of R15: when the number of aperiodic CSI trigger states configured by RRC exceeds the number that can be indicated by CSI trigger field in DCI, MAC CE is used to activate some of the states to be triggered through DCI. In this way, whether to use MAC CE for activation completely depends on the configuration on the network side. In addition, if the measurement of aperiodic CSI is based on aperiodic measurement resources (such as aperiodic CSI−RS), the corresponding trigger signaling also indicates the transmission of the aperiodic RS used for measurement and the reporting of aperiodic CSI (as shown in Fig. 8.3).

On the basis of the PUSCH-based aperiodic CSI reporting in NR that is similar to that in LTE, in the RAN1#AdHoc1709 meeting, 3GPP agreed to support DCI-triggered PUCCH-based aperiodic CSI reporting [6]. Some companies thought the PUCCH-based method required high implementation complexity and great standardization effort, and thus it is difficult to complete it within R15 time window. In the RAN1#90bis meeting, three candidate trigger methods were discussed for PUCCH-based aperiodic CSI reporting [7]: triggered by DCI scheduling PDSCH, triggered by DCI scheduling PUSCH (RRC configures PUCCH or PUSCH), and triggered by DCI scheduling PUSCH (DCI configures PUCCH or PUSCH), and conclusion was reached. During the email discussion after the meeting, there was still great divergence among

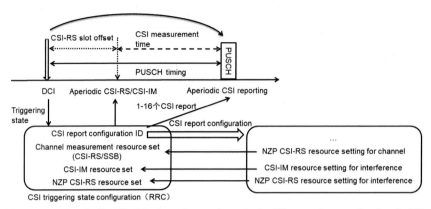

Figure 8.3 Aperiodic CSI–RS triggering and aperiodic CSI reporting mechanism in NR.

companies on which DCI is used to trigger the PUCCH-based CSI reporting. In the end, the solution of PUCCH-based aperiodic reporting was not supported in NR standardization.

8.1.1.4 Enhancement on content of CSI reporting

In addition to the CSI reporting method, the NR has also enhanced the content of CSI reporting so that the CSI feedback in the NR can support more functions. The CSI report in the LTE system can only support CSI for PDSCH scheduling, such as CSI–RS resource indicator (CRI), rank indicator (RI), precoding matrix indicator (PMI), and channel quality indicator (CQI). The NR introduces beam management-related content in CSI, such as CRI/SSB index and corresponding RS reception power (RSRP). The network side can determine the downlink beam according to the feedback from the UE. Furthermore, a CSI reporting configuration in NR can also correspond to a CSI report with empty content (i.e., the UE does not need to perform actual CSI reporting). Such a CSI reporting configuration can be used to trigger aperiodic tracking RS (TRS) used for time-frequency synchronization or aperiodic CSI–RS used in downlink beam management. The UE does not need to perform reporting based on these RSs (Table 8.1).

8.1.1.5 Enhancement of reciprocity-based CSI

In the LTE system, the UE can report the corresponding CQI based on nonprecoded CSI–RS ports and a predefined transmission scheme. Other CSI information is derived by the network side based on channel reciprocity. In order to better support downlink transmission based on channel

Table 8.1 CSI reporting contents and corresponding application scenarios in NR.

Report content	Objective	Applicability	Notes
"cri–rI–PMI–CQI"	CSI reporting based on PMI	Type I/II codebook	It can support CSI–RS–based beam selection and CSI reporting, similar to LTE class A/B CSI reporting.
"cri–rI–LI–PMI–CQI"	CSI reporting based on PMI	Type I/II codebook	The strongest transport layer indicator LI is added to determine the transport layer associated with DL PTRS.
"cri–rI–i1"	Partial CSI report	Type I codebook	Only wideband channel information is reported, which can be used together with other reported content.
"cri–rI–i1-CQI"	CSI report for semiopen-loop transmission	Type I codebook	Only W_1 is reported and W_2 is not reported. The UE randomly selects one W_2 corresponding to W_1 on each PRG to calculate CQI.
"cri–rI–CQI"	No-PMI-based CSI reporting	indicated CSI–RS port	RI and CQI are estimated based on CSI–RS ports indicated for each rank from network devices.
"cri–rSRP"	Tx beam management	RSRP measurement	RSRP measurement and reporting based on CSI–RS resources (with different beams).
"ssb-Index–rSRP"	Tx beam management	RSRP measurement	RSRP measurement and reporting based on SSB (with different beams).
"none"	TRS or downlink receive beam management	CSI reporting based on PUSCH	It cannot be configured for CSI reporting on PUCCH.

reciprocity, at the RAN1#88bis meeting, a non–PMI–based feedback solution was agreed on for the case that the network can obtain complete downlink channel information from uplink measurement. In this solution, the CSI contains RI and CQI, where CQI is calculated based on a codebook, and precoding information is derived based on channel reciprocity and the reported RI. Regarding the codebook assumed for CQI feedback, three candidate schemes were agreed on at the RAN1#90 meeting [8]: using port selection codebook, using columns of an identity matrix as the codebook, or using the existing codebook. In the next RAN1 meeting, the third scheme was first excluded, and the first two schemes further evolved into two port selection methods:

- Method 1: The CQI is calculated based on the port selection codebook, and each column is used to select a CSI–RS port corresponding to a layer.
- Method 2: Port index indicated by the network is used to select part of CSI–RS ports from the CSI–RS ports in a CSI–RS resource for RI/CQI calculation. The network indicates the CSI–RS ports used for measurement for each rank.

At the RAN1#90bis meeting, the latter was agreed upon. The network first determines the candidate precoding vectors based on the channel reciprocity, and CSI–RS ports are precoded based on the determined precoding vectors. Then the network informs the UE with the CSI–RS ports for measurement corresponding to different ranks in advance. The UE reports RI and CQI based on measuring the indicated CSI–RS ports, where the UE performs CQI calculation based on the CSI–RS ports corresponding to the reported rank. The CSI feedback mechanism for the case of partial channel reciprocity (i.e., the UE has fewer transmit antennas than reception antennas, and the network can only obtain part of the downlink channel information) was also discussed during the standardization process. Companies performed a lot of evaluation based on the candidate feedback mechanisms for that, but no scheme was finally agreed on due to the large divergence among companies.

8.1.2 R15 Type I codebook

R15 supports two types of codebooks: Type I codebook and Type II codebook. R16 further supports the enhanced eType II codebook. The design of Type I codebook basically follows the codebook design of LTE, and it supports normal spatial-resolution and CSI accuracy. Type II and eType II codebooks adopt the design idea of eigenvector quantization and eigenvector linear weighting to support higher spatial-resolution and CSI accuracy. This section will focus on the design of the Type I codebook, while the design of the Type II and eType II codebooks will be discussed in the following chapters.

At the initial stage of the NR discussion, it was agreed that the NR Type I codebook could reuse the two-stage codebook design in the LTE system (i.e., $W = W_1 W_2$,) where W_1 is used to report a beam (or beam group), and W_2 is used to report at least one of the following information: the beam selected from the beam group, the weighting coefficient between the beams, and the phase between polarization directions. Two issues need to be solved for designs based on such a scheme: whether the beam is selected only based on W_1 (number of beams in a beam set $L = 1$) or based on both W_1 and W_2 as LTE ($L = 4$) and whether a single beam or a linear weighting of multiple beams is applied to a layer. It should be noted that the beams used in codebook design are generally digital beams that are mapped to CSI–RS antenna ports (i.e., the beams formed by digital precoding). They are different from the analog beams discussed in other chapters.

At the RAN1#89 meeting, more than 30 companies jointly presented a package proposal for NR codebook design [9], which provided most of the designs for Type I and Type II codebooks, including the solutions for the above two issues. Although some details of the design were still controversial and not acceptable to some companies, the joint proposal was finally approved due to majority supporting. According to that proposal, with the exception of the 2-port codebook that reuses the design of LTE, all other codewords in Type I codebook can be expressed as: $W = W_1 W_2$, where $W_1 = \begin{bmatrix} B & 0 \\ 0 & B \end{bmatrix}$ and $B = [b_0, b_1 \ldots, b_{L-1}]$ corresponds to L oversampled DFT beams (they can be horizontal and vertical two-dimensional beams). Although adopting similar expressions, the NR introduces many enhancements in the Type I codebook compared to LTE:

- When rank = 1 or 2, the network can configure either $L = 1$ or $L = 4$. If $L = 1$, W_2 is used to only reports the phase between two polarizations. If $L = 4$, W_2 is used to select a beam from the beam group corresponding to W_1 and also report the interpolarization phase. The definition of a beam group in NR reuses one of the three definitions of beam groups in LTE, as shown in Fig. 8.4.

Figure 8.4 Pattern of beam group supported by $L = 4$ (*left*: horizontal port, *right*: 2D port).

- When rank $= 3$ or 4, only $L = 1$ is supported. The LTE codebook design is reused when the number of ports is less than 16, while the phase between orthogonal beams is additionally reported for 16 and 32 ports.
- When rank is greater than 4, only $L = 1$ is supported and fixed orthogonal beams are used.
- Based on the single-panel codebook, the multipanel codebook is designed with an additional reporting interpanel phase (which can be wideband or subband). A single-panel codebook is adopted for each panel.
- The codebook subset restriction (CSR) is introduced for the Type I codebook. The CSR can be performed for each DFT beam and each rank, and the PMI corresponding to the restricted beam is not reported by the UE.

Based on the single-panel codebook, CSI reporting for semiopen-loop transmission is also supported in NR, where the UE only reports CRI/RI and beam information (W_1), while W_2 is not reported. The UE assumes that network adopts the beam(s) corresponding to W_1 for downlink transmission. There are many candidate schemes for the precoding assumption that is used by the UE to calculate CQI, which are diversity transmission (SFBC, similar to CQI assumption used in LTE transmission mode 7) and codeword polling and random selection of codewords. At the RAN1AdHoc1709 meeting, it was agreed that the UE assumes the codeword used in each precoding resource group (PRG) is randomly selected from the multiple W_2 corresponding to the reported W_1. The codewords for random selection can be further indicated via the CSR configured by the network.

8.1.3 R15 Type II codebook

The previous section introduced the Type I codebook design in NR. It can meet the requirement of low-precision CSI feedback with lower feedback overhead for SU—MIMO. However, in some application scenarios such as MU–MIMO, CSI feedback with higher channel precision is desired. For this reason, RAN1# 86bis agreed that the Type II codebook can support high-precision CSI in NR [10].

The main differences between Type I and Type II codebooks are:

- The Type I codebook is mainly used for SU–MIMO and can support high-rank transmission. The Type II codebook is mainly for MU–MIMO and rank is generally low. Specifically, only rank = 1/2 is supported in the Type II codebook for low overhead.
- The Type I codebook reports only one beam per precoder, while the Type II codebook supports reporting combinations of multiple beams.
- The power of the Type I codebook ports is constant, while power on each port of the Type II codebook varies due to superposition.
- The layers in the Type I codebook are orthogonal to each layer, while the Type II codebook does not have such a constraint.
- Only phase information is reported on the subband in the Type I codebook, while the amplitude coefficient of each subband can be reported in the Type II codebook through the mode of wideband + subband.
- The feedback payload in the Type I codebook is low, only dozens of bits, which can be obtained by exhaustive searching. However, the feedback of the Type II codebook requires hundreds of bits (~ 500 bits) and the weighting coefficients are computed by minimizing the mean square error and then quantization.
- The feedback payload of the Type I codebook is fixed, while the feedback payload of the Type II codebook varies with the channel state.

8.1.3.1 Structure of the Type II codebook

At the RAN1#AdHoc1701 meeting, it was agreed to support high-precision CSI reporting through a linear combination of multiple beams [11]. The spatial-domain (SD) channel information is projected onto a set of bases, and then the UE reports dominated components and the weighting coefficients. There exist three different candidate schemes: channel correlation matrix feedback, precoding matrix feedback, and hybrid feedback. The method of precoding matrix feedback extends the design of LTE. The UE recommends reporting precoder and RI/CQI information and determines the transmission rate associated with the precoder. In the method of correlation matrix feedback, the UE reports the long-term and wideband channel covariance matrix. The method of hybrid feedback is

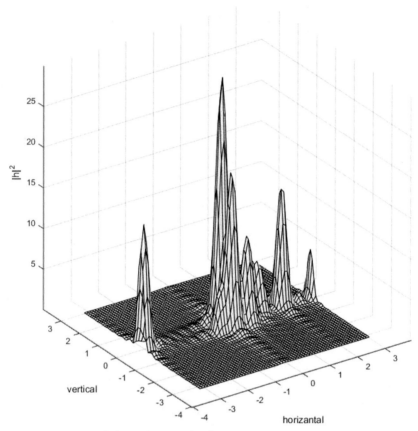

Figure 8.5 Sparsity of channel in angular domain.

similar to Class B in LTE, and the CSI−RS port with beam-forming is selected for feedback.

The basic scheme of linear combining [12−15] is shown in Fig. 8.5. The principle is to transform the spatial-channel into the angle domain, and the combining coefficient represents the amplitude and phase per component. On the one hand, the radio channel itself is sparse in space, which means only few directions have energy. When spatial-sampling rate increases with more antenna, only a few coefficients have nonzero energy. The UE only needs to feed back a few nonzero components, which greatly reduces the reporting overhead. On the other hand, each precoder (e.g., eigenvectors) can also be expressed as a linear combination of chan-nel spatial-vectors (Fig. 8.6).

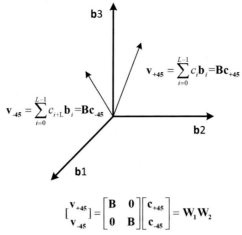

$$\left[\begin{matrix} \mathbf{v}_{+45} \\ \mathbf{v}_{-45} \end{matrix}\right] = \left[\begin{matrix} \mathbf{B} & 0 \\ 0 & \mathbf{B} \end{matrix}\right]\left[\begin{matrix} \mathbf{c}_{+45} \\ \mathbf{c}_{-45} \end{matrix}\right] = \mathbf{W}_1\mathbf{W}_2$$

Figure 8.6 Schematic diagram of linear combination.

Regarding the codebook structure, most companies agreed to adopt the dual codebook structure for the Type II codebook [12]:

$$W = W_1 W_2 \tag{8.1}$$

At the RAN1#AdHoc1701 meeting, the candidate schemes for the Type II codebook were agreed on. Similar as Type I codebook, two models are discussed. The first model is:

$$W_1 = \left[\begin{matrix} B_1 & B_2 \\ B_1 & -B_2 \end{matrix}\right] \tag{8.2}$$

And another one is (Fig. 8.6):

$$W_1 = \left[\begin{matrix} B & 0 \\ 0 & B \end{matrix}\right] \tag{8.3}$$

The above two alternatives are actually equivalent. There exist two options for the selection of vector B. The first one ensures B to be a set of orthogonal vectors, while the second option allows B to be nonorthogonal. Finally, at the RAN1#89 meeting [16], most companies agree to use the same beam for both polarizations, thus W_1 has a block diagonal structure.

In order to improve the feedback accuracy, NR adopts four times as much oversampled 2D−DFT vectors (Fig. 8.7) to quantize the beam in both horizontal and vertical directions. Similar to the codebook

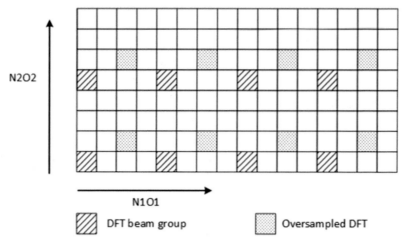

Figure 8.7 Diagram of beam selection.

Table 8.2 Codebook configuration of Type II codebook.

CSI−RS port	(N_1, N_2)	(O_1, O_2)
4	(2,1)	(4,-)
8	(2,2)	(4,4)
	(4,1)	(4,-)
12	(3,2)	(4,4)
	(6,1)	(4,-)
16	(4,2)	(4,4)
	(8,1)	(4,-)
24	(6,2), (4,3)	(4,4)
	(12,1)	(4,-)
32	(8,2), (4,4)	(4,4)
	(16,1)	(4,-)

introduced in LTE R14, the orthogonality of L selected beams is considered, where $L \leq N_1 N_2$. The number of CSI−RS ports supported by the Type II is shown in Table 8.2.

In the above expression, W_2 is the combined coefficients corresponding to L beams on the subband, including the amplitude and phase. The coefficients per-layer and per-polarization direction are selected independently. The codebook can be expressed as:

$$\tilde{w}_{r,l} = \sum_{i=0}^{L-1} b_{k_1^{(i)} k_2^{(i)}} \times p_{r,l,i}^{(WB)} \times p_{r,l,i}^{(SB)} \times c_{r,l,i} \qquad (8.4)$$

where $b_{k_1^{(i)} k_2^{(i)}}$ is an oversampled DFT vector, which is determined by the W_1 vector; $p_{r,l,i}^{(WB)}$ is the wideband amplitude of the ith beam in the polarization r and layer l; $p_{r,l,i}^{(SB)}$ is the subband amplitude of the ith beam in the polarization r and layer l; $c_{r,l,i}$ is the phase of the ith beam in the polarization r and layer l.

8.1.3.2 Quantization
During the 3GPP discussion, some companies believed that subband reporting can bring better trade-off between overhead and performance. Wideband reporting has low overhead while its performance is slightly worse than that of the subband reporting. In subband reporting, subband amplitude is reported in differential manner [17]. The feedback payload increases but the system performance is also improved. Finally, the NR decided to support both amplitude reporting scheme through higher layer configuration [16]:

- In wideband reporting mode, the UE does not report the differential amplitude on the subband (i.e., $p_{r,l,i}^{(SB)} = 1$.).

- For wideband + subband amplitude reporting, the UE reports differential amplitude on subband with 1bit quantization, namely $\{1, \sqrt{0.5}\}$, while wideband amplitude adopts 3-bit quantization and uniform quantizer with 3 dB step size to facilitate implementation, $\{1, \sqrt{0.5}, \sqrt{0.25}, \sqrt{0.125}, \sqrt{0.0625}, \sqrt{0.0313}, \sqrt{0.0156}, 0\}$.

The phase of each coefficient can be quantized by QPSK and/or 8PSK, which is determined by high-level configuration parameters. For wideband + subband amplitude reporting, the Type II codebook adopts nonuniform phase quantization [18]. High-precision (8PSK) phase quantization is used for strong beams while low-precision (QPSK) phase quantization is used on weak beams in order to optimize the trade-off between overhead and performance. If $L = \{2, 3, 4\}$, the first $K = \{4, 4, 6\}$ beams are quantized with 8PSK. The precoding vector only needs the differential phase and amplitude between different ports. Thus the dimension is $2L$-1. The Type II codebook takes the beam with the strongest wideband power as the reference "1", which is indicated by $\log_2(2L)$) bits. The UE reports the amplitude and phase of remaining $2L$-1 beam [17]. Similar to the UE selective report subband coding, the Type II codebook utilize a combination number to report L selected beams $(\log_2 C_{N1N2}^L$ bit), which can reduce feedback overhead compared with the method bitmap indicator but at cost of certain complexity [19].

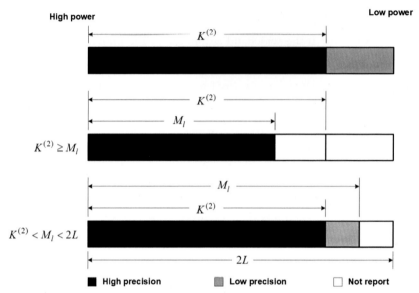

Figure 8.8 W_1 Report the number of nonzero coefficients.

Because L is a high-layer parameter and the actual number of beams selected may be less than L, the wideband amplitude needs to include zero elements [20] to reduce the feedback overhead for the subband [21]. The wideband amplitudes in the two polarization directions are selected independently (allowed to be zero). In order for the network to determine the information length, the UE reports the number of wideband nonzero coefficients [22] (i.e., the number of L beams whose power is greater than zero). As shown in Fig. 8.8, M_l is the number of reported nonzero coefficients and $K^{(2)}$ is the number of 8PSK quantized beams.

8.1.3.3 Port selection codebook

If the network knows partial downlink channel information, for example, the direction of beam that is obtained through the UL/DL reciprocity, the UE can measure the beam-formed CSI–RS. In this case, the UE only needs to measure a few ports, and that can reduce complexity of calculating, storage, and also the CSI feedback overhead. On the basis of the Type II codebook, the W_1 of port selection codebook is [23]:

$$W_1 = \begin{bmatrix} E & 0 \\ 0 & E \end{bmatrix}$$

The matrix E is expressed as:

$$E = \left[e_{\text{mod}\left(md,\frac{X}{2}\right)} \; e_{\text{mod}\left(md+1,\frac{X}{2}\right)} \; \cdots \; e_{\text{mod}\left(md+L-1,\frac{X}{2}\right)} \right] \tag{8.5}$$

where X is the number of CSI−RS ports. In this method, the candidate ports are divided into d groups (d is high-layer configuration) for the UE to select continuous L ports. Each column of E has only one (i.e., e_i represents the position of port). In E, L elements are one while the rest are zero, which is used to indicate the port selection.

8.1.3.4 CSI omission

One major issue of the Type II codebook is that the overhead is too high. As shown in Table 8.3, the feedback overhead of the Type II codebook is proportional to the rank. In some configurations, the payload is more than 500 bits, and the overhead of different configurations varies greatly. When the network side allocates PUSCH resources for CSI reporting, it is difficult to estimate the overhead accurately. Resources can only be allocated according to the maximum overhead, which will cause a waste of resources.

In order to solve this issue, NR considers some methods to reduce resource overhead. Since channel in frequency domain are correlated, the CSI of adjacent subbands are generally similar. Downsampling in subband can reduce overhead significantly. When the PUSCH resource allocation is insufficient, half of the subbands can be dropped, thereby reducing the overhead by about half, while ensuring there is no obvious distortion of the CSI. Specifically, CSI on the subband with even index are reported with high priority, while the CSI of odd subband may be omitted [24] (Fig. 8.9).

8.2 R16 codebook enhancement

Compared with the traditional codebook, the Type II codebook can significantly improve the accuracy of the channel information, thereby improving the performance of downlink transmission, especially for downlink multiuser transmission [25,26]. However, the feedback overhead of the Type II codebook is very high. For example, when $L = 4$, rank $= 2$ and the number of subbands is 10, the total feedback overhead reaches 584 bits. At the same time, the Type II codebook can only support single-layer and two-layer transmission due to high feedback cost,

Table 8.3 Feedback cost of Type II codebook.

Beam number	Over-sampling	Beam selection	Reference beam	Wideband amplitude	Wideband overhead	Subband amplitude	Subband phase	Total
Rank1 overhead (Number of bits)								
2	4	7	2	9	22	3	9	142
3	4	10	3	15	32	3	13	192
4	4	11	3	21	39	5	19	279
Rank2 overhead (Number of bits)								
2	4	7	4	18	33	6	18	273
3	4	10	6	30	50	6	26	370
4	4	11	6	42	63	10	38	543

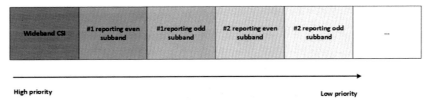

| Wideband CSI | #1 reporting even subband | #1reporting odd subband | #2 reporting even subband | #2 reporting odd subband | ... |

High priority   **Low priority**

Figure 8.9 CSI priority.

which limits its application scenarios. In order to reduce the feedback signaling overhead of the Type II codebook and extend it to higher rank transmission, NR introduces an enhanced Type II codebook in R16, called the eType II codebook.

8.2.1 Overview of the eType II codebook

The Type II codebook compresses the SD of eigenvectors through L DFT beams, but does not compress the frequency-domain (FD) part. The overhead of the Type II codebook is mainly contributed by the coefficient matrix $W2$, so the overhead of the Type II codebook can be reduced by compressing the linear combination coefficient of $W2$ [27]. Therefore the eType II codebook considers two methods to reduce the overhead: firstly, compressing the FD complexity, and then selectively reporting some linear combination coefficients to compress the correlation of coefficient matrix.

In order to reduce the feedback overhead of the Type II codebook, two compression schemes were proposed at the RAN1#94bis meeting: the method of FD compression (FDC) and the method of time-domain compression (TDC). The key idea of both schemes is to reduce the feedback overhead by using FD correlation or time-domain sparsity. The FDC scheme uses the correlation of precoding coefficients between adjacent subbands and introduces a set of FD basis vectors to compress the FD complexity. The TDC scheme uses DFT or IDFT to convert the coefficients of the subbands into coefficients in the time-domain. If the DFT vectors are selected as the FD basis vectors, the FDC and TDC compression schemes are equivalent. Therefore at the RAN1#95 meeting, it was agreed to use the DFT vector as the basis vector for FDC matrix. Thus the FDC and TDC schemes were combined into one scheme [28−30], which was referred to as the FDC scheme.

At the RAN1#95 meeting, some companies also proposed another candidate compression scheme in order to reduce the FD feedback overhead, which is a compression scheme based on singular value decomposition (SVD) [31]. This scheme uses SVD of W_2 to obtain W_f and the compressed $\tilde{W}_2$. Since this SVD scheme does not use a predefined set of basis vectors, the UE needs to dynamically report the basis vectors corresponding to W_f [32]. Although this scheme can capture the maximum signal energy and obtain the best compression effect, it is more sensitive to errors. Furthermore, the SVD scheme requires a relatively large feedback overhead for reporting the W_f of the subband. Compared with the aforementioned schemes, the SVD scheme has advantage when the feedback overhead is greater. Finally, the 3GPP adopted the FDC scheme as the FDC scheme.

According to the FDC scheme, the eType II codebook can be expressed as $W = W_1\tilde{W}_2W_f^H$. The UE needs to report $W_1 = \{b_i\}_{i=0}^{L-1}$, $W_f = \{f_k\}_{k=0}^{M-1}$ and linear combination coefficients $\tilde{W}_2 = \{c_{i,f}\}, i = 0, 1, \ldots, L-1, f = 0, 1, \ldots, M-1$, where

- The W has N rows and N_3 columns, where N is the number of CSI−RS ports, and N_3 is the number of FD units.
- W_1 reuse the Type II codebook design and each polarization group contains L beams (i.e., $W_1 = \begin{bmatrix} b_0 \ldots b_{L-1} & 0 \\ 0 & b_0 \ldots b_{L-1} \end{bmatrix}$.)
- Similar to the design of the Type II codebook, the $\tilde{W}_2$ matrix contains all the $2L \times M$ coefficients of linear combination, where M is the number of FD basis vectors.
- W_f is the matrix of DFT basis vectors (i.e., $W_f = \begin{bmatrix} f_0 \ldots f_{M-1} \end{bmatrix}$.)

For example, if rank = 1, the precoding matrix of the eType II codebook can be expressed as shown in Fig. 8.10 [33].

Fig. 8.11 shows the evolution from the Type II codebook to the eType II codebook. Compared with the Type II codebook, the eType II codebook adds a new part W_f, where H is the eigenvector matrix, and the kth column of H is the channel eigenvector of the kth subband [27].

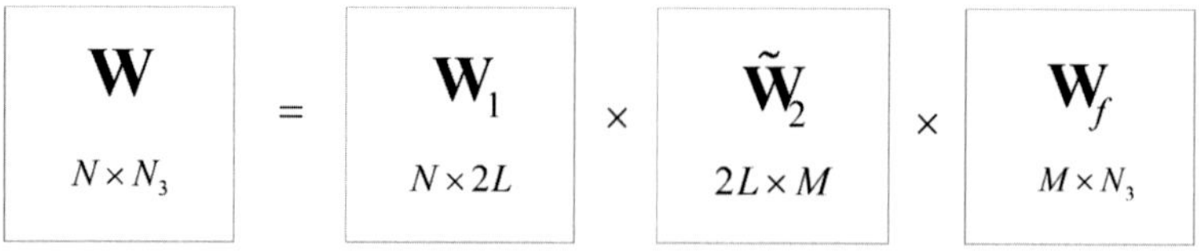

$$W \quad = \quad W_1 \quad \times \quad \tilde{W}_2 \quad \times \quad W_f$$
$$N \times N_3 \qquad N \times 2L \qquad 2L \times M \qquad M \times N_3$$

Figure 8.10 General form of R16 eType II precoding matrix.

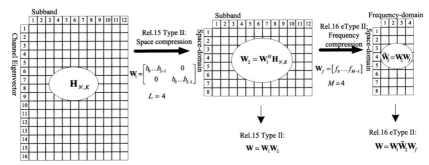

Figure 8.11 Schematic diagram of the evolution process from R15 to R16 Type II.

After determining the general form of the eType II codebook, the design of W_1, $\tilde{W}_2$, W_f^H will be discussed in next sections. The design of W_1 is the same as that of the Type II codebook, so this section will not repeat the discussion on it. The design of $\tilde{W}_2$ and W_f for rank $= 1$ will be first introduced, and then the case of rank > 1.

8.2.2 Frequency-domain matrix design

The critical part of W_f design is the selection of DFT basis vector (i.e., how to select M DFT basis vectors from N_3 FD units, where $N_3 = N_{sb} \times R$, $M = p \times \frac{N_3}{R}$, N_{sb} is the number of subbands, R indicates the number of FD units contained in each subband, and p is configured by high-level signaling to determine the number of DFT basis vectors).

For rank $= 1$, 3GPP considered the following two candidate schemes for the selection of DFT basis vectors:

- Common basis vector [34]: $2L$ beams select the same DFT basis vector, where $W_f = [f_0 \ldots f_{M-1}]$, and M basis vectors are dynamically selected, as shown in Fig. 8.12.
- Independent basis vector [35]: each beam independently selects the DFT basis vector, $W_f = [W_f(0), \ldots, W_f(2L-1)]$ where $W_f(i) = [f_{k_{i,0}} f_{k_{i,1}} \cdots f_{k_{i,M_i-1}}]$, $i \in \{0, 1, \ldots, 2L-1\}$, as shown in Fig. 8.13.

For the above two schemes, companies conducted a lot of simulations and found that the implementation of the scheme of common basis vector is simpler and the overhead is relatively small [36]. Therefore it was agreed to adopt the scheme of common basis vector for rank $= 1$ at the RAN1#AH1901 meeting.

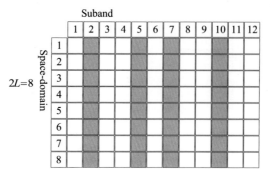

Figure 8.12 2L beams using the same DFT basis vector.

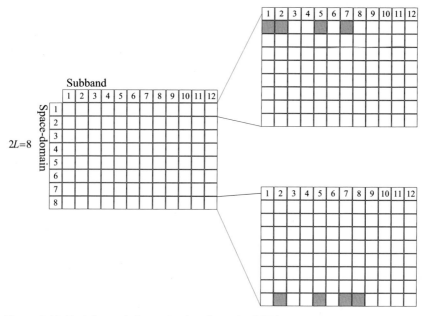

Figure 8.13 Each beam independently selects the DFT basis vector.

In addition, the companies also found that when selecting M DFT basis vectors in N_3 FD units, $\left\lceil \log_2 \binom{N_3 - 1}{M - 1} \right\rceil$ bits need to be fed back. If N_3 increases, the feedback bit will also increase. In order to further reduce the feedback overhead, a one-stage scheme and a two-stage window scheme that are suitable for different scenarios were agreed on at the RAN1 97 meeting [37]. When $N_3 \leq 19$, the one-step scheme is adopted,

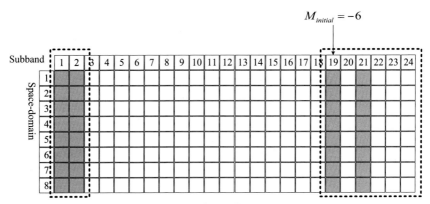

Figure 8.14 $N_3 > 19$, the two-stage window scheme.

as shown in Fig. 8.12. When $N_3 > 19$, the two-step windowing scheme is adopted. A window with a length of $2M$ is configured through high-level signaling, and the UE reports $M_{initial}$ to indicate the starting position of the window, and selectively reports M DFT basis vectors within that window, where $M_{initial} \in \{-2M + 1, -2M + 2, \ldots, 0\}$. An example is illustrated in Fig. 8.14. The black virtual frame is a window with a length of 2M, and the gray grid is the selected M DFT basis vectors.

8.2.3 Design of coefficient matrix

The W_2 matrix of the Type II codebook is a coefficient matrix containing $K = 2LM$ linear combinations, and each row contains related linear combination coefficients. Therefore a parameter K_0 is introduced in the eType II codebook in order to further reduce the overhead of the eType II codebook, where $K_0 = \beta \times 2LM$. A UE can select up to K_0 nonzero coefficients for reporting from a set of $2LM$ linear combination coefficients, where β is an RRC configuration parameter.

Regarding how to determine K_0, RAN1 discussed two schemes [38,39]:
- Unrestricted subset selection: K_0 nonzero coefficients are freely selected in the set of size $2LM$. An example is shown in Fig. 8.15 with $L = 4$, $M = 4$, $\beta = 0.25$.
- Polarization-common subset selection: the same nonzero coefficients are selected in both polarization directions. An example of this scheme is shown in Fig. 8.16.

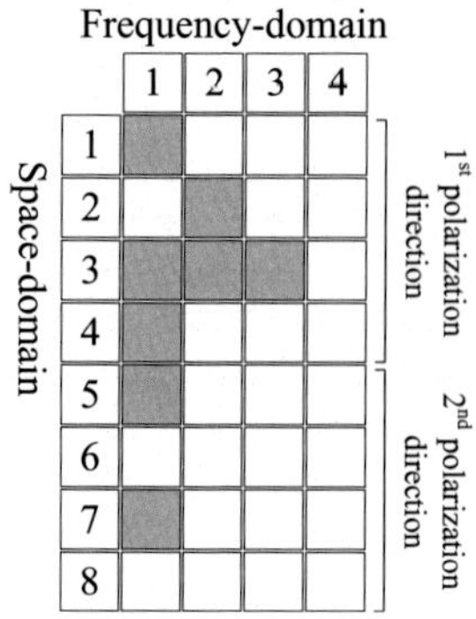

Figure 8.15 Unrestricted subset selection.

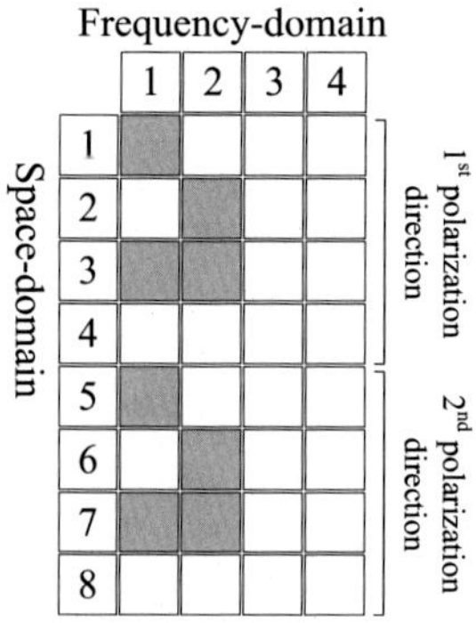

Figure 8.16 Polarization-common subset selection.

Based on the simulation evaluations, companies found that these two schemes do not have significant difference in terms of performance [40−42]. Although the scheme of common subset selection only needs a bitmap of size LM to report the exact location of zero or nonzero coefficients, it may happen that the coefficients in one polarization direction are nonzero while the same coefficients in the other polarization direction are zero. In order to avoid such a situation, at the RAN1#96 meeting, the method of unrestricted subset selection was adopted for K_0 selection. As shown in Fig. 8.17, the positions marked with 1 are the nonzero coefficients, while the positions marked with 0 are the zero coefficients.

After indicating the position of the nonzero coefficient, the UE needs to report the amplitude and phase of the corresponding nonzero coefficient. Three schemes were mainly considered in NR for the quantization of the nonzero coefficient. In the following discussion, the linear

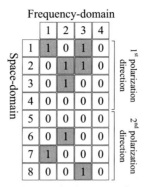

Figure 8.17 The 2*LM* bitmap indicates the location of nonzero coefficients.

combination coefficients related to beam $i \in \{0, 1, \ldots, L-1\}$ and FD units $f \in \{0, 1, \ldots, M-1\}$ are denoted as $c_{i,f}$, and the strongest coefficient is denoted as $c_{i*,f*}$.

The major features of quantization scheme 1 (which is similar to R15 Type II W_2) are:

- The position of the strongest linear combination coefficient $(i*, f*)$ is indicated by $\log_2 K_{NZ}$ bits, where K_{NZ} is the actual number of non-zero coefficients reported. The strongest coefficient $c_{i*,f*} = 1$, so the UE does not need to report its amplitude and phase.
- For all the other linear combination coefficients $\{c_{i,f}, (i,f) \neq (i*,f*)\}$: the amplitude is quantized with three bits, and the phase is quantized with three bits (8PSK) or four bits (16PSK). The table of amplitude 3-bit quantization set is the same as that in R15.

The major features of quantization scheme 2 as shown in Fig. 8.18 [43] are:

- The position of the strongest linear combination coefficient $(i*, f*)$ is indicated by $\log_2 K_{NZ}$ bits. The strongest coefficient $c_{i*,f*} = 1$, so the UE does not need to report its amplitude and phase.
- Polarization-direction reference amplitude $p_{ref}(i,f)$: the reference amplitude of the polarization direction that contains the strongest coefficient is not reported. While in another polarization direction, the reference amplitude is quantized with four bits with respect to the strongest coefficient, and the quantization set table is $\left\{ 1, \left(\frac{1}{2}\right)^{\frac{1}{4}}, \left(\frac{1}{4}\right)^{\frac{1}{4}}, \left(\frac{1}{8}\right)^{\frac{1}{4}}, \ldots, \left(\frac{1}{2^{14}}\right)^{\frac{1}{4}}, 0 \right\}.$
- For all the other coefficient $\{c_{i,f}, (i,f) \neq (i*,f*)\}$: in each polarization direction, the differential amplitude of the nonzero coefficient is

quantized with three bits with respect to the reference amplitude in that polarization direction. The quantization set table is $\left\{ 1, \frac{1}{\sqrt{2}}, \frac{1}{2}, \frac{1}{2\sqrt{2}}, \frac{1}{4}, \frac{1}{4\sqrt{2}}, \frac{1}{8}, \frac{1}{8\sqrt{2}} \right\}$. The final quantized amplitude is the multiplication of the reference amplitude and the differential amplitude: $p_{i,f} = p_{ref}(i,f) \times p_{diff}(i,f)$.

- The phase of each coefficient is quantized with three bits (8PSK) or four bits (16PSK).

The quantization scheme 3 [44] has the following major features:

- The position of the strongest linear combination coefficient $(i*, f*)$ is indicated by $\log_2 K_{NZ}$ bits, and the strongest coefficient $c_{i*,f*} = 1$, so the UE does not report its amplitude and phase.
- For $\{ c_{i,f*}, i \neq i* \}$: the amplitude is quantized with four bits and the phase is quantized with 16PSK. The quantization set table is $\left\{ 1, \left(\frac{1}{2}\right)^{\frac{1}{4}}, \left(\frac{1}{4}\right)^{\frac{1}{4}}, \left(\frac{1}{8}\right)^{\frac{1}{4}}, \ldots, \left(\frac{1}{2^{14}}\right)^{\frac{1}{4}}, 0 \right\}$.
- For $\{ c_{i,f}, f \neq f* \}$: the amplitude is quantized with three bits and the phase is quantized with 8PSK or 16PSK. The quantization set table is $\left\{ 1, \frac{1}{\sqrt{2}}, \frac{1}{2}, \frac{1}{2\sqrt{2}}, \frac{1}{4}, \frac{1}{4\sqrt{2}}, \frac{1}{8}, \frac{1}{8\sqrt{2}} \right\}$.

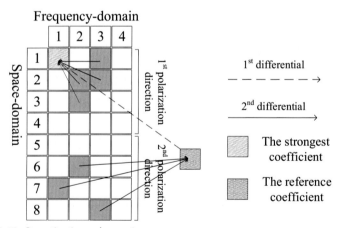

Figure 8.18 Quantization scheme 2.

Companies conducted a lot of simulation evaluations based on above schemes [40,45−47], and finally agreed on to adopt quantization scheme 2 at the RAN1#96 meeting.

According to the design in quantization scheme 2, the amplitude of the nonzero coefficient in $\tilde{W}_2$ can be expressed as the multiplication of the reference amplitude and the differential amplitude. Assuming that the strongest coefficient is located in the first polarization direction (i.e., $c_{1,1}$ in the example shown in Fig. 8.1), the reference amplitude of its polarization direction is $p_0^{(1)} = 1$. The reference amplitude of the second polarization direction $p_1^{(1)}$ is quantized with four bits with respect to the strongest coefficient, as shown by the dotted line in the figure. If rank $= 1$, the specific expression form of $\tilde{W}_2$ is:

$$\tilde{W}_2 = \begin{bmatrix} \sum_{i=0}^{L-1}\sum_{f=0}^{M-1} c_{i,f} \\ \sum_{i=0}^{L-1}\sum_{f=0}^{M-1} c_{i+L,f} \end{bmatrix} = \begin{bmatrix} \sum_{i=0}^{L-1} p_0^{(1)} \sum_{f=0}^{M-1} p_{i,f}^{(2)} \varphi_{i,f} \\ \sum_{i=0}^{L-1} p_1^{(1)} \sum_{f=0}^{M-1} p_{i+L,f}^{(2)} \varphi_{i+L,f} \end{bmatrix}, \qquad (8.6)$$

where $c_{i,f}$ is the linear combination coefficient, $p_0^{(1)}$ is the first polarization-direction reference amplitude, $p_1^{(1)}$ is the second polarization-direction reference amplitude, $p_{i,f}^{(2)}$ is the differential amplitude, and $\varphi_{i,f}$ is the phase of the nonzero coefficient.

8.2.4 Codebook design for rank $= 2$

For rank $= 2$, RAN1 further discussed the selection methods of SD subset, FD subset, and coefficient subset. The SD subset selection means selecting L spatial-domain DFT vectors in the beam set of $N_1 N_2$, the FD subset selection means selecting M DFT basis vectors in N_3 FD units, and coefficient subset selection means selecting K_{NZ} nonzero coefficients from the $2LM$ linear combination coefficient set. Although some simulation results show that it can improve the performance by selecting SD subsets independently in each layer, R16 still uses the same beam in different layers, which are consistent with R15 due to the overhead and complexity. Three different schemes for selecting FD subset and coefficient subset were considered:

- The selection of FD subset and coefficient subset is common to all the layers.
- FD subset selection is common to all the layers, but the coefficient subset selection is independent of each layer.
- The selection of FD subset and coefficient subset is independent for each layer.

For rank = 2, the independent selection of FD subsets and coefficient subsets for each layer can provide sufficient configuration flexibility, and can provide considerable performance gain without increasing the feedback overhead by too much. Considering the tradeoff between performance and overhead, the independent selection of the FD subset and coefficient subset scheme of each layer was adopted at the RAN1#96 meeting [48−50]. K_0 of the two layers is the same, where $K_0 = \beta \times 2LM$ (β is configured by RRC signaling), and the number of nonzero coefficients reported of each layer cannot exceed K_0, (i.e., $K_l^{NZ} \le K_0$, $l = 1, 2$).

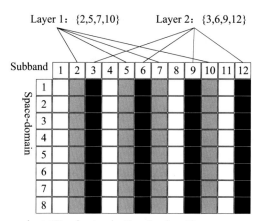

Figure 8.19 Independent FD subset selection (rank = 2).

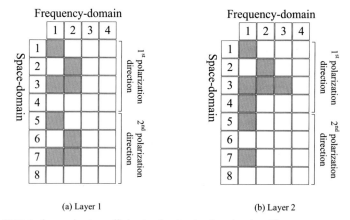

(a) Layer 1   (b) Layer 2

Figure 8.20 Independent coefficient subset selection (rank = 2).

8.2.5 Codebook design for high rank

The low-rank eType II codebook design was introduced is previous sections, and can significantly reduce the overhead compared to the Type II codebook. In order to provide better performance in scenarios with good channel quality, the eType II codebook introduces enhancements for high rank and extends the design for rank $= 1/2$ to rank $= 3/4$. The method of SD subset selection, FD subset selection, and coefficient subset selection of high rank follow the same principle as for rank $= 2$. SD subset selection is the same for each layer, while FD subset selection and coefficient subset selection are independent in each layer.

Since the overhead of the codebook is proportional to the number of nonzero coefficients and quantization parameters, so directly extending the design of rank $= 1/2$ to rank $= 3/4$ will bring a significant increase in the overhead. Considering that the purpose of introducing the eType II codebook in R16 is to reduce the overhead of the Type II codebook, it was decided at the RAN1#96bis meeting that extending the design to high rank will not increase the feedback overhead. Specifically, the actual number of nonzero coefficients reported for each layer will not be greater than K_0, (i.e., $K_l^{NZ} \leq K_0$,) and the total number of nonzero coefficients reported for all layers cannot be greater than $2K_0$, (i.e., $\sum_{l=1}^{RI} K_l^{NZ} \leq 2K_0$) [51,52]. In order to meet the above two restrictions, after considering the complexity and performance, the final determined parameter combination (L, P, β) is shown in Table 8.4 [53].

Table 8.4 Parameter combination of eType II codebook.

paramCombination – r16	L	p_v		β
		$v \in \{1, 2\}$	$v \in \{3, 4\}$	
1	2	¼	1/8	¼
2	2	¼	1/8	½
3	4	¼	1/8	¼
4	4	¼	1/8	½
5	4	¼	¼	¾
6	4	½	¼	½
7	6	¼	—	½
8	6	¼	—	¾

Table 8.5 Expression of eType II codebook.

Layer	W
$v = 1$	$W^{(1)} = W^1$
$v = 2$	$W^{(2)} = \frac{1}{\sqrt{2}}\left[W^1 W^2\right]$
$v = 3$	$W^{(3)} = \frac{1}{\sqrt{3}}\left[W^1 W^2 W^3\right]$
$v = 4$	$W^{(4)} = \frac{1}{\sqrt{4}}\left[W^1 W^2 W^3 W^4\right]$

8.2.6 eType II codebook expression

The eType II codebook can be expressed by Table 8.5.where

$$
W^l = W_1 \tilde{W}_2^l \left(W_f^l\right)^H = \frac{1}{\sqrt{N_1 N_2 \gamma_{t,l}}}
\begin{bmatrix}
\sum_{i=0}^{L-1} v_{m_1^{(i)},m_2^{(i)}} p_{l,0}^{(1)} \sum_{f=0}^{M_v-1} y_{t,l}^{(f)} p_{l,i,f}^{(2)} \varphi_{l,i,f} \\
\sum_{i=0}^{L-1} v_{m_1^{(i)},m_2^{(i)}} p_{l,1}^{(1)} \sum_{f=0}^{M_v-1} y_{t,l}^{(f)} p_{l,i+L,f}^{(2)} \varphi_{l,i+L,f}
\end{bmatrix},
$$

$l = 1, \ldots, v$, v is the layer number, $i = 0, 1, \ldots, L-1, f = 0, 1, \ldots, M_v - 1$,

$t = 0, \ldots, N_3 - 1$, $\gamma_{t,l} = \sum_{i=0}^{2L-1} \left(p_{l,\frac{i}{L}}^{(1)}\right)^2 \left|\sum_{f=0}^{M_v-1} y_{t,l}^{(f)} p_{l,i,f}^{(2)} \varphi_{l,i,f}\right|^2$. $v_{m_1^{(i)},m_2^{(i)}}$ is the 2D-

DFT beam. $y_{t,l} = \left[y_{t,l}^{(0)} y_{t,l}^{(1)} \ldots y_{t,l}^{(M_v-1)}\right]$ is the M_v DFT basis vectors of the

layer l. $p_{l,0}^{(1)}$ is the reference amplitude of the layer l in the first polarization

direction and $p_{l,1}^{(1)}$ is the reference amplitude of the layer l in the second

polarization direction. $p_{l,i,f}^{(2)}$ is the differential amplitude of the layer l, $\varphi_{l,i,f}$

is the phase of the nonzero coefficient of layer l. For all the zero coeffi-

cients, $p_{l,i,f}^{(2)} = 0$, $\varphi_{l,i,f} = 0$.

8.3 Beam management

With the rapid development of mobile communication technology and
the wide deployment of mobile network, radio spectrum resources are
becoming increasingly scarce since the low-frequency band with low
pathloss and good coverage is almost used up. In order to offer higher
transmission rate and larger system capacity, the new generation of mobile
communication tends to exploit middle-frequency band and high-
frequency band, such as the middle-frequency band of 3.5–6 GHz and

the millimeter wave band. The middle and high-frequency bands have the following major characteristics (Fig. 8.18):

- Due to the relatively rich frequency resource in middle-frequency and high-frequency bands, it is easy to allocate continuous frequency resource with larger bandwidth, which is beneficial for the commercial deployment and UE experience.
- Due to higher frequency, the larger pathloss will lead to a smaller coverage. Therefore better isolation in the space dimension can be achieved and it can facilitate the deployment of ultradense network.
- Due to the high-frequency band, the corresponding antenna and other hardware modules are small. Thus a larger number of antennas can be used for NR, which is conducive to the realization of the massive MIMO technology (Fig. 8.19).

As can be seen from the characteristics described above, the high frequency leads to the problem of limited coverage. Furthermore, the large number of antennas brings the problem of huge complexity and cost of implementation. In order to effectively solve these two problems, Analog Beamforming technology is adopted in NR, which can enhance network coverage and reduce the implementation complexity of devices (Fig. 8.20).

8.3.1 Overview of analog beam-forming

The basic principle of analog beam-forming technology is to adjust the phase/amplitude of the output of each antenna element (e.g., through the phase shifter), so that a group of antennas can form the beam in different directions and improve cell coverage through beam sweeping. By using phase shifter to form beams in different directions, the high complexity of traditional MIMO schemes when applying in a wide bandwidth can be avoided, where the traditional MIMO schemes can also be referred to as digital beam-forming.

An example of analog beam-forming is shown as Fig. 8.21. The left is a traditional LTE or an NR system without analog beam-forming, and the right is an NR system using analog beam-forming:

- As shown in the left of Fig. 8.21, the LTE/NR network uses a wide beam to cover the entire cell. Users 1−5 can receive the network signal at any time.
- As shown in the right of Fig. 8.21, the network uses narrow beams (such as beams 1−4 in the figure), and uses different beams to cover different areas within the cell at different time. For example, at time 1,

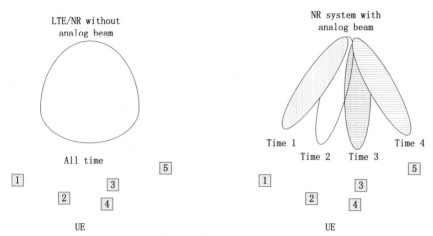

Figure 8.21 The left is a traditional LTE/NR system without analog beam-forming and the right is an NR system using analog beam-forming.

the NR network covers the area around user 1 through beam 1; at time 2, the NR network covers the area around user 2 through beam 2; at time 3, the NR network side covers the area around user 3 and user 4; at time 4, NR network side covers the area around user 5 through beam 4.

In the right of Fig. 8.21, since the network uses narrow beams, the transmission power can be more concentrated. Therefore it can provide a better coverage for a UE with a longer distance. At the same time, since the analog beams are narrow and each beam can only cover a portion of the cell, the analog beam-forming technology is "exchanging time for space" in some sense.

Analog beam-forming can be used not only in the network equipment, but also in UEs. It is generally believed in the industry that in 2Ghz-6Ghz band, network equipment can choose to use analog beam-forming technology, while the UE still uses traditional omni-directional antenna(s). In the millimeter wave band, both network equipment and UEs usually adopt analog beam-forming. The analog beam can be used for both transmitting signals (called transmit beams or Tx beams) and receiving signals (called receive beams or Rx beams).

The main differences between the analog beam-forming and the digital beam-forming (i.e., traditional MIMO schemes) are:

- In analog beam-forming technology, the same analog beam is applied to all the PRBs in a frequency band.

- In digital beam-forming technology, different subbands or even different PRBs can use different precoders.

In order to effectively support the analog beam-forming technology, the NR system has designed the mechanisms of analog beam measurement, analog beam selection, and analog beam indication, which are referred to as beam management.

In this chapter, we will focus on the main procedures of beam management for downlink and uplink transmission. Since the analog beam is narrow and easily blocked, the communication link may suffer severe degradation of quality. How to effectively improve the reliability will also be introduced in this chapter.

8.3.2 Basic procedures of downlink beam management

From the perspective of downlink transmission, beam management needs to solve two basic issues:

- If both network and UE adopt analog beams, it is necessary to align the transmit beam with the receive beam to form a beam pair in order to obtain a reliable communication link with good quality. Therefore how to determine one or more beam pairs is a basic problem in beam management.
- When the network selects a transmit beam for DL data transmission, the UE should know which is the best receive beam for the reception of the DL data.

The first problem is related to the transmit beam of the network and the receive beam of the UE:

- From the perspective of network, it should determine transmit beam(s) that is suitable for DL data transmission. Generally, such a decision in the network relies on the measurement and reporting of downlink transmit beam from the UE.
- From the perspective of the UE, it should determine which downlink transmit beam(s) can offer good quality for downlink transmission. At the same time, for a downlink transmit beam, a UE capable of supporting millimeter wave also needs to determine which receive beam is suitable for the reception, which also depends on the beam measurement at the UE side.

The second problem is related to how the network notifies the UE with DL beam-related information, which can help the UE to determine the corresponding downlink receive beam.

In this section, we will introduce the basic principles for determining downlink transmit and receive beam pair. The function of downlink beam measurement/reporting and beam indication will be described in Sections 8.3.3 and 8.3.4, respectively.

The procedures to pair a transmit beam and receive beam for downlink transmission generally involve the following three processes (P1, P2, and P3, respectively) [54]:

- P1: A coarse alignment of downlink transmit beam and receive beam.
- P2: Refinement of downlink transmit beam at the network side.
- P3: Refinement of downlink receive beam at the UE side.

Fig. 8.22 shows the coarse pairing process (P1) of downlink transmit and receive beams. It can be implemented in various ways in a practical system. For example, during the initial access process, coarse alignment can be achieved through the four-step random access process. After the initial access process, a coarse beam alignment has been established between the network and UE, which can support the subsequent data transmission. For the coarse alignment of downlink beams, if the transmit beam and receive beam are narrow, it will take a long time to complete the alignment and thus a large delay is caused to the system. Therefore in

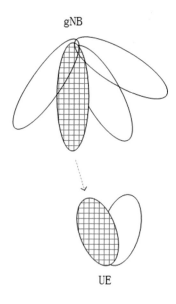

gNB

UE

P1

Figure 8.22 Coarse a of downlink transmit and receive beams (P1).

order to quickly complete the coarse pairing of downlink beams, the corresponding transmit and receive beams may be implemented with relatively wide beams. The established beam pairing can still offer good performance, although it may not be the optimal performance that can be achieved with narrow beams.

On the basis of P1 coarse pairing, NR also supports to further refine transmit beam and receive beam (corresponding to P2 and P3 process, respectively), where the refined beam(s) is expected to further improve transmission performance. P2 is the process to refinement downlink transmit beam at the network side which is illustrated in Fig. 8.23.

- Based on the P1 process, a coarse pairing between downlink transmit beam 2 and downlink receive beam A is established.
- The network sends RSs through three narrow beams 2−1, 2−2, 2−3 to facilitate the refinement of transmit beam 2.
- The UE uses receive beam A to measure the RSs transmitted on beam 2−1, 2−2, and 2−3, respectively, and the corresponding metric is L1−RSRP.
- Based on the L1−RSRP measurement results, the UE transmits reporting to the network with the recommended DL transmit beams and their associated L1−RSRP measurement results.

P3 is the process to refinement downlink transmit beam at the network side which is illustrated in Fig. 8.24.

- Based on the P1 process, a coarse pairing between downlink transmit beam 2 and downlink receive beam A is established.

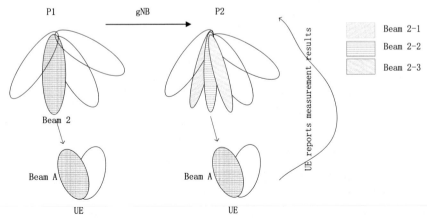

Figure 8.23 Refinement of downlink transmit beam(s) at network side (P2).

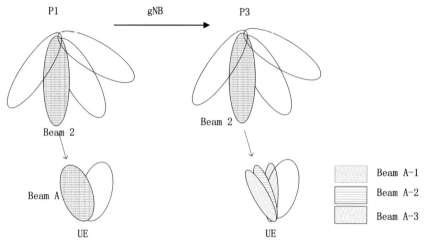

Figure 8.24 Refinement of downlink receive beam at the UE side (P3).

- The network sends multiple RSs to beam 2 to facilitate the refinement of the corresponding receive beam.
- The UE uses three narrower beams (A−1, A−2, and A−3) to receive and measure the signals transmitted on beam 2.
- According to the measurement results, the UE can determine the best narrow receive beam corresponding to transmit beam 2. For the P3 process, the UE does not need to report the measurement results to the network on which the narrow receive beam is selected. The selection of receive beam is transparent to the network.

The P1, P2, and P3 processes are not explicitly defined in the NR specification. However, the above processes can be completed through the functional components of the NR specification. The corresponding functional components are introduced in the following sections on downlink beam measurement and reporting, and downlink beam indication.

8.3.3 Downlink beam measurement and reporting

This section discusses how to solve the following three problems:
- What is used for the downlink beam measurement?
- What is the metric of the beam measurement?
- How to report measurement results?

8.3.3.1 Reference signals for downlink beam measurement

A beam is an objective physical entity, but logical concepts rather than physical entities are generally used in NR specifications, especially in the physical layer specifications. Therefore the measurement of a beam is realized by measuring the RSs transmitted through the beam.

In a NR system, a larger number of measurement-related functionalities (e.g., mobility measurement, CSI measurement, etc.) are based on CSI−RS/SSB signals. For downlink beam management, the NR system also supports beam measurement and reporting based on CSI−RS and SS/PBCH (SSB) [55−57]. For simplicity of discussion, the beam will be described indirectly by the CSI−RS signal or SSB transmitted on the beam in the remaining parts of this chapter.

During the standardization discussions, two different types of CSI−RS were proposed for the downlink beam management [58]:

- UE-specific CSI−RS
- Cell-specific CSI−RS

For the NR system, a basic design principle is to configure dedicated RSs for each the UE (namely UE-specific RS) and avoid or reduce the cell-level (i.e., cell-specific) RSs. The main advantage of that is the network can flexibly configure the RS for each UE according to the particular requirements and it is beneficial for power-saving and overhead reduction of reference signals. Due to that consideration, most companies prefer to use UE-specific CSI−RS signal for beam management. However, some companies prefer to use cell-specific CSI−RS for beam management [55]. The main advantages of cell-specific CSI−RS are as follows:

- It can provide better performance during the initial access. Based on cell-specific CSI−RS, the UE can select a better transmit beam for DL data transmission and improve the quality of the link during the initial access process. However, the UE-specific CSI−RS can only be configured after the UE enters the RRC-connected state.
- It can provide Lower cost of time-frequency resources. The transmit beams for control channels are usually wider than that of data channels since they need to cover more users and require higher reliability. Thus from the perspective of resource utilization, it is more efficient for multiple UEs to use the same cell-specific CSI−RS to measure the same transmit beam.
- It has lower signaling overhead of configuration. For a UE-specific CSI−RS, the network needs to send dedicated configuration signaling to each individual UE, whereas the configuration of cell-specific CSI−RS can be broadcasted to all UEs through one signal. Thus the overall signaling overhead of cell-specific CSI−RS is lower.

As mentioned above, one principle for designing RS in NR is to reduce the use of cell-specific RSs. The idea of using cell-specific

CSI—RS for beam management is inconsistent with this principle, so it is opposed by many companies. After several discussions in RAN1 meetings, the final decision is that NR only support UE-specific CSI—RS and SSB for downlink beam measurement. According to the specific configuration provided by the network, a UE may need to measure UE-specific CSI—RS, SSB, or both UE-specific CSI—RS and SSB.

8.3.3.2 Measurement quantity

In NR, commonly used measurements can be divided into two categories:

- L3 measurement: layer 3 (L3) measurement is generally used in mobility management measurement and path loss measurement. L3 measurement generally requires L3 filtering operation on multiple measurement samples, which leads to large delay.
- L1 measurement: L1 measurement is directly processed in the physical layer without L3 filtering and has a small latency. One typical example is the CSI measure.

Beam management is mainly performed in the physical layer and the L1 measurement can effectively reduce the latency of the beam management procedure. There exist two options on the measurement metric (i.e., measurement quantity) of downlink beam measurement:

- L1—RSRP: layer 1-RS receiving power.
- L1—SINR: layer 1-signal to interference to noise ratio.

Compared to L1—SINR, the advantages of using L1—RSRP as measurement quantity are as follows:

- The measurement of L1—RSRP is simple and has low complexity, and can be easily implemented in the UE.
- The L1—RSRP mainly reflects the channel quality corresponding to the downlink beam, and its variation is relatively slow, which can facilitate selecting relatively reliable downlink beam. On the contrary, interference changes quickly and the value of L1—SINR has greater fluctuation, which would cause frequent beam switching.
- In the framework of MIMO transmission, CSI feedback already takes into account the interference characteristics. Combining the L1—RSRP beam management and CSI feedback can achieve the same purpose as L1—SINR.
- Compared with L1—RSRP, the additional gain of L1—SINR is not obvious, and some companies found in some cases it may even cause performance loss [59,60].

The advantages of using L1−SINR as beam measurement quantity are as follows:

- The reception quality of downlink beam depends on both of the corresponding channel quality and the interference. L1−RSRP does not fully reflect the quality of the transmission link corresponding to the downlink transmit beam [61]. The L1−SINR can better consider the mutual interference between different downlink transmit beams.
- The combination of L1−RSRP measurement and CSI feedback requires two steps: beam management and CSI measurement feedback. While L1−SINR-based beam management can complete user pairing in one process.
- The L1−SINR-based solution can assist the network to coordinate the downlink transmit beams in different cells or different TRPs, thereby improving the overall performance of the system.

After lengthy technical discussion among companies, only L1−RSRP is finally supported for Rel-15 downlink beam measurement. Later on, L1−SINR is introduced as a new measurement metric during the beam management enhancement of R16 NR. Therefore in NR R16 if a UE supports the capability of L1−SINR-based downlink beam management, the network can configure the UE to perform beam measurement based on either L1−RSRP or L1−SINR.

8.3.3.3 Beam reporting

According to the measurement results, the UE can report the information of $K \geq 1$ beams to the network. Each information includes the recommended beam (e.g., CRI, SSB index) and the corresponding L1−RSRP result. When $K > 1$, the absolute RSRP value is reported for the largest L1−RSRP, while the rest $K-1$ L1−RSRP values are reported in the form of differential values with respect to the largest L1−RSRP value [62]. In the L1−SINR measurement and report, the same differential reporting mechanism is adopted.

Based whether a UE can receive the data on K downlink beams simultaneously, the reporting methods of beam measurement results can be divided into two categories:
- Nongroup-based reporting
- Group-based reporting [63]

In the nongroup-based reporting, the UE chooses to report the measurement information of K RSs according to the measurement results of N RSs. The value of K is configured by the network, and can be 1, 2, 3, or 4. Here, K RSs actually correspond to K beams. The network cannot transmit signals on multiple downlink beams simultaneously, because the UE cannot receive signals from multiple beams at the same time. The UE can select those K from N RSs according to its own implementation algorithm. For example, the UE can choose the strongest K L1−RSRP values, or select K by considering the arrival direction of different beams (i.e., considering the spatial correlation between different RSs). That is not specified in the NR specification and is up to UE implementation.

Group-based reporting requires the UE to have the capability to receive multiple downlink transmit beams simultaneously. If the network configures UE group-based reporting, the UE reports the measurement results of $K = 2$ beams according to the measurement results of N RSs. Each reporting information includes the indicator of the selected RS corresponding to a recommended beam (e.g., CSI−RS resource identification, SSB number) and the corresponding L1−RSRP measurement. The network can transmit data to the UE on those reported $K = 2$ beams simultaneously.

During the discussion of NR standardization, group-based reporting refers to beam reporting where the reported beams belong to one group or different groups. Let's assume there are G groups, and the UE reports the corresponding $M_i(i = 1,\ldots, G)$ downlink transmit beam(s) for each group of transmit beams. Two basic schemes for group-based reporting were discussed during the NR standardization [64]:

- Scheme 1: The UE can simultaneously receive the signals transmitted on multiple downlink transmit beams that are corresponding to the same group. However, the UE cannot receive the signals transmitted on multiple downlink transmit beams in different groups at the same time.
- Scheme 2: The UE can simultaneously receive the signals transmitted on multiple downlink transmit beams that are corresponding to different groups. But the UE cannot receive the signals transmitted on multiple different downlink transmit beams in the same group at the same time.

The difference between Scheme 1 and Scheme 2 is reflected in the fact that group corresponds to different physical entities in the implementation of products.

- In Scheme 1, a group corresponds to a set of receive beams at the UE side that can be used at the same time, while the receive beams corresponding to different groups cannot be used at the same time.

Therefore all downlink transmit beams corresponding to a group can be received by the UE at the same time.

- In Scheme 2, one group corresponds to one antenna panel at the UE side. When the UE uses multiple antenna panels for reception, multiple antenna panels can simultaneously receive the signals from different downlink transmit beams where each panel can be aligned with one downlink transmit beam. For a UE with a single antenna panel, only one receive beam is active at each time, so it cannot receive signals from multiple downlink transmit beams simultaneously.

Scheme 1 and Scheme 2 have their own advantages and disadvantages in different scenarios. Some examples are shown in Figs. 8.25 and 8.26 to illustrate some scenarios where Scheme 1 or Scheme 2 can work better than the other one. In the examples, two TRPs are used for downlink transmission (or two antenna panels of the same TRP). The UE has two antenna panels, so it can support two active receive beams in two directions at the same time.

In Fig. 8.25, transmit beams 1, 2, A, and B correspond to receive beams X1, X2, YA, YB, respectively. Receive beams X1 and X2 are on the first antenna panel and only one of them can be used at a time. The receive beams YA and YB are also on the second antenna panel and only one of them can be used at a time. Each antenna panel can have one active receive beam at each particular moment. Thus only one of four

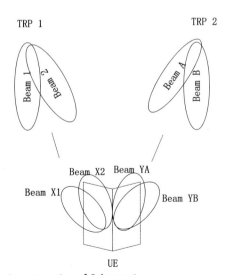

Figure 8.25 Application scenarios of Scheme 1.

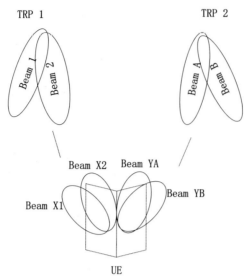

Figure 8.26 Application scenarios of Scheme 2.

combinations can be used for simultaneous reception at any time: {beam X1, beam YA}, {beam X1, beam YB}, {beam X2, beam YA}, and {beam X2, beam YB}.

Assume the transmission on beam 2 and beam A will lead to large mutual interference at the UE side. Therefore beam 2 and beam A are not suitable for simultaneous DL transmission. If they are used for downlink transmission simultaneously, the transmission performance will be degraded due to the large mutual interference between them. In this scenario, if Scheme 1 is applied, the UE can report three groups:{beam1, beamA}, {beam1, beamB}, and {beam2, beamB}. After receiving the beam report from the UE, the network can transmit signals on the two beams in any of those three reported groups, and the UE can use the corresponding receive beams to receive signals simultaneously. On the contrary, if method 2 is adopted, the UE can report two groups, and some potential options for these two groups can be:

- The two groups are {beam1} and {beam B}, which correspond to $G = 2$, $M_1 = 1$, $M_2 = 1$. According to such a reporting, the network is not aware of that it can transmit from beam 2 and beam B at the same time, which will negatively affect the load balancing and scheduling optimization of the network.
- The two groups are {beam 1, beam 2} and {beam B}, which correspond to $G = 2$, $M_1 = 2$, $M_2 = 1$. With this reporting, the network

does not know that it can transmit from beam 1 and beam A. That will also negatively affect the load-balancing and scheduling optimization of the network.

- The two groups are {beam 1} and {beam A, beam B}, which correspond to $G = 2$, $M_1 = 1$, $M_2 = 2$. In this case, the network does not know that it can transmit from beam 2 and beam B at the same time. Thus the load balancing and scheduling optimization of the network can be impacted.
- The two groups are {beam 1, beam 2} and {beam A, beam B}, which correspond to $G = 2$, $M_1 = 2$, $M_2 = 2$. In this case, if the network transmits data from beam 2 and beam A to the UE simultaneously based on to the reporting, the transmission on these two transmit beams will suffer large mutual interference, thus impairing the performance.

In the scenario shown in Fig. 8.25, Scheme 1 can better support the UE to select and report the beams that are more suitable for simultaneous transmission so that the interference between two simultaneous beams can be well controlled. However, Scheme 2 has to consider the tradeoff between network scheduling flexibility and performance degradation caused by mutual interference between two transmit beams. Scheme 2 might have to sacrifice some aspect, for example reducing network scheduling flexibility or tolerating interference between simultaneous transmit beams.

Fig. 8.26 is an example of scenarios where Scheme 2 has advantages. The scenario shown in Fig. 8.26 is similar to that in Fig. 8.25 and the difference is that the signals of beam 2 and beam A do not suffer large mutual interference at the UE side. Thus beam 2 and beam A can be used for transmission simultaneously. In Scheme 2, the UE can report two groups, which are {beam 1, beam 2} and {beam A, beam B}. After receiving the report from the UE, the network can choose to send data through any of the following beam combinations according to the situation:

- Beam 1, beam A
- Beam 1, beam B
- Beam 2, beam A
- Beam 2, beam B

In contrast, in order to provide the same scheduling flexibility for downlink transmission, the UE should report $G = 4$ groups in Scheme 1, which are {beam1, beam A}, {beam1, beam B}, {beam2, beam A}, and {beam2, beam B}. We can see that in the scenario of Fig. 8.26, in order to achieve the same effect, Scheme 1 nccds $G = 4$ groups with 2 beams ($M_1 = M_2 = M_3 = M_4 = 2$) in each reported group, while Scheme 2 only needs $G = 2$ groups with 2 beams ($M_1 = M_2 = 2$) in each group. Therefore the feedback overhead of reporting Scheme 2 is smaller than that of the reporting Scheme 1.

Since both schemes have their own advantages and disadvantages in different scenarios, it is still difficult to reach agreement after intensive and lengthy discussion. In order to promote the progress, some companies proposed new solutions that are compatible with both schemes, which are mainly belong to two different categories:

- Category 1: these two schemes are merged into a new scheme. Through different parameter configuration, the new scheme can be downgraded to Scheme 1 or Scheme 2.
- Category 2: both Scheme 1 and Scheme 2 are supported in the NR. The network configures Scheme 1 or Scheme 2 for reporting through RRC signaling.

The above discussion did not make any substantial progress until the end of R15, and finally a compromise conclusion was reached, which is, if group-based reporting is configured, the UE reports two beams that can be received simultaneously by the UE. That can be considered as a special case of both schemes [63]:

- For Scheme 1: $G = 1$, $M_1 = 2$. From the perspective of Scheme 1, the UE reports one group with two beams.
- For Scheme 2: $G = 2$, $M_1 = M_2 = 1$. From the perspective of Scheme 2, the UE reports two groups and each group contains only one beam.

8.3.4 Downlink beam indication

In the NR system, PDSCH and PDCCH can use their respective beams, which provides more flexibility for optimization in practical commercial systems. For example, one actual network deployment can apply a wider beam for transmitting PDCCH while a narrower beam for transmitting PDSCH. From a technical point of view, the reasons why PDSCH and PDCCH may use different beams are as follows:

- PDCCH and PDSCH generally require different link quality. PDCCH requires high reliability for a single transmission, but does not require high transmission data rate. On the contrary, PDSCH requires high data rate, while the transmission reliability can be improved through HARQ retransmissions.

- The users severed by PDCCH and PDSCH may be different. For example, some PDCCH serves a group of UEs (called group common PDCCH), while a PDSCH is usually for a certain UE. Thus the requirements for transmit beam will be different.

In NR, the beam indication is realized through the configuration and indication of TCI state(s), which is based on the concept of quasi colocation (QCL).

The network can indicate a corresponding TCI state for a downlink signal or downlink channel, which contains the corresponding QCL RS (s) for the target downlink signal or target downlink channel. Based on the larger-scale parameter(s) derived from the indicated QCL RS, the UE can optimize the receivers for the reception of the target downlink signal or the downlink channel. A TCI state can include the following configurations:

- A TCI state ID that is used to identify a TCI state
- QCL information 1
- QCL information 2
  Among them, QCL information contains the following content:
- Configuration of QCL type, which can be one of QCL-typeA, QCL-typeB, QCL-typeC, or QCL-typeD.
- Configuration of QCL RS, which includes the cell ID, BWP ID, and RS identification (CSI−RS resource index or SSB index).

In QCL information 1 and QCL information 2, at least one QCL type of QCL information must be QCL-typeA, QCL-typeB or QCL-typeC, and the QCL type of the other QCL information (if configured) must be QCL-typeD.

In NR specification, the following QCL types are defined:

- "QCL type A": {Doppler shift, Doppler spread, average delay, delay spread}
- "QCL TypeB": {Doppler shift, Doppler spread}
- "QCL typeC": {Doppler shift, average delay}
- "QCL typeD": {spatial RX parameter}

QCL-typeA, QCL-typeB, and QCL-typeC correspond to different large-scale parameters of wireless channel. If a PDSCH channel and a RS X are quasi colocated with respect to the QCL-typeA, the UE may assume that the four large-scale parameters {Doppler frequency offset, Doppler spread, average delay, delay spread} of the wireless channel corresponding to the PDSCH and RS X are the same. Therefore the UE can estimate the Doppler frequency offset, Doppler spread, average delay, and delay spread through measuring the RS X, and then apply these estimated parameters to optimize the channel estimation for PDSCH demodulation RS (DMRS), and thus improve the reception performance of PDSCH.

QCL-typeD provides information for a UE to determine receive beam for receiving downlink signals or channels that may be transmitted through different transmit beams. If a PDSCH channel and RS X are quasi colocated with respect to QCL-typeD, the UE may consider to apply the same receive beam to receive the RS X and PDSCH transmission. For example, through beam management process explained in previous sections, a UE determines that for a RS X, the best receive beam is receive beam A. If the network indicates that a PDSCH channel and RS X are quasi colocated with respect to QCL-typeD, the UE uses the same receive beam A to receive PDSCH. In low-frequency band (e.g., less than 6 GHz), the UE generally does not implement analog beam, but use omni-directional antenna for reception. Thus the indication of QCL-typeD is not necessary in this scenario.

TCI state can be indicated by RRC signaling, MAC CE, or DCI signaling. The main advantages and disadvantages of these three methods are summarized in Table 8.6 [65].

As can be seen from Table 8.6, different signaling indication schemes have different tradeoffs between latency, reliability, and overhead. PDCCH requires high transmission reliability. If a UE does not to receive the beam indication for PDCCH correctly, the UE could miss the DCI signaling and then the associated data transmission, which greatly affects the system performance. Therefore RRC signaling and MAC CE signaling are suitable for beam indication of PDCCH while DCI signaling is not. On the contrary, PDSCH can use HARQ retransmission to improve the transmission reliability. Furthermore, in some scenarios, the network might need to switch beams quickly to achieve good load balancing among different beams based on the user distribution and actual traffic/service requirements. Therefore it was finally decided that the DCI signaling with low latency and low overhead be used to indicate the beam of PDSCH.

As defined in NR specification, the signaling of RRC + MAC CE is used to indicate TCI state for PDCCH. An example of TCI state indication for PDCCH is shown in Fig. 8.27.

Table 8.6 Different signaling methods.

	Advantages	Disadvantages
RRC signaling	• Higher reliability than MAC CE signaling and DCI signaling (RRC: about 10^{-6}, MAC CE: about 10^{-3}, DCI: about 10^{-2}). • The RRC signaling has corresponding ACK/NACK feedback.	• High signaling processing delay • More signaling overhead
MAC CE signaling	• Higher reliability than DCI signaling. • There is corresponding ACK/NACK feedback. • •The delay of signaling is lower than RRC signaling.	• The delay is larger than DCI signaling. • The signaling overhead is larger than DCI signaling.
DCI signaling	• Low latency • Low signaling overhead	• The reliability is worse than MAC CE signaling and RRC signaling.

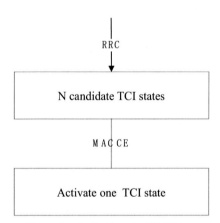

Figure 8.27 TCI state configuration/indication for PDCCH.

- The network first configures N TCI states for a CORESET via RRC signaling.
- If $N = 1$, the configured TCI state is used for the reception of PDCCH related with that CORESET.
- If $N > 1$, the network uses MAC CE signaling to activate one of the configured TCI states for the reception of PDCCH related with that CORESET.

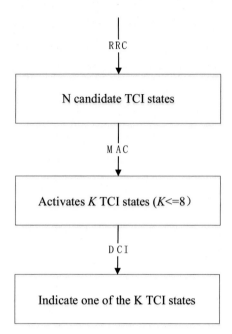

Figure 8.28 TCI state configuration and indication method of PDSCH.

For PDSCH channel, NR adopts the signalling method of RRC + MAC CE + DCI to indicate a TCI state and the procedure of the method is illustrated in Fig. 8.28.

- The network first configures a set of TCI states through RRC signaling.
- The network then uses MAC CE signaling to activate K TCI states ($K \leq 8$) from this group of TCI states.
- When a PDCCH schedules a PDSCH, the scheduling DCI indicates one of those K activated TCI states for the scheduled PDSCH transmission.

The above description is for a single TRP transmission scenario. The last section in this chapter will describe the multiple TRP transmissions, where a DCI can indicate two TCI states for PDSCH transmission. In addition, in R15, the activation of TCI state via MAC CE is per carrier. In R16, in order to reduce the signaling overhead, a new MAC CE format is introduced to activate TCI states for a group of carriers.

8.3.5 Basic procedures of uplink beam management

In the low-frequency band (e.g., less than 6 GHz), the UE does not use analog transmit beams. Therefore uplink beam management is only used for high-frequency band, such as in millimeter wave band. The main purpose

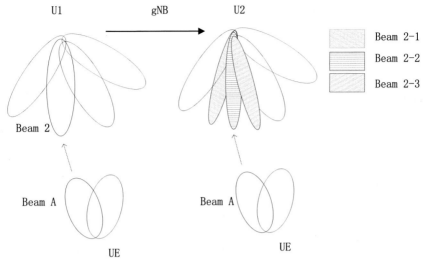

Figure 8.29 U2 process.

and problems in designing uplink beam management are similar to those in downlink beam management. Correspondingly, uplink beam management also has three main processes, which are denoted as U1, U2, and U3 (corresponding to P1, P2, and P3 in downlink beam management) [66]:

- U1: Coarse pairing of uplink transmit beam and receive beam.
- U2: Refining uplink receive beam at the network side.
- U3: Refining uplink transmit beam at the UE side.

Similar to the P1 process, the U1 process can be performed during the initial access procedure. In order to complete the coarse pairing between the uplink transmit beam and the uplink receive beam with low latency, the uplink transmit beam and receive beam may be relatively wide. The beam pairing determined during initial access procedure generally can provide reasonable performance, but not the optimal performance. In order to further improve the performance, on the basis of coarse pairing, beam refinement can be performed for uplink receive beam at the network side and uplink transmit beam at the UE side, through U2 and U3 processes, respectively. Fig. 8.29 illustrated the P1 process, in which the procedures are as follows:

- Based on the U1 process, the coarse pairing between uplink transmit beam A of the UE and uplink receive beam 2 of the network side is established.
- In order to refine the receive beam, the network tells the UE to send multiple RSs on beam A.

- The network applies three different narrower beams 2–1, 2–2, and 2–3 to receive and measure the RSs transmitted through Beam A
- Based on the measurement results, the network can determine the best narrow beam for receiving the uplink signals or channel transmitted through transmit beam A.

Fig. 8.30 illustrates one example for the U3 process, and the specific steps are as follows:

- Based on the U1 process, coarse paring between uplink transmit beam of the UE and uplink receive beam 2 of the network is established.
- In order to refine the uplink transmit beam of the UE, the network tells the UE to transmit multiple RSs through three narrower beams A-1, A-2, and A-3. In fact, how to formulate those three narrower beams is totally up to UE implementation and the network is not aware of the association between RSs and uplink transmit beams.
- The network then uses receive beam 2 to receive and measure the RSs transmitted on transmit beams A-1, A-2, and A-3, respectively. Based on the measurement results, the network can determine the best uplink transmit beam for uplink transmission.
- The network then indicates the UE to apply the selected transmit narrow beam on uplink transmission. In fact, what the network indicates to the UE is the ID of one of those RSs. The UE can derive the corresponding transmit beam since the UE is aware of the mapping between transmit beam and RS.

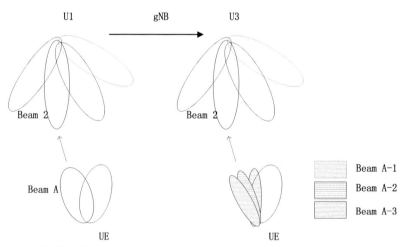

Figure 8.30 The U3 process.

8.3.6 Uplink beam measurement

As mentioned above, downlink beam measurement is based on CSI−RS and SSB signals. Correspondingly, uplink beam management is based on the sounding RS (SRS) signal. For example:

- In the U2 process shown in Fig. 8.29: the network configures three SRS resources for the UE, and the UE uses the same uplink transmit beam A to send those three SRS resources. The network receives and measures them with three different receive beams (2−1, 2−2, and 2−3) to determine the receive narrow beam with the best performance.

- In the U3 process shown in Fig. 8.30: the network configures three SRS resources for the UE, and the UE uses three different uplink transmit narrow beams (A-1, A-2, and A-3) to transmit those three SRS resources, respectively. The network receives and measures them with beam 2 to determine which one has the best performance.

Based on the measurement results, the network determines which uplink transmit beam is used for transmission. The UE only needs to know the uplink transmit beam that will be used for uplink transmission, but does not need to know how the network measures and determines the best uplink transmit beam. Therefore the measurement and selection of uplink beam is up to network implementation and is not specified in the NR protocol.

8.3.7 Uplink beam indication

Before discussing the uplink beam indication, we need to firstly explain the concept of beam correspondence. When there exists beam correspondence in the system, a UE can determine its own uplink transmit beam based on the downlink receive beam, or determine its own downlink receive beam based on the uplink transmit beam. Specifically, if receive beam A of the UE is the better or optimal choice for receiving downlink signals, the UE can determine that the corresponding uplink transmit beam A′ is also the better or optimal uplink transmit beam.

If the beam correspondence exists at the UE side, when the network indicates a downlink RS X corresponding to a certain downlink transmission beam, the UE can derive an uplink transmit beam A′ that corresponds to the receive beam A used in the reception of RS X. For the UE with beam correspondence, the system can signal the downlink signal X to directly indicate the uplink transmission beam A′ of the UE.

In the NR system, the network can indicate the uplink transmission beam in two ways:

- Case 1: the network configures multiple SRS resources, and the UE applies different uplink transmission beams on different SRS resources. For an uplink transmission, the network can inform the UE with an uplink transmit beam by indicating an SRS resource S1, and the UE will apply the same uplink transmit beam of the SRS resource S1 for this uplink transmission. Case 1 is applicable to any UE with or without valid beam correspondence.

- Case 2: if the beam correspondence of a UE is valid, the UE can derive the best receive beam and then determines the corresponding uplink transmit beam. The advantage of Case 2 is that the uplink beam management measurement is not necessary. The network can determine the uplink transmit beam and the uplink receive beam according to the downlink beam management process, thus reducing the total resource overhead and latency of the beam management process.

In NR, the abovementioned Case 1 and Case 2 are realized with the concept of spatial-relation information. The parameter of spatial-relationship information contains the information of a source RS, which can be one of the SRS resource ID (corresponding to Case 1), CSI−RS resource index (corresponding to Case 2) and SSB index (corresponding to Case 2). The network can provide spatial-relation information for uplink channel or uplink signal. If the source RS in the spatial-relation information is X, the UE determines the uplink transmission beam based on the source signal X according to the method described in Case 1 or Case 2. For each uplink transmission, the uplink beam indication method specified in NR is described as follows:

- For SRS transmission, RRC signaling is used to indicate uplink transmit beam. Specifically, the configuration of each SRS resource contains a spatial-relation information that indicates an uplink transmit beam. For semipersistent SRS, the network uses MAC CE to indicate the corresponding spatial-relation information when the SRS transmission is activated. In R16, NR supports using MAC CE to update the spatial-relation information for an aperiodic SRS.

- For PUCCH, a two-step indication procedure of RRC signaling + MAC CE signaling is adopted. The network first configures K spatial-relation information for a PUCCH resource. If

$K = 1$, the UE determines the uplink transmit beam of PUCCH resource based on the source RS indicated in that spatial–relationship information; if $K > 1$, the network activates one of K spatial–relationship information through MAC CE, and then the UE determines the uplink transmit beam for PUCCH resource based on the source RS in the activated spatial–relationship information. In R15, MAC CE-based spatial–relationship information indication is per PUCCH resource and the number of PUCCH resource could be large. Thus beam indication for PUCCH could cause a large overhead of signaling. In order to reduce the signaling overhead of beam indication for PUCCH, a new mechanism is introduced in R16, where one MAC CE format can update/activate the spatial–relationship information for a group of PUCCH resources.

- Different uplink beam indication methods are applied to different types of PUSCH transmission. For example, for a dynamically scheduled PUSCH based on codebook transmission, if only one SRS resource is configured for PUSCH transmission, the UE applies the same transmit beam of that SRS resource for PUSCH transmission; if two SRS resources are configured for PUSCH transmission, the network indicates one of the SRS resources through DCI, and the UE applies the same transmit beam of this indicated SRS resource on the PUSCH transmission.

During the discussion of MIMO enhancement in R16, some companies thought that R15 design is too flexible and thus causes excessive RRC signaling overhead in some typical network deployment. Therefore a series of mechanisms for determining the default spatial–relationship information were proposed. For uplink channels or signals (including PUCCH resources and SRS resources) that are not provided with spatial–relationship information, the corresponding uplink transmit beam is determined according to some specific rules. Those specific rules are mainly used to reduce beam indication control signaling overhead. The rationale behind these rules is quite straightforward and we will not introduce them in detail here.

8.4 Beam failure recovery on primary cell(s)

As mentioned above, the analog beam is mainly used in high-frequency bands such as millimeter wave bands. Due to the large penetration loss of high-frequency electromagnetic wave and the narrower beamwidth of analog beam, the communication link will be easily blocked, resulting in poor communication quality and even service interruption.

In order to improve the robustness of transmission through analog beams in high-frequency bands, the following two different strategies can be considered:

- Active strategy: multiple beams are used to transmit data simultaneously. The beams point to different directions and are unlikely blocked at the same time. Thus the reliability of transmission can be improved. The group based reporting described in previous sections can support simultaneous transmission of two downlink beams. For a UE not capable of receiving multiple simultaneous beams, a method of alternate transmission can be applied on those beams.
- Passive strategy: when the quality of current beam transmission becomes poor, the UE searches for a new beam with good link quality, and then informs the network to recovery high-quality reliable communication link with that new beam. This method is called the BFR mechanism.

In the NR system, the recovery process of beam failure is designed for downlink transmit beam, and does not consider the blockage of uplink transmit beam. The main reason is that if the downlink communication quality is good, the network can still send commands to indicate the UE to switch to a better uplink transmit beam for uplink transmission; but if the downlink communication quality is poor, the UE may not be able to receive the control signaling from the network, thus the network is not able to communicate with the UE to determine the new beam pairing.

8.4.1 Basic procedure of BFR

In NR R15, the BFR mechanism is designed only for Pcell and PScell. The main functional components can be divided into four parts [67]:
- Beam failure detection (BFD)
- New beam identification (NBI)
- BFR request (BFRQ)
- Response from the network

The UE first measures PDCCH to determine the link quality corresponding to the downlink transmit beam. If the corresponding link quality is worse than a threshold, the UE can claim that the downlink beam failure happens. The UE also measures a set of candidate beams and select one beam with good link quality, which is referred to as the new beam.

Then the UE notifies the network with the beam failure event and the new beam through the message of BFRQ. After receiving the BFRQ message sent by the UE, the network can send the PDCCH with the new beam reported through the BFRQ. If a PDCCH is detected correctly on the new beam, the UE can conclude that the response from the network side corresponding to the BFRQ is received. At this point, the BFR process is successfully completed. Fig. 8.31 illustrates the complete process of BFR.

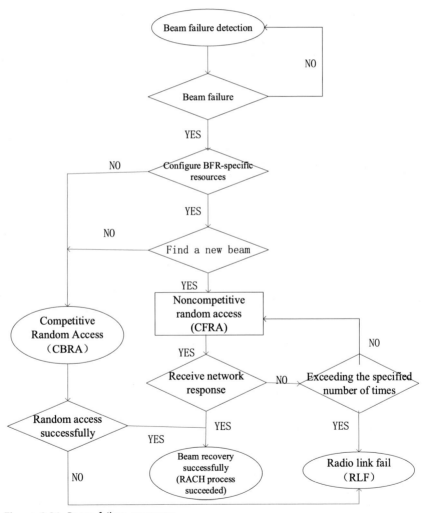

Figure 8.31 Beam failure recovery.

In the following sections, for simplicity of description, we will refer to both PCell and PScell as the primary cell.

8.4.2 Beam failure detection

From the perspective of physical layer, PDCCH and PDSCH are the most frequently used channels in downlink transmission. Therefore the link quality of these two channels should be guaranteed. The first question is: should the BFD be based on measuring PDCCH channel quality or measuring PDSCH channel quality? In NR, DCI can dynamically schedule PDSCH to transmit from different transmit beams. Thus if the link quality of PDCCH transmission is good enough, the network can always use PDCCH to indicate a transmit beam with good quality for downlink data transmission, and the communication link of PDSCH will be maintained. For this reason, in R15, NR BFD is based on the channel quality of the PDCCH.

The second question for BFD is what metric is used to determine the link quality? During the discussions of NR standardization, two options were proposed for the measurement metric [68,69]:

- Hypothetical PDCCH BLER: Since the real BLER of PDCCH transmission cannot be directly calculated, the UE may calculate the an artificial BLER by mapping the measured SINR to a value of BLER. This artificial BLER is referred to as hypothetical PDCCH BLER. The main reason for supporting the hypothetical BLER is that the beam failure is similar to the radio link failure (RLF) and a unified design will be applied for BFD and RLF detection. Therefore BFD should be based on hypothetical BLER that is used for the detection of RLF. Furthermore, the L1−RSRP measurement cannot reflect the real link quality since it does not take into account the interference and noise.

- L1−RSRP: Since L1−RSRP is used in downlink beam management, many companies support to use L1−RSRP in BFD in order to keep the consistency between the two related functionalities. Using the same metric in beam management and BFD can reduce the complexity of UE implementation and also avoid the ping-pong effect. Using hypothetical BLER in BFD could cause the ping-pong effect due to the fact that one beam could have a good value for one measurement metric but a bad value for another measurement metric since L1−RSRP and the hypothetical BLER don't have one-to-one mapping. For example, in the downlink beam management, a UE reports a certain beam with good quality based on L1−RSRP measurement, but the UE may claim beam failure on the same beam based on bad hypothetical PDCCH BLER measurement.

These two metrics were discussed for a long time by the 3GPP. Finally, in order to complete the standardization of BFR mechanism on time, companies agreed to support using the hypothetical PDCCH BLER as the measurement metric for BFD.

The third question for BFD is: what signal should be measured? In order to obtain a reliable measurement of PDCCH BLER in time, the signal to be measured needs to meet the following requirements [70]:

- It will be periodic signal: only the periodic signal can ensure that the UE can estimate the corresponding link quality in time.
- It can represent the quality of PDCCH channel.

Based on these two requirements, the following signals were proposed for BFD during the 3GPP discussion:

- PDCCH DMRS: it can provide accurate estimate of the link quality of PDCCH channel. However, PDCCH and its corresponding DMRS are not periodically transmitted and their transmission are based on the scheduling requirement. Therefore the UE is not able to measure the quality of PDCCH channel in time in some cases.
- SSB: The companies hope to design a common solution that can be applied to various deployment scenarios of NR, including possible new scenarios in the future. However, BFD based on SSB is not applicable in some cases. For example, in the multi-TRP scenarios, a SSB may transmitted from multiple TRPs, but a PDCCH may only be transmitted from one TRP. In this case, the beams used for the PDCCH and SSB are different.
- Periodic CSI−RS: if a periodic CSI−RS and PDCCH use the same transmit beam, the CSI−RS can represent the channel quality of the PDCCH. Therefore the NR adopts that the UE detects the beam failure based on measuring a periodic CSI−RS corresponding to the PDCCH.

The DMRS of PDCCH only supports a single port and CSI−RS used in RLF is also a single port RS. Therefore it is straight forward to conclude that only periodic CSI−RS with a single port can be used for BFD. Specifically, the UE can determine the periodic CSI−RS resource (s) for BFD in two different ways:

- Explicit way: the network explicitly configures a set of periodic CSI−RS resources for BFD through RRC signaling.

- Implicit way: if the network does not provide the abovementioned explicit configuration, the UE can derive a set of periodic CSI−RS resources from the active TCI state(s) of PDCCH for BFD. In R15, a UE is required to measure up to two periodic CSI−RS resources for BFD. In the next section BFR on secondary cell(s) (Scell), the potential problems with this limitation, and some potential solutions will be discussed in detail.

As mentioned earlier, each CORESET is provided with the corresponding transmit beam. Since a UE can be configured with multiple CORESETs, multiple transmit beams would be used for PDCCH transmission. That leads to the fourth question of BFD: how many PDDCH beams need to meet failure before we can claim beam failure? The RAN1 discussion focused the following two options regarding this question [71]:

- Option 1: When some of the PDCCH beams suffer poor link quality, the UE can consider that beam failure occurs. The proponents of Option 1 believe that when the quality of some PDCCH beams is poor, the BFR mechanism can be used to quickly reestablish the link and restore the reliable transmissions, which can effectively reduce the latency of the entire BFR procedure and also reduce the probability of service interruption. In contrast, if Option 2 is adopted, the BFR procedure is triggered only when all the beam of PDCCH are of poor link quality, and thus the latency of the entire BFR procedure will be enlarged. That will cause problems such as discontinuity of user service quality.

- Option 2: Beam failure is claimed only when all the PDCCH beam(s) suffer poor link quality. The supporting companies believe that when only a subset of the PDCCH beams is of poor quality, the network can still resolve the issue by switching the beam such as configuring new beam through RRC signaling or activating new beam through MAC CE signaling. Furthermore, the proponents of Option 2 think Option 1 will trigger beam recovery process frequently, thereby increasing the overhead of the system.

During the 3GPP discussions, it was agreed to study and design the corresponding mechanisms for both options. However, NR only completes the BFR procedure based on Option 2 due to the limited time budget.

In the NR protocol, the physical layer and MAC layer need to coordinate to determine the beam failure. The specific steps are as follows:

- The physical layer measures the hypothetical BLER of each beam corresponding to the PDCCH. If the hypothetical BLER of all beams are worse than a given threshold, the physical layer generates a beam failure instance (BFI), and then reports it to the MAC layer. The physical layer needs to report BFI periodically to the MAC layer. If there is no report for a certain time, the MAC layer assumes there is no BFI in physical layer.
- MAC layer maintains a timer (*beamFailureDetectionTimer*) and a counter (*BFI_COUNTER*) for BFD. When receiving a BFI report, the MAC layer starts or restarts the BFD timer, and increase the beam failure counter count by 1. If the BFD timer expires before a new BFI is received, the MAC layer resets the counter to 0. The reason for that is the detection of beam failure is based on continuous BFI reporting. If the beam failure counter reaches a value that is configured by network, the MAC layer can claim that the beam failure has occurred and will initiate the random access process on the corresponding primary cell.

8.4.3 New beam identification

If the UE only knows that a beam failure has occurred, it still cannot quickly reestablish a new link with the network. The UE also needs to know which beam with good link quality can be used for the link recovery. If the UE is requested to blindly search to find a good downlink transmit beam, it will cause high UE implementation complexity and large latency. Therefore in order to assist the UE to identify downlink transmit beam with good link quality, the network configures a set of RSs (such as CSI−RS signal and/or SSB) in advance. In fact, each RS corresponds to a candidate downlink transmit beam. Thus the network equivalently provides a set of candidate downlink transmit beams to the UE. The UE determines a new beam based on the L1−RSRP measurement. The network configures a threshold of L1−RSRP in advance, and the UE selects a new beam whose L1−RSRP is greater than the threshold.

The starting time of the new beam selection process is not specified in the NR specification, so it is totally up to UE implementation. For example, the UE can start the new beam selection process when the beam failure occurs. The advantage of such an implementation is that the UE needs to search new beam only when the beam failure happens and the required processing at the UE side is reduced, which is beneficial from the perspective of power consumption, and the disadvantage is that it may introduce large latency. Another possible UE implementation is to start the new beam selection process before the beam failure occurs. The main advantage is the latency is reduced, because the UE can send BFRQ to notify NW the selected new beam immediately after the beam failure occurs. The disadvantage is that the UE has to do more measurements on the candidate beams and will consume additional power.

8.4.4 Beam failure recovery request

After beam failure happens, the UE can notify the event of beam failure and report the selected new beam to the network. This report process is referred to as BFR request (BFRQ). Three different schemes were discussed by the 3GPP on the design of BFRQ and the advantages and disadvantages of each of them are summarized in Table 8.7 [67].

The PRACH-like scheme adopts a similar structure of PRACH, but some parameters are adjusted, such as using a shorter length. During the

Table 8.7 Comparison of different BFRQ transmission schemes.

Schemes	Advantages	Disadvantages
PRACH-based	• Reuse the existing PRACH signal, and has little impact on the specification • High transmission reliability	• The length of PRACH is large, which causes high overhead of time and frequency resources. • Using PRACH for BFRQ may reduce PRACH capability for other applications. • It can only carry little information.
PUCCH-based	• Can carry large amount of information • Low overhead of time and frequency resources	• A new PUCCH format is needed, which has a large impact on the specification • Multiplexing with existing PUCCH formats will cause high complexity.
PRACH-like	• Low overhead of time and frequency resources • Short transmission time	• Introducing a new signal has a large impact on the specification • The performance of a new signal has not been verified

discussion in 3GPP, each of those three schemes are supported by many companies, especially the PRACH-based scheme and PUCCH-based scheme.

Finally, considering the complexity of the scheme and time available for NR tenderization work, the PRACH-based scheme is adopted for BFRQ transmission. The final design adopted in NR specification is: when the beam failure occurs, the UE triggers a random access process and the selected new beam is indicated through the transmission of RACH Msg1 to the network, which also implicitly indicates the occurrence of beam failure. Specifically, the network preconfigures a set of candidate beams (such as CSI−RS signal and/or SSB) for the UE, and each candidate beam is associated with a dedicated PRACH preamble/resource. When a UE selects a new beam, the UE transmits the PRACH that is associated the selected new beam. After detecting that PRACH transmission successfully, the network determines that the beam failure occurs at the UE and also determines the new beam recommended by the UE.

The abovementioned random access process is based on the dedicated PRACH that are preconfigured by the network, which is a contention-free random access. In a practical system, a UE may not perform contention-free random access for BFRQ in various cases. The first case is that the network does not configure RS for NBI and corresponding PRACH resources to the UE. In this case, the UE does not have any candidate beams to measure. The second case is the UE cannot find a new beam with satisfactory link quality since the L1−RSRP measurement of all the candidate beams is lower than the threshold that is configured by network.

When any one of the above cases happens, the UE will initiate the existing contention-based random access process to reestablish the connection with the network based on the SSB signal quality of that cell. Such RACH transmission is not based on dedicated PRACH resource for beam failure requests of that UE. Therefore after receiving the RACH Msg1, the network does not know is the purpose of the RACH transmission. In R16, further enhancement is introduced for the BFR mechanism based on contention-based random access. The UE can transmit a MAC CE dedicated to BFR information in Msg3 or MsgA to indicate that the random access is triggered by beam failure.

8.4.5 Response from network

After sending Msg1 for BFR, the UE starts a window (*ra−ResponseWindow*) to monitor PDCCH in order to detect the response to the BFRQ (i.e., response from network for the corresponding random access). If the contention-free random access is used for BFRQ transmission, the UE monitors the response in a dedicated search space on the new beam. For that, the network configures a dedicated CORESET and one associated search space in advance, and this CORESET is dedicated for BFR and is not associated with any other search space. The detailed operations in the UE are as follows:

- If the UE detects a DCI on the new beam within the time window of ra−ResponseWindow, the beam recovery is completed successfully. Given that, the network and the UE can continue the communication through the new beam. Later on, the network can perform normal beam management procedure, such as indicating the UE to perform beam measurement and beam switch.
- If the UE does not detect any DCI on the new beam within the window of ra−ResponseWindow, the UE can retransmit the BFRQ. This process can be repeated until the beam recovery is successfully completed or the number of retransmissions of BFRQ exceeds a threshold configured by the network.

If contention-based random access is used for BFRQ transmission, the network does not configure the dedicated search space. Instead, the UE monitors PDCCH in the common search space for the response to the BFRQ. The mechanism of monitoring network response and retransmission of BFRQ (random access) is the same as that in the existing contention-based random access.

8.5 Beam failure recovery on secondary cell(s)

In carrier aggregation (CA), both the primary cell and the Scell are configured to a UE. In previous sections, we have introduced the BFR procedure for the primary cell(s). In NR R15, some companies proposed to extend the design of the BFR procedure to the Scell(s). They suggested that the fast recovery after the beam failure on the Scell(s) is very important due to the following reasons:

- If the BFR function is supported for Scell, high-quality radio links can be reestablished for a Scell when beam failure happens. Otherwise, the beam failure could lead to a deactivation of the Scell, which will bring a large latency, thereby affecting the data rate and user experience.
- In typical NR deployment scenarios, the primary cell is generally configured in a low-frequency band, while some Scell(s) is generally

configured in a high-frequency band (such as millimeter wave band). Thus the blockage of the PDCCH link is more likely to occur in some Scells.

However, the opposing companies thought that it is not worthwhile to design an additional BFR for the Scell [72], and the specific reasons are as follows:

- For a UE configured with CA, as long as the link quality of the primary cell is good enough, the reliability and continuity of data transmission can be guaranteed. When the beam quality on the Scell is poor, the network can transmit the relevant configuration or indications through the primary cell to indicate the UE to perform corresponding beam management process, and switch to the beam(s) with good link quality, without deactivating the Scell. Therefore there is no need to design an additional BFR procedure for SCell.
- Performing beam failure procedures additionally on the Scell will unnecessarily increase the complexity of the UE implementation.
- If the same BFR mechanism of that of the primary cell is reused, the contention-based PRACH needs to be reserved for the Scell, which causes additional resource waste. On the other hand, if a different design is adopted, the system would have to support two different BFR mechanisms, which will increase the complexity of the specification, and also the implementation complexity at both network side and UE side.
- The various aspects involved in the procedure to be designed are complicated and there is no enough time for R15 standardization.

Due to the stalemate between the two sides, at the late stage of R15, different working groups reached the opposite conclusions. RAN1 agreed to design a BFR mechanism for SCell [73], while RAN2 suggested to study it in the subsequent version (e.g., R16). Finally, the BFR mechanism for SCell is developed in R16.

Regarding the design of BFR mechanism on Scell(s), a basic principle at the beginning of standardization work is to reuse the design of existing BFR of primary cell as much as possible. Therefore the BFR on SCell also has the following four functional components:

- BFD
- NBI
- BFR request
- Response from network

Among them, the components of BFD and new beam selection are basically similar to the counterparts in R15 primary cell BFR mechanism;

the other two components have big changes, which will be introduced in detail later.

8.5.1 Beam failure detection

The BFD on each Scell is performed independently and that is related with the process on primary cell or any other SCell. Similar to the existing mechanism in NR R15, the BFD on a Scell is also based on the single-port periodic CSI−RS resource(s). The periodic CSI−RS resource (s) can be explicitly configured by network or implicitly derived from the TCI state(s) of CORESETs of PDCCH by the UE. These periodic CSI−RS resources for BFD are referred to as BFD RS for simplicity. The other aspects of BFD process of Scell (such as threshold configuration, beam failure determination, etc.) also reuse the design of NR R15.

During the design of BFD for Scell(s), a controversial issue is the maximum number of CSI−RS resources (namely the active beams of PDCCH) that a UE can measure for BFD [74]:

- Option 1: Up to two CSI−RS resources can be used for BFD on one BWP. If the BFD RS is determined with implicit method, it is up to the UE to select two BFD RSs based on the PDCCH beams when PDCCH uses more than two active beams.
- Option 2: The maximum number of CSI−RS resources configured for BFD on one BWP is two. For the implicit method of BFD RS determination, some rule(s) will be specified in the NR specification for selecting two PDCCH beams for the BFD when PDCCH uses more than two active beams.
- Option 3: Up to three CSI−RS resources can be used for BFD on one BWP

The same problem for the primary cell BFR was actually also discussed in R15. A R15 UE can be configured with up to three CORESETs on a BWP through RRC signaling. When designing the BFR mechanism of primary cell in R15, RAN1 assumed that one dedicated CORESET (associated with a dedicated search space) is always configured for the UE to monitor the response of the BFRQ and this dedicated CORESET is not used for PDCCH transmission when the beam failure does not occur. Based on this assumption, each of the other two CORESETs has its own TCI state, and thus there are up to two active downlink transmit beams for PDCCH. Therefore it was agreed that the UE can measure up to two CSI−RS resources for BFD. At the late stage of R15 standardization, after combining the conclusions made in working groups RAN1 and RAN2, companies found that there might exist more than two active PDCCH beams on one BWP. The reason is the configuration of that dedicated search space for BFR (and the corresponding dedicate CORESET) is optional, according to the design in RAN2, and thus there might not be a dedicated CORESET for the BFR in some scenarios. The PDCCH can have up to three active downlink transmit beams in these scenarios. As discussed in previous sections, when no dedicated CORESET is configured for BFR, the UE performs a four-step random access process when beam failure occurs. In order to address this issue in R15, some companies proposed to modify the previous conclusions by adopting Option 2 or Option 3. However, at that time, R15 was close to the completion, and most companies did not agree to work on additional optimization for BFR. Thus at the end, the original conclusion was kept. When starting the study of BFR for Scell(s) in R16, this issue was raised again and the abovementioned three options were discussed once more. The technical reasons for supporting each of those three options are summarized here.

The main reasons for supporting Option 1 (i.e., R15 design) are as follows:

- If different designs are adopted for the BFD on Scell(s) and primary cell(s), the consistency of system design would be destroyed.
- Increasing the maximum number of CSI−RS resources for beam failure detection will increase the UE complexity and power consumption.
- The network can explicitly provide the UE with BFD signals via RRC signaling, which can ensure the same understanding between the network and the UE. The additional rules suggested by Option 2 for implicit method will unnecessarily increase the complexity but not bring obvious benefits.

The main reasons for supporting Option 2 are as follows:

- In the implicit method of the BFD RS determination, if Option 1 is adopted, the network does not know which two beams are chosen by the UE for BFD when the number of active PDCCH transmit beams is greater than 2. That would cause inconsistent understanding between the network and the UE, and could affect the performance of the system. Therefore it is necessary to specify some rules to select two beams for BFD in order to maintain the same understanding of the network and the UE.
- In R16, a UE can be configured with up to five CORESETs if multi-TRP transmission is configured. Thus the maximum number of transmit

beams used by PDCCH could be five. The same problem still exists if the maximum number of BFD signals is set to two as suggested by Option 1 or three as suggested by Option 3.

The main reasons for supporting Option 3 are as follows:

- On the one hand, Option 3 can avoid the complicated rules proposed in Option 2. On the other hand, it can take into account all three active PDCCH beams in BFD.
- Increasing the maximum number to three has limited impact on the UE complexity. Furthermore, a UE can control the overall complexity by reporting UE capability, such as the maximum number of Scells supported for BFR at the same time.
- The standardization work in R16 does not consider the BFR in multi-TRP transmission, which is not in the scope of design and can be considered in future release(s).

In the process of standardization, most companies acknowledged that the R15 mechanism (i.e., Option 1) has some restrictions and some enhancement or optimization might be needed. However, regarding the specific enhancement, companies were mainly divided into two groups supporting either Option 2 or Option 3. But they could not reach any compromised solution to move forward. As a result, neither of these two options were agreed on. Finally, the same design of NR R15 is used in R16. For the implicit method of determining BFD RS, if there are more than two active PDCCH transmit beams, the UE autonomously selects two downlink transmit beams for BFD.

Just as with R15 primary cell BFD, the determination of beam failure for Scell(s) is also completed by physical layer and MAC layer.

8.5.2 New beam identification

The NBI for BFR on Scell basically reuses the existing design of R15. The difference is that the RSs (including CSI−RS resources and/or SSB) for NBI must be configured to the UE, while those RSs can be configured or not in R15. The BFR mechanism in R15 is aimed at the primary cell. Therefore if the network does not configure RSs for NBI, the UE can initiate a four-step contention-based random access process to reestablish the link. In contrast, the BFR mechanism in R16 is aimed at the Scell, which does not have a contention-based PRACH. The Scell itself cannot initiate a contention-based random access.

- If supporting to initiate a contention-based random access for SCell BFR on the primary cell, the solution will have to involve the interoperation or priority determination between the R16 new BFR procedure and the R15 primary cell BFR procedure and the specification design will be more complicated.
- The four-step contention-based random access has relatively large latency and costs more uplink resources.

Due to the above reasons, it is stipulated in R16 that for the BFR of Scell, the network will always configure RS for NBI.

8.5.3 Beam failure recovery request

In R15, the beam recovery procedure is for the primary cell, and the PRACH is used to transmit BFRQ. Compared with the primary cell, the Scell has its own characteristics. For example, a Scell may have only downlink carrier but no uplink carrier. During the R16 discussions, companies have proposed various design on the BFRQ transmission of the Scell, mainly including the following three schemes [75]:

- Scheme 1: Use PRACH to transmit BFRQ (similar to R15 design).
- Scheme 2: Use PUCCH to transmit BFRQ.
- Scheme 3: Use MAC CE signaling to transmit BFRQ.

The advantages and disadvantages of each scheme are summarized in Table 8.8.

Among them, relatively few companies support Scheme 1, and most of the companies support either Scheme 2 or Scheme 3. It was very hard for each side to give up the position and accept the other scheme. Finally, a compromised solution combining PUCCH and MAC CE (namely BFR MAC CE) was agreed on, which mainly includes the following processes:

- When beam failure occurs in a SCell, if there exist available uplink resource grant, the UE transmit BFR MAC CE in PUSCH transmission to report the beam failure of the SCell and a selected new beam for the Scell.
- If there does not exist uplink resource grant, the UE can send SCell BFR SR in PUCCH to request uplink grant for sending BFR MAC CE.
- If the UE does not have available uplink resources for transmitting BFR MAC CE and the network does not provide SR configuration, the UE can trigger random access process to request uplink resources based on as normal R15 mechanism.

Table 8.8 Comparison of BFRQ schemes for different secondary cells.

Scheme	Advantages	Disadvantages
PRACH	• Reuse the existing mechanism of R15 to avoid adopting different solutions for similar problems	• PRACH consumes a lot of resources • Unlike the primary cell, the use of PRACH resources in the Scell is relatively limited (e.g., contention-based PRACH are not available) • Because of the existence of the primary cell, other more flexible and efficient mechanisms can be used (such as Scheme 2 and Scheme 3), and the mechanism adopted for the primary cell is not required
PUCCH	• The overall latency is small • Low overhead • PUCCH is more flexible than PRACH • A good compromise between flexibility and performance	• The Scell may not have uplink carrier, or no PUCCH is configured in the uplink carrier. Thus PUCCH resources in other cells has to be used. • To maintain low latency, a dedicated PUCCH resources is needed. Even when the beam failure does not occur, that PUCCH resource has to be reserved and cannot be used for other purpose, which result in resources waste. • PUCCH-based method can carry more information payload than PRACH, but it is less flexible than MAC CE-based method. • The priority rule or multiplexing with other PUCCH resources will complicate the system design.

(Continued)

Table 8.8 (Continued)

Scheme	Advantages	Disadvantages
MAC CE	• The MAC CE can easily carry more information bits. Compared with PUCCH format, the MAC CE format is simple in design. • The MAC CE can be reliably transmitted on the primary cell regardless of whether a Scell has uplink carrier or not. • The MAC CE is transmitted through PUSCH. So the overall procedure of beam recovery is simple and the additional standardization effort is small.	• The MAC CE signaling will bring more latency than PUCCH-based method.

Similar to R15 NBI, the new beam selection process on the Scell may not find a beam that can meet the threshold configured by the network. In R15, if that happens, the UE may start a four-step contention-based random access process. But in R16, a MAC CE is used for BFRQ reporting, which is more flexible and can carry more information than the design in R15. Companies proposed different reporting schemes for the Scell for the case when the UE does not find a new beam:

- Scheme 1: the UE reports the identification of the Scell where beam failure happens and a special state to indicate that no new beam is founded for this Scell.
- Scheme 2: the UE reports the identification of the Scell where beam failure happens and the best candidate beam and its L1−RSRP value.
- Scheme 3: the UE only reports the identification of the Scell that encounters beam failure.

All three schemes support to report the identification of the Scell(s) where beam failure happens. The difference lies in whether to report other additional information. The main technical motivation of Scheme 1 is that the network will be notified that none of new beams can meet the threshold, so that the network can make better decisions accordingly. Furthermore, the Scheme 1 can support to use the same MAC CE format for the cases of reporting new beam or not finding a new beam. The main motivation of Scheme 2 is to report the beam with the best L1—RSRP to the network when no new beam is found, which can assist the network to make better judgment. The main technical reason for opposing Scheme 2 is that the network can flexibly configure different thresholds for NBI according to requirements, and the configuration of the network should be respected. Therefore the UE will not report a beam that does not meet the requirement. Moreover, introducing different reporting contents generally requires different MAC CE reporting formats, which results in different MAC CE formats and unnecessarily increases the system complexity. The main advantage of Scheme 3 is that it can theatrically reduce the resources used for MAC CE transmission. However, many companies think that it is hard to achieve that goal in practice. Moreover, it requires different reporting contents in MAC CE and thus requires to support different MAC CE formats, which will complicate the system design. Overall, the Scheme 1 is more intuitive and the design is simple. Therefore it is adopted in 3GPP finally. The final design adopted in the NR specification is that, if the UE does not find a new beam with link quality better than the configured threshold, BFR MAC CE should inform the network with that.

In addition, a UE may be configured with multiple Scells and each Scell can perform its own BFR procedure independently. A R16 UE can report the maximum number of Scells it can support BFR procedures simultaneously through UE capability reporting. If beam failure happens in multiple Scells, their BFRQ can be transmitted in one single BFR MAC CE, so as to save UL resource overhead and reduce the latency. If the available uplink resources are not sufficient to transmit all the BFRQs, a truncated BFR MAC CE can be used to report on identification of the Scell(s) with beam failure. After receiving the truncated MAC CE, the network knows which cell(s) encounters beam failure, and also knows that the UE does not have enough uplink resources for reporting detailed information of SCell beam failure. It is allowed to transmit the BFR MAC CE report in either the primary cell or any Scell, for which the NR specification does not define any restrictions. It can be determined based on the available uplink resources at that time.

8.5.4 Response from network

The MAC CE signaling is transmitted through PUSCH. Whether the network receives the MAC CE signaling successfully can be determined through the HARQ mechanism of PUSCH. As specified in R15, if a UE receives an uplink grant for scheduling new data corresponding to the same HARQ process number, the UE can assume that the previous PUSCH transmission has been correctly received by the network. For the Scell BFRQ transmission, the same mechanism is used to determine whether the network has received the MAC CE correctly.

On the abovementioned mechanism for determining the response of network, some companies suggested some further optimization for the reasons that the network may not schedule new uplink data transmission for a period of time in some scenarios. During that period, the UE cannot determine whether the network has correctly received the BFR MAC CE. The opposing companies believed that the network can make a wise choice and no additional enhancement is needed. On one hand, the network can to decide when to send the response to the UE according to its scheduling. On the other hand, even if there is no new uplink data transmission, the network can still grant a new PUSCH transmission if the network thinks a response is needed. Finally, no additional enhancement was introduced on this.

8.6 Multi-TRP cooperative transmission

In order to improve the performance of cell edge users and provide a more balanced quality of service in the coverage area, a multi-TRP cooperative transmission scheme is introduced in NR. Multi-TRP cooperative transmission can improve the throughput of edge users and improve the performance of edge users through noncoherent joint transmission (NC−JT) or diversity transmission between multiple TRPs, thereby better supporting the eMBB service and URLLC service. In NR R16, the cooperative transmission is limited to two TRPs and thus the following discussion in this section will be based on the assumptions of coordination between two TRPs.

8.6.1 Basic principles

Based on mapping relationship between the layers of transmitted data and the transmission TRPs, the multi-TRP cooperative transmission schemes can be divided into two types: coherent transmission and noncoherent transmission.
- In coherent joint transmission, data transmitted by multiple transmission points are jointly beamformed, and the precoding matrix (relative

phase) applied on different transmission points is coordinated so that the data in the same layer can be coherently combined at the UE. In other words, the subarrays of multiple transmission points are modeled as a higher-dimensional antenna array to obtain higher beam-forming gain. The coherent transmission scheme has high requirements for synchronization and cooperation between transmission points. In the practical deployment environment of NR, the cooperation performance between transmission points is easily affected by some nonideal factors such as frequency deviation. And the differences in physical location and large-scale parameters such as path loss also negatively affect the performance of joint beam-forming. Therefore it is difficult to guarantee the performance gain of coherent transmission in practical systems.

- Compared with coherent transmission, noncoherent transmission (NC–JT) does not require joint beam-forming among multiple transmission points. Each transmission point can independently precode its own data without coordinating the relative phase with other TRPs. The influences of the abovementioned nondeal factors are relatively small. Therefore the NR system has studied and standardized the noncoherent transmission as an important technical feature to improve the performance of cell edge users.

Considering the various application of multi-TRP transmission under different backhaul capabilities and service requirements, the NR mainly studies the following multi-TRP cooperative transmission schemes (as shown in Fig. 8.32):

- The NC–JT transmission scheme based on a single DCI is mainly used to improve the data rate of eMBB service. In a single DCI-ß-based MTRP enhancement, a single PDCCH is used to schedule a

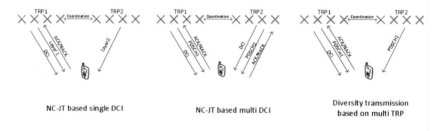

NC-JT based single DCI   NC-JT based multi DCI   Diversity transmission based on multi TRP

Figure 8.32 three transmission schemes of multi-TRP.

PDSCH that can be transmitted from multiple TRPs. In each PDSCH, each transmission layer can be transmitted from one TRP, and different layers can be mapped to different TRPs. Due to the requirement for dynamic and rapid collaboration between TRPs, this transmission scheme is only suitable for the deployment scenarios with ideal backhaul.

- The NC–JT transmission scheme based on multi-DCI is also mainly used to improve the data rate of eMBB service. In the multi-DCI based multi-TRP transmission, each transmission point schedules its own PDSCH transmission through an independent PDCCH. In this case, close cooperation and frequent signaling and data exchange between TRPs are not required. Thus this scheme can be deployed in the nonideal backhaul scenarios.
- The multi-TRP-based diversity transmission scheme is mainly used to improve the transmission reliability of URLLC services, in which multiple TRPs repeatedly transmits the same data in order to improve the transmission reliability, thereby better supporting the URLLC service.

8.6.2 NC–JT transmission based on a single DCI

In the single DCI–based NC–JT transmission, different layers of a PDSCH are transmitted in parallel from multiple transmission points on the same time-frequency resources, which can effectively improve the edge spectrum efficiency. The signaling overhead of a single PDCCH-based transmission is smaller than that in the multi-PDCCH-based scheme. The research and standardization work on a single DCI-based MTRP enhancement mainly includes: codeword mapping scheme, DMRS ports allocation indication scheme, and TCI states indication and mapping scheme.

8.6.2.1 Codeword mapping

In R15, when the number of layers (i.e., the number of MIMO layers) is less than four, single codeword transmission is used while two codewords are used if the number of layers is greater than four. The multi-TRP cooperation is mainly used to improve the performance of cell edge users and four receiving antennas will be the popular configuration in NR UEs. Therefore multi-TRP cooperative transmission mainly considers data transmission with up to four layers. If the existing codeword mapping rules defined in R15 are reused here, multiple TRPs have to share the same codeword in one PDSCH transmission. That would result in one problem that each TRP is not allowed to adjust the PDSCH MCS, respectively, according to its own link quality because one codeword can only have one corresponding MCS according to specification in NR protocol. However, adjusting the existing codeword mapping scheme [76,77] will lead to significant changes in the specification, and the simulation results given by some companies [78] show that the using a single codeword does not bring significant performance loss. Thus it is decided that the codeword mapping scheme in single DCI-based NC–JT transmission still follows the rules specified in R15.

8.6.2.2 DMRS port indication

The channels between different TRPs and the UE have different large-scale channel properties. In order to ensure the orthogonality between the DMRS ports within the same CDM group, those DMRS ports are required to be QCLed. Therefore in the design of DMRS port allocation for multi-TRP cooperative transmission, it is necessary to allocate at least two CDM groups, where each CDM group is used for data transmission of one TRP. As specified in the TS 38.212 of R15, the DMRS port indication table lists the DMRS port indication of different CDM groups. Different DMRS configurations correspond to different layer combinations for multi-TRPs. Here are few examples of layer combinations for multi-TRPs (only no more than four layers are considered here):

- 1 + 1: each TRP transmits one layer.
- 2 + 1: two layers are transmitted from the first TRP and one layer is transmitted from the second TRP.
- 2 + 2: two layers are transmitted from two TRPs, respectively.

In order to support more flexible scheduling, R16 introduces a new transport layer combination {1 + 2}, which can be achieved by introducing DMRS port {0,2,3}. It can support that one layer is transmitted from the first TRP and two layers are transmitted from the second TRP.

For the case of rank = 4, some companies [79] proposed to introduce the configuration of one DMRS port in one CDM group and three DMRS ports in another CDM group to support {1 + 3} or {3 + 1} layer combination. However, multi-TRP is mainly aimed at cell edge users and the signal strength difference from users to two cooperative TRPs will not be very large. As a result, the numbers of layers transmitted from two cooperative TRPs are expected to be similar. Based on this assumption, some companies [80] simulated the case that the total number of rank is no more than four and the layer combination can be 1 and 3. The results show that the performance of such transmission configuration is not significantly improved compared with that of not allowing this configuration. Therefore NR does not support the layer combination of $\{1+3\}$ and $\{3+1\}$.

In addition to the rank combination, NR also discusses whether to support both NC—JT and MU—MIMO at the same time. Some companies [81,82] believe that the transmission of NC—JT is mainly aimed at improving the performance of edge users while MU—MIMO is mainly used to increase the throughput of central users. Applying MU—MIMO on edge users may offset the gain of NC—JT. In addition, the NC—JT transmission is expected to provide system performance gain only when the system resource utilization rate is low, but the typical scenario of MU—MIMO performance improvement is high resource utilization. It is rare that both of these two technical functions are applicable. At the same time, some companies [83] conducted simulation evaluation for multi-TRP multiuser transmission based on a single DCI scheme, and no obvious performance gain is observed. Therefore NR does not support this feature finally.

8.6.2.3 TCI state indication

Due to the different physical location of different TRPs, the channel large-scale characteristics corresponding to each TRP have obvious differences. Therefore in the joint transmission of multiple TRPs, it is necessary to indicate the QCL information corresponding to each TRP. In R15, one code point of TCI field in DCI only corresponds to one TCI state. In order to support multi-TRP-based transmission, R16 enhances the MAC—CE signaling for PDSCH TCI state activation and one code point of TCI field in DCI can be mapped to up to two TCI states. If the TCI field in a DCI indicates two TCI states, the data associated with the first TCI state will be transmitted using the DMRS port(s) indicated in the first CDM group, and the data associated with the second TCI state will be transmitted using the DMRS port(s) indicated in the second CDM group.

The configuration and indication of TCI state includes three steps: RRC configuration, MAC—CE activation, and DCI indication:

- The RRC configures M TCI states for the UE through RRC parameter PDSCH-Config. The value of M is determined by the UE capability and the maximum value of M can be 128.
- MAC—CE activates up to eight TCI state groups and maps each of them to one codepoint of the 3-bit TCI field in the DCI. Each TCI

state group activated by the MAC−CE can contain one or two TCI states. If the higher-layer configuration indicates that TCI field is present in a DCI, DCI format 1_1 can indicate a TCI state group from the TCI state group activated by MAC. If the higher-layer configuration does not indicate that TCI is present in DCI or the data is scheduled by DCI format 1_0, the TCI state indication field is not included in the DCI.

- If at least one TCI state group activated by MAC CE contains two TCI states, and the time interval between DCI and scheduled PDSCH is less than the threshold *timedurationForQCL* that is reported by the UE, the UE uses the TCI state group with the lowest index among the TCI state groups containing two TCI states activated by MAC−CE to receive the data.

8.6.3 NC−JT transmission based on multi-DCI

Because the backhaul capacity of multiple TRPs may be limited, there will be a large delay in the interaction between TRPs and thus it is impossible to schedule multiple TRPs through a single DCI. Furthermore, the channel conditions between different transmission points and the UE are independent and performance gain can be expected through separate scheduling and link adaptation from each TRP. Therefore multi-TRP transmission scheme based on multi-DCI is introduced and specified in R16. Although the noncoherent transmission based on multi-DCI is mainly introduced for scenarios of nonideal backhaul, this scheme can also be used in the case of ideal backhaul. For HARQ−ACK feedback of PDSCHs from different TRPs, the network can configure separate feedback or joint feedback based on the actual backhaul capacity between transmission points.

8.6.3.1 PDCCH enhancement

Since each TRP schedules its own PDSCHs through its own PDCCH, the UE needs to monitor PDCCHs from different TRPs separately. In NR, different TRPs are distinguished by different CORESETs (i.e., each CORESET only corresponds to one TRP). Different TRPs can use different CORESET groups to transmit PDCCHs. Specifically, CORESETs in different CORESET groups are distinguished by different CORESET group indexes (*CORESETPoolIndex*). For a CORESET without a *CORESETPoolIndex* configured, the UE assumes its *CORESETPoolIndex* is 0 by default. With this parameter, the UE can assume that all the data scheduled by PDCCH in CORESETs with the same *CORESETPoolIndex* comes from the same TRP. Compared with R15, since the CORESET allocated by each TRP is independent, the network needs to configure more CORESETs to support the independent scheduling from two TRPs. In R16, a UE can be provided with up to five CORESETs (three CORESETs can be configured for R15) on each BWP, and the specific number of CORESETs that can be configured depends on the capability of the UE.

The network can configure up to ten search space sets for each UE per BWP, and the monitoring periods of different search spaces can be different. Due to the different monitoring periodicity, the network can configure the PDCCH candidates in each search space according to maximum UE capability. An issue related with blind decoding arises when the PDCCH monitoring occasions coincide. For example, when the UE needs to monitor both the common search space and UE-specific search space at the same time, the number of PDCCH that the UE needs to blindly decode may exceed the UE capability. In the specification, overbooking rule is specified in R15. When the number of PDCCHs that the UE needs to blindly decode exceeds the UE capability, some SS sets will be dropped according the specified rules. In the cells supporting multi-DCI-based transmission, the maximum number of blindly decoding PDCCH candidates and nonoverlapped CCEs per slot for a DL BWP will not be greater than the R15 limits. If maximum number of blind detection/CCE that a UE is supposed to monitor over CORESETs with the same *CORESETPoolIndex* value is the same as the maximum number of BD/CCE over all configured CORESETs, R15 overbooking rule is followed. Otherwise, overbooking is only applicable for USS sets associated with the CORESETs that are configured with *CORESETPoolIndex* being 0.

8.6.3.2 PDSCH enhancement

When each transmission point schedules the corresponding PDSCH through its own PDCCH, two PDSCHs from two different TRPs might be fully overlapped, partially overlapped, or not overlapped in time-frequency resources. Different cases of overlapping have different impact on the performance of cooperative transmission and UE complexity [84–87]:

- If the resources of different PDSCHs scheduled by different TRPs are nonoverlapped, the processing complexity of the UE can be reduced from the perspective of UE implementation, and there does not exist additional interference between PDSCHs. However, in order to ensure the nonoverlapping resource allocation, the TRPs have to coordinate with each other in advance, which requires real-time data interaction or semistatic resource coordination between transmission points.

- If the resources of PDSCHs from different transmission points are fully overlapped, it can enhance the spectrum efficiency of the system under ideal circumstances. However, considering the differences in the channel conditions of different transmission points, this transmission method may also lose the gain of frequency selective scheduling. From the perspective of UE implementation, the statistical characteristics of interference between PDSCHs are relatively stable within the allocated resources, so it is more convenient to implement the interference estimation and interference mitigation algorithms. In order to ensure that the resources of two PDSCHs are completely overlapped, dynamic coordination of the available resources is necessary, which also requires high real-time data interaction between transmission points.
- If the resources of PDSCHs are partially overlapped, the interference between the PDSCHs will be different on different part of the allocated resources, which will bring extra complexity to channel estimation and interference suppression of the UE.

Considering the above factors, if the resource allocation is not restricted, the base station can have full scheduling flexibility, and the requirement for backhaul can be relaxed. But the implementation complexity of the UE will be significantly increased. Therefore whether to support all of the cases of no-overlapping/partial-overlapping and fully overlapping resources becomes a hotspot problem between terminal/chip manufacturer and network manufacturer during the discussions on multi-DCI-based transmission. Finally, after several rounds of extensive discussion on UE capability [88], it was decided that partial overlapping and full overlapping would be supported as optional UE capabilities and the UE can report it to the network.

In order to further reduce the implementation complexity of the UE and also mitigate the interference between PDSCHs, the following restrictions and enhancements are specified in NR [89]:

- The configuration of PDSCH DMRS, including the configuration type, symbol location and number of symbols of the front loaded DMRS and additional DMRS, will be the same in each PDSCH. The

UE is not expected to have more than one TCI state for the DMRS ports within one CDM group. In this way, the DMRS of different PDSCHs are mapped to different CDM groups, which results in no interference between DMRS of different PDSCHs.

- If the allocated resources of two PDSCHs are fully overlapped or partially overlapped, the network scheduling will ensure that the data of one PDSCH is not mapped on the REs that are used by the DMRS of the other PDSCH.

- The UE only receives data from two TRPs simultaneously within the same BWP and the same subcarrier spacing. Every transmission point can indicate BWP switching through its own DCI. However, since the UE can only receive and transmit on one BWP at any time in a component carrier, if different TRPs indicate different BWPs at the same time, the UE is not able to receive and transmit to these two BWPs at the same time.

- The PDSCHs scheduled by PDCCHs in CORESETs with different CORESETPoolIndex values will use different PDSCH scrambling sequences. If the scrambling IDs of the PDSCHs transmitted by different TRPs are the same, the scrambling sequences of PDSCH transmitted by different TRPs are also the same. In this way, there will be persistent interference between PDSCHs from different TRPs when the resources are fully overlapped or partially overlapped. In order to randomize the interference between PDSCHs from different TRPs, the network can configure two scrambling IDs through higher level signaling and associate them with different *CORESETPoolIndex* values.

In NR, in addition to considering the interference between PDSCHs scheduled by multiple DCI, we also need to take into account the confliction between NR PDSCH and LTE CRS transmitted by adjacent cells. If NR and LTE are deployed in the same frequency band, the collision between NR PDSCH and LTE CRS will be avoided by rate matching of PDSCH RE. Specifically, in the multi-TRP cooperative transmission based on multi-DCI, the network can configure up to three CRS patterns for two TRPs. If the interference between the PDSCH transmitted by one TRP and the CRS transmitted by another TRP is not significant, the network can independently match the CRS pattern of each TRP and the rate matching is done based on the CRS pattern configured by the TRP itself, which can reduce the impact on system throughput. In this case, CRS patterns associated with each

CORESETPoolIndex are applied to the PDSCH scheduled with a DCI detected on a CORESET with the same *CORESETPoolIndex*. This feature requires separate UE capability signaling. If a UE does not support this feature, the UE will rate match PDSCH around the configured CRS patterns from multiple TRPs (Fig. 8.33).

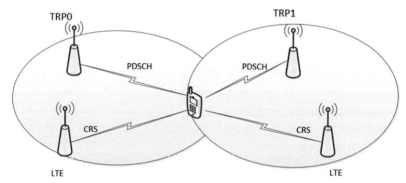

Figure 8.33 The UE is interfered with by two CRS at the same time.

8.6.3.3 HARQ—ACK enhancement

Two HARQ—ACK feedback modes are supported for multi-DCI-based NC—JT transmission: one is separate ACK/NACK feedback and the other is joint ACK/NACK feedback. RRC signaling is used to switch between separate feedback and joint feedback if two feedback modes can be supported by UE capability. In these two feedback modes, both dynamic and semistatic codebook of HARQ—ACK need to be considered. For separate feedback mode, one more issue is how to ensure the PUCCH resources for corresponding to different TRP are not overlapped in time domain.

• Separate HARQ—ACK feedback

In separate HARQ—ACK feedback mode, the UE uses separate PUCCH resources to feedback HARQ—ACK for PDSCHs that are received from different TRPs, as shown in Fig. 8.34.

In the semistatic codebook of R15, the size of HARQ—ACK codebook is predefined or determined by the parameters configured by RRC. For example, the HARQ—ACK codebook is determined based on a set of occasions for candidate PDSCH reception that is configured through the high-level parameters. In the semistatic codebook, HARQ—ACK bits are concatenated according to the order of the reception occasions and

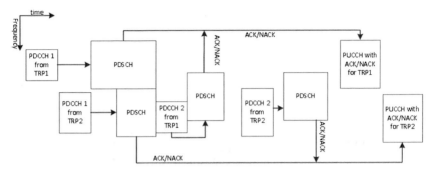

Figure 8.34 independent HARQ—ACK feedback.

the serving cell index. In the dynamic HARQ—ACK codebook of R15, the size of HARQ—ACK codebook changes dynamically with the actual PDSCH scheduling. The HARQ—ACK bits for each downlink transmission are concatenated according to the C—DAI in DCI. In multi-DCI-based multi-TRP transmission, if the separate HARQ—ACK feedback mode is configured, the HARQ codebook of each TRP is still generated independently according to R15 rules. Semistatic codebook is generated based on the candidate PDSCHs scheduled by all the PDCCHs associated with the same *CORESETPoolIndex* value. If the dynamic codebook is configured, the DAI corresponding to each *CORESETPoolIndex* value is counted independently, and the UE generates the ACK/NACK codebooks for PDSCHs which are scheduled by PDCCHs with the same *CORESETPoolIndex*.

If separate HARQ—ACK feedback mode is configured, the PUCCH resources indicated by the PRI in two different DCIs from two TRPs will not overlap. To solve this problem, some companies [90,91] proposed to configure two TDMed PUCCH resource groups over PUCCH resource set and each TRP is associated with one PUCCH resource group. The drawback of this proposal is that when the traffic loads of two TRPs are not balanced, such an explicit grouping scheme will impair PUCCH resource utilization efficiency. Furthermore, the resource allocation of CSI and SR should also be considered. Another proposed scheme [92] is that instead of explicitly grouping the PUCCH resources, the system can realize the TDMed PUCCH resource allocation through network scheduling, which is transparent to the UE. The other proposal [93] is that the network configures PUCCH resources between TRPs and PUCCH resources can overlap in time domain. After considering all the potential influencing factors, NR did not adopt the method of explicit PUCCH grouping [94]. Instead, it is up to network implementation to avoid the confliction between uplink signals of different TRPs.

- Joint HARQ—ACK feedback

In a scenario with ideal backhaul, the network can configure joint HARQ—ACK feedback mode, in which the HARQ—ACK feedback of two TRPs is reported to one TRP. When the PUCCH resources carrying HARQ—ACK of two TRPs are allocated in the same slot, the HARQ—ACK bits of PDSCH of the two TRPs are concatenated together in one HARQ—ACK codebook carried in one PUCCH resource. This feedback mechanism completely reuses the HARQ—ACK feedback methods specified in R15. If the PUCCH resources of two TRPs are configured in different slots, the UE still feeds back the HARQ—ACK codebooks of each TRP separately. The specific feedback mechanism is shown in Fig. 8.35.

If joint feedback mode is configured, the receiving time of PDSCH, the serving cell index and the associated *CORESETPoolIndex* (corresponding to TRP) are sorted in the semistatic codebook. In R15, the HARQ—ACK information bits are concatenated in ascending order of PDSCH reception occasions and then serving cell index. In order to be compatible with the sorting rules defined in R15, if the network provides different *CORESETPoolIndex*, the feedback of semistatic codebook is generated according to the PDSCH reception occasion index at first and then serving cell index and last the *CORESETPoolIndex* value, as shown in Fig. 8.36.

Two TRPs can transmit different PDSCHs to a UE simultaneously. How the DAI is generated determines the order of HARQ—ACK bits in

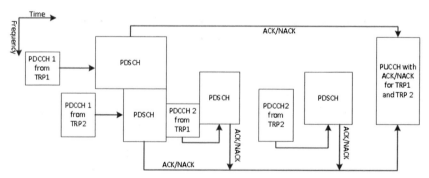

Figure 8.35 joint HARQ—ACK feedback.

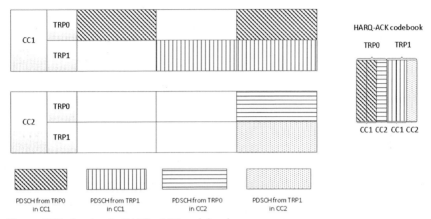

Figure 8.36 Semistatic HARQ—ACK codebook.

a dynamic codebook. Regarding the counting rules of DAI, NR discussed two schemes [95—97]: independent counting within each TRP and joint counting of two TRPs. In the independent counting mode, at the same receiving occasion, the DAI of each TRP is accumulated with the downlink data of each TRP. The HARQ—ACK codebook of each TRP can be generated separately and then the joint dynamic codebook is generated by concatenating the HARQ—ACK codebooks of two TRPs. In the joint counting mode, the same receiving occasion is accumulated with the scheduled downlink data according to the ascending order of *CORESETPoolIndex*. The dynamic codebook is generated in ascending order of DAI. Considering that one missed DCI from one TRP can be determined by the DAI of the other TRP if the joint counting method is adopted, joint counting method is adopted in NR finally. In joint feedback mode, the PUCCH resource used for joint feedback is the PUCCH resource indicated in the last DCI, which follows the designs specified in R15. However, because the DCIs sent by different TRPs may occupy the same time-domain resources, some rules are needed for determining the last DCI in those scenarios. Specifically, when a UE is configured with multiple *CORESETPoolIndex* values, the last DCI is first determined according to the ascending order of PDCCH monitoring occasion; if there are PDCCHs of multiple serving cells at the same monitoring occasion, the last DCI is determined according to the ascending order of serving cell index; if there are multiple PDCCHs of the same serving cell from different TRPs at the same time, the last DCI is determined according to the ascending order of the associated *CORESETPoolIndex* value.

8.6.4 Diversity transmission based on multi-TRP

In addition to the enhancements for eMBB services, the multi-TRP scheme of R16 also considers the enhancements for URLLC services. Since the channel propagation characteristics between different TRPs and the UE are relatively independent, a diversity transmission using multiple TRPs would improve the reliability of data transmission, in which multiple TRPs repeat the same data transmission in spatial, time, or frequency domain. Considering that the URLLC enhancement scheme is mainly for the scenarios with ideal backhaul, the NR designed the diversity transmission only based on the single DCI-based multi-TRP transmission scheme. Specifically, the NR considers four diversity transmission schemes: spatial-division multiplexing (SDM), frequency division multiplexing (FDM), intraslot time division multiplexing (TDM), and interslot TDM [98−100]:

- SDM scheme: Different TRPs transmit PDSCH with different TCI states and different DMRS ports on the same time-frequency resources. Multiple layers of PDSCH are transmitted in the same resources and the resource utilization of SDM scheme is relatively high. However, there exist interlayer interference.
- FDM scheme: Different TRPs use nonoverlapping FD resources on the same OFDM symbol to repeatedly transmit the same data. Because the transmission resources occupied by different TRPs are not overlapped in the frequency domain, they do cause no interference to each other. But the frequency resource utilization rate in FDM scheme is low.
- Intraslot TDM scheme: In a slot, different TRPs use different OFDM symbols on the same FD resource for repeated data transmission. Because the available time-domain resources in a slot are relatively small, the number of repetitions is limited. But the repetition can be completed in a very short time. The intraslot TDM scheme is suitable for the URLLC services that require high reliability and low latency, and have relatively small packets.
- Interslot TDM scheme: different TRPs use different slots on the same frequency domain for repeated data transmission. This scheme can provide high reliability. However, since the repeated transmissions span multiple slots, this scheme is only suitable for URLLC services that are not sensitive to delay.

Among them, the schemes of FDM, intraslot TDM and interslot TDM can be switched through RRC signaling. Once one of these three schemes is configured through RRC signaling, the switching between it and the R15 nonrepetition transmission scheme or SDM repetition scheme is dynamically realized through DCI.

8.6.4.1 Spatial-division multiplexing schemes

NR discussed three different solutions for SDM schemes:

*   Scheme 1A (Fig. 8.37): two groups of layers are generated for the same transmission block (TB) and different group of layers are sent through different transmission points. Each transmission point uses a different set of DMRS ports.
*   Scheme 1B: two different transmission points transmit the same TB with two independent RV versions. Each transmission point uses a different set of DMRS ports. The RV versions of data transmitted by two transmission points can be the same or different.
*   Scheme 1C: two transmission points use the same set of DMRS ports to transmit the same data with the same RV version.

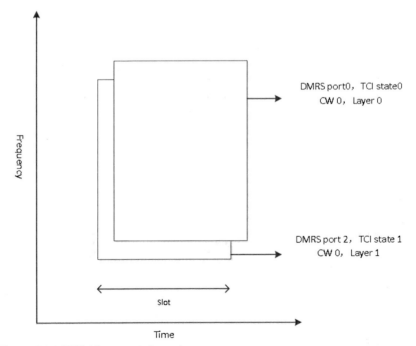

Figure 8.37 SDM 1A transmission scheme.

Among the three schemes, Scheme 1A does not need any additional standardization work compared with the aforementioned single DCI-based scheme. In Scheme 1B, if those two RV can be self-decoded, the performance is more robust, especially when one of the two TRPs is blocked. But in fact, Scheme 1A can also support the self-decoding of each TRP when the bit rate is low. According to the performance evaluation provided by some companies [101,102], the gain of Scheme 1C is not obvious compared with SFN transmission, which can be implemented without specification support. Taking into account the impact on the NR protocol, NR finally adopted SDM Scheme 1a which implicitly supports the multi-TRP-based diversity transmission through a single DCI-based scheme.

8.6.4.2 Frequency division multiplexing scheme
According to whether data transmitted in different FD resources use the same RV version(s), the FDM scheme can be divided into two solutions:

- Scheme 2A: Single codeword with one RV is mapped in frequency resource corresponding to all indicated TCI states. Each TRP transmits only part of the information bits in the codeword.
- Scheme 2B: Two independent RV versions of a single codeword are mapped to FD resources corresponding to two different TCI states, respectively. Different TRPs transmit different RV of the same TB on nonoverlapping FD resources. These two RV versions can be the same or different.

Because the FD resources occupied by different TRPs do not overlap with each other, the PDSCHs transmitted by different TRPs (i.e., associated with different TCI states) can use the same DMRS port(s). Soft combining of PDSCHs with different RVs can improve the performance of Scheme 2B. The UE capability on soft combing is specified in NR and a UE can report to the network whether the UE supports soft combining. If the UE does not support this capability, the performance of the scheme will be significantly reduced. Fig. 8.38 shows Scheme 2A and Scheme 2B.

Schemes 2A/2B need to address the issue of frequency resource partition between two TRPs and also the issue of mapping between TCI states and FD resources. The diversity transmission of Scheme 2A/2B is still based on a single DCI-based transmission (i.e., the PDSCHs transmitted from two different TRPs are scheduled through a single DCI). Trying not to increase the DCI signaling overhead, the NR only supports comb-like frequency resource partition between two TRPs. The specific

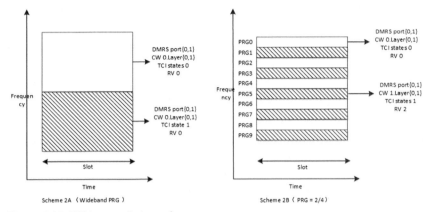

Figure 8.38 FDM transmission schemes.

resource partition method depends on the precoding granularity of the network configuration:

- When the precoding granularity is wideband PRG, the first half of the allocated PRBs indicated by DCI are assigned to the transmission associated with first TCI state (i.e., the first TRP) and the remaining PRBs are assigned to the transmission associated with the second TCI state (i.e., the second TRP).
- For PRG size two or four, the even number of PRGs in the FDRA are assigned to the transmission associated with the first TCI state, and the odd PRGs are assigned to the transmission associated with the second TCI state.
- When the numbers of PRBs allocated to two TCI states are different, the TBS determination is calculated with the number of PRBs associated with the first TCI state in the TCI code point.

For determining the RVs of the PDSCHs sent by two TRPs in Scheme 2B, NR defines four sets of RV sequence candidates. Each RV sequence candidate contains two RV values the PDSCH transmissions from two different TRPs. The specific RV definitions are shown in Table 8.9, where RV_{id} is the index of RV sequence. The DCI indicates the value of RV_{id} and then the UE can determine the corresponding RV1 and RV2 accordingly.

8.6.4.3 Intraslot time division multiplexing scheme

In the TDM scheme, the time-domain resources occupied by transmission from different TRPs are not overlapped. All the transmission occasions in

Table 8.9 Indication of RV in Scheme 2B.

RV$_{id}$	RV1	RV2
0	0	2
1	2	3
2	3	1
3	1	0

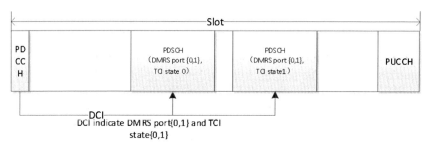

Figure 8.39 TDM transmission scheme in time slot.

one PDSCH use a common MCS with the same DMRS port(s). For the number of repetitions within the single slot, various maximum repetition numbers such as two, four, and seven have been proposed during 3GPP discussions. Transmission schemes with more than two repetitions do not provide more diversity gain than the transmission with only two repetitions, but increase the DMRS overhead. As a result, the maximum repetition number for intraslot TDM scheme is decided to be two. The specific number of repeated transmissions is implicitly indicated by the number of TCI states indicated by the DCI. If the number of TCI states is two, the number of repetitions is two, otherwise it is one. The intraslot TDM scheme use the same the RV indication method as the Scheme 2B and one of those four predefined RV combinations is indicated for each scheduled PDSCH.

In the intraslot TDM scheme, the starting symbol and duration of the first transmission occasion are indicated by SLIV field in DCI. The starting of the second transmission occasion is K symbols after the last symbol of the first transmission occasion whereas K can be configured by RRC or 0 if not configured, and the duration is the same as the first transmission occasion (Fig. 8.39).

8.6.4.4 Interslot time division multiplexing scheme

In the interslot TDM scheme, the number of PDSCH repetitions is indicated through the SLIV field in DCI, which also indicates the time-domain resource allocation of the PDSCH. The network can indicate one or two TCI states for PDSCH transmission in the interslot TDM scheme. Each TCI state corresponds to the transmission opportunities in one or more time slots. The same group of DMRS ports are used in the time-domain resources corresponding to all the indicated TCI states and all the DMRS ports belong to the same CDM group. For one PDSCH in interslot TDM scheme, all the PDSCH transmission occasions in different slots occupy the same FD resources, which are indicated in DCI (Fig. 8.40).

For a PDSCH with interslot repetitions, the UE needs to determine the TCI state and RV for each PDSCH transmission occasion in each slot. Regarding the case when DCI indicates two TCI states, NR discussed two different patterns of mapping between TCI states and transmission occasions: cyclic mapping and sequential mapping. In the method of cyclic mapping, two TCI states are mapped alternatively on the transmission occasions in different slot, which can improve time-domain diversity. In the method of sequential mapping, the first TCI state is associated with the first half of the transmission occasions and the second TCI state are mapped to the remaining transmission occasion when the repetition number is four. If the number of transmission occasions is greater than four, the sequential mapping is applied on every four consecutive transmission occasions. The sequential mapping can reduce the number of receiving beam switching at the UE side, thereby reducing the UE complexity. Considering both mapping methods have their own advantages and disadvantages, the NR finally agreed on to adopt both options and use RRC signaling to switch between these two methods.

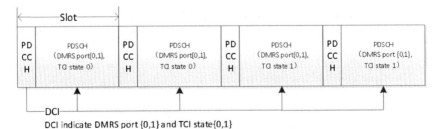

Figure 8.40 Interslot TDM transmission scheme.

8.7 Summary

After the continuous enhancement over multiple releases, large-scale antenna technology has become very mature in LTE systems. On the basis of inheriting the multiantenna technology of the LTE and introducing necessary MIMO enhancements for new NR services, multiantenna technology in NR can achieve higher spectrum efficiency and transmission reliability. For example, by enhancing the CSI feedback and codebook, the network can obtain more accurate downlink channel information for more precise scheduling; by introducing the mechanism of beam management and BFR, analog beam-forming can be widely used in millimeter wave band to provide considerable beam-forming gain to defeat the energy loss; and multi-TRP transmission can improve the throughput and transmission reliability of cell edge users. With the emergence of new application scenarios, the multiantenna technology will be further improved in the subsequent releases.

References

[1] R1−1613175. WF on CSI framework for NR, Samsung, AT&T, NTT DOCOMO, 3GPP RAN1#87, Reno, United States, 14th−18th November 2016.
[2] R1−1701292. WF on CSI framework for NR, Samsung, Ericsson, LG Electronics, NTT DOCOMO, ZTE, ZTE Microelectronics, KT Corporation, Huawei, HiSilicon, Intel Corporation, NR Ad-Hoc, Spokane, United States, 16th−20th January 2017.
[3] R1−1706927. Channel and interference measurement for CSI acquisition, Huawei, HiSilicon, 3GPP RAN1#89, Hangzhou, P.R. China, 15th−19th May 2017.
[4] R1−1708455. On CSI measurement for, NR NTT DOCOMO, INC, 3GPP RAN1#89, Hangzhou, P.R. China, 15th−19th May 2017.
[5] R1−1709295. WF on NZP CSI−RS for interference measurement, Huawei, HiSilicon, Xinwei, MediaTek, AT&T, CeWiT, Intel, ZTE, ZTE Microelectronics, IITH, IITM, China Unicom, Tejas Networks, Softbank, Qualcomm, LGE, Ericsson, KDDI, Deutsche Telekom, Mitsubishi Electric, InterDigital, NEC, SONY, Spreadtrum, China Telecom, CATR, SHARP, 3GPP RAN1#89, Hangzhou, P.R. China, 15th−19th May 2017.
[6] R1−1716901. WF for open issues on CSI reporting, Samsung, Ericsson, Huawei, HiSilicon, ZTE, Sanechips, Mediatek, NTT DOCOMO, Nokia, Nokia Shanghai Bell, KDDI, Vodafone, CEWiT, IITH, IITM, Tejas Networks, Verizon, Deutsche Telekom, Softbank, CHTTL, NEC, WILUS, Sharp, China Unicom, ITL, KRRI, CMCC, ASTRI, KT Corporation, BT, Sprint, LG Electronics, AT&T, NR Ad-Hoc, Nagoya, Japan, 18th−21st September 2017.
[7] R1−1719142. Offline session notes CSI reporting (AI 7.2.2.2), Ericsson, 3GPP RAN1#90bis, Prague, CZ, 9th−13th October 2017.
[8] R1−1714907. Way forward on reciprocity based CSI, ZTE, Ericsson, Samsung, LG Electronics, Nokia, NSB, 3GPP RAN1#90, Prague, Czech Republic, 21st−25th August 2017.

[9] R1−1709232. WF on Type I and II CSI codebooks, Samsung, Ericsson, Huawei, HiSilicon, NTT DOCOMO, Intel Corporation, CATT, ZTE, Nokia, Alc atel-Lucent Shanghai Bell, AT&T, BT, CATR, China Telecom, CHTTL, Deutsche Telekom, Fujitsu, Interdigital, KDDI, Mitsubishi Electric, NEC, OPPO, Reliance Jio, SK Telecom, Sharp, Sprint, Verizon, Xiaomi, Xinwei, CEWiT, IITH, Tejas Networks, IITM, 3GPP RAN1#89, Hangzhou, P.R. China, 15th−19th May 2017.

[10] Final Report of 3GPP TSG RAN WG1 #86bis v1.0.0. Lisbon, Portugal, 10th−14th October 2016.

[11] R1−1701553. Final report of 3GPP TSG RAN WG1 #AH1_NR v1.0.0, Spokane, United States, 16th−20th January 2017.

[12] R1−1700752. Type II CSI feedback, Ericsson, Spokane, WA, United States, 16th−20th January, 2017.

[13] R1−1609012. Linear combination W1 codebook, Samsung, Lisbon, Portugal 10th−14th October 2016.

[14] R1−1609013. Linear combination W2 codebook, Samsung, Lisbon, Portugal 10th−14th October 2016.

[15] R1−1700415. Design for Type II feedback, Huawei, HiSilicon, Spokane, United States, 16th−20th January 2017.

[16] R1−1709232. WF on Type I and II CSI codebooks, Hangzhou, China, 15th−19th May 2017.

[17] R1−1705899. Type II CSI feedback, Ericsson, Spokane, United States, 3rd−7th April 2017.

[18] R1−1705076. Design for Type II Feedback, Huawei, HiSilicon, Spokane, United States, 3rd−7th April 2017.

[19] R1−1713590. Remaining details of Type I and Type II CSI codebooks, Samsung, Prague, P. R. Czechia, 21th−25th August 2017.

[20] R1−1708688. Codebook design for Type II CSI feedback, Ericsson, Hangzhou, China, 15th−19th May, 2017.

[21] R1−1716505. Reduced PMI payload in the NR Type II codebooks, Nokia, Nokia Shanghai Bell, Nagoya, Japan, 18th−21st September 2017.

[22] R1−1716349. On CSI reporting, Ericsson, Nagoya, Japan, 18th−21st September 2017.

[23] R1−1707127. Type II CSI feedback based on linear combination, ZTE, Hangzhou, China, 15th−19th May 2017.

[24] R1−1718886. WF on omission rules for partial Part 2, Prague, Czeck Republic, 9th−13th October 2017.

[25] Final report of 3GPP TSG RAN WG1 #94bis. Chengdu, China, 8th−12th October 2018.

[26] R1−1811276. CSI enhancement for MU-MIMO support, Qualcomm Incorporated, 3GPP RAN1#94bis, Chengdu, China, 8th−12th October 2018.

[27] R1−1810884. CSI enhancement for MU-MIMO, Samsung, 3GPP RAN1#94bis, Chengdu, China, 8th−12th October 2018.

[28] R1−1811654. Summary of CSI enhancement for MU-MIMO support, Samsung, 3GPP RAN1#94bis, Chengdu, China, 8th−12th October 2018.

[29] Final report of 3GPP TSG RAN WG1 #95. Spokane, United States, 12th−16th November 2018.

[30] R1−1812242. Discussion on CSI enhancement, Huawei, HiSilicon, 3GPP RAN1#95, Spokane, United States, 12th−16th November 2018.

[31] R1−1813913. CSI enhancements for MU-MIMO support, ZTE, 3GPP RAN1#95, Spokane, United States, 12th−16th November 2018.

[32] R1−1813002. Summary of CSI enhancement for MU-MIMO, Samsung, 3GPP RAN1#95, Spokane, United States, 12th−16th November 2018.

[33] R1−1813441. CSI enhancement for MU-MIMO support, Qualcomm Incorporated, 3GPP RAN1#95, Spokane, United States, 12th−16th November 2018.

[34] R1−1901276. Samsung, CSI enhancement for MU-MIMO, 3GPP RAN1#AH1091, Taipei, Taiwan, 21st−25th January 2019.

[35] R1−1900904. CSI enhancement for MU-MIMO support, Qualcomm Incorporated, 3GPP RAN1#AH1091, Taipei, Taiwan, 21st−25th January 2019.

[36] R1−1900265. Enhancements on overhead reduction for type II CSI feedback, OPPO, 3GPP RAN1#AH1091, Taipei, Taiwan, 21st−25th January 2019.

[37] Final report of 3GPP TSG RAN WG1 #97. Reno, United States, May 13−17, 2019.

[38] Final report of 3GPP TSG RAN WG1 #AH_1901 v1.0.0. Taipei, Taiwan, 21st−25th January 2019.

[39] R1−1901075. Summary of CSI enhancement for MU-MIMO, Samsung, 3GPP RAN1#AH1091, Taipei, Taiwan, 21st−25th January 2019.

[40] R1−1902700. Discussion on overhead reduction for type II CSI feedback, OPPO, 3GPP RAN1#96, Athens, Greece, 25th February−1st March 2019.

[41] R1−1902123. Enhancements on Type-II CSI reporting, Fraunhofer IIS, Fraunhofer HHI, 3GPP RAN1#96, Athens, Greece, 25th February−1st March 2019.

[42] R1−1901701. Further discussion on type II CSI compression and feedback parameters, vivo, 3GPP RAN1#96, Athens, Greece, 25th February−1st March 2019.

[43] R1−1903343. CSI enhancements for MU-MIMO support, ZTE, 3GPP RAN1#96, Athens, Greece, 25th February−1st March 2019.

[44] R1−1900690. Nokia, CSI enhancements for MU-MIMO, 3GPP RAN1#AH1091, Taipei, Taiwan, 21st−25th January 2019.

[45] R1−1901566. Discussion on CSI enhancement, Huawei, HiSilicon, 3GPP RAN1#96, Athens, Greece, 25th February−1st March 2019.

[46] R1−1903038. On CSI enhancements for MU-MIMO, Ericsson, 3GPP RAN1#96, Athens, Greece, 25th February−1st March 2019.

[47] R1−1902501. Type II CSI feedback compression, Intel Corporation, 3GPP RAN1#96, Athens, Greece, 25th February−1st March 2019.

[48] Final report of 3GPP TSG RAN WG1 #96. Reno, United States, May 13−17, 2019.

[49] R1−1902304. Summary of CSI enhancement for MU-MIMO, Samsung, 3GPP RAN1#96, Athens, Greece, 25th February−1st March 2019.

[50] R1−1902018. Discussions on Type II CSI enhancement, CATT, 3GPP RAN1#96, Athens, Greece, 25th February−1st March 2019.

[51] Final report of 3GPP TSG RAN WG1 #96 bis. Xi'an, China, 12th−16th April 2019.

[52] R1−1905724. Feature lead summary for MU-MIMO CSI Tuesday offline session, Samsung, 3GPP RAN1#96bis, Xi'an, China, 12th−16th April 2019.

[53] Final report of 3GPP TSG RAN WG1 #98. Prague, Czech Republic, 26−30 August 2019.

[54] Final report of 3GPP TSG RAN WG1 #86. Lisbon, Portugal, 10th−14th October 2016.

[55] R1−1707953. Downlink beam management details, Samsung, 3GPP RAN1#89, Hangzhou, China, 15th−19th May 2017.

[56] Final report of 3GPP TSG RAN WG1 #88bis.

[57] R1−1706733. WF on use of SS blocks in beam management, Qualcomm, LG, AT&T, Ericsson, Xinwei, Oppo, IITH, CEWiT, Tejas Networks, IITM, ZTE, 3GPP TSG RAN1 #88bis, Spokane, United States, 3rd−7th April 2017.

[58] R1−1706457. WF on beam measurement RS, Samsung, 3GPP RAN1 Meeting #88bis, Spokane, United States, 3rd−7th April 2017.

[59] R1−1901084. Evaluation on SINR metrics for beam selection, Samsung, 3GPP RAN1 AH-1901, Taipei, Taiwan, 21st−25th January 2019.

[60] R1−1901204. Performance of beam selection based on L1-SINR, Ericsson, 3GPP TSG RAN1 Ad-Hoc Meeting 1901, Taipei, Taiwan, 21st−25th January 2019.

[61] R1−1902503. On beam management enhancement, Intel, 3GPP RAN1 #96, Athens, Greece, 25th February−1st March 2019.

[62] Final report of 3GPP TSG RAN WG1 #92.

[63] R1−1700122. Group based beam management, ZTE, ZTE Microelectronic, 3GPP RAN1 NR Ad-Hoc Meeting, Spokane, United States, 16th−20th January 2017.

[64] R1−1710183. Discussion on DL beam management, ZTE, 3GPP RAN1 NR Ad-Hoc#2, Qingdao, P.R. China, 27th−30th June 2017.

[65] R1−1705342. DL beam management details, Samsung, 3GPP TSG RAN1#88bis, Spokane, United States, 3rd−7th April 2017.

[66] Final report of 3GPP TSG RAN WG1 #86bis. Lisbon, Portugal, 10th−14th October 2016.

[67] Final report of 3GPP TSG RAN WG1 #88bis. Spokane, United States, 3rd−7th April 2017.

[68] R1−1717606. Beam failure recovery, Samsung, 3GPP RAN1# 90bis, Prague, CZ, 9th−13th October 2017.

[69] R1−1718434. Basic beam recovery, Ericsson, 3GPP RAN1#90bis, Prague, CZ, 9th−13th October 2017.

[70] R1−1705893. Beam failure detection and beam recovery actions, Ericosson, 3GPP RAN1#88bis, Spokane, United States, 3rd−7th April 2017.

[71] R1−1715012. Offline discussion on beam recovery mechanism, MediaTek, 3GPP RAN1#90, Prague, Czech, 21th−25th August 2017.

[72] R1−1807661. Summary 1 on remaing issues on beam failure recovery, MediaTek, 3GPP RAN1#93, Busan, Korea, 21th−25th May 2018.

[73] R1−1807725. DRAFT reply LS on beam failure recovery, RAN1, 3GPP RAN1 #93, Busan, Korea, 21th−25th May 2018.

[74] R1−1911549. Feature lead summary 3 on SCell BFR and L1-SINR, Apple, 3GPP RAN1#98bis Chongqing, China, 14th−20th October 2019.

[75] R1−1903650. Summary on SCell BFR and L1-SINR, Intel, 3GPP RAN1 #96, Athens, Greece, 25th February−1st March 2019.

[76] R1−1906029. Enhancements on multi-TRP/panel transmission, Huawei, HiSilicon, 3GPP RAN1#97, Reno, United States, 13th−17th May 2019.

[77] R1−1906345. On multi-TRP/panel transmission, CATT, 3GPP RAN1#97, Reno, United States, 13rd−17th May 2019.

[78] R1−1905513. On multi-TRP/panel transmission, Ericsson, 3GPP RAN1#96b, Xi'an, China, 8th−12th April, 2019.

[79] R1−1906738. Discussion on DMRS port indication for NCJT, LG electronics, 3GPP RAN1#97, Reno, United States, 13rd−17th May 2019.

[80] R1−1905166. NC−JT performance with layer restriction between TRPs, Ericsson, 3GPP RAN1#96bis, Xi'an, China, 8th−12th April 2019.

[81] R1−1907289. Multi-TRP enhancements, Qualcomm Incorporated, 3GPP RAN1#97, Reno, United States, 13rd-17th May 2019.

[82] R1−1909465. On multi-TRP and multi-panel, Ericsson, 3GPP RAN1#98, Prague, Czech Republic, 26th−30th August 2019.

[83] R1−1908501. Enhancements on multi-TRP/panel transmission, Samsung, 3GPP RAN1#98, Prague, Czech Republic, 26th−30th August 2019.

[84] R1−1901567. Enhancements on multi-TRP/panel transmission, Huawei, HiSilicon, 3GPP RAN1#96, Athens, Greece, 25th February−1st March 2019.

[85] R1−1902019. Consideration on multi-TRP/panel transmission, CATT, 3GPP RAN1#96, Athens, Greece, 25th February−1st March 2019.

[86] R1−1902091. Enhancements on multi-TRP/panel transmission, LG Electronics, 3GPP RAN1#96, Athens, Greece, 25th February−1st March 2019.

[87] R1−1902502. On multi-TRP/panel transmission, Intel Corporation, 3GPP RAN1#96, Athens, Greece, 25th February−1st March 2019.

[88] R1−2005110. RAN1 UE features list for R16 NR updated after RAN1#101-e, Moderators, 3GPP RAN1#101-e, e-Meeting, 25th−June 5th May 2020.

[89] Chairman's notes RAN1#96 final, 3GPP RAN1#96, Athens, Greece, 25th February−1st March 2019.

[90] R1−1904013. Enhancements on multi-TRP and multi-panel transmission, ZTE, 3GPP RAN1#96b, Xi'an, China, 8th−12th April 2019.

[91] R1−1905026. Multi-TRP enhancements, Qualcomm Incorporated, 3GPP RAN1#96b, Xi'an, China, 8th−12th April 2019.

[92] R1−1906029. Enhancements on multi-TRP enhancements, Huawei, HiSilicon, 3GPP RAN1#97, Reno, United States, 13rd−17th May 2019.

[93] R1−1906274. Discussion of multi-TRP enhancements, Lenovo, Motorola Mobility, 3GPP RAN1#97, Reno, United States, 13rd−17th May 2019.

[94] Final_Minutes_report_RAN1#98b_v200. 3GPP RAN1#98b, Chongqing, China, 14th−20th October 2019.

[95] R1−1910073. Enhancements on multi-TRP enhancements, Huawei, HiSilicon, 3GPP RAN1#98b, Chongqing, China, 14th−20th October 2019.

[96] R1−1910865. Remaining issues for mTRP, Ericsson, 3GPP RAN1#98b, Chongqing, China, 14th−20th October 2019.

[97] R1−1911126. Multi-TRP enhancements, Qualcomm, 3GPP RAN1#98b, Chongqing, China, 14th−20th October 2019.

[98] R1−1900017. Enhancements on multi-TRP, Huawei, HiSilicon, 3GPP RAN1 AH-1901, Taipei, 21th−25th January 2019.

[99] R1−1900728. On multi-mRP and multi-panel, Ericsson, 3GPP RAN1 AH-1901, Taipei, 21th−25th January 2019.

[100] Final_Minutes_report_RAN1#AH_1901_v100, 3GPP RAN1 AH-1901, Taipei, 21th−25th January 2019.

[101] R1−1812256. Enhancements on multi-TRP and multi-panel transmission, ZTE, 3GPP RAN1#95, Spokane, United States, 12th−16th November 2018.

[102] R1−1813698. Evaluation results for multi-TRP/panel transmission with higher reliability/robustness, Huawei, HiSilicon, 3GPP RAN1#95, Spokane, United States, 12th−16th November 2018.

Further reading

Final report of 3GPP TSG RAN WG1 #93, 2018 Final report of 3GPP TSG RAN WG1 #93, Busan, Korea, 21st−25th May 2018.

R1−1903501, 2019 R1−1903501. Summary of offline email discussion on MU-MIMO CSI, Samsung, 3GPP RAN1#96, Athens, Greece, 25th February−1st March 2019.

CHAPTER 9

5G radio-frequency design

Jinqiang Xing, Zhi Zhang, Qifei Liu, Wenhao Zhan, Shuai Shao and Kevin Lin

9.1 New frequency and new bands

The radio-frequency (RF) spectrum is the first topic to be considered in the design of a 5G NR system. The radio-frequency spectrum and its associated electromagnetic wave propagation characteristics, to a great extent, determine the key technical features of 5G NR wireless communication. At present, 2G, 3G, and 4G mobile communication systems occupy the low-frequency spectrum mostly below 3 GHz, and in the same frequency range there is little spectrum available for the 5G new radio (NR) system. In order to deliver a much higher data rate transmission for the next generation of mobile communication system, at the beginning of the 5G NR system design, the target radio-frequency spectrum was set to be the high-frequency spectrum with much more spectrum available, such as the millimeter wave spectrum above 3 GHz.

9.1.1 Spectrum definition

Because there are significant differences in RF and antenna designs between the millimeter-wave spectrum and the low-frequency spectrum (below 7.125 GHz), the 3GPP divides the frequency spectrum into frequency range 1 (FR1) and frequency range 2 (FR2) as shown in Table 9.1.

9.1.1.1 New band definition

The FR1 spectrum includes all the existing frequency bands occupied by the current 2G, 3G, and 4G mobile communication systems, as well as some newly defined frequency bands for 5G. It is expected that some frequency bands in FR1 might be used for global roaming as shown in Table 9.2.

Table 9.1 Frequency-range definition.

	Frequency range
FR1	410−7125 MHz
FR2	24,250−52,600 MHz

5G NR and Enhancements
DOI: https://doi.org/10.1016/B978-0-323-91060-6.00009-X

Table 9.2 Typical FR1 bands.

Band	UL	DL	Mode
n77	3300−4200 MHz	3300−4200 MHz	TDD
n78	3300−3800 MHz	3300−3800 MHz	TDD
n79	4400−5000 MHz	4400−5000 MHz	TDD

TDD, Time division duplex.

For the above spectrum bands, demand for the n79 band is mainly in China and Japan, as such it is less likely to be used as a global roaming band. The n77 and n78 bands are the most promising bands for global roaming in 5G NR.

In the process of formulating the 5G NR standards, the division of the n77 and n78 bands was made to be compatible with the needs and regulatory requirements of different regions and countries. At the beginning of the standardization process, the spectrum allocation in China and Europe was mainly concentrated in the range from 3.3 GHz to 3.8 GHz, while Japan targeted 3.3 GHz to 4.2 GHz for 5G use. Therefore in order to achieve a unified frequency band for global roaming and to also rely on economies of scale to reduce equipment costs, operators in Japan intended to combine the spectrum range from 3.3 GHz to 4.2 GHz with the spectrum range from 3.3 GHz to 3.8 GHz and define it as one single-frequency band. However, Japan has restrictions on terminal transmission power to protect radar and satellite services (i.e., high-power terminals cannot be used for transmission). On the contrary, high-power UE is a very important feature to improve uplink coverage in China and Europe. In view of this, the n77 and n78 bands are defined separately in the 3.5 GHz spectrum.

At present, there are a few radio systems deployed in FR2, which makes it possible to utilize a larger bandwidth for 5G NR to achieve a higher throughput than FR1. According to the definition of Shannon channel capacity, the channel capacity is proportional to the channel bandwidth when the Signal-Noise Ratio (SNR) is constant. The direct benefit of increasing bandwidth is the increase of terminal peak data rate, but it is difficult to find spectrum large enough for NR in FR1. Therefore the use of millimeter wave spectrum [which refers to the millimeter wave band above 24 GHz in 3rd Generation Partnership Project (3GPP) Release 15] is highly anticipated by the industry due to its large available bandwidth. For the 5G NR system, the maximum channel bandwidth is increased to 100 MHz in FR1 (below 7.125 GHz in Release 15) and 400 MHz in FR2.

For the allocation of frequency spectrum in 3GPP Release 15, FR2 defines four frequency bands, n257, n258, n260, and n261, according to the needs of different countries/regions. In Release 16, the n259 band, namely 39.5−43.5 GHz is newly added (Table 9.3).

9.1.1.2 Refarming of existing bands

Many frequency bands are defined in the 2G, 3G, and 4G era for their own use. But in fact, with the deployment and application of a new communication system, users of the old communication system will gradually migrate to the new communication system, resulting in a gradual reduction in the number of remaining users still in operation using the old communication system. As such, the value of the old communication systems is also gradually reduced. For the occupied frequency bands where the utilization is not high anymore, operators will gradually refarm them based on the actual demand and reutilize them to deploy 5G systems.

In the process of standardization, the frequency-band refarming is often done much earlier than the withdrawal of the old communication system and the introduction of the new communication system. Thus it is ensured that at some point in the future, when the operator decides to replace the old communication system with a new communication system, the spectrum is ready and can be used directly. Based on this situation, most of the existing 2G, 3G, and 4G bands are refarmed in Release 15. As such, the definition of a frequency band for NR is generally consistent with the existing frequency band in the frequency range, but the subcarrier spacing (SCS), channel raster, and other system parameters may be different.

The typical refarming band is B41 (2496−2690 MHz), and it has been deployed in China, Japan, and the United States. There is large amount

Table 9.3 FR2 band definitions.

Band	UL/DL			Mode
n257	26,500 MHz	−	29,500 MHz	TDD
n258	24,250 MHz	−	27,500 MHz	TDD
n259	39,500 MHz	−	43,500 MHz	TDD
n260	37,000 MHz	−	40,000 MHz	TDD
n261	27,500 MHz	−	28,350 MHz	TDD

of frequency spectrum that can be used by the NR system even if the existing long term evolution (LTE) network is still deployed. Therefore B41 is naturally refarmed to the n41 band.

For most of the refarming frequency bands, the 100 kHz channel raster from LTE is reused for NR in order to keep alignment between them. But the n41 band is different as its channel raster is based on the assigned SCS (15 or 30 kHz), which can potentially provide a higher system spectrum efficiency, but it also leads to incompatibility with the LTE system. In order to avoid the interference between different systems in the same frequency band, it is necessary to have a certain protection frequency between LTE and NR systems. Later when the 5G spectrum allocation in each region becomes clearer, this problem will gradually emerge and make it difficult for operators to carry out accurate dynamic spectrum sharing of LTE and NR systems at the RB level. Finally, 3GPP has to define n90 band that has same frequency as n41 to solve the problem of coexistence between the NR system and LTE system.

9.1.2 Band combination

There are currently already a large number of band combinations in the NR, including CA, DC, EN−DC, and so on. With frequency-band combination, new frequency bands or the refarming of frequency bands mentioned above can be combined according to the actual demand to achieve higher peak data rates.

9.1.2.1 CA band combination

NR CA in terms of combining different parts of frequency spectrum basically follows a mechanism similar to LTE, including intraband continuous CA, intraband noncontinuous CA, interband CA, and so on. For the NR system, a variety of CA bandwidth classes have been defined to cater to the different bandwidth aggregation demands in actual network deployment. But in the NR, the bandwidth combination is more complex and a new CA bandwidth fallback group is defined. Bandwidth classes in the same CA fallback group can fall back from high-bandwidth class to low-bandwidth class. In the actual network, the base station configures a number of carriers and aggregates those bandwidths for the terminal according to the needs. Tables 9.4 and 9.5 list the CA bandwidth classes defined in Release 15 for FR1 and FR2, respectively.

Table 9.4 NR FR1 CA bandwidth class.

NR CA bandwidth class	Aggregated channel bandwidth	Number of contiguous CC	Fallback group
A	$BWChannel \leq BWChannel,max$	1	1, 2
B	$20\ MHz \leq BWChannel\_CA \leq 100\ MHz$	2	2
C	$100\ MHz < BWChannel\_CA \leq 2\ x$ $BWChannel,max$	2	1
D	$200\ MHz < BWChannel\_CA \leq 3\ x$ $BWChannel,max$	3	
E	$300\ MHz < BWChannel\_CA \leq 4\ x$ $BWChannel,max$	4	
G	$100\ MHz < BWChannel\_CA \leq 150\ MHz$	3	2
H	$150\ MHz < BWChannel\_CA \leq 200\ MHz$	4	
I	$200\ MHz < BWChannel\_CA \leq 250\ MHz$	5	
J	$250\ MHz < BWChannel\_CA \leq 300\ MHz$	6	
K	$300\ MHz < BWChannel\_CA \leq 350\ MHz$	7	
L	$350\ MHz < BWChannel\_CA \leq 400\ MHz$	8	

BW, Bandwidth; *CA*, carrier aggregation; *CC*, component carrier; *DL*, downlink; *EIRP*, effective isotropic radiated power; *FCC*, Federal Communications Commission; *UE*, user equipment; *IEC*, International Electro Technical Commission; *ISO*, International Organization for Standardization; *MIMO*, multiple input multiple output; *OTA*, over the air; *UL*, uplink.

9.1.2.2 EN−DC band combination

As introduced in a previous chapter, 3GPP defines two network structures, namely SA and NSA, according to the difference in base stations and core networks. In the NSA network, the terminal needs to maintain the connection with the LTE and NR base stations at the same time.

According to whether the LTE band and NR band are the same, the EN−DC band combinations is divided into intraband EN−DC and interband EN−DC. In the intraband EN−DC, it is further divided into intraband contiguous EN−DC and intraband noncontiguous EN−DC according to whether the LTE carrier and NR carrier are contiguous or not. For example, DC_3_n78 is inter band EN−DC, DC_(n)41 is intraband contiguous EN−DC, and DC_41_n41 is intraband noncontinuous EN−DC.

9.1.2.3 Bandwidth combination set

A bandwidth combination set is used to describe the bandwidth combinations that can be supported by different frequency bands in a combination of CA, DC, EN−DC, or NE−DC. The UE can clearly inform the base station of the bandwidth combination it supports by reporting the

Table 9.5 NR FR2 CA bandwidth class.

NR CA bandwidth class	Aggregated channel bandwidth	Number of contiguous CC	Fallback group
A	BWChannel $\leq$ 400 MHz	1	1, 2, 3, 4
B	400 MHz < BWChannel_CA $\leq$ 800 MHz	2	1
C	800 MHz < BWChannel_CA $\leq$ 1200 MHz	3	
D	200 MHz < BWChannel_CA $\leq$ 400 MHz	2	2
E	400 MHz < BWChannel_CA $\leq$ 600 MHz	3	
F	600 MHz < BWChannel_CA $\leq$ 800 MHz	4	
G	100 MHz < BWChannel_CA $\leq$ 200 MHz	2	3
H	200 MHz < BWChannel_CA $\leq$ 300 MHz	3	
I	300 MHz < BWChannel_CA $\leq$ 400 MHz	4	
J	400 MHz < BWChannel_CA $\leq$ 500 MHz	5	
K	500 MHz < BWChannel_CA $\leq$ 600 MHz	6	
L	600 MHz < BWChannel_CA $\leq$ 700 MHz	7	
M	700 MHz < BWChannel_CA $\leq$ 800 MHz	8	
O	100 MHz $\leq$ BWChannel_CA $\leq$ 200 MHz	2	4
P	150 MHz $\leq$ BWChannel_CA $\leq$ 300 MHz	3	
Q	200 MHz $\leq$ BWChannel_CA $\leq$ 400 MHz	4	

bandwidth combination set, and then it is used for the base station to configure the bandwidth for different UEs according to its capabilities.

Taking EN—DC as an example, Table 9.6 shows the optional bandwidth combination set supported by the UE under the intraband contiguous EN—DC between LTE B41 and NR n41.

9.2 FR1 UE radio-frequency

Generally speaking, the RF technology for 5G in FR1 is similar to LTE, and still uses the conduct-based requirements. However, NR has larger channel bandwidth and more complex EN—DC structure than LTE. The

Table 9.6 EN−DC bandwidth combination sets.

DL EN−DC Configuration	UL EN−DC Configuration	Component carriers in order of increasing carrier frequency (MHz)			Maximum aggregated channel BW (MHz)	Bandwidth combination set
		E−UTRA Channel BW	NR Channel BW	E−UTRA Channel BW		
DC_(n)41AA	DC_(n)41AA	20	40, 60, 80,100 40, 60, 80,100	20	120	0
		20	40, 50, 60, 80, 100 40, 50, 60, 80,100	20	120	1

Table 9.7 UE-power class.

Power class	PC 1	PC 1.5	PC 2	PC 3
Max output power (dBm)	31	29	26	23

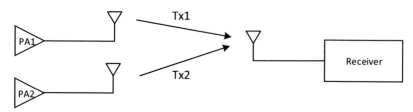

Figure 9.1 Transmit diversity.

following sections mainly discuss the key features and aspects of 5G NR in FR1 (i.e., high-power UE, reference sensitivity, and interference).

9.2.1 High-power UE

9.2.1.1 UE-power class

The UE-power class is used to describe the maximum transmit power capability of the UE. Different power classes are defined for UEs with different power capabilities in the 3GPP standard, as shown in Table 9.7. Additionally, power tolerance is usually introduced considering the accuracy of the terminal transmit power accuracy. The power tolerance level is different for different frequency bands; usually the upper and lower tolerance is +2 dB/−3dB. For some frequency bands, the lower limit may be −2.5 dB or −2dB, which is defined in detail in 3GPP TS38.101−1.

In the above UE-power class table, PC3 is usually considered as the normal and most commonly used power class, and the power higher than PC3 is usually called the high-power class.

In addition, power class is also defined for band combinations like CA and EN−DC as the maximum transmit power that can be achieved when multiple carriers transmit.

The power class of a UE will be reported to the network together with other wireless capabilities during the initial access to the network. If the UE does not report the power-class capability, then the default power class is used.

9.2.1.2 Transmit diversity

Transmit diversity has been widely used since LTE, but it is not defined in the 3GPP NR physical layer protocol and it is up to UE implementation to decide; in other words, it is transparent to the network. The typical UE implementation is shown in Fig. 9.1. Of course, the complete transparency of the 3GPP standard is difficult to achieve, especially when defining UE RF requirements.

A typical example is that the UE achieves the power level of PC2 by using two PC3 PAs. Then for this UE, the PC2 can be achieved under UL MIMO, while the power-class capability under a single-antenna port operation depends on whether the UE supports transmit diversity or not. For UEs that support transmit diversity, PC2 can still be achieved. However, for a UE that does not support transmit diversity, its max output power level is actually PC3 when the network configures the UE for single-antenna port transmission, due to only one PA working.

With this example, it can be seen that although transmit diversity is transparent in the physical layer standard, it has a nonnegligible impact on the definition of RF requirements and subsequent conformance testing. Especially in the 3GPP Release 15 standard, the UE will only report one power class to the network, but there is no clear definition of how to report the power class when the maximum power capability is different between single-antenna port and two-antenna port operation.

At the beginning of the UE RF requirement definition, the whole RF requirement is divided into RF requirements under the single-antenna port and UL MIMO. In the single-antenna port, UE works with one physical transmitting antenna or two physical antennas with transmit diversity. However, at the beginning of Release 15, transmit diversity did not draw the group's attention at 3GPP, so the RF requirements were defined with the assumption of one physical antenna. After that, the discussion about transmit diversity became more and more at the end of Release 15 and triggered even more discussions after Release 15 was frozen. The main problem is how to avoid big changes to the Release 15 standard, and at the same time still compatible with different UE implementations like UE achieve PC2 through two PC3 PAs.

9.2.1.3 High-power UE and specific absorption rate solutions

As mentioned earlier, high-power UE transmits higher power to improve uplink coverage, but at the same time it may also cause some associated problems, such as safety impact to human body due to higher electromagnetic radiation. Standard organizations such as ISO, IEC, and FCC have defined their Specific Absorption Rate (SAR) requirements. As such, all kinds of wireless transmitting terminal equipment need to comply with the requirements.

SAR is the average value of electromagnetic radiation absorbed by the human body over a period of time when a UE transmits and maintains its maximum power level, as such SAR is closely related to the transmission power and time duration. The higher the transmission power and the longer the transmission time, the greater the possibility of SAR exceeding the standard. Consequently, high-power UEs such as PC1, PC1.5, and PC2 are facing challenges in meeting the SAR requirements, so it is necessary to consider how to resolve this issue for high-power UEs.

LTE high-power UEs are able to meet the SAR requirements by restricting the uplink and downlink configuration in a cell. Since the max transmit power of a PC2 UE is doubled, then the corresponding transmission time should be reduced by 50%. Based on this emission control mechanism, LTE PC2 high-power UEs are only suitable for uplink and downlink configurations with less than 50% UL transmission time allocation (i.e., uplink and downlink configurations 0 and 6 in Table 9.8 are naturally excluded). When the uplink and downlink configuration of the network is 0 or 6, the high-power UE falls back to PC3.

Following the above SAR solution idea for LTE high-power UEs, the NR tried to limit the uplink and downlink configuration for SA high-power UEs at the initial stage. But unlike LTE, the uplink and downlink configurations in NR can be flexibly scheduled by the network, and there is no fixed uplink and downlink configuration as in LTE. As shown in Table 9.9, the NR network can flexibly schedule UEs for uplink transmission or downlink

Table 9.8 LTE UL and DL configuration.

LTE Uplink-downlink configuration	Subframe number									
	0	1	2	3	4	5	6	7	8	9
0	D	S	U	U	U	D	S	U	U	U
1	D	S	U	U	D	D	S	U	U	D
2	D	S	U	D	D	D	S	U	D	D
3	D	S	U	U	U	D	D	D	D	D
4	D	S	U	U	D	D	D	D	D	D
5	D	S	U	D	D	D	D	D	D	D
6	D	S	U	U	U	D	S	U	U	D

Table 9.9 New radio UL and DL configuration.

NR Slot Format	Symbol number in a slot													
	0	1	2	3	4	5	6	7	8	9	10	11	12	13
0	D	D	D	D	D	D	D	D	D	D	D	D	D	D
1	U	U	U	U	U	U	U	U	U	U	U	U	U	U
2	X	X	X	X	X	X	X	X	X	X	X	X	X	X
3	D	D	D	D	D	D	D	D	D	D	D	D	D	X
4	D	D	D	D	D	D	D	D	D	D	D	D	X	X
5	D	D	D	D	D	D	D	D	D	D	D	X	X	X
56	D	X	U	U	U	U	U	D	X	U	U	U	U	U
57	D	D	D	D	X	X	U	D	D	D	D	X	X	U
58	D	D	X	X	U	U	U	D	D	X	X	U	U	U
59	D	X	X	U	U	U	U	D	X	X	U	U	U	U
60	D	X	X	X	X	X	U	D	X	X	X	X	X	U
61	D	D	X	X	X	X	U	D	D	X	X	X	X	U
62−255	Reserved													

reception on the X symbol. While the operators want to retain the scheduling flexibility, UE manufacturers prefer to ensure the security of the terminal, and hence different views are expressed on the use of X in calculating the uplink duty cycle. As such, it is difficult to reach an agreement between operators and terminal manufacturers in excluding some parts of the uplink and downlink configurations.

Finally, a solution is introduced in 3GPP based on the UE reporting of maximum uplink duty cycle in Release 15. The mechanism of this scheme is not to exclude any uplink and downlink configuration, but to allow a high-power UE to fall back to PC3 when it detects that the uplink configuration of network scheduling exceeds its maximum uplink duty cycle. However, the price is that the UE needs to keep track of a statistic proportion of the uplink transmission time in real-time, which might have an impact on the power consumption of the UE.

The above SAR solution for SA is widely used to solve other high-power UE in scenarios like EN−DC LTE TDD band + NR TDD band high-power UE and EN−DC LTE FDD band + NR TDD band high-power UE. However, the difference is that an EN−DC UE needs to consider its transmission power in both the LTE and NR bands.

For a UE which is PC2 under LTE TDD + NR TDD band combination, usually it has the ability to achieve 23dBm in both the LTE TDD band

Table 9.10 EN−DC LTE TDD band + NR TDD band duty cycle capability.

LTE UL and DL configuration	0	1	2	3	4	5	6
NR max Uplink duty cycle capability	Capability 0	Capability 1	Capability 2	Capability 3	Capability 4	Capability 5	Capability 6

and NR TDD band, respectively. When the UE tries to connect to the LTE cell during the initial access, it will first read the system broadcast message and acquire the uplink and downlink configuration information of the cell, then further obtain the maximum available transmission time in the NR TDD band to meet the SAR requirement that is associated with the uplink and downlink configuration of the LTE TDD band (Table 9.10).

9.2.2 Reference sensitivity

Reference sensitivity is the most basic and important receiver requirement for a mobile communication UE. The definition of reference sensitivity requirement for FR1 follows the existing sensitivity calculation method. More specifically, it is determined on the basis of taking the thermal noise level as the baseline, taking into account of UE demodulation threshold, UE receiver noise figure, and UE multiantenna diversity gain, and then further reserving a certain implementation margin. The specific calculation formula is as follows:

$$\text{REFSENS (dBm)} = -174 + 10 * \lg(\text{Rx BW}) + 10 * \lg(\text{SU}) \\ + \text{NF} - \text{diversity gain} + \text{SNR} + \text{IM} \quad (9.1)$$

In the above calculation formula (9.1), Rx BW is the receiving bandwidth of the UE; Spectrum Utilization (SU) is the full RB spectrum efficiency; NF (Noise Figure) is related to the receiver noise figure for a given frequency band; Diversity gain is the diversity reception gain of the UE with two antennas; SNR is the baseband demodulation threshold; and Implementation Margin (IM) is the UE implementation margin.

With the above calculation formula (Eq. 9.1), the UE reference sensitivity requirement under different bandwidths in each frequency band can be obtained. Table 9.11 shows the reference sensitivity of the two−antenna receiver. For the reference sensitivity requirement of a four-antenna receiver, the sensitivity gain caused by the additional two receiving antennas can be further considered, as shown in Table 9.12.

Table 9.11 Reference sensitivity for UEs with two antennas.

Band	SCS (kHz)	10 MHz (dBm)	15 MHz (dBm)	20 MHz (dBm)	40 MHz (dBm)	50 MHz (dBm)	60 MHz (dBm)	80 MHz (dBm)	90 MHz (dBm)	100 MHz (dBm)
n41	15	−94.8	−93.0	−91.8	−88.6	−87.6				
	30	−95.1	−93.1	−92.0	−88.7	−87.7	−86.9	−85.6	−85.1	−84.7
	60	−95.5	−93.4	−92.2	−88.9	−87.8	−87.1	−85.6	−85.1	−84.7
n77	15	−95.3	−93.5	−92.2	−89.1	−88.1				
	30	−95.6	−93.6	−92.4	−89.2	−88.2	−87.4	−86.1	−85.6	−85.1
	60	−96.0	−93.9	−92.6	−89.4	−88.3	−87.5	−86.2	−85.7	−85.2
n78	15	−95.8	−94.0	−92.7	−89.6	−88.6				
	30	−96.1	−94.1	−92.9	−89.7	−88.7	−87.9	−86.6	−86.1	−85.6
	60	−96.5	−94.4	−93.1	−89.9	−88.8	−88.0	−86.7	−86.2	−85.7
n79	15				−89.6	−88.6				
	30				−89.7	−88.7	−87.9	−86.6		−85.6
	60				−89.9	−88.8	−88.0	−86.7		−85.7

Table 9.12 Diversity gain of four antennas comparing to two antennas.

Band	$\Delta R_{IB,4R}$ (dB)
n1, n2, n3, n40, n7, n34, n38, n39, n41, n66, n70	-2.7
n77, n78, n79	-2.2

9.2.3 Interference

Receiver sensitivity degradation means that the receiver is impacted by interference or noise, which results in a certain degradation to the sensitivity of the receiver. There are many factors in the NR that will cause sensitivity degradation like the harmonics or intermodulation interference in EN–DC or interband CA. In the following, the EN–DC is taken as an example to give a brief introduction.

Usually, the interference in the UE mainly comes from the nonlinearity of the RF front-end components. Nonlinear components can be divided into two categories (i.e., passive and active), and the harmonics and intermodulation interference generated by passive components are generally weaker than those of active components. PA is the main source of nonlinearity in active components.

The following is the Taylor series expansion describing the input and output signals of nonlinear devices:

$$y = f(v) = a_0 + a_1 v + a_2 v_2 + a_3 v_3 + a_4 v_4 + a_5 v_5 + \ldots \qquad (9.2)$$

where v is the input signal and y is the output signal.

When the input is a single-tone signal cos(wt), the output signal contains high-order harmonic components such as 2 wt, 3 wt, and so on. For example, harmonic interference is caused when harmonics fall into the receiving frequency band. The interference often occurs in the scenario where low-frequency transmission and high-frequency reception are carried out at the same time.

When the input signal contains multiple frequency components, the output signal will contain the intermodulation products of these frequency components. Taking the input of two frequency components, cos(W_1t) and cos(W_2t), as an example, the output signal will include a second-order intermodulation ($W_1 \pm W_2$), a third-order intermodulation ($2W_1 \pm W_2$, $W_1 \pm 2W_2$), and so on. If the intermodulation product falls into the receiving band, it will cause intermodulation interference. The interference often occurs in the scenario where high and low frequencies coexist, or the external signal is injected into the UE transmission chain and so on.

Take the interference between B3 and n77 as an example and illustrated in Fig. 9.2. The second harmonic of B3 uplink will cause second harmonic interference to the downlink in n77. The second-order intermodulation products of B3 uplink and n77 uplink will interfere with B3 downlink reception [1].

In the NR, the above harmonics and intermodulation interference have a serious impact on the UE receiver performance; in particular, the second harmonic and second-order intermodulation products may cause the sensitivity degradation in tens of dB. The frequency bands at around 3.5 GHz, such as n77 and n78, which are newly introduced in NR, are expected to become the main frequency bands for global roaming. The half of this frequency band is about 1.8 GHz, which is the midfrequency range of LTE. In other words, under EN−DC, there may be strong second harmonic or second-order intermodulation interference when the LTE intermediate band and the NR 3.5 GHz band are working/combined together at the same time in a UE. Therefore the intermodulation and harmonic interference under the EN−DC operation have become an important topic during the process of 3GPP standardization. Finally, the maximum sensitivity degradation value is defined according to the interference condition.

Furthermore, in addition to the above harmonics and intermodulation interference caused by the reverse coupling through the UE transmitting and receiving link, the interference directly leaked into another branch through the terminal PCB (printed circuit board) in some specific RF designs has also become a factor that cannot be ignored.

Fig. 9.3 is a diagram of the second harmonic interference of B3 to a 3.5 GHz receiving chain. Part of the second harmonic passes through the B3 transmitting chain and goes into the 3.5 GHz receiving chain; another part of the second harmonic of B3 PA directly leak into the 3.5 GHz receiving chain through the PCB. Among these two kinds of interference, the interference directly leaking through the PCB has become an important interference source.

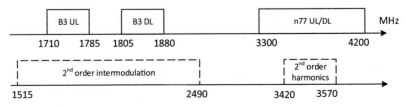

Figure 9.2 Harmonics and IMD interference.

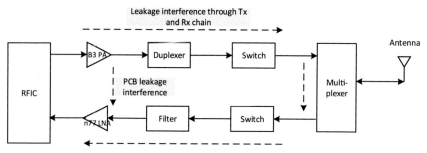

Figure 9.3 Interference sources inside the UE.

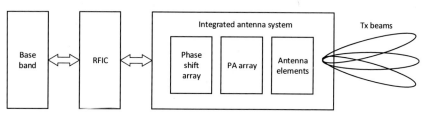

Figure 9.4 FR2 UE structure.

9.3 FR2 radio-frequency and antenna technology

Both the UE RF and antenna technology for operating in FR2 are very different compared to FR1, in all areas including the UE RF, antenna architecture, UE requirements, and performance testing. This section will briefly introduce the UE RF, antenna technology, and its corresponding requirements for FR2.

9.3.1 UE radio-frequency and antenna architecture

The RF front-end and antenna array are typically integrated in FR2. As shown in Fig. 9.4, a series of phase shifters and power amplifiers are implemented before the antenna array to amplify the signals and to also form beams.

The RF antenna connectors used in FR1 are no longer available or no longer needed in FR2 due to integrated design of RF front-end and antenna array. This leads to testing of RF requirements under OTA environment, which is different from the conducted tests used for FR1.

9.3.2 Power class

For the emission requirements based on OTA, the power–class definition needs to be clearly defined first. In FR1, the power class is defined for the

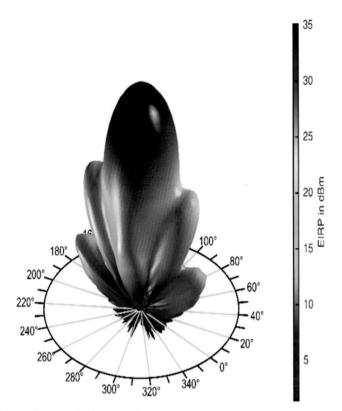

Figure 9.5 Typical FR2 UE transmitting beam.

maximum transmission power capability of the UE, and this basic principle is also applied to the definition of FR2 power classes.

9.3.2.1 Requirement definition

A typical transmit beam of a FR2 UE is shown in Fig. 9.5. The transmit beam of the UE has strong directivity. In the case of the same total transmission power, the narrower the transmit beam, the stronger the peak signal, and of course, the weaker it is in other directions. However, considering the large spatial propagation path loss of FR2 signal, it is necessary to use a communication method based on narrow transmit beam to concentrate energy to overcome the spatial path loss. Therefore defining the minimum peak EIRP of the strongest beam is needed.

Besides the minimum peak EIRP requirement, the mobility requirement for the FR2 UE also needs to be considered. As such, the UE needs to ensure the signal strength in directions other than the strongest beam. In order to quantify this requirement, the 3GPP takes spherical samples of transmitting signal and derives the Cumulative Distribution Function (CDF). Then the UE needs to fulfill the requirement for a certain percentage of the CDF curve, which is called the spherical coverage requirement.

Similar to FR1, the regulatory requirements of various countries and regions need to be additionally considered in the 3GPP. When discussing the requirements in 3GPP, with the exception of the United States, there are not many RF requirements for FR2 in the world. Therefore FR2 requirements in the 3GPP are largely based on US regulations, including the maximum peak EIRP of the strongest transmit beam and the maximum TRP. In the subsequent Release 17 standards as part of 3GPP evolution, other regional regulations like Japan are also been adopted.

To sum up, the power class of the millimeter wave terminal is ultimately a combination of multiple requirements, namely:

- Minimum peak EIRP
- Maximum peak EIRP
- Maximum TRP
- Spherical coverage

9.3.2.2 UE types

After the above power-class metrics are determined, the next step is to define the power-class requirements for the UE. In FR1, the definition of a power class is relatively simple and is based on the UE's maximum transmit power. However, the definition of the UE power class in FR2 is a combination of the aforementioned four metrics, which is more complex.

When the 3GPP discussed how to distinguish UEs with different transmission power capabilities, it was found that the types of UEs involved in FR2 can be very different, including Fix Wireless Access (FWA) UEs, vehicle UEs, handheld UEs, high-power nonhandheld UEs, etc. In view of different application scenarios, use cases, and UE types, the 3GPP finally defined the power class for each UE type. The UE power-class definition in FR2 not only distinguishes the transmission power capability of a UE, but it also actually represents a UE type. This is different from the UE power-class definition in FR1.

9.3.2.3 Peak EIRP requirement

When defining requirements, many factors should be considered such as the needs of network deployment, limitations in UE implementation, and so on.

The peak EIRP requirement is calculated based on the conduct power as:

$$\text{peak EIRP} = \text{conduct power} + \text{antenna gain} - \text{implementation loss}$$

$$(9.3)$$

Out of these parameters, the determination of the conduct power should take into account the multiple PAs incorporated inside of FR2 UE, such that the derivation needs to compute the total transmission power from the multiple PAs; for the antenna gain, it needs to consider both the single-antenna element gain and the total gain of antenna arrays. Here, both the conduct power and the antenna gain can be determined in principle based on calculation. However, for the implementation loss many different factors should be considered and discussed.

The implementation loss generally includes the loss due to RF components mismatch, transmission line loss, imperfect beamforming loss, and antenna integration loss caused by peripheral components. The specific numerical range of these implementation losses depends on the UE design. Finally, the total agreeable implementation loss in the 3GPP is about 7 dB. In the end, the minimum peak EIRP requirement is therefore defined as 22.4 dBm for smartphones. The definition of peak EIRP requirements for other UE types are specified in a similar approach, but the requirement values are different due to the difference in UE types and implementation limitations.

9.3.2.4 Spherical coverage requirement

Compared with peak EIRP, the discussion of spherical coverage requirement is more complex and controversial in the 3GPP. The definition of spherical coverage requirement is to ensure that the UE is able to achieve high signal strength in all directions in order to fulfill the mobility needs of the UE. This requirement will be used as a key reference for operators when deploying networks. If the UE cannot reach a high spherical coverage level, then the mobility performance of the UE within the network cannot be guaranteed. However, from the UE implementation point of view, it is challenging to ensure strong transmission power in all directions.

A typical problem is the impact to the UE transmission power caused by the display screen of the device. As shown in Fig. 9.6, typically, the phone screen has a metal back panel that supports the phone screen and also acts as a reflector for the signal. The antennas can achieve omnidirectional coverage without a mobile phone screen, but nearly half of the

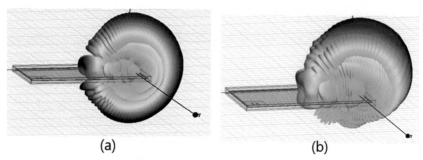

(a)   (b)

Figure 9.6 Screen impact to UE transmit power.

Table 9.13 PC1 for fix wireless access UE.

	Minimum peak EIRP (dBm)	Maximum peak EIRP (dBm)	Spherical coverage @85% CDF (dBm)	Maximum TRP (dBm)
n257	40	55	32	35
n258	40	55	32	35
n260	38	55	30	35
n261	40	55	32	35

signal is blocked when there is a screen, resulting in a sharp reduction of the radiation surface [2].

In addition, the number and location of the antenna array also have a significant impact on the spherical coverage. The greater number of antenna arrays used in a UE, the better the spherical coverage can be achieved. Peripheral components in the UE such as the display screen need to be considered when designing the antennas.

In view of the fact that the radiation effect of the FR2 UE antenna array is affected by many factors in the implementation, the spherical coverage requirement is obtained through simulation of the real UE.

The final definition of power class for each type of UE is shown in Tables 9.13—9.16:

9.3.2.5 Multiband impact

Compared with FR1, the one outstanding feature of FR2 frequency bands is the extra wide bandwidth that is available (even more than 3 GHz), as shown in Fig. 9.7.

For the 28 GHz spectrum, the total frequency bandwidth of the band group (including n257, n258, and n261) is 5.25 GHz, and these three bands

Table 9.14 PC2 for vehicular UE.

	Minimum peak EIRP (dBm)	Maximum peak EIRP (dBm)	Spherical coverage @60% CDF (dBm)	Maximum TRP (dBm)
n257	29	43	18	23
n258	29	43	18	23
n261	29	43	18	23

Table 9.15 PC3 for handheld UE.

	Minimum peak EIRP (dBm)	Maximum peak EIRP (dBm)	Spherical coverage @50% CDF (dBm)	Maximum TRP (dBm)
n257	22.4	43	11.5	23
n258	22.4	43	11.5	23
n260	20.6	43	8	23
n261	22.4	43	11.5	23

Table 9.16 PC4 for high-power nonhandheld UE.

	Minimum peak EIRP (dBm)	Maximum peak EIRP (dBm)	Spherical coverage @20% CDF (dBm)	Maximum TRP (dBm)
n257	34	43	25	23
n258	34	43	25	23
n260	31	43	19	23
n261	34	43	25	23

Figure 9.7 FR2 band frequency range.

are usually implemented with the same antenna array in a UE. If it is further considered that the n260 band and the 28 GHz band group are implemented with the same antenna array, the total bandwidth reaches 15.75 GHz. In the antenna design, in order to support such a wide frequency range, a certain compromise in performance should be made.

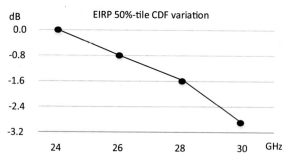

Figure 9.8 Relative antenna performance from 24 GHz to 30 GHz.

Table 9.17 PC3 UE multiband relaxation.

Supported band	Peak EIRP total relaxation ($\sum MB_P$ (dB))	Spherical coverage total relaxation ($\sum MB_S$ (dB))
n257, n258	≤ 1.3	≤ 1.25
n257, n260	≤ 1.0	≤ 0.75
n258, n260		
n257, n261	0.0	0.0
n258, n261	≤ 1.0	≤ 1.25
n260, n261	0.0	≤ 0.75
n257, n258, n260	≤ 1.7	≤ 1.75
n257, n258, n261		
n257, n258, n260, n261		
n257, n260, n261	≤ 0.5	≤ 1.25
n258, n260, n261	≤ 1.5	≤ 1.25

Fig. 9.8 is an example of the antenna performance impact of an antenna array in the 28 GHz band group and the n260 band, where when the antenna is optimized according to 24 GHz; its spherical coverage loss at 30 GHz reaches nearly 3 dB [3].

Therefore during 3GPP discussions, it was finally decided that on the basis of defining the single-band requirement, an additional multiband relaxation would be introduced for UEs that support multiband to give the UE design more freedom to meet the market needs of different countries and regions. Specific multiband relaxation requirements are shown in Table 9.17 [4].

9.3.3 Reference sensitivity

Similar to FR1, reference sensitivity is also the core requirement of a FR2 receiver. This reference sensitivity requirement mainly includes two parameters; one is the peak Equivalent Isotropic Sensitivity (EIS) and the other is the EIS spherical coverage.

9.3.3.1 Peak equivalent isotropic sensitivity

Similar to FR1, the peak EIS can be calculated according to the following formula:

$$\text{Peak EIS} = -174 + 10\lg(\text{Rx BW}) + 10\lg(\text{SU}) + \text{NF} + \text{SNR} - \text{Ant.gain} + \text{ILs} \tag{9.4}$$

In the formula, Rx BW is the receiving bandwidth of the UE; Spectrum Utilization (SU) is the spectrum efficiency under full RB; NF is the noise figure of the UE receiver associated with the frequency band; SNR is the baseband demodulation threshold of the UE; Ant.gain is the antenna array gain of the terminal; and ILs (Insertion Losses) is the insertion loss of the UE, which includes the transmission line loss, mismatch loss, the loss of antenna integration into the UE, and an implementation margin reserved for production.

We use the 50 MHz bandwidth in the 28 GHz band as an example of how to calculate the sensitivity. It should be noted that some of the following parameters, such as antenna gain and ILs, do not have a unified standard value, but are based on different UE implementation and varies according to different supplier (Tables 9.18–9.23).

Table 9.18 50 MHz CBW reference sensitivity at 28 GHz calculation example.

kTB/Hz [dBm]	−174
10lg(Rx BW) [dB]	76.77
Antenna array gain [dB]	9
SNR [dB]	−1
NF [dB]	10
ILs [dB]	8.93
Peak EIS [dBm]	−88.3

Table 9.19 FR2 reference sensitivity in each band.

Band	Peak EIS (dBm)			
	50 MHz	100 MHz	200 MHz	400 MHz
n257	−88.3	−85.3	−82.3	−79.3
n258	−88.3	−85.3	−82.3	−79.3
n260	−85.7	−82.7	−79.7	−76.7
n261	−88.3	−85.3	−82.3	−79.3

Table 9.20 FR2 PC1 FWA UE reference sensitivity.

Band	EIS @ 85th percentile CCDF (dBm)/Channel BW			
	50 MHz	100 MHz	200 MHz	400 MHz
n257	−89.5	−86.5	−83.5	−80.5
n258	−89.5	−86.5	−83.5	−80.5
n260	−86.5	−83.5	−80.5	−77.5
n261	−89.5	−86.5	−83.5	−80.5

Table 9.21 FR2 PC2 Vehicular UE reference sensitivity.

Band	EIS @ 60th percentile CCDF (dBm)/Channel BW			
	50 MHz	100 MHz	200 MHz	400 MHz
n257	−81.0	−78.0	−75.0	−72.0
n258	−81.0	−78.0	−75.0	−72.0
n261	−81.0	−78.0	−75.0	−72.0

Table 9.22 FR2 PC3 Handheld UE reference sensitivity.

Band	EIS @ 50th percentile CCDF (dBm)/Channel BW			
	50 MHz	100 MHz	200 MHz	400 MHz
n257	−77.4	−74.4	−71.4	−68.4
n258	−77.4	−74.4	−71.4	−68.4
n260	−73.1	−70.1	−67.1	−64.1
n261	−77.4	−74.4	−71.4	−68.4

Table 9.23 FR2 PC4 High-power nonhandheld UE reference sensitivity.

Band	EIS @ 20th percentile CCDF (dBm)/Channel BW			
	50 MHz	100 MHz	200 MHz	400 MHz
n257	−88.0	−85.0	−82.0	−79.0
n258	−88.0	−85.0	−82.0	−79.0
n260	−83.0	−80.0	−77.0	−74.0
n261	−88.0	−85.0	−82.0	−79.0

In Release 15, the reference sensitivity requirement for different bandwidths in each frequency band defined for the handheld UEs (power class 3) is as follows:

9.3.3.2 Equivalent isotropic sensitivity spherical coverage

The receiver sensitivity spherical coverage is defined as EIS spherical coverage, which is similar to the definition of the transmitter EIRP spherical coverage requirement. The EIS spherical coverage uses a method of sampling the receiver sensitivity in all directions of the UE, and using the statistics to obtain the EIS CCDF (Complementary Cumulative Distribution Function) in order to define the spherical coverage capability of the UE.

In the requirement definition, taking the handheld UE (power class 3) as an example, the EIS spherical coverage defines the sensitivity level at the 50 percentile of the CCDF. In addition, in order to simplify the complexity, the EIS spherical coverage requirement reuses the simulation results of EIRP CDF. That is, the power difference from the peak EIRP to the 50 percentile of the EIRP CDF is used as the sensitivity difference from peak EIS to 50 percentile of the EIS CCDF. As a result, the EIS spherical coverage requirement is obtained. The spherical coverage for each UE type is captured in the following tables Tables 9.20–9.23:

9.3.3.3 Multiband relaxation

Similar to the UE transmission power definition, it is difficult to optimize the receiver sensitivity for each frequency band when the UE supports multiple FR2 bands, which leads to the relaxation of the requirements compared with the single-frequency band as such. In order to simplify the complexity of the standard discussion, the peak EIS and EIS spherical coverage values for the multiband relaxation of the receiver sensitivity reuse the values from the transmission power multiband relaxation, which can be found in the relaxation definitions of peak and spherical coverage values given in Section 9.3.2.

9.3.3.4 Interference

The harmonics and intermodulation interference between different FR1 frequencies in the UE are extensively discussed in 3GPP. However, the operating frequencies of the FR2 system are far different from the existing FR1 communication system, as it is difficult for the transmitted signal and its harmonics of the existing system to fall directly into a FR2 band and cause interference.

However, when the UE upconverts a transmit signal to a FR2 band, it usually first upconverts the transmit signal to about 10 GHz, and then further upconverts to the target FR2 band such as 28 GHz or 39 GHz. When there are other communication systems operating at the same time in the terminal, such as LTE or GPS, if the isolation is insufficient, then potentially there will be interference among these systems, resulting in performance degradation.

9.3.4 Beam correspondence

Beam correspondence is an important Release15 MIMO feature that a FR2 UE needs to support. Its basic meaning is that the UE needs to determine an uplink transmit beam according to the direction of the downlink beam. This feature enables the UE to quickly select a transmit beam and complete the transmission of the uplink signal, thus reducing the time delay. If the UE does not have this capability, it will cause the UE to use the time-consuming beam sweeping process to complete the optimal transmit beam selection with the assistance of the base station. Therefore beam correspondence has become a feature that must be supported by the FR2 UE.

For the requirement definition aspect, beam correspondence reuses the already defined peak EIRP requirement and spherical coverage requirement, and further defines a relaxation value, namely a beam correspondence tolerance requirement. The beam correspondence tolerance requirement is defined as follows:
- Fix the downlink beam direction of the base station; the UE selects an optimal transmit beam using the uplink beam sweeping process assisted by the base station, and measures the peak $EIRP_1$ of the transmit beam.
- Fix the downlink beam direction of the base station; the UE selects a transmit beam using the autonomous beam correspondence, and measures the peak $EIRP_2$ of the transmit beam.
- In each downlink beam direction of the base station, a statistical CDF curve of the difference between peak $EIRP_1$ and peak $EIRP_2$ is obtained. The 85 percentile position of the CDF curve is taken as the beam correspondence tolerance requirement.

The beam correspondence tolerance requirement is only defined for handheld UEs (power class 3) in the Release 15 specification (i.e., the 85th percentile position of the peak EIRP difference CDF curve should not be higher than the 3 dB). It should be noted that the above peak EIRP difference CDF curve does not include all the beam directions in the statistics, but selects the beam direction in which the peak $EIRP_1$ meets the EIRP spherical coverage requirement.

The use of the three requirements (peak EIRP, spherical coverage, and beam correspondence tolerance) follows these principles:

- If the UE meets the peak EIRP and spherical coverage requirements of this frequency band by using the autonomous beam selection, it is considered that the UE is able to fulfill the beam correspondence requirements. There is no need to verify the beam correspondence tolerance requirement.
- If the UE can only meet the peak EIRP and spherical coverage requirement of the frequency band when the base station assists in uplink beam selection, then the UE needs to further fulfill the beam correspondence tolerance requirement.

9.3.5 Max permissible emission

Max Permissible Emission (MPE) is a requirement for electromagnetic radiation of a FR2 UE from the perspective of human body safety. At present, this requirement is mainly developed by standards organizations such as FCC and ICNIRP that specify the average maximum radiation power density of the UE in a certain direction, and the limitation of this requirement needs to be considered in the definition of 3GPP standards.

Taking FCC and ICNIRP as examples, MPE requirements are still under development, as shown in Table 9.24. The calculation of the average radiation power density includes not only the average in area, but also the average in time, for which the FCC has a shorter time for about several seconds in the 28 GHz and 39 GHz bands, while the ICNIRP takes the average in minutes.

9.3.5.1 Max permissible emission consideration in release 15

The MPE requirement is defined to limit the transmission power of the UE. Taking the handheld UE (power class 3) as an example, the peak EIRP is required to be greater than 22.4 dBm and less than 43 dBm, while the TRP needs to be less than 23 dBm. Under this maximum power condition, it is

Table 9.24 Max permissible emission requirements.

Frequency range (GHz)	FCC >6 GHz	ICNIRP >6 GHz
Power density (W/m^2) Area	10 4 cm^2	$55*f^{-0.177}$ 4 cm^2 (6 GHz $\sim$ 30 GHz) 1 cm^2 ($>$ 30 GHz)

very difficult for the UE to keep on transmitting when it is close to the human body in order to fulfill the MPE requirement.

The available UE transmission time is calculated theoretically in Ref. [5]. The proportion of the maximum available transmission time is 28.8% if the UE is to meet the above FCC index in 30 GHz band, and 91.2% if the UE is to meet the above ICNIRP index in 30 GHz band. The maximum available transmission time is only 43.7% in 40 GHz band to meet above FCC and ICNIRP requirement.

It is also analyzed in Ref. [6] that when the terminal is operating in the 28 GHz band, in order to meet the above FCC and ICNIRP MPE requirements, the maximum uplink time available for transmission is reduced to an astonishing 2%.

When the UL transmission time scheduled by the base station exceeds the UE capability, the UE has to do power fallback to meet the MPE requirement. Therefore in order to ensure that the UE is able to transmit at the maximum power, the base station needs to ensure that the uplink transmission time is lower than a certain threshold, which is related to the UE capability. Therefore the maximum uplink transmission capability (maxUplinkDutyCycle-FR2) reported by the UE is introduced into the Release 15 specification to assist the base station with scheduling, while ensuring that the UE can meet the MPE requirements. This capability is optionally reported, and the values include (15%, 20%, 25%, 30%, 40%, 50%,..., 100%).

The above maximum uplink duty cycle capability does not restrict the scheduling from the base station. When the actual uplink duty cycle scheduled by the base station exceeds this capability, the UE will perform power back-off. In some cases, the power back-off may be as high as 20 dB and cause the radio-link failure, for which it is optimized in Release 16.

9.3.5.2 R16 radio-link failure optimization

In Release 16, there has been an extensive discussion on how to solve the above potential radio-link failure problem, including a solution based on beam selection and switching in the physical layer, etc. But in the end, there was no common understanding of the actual gain and it was not standardized.

When the UE detects that it may face excessive MPE and is about to perform power back-off, the UE reports its power back-off value for the base station to take certain measures to avoid radio-link failure. Finally,

3GPP agreed to allow UE reporting the power back-off values (P-MPR) caused by MPE to base station as early warning information.

9.4 New radio test technology

Testing is an important link in quality work, and it plays a key role in mobile communication systems. The terminal must pass numerous tests and verifications before being able to access the system. Accurate and comprehensive terminal testing can guarantee the functions and performance of the terminal, and ensure that the terminal is able to operate normally in all kinds of network environments.

5G is a brand-new system that includes two network architectures, namely NSA and SA, and a variety of communication features, which greatly increase the workload of 5G testing. Moreover, radio communication using millimeter waves was also introduced in 5G, and millimeter wave testing is a new topic at the 3GPP. In particular, millimeter wave radio-frequency testing methods will be completely different from traditional radio-frequency testing. In addition, the introduction of important 5G features such as HPUE, UL MIMO, and CA has created greater challenges for terminal testing. Therefore the testing of 5G terminals is more complicated than that of previous generations of mobile communication systems.

The 3GPP NR conformance test standards have been studied and formulated since 2017, and the development of test cases for the core specification of the Release 15 version was still not complete as of mid-2020. A large number of test cases, new test methods, and the introduction of key 5G features are all important factors that affect the development of 5G test standards. In addition to developing test cases that ensure the terminal is able to meet the functional and performance requirements, how to improve the test efficiency is also a problem that must be considered when the standard is formulated.

9.4.1 SA FR1 radio-frequency test

The RF test for 5G SA in the FR1 still uses the traditional conduction test method, so it has little change compared with LTE, and the test cases basically refer to the LTE specification. New features such as HPUE, UL MIMO, and four-antenna reception introduce din Release 15 version of the NR have had a significant impact on the design of RF test cases.

9.4.1.1 Maximum transmit power test

The maximum output power test is a very important RF test. Generally speaking, the maximum output power of the terminal is always considered during network planning and deployment. Therefore the maximum output power of the terminal directly affects the communication capability of the terminal. If the maximum output power of the terminal is too small, the coverage area of the base station will be reduced, but if the maximum output power of the terminal is too large, it may cause interference to other communication systems and equipment. As mentioned in the previous sections of this chapter, NR FR1 introduced an important feature of HPUE at the beginning of its design, and defined the terminal capabilities to support power class 2 for frequency bands such as n41/n77/n78/n79. However, when designing the terminal, HPUE may be implemented in different ways. For example, the terminal can achieve the requirements of HPUE through a single antenna emitting 26 dBm or dual antennas each emitting 23 dBm. Different implementation methods will require different test methods. Therefore different terminal implementation methods need to be considered when designing the maximum output power and related test cases, so that the test can fulfill the intended output requirements for all terminals. Considering that in the actual network deployment, in order to reduce intercell interference and for other reasons, the base station may limit the terminal maximum output power in the cell to 23 dBm. Consequently, it is required that the maximum output power of HPUE needs to be further tested when it falls back and operates at 23 dBm.

9.4.1.2 Selection of test points

The selection of test points is an important step in the design of test cases. Choosing appropriate test points can make the tests more accurate and efficient. For NR testing, the selection of test points needs to consider factors such as test environment, frequency, bandwidth, SCS, and uplink and downlink configurations, and select the most representative/commonly used configuration to make the test cover as many scenarios as possible. At the same time, the time duration required to conduct the test should be also taken into account, and test points that are of a low priority or reusable can be appropriately removed to improve test efficiency. Since the radio-frequency tests of LTE are relatively mature, and the RF test method for NR SA FR1 has little change from LTE, the test points of LTE are usually used as a reference when selecting the test points for NR

SA FR1 test cases. When combined with the new features of NR, the most suitable test points are then selected.

For the individual test case, there may be scenarios where no test point can be found to fulfill the test requirements, such as the maximum output power test for UL MIMO. Under UL MIMO, the terminal is only able to use the CP−OFDM waveform for transmission, but the maximum power reduction (MPR) is configured for the CP−OFDM waveform in the Release 15 specification, and hence it is not possible to obtain test points to carry out the maximum output power test under UL MIMO without the MPR during the test case design. The main reason for the problem of not having suitable test points is that the test requirements were not fully considered in the design of the core requirement. The most serious consequence is that the test cannot be carried out at all, and subsequently the corresponding requirement cannot be fully verified, which will have a potential impact on the system. Therefore in the process of formulating standards and requirements, one should always consider the needs of the testing. If the defined requirements and functions cannot be tested or verified, the capabilities of the terminal and the network performance will not be guaranteed.

9.4.2 SA FR2 radio-frequency test

The millimeter wave spectrum is a newly introduced frequency spectrum for 5G. Compared with the traditional low-frequency spectrum, the millimeter wave has a shorter wavelength. If a conductive test is used, a large insertion loss will occur through the wire in the test setup, which will seriously affect the accuracy of the test. Moreover, the RF front-end and antennas of the millimeter wave terminal mostly adopt an integrated design, and there is no test port between the antenna and the RF front-end such that it can be connected to the test instrument. This directly causes the millimeter wave test to no longer use the traditional conduction test scheme, and instead uses the OTA testing method.

9.4.2.1 FR2 OTA test methods

At present, the 3GPP has defined mainly three OTA test schemes: Direct far field (DFF), Indirect far field (IFF), and Near field far field transform (NFTF) [5−12].

9.4.2.1.1 Direct far-field method

The DFF method is a more traditional OTA test scheme. When the DFF method is used for OTA testing, the quiet zone of the test needs to satisfy the far-field conditions. Specifically, the distance between the terminal being tested and the test antenna must ensure $R > \frac{2D^2}{\lambda}$, where D is the diagonal of the tested terminal. Due to the short wavelength of the millimeter wave, when the size of the measured terminal is large, the far-field distance of the test will be very large. The increase in the test distance will result in a further increase in the size of the chamber, which significantly increases the cost of the OTA test system. Moreover, the increase of the test distance will cause the signal propagation loss to further increase, and thus it affects the dynamic range of the system and the accuracy of the test. Therefore the DFF method is more suitable for small-sized terminal under test when performing FR2 RF testing.

9.4.2.1.2 Indirect far-field method

The IFF method can generate plane waves through indirect methods such as reflective surfaces and antenna arrays to achieve the required test conditions for the far field. The specification mainly defines a Compact Antenna Test Range based on a metal reflecting surface. As shown in Fig. 9.9B, the spherical wave emitted by the feed antenna can produce a plane wave after being reflected by the metal parabolic surface, forming a quiet zone with stable amplitude and phase in a short distance. The IFF method does not need to meet the distance requirements of the far field. The size of the quiet zone mainly depends on the size of the reflecting surface. Therefore the size of the chamber can be small, and can fulfill the test requirements of the larger-sized terminal. The size of the reflecting surface and the manufacturing process to a large extent determine the cost and accuracy of the test system.

(a) Direct far field

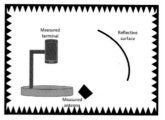

(b) Indirect far field

Figure 9.9 FR2 OTA test method.

9.4.2.1.3 Near-field far-field transform method

The principle of the NFTF method is to sample the radiation amplitude and phase information of the measured terminal in the near-field area, and then calculate the far-field radiation value through Fourier spectrum transformation. This test solution requires the smallest size of the chamber, the conversion accuracy depends on the spatial resolution of the sampling, the test is time-consuming, and it is only suitable for a few RF tests such as EIRP, TRP, and spurious emission.

9.4.2.2 Applicability of FR2 OTA test method

The uncertainty of the test system directly reflects the accuracy of the test. If the test uncertainty is too large, it will cause test results to be very unreliable, seriously affecting the verification of the terminal function and performance, and it can even cause the test to become meaningless. Compared with the low-frequency conduction test, the millimeter-wave OTA test system is much more complicated as there are more factors that affect the uncertainty of the test, such as the placement of test terminal, test distance, test grid, etc. These also cause the uncertainty of the millimeter wave OTA test to be much greater than the low-frequency conduction test.

Calculation methods and values of the test uncertainty for different test schemes are different. For the measured terminal with a diameter of less than 5 cm, the direct far-field method is used as a benchmark to define the test uncertainty. For the measured terminal with a diameter greater than 5 cm, if the direct far-field method and a larger chamber are used, the measurement uncertainty can be significantly improved due to the increase in the test distance. Therefore the specification definition uses the indirect far-field method as the benchmark to define the measurement uncertainty. If the test uncertainty of other test methods is less than the benchmark test uncertainty, the test program can also be applied to conformance testing (Table 9.25).

9.4.2.3 FR2 OTA measurement grid

For EIRP, EIS, TRP, and other tests, the radiation value of the test terminal on a closed spherical surface needs to be obtained. For the test system, only one location point on the spherical surface can be tested at a time, so multiple points on the closed spherical surface sampling need to be collected. If the sampling points are too dense, the test time will be significantly increased. If the sampling points are too sparse, the test will

Table 9.25 Applicability of FR2 OTA test method.

Test terminal antenna configuration	Indirect far field	Indirect far field	Near-field far-field transform
An antenna array with a diameter of less than 5 cm	Applicable	Applicable	Applicable
Multiple uncorrelated antenna arrays with a diameter of less than 5 cm	Applicable	Applicable	Applicable
Any size phase-correlated antenna array	Not Applicable	Applicable	Not Applicable

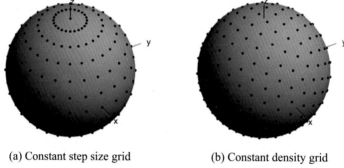

(a) Constant step size grid   (b) Constant density grid

Figure 9.10 FR2 OTA measurement grid [5−10].

not be accurate enough and the uncertainty of the test will increase. Therefore choosing a suitable measurement grid is very important in OTA testing. The 3GPP defines two types of measurement grids, which are Constant step size grid and Constant density grid, as shown in Fig. 9.10. When the number of sampling points is the same, the measurement uncertainty of the Constant density grid is smaller, but the Constant step size grid has lower requirements for the test turntable. Both types of test grids can fulfill the needs of millimeter wave OTA testing. Which type of grid to choose depends mainly on the design of the test system. For saving test time, the measurement grid with as few sampling points as possible can be selected while ensuring that the test uncertainty still meets the requirements.

The FR2 OTA test needs to obtain the beam peak direction information of the measured terminal. The benchmark beam peak direction search scheme uses a relatively fine measurement grid to measure the measured terminal, so as to find the direction of the beam peak. This means that even when the EIPR/EIS of a certain area is known to be low, it is still required to measure all the grid points in this area. Obviously, the measurement of these grid points is not necessary and will only increase the overall test time. In order to improve the efficiency of beam peak direction search, the specification defines a method of coarse and fine measurement grids [5−10].

When using a coarse and fine measurement grid, first select a coarse measurement grid with a larger step size or a lower density for measurement, and find the area where the beam peak direction may exist, as shown in Fig. 9.11(a). Then use a fine measurement grid in these areas, as shown in Fig. 9.11(b), and perform the measurement again. The point where the measured EIPR/EIS is the largest is the direction of the beam peak. The coarse and fine measurement grids can significantly reduce the sampling points of the test and reduce the test time without affecting the accuracy of the test.

9.4.2.4 FR2 OTA test limitations

Compared to the more mature RF tests for FR1, the research and system development of RF tests for FR2 is still in its infancy due to the influencing factors such as test environment and system. For some traditional RF test cases, the millimeter wave OTA system has a problem that cannot be

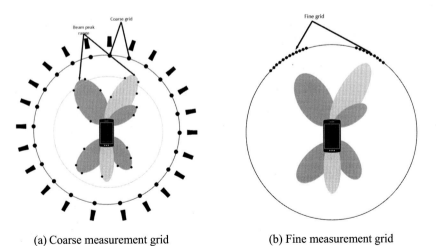

(a) Coarse measurement grid   (b) Fine measurement grid

Figure 9.11 Coarse and fine measurement grids method.

tested. Taking the maximum input level test as an example, since the propagation loss for the millimeter wave is relatively large, the maximum input level value defined by the core specification cannot be reached under the existing test system hardware conditions, so the maximum input level index cannot be verified.

For test items with lower terminal transmit power, since the bandwidth of the millimeter wave band is larger, the power spectral density of the terminal will be lower and the OTA test noise will be larger, and as such the SNR of the signal to the test antenna will be very low. Then the uncertainty of the test will be increased significantly as well, making the test impossible. For such test cases, in order to enable the tests to be carried out smoothly, the standard will appropriately exclude points with larger channel bandwidths when selecting the test points to improve the signal-to-noise ratio. However, even if a smaller bandwidth is used for test cases with very low power, a more ideal signal-to-noise ratio cannot be achieved. For such cases, relaxing the test requirements for the tests seems to be the only solution at present.

Due to factors such as the capabilities of the test system, the overall test uncertainty of the FR2 RF test is still relatively high. For some of the FR2 test cases, especially the receiver tests, there are many test points and sampling points, so a long time period is needed to perform the tests. The FR2 test system and test cases still have many areas that can be optimized. How to reduce test uncertainty and improve test accuracy and test efficiency are still the focus of follow-up research on the FR2 RF test specification work.

9.4.3 EN–DC radio-frequency test

For the EN–DC RF test, the specification defines a new test method, namely the LTE anchor agnostic approach [5–11]. This test method uses LTE as the anchor to ensure the link between the terminal and test system is always maintained and then verifies the RF requirement on the NR carrier. Generally speaking, if the LTE carrier does not interfere with the NR carrier, the LTE anchor agnostic method can be used for RF testing. When the LTE anchor agnostic method is used, the specification defines that for the EN–DC band combination of the same NR band, any LTE band can be selected as the anchor point for testing, and the EN–DC combination of the same NR band only needs to be tested once. For the EN–DC combination with NR in the FR2 frequency band, the OTA method is also required for testing, and most test cases can be tested using

LTE as the anchor. There are many combinations of EN−DC frequency bands, and the LTE anchor agnostic approach can greatly improve test efficiency, reduce test volume, and avoid repeated tests.

Although EN−DC has two links, LTE and NR, the specification only defines the total maximum transmit power requirement of the terminal. For EN−DC terminals, there are two states: simultaneous transmission of LTE and NR links and separate transmission of LTE and NR links. When defining the test specification, it is necessary to consider if both states can be tested.

- When LTE and NR links are transmitted separately, they need to fulfill the maximum transmission power requirements of the LTE single carrier and NR single carrier independently.

- For the case when LTE and NR links are transmitting at the same time and the terminal that supports dynamic power sharing, if the maximum transmission power of the LTE and NR link is not limited during the test and the test equipment schedules the terminal to transmit at the maximum output power, the terminal may allocate most of the power to the LTE link to maintain the connection. As such the NR link may be disconnected since the remaining available power is insufficient to fulfill the requirements for simultaneous LTE and NR transmission. Therefore for the simultaneous transmission of LTE and NR, the specification defines the maximum transmission powers of the LTE and NR links, respectively. For the example of a PC3 terminal, the maximum transmission power of both LTE and NR is set to 20 dBm to ensure that the terminal is at the maximum transmission power. Both LTE and NR links can guarantee a certain transmit power to ensure that the test can be carried out smoothly.

In the actual network, it is only a part of the scenario that the transmission power of the terminal is limited by the instruction from the base station. Therefore setting the maximum transmission power of the LTE and NR links does not reflect all application scenarios. For the conformance test specification, it is difficult to define the test for all scenarios. How to more accurately and efficiently ensure that the function and performance of the terminal meet the needs of the network is the purpose of the specification definition.

9.4.4 MIMO OTA Test

As mentioned in previous sections, MIMO is an effective way to improve the system capacity, and it is also one of the most important key technologies of 5G. Since MIMO uses spatial multiplexing technology, the UE MIMO performance should be verified over the air interface (i.e., OTA mode). The MIMO OTA test is derived from SISO (single in, single out) OTA. SISO OTA is designed to test the ultimate access capability of the UE, including transmit power (TRP) and receiver sensitivity (TRS). Obviously, SISO OTA testing cannot reflect the real performance of the UE under the condition of multiple inputs and multiple outputs. In order to verify the MIMO receiving performance of the UE, referring to the SISO OTA test method, on the basis of the constructed "quasifree space" environment in the anechoic chamber, and a space channel environment conforming to the propagation characteristics of a certain channel is superimposed. The UE throughput test under this controllable and reproducible MIMO space wireless environment is called the MIMO OTA test.

The research and development of the Release 16 NR MIMO OTA test method were basically completed within only 15 months, including Multi-Probe Anechoic Chamber (MPAC) and Radiation Two Stage for FR1, and 3D−MPAC for FR2. At the same time, the channel model for the MIMO OTA test was also determined. The standardization of the NR MIMO OTA testing was actually a product that combines technical research and industrial experience accumulated over 8 years since the 4G phase, including verification method of test zone performance test, multi probe calibration and compensation method, channel model verification method, throughput test comparison based on SNR and Reference Signal Reception Power, MIMO channel model selection, and so on. Regardless of 4G or 5G, MIMO OTA testing standardization work not only needs to study the above technical problems and solutions, but also needs to further consider how to simplify the test system complexity and shorten the test time and other cost factors, such as the number of probes, polarization direction, the number of channel models, DUT test attitude, etc. Thus this extremely complex system level test is based on solid engineering implementation and incorporates value from industrial promotion.

9.4.4.1 Channel model for NR MIMO OTA
In order to verify the performance of NR MIMO OTA effectively and efficiently, the most fundamental and urgent problem in NR MIMO OTA research is the selection of channel model. The effectiveness lies in that the selected channel model is able to be used for clearly distinguishing

Table 9.26 Channel Model for NR MIMO OTA.

	Channel Model 1	Channel Model 2
FR1	UMi CDL−A	UMa CDL−C
FR2	InO CDL−A	UMi CDL−C

Note: *CDL*, Clustered delay line; *InO*, indoor office; *UMa*, urban macro; *UMi*, urban micro.

Table 9.27 RMS Elevation spread with different CDL model and BS antenna combinations.

	Elevation Spread [deg]	
	CDL−A	CDL−C
Directive BS w/o beam	8.2	7.0
8 × 8 URA, 1 strongest beams	5.4	5.3
8 × 8 URA, 2 strongest beams	5.4	5.5
8 × 8 URA, 4 strongest beams	5.6	7.3

Note: *URA*, Uniform rectangular array.

the MIMO performance of different UEs; and the efficiency is reflected in covering the most typical application scenarios with the least number of channel models. During the RAN4#92 meeting in August 2019, all companies reached an agreement on the selection of channel model [13], as shown in Table 9.26. This conclusion greatly simplifies the workload of the follow-up research.

9.4.4.2 Chamber layout for NR FR1 MIMO OTA

After resolving the channel model selection problem, the next most important and complex problem to be solved is the number and distribution of probes. This issue was raised and discussed for many rounds in the early stages of the project.

As shown in Table 9.27, the channel model has a certain energy distribution in the vertical direction [14]. Based on this, the 3GPP discussed the 3D measurement probe scheme in the early stage of standardization.

However, after only two meetings, the 3D scheme was abandoned and the 2D scheme returned to the center of discussion. The reason is that the number of probes needed in the 3D scheme is too large. For example, in the case when the base station antennas do not use beamforming, UMA CDL−C channel model is adopted, UE operating frequency is 7.25 GHz, and UE antenna spacing is 0.3 m, and the required

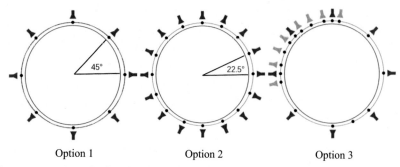

Option 1   Option 2   Option 3

Figure 9.12 2D probe layout options for FR1 MIMO OTA.

number of probes reaches as many as 429. This not only means that 429 dual polarization probes [15] need to be configured at the designated position in the full anechoic chamber, but also 858 channels of channel simulation equipment are required behind the probes. This is obviously not a solution suitable for industrial implementation.

For the 2D scheme, three alternative methods were the main focus, as shown in Fig. 9.12. The probe layout of Option 1 follows the scheme of LTE MIMO OTA (i.e., eight dual polarization probes are evenly distributed on the horizontal loop). In Option 2, the probe density is doubled based on Option 1 (i.e., 16 dual polarization probes are evenly distributed on the horizontal ring). In Option 3, eight additional dual polarization probes are added to scheme 1 to construct the 5G FR1 MIMO OTA channel environment.

The advantage of Option 1 is that it can fully reuse the hardware of LTE MIMO OTA chamber, and the cost of test environment reconstruction is the lowest. However, due to the large error in the spatial correlation coefficient of the channel model constructed by this option (the error is more than 0.25) [16], it is excluded first.

The main advantage of Option 2 is that the cost of the hardware transformation of the chamber is low. Many built chambers have been equipped with or have the conditions to upgrade to 16 evenly distributed probes. Moreover, the uniform distribution of probes is beneficial to adapt to other channel models other than the selected channel models listed above, which makes the chamber more scalable.

The advantage of Option 3 is that it is compatible with LTE MIMO OTA chambers and better matches the selected NR MIMO OTA channel models of UMi CDL−A and UMa CDL−C listed above.

Comparing Option 2 and Option 3, their error of reconstructed channel model does not differ significantly; in addition, uniform distributed probes have obvious scalability advantages, which makes the valuable chambers applicable to more scenarios. Considering the technical index, cost, and the expansibility of the options, Option 2 is finally chosen as the probe layout of the chamber.

9.4.4.3 Chamber layout for NR FR2 MIMO OTA

For the deployment scheme of FR2 probe, it is relatively simple to produce the scheme since the issue of backward compatibility does not need to be taken into account. First of all, due to the strong directionality of the mmW array, the probe layout used is no longer a ring or a sphere, but a 3D sector covering a certain area. Through the simulation of the channel model with different chamber layouts, the scheme of using six dual-polarization probes was finally determined [17,18]. The probe positions are shown in Table 9.28 and Fig. 9.13, and the relative position relationship between the DUT and the test probes are shown in Fig. 9.14.

In June 2020, the work item phase of the standardization project for NR MIMO OTA testing started, and the task will specifically discuss the definition of MIMO OTA performance metrics, etc.

Table 9.28 FR2 3D MPAC Probe Locations.

Probe Number	Theta/ZoA [deg]	Phi/AoA [deg]
1	90	75
2	85	85
3	85	55
4	85	95
5	95	95
6	90	105

Note: *AoA*, Azimuth angle of arrival; *ZoA*, zenith angle of arrival.

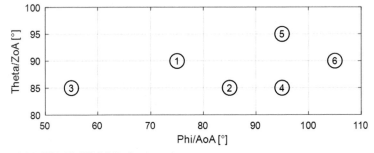

Figure 9.13 FR2 3D MPAC Probe Locations

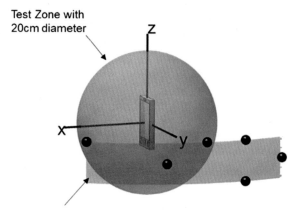

Test Zone with
20cm diameter

Sector with 6 probes at fixed
min. radius from centre of test
zone of 0.75m

Figure 9.14 3D MPAC Probe Layout for NR FR2 MIMO OTA testing.

9.5 New radio RF design and challenges

9.5.1 NR RF Front-end

5G NR brings unprecedented challenges to the implementation of UE RF components. There are a few key aspects that need to be considered and addressed during the process of designing the UE. These aspects are:

- NR supports up to 100 MHz channel bandwidth in sub-6GHz frequency spectrum
- 256QAM (de)modulation in uplink and downlink
- HPUE with 26 dBm max power
- Interference between NR and LTE caused by EN−DC
- Coexistence of NR, WiFi and others
- Multiantenna design and switching time required (SRS)

In addition to the new features and functionalities above that would cause significant impact to the UE RF design for supporting 5G NR, a UE also needs to support LTE, UMTS, GSM, and other cellular technologies, as well as WiFi, Bluetooth, GNSS, Near Field Communication (NFC), and other wireless communication modules. As shown in Fig. 9.15, the RF architecture for the terminal has reached unprecedented complexity. Furthermore, it will be more challenging to take into account the space constraints and overall power consumption while ensuring the RF performance.

In order to overcome the challenges above, it is necessary to improve the selection of RF components and designs of RF architecture. For example,

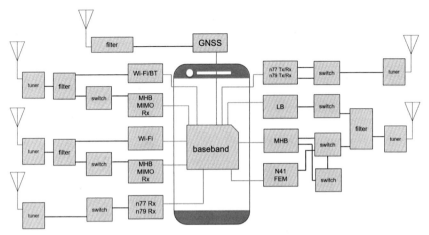

Figure 9.15 UE with NR capability.

using an envelope-tracking chip to support larger bandwidth and front-end components such as PA and switch with better linearity to reduce interference.

The RF front-end of a UE-supporting NR needs to be transformed from discrete components to integrated modules. As shown in Fig. 9.16, the use of LPMAiD module increases the flexibility of UE RF front-end design. LNA bank, MMPA, and duplexer are integrated in the module. Compared with discrete approach, integrated design is able to effectively reduce the insertion loss and noise figure, therefore improving RF performance. In addition, due to the difference of multiantenna performance and the needs of EN−DC, the integrated approach reduces the use of multiplexer and is able to fulfill the demand of flexible antenna switching in the RF front-end.

9.5.2 Interference and Coexistence

Interference or desensitization is a common problem in UE terminal design. The interference caused by NR has been described in the RF standards in Section 9.2.3. This section focuses on the interference between LTE and NR under the EN−DC architecture, as well as coexistence between NR and WiFi.

9.5.2.1 EN−DC Interference

In the EN−DC structure, since the UE has to perform transmission and reception of 4G and 5G signals simultaneously, this may cause mutual interference between the two system operations for the UE, such as

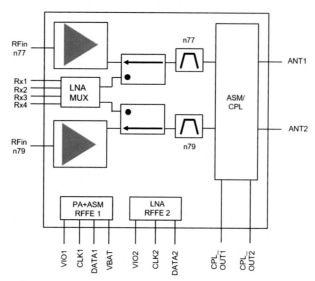

Figure 9.16 schematic diagram of NR RF front-end module [5–20].

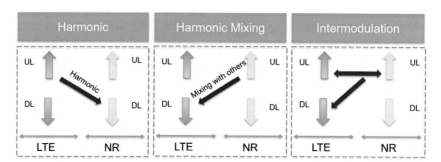

Figure 9.17 Common interference problems between 4G LTE and 5G NR.

harmonic interference, harmonic mixing, and intermodulation interference (Fig. 9.17).

In the harmonic interference, such as the combination of B3 and n78, the second harmonic of B3 in uplink may fall into n78 band, causing desensitization and affecting the sensitivity of the UE receiver. There are several causes of the interference.

- Second harmonic due to the nonlinearity of B3 PA
- Second harmonic due to the nonlinearity of the RF switch in B3 transmitter
- Second harmonic due to the nonlinearity of the RF switch in n78 receiving path

- Antenna and its close-by nonlinear components, such as transient voltage suppression diode and antenna tuner

The first step to solve the interference problem is to locate the source of the interference. As listed above, there may be several causes. The interference source can be located through comparative tests and then the problem can be solved. In order to solve the harmonic interference of B3 to n78, the main methods can be increasing the isolation of B3 and n78 physical antennas, adding low-pass filter in B3 transmission channel, adding high-pass filter in n78 receiving channel, using PA with good linearity, and devices with high linearity near the antenna.

High-order intermodulation is also a difficult problem to resolve. Taking the B1 + n3 combination as an example, B1 receiver sensitivity might be degraded due to the third-order intermodulation interference caused by the second harmonic of B1 uplink and n3. The second harmonic may still be caused by nonlinear devices such as PA and switches on the RF path. Therefore in the design of NR UE, we need to pay extra attention to the linearity of the components.

9.5.2.2 Coexistence of NR and WiFi

Coexistence of cellular radio and other wireless systems is also a persisting problem that needs to be solved in UE design. In a NR terminal, the spectrum of new NR band n41 is close to the WiFi 2.4G band, so NR transmission will cause significant interference to WiFi reception. If the 100 MHz channel bandwidth is being used in n41, its side lobe will fall directly into the WiFi 2.4 GHz band under the maximum transmission power, resulting in serious deterioration of sensitivity. In addition, the second harmonic of n41 can fall directly into WiFi 5 GHz band, which could also cause serious sensitivity deterioration.

The most straightforward method to resolve the problem is to apply time sharing between WiFi and NR. However, this method will reduce the usage time of each wireless connectivity, thus affecting the wireless transmission throughput and negatively affecting the user experience.

From the perspective of hardware design, there are several solutions:

- increase the isolation between n41 transmitter and WiFi 2.4G antenna
- add a filter in n41 transmission path to reduce the out of band noise emission, as shown in Fig. 9.18
- add a filter in the WiFi receiving path
- reduce the spurious emission of n41 RF front-end by optimizing the internal structure of the device

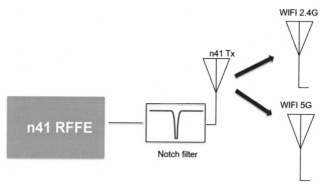

Figure 9.18 NR n41 and WiFi interference solution.

For the methods of reducing out-of-band leakage by adding new hardware, the insertion loss of the link may be increased, which is also a factor that should be considered (Fig. 9.18).

9.5.3 Design of SRS RF front-end

SRS switching requires a UE to transmit reference signals on all receiving antennas for MIMO channel condition estimation and further enhancing the downlink throughput. SRS switching requires a UE to switch physical antennas quickly. In EN—DC, the UE needs to take both LTE and NR antennas into consideration. Therefore SRS switching poses a great challenge to the design of antenna structure and the switching mechanism.

In EN—DC, the NR and LTE may share the same physical antennas. When the UE needs to transmit SRS signal, the mapping relationship between RF ports and antennas will be changed. LTE service may get interrupted due to the loss of antenna connection. Ensuring the continuation of LTE service while transmitting SRS signals is a factor that should be considered in RF front-end design. Refs. [5—21] propose an antenna-switching architecture to realize SRS signal transmission under the condition of ensuring LTE communication.

As shown in Fig. 9.19, the NR transmitter and LTE transceiver can be connected to physical antenna ANT0 and ANT1 by controlling the antenna switches. In (a), connect port 1—3 in switch 1, connect port 2—6 in switch 2, and use port 1 in switch 3 to realize the connection of NR transmitter with antenna ANT0 and LTE transceiver with ANT1.

In (b), connect port 1—4 in switch 1, connect port 1—6, 2—5 in switch 2, and use port 2 in switch 3 to realize the connection of NR

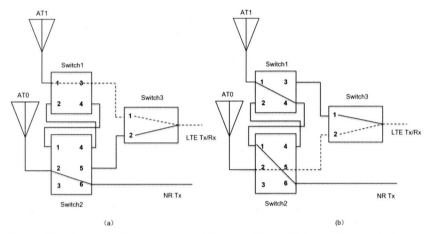

Figure 9.19 Correspondence between NR transmitter, LTE transceiver, and physical antenna in SRS signal transmission. (A) NR transmitter is connected to physical antenna ANT0, and LTE transceiver is connected to physical antenna ANT1. (B) NR transmitter is connected to physical antenna ANT1 and LTE transceiver is connected to physical antenna ANT0.

transmitter to antenna ANT1 and LTE transceiver to antenna ANT0. By doing so, LTE continues to function while the SRS signal is transmitted.

In the RF design, we need to consider the performance of UE antennas, UE form factor and mechanical structure, and cost constraints, and select the optimal RF and antenna architecture accordingly. For different EN−DC combinations, UE design requires different RF and antenna architectures, and hence there is always no best solution available. As such, tradeoffs must be taken based on the choice of RF performance, UE form factor, and overall cost.

9.5.4 Other new radio challenges

9.5.4.1 Dual SIM card support

5G NR also brings new challenges for UEs that support dual SIM cards. In the 4G era, many UEs adopt a dual SIM dual standby mode. In such mode, SIM 2 will frame SIM 1 when monitoring system paging messages. Therefore if a large data is being exchanged on the SIM 1 connection, such as gaming and videos, the user will experience obvious jamming problems.

In the 5G era, UE dual SIM will most likely adopt a dual-receive dual SIM dual standby mode (Dr−DSDS). In terms of hardware support, separate RF receiving channels for both SIM cards are provided for operating in this mode, so the jamming problem will not occur and the user experience will be improved. Another upgrade to this technology is Dr–DSDS + TX sharing. That is, the sharing of RF transmission channels by both SIM cards is supported based on the Dr–DSDS framework to allow a user playing games over 5G while using VoLTE in 4G.

At the time of writing this book, 5G dual SIM UEs mainly adopt 5G + 4G mode. More specifically, one SIM card works on 5G and the other SIM works on 4G. 5G can be in NSA or SA network operating mode. For 5G + 5G dual SIM card terminal, SA + SA or NSA + SA modes are mainly used.

Careful analysis of the RF and antenna architecture is required for a UE supporting 5G + 4G dual SIM card operation mode. Due to limitations caused by the UE form factor, mechanical structure, and communication chipset, it is necessary to consider tradeoffs between five-antenna and six-antenna designs. Regardless of UEs operating in NSA + LTE or SA + LTE mode, the working principle is that the NR path does not need to be reconfigured when SIM 1 and SIM 2 switches their operating mode. In order to achieve this, the switch control in the RF front-end component also needs to support MIPI2.1 to realize the function of "hardware mask write."

9.5.4.2 Radio-frequency front-end switch

For the usage of RF components, it is not only required to increase the quantity of 5G UEs, but also set higher standards for the quality. Switches are widely used in a RF system, which can be used externally or integrated into other modules. A few factors need to be considered when it comes to the selection of switches for 5G UEs. These may include:

- isolation
- linearity (avoiding interference and coexistence problems)
- insertion loss
- switching time required (SRS)
- MIPI register control (whether hardware mask write is supported or not)
- VDD/VIO voltage
- GPIO/MIPI control
- P2P resources

9.6 Summary

This chapter started with a discussion of RF technology for the radio spectrum, introduced the new 5G spectrum, and defined the new frequency band/band combinations, and further described RF technology for FR1 and FR2, NR test methodologies, RF implementations and challenges, etc. In the section on RF technology for FR1, the key techniques, requirements, and specifications for high-power UE, receiver sensitivity, and mutual interference were introduced. In the FR2 RF and antenna technology section, RF antenna architecture, power-level definition, sensitivity, beam correspondence, MPE, and other technologies and standards that are different from FR1 were discussed. In the section on NR testing technology, we focused on testing technology for evaluating performance in SA and NSA deployments, and discussed the development of test methodologies for evaluating MIMO OTA performance. In the section on RF implementation and challenges for NR, new problems identified in the NR and their corresponding solutions were discussed from the perspective of UE RF design.

References

[1] Xing Jinqiang, Research of LTE and 5G NR UE interference, mobile communications, February, 2018.
[2] R4−1711036, Consideration of EIRP spherical coverage requirement, Samsung, 3GPP RAN4#84-Bis, Dubrovnik, Croatia, 09−13 October, 2017.
[3] RP-180933, Extending FR2 spherical coverage requirement to multi-band UEs, Apple Inc, 3GPP RAN#80, La Jolla, USA, 11−14 June, 2018.
[4] 3GPP TS 380.101−2 V150.8.0, NR; UE radio transmission and reception; Part 2: Range 2 Standalone (Release 15), 2019−12.
[5] R4−1900253, Discussion on FR2 UE MPE remaining issues, OPPO, 3GPP RAN4#90, Athens, GR, 25 Feb - 1 Mar, 2019.
[6] R4−1900440, P-MPR and maxULDutycycle limit parameters, Qualcomm Incorporated, 3GPP RAN4#90, Athens, GR, 25 Feb−1 Mar, 2019.
[7] R4−1814719, Update on RF EMF regulations of relevance for handheld devices operating in the FR2 bands, Ericsson, Sony, 3GPP RAN4#89, Spokane, WA, United States, 12 November−16 November, 2018.
[8] R4−1908820, Mitigating Radio Link Failures due to MPE on FR2, Nokia, Nokia Shanghai Bell, 3GPP RAN4#92, Ljubljana, Slovenia, August 26st−30th, 2019.
[9] 3GPP TS 38.521−1 V16.3.0, NR; User Equipment (UE) conformance specification; Radio transmission and reception; Part 1: Range 1 standalone (Release 16), 2020.
[10] 3GPP TS 38.521−2 V16.3.0, NR; User Equipment (UE) conformance specification; Radio transmission and reception; Part 2: Range 2 standalone (Release 16), 2020.
[11] 3GPP TS 38.521−3 V16.3.0, NR; User Equipment (UE) conformance specification; Radio transmission and reception; Part 3: Range 1 and Range 2 Interworking operation with other radios (Release 16), 2020.
[12] 3GPP TS 38.810 V16.5.0, NR; Study on test methods (Release 16) 2020.
[13] R4−1910609, WF on NR MIMO OTA, CAICT, 3GPP RAN WG4 Meeting #92, Ljubljana, Slovenia, 26−30 Aug, 2019.

[14] R4—1814833, 2D versus 3D MPAC Probe Configuration for FR1 CDL channel models, Keysight Technologies, 3GPP RAN WG4 Meeting #89, Spokane, USA, 12—16 November, 2018.

[15] R4—1900498, 3D MPAC System Proposal for FR1 NR MIMO OTA Testing, Keysight Technologies, 3GPP RAN WG4 Meeting #90, Athens, Greece, 25 Feb—1 Mar, 2019.

[16] R4—1909728, System implementation of FR1 2D MPAC, Keysight Technologies, 3GPP RAN WG4 #92, Ljubljana, SI, 26—30 Aug, 2019.

[17] R4—2002471, WF on finalizing FR2 MIMO OTA, CAICT, Keysight, 3GPP RAN WG4 Meeting #94-e, Electronic Meeting, Feb 24th—Mar 6th 2020.

[18] R4—2004718, 3D MPAC Probe Configuration Proposal, Spirent Communications, Keysight Technologies, 3GPP RAN WG4 Meeting #94Bis, e-meeting, April 20th — April 30th, 2020.

[19] 3GPP TR 38.827 V1.4.0, Study on radiated metrics and test methodology for the verification of multi-antenna reception performance of NR User Equipment (UE) (Release 16), 2020.

[20] Skyworks, 5G new radio solutions: revolutionary applications here sooner than you think, (white paper), 2018.

[21] Natarajan, Vimal; Daugherty, John; Black, Gregory; and Burgess, Eddie, Multi-antenna switch control in 5G, Technical Disclosure Commons, 2019.

CHAPTER 10

User plane protocol design

Cong Shi, Xin You and Xue Lin

10.1 Overview

The user plane refers to the protocol stack and procedures for transmitting UE data, corresponding to the control plane that takes charge of control signaling transmission. The focus of this chapter is the introduction of the user plane protocol and related procedures; more details on the control plane can be found in Chapter 11, Control Plan Design. Fig. 10.1 [1] shows the protocol stack for the user plane.

Compared with LTE, a new protocol sublayer, called the Service Data Adaptation Protocol (SDAP) sublayer, is added to 5G NR user plane. The protocol stack below the SDAP inherits the structure in LTE. The main functions of each sublayer of the NR layer 2 protocol stack (i.e., the protocol layers above PHY) are summarized as follows:

- SDAP sublayer: mapping between a Quality of Service (QoS) flow and a Data Radio Bearer (DRB).
- Packet Data Convergence Protocol (PDCP) sublayer: ciphering and deciphering, integrity protection, header compression, maintenance of PDCP sequence numbers, reordering and in-order delivery, etc. Different form LTE, out-of-order delivery can be supported in NR PDCP based on the network configuration. In addition, in order to improve the reliability of data packet transmission, NR PDCP also supports functions such as duplication, which is introduced in detail in Chapter 16, URLLC in High Layer.
- Radio Link Control (RLC) sublayer: segmentation and reassembly of RLC Service Data Unit (SDU), error detection, etc. Compared with LTE, NR RLC removes the function of concatenation.
- Medium Access Control (MAC) sublayer: mapping between logical channels and transport channels, multiplexing and demultiplexing of MAC SDUs, uplink and downlink scheduling, random access procedure, etc. Compared with LTE, NR MAC introduces some new features, such as BWP activation/deactivation, beam failure recovery, and so on.

5G NR and Enhancements
DOI: https://doi.org/10.1016/B978-0-323-91060-6.00010-6

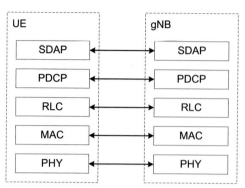

Figure 10.1 User plane protocol stack.

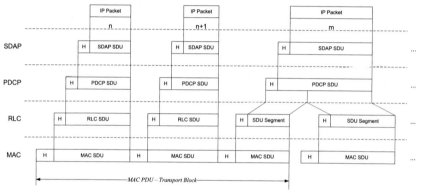

Figure 10.2 Data flow example.

Let's take UE data transmission as an example to introduce the work-flow of the user plane. The user data in the form of QoS flows (refer to Chapter 13: QoS Control, for more details) first arrives at the SDAP sub-layer. The SDAP sublayer is responsible for mapping the data from different QoS flows to different DRBs and generating SDAP Packet Data Units (PDUs) marked with QoS flow identification according to the network configuration. SDAP delivers the SDAP PDUs to the PDCP sublayer where header compression, encryption, integrity protection, etc., are performed to generate PDCP PDUs submitted to the RLC sublayer. The RLC sublayer processes the segmentation, retransmission, etc., of RLC SDUs based on the configured RLC mode. In NR RLC, the function of concatenation is removed. The MAC layer multiplexes the data from each logical channel into a MAC PDU, which is also known as a transport block. A MAC PDU could consist of multiple RLC SDUs or RLC SDU segmentations, which may come from different or the same logical channels. Fig. 10.2 [1] illustrates a typical Layer 2 data flow. As can be seen from the figure, for a certain protocol layer, the data received from the upper layer is referred to as an SDU, and the data generated by adding the corresponding protocol layer header after processed by the protocol layer is referred to as a PDU.

The 5G NR user plane is designed to support various requirements of QoS, data rates, reliability, and delay [2]. Some enhancements are made in the following aspects:

- Enhancement of MAC [3]: A Scheduling Request (SR) configuration based on logical channel is introduced, so that the UE can trigger the corresponding SR according to different service types, thus enabling the network to schedule the uplink resources appropriately. In addition, configured grant (CG) Type1 can be provided by the RRC, which is activated once configured and aims to achieve fast data transmission. Regarding the multiplexing and assembly of MAC PDU, logical channel priority (LCP) should be subject to the attributes of physical layer resources, so as to better meet the QoS requirements. Furthermore, the enhanced interleaving format of MAC PDU is beneficial to the fast data processing at the terminals.
- Enhancement of PDCP and RLC [4,5]: The majority of functions of PDCP and RLC inherit LTE. In order to support faster data processing, two main enhancements are made at PDCP. On the one hand, the RLC sublayer removes the concatenation function and only retains the data packet segmentation function. The purpose is to enable the terminal to prepare the data in advance without receiving the indication of the physical layer resource. On the other hand, the reordering function at RLC in LTE is removed; instead, the reordering and in-order delivery are all handed over to PDCP. Out-of-order delivery can also be supported based on the network configuration. In addition, duplication is introduced to improve the reliability of data transmission, which is described in Chapter 16, URLLC in High Layer.
- Another two important enhancments can also be found in this capture: the one is two-step Ransom Access Channel (RACH) procdure which aims to support lower delay. The other is the new feature of PDCP due to introduction of mobility enhancement based on Dual Active Protocol Stack (DAPS).

In the following sections, we introduce the enhancements of each protocol sublayer from SDAP to MAC.

10.2 Service data adaptation protocol

SDAP is a newly added protocol sublayer in the NR user plane [6]. The data is carried on per DRB basis in air interface transmission, while a more refined QoS flow design is introduced in the NR core network. Therefore the main function of the SDAP sublayer is to perform the mapping between QoS flows and DRBs according to the rules configured by the network. One or more QoS flows can be mapped onto one DRB, while one QoS flow can be mapped onto one DRB. Fig. 10.3 [6] illustrates the structure for the SDAP sublayer.

As can be seen from the figure, the UE can support the configuration of multiple SDAP entities, and each SDAP entity corresponds to a PDU session. A PDU session (see Chapter 13: Quality of Service Control, for details) corresponds to the data of one or more QoS flows. The main functions of the SDAP sublayer include two aspects: one is to transfer user plane data, and the other is to ensure in-order delivery.

For uplink data transmission, when the UE receives the SDAP SDU from the upper layer, it maps the data of different QoS flows to the corresponding DRBs according to the stored mapping rule. QoS flows would be mapped to a default DRB if this mapping rule is not available. After determining the mapping rule, the UE generates an SDAP PDU based on the network configuration and submits it to the lower layer. The SDAP PDU may or may not carry the SDAP header according to the network configuration.

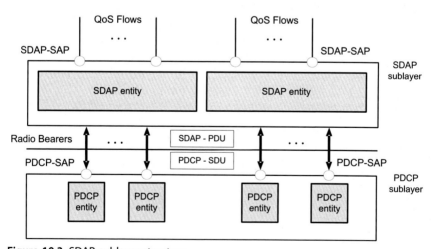

Figure 10.3 SDAP sublayer structure.

For downlink data reception, when the UE receives an SDAP PDU from the lower layer, the SDAP sublayer performs different processing according to whether the SDAP PDU is configured with a SDAP header. If the data is not configured with a SDAP header, it can be directly delivered to the upper layer. Otherwise, the SDAP sublayer needs to remove the SDAP header before delivering the data to the upper layer. The processing is based on the information carried in the packet header, such as indicating whether the reflective mapping (see more details in Chapter 13: Quality of Service Control) is activated. The reflective mapping is used to store the QoS flow to the DRB mapping of the downlink SDAP data PDU as the QoS flow to DRB mapping rule for the uplink, thus saving the signaling overhead to configure the mapping rule.

For the uplink, the change of mapping rule may mean the DRB is subject to the new rule is received before than the old one. This may lead to out-of-order delivery since the SDAP sublayer is not capable of reordering. In order to solve this problem, End-Marker based mechanism is supported in the SDAP sublayer. The mechanism enables the SDAP sublayer on the UE side to carry an End-Marker in the old DRB when a QoS flow is remapped, so that the SDAP on the receiving side does not deliver the new DRB to the upper layer before receiving the End-Marker from the old DRB, thus ensuring in-order delivery.

For downlink, the mapping between QoS flows and DRBs is determined by network implementation. After the UE receives the SDAP PDU, it can restore the SDAP PDU to the SDAP SDU and deliver it to the upper layer.

10.3 Packet data convergence protocol

For uplink, the PDCP sublayer is mainly responsible for processing the PDCP SDU received from the SDAP sublayer and generating PDCP PDU submitted to the RLC sublayer. For downlink, the PDCP layer is mainly responsible for receiving PDCP PDU submitted from the RLC sublayer and delivering them to SDAP sublayer after removing the PDCP header. The PDCP and radio bearer follow one-to-one mapping; that is, each radio bearer (including SRB and DRB, excepting for SRB0) is associated with one PDCP entity. Most functions of NR PDCP are similar to LTE, mainly including:
- Maintenance of PDCP SNs
- Header compression and decompression
- Ciphering and deciphering, integrity protection

- Timer based SDU discard
- For split bearers and DAPS bearer, routing
- Duplication
- Reordering and in-order delivery

The PDCP sublayer is similar to LTE in the data transfer process. One improvement is that NR PDCP adopts the COUNT-based method in local variable maintenance and condition comparison. This enhancement can greatly improve the readability of the protocol [7]. The COUNT value with a fixed size of 32 bits consists of SN and a hyperframe number. It should be noted that the header of PDCP PDU still only contains SN instead of COUNT value, so it does not increase the overhead of air interface transmission.

Specifically, for uplink transmission, the transmitting PDCP entity will associate the COUNT value corresponding to TX_NEXT, which is initially set to 0. Each time a new PDCP PDU is generated, the SN included in the PDCP header is set to the value of TX_NEXT, and the TX_NEXT is added by 1. The transmitting PDCP entity performs header compression, integrity protection, and ciphering on the PDCP SDU according to the network configuration [5]. It should be noted that the header compression function of NR PDCP is not applicable to the header of SDAP.

For downlink reception, the receiving PDCP entity maintains a receiving window according to the COUNT value, and the window is maintained by the following local variables:

- RX_NEXT: COUNT value of the next expected PDCP SDU.
- RX_DELIV: COUNT value corresponding to the next PDCP SDU expected to be delivered to the upper layer, which determines the lower boundary of the receiving window.
- RX_REORD: COUNT value corresponding to PDCP PDU that triggers the reordering timer.

During the standardization stage, two potential mechanisms for the receiving window were discussed [8]: PULL window and PUSH window. In brief, the PULL window uses local variables to maintain the upper bound of a receiving window, while the lower bound is the upper bound minus the window length. The PUSH window uses local variables to maintain the lower bound of a receiving window, while the upper bound is the lower bound plus the window length. Essentially, these two window mechanisms can both work. Finally, the PUSH window-based mechanism is adopted, because it is simpler from the perspective of the protocol description. Based on the PUSH window mechanism, the receiving PDCP entity performs the following steps for the received PDCP PDU:

- The SN of the PDCP PDU is mapped to the COUNT value; when determining the COUNT value, it is necessary to calculate the hyper-frame number of the received PDCP PDU, denoted by RCVD_HFN. The COUNT value is [RCVD_ HFN, RCVD_ SN].
- The receiving PDCP entity decides whether to discard the PDCP PDU according to the calculated COUNT value. The conditions for discarding a PDCP PDU are as follows:
  - Integrity verification fails.
  - PDCP PDU is outside the PUSH window.
  - The PDCP PDU has been received before.
- If the PDCP PDU is not discarded, the receiving PDCP entity will store the resulting PDCP SDU in the reception buffer, and update the local variables according to the COUNT value of the received packets. The conditions are divided into the following categories, as shown in Fig. 10.4.
  - When COUNT value is in Case 1 and 4, the PDCP PDU is discarded.
  - When COUNT value is in Case 2 (i.e., it is within the PUSH window but less than RX_NEXT), the local RX_DELIV value is updated accordingly, and the stored PDCP SDU is delivered to the upper layer.
  - When COUNT value is in Case 3, the RX_NEXT value is updated.

It should be noted that the PDCP sublayer can also be configured with out-of-order delivery. The receiving PDCP entity can directly deliver the generated PDCP SDU to the upper layer without waiting for any unreceived data packet, which means the transmission delay can be further reduced.

Duplication is also supported in NR PDCP. Simply speaking, PDCP PDU can be duplicated into two identical copies according to the configuration and activation indication from the network. The copies are

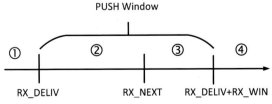

Figure 10.4 The COUNT value of a received PDCP PDU.

submitted to different RLC entities. More details on duplication are described in Chapter 16, URLLC in High Layer.

In addition, the mobility enhancement based on DAPS handover also has an impact on the PDCP protocol. Refer to Chapter 11, Control Plan Design, for a detailed introduction of DAPS handover, since this chapter only introduces the impact of DAPS handover on the PDCP sublayer. Specifically, DAPS handover means the UE initiates a random access procedure to the target cell while maintaining connection with the source cell during the handover execution period. The main impacts of this design on the PDCP sublayer are as follows.

First, for a DRB configured with DAPS, the PDCP entity of the DRB is configured with separate security functions and header compression protocols. One is for the data processing in the source cell and the other is for the data processing in the target cell. The DAPS PDCP entity determines the header compression and security algorithm to be used in data processing based on whether the data is for the source cell or the target cell.

Second, NR PDCP introduces PDCP entity reconfiguration. Specifically, the transfer between the PDCP entity and DAPS PDCP entity is defined as PDCP reconfiguration. When the upper layer indicates the reconfiguration of the PDCP entity (PDCP entity is reconfigured to the DAPS PDCP entity), the UE establishes a corresponding ciphering function, integrity protection function, and header compression protocol for the DRB based on the ciphering algorithm, integrity protection algorithm, and key and header compression configuration provided by the upper layer. That is, when the DAPS handover command is received, the UE adds the header compression protocol and corresponding security functions to the target cell. Conversely, when the upper layer reconfigures the DAPS PDCP entity to the PDCP entity, the UE releases the ciphering function, integrity protection function, and header compression protocol corresponding to the cell to be released. For example, if the RRC layer instructs to reconfigure the PDCP when the DAPS handover completes and releases the source cell, the UE deletes the header compression protocol and security function corresponding to the source cell. If the RRC layer instructs to reconfigure PDCP when the DAPS handover fails and returns to the source cell, the UE releases the header compression protocol and security function corresponding to the target cell.

Third, DAPS handover would impact the trigger condition of the PDCP sublayer status report. The PDCP status report is mainly used to inform the network side of the current reception status of downlink data, and is used for effective data retransmission or new transmission from the network side. The existing PDCP status report is triggered at the time of PDCP reestablishment or PDCP data recovery, and only for AM DRB. For DAPS handover, the source cell can forward data to the target cell during the handover preparation period, and the UE still keeps sending and receiving data with the source cell during the DAPS handover execution. As a result, when the UE successfully establishes a connection with the target cell and starts data interaction, the target cell may send redundant data to the UE, which leads to extra overhead. The same situation happens when releasing the source cell. Thus for AM DRB, DAPS handover introduces a new trigger condition of PDCP status report (i.e., when the upper layer requests an uplink data switching or releasing the source cell, the PDCP status report is triggered). For UM DRB, the status report is only triggered when the upper layer requests an uplink data switching since UM DRB is delay sensitive and does not support retransmission (the PDCP status report is mainly to avoid the redundant data transmission).

10.4 Radio link control

Similar to LTE RLC, the NR RLC sublayer supports three types of transmission modes: Transparent Mode (TM), Unacknowledged Mode (UM), and Acknowledged Mode (AM). The features of these three RLC modes are summarized as follows:

- RLC TM: when RLC is in TM mode, the RLC sublayer directly submits an RLC SDU without any modification to the lower layer or delivers an RLC PDU without any modification to the upper layer. TM mode is applicable to logical channels for broadcast, public control, and paging.
- RLC UM: when RLC is in UM mode, the RLC sublayer processes RLC SDU, including segmentation and adding headers. The UM is generally suitable for logical channels with lower requirements in reliability but sensitive to delay, such as logical channels for voice service.
- RLC AM: when RLC is in AM mode, the RLC sublayer has the function of UM and can also support data reception feedback. AM is suitable for logic channels requiring high reliability.

The functions of NR RLC are summarized as follows [4]:
- Data transfer
- ARQ-based error correction (AM only)

- Supports RLC SDU segmentation (AM and UM) as well as resegmentation (AM only)
- RLC SDU discard (AM and UM) and other functions

The data transfer procedure would be different if the RLC is configured with different modes. The following are some enhancements of NR RLC over LTE RLC.

The first enhancement is to remove the data packet concatenation in transmitting RLC entity. In LTE, the RLC sublayer supports data packet concatenation, which means the RLC sublayer concatenates PDCP PDU into an RLC PDU according to the resource size indicated by the lower layer. Then the MAC sublayer assembles the concatenated RLC PDUs into a MAC PDU. This means that the size of uplink scheduling resources needs to be known in advance for the concatenation of data packets. This processing would cause extra processing delay when generating a MAC PDU.

In order to optimize the data processing delay, NR RLC removes the concatenation function, which means that an RLC PDU at most contains one SDU or one SDU segment. In this way, the RLC sublayer can prepare the data packet in advance without considering the physical resource indication, and the MAC sublayer can also generate the corresponding MAC subheader. The only processing delay that needs to be considered is that when a given physical resource cannot completely contain the generated RLC PDU, the RLC sublayer still needs to perform the segmentation or resegmentation. However, the delay is still greatly reduced compared with legacy LTE RLC.

Another enhancement of NR RLC is that the in-order delivery of SDU is no longer supported in receiving RLC entity. That is to say, from the perspective of the RLC sublayer, if a complete RLC PDU is received from the lower layer, the RLC sublayer can directly deliver it to the upper layer after removing the RLC PDU header, regardless of whether the RLC PDU with the previous serial number has been received [9]. This purpose is mainly to reduce the processing delay of data packets. If an SDU segment is included in the RLC PDU, the receiving RLC entity can not deliver the SDU segment to the upper layer, but needs to wait until other segments are received.

10.5 Medium access control

The structure of the MAC sublayer is shown in Fig. 10.5.

Most functions of the NR MAC sublayer follow LTE design, such as uplink and downlink data transfer, random access, discontinuous

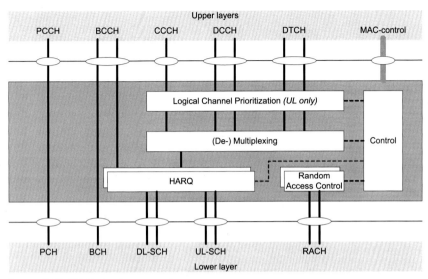

Figure 10.5 MAC structure overview.

reception, etc. But some specific features and enhancements are introduced into NR MAC according to NR requirements.

For random access, the four-step contention-based random access procedure in NR MAC inherits LTE. Some improvements are made to support beam management. Furthermore, a two-step random access procedure is introduced to further reduce the delay and signaling overhead.

In terms of data transmission, NR MAC supports both dynamic scheduling and nondynamic scheduling. In order to better support the URLLC service, NR MAC optimizes the procedure of SR and Buffer Status Report (BSR). In addition, the multiplexing and assembly based on LCP has been optimized accordingly, so that specific logical channel data can be multiplexed to more suitable uplink resources for transmission. For nondynamic scheduling, Semi-persistent Scheduling (SPS) supported by LTE is also reused in NR, except that NR introduces a new uplink configuration resource referred to as CG Type1 and the uplink SPS supported by LTE is referred to as CG Type2.

In terms of power saving, the NR Discontinuous Reception (DRX) mechanism in the time domain is basically the same as that in LTE. In the subsequent evolution, the wake-up mechanism is introduced in order to better save power, which has an influence on DRX (see more details in Chapter 9: 5G RF Design). In the frequency domain, the most significant feature of NR is the introduction of BWP, and the NR MAC sublayer

Table 10.1 Features of medium access control (MAC).

Function	LTE MAC	NR MAC
Random access	Four-step contention-based random accessFour-step contention-free random access	Four-step contention-based random accessFour-step contention-free random accessTwo-step contention-based random accessTwo-step contention-free random access
DL data transmission	HARQ-based downlink transmission	HARQ-based downlink transmission
UL data transmission	Scheduling requestBuffer status reportLogical channel prioritization	Enhanced scheduling requestEnhanced buffer status reportEnhanced logical channel prioritization
Semi-persistent scheduling	DL SPSUL SPS	DL SPSCG type1CG type2
DRX	DRX	DRX
MAC PDU	UL/DL MAC PDU	Enhanced UL/DL MAC PDU
NR specific		BWPRACH-based BFR

supports the BWP activation and deactivation mechanism, which is described in detail in Chapter 5, 5G Flexible Scheduling.

NR MAC also made some enhancements in MAC PDU format, which is embodied in support of the so-called interleaved MAC PDU format to improve data processing efficiency.

Finally, the NR MAC sublayer supports new features, such as beam failure recovery and activation/deactivation of PDCP duplication.

Some enhanced features of NR MAC are summarized in (Table 10.1).

Some features of the NR MAC sublayer are described in detail in the following.

10.5.1 Random access procedure

Random access is a basic procedure in the MAC sublayer. NR MAC follows LTE's four-step contention-based random access (four-step CBRA) procedure and contention-free random access procedure (CFRA). In the NR, due to the introduction of beam operation, preamble resources are associated with reference signals, such as SSB or CSI—RS, so that beam management can be performed based on RACH. Simply speaking, when sending a preamble, the UE measures the corresponding reference signal and select a reference signal whose signal quality meets the conditions, and uses the random access preamble resource associated with this reference signal to transmit the corresponding preamble. With this association, the network can know which reference channel is better from the UE side when it receives the preamble sent by the UE, and send downlink data with the beam direction corresponding to this reference signal. Refer to Chapter 8, MIMO Enhancement and beam Management, for the detailed procedure. In addition, the NR introduces some new random access trigger events, such as random access for beam failure recovery and requesting system information.

In the evolution of standards, in order to further optimize the random access procedure, a two-step contention-based random access (two-step CBRA) procedure and two-step contention-free random access (two-step CFRA) procedures were introduced, aimed at reducing the delay and signaling overhead. In addition, considering that the unlicensed spectrum would be supported in NR, a two-step random access procedure can further reduce the probability of channel preemption, thus improving the spectrum utilization. For a detailed description of the unlicensed spectrum, refer to Chapter 18, 5G NR in Unlicensed Spectrum. Taking the CBRA procedure as an example, the four-step CBRA and two-step CBRA are shown in Fig. 10.6.

Four-step CBRA requires four signaling interactions.

- Step 1: UE selects random access resources and transmits preamble, which is called as message 1 (Msg1). Before sending Msg1, the UE needs to measure the quality of the reference signal to select a relatively good reference signal and corresponding random access resources and preamble.
- Step 2: The UE receives Random Access Response (RAR) sent by the network within a preconfigured receiving window, which is called message 2 (Msg2). The RAR includes timing advance command, uplink grant, and TC—RNTI for subsequent uplink data transmission.
- Step 3: the UE performs uplink transmission according to the scheduling information in the RAR [i.e., the transmission of message 3 (Msg3)]. The UE identifier for subsequent contention resolution will be carried in Msg3. This identifier will be different according to the RRC state of the UE. UEs in RRC connected state will carry

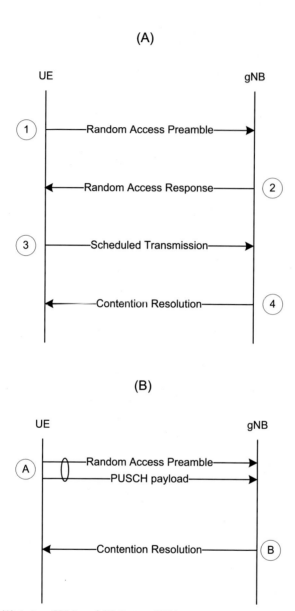

Figure 10.6 (A) 4-step CBRA and (B) 2-step CBRA.

C-RNTI in Msg3, while UEs in RRC idle state and inactive state will carry a UE identity generated by th RRC layer in Msg3. No matter what form of identifier, this identifier can help the network identify the UE.

- Step 4: after sending Msg3, the UE will receive the contention resolution message sent by the network. If the network can successfully receive the Msg3 sent by the UE, the network has already identified the UE; that is to say, the contention is resolved from the network side. If the UE can detect the contention resolution identification in message 4 (Msg4), the contention is also resolved from the UE side.

On the basis of four-step CBRA, the NR further introduces two-step CBRA, which contains only two signaling interactions. To be specific, the first message is called message A (MsgA), which consists of the preamble transmitted on random access resources and the payload transmitted on PUSCH, corresponding to Msg1 and Msg3 in four-step CBRA. The second message is called message B (MsgB), which corresponds to Msg2 and Msg4 in four-step CBRA. MsgB is described in detail later.

The trigger events of four-step CBRA are also applicable to two-step CBRA. Therefore if a certain event triggers a CBRA procedure while the UE is configured with two types of CBRA resources, the UE needs to know which random access type to choose. There are two main schemes for type selection in the process of standardization [10]. One scheme is based on the radio link quality. The network configures the UE with the threshold for determining the radio link quality, and the UE selects two-step CBRA if the measured RSRP meets the threshold. With this criteria, only when the channel quality is good enough can the UE try to use two-step CBRA. The purpose is to improve the successful reception probability of MsgA. The disadvantage of this scheme is that when a large amount of UEs meet the threshold, they suffer serious contention under two-step CBRA. The second scheme is a random number-based selection scheme. The network broadcasts a load factor to the UEs according to the configuration of random access resources, and the UEs compare the generated random number with the load factor to determine the random access type. Load balance between the two types can be achieved with the second scheme. Considering the feedback from the physical layer and the simplicity of the scheme, the wireless quality-based method is finally determined as the type selection criteria.

After the UE selects the two-step random access process and transmits the MsgA, it needs to monitor MsgB within the configured window. Referring to four-step CBRA, UEs in different RRC connection states will have different behaviors of MsgB monitoring. Generally speaking, when the UE is in the RRC connection state (i.e., when the UE carries C—RNTI in MsgA), the UE monitors the PDCCH scrambled by

C—RNTI and the PDCCH scrambled by MsgB—RNTI. When the UE is in the RRC idle or inactive state (i.e., when the UE carries the RRC message as an identification in MsgA), the UE monitors the PDCCH scrambled by MsgB—RNTI. MsgB—RNTI is calculated based on the time-frequency information of random access resources selected by UE, which reused the design of RA—RNTI. Considering that two-step random access and four-step random access would share the same PRACH occasions, an offset is added in MsgB—RNTI calculation in order to avoid confusion.

The MsgB message, as mentioned above, corresponds to Msg2 and Msg4 in four-step CBRA. Therefore the design of MsgB needs to consider the functions of Msg2 and Msg4. On the one hand, MsgB should support the contention resolution, such as including the contention resolution identification of the corresponding UE. On the other hand, MsgB should also provide the function as Msg2, such as backoff indication and the contents in RAR. When the network is able to successfully decode the contents of MsgA (preamble and payload of MsgA), the network can send a contention resolution message through MsgB, which corresponds to the function of Msg4. If the network only detects the preamble of MsgA but does not decode the MsgA payload successfully, the network does not recognize the UE. In this case, the network can still send a fallback indication (corresponding to the function of Msg2) through the MsgB to indicate the UE to send Msg3. This procedure is defined as fallback, as shown in Fig. 10.7.

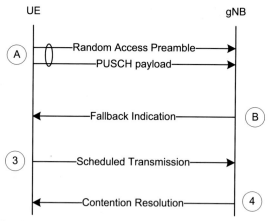

Figure 10.7 Fallback for two-step CBRA.

If the access is still not completed after Msg3 transmission, the UE can go back to attempt MsgA. The network can configure the maximum number of MsgA transmission. If the random access procedure with two-step RA type is not completed after the maximum transmission times, the UE can switch to CBRA with four-step RA type.

10.5.2 Data transmission procedure

The downlink data transmission process basically follows the design of LTE, and dynamic scheduling based on C−RNTI and nondynamic scheduling based on RRC configuration are also supported in NR MAC. Since the scheduling algorithm of downlink scheduling depends on the implementation of the base station, it is not introduced here.

For uplink data transmission, the NR supports dynamic scheduling and nondynamic scheduling. For uplink dynamic scheduling, NR MAC supports SR and BSR, and made some enhancements compared with LTE.

First of all, for SR, the function is to request for dynamic scheduling resources when the UE has uplink data to transmit. In LTE, SR can only inform the network whether the UE has data to transmit. When the network receives SR sent by the UE, it can only schedule one uplink resource based on implementation. In the NR, because the UE supports various types of services, the attributes of physical layer resources supported by the NR are different. Some physical resources may be more suitable for transmitting delay-sensitive services, while some are more suitable for transmitting high-throughput services. Therefore in order to let the network know the data type the UE wants to transmit the first time, the NR enhances SR for more targeted scheduling. Different logical channels can be mapped to different SR configurations according to the RRC configuration, so that when a certain logical channel triggers SR transmission, the UE can use corresponding configuration resources to transmit the SR. The network side can deduce which logical channel or channels the received SR corresponds to according to the relationship between logical channels and SR configuration, so that more suitable physical resources can be scheduled for the UE.

The function of BSR is similar to LTE. It is used to report the current buffer status of the UE to the network, so that the network can further schedule uplink resources. NR further increases the number of logical channel groups to 8 on the basis of LTE. The main purpose is to support precise logical channel BSR to facilitate more accurate network scheduling. With the increase of the number of logical channel groups, the format of BSR MAC CE has changed. The NR mainly supports long and short BSR MAC CE formats. The long MAC CE format can be divided into Long BSR MAC CE with variable length and Long Truncated BSR MAC CE with variable length. The format of short MAC CE can be divided into Short BSR MAC CE and Short Truncated BSR MAC CE. Truncated BSR MAC CE is a newly introduced BSR MAC CE type in NR, which is mainly used for the UE to let the network know that some data of logical channel groups are not put in the resources for reporting when the uplink resources are insufficient and the UE has more than one buffered data to report. Different BSR MAC CE formats are mainly used in different scenarios, which are similar to LTE.

When the UE obtains uplink transmission resources, it needs to assemble corresponding MAC PDU according to the size of the uplink resources. An LCP-based multiplexing and assembly procedure is adopted in the NR MAC sublayer. Compared with LTE, this process has been optimized. In LTE, for uplink transmission, the MAC sublayer will decide the order and size of allocating resources to each logical channel according to parameters such as the priority allocated to logical channels by RRC layer. There is no difference in physical transmission characteristics of a given grant resource, such as subcarrier spacing and other parameters, and the LCP process is the same for all logical channels of data to be transmitted. That is to say, for a given resource, if the size of the resource is large enough, theoretically all data of logical channels can be transmitted on the resource. In NR, in order to support services with different QoS requirements, data of different logical channels needs to be transmitted on uplink resources with specific physical transmission attributes. For example, URLLC services need to meet their delay requirements when the subcarrier spacing is large enough and the duration of physical transmission channel (PUSCH) is short enough. In order to distinguish different resource attributes, the LCP process needs to be enhanced. At the study stage of NR, some conclusions were formed, and it was believed that the network needs to support some control mode so that the data of different logical channels can be mapped to physical resources with different attributes.

In order to achieve the above purpose, in the NR standardization stage, a mechanism based on Transmission Profile is discussed to limit the transmission of logical channel data in specific physical resources [11]. The RRC layer configures one or more transmission characteristics for the

UE, and each characteristic is associated with a unique identifier and is mapped to a set of physical layer parameters. The RRC layer also configures one or more transmission feature identifiers for the logical channel, indicating that the logical channel can only be transmitted on authorized resources with corresponding physical layer parameters. When the UE obtains the uplink resource, the UE selects the logical channels that can match the transmission characteristic identifier to perform LCP. Considering the complexity of implementation and forward compatibility, the way to associate a series of physical resource parameters to a feature identifier was not agreed on. Thus a simpler logical channel selection scheme was specified, in which each logical channel is directly configured with a series of physical layer resource parameters as follows:

- *allowedSCS-List*: set the allowed subcarrier spacing for transmission. The data can only be transmitted on the resources with the allowed subcarrier spacing.
- *maxPUSCH-Duration*: this parameter specifies that the data of this logical channel can only be transmitted on PUSCH resources smaller than this parameter.
- *configureGrantType1Allowed*: this parameter specifies whether the logical channel is allowed to use CG type1 resources for transmission.
- *allowedServingCells*: this parameter specifies the allowed serving cells for the data of this logical channel.

Upon receiving an uplink resource, the MAC sublayer determines the physical attributes of the grant, such as the subcarrier interval, the PUSCH duration, and so on, according to the related indications or configured parameters of this uplink resource. After determining the attributes, the MAC sublayer filters out the logical channels with configured parameters matching the physical attributes of the grant. For these selected logical channels, the MAC sublayer multiplexes the data based on the LCP procedure.

The aforementioned uplink transmission resources can be dynamically scheduled by the network, or preconfigured by the RRC layer. For non-dynamically scheduled transmission resources, the NR reuses the uplink SPS in LTE, which is referred to as the CG Type2 in NR. For the CG Type2, the RRC layer configures a set of parameters such as the period, the number of HARQ processes, etc., and the resource is activated or deactivated by PDCCH. In addition, in the NR, in order to better support services with ultralow latency requirements, the CG Type1 is introduced, which can be activated and used once it is configured. Specifically, the RRC layer provides related configuration parameters such as period, time offset, frequency domain location, etc. When the UE receives the RRC configuration, the resource is activated, which can reduce the time delay caused by activation.

10.5.3 Medium access control packet data units format

Compared with LTE, the format of NR MAC PDU has also been enhanced. As early as the NR research stage, it was decided that the format of NR MAC PDU needs to adopt an interleaving structure. The MAC subheaders contained in NR MAC PDU are close to their corresponding payloads, which is different from LTE. In LTE, all the payload subheaders are put at the front of the whole MAC PDU [12−14]. The advantage brought by this interleaving structure of NR MAC PDU is that the receiver can adopt a similar "pipeline" processing mode when processing MAC PDU. The MAC subheaders and their corresponding payload can be treated together, instead of waiting for all the subheaders and loads in MAC PDU to be treated as a complete MAC PDU. This can effectively reduce the processing delay.

After deciding the interleaving structure of MAC PDU, another issue is the placement of MAC subheaders [12,13]. One view is that the MAC subheader should be placed in front of the corresponding payload. The main reason is that it can speed up the processing at the receiver by processing the MAC subPDUs one by one from front to back. The main reason for placing the MAC subheader behind the corresponding payload is that the receiver can process the MAC CE from the back to the front. Considering that the MAC CE is generally placed behind, the scheme can help the receiver to process the MAC CE quickly. Finally, after discussion, it is considered that the benefits brought by the scheme of putting the MAC subheader behind the corresponding payload are not as obvious as those brought before, and would increase the processing complexity at the receiver, so it was finally decided to put the MAC subheader before the corresponding payload.

There are also different views on the location of the MAC CE in the whole MAC PDU. According to one view, the MAC CE is placed at the front of downlink MAC PDU, and is placed at the back of uplink MAC PDU [12]. Another view is that the MAC CE positions of uplink and downlink MAC PDU should be unified and placed at the end [13,14]. Generally speaking, the reasonable way should be to let the receiver process the control information first as much as possible. For the downlink, the MAC CE can be generated before scheduling, so it can be put at the front. However, for the uplink, some MAC CEs cannot be generated in advance, but need to wait until the uplink resources are available. Forcing these MAC CEs to be placed at the front of the whole MAC PDU will slow down the generation speed of the MAC PDU. Therefore in the MAC sublayer, the MAC CE of the downlink MAC PDU is placed at the front, while the position of the uplink MAC CE is placed at the back of the MAC PDU.

10.6 Summary

This chapter mainly introduced the user plane protocol stack and related procedures. According to the protocol stack from top to bottom, it introduced SDAP-related functions, PDCP sublayer data transfer procedures, the impacts of DAPS on PDCP, and the enhancements of RLC. Finally, some specific features of the MAC sublayer compared with LTE were introduced.

References

[1] 3GPP TS 38.300 V15.8.0 (2019−12). NR and NG-RAN overall description; stage 2 (release 15).
[2] 3GPP TS 38.913 V15.8.0 (2019−12). Study on scenarios and requirements for next generation access technologies (release 15).
[3] 3GPP TS 38.321 V15.8.0 (2019−12). Medium Access Control (MAC) protocol specification (release 15).
[4] 3GPP TS 38.322 V15.5.0 (2019−03). Radio Link Control (RLC) protocol specification (release 15).
[5] 3GPP TS 38.323 V15.6.0 (2019−06). Packet Data Convergence Protocol (PDCP) specification (release 15).
[6] 3GPP TS 37.324 V15.1.0 (2018−09). Service Data Adaptation Protocol (SDAP) specification (Release 15).
[7] R2−1702744 PDCP TS design principles. Ericsson.
[8] R2−1706869 E-mail discussion summary of PDCP receive operation. LG Electronics Inc.
[9] R2−166897 Reordering in NR. Intel Corporation.
[10] R2−1906308 email discussion report: Procedures and mgsB content. ZTE Corporation.
[11] R2−1702871 Logical Channel Prioritization for NR. InterDigital Communications.
[12] R2−1702899 MAC PDU encoding principles. Nokia, Alcatel-Lucent Shanghai Bell.
[13] R2−1703511 Placement of MAC CEs in the MAC PDU. LG Electronics Inc.
[14] R2−1702597 MAC PDU format. Huawei, HiSilicon.

CHAPTER 11

Control plan design

Zhongda Du, Shukun Wang, Haitao Li, Xin You and Yongsheng Shi

11.1 System information broadcast

5G NR system information is in general similar to 4G LTE in terms of content, broadcast, update, acquisition methods, and validity. But the 5G NR also introduces some new mechanisms such as on-demand SI requests so that the UE can "ask for" system information messages.

11.1.1 Content of system information

Similar to the LTE the content of 5G NR system information is also defined in the form of message SIBs (System Information Blocks). It can be divided into main message block (MIB); SIB1 (also known as RMSI); and SIBn (n = $2 \sim 14$). With the exception that MIB and SIB1 are single RRC messages, different SIBn can be multiplexed into one RRC message at the RRC layer, which is called Other System Information (OSI). The specific SIBn contained in an OSI is scheduled in SIB1.

During the initial access, after obtaining time-frequency domain synchronization the first action is to acquire MIB. The main function of the parameters included in the MIB is simply to let the UE know whether the current cell is allowed to camp on and whether it is broadcasting SIB1, and the configuration information of the downlink control channel of SIB1. If the MIB indicates that SIB1 is not broadcast in the current cell, the UE is not allowed to access the cell. If SIB1 is broadcast, then the UE will acquire SIB1 further. A special case is that the UE will acquire SIB1 after acquisition of MIB and the completion of handover procedure according to the random access information received in handover command.

There are two other important information elements in MIB related to cell selection and reselection. The *CellBarred* information element indicates whether the current cell is barred. If so, the UE in the RRC_IDLE or RRC_INACTIVE state cannot camp on such a cell. This is because some cells in the network may not be suitable for the UE to camp on, such as the cells on the SCG (secondary cell group) node in the EN−DC

5G NR and Enhancements
DOI: https://doi.org/10.1016/B978-0-323-91060-6.00011-8

(LTE NR dual connectivity) architecture. The SCG may be configured to the UE only after the UE has established an RRC connection. Letting the UE know that the cell is barred from accessing allows the UE to skip the subsequent process of cell camping to save energy consumption. The *intraFreqReselection* information element related to cell selection/reselection indicates whether the UE is allowed to select or reselect to the second best cell on the current frequency layer if the best cell on the current frequency layer is not allowed to camp. If the value of this cell is "*not allowed*," the UE needs to select/reselect to other frequencies; otherwise it is fine. Generally, if there is only one frequency in the network, this value will be set to "*allowed*" because there is no choice; otherwise, it will always be set to "*not allowed*." The disadvantage of the UE working in the suboptimal cell is that it will suffer from the interference of the best cell. The effective time for barring is 300 seconds. This means that the UE needs to recheck whether this restriction still exists after 300 seconds, until the UE is allowed to select or reselect the cell.

After the UE acquires the MIB, it will continue to acquire SIB1 in general. SIB1 mainly contains the following types of information:

- Cell selection parameters related to the current cell
- General access control parameters
- The configuration parameters of the common physical channel related to the initial access process
- Configuration parameters for system information request
- OSI scheduling information and area validity information
- Other parameters, such as whether to support emergency calls, etc.

For the specific details of the unified access control parameters, refer to Section 11.3.1. For the specific details of the configuration parameters of the common physical channel related to the initial access process, refer to Chapter 6, NR Initial Access. Refer to Section 11.1.2 for details of system information request configuration parameters and OSI scheduling information, and refer to Section 11.1.3 for the validity of OSI.

Table 11.1 lists the SIB content.

SIB2/3/4/5 is related to cell reselection and details can be found in Section 11.4.2. SIB6/7/8 broadcast public safety-related messages. The broadcast and update methods of these three SIBs are different from other OSIs. For details, refer to Section 11.1.2. SIB9 provides a globally synchronized time which can be used, for example, to initialize GPS or correct the internal clock of the UE.

Table 11.1 SIB content.

SIB	SIB content
SIB2	Common parameters for intrafrequency or interfrequency reselection or reselection between LTE and NR, and other parameters other than neighboring cells required for intrafrequency cell reselection
SIB3	Configuration parameters of neighboring cells required for intrafrequency reselection
SIB4	Neighboring cell and other frequency configuration parameters required for interfrequency cell reselection
SIB5	LTE frequency and neighboring cell configuration parameters required for reselection to LTE cell
SIB6	ETWS (Earthquake and Tsunami Warning System) main notification message
SIB7	ETWS second notification message
SIB8	CMAS (commercial mobile alert system) notification message
SIB9	GPS (Global Positioning System) and UTC (Coordinated universal Time) time information
SIB10	HRNN (list of readable network names for private networks)
SIB11	Premeasurement configuration information in RRC_IDLE and RRC_INACTIVE state
SIB12	Sidelink communication configuration parameters
SIB13	LTE sidelink communication configuration parameters (LTE system information block 21)
SIB14	LTE sidelink communication configuration parameters (LTE system information block 26)

11.1.2 Broadcast and update of system information

MIB is broadcast through the BCCH (Broadcast Logical Channel) channel mapped to the BCH (Broadcast Transport Channel) channel. SIB1 and OSI are broadcast through the BCCH channel mapped to the DL−SCH (Downlink Shared Channel) channel. The layer 2 protocol on the user plane including PDCP (Packet Data Convergence Protocol), RLC (Radio Link Control Protocol), and MAC (Media Access Control) layer is transparent. It means that the RRC message after Abstract Syntax Notation 1 encoding is directly delivered to the physical layer for processing.

The broadcast periods of MIB and SIB1 are 80 and 160 ms, respectively. MIB and SIB1 will be repeated in their respective cycles. Refer to Chapter 6, NR Initial Access, for details. The scheduling information of the OSI message is contained in the SIB1. And the scheduling scheme is similar to the LTE system (i.e., the method of period plus broadcast window) is adopted. Fig. 11.1 is an example of system information scheduling.

Each small box in the figure represents a broadcast window. The minimum window value is five time slots while the maximum is 1280 time slots. The size of the window is related to the subcarrier spacing of the carrier and the bandwidth of the system. The broadcast window of the FR1 carrier is generally larger than the broadcast window of the FR2. The broadcast windows of all OSIs in a cell are the same. They are arranged in the order of the OSI scheduling in SIB1. The OSI scheduling cycle ranges from 80 to 5120 ms and increases by a power of 2.

Generally speaking an OSI can contain one or more SIBn. But when the content contained in the SIB is relatively large, an OSI may only include one segment of a SIBn. This applies to SIB6 and SIB7 for broadcasting ETWS and SIB8 for broadcasting CMAS. This is because the maximum size of system information in 5G NR cannot exceed 372 bytes. The notification messages of ETWS and CMAS generally exceed this limit. Thus these SIBs are segmented in the RRC layer. After receiving the SIB segments, the UE needs to assemble them in the RRC layer before it can get the complete ETWS or CMAS notification message.

When a network plans to update system information, it will notify the UE through a paging message. The UE that receives the paging message will generally obtain new system information (except for SIB6, SIB7, and SIB8) in the next modification period. A modification period is an integer multiple of the default paging cycle. Unlike the LTE system, the paging message that triggers the system information update is carried on a PDCCH, which is called the short message. The reason for this is that the message itself needs to contain very limited content—currently only 2 bits. In addition, the UE does not need to always decode the PDSCH

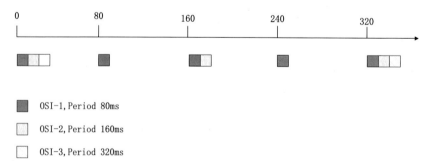

Figure 11.1 OSI scheduling diagram.

like receiving other types of paging messages, thereby saving the processing resources and power consumption of the UE.

One bit in the short message is used to indicate *systemModification*, which is used to update SIBn except for SIB6, SIB7, and SIB8. If the other bit *etwsAndCmasIndication* is 1, the UE will try to receive a new SIB6, SIB7, or SIB8 in the current modification period. This is because the contents of this system information are used to broadcast public safety-related information such as earthquakes, tsunamis, etc. The base station will broadcast these messages once it receives them. Otherwise unnecessary delay will be introduced.

From the perspective of the UE, the behavior of monitoring short messages is related to the RRC state of the UE. In the RRC_IDLE or RRC_INACITVE state, the UE monitors its own paging occasion. In the RRC_CONNECTED state, the UE monitors the short message at any paging occasion in every modification period. However, the specification also requires the UE to monitor short messages at any paging occasion at least once every default paging cycle to receive short messages intended to update ETWS or CMAS. Thus the UE supporting ETWS or CMAS will monitor short messages every default paging cycle in the RRC_CONNECTED state.

In addition to periodic broadcasting, SIB1 and SIBn can also be sent to the UE through dedicated signaling. The reason behind is that the BWP where the UE is currently activated may not necessarily be configured with a PDCCH (Physical Downlink Control Channel) channel for receiving system information or paging. In this case, the UE cannot directly receive the broadcast SIB1 or SIBn. In addition, the system information of other serving cells except PCell is also sent to the UE through dedicated signaling if the UE is configured with carrier aggregation or DC.

The NR UE will find that OSI may not be broadcasting, even if this system information is scheduled in SIB1! This requires the on-demand broadcast mechanism unique to 5G NR. Refer to Section 11.1.3 for details. Fig. 11.2 shows an illustration in the 3GPP Phase 2 protocol TS 38.300 [1], including the transmission methods of the above three pieces of system information.

11.1.3 Acquisition and validity of system information

From the perspective of the UE, the basic principle of acquiring system information is that if there is no stored system information or the stored

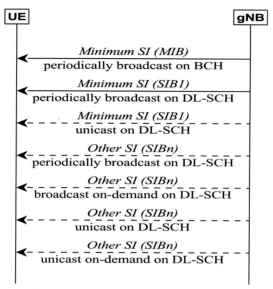

Figure 11.2 System information acquisition process.

system information is invalid, the UE needs to obtain or reacquire the system information. Then how does the UE judge whether a certain SIB is valid?

In terms of time validity, the methods of 5G NR and LTE systems are the same. Each SIBn has a 5-bit long tag (value tag), ranging from 0 to 31. The initial value of the SIBn tag is 0, and it is increased by 1 every time it is updated. The 5G NR also stipulates that the longest time limit for a stored system information is 3 hours. The combination of these two conditions requires the network to update system information no more than 32 times within 3 hours. Otherwise, the UE may mistakenly think that the updated system information is valid and not reacquire it, resulting in inconsistency between the network and the UE. This may happen to UEs that come back after leaving a certain cell for a period of time (less than 3 hours).

The 5G NR also introduces area validity. There is a parameter called system information area identifier (*systemInformationAreaID*) in the scheduling message of SIB1. In the scheduling information of each SIBn it will be marked whether it is valid in the area specified by this area identifier. If the area identifier is not applicable for one SIBn then the valid area of this SIBn is the current cell.

As shown in Fig. 11.3, the cells configured with the same area identifier (corresponding to the color in the figure) usually form a coherent area. Some SIBn is valid in this area. The reason for introducing area validity is that certain system information is often the same among different cells. For example, SIB2, SIB3, and SIB4 are used for cell reselection. If there is no cell-specific information such as blacklist, neighboring cells are configured with frequency as the granularity. For neighboring cells on the same frequency, the related parameters are likely to be the same in a certain area. In this case, after the UE obtains these SIBs on a certain cell in the area, it does not need to acquire the same SIBs in other neighboring cells again. The area validity setting is configured per SIB instead of OSI. The reason is that the mapping of SIB to OSI in different cells may be different. And it is the content of the system information that determines whether the area is valid, but not the way of scheduling. Table 11.2 shows the attributes related to the validity of system information.

Going back to the question of SIB validity raised in Section 11.1.3, the answer is that if the value tag of the one SIBn in its valid area has not changed within 3 hours after acquiring the SIB1 of a certain cell then this SIBn is valid, otherwise it is invalid. Once the UE finds that a certain SIBn is invalid, it needs to obtain the SIBn being broadcast through the scheduling information of SIB1. If the broadcast status of this SIBn in SIB1 is "*nonbroadcast*," then the UE needs to obtain it through an on-demand broadcast mechanism. The introduction of time validity is to avoid the inconsistency of system parameters between the network and the UE, and the introduction of area validity is mainly to save power consumption of the UE.

The main reason for introducing on-demand broadcasting is to save network energy. MIB and SIB1 are essential SIBs so they must be

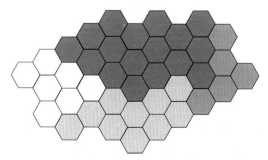

Figure 11.3 Schematic diagram of the validity of the system information area.

Table 11.2 System information validity parameters.

Parameters of SI	MIB	SIB1	SIBn
Validity time (h)	3	3	3
Value tag	N/A	N/A	$0 \sim 31$
Validity Area	Current Cell	Current Cell	Current Cell or Validity Area

broadcast periodically. However, other SIBs, such as SIBs for cell selection and reselection, do not need to be broadcast when there is no UE camping on the cell. This happens every day (e.g., late at night in commercial districts of a big city).

In the 5G NR system, on-demand broadcasting is based on the random access procedure, and there are two ways. The first way is to use the first message (i.e., PRACH preamble to indicate which OSI the UE wants to ask for). If there is only one preamble resource configured in SIB1 for the system information request the network will interpret that the UE wants all OSIs identified as "*nonbroadcast*" in SIB1 once it receives it. If there is more than one preamble resource configured in SIB1, then there is a one-to-one mapping between the preamble and the OSI identified as "*nonbroadcast*" according to the order of the preamble configuration and the scheduling. The network can judge according to the received preamble which SIB the UE wants. In order to prevent congestion of the random access channel, in the latter case, different transmission timings are arranged in different RACH association periods. The maximum transmission period is 16 RACH association periods.

The second way is to indicate the SIB that the UE wants to acquire through an RRC message. This message is called *RRCSystemInfoRequest*. In Rel-15 and Rel-16 versions, a maximum of 32 OSIs can be requested. This RRC message can be sent to the network through the third message (Message 3) of the random access procedure.

After receiving the request from the UE, the network starts to broadcast system information. From the perspective of the UE, after sending the request, system information acquisition can be started immediately in the current modification period. The on-demand request is only applicable to UEs in RRC_IDLE and RRC_INACTIVE states in R15. In R16, the UE can also trigger this process when it is in the RRC_CONNECTED state. In this case, the network sends the system information required by the UE through dedicated signaling.

11.2 Paging

There are three kinds of paging in NR systems: paging initiated by core network, paging initiated by gNB, and system information update notification initiated by gNB.

Paging initiated by the core network is linked to mobility management of NAS (Nonaccess Stratum) layer under RRC_IDLE state [i.e., the paging area is the current registered tracking area (TA) of the UE]. The set of tracking area is a compromise between the paging load and the tracking area update load. This is because the larger the tracking area is, the less frequent the tracking area update, but the higher the system paging load. Paging initiated by the core network is usually only for UEs in the RRC_IDLE state. A UE in the RRC_INACTIVE state receives a page initiated by the gNB. This is because in the RRC_INACTIVE state, the NR introduces a concept similar to the tracking area, namely the RAN Notification Area. The UE in RRC_INACTIVE state only needs to inform the network of new RNA through the notification area update process (RNA Update) when it moves across RNA. When new downlink data or signaling (such as NAS signaling) needs to be sent to the UE, the paging will be triggered. The initiator of this paging process is the anchor gNB (i.e., the gNB that drives the UE to enter the RRC_INACTIVE state through the RRCRelease message). Since the cells contained within one RNA may cover multiple gNBs, paging initiated by gNB also needs to be forwarded through the Xn interface.

The two paging mechanisms are not completely independent. The core network can provide some assistance information to gNB to determine the size of RNA. Moreover, the paging initiated by the core network is a fallback scheme for the paging initiated by gNB. When the gNB does not receive a paging response from the UE after initiating the paging, it will consider that the synchronization between the UE and the network has lost at the RNA level, and the gNB will therefore notify the core network. Then the core network triggers the paging process within the TA range. If the UE receives the paging message from the core network in the RRC_INACTIVE state, it will enter the RRC_IDLE state first and then respond to the paging message. Paging messages triggered by the core network and gNB can be distinguished by

UE identifiers contained in the paging messages. The paging message including I—RNTI (radio network temporary identifier in RRC_INACTIVE state) is a gNB-triggered paging message, and the paging message including ng-5G-S-TMSI (5G system temporary mobile description identifier) is a core network-triggered paging message.

The update of system information is called the Short Message in RRC protocol, which is actually RRC information contained in the PDCCH channel. It contains two bits. The first bit indicates that the update of OSI except ETWS/CMAS messages (SIB6, SIB7, SIB8) is triggered, and the second bit indicates that the notification of ETWS/CMAS messages (SIB6, SIB7, SIB8) is triggered. Refer to the first section of this chapter for the process of updating system information. The UE in any RRC state may receive this Short Message to update the system information.

The PDCCH channel of the above three paging messages is scrambled with the paging-specific identifier (i.e., P—RNTI, Paging Radio Network Temporary Identifier). The paging message initiated by the core network and the paging message initiated by the gNB may contain multiple paging records, which are for different UEs (Table 11.3).

The transmission of paging message is described in Section 7.1 of Ref. [2] in detail. The basic principle is the same for paging initiated by core network and by gNB. From the network point of view, it will be sent on each PO (paging occasion) in the system information modification period. The UE in RRC_IDLE and RRC_INACTIVE will only monitor their own POs. The UE in RRC_CONNECTED may also monitor the PO not associated with its own UE ID to avoid colliding with other dedicated downlink data as much as possible.

Before introducing specific paging mechanism, we need to introduce two basic concepts: PF (Paging Frame) and PO (Paging Occasion). In the LTE

Table 11.3 Paging message comparison.

Paging	Paging range	RRC state	Identification of convolution	UE identity
CN-level paging	TA	RRC_IDLE	P—RNTI	ng–5G–S-TMSI
gNB-level paging	RNA	RRC_INACTIVE	P—RNTI	I—RNTI
Short Message	Cell	RRC_IDLE, RRC_INACTIVE, RRC_CONNECTED	P—RNTI	N/A

system, the definition of PF and PO is very simple. PF is a radio frame containing PO in a DRX cycle while PO is a subframe in PF where paging messages can be transmitted. In the NR, if the PDCCH monitoring opportunity (PMO) is defined by the paging search space in SIB1, then the definition of PF can be consistent with that of LTE. But the definition of PO becomes a slot containing multiple PMOs. The number of PMO is equal to the number of SSBs actually sent in the cell. When the PMO is defined by search space #0 in MIB, the definition of PO can follow the new definition, but PF is actually the reference radio frame associated with the PO. This reference radio frame actually contains SSB burst. The timing relationship between the reference radio frame and the radio frame of the associated PO (as well as the slot and OFDM symbol) is fixed so the UE can accurately locate the PO according to the reference radio frame.

The formula for determining PF is defined in the protocol, which is applicable to the above two cases:

$$(SFN + PF\_offset) \bmod T = (T \text{ div } N) * (UE\_ID \bmod N) \quad (11.1)$$

The meanings of the parameters are as follows:
- SFN: the frame number of the radio frame where the PF is located
- PF_offset: radio frame offset
- T: paging cycle
- N: the number of PF in one paging cycle
- UE_ID: the ID of UE, equal to 5g-s-tmsi mod 1024

This formula has one more parameter (i.e., PF_offset compared to LTE). If the paging search space is configured by SIB 1, this PF_offset is not needed. If the search space is configured by MIB, then the search space is the same as that of the UE to obtain SIB1 which bears fixed relative position of SSB in the time and frequency domain in the cell with a total of three patterns. In pattern 1, the search space is after several OFDM symbols of SSB and the search space must appear in an even frame. In this case PF_offset is always equal to 0. In patterns 2 and 3, the search space is in the same slot as SSB, so PF is the radio frame where SSB is located. While the NR system allows SSB to be located in any radio frame, SSB cycles can be {5, 10, 20, 40, 80, 160 ms}. When the period is greater than 10 ms, the radio frame where SSB is located can be any radio frame that meets the following condition:

$$SFN \bmod (P/10) = SSB\_offset \quad (11.2)$$

where P is the period of the SSB and the range of SSB_offset is $0 \sim (P/10) - 1$. For example, when p = 40 ms, the formula becomes SFN mod 4 = SSB_ Offset, where SSB_offset = 0, 1, 2, 3.

In the above formula for calculating PF the value range of period T is {320, 640, 1280, 2560 ms}, while N is a constant that can always be divisible by T. Note that T/N is always an even number. So if there is no PF_offset in the formula, PF must be in an even number of radio frames. This contradicts with the requirements for PF in modes 2 and 3. In order to meet the flexibility of PF setting in NR system, PF_offset is added to the formula parameter. Mathematically, PF_offset and SSB_offset are actually consistent.

The formula to calculate PO is as follows:

$$i\_s = \text{floor}\,(UE\_ID/N)\ \text{mod}\ Ns \qquad (11.3)$$

Among them, UE_ID and N are consistent with the parameters in the formula for determining PF. Ns is the number of PO in a PF, which can be 1, 2, and 4.

If the search space is defined in MIB, then Ns = 1 in pattern 1 and Ns = 1 or 2 in pattern 2 and 3. If the search space is defined by the paging search space in SIB1, then Ns can be 1, 2, and 4.

Two more parameters are to define PO. The first parameter is used to define the number of PMO corresponding to an SSB to increase paging opportunities for NR−U cells. The second parameter is used to define the sequence number of PMO at the beginning of each PO in a PF, which can be called PO_start. This parameter only applies when the search space is defined by the paging search space in SIB1. In this configuration, all PMOs in the PF are sorted by time. When there is no PO_start is used, the PMOs owned by each PO are arranged in order. The index of PMO corresponding to the first PO is $0 \sim (S - 1)$ PMO, the index of PMO corresponding to the second PO is $S \sim (2S - 1)$, and so on, where S is the actual number of actually transmitted SSB burst in the cell. One of the main problems of this method is that the PMOs corresponding to the PO are always listed from beginning continuously among all PMOs in PF, which makes PMO unevenly distributed in the time domain. Receiving paging usually triggers the random access process and thus will also lead to the uneven use of PRACH resources. In order to overcome this problem, PO_start is introduced (paging opportunity start symbol) so that the PMO at the beginning of each PO can be any one of all PMOs in the PF.

11.3 RRC connection control

11.3.1 Access control

Before a call is initiated by the UE, access control must be carried out first. There are two kinds of access control. One is executed by the UE itself, which is called UAC (unified access control) in the NR system, and the other is done by the base station according to "RRC establishment cause" in the RRC setup request message. When the NAS layer of the UE initiates a call, it will get a "RRC establishment cause" according to the access ID and access category. For the specific mapping table, refer to table 4.5.6.1 in Ref. [3]. The "RRC establishment cause" is encoded by the RRC layer and sent to gNB in the RRCSetupRequest message. How to control the access according to "RRC establishment cause" is the internal algorithm of gNB. If gNB accepts the RRCSetupRequest message initiated by the UE, it will respond with a RRCSetup message otherwise it will send a RRCReject message to reject. See Section 11.3.2 for details.

In this section, we will focus on the UAC mechanism. First of all, the concept of access identity and access category should be clarified. The access identity is similar to the concept of access class. Currently there are 16 $(0 \sim 15)$ standardized access IDs in NR, among which $3-10$ are undefined parts. Refer to table 4.5.2.1 in Ref. [3] for more information. The access category represents the service attribute of the call initiated by the UE. It can be found from table 4.5.2.2 in Ref. [3] that $0 \sim 7$ are standardized access categories while $32 \sim 63$ are operator-defined access categories. The rest are undefined access categories.

The SIB1 of the NR cell defines specific access control parameters, among which the key parameters are called "uac-BarringFactor," "uac-BarringTimer," and "uac-BarringForAccessIdentity." The first two parameters have a mapping relationship with the access category. Therefore when the UE performs the UAC process in a cell, it must first obtain the SIB1 of the cell.

The process of executing UAC in the AS layer of the UE can be divided into three steps. Step 1 is to determine whether the "green lane" or "red lane" can be directly accessed according to the access category given by the NAS layer. Access category 0 indicates the process of paging response initiated by the UE. This access class will be reflected in the "RRC establishment cause." And it is unconditionally accepted. Access category 2 represents an emergency call. Only when the UE is in the state of being rejected to establish RRC connection by the network (i.e., when t302 is running), the UE needs to carry out UAC process for this access category. In other words, when the network is particularly busy, emergency calls also need to go through the UAC control process. In addition to the above situation, if the UE finds that the current network does not broadcast any relevant UAC parameters, it will also consider that it can directly enter the "green lane." Entering the "red lane" means being rejected directly. When t302 is running, all access category will be blocked except access category 0 and 2. When an access class fails to pass the UAC, the UE will start timer t390 for this access category (its time duration is defined in the UAC barringtimer parameter). When the same access category is initiated again while t390 is running, it will be rejected directly. In addition to the "green lane" or "red lane" situation mentioned above, other access categories will enter the "yellow lane" and the process of step 2 will be executed.

In step 2, the access identity is to be verified. The lane corresponding to access ID 1, 2, 11 ~ 15 is defined in the parameter uac-BarringForAccessIdentity. This parameter is actually a bitmap. The access ID set to 1 can directly enter the "green lane," while the access ID set to 0 will directly enter the "red lane." Only the access ID 0 has no corresponding bitmap because it needs to be judged in step 3.

In step 3, the unique control parameters "uac-BarringFactor" and "uac-BarringTimer" should be determined according to the access category. Then the AS layer generates a random number between 0 and 1. If the generated random number is less than the corresponding uac-BarringFactor, it is considered that the UAC control has been passed. Otherwise the timer t390 will be started whose length is (0.7 + 0.6 * uac-BarringTimer).

11.3.2 RRC connection control

Before introducing RRC connection control it is worth introducing the RRC_INACTIVE state in NR systems. The RRC_INACTIVE state intends to save power and to reduce the access latency of control plane. In the RRC_CONNECTED state whether to transmit or receive data mainly depends on the current traffic model. For some services, such as "wechat" applications, the interval between packets will be relative longer. In this case, to keep the UE in the RRC_CONNECTED state all the time is not helpful to the user experience. In order to maintain RRC connection the UE needs to continuously measure the current serving cell and neighboring cells, to maintain radio link in the current cell and avoid dropping calls during across the cells. Transmission of measurement report requires UE's hardware to be active. In the RRC_CONNECTED state CDRX (discontinuous reception) can save battery consumption to a certain extent but it cannot solve the problem completely. One option is to release the RRC connection and then establish the RRC connection upon arrival of data. The problem with this approach is that the user experience will be poor because of long latency of control plane. The RRC connection establishment in idle state requires a whole set of call establishment processes, which usually takes tens of milliseconds or even longer. Another problem is that frequent RRC connection release and establishment can lead to a lot of control signaling. When the network is busy such signaling will lead to the so-called signaling storm. It will impact on the stability of core network operation. RRC_INACTIVE can be considered as a compromise between these two alternatives.

As shown in Fig. 11.4 in the RRC_ INACTIVE state the UE maintains the context of NAS layer and radio-bearer configuration of the AS layer but suspends all SRB and DRB and releases semistatic uplink radio resources such as PUCCH/SRS resources. In addition to maintaining the radio bearer of the AS layer gNB will retain the NG interface and the UE-related context.

When the UE needs to receive and transmit data the UE usually needs to enter the RRC_CONNECTED state. From the RRC_INACTIVE state the establishment process of NG connection and NAS connection can be saved, hence the delay of the control plane can be reduced to 10 Ms. In Rel-17 small data transmission in the RRC_INACTIVE state is introduced to avoid entering the RRC_CONNECTED state. Together with two-step RACH (two-step random access process) this scheme can not only shorten the control plane delay but also save the signaling to save power consumption. The power-saving function of the RRC_INACTIVE state is mainly due to the fact that the UE need only

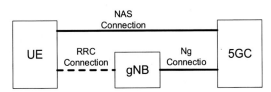

Figure 11.4 RRC_INACTIVE state.

follow mobility management almost identical to the RRC_IDLE state instead of the RRC_CONNECTED state. Thus unnecessary RRM (radio resource management) measurement and signaling overhead can be saved.

Except for compliance with the mobility management and measurement rules similar to the RRC_IDLE state, the UE also needs to execute an RNA update procedure when crossing the RNA or periodically in order to inform the network of the current RNA. When the UE moves within the RNA only a periodic RNA update process is necessary.

Fig. 11.5 shows the qualitative comparison of different RRC states on key indicators.

RRC connection control mainly includes two parts. One part is the maintenance of the RRC connection during the RRC state transition. The other part refers to the RRC_CONNECTED maintenance of the radio link including radio–link failure and RRC reestablishment. For DC the radio link of MCG and SCG can be maintained separately (Fig. 11.6).

In this state machine RRC_INACTIVE state fallback to RRC_IDLE state is a rare case. It may occur during the transition from RRC_INACTIVE to RRC_CONNECTED state. When gNB receives RRCResumeRquest message, gNB will respond with RRCRelease to drive the UE into RRC_IDLE state.

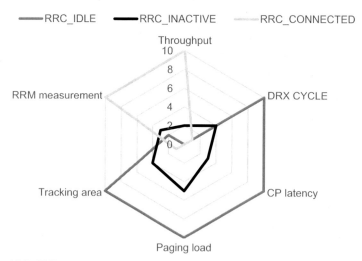

Figure 11.5 RRC state comparison.

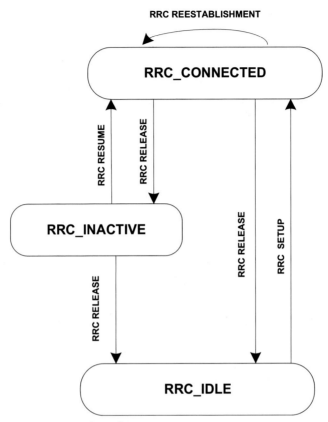

Figure 11.6 RRC State machine diagram.

The following chapters focus on RRC connection establishment and release, RRC state transition between RRC_INACTIVE state and RRC_CONNECTED, RRC reestablishment, and so on.

There are two handshakes in the RRC connection process as shown in figure 5.3.3.1-1 in Ref. [4] (Fig. 11.7).

The RRCSetupRequest message contains the UE ID and "RRC establishment cause." If the UE has been registered in the currently selected PLMN, the UE ID is the temporary identity of the NAS layer called "ng-5G-S-TMSI." The full length of the "ng-5G-S-TMSI" is 48 bits. The RRCSetupRequest message is transmitted in message 3 of the RACH procedure. Message 3 size is limited without the help of the ARQ mechanism in the RLC layer. When message 3 is used to carry RRCSetupRequest messages it is limited to 56 bits. Without 8-bit MAC

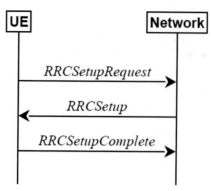

Figure 11.7 RRC connection establishment procedure.

header overhead the size of RRCSetupRequest message is limited to 48 bits. In order to include the 4-bit "RRC establishment cause" in the same message, only 39 bits of the low bit of ng-5G-S-TMSI can be included in the message. One bit is reserved for future extension. The nine MSB bits of ng-5g-s-tmsi will be sent to the gNB via RRCSetupComplete message. The reason to include 39 bits of ng-5G-S-TMSI is to ensure the uniqueness of the UE ID in the message as much as possible. It is important is that this message will be used as the basis for contention resolution in the RACH process of the MAC layer. ng-5G-S-TMSI is unique because the UE ID is uniformly assigned by 5GC. When the UE is not registered in the current PLMN, it can only select a 39-bit random number. In theory, the random number generated by the UE may be identical with ng-5G-S-TMSI of a UE, resulting in the failure of the RACH process. But the probability is very low because it will occur only when two UEs with the same UE ID initiate the RACH process at the same time.

The RRCSetup message contains the radio-bearer configuration of SRB1 as well as the configuration parameters of the MAC layer and PHY layer. After configuring SRB1 the UE will send the RRCSetupComplete message on SRB1. SRB1 is sent with RLC AM mode so its size has no special limit. There are three parts in the RRCSetupComplete message. The first part is the UE ID information. When the RRCSetupRequest message contains 39 bits of ng-5G-S-TMSI, then the UE ID is the remaining nine bits. Otherwise, the UE ID contains the complete 48 bit ng-5G-S-TMSI. The second part is the NAS message. The third part is the routing information needed by the gNB, including the UE registered network slice list, selected PLMN, registered AMF, and so on. The gNB uses this information to ensure that the NAS message as well as the complete ng-5G-S-TMSI are routed to the appropriate core network nodes through the initial UE message of the NG interface.

If the gNB does not accept the current call when the network is congested after checking the "RRC establishment cause," it will reject the UE through the RRCReject message. The RRC connection establishment process fails.

When the UE is in RRC_CONNECTED, gNB can drive the UE into RRC_INACTIVE state through RRCRelease message. There is a parameter called "SuspendConfig" in this RRCRelease message which contains three parts. The first part is the new ID assigned to the UE (i.e., I−RNTI). In addition to the 40-bit complete I−RNTI (called full I−RNTI), there is also a 24-bit short I−RNTI (called short I−RNTI). The second part is the RRC_INACTIVE configuration parameters related to mobility management including the length of the dedicated paging cycle, notification area, and timer (T380) for periodic RNA update procedure. These parameters are described in the first section of this chapter. The third part is the security-related parameters, namely NCC parameters. The usage of UE IDs and NCC will be described in detail in the introduction to the RRC resume procedure.

The UE may initiate the RRC resume procedure due to paging or RNA update procedure or originate a data service. The flowchart of this procedure can be found in Ref. [4] and Fig. 11.8.

Before sending RRCResumeRequest/RRCResumeRequest1 the UE also needs to perform the access control procedure. The determination of access category is unique. If the UE is to respond to the paging of Gnb, the access category is set to 0. If it is due to RNA update, the access category is 8. Others are consistent with the UAC mechanism described in this chapter.

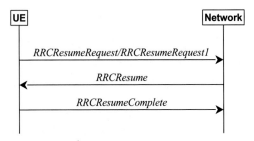

Figure 11.8 RRC Resume procedure.

The difference between RRCResumeRequest and RRCResumeRequest1 is the length of the contained I−RNTI. Similar to RRCSetupRequest, the size of these two messages is limited. The RRCResumeRequest message is limited to 48 bits. Because the message also needs to contain 16 bits of Short-MAC-I and RRC resume only 24 bits of I−RNTI (short I−RNTI) are included in the message. The drawback of not having a complete I−RNTI is that the determination of anchor gNB based on short I−RNTI may have some ambiguity. In order to solve this problem, for example in a small coverage cell, the UE will send a RRCResumeRequest1 containing 40 bit I−RNTI (full I−RNTI). There is a parameter in the SIB1 of the cell to indicate whether the UE is allowed to send RRCResumeRequest1.

Both messages contain Short-MAC-I for security authentication and UE context identification. Short-MAC-I is actually a 16 bit MAC-I. The parameters to calculate the Short-MAC-I of RRCResumeRequest or RRCResumeRequest1 are {source PCI (physical cell ID), target cell ID (target cell ID), and source C-RNTI (source cell allocated C-RNTI)}. Then the old integrity protection security key and algorithm used by the UE in the anchor gNB are used. Assuming that the current serving gNB is the anchor gNB, then the gNB will calculate and verify the above calculation parameters according to the same security key and algorithm. If the integrity verification is passed, the gNB will respond with the RRCResume message. The intention of the RRC resume message is to recover SRB2 and all DRBs and configure the radio parameters of MAC and PHY layers (SRB1 is considered to have recovered when it transmits the RRCResumeRequest message). The RRCResume message is sent to the UE through SRB1. Later the UE will reply with the RRCResumeComplete message on the SRB1. At this point, the UE enters the RRC _CONNECTED state.

The K_{gNB} in Table 11.4 is calculated by horizontal or vertical security key based on the old K_{gNB} or NH corresponding to NCC. The PCI and frequency information of the current serving cell also need to be input during the calculation. For details, refer to Section 6.9.2.1.1 of the protocol 33.501.

If the current serving gNB is not the anchor gNB, then the current gNB needs to forward the received information and information of the serving gNB including PCI and allocated C−RNTI to the anchor gNB. If the anchor gNB completes the Short-MAC-I verification according to the forwarded information, it will forward the stored UE AS context, including the security context, to the serving gNB. Then the serving gNB will continue with the subsequent process. Fig. 11.9 refers to

Table 11.4 RRCresume-related messages.

RRC message	SRB	Key	Security protection
RRCResumeRequest (1)	SRB0	No	No
RRCResume	SRB1	K_{gNB} or NH addressed by NCC	Integrity protection and encryption
RRCResumeComplete	SRB1	K_{gNB} or NH addressed by NCC	Integrity protection and encryption

Figure 11.9 UE context retrieve procedure in Xn interface.

Fig. 8.2.4.2-1 of Ref. [5] and identifies the process of exchanging UE context between old NG—RAN node and new NG—RAN node by the RETRIEVE UE CONTEXT REQUEST message and RETRIEVE UE CONTEXT RESPONSE message.

Different abnormal processes may occur in the RRC resume procedure. Table 11.5 lists various possible situation.

The trigger of the RRC reestablishment in NR is similar to that of LTE.

It can be seen from Table 11.6 that the RRC Reestablishment procedure in the NR has better security protection for RRCRestablishment messages. The configuration of SRB1 depends on the default configuration. The advantage is that the message is very simple. However, due to the lack of necessary PUCCH resource configuration, the gNB must send a UL grant to the UE at an appropriate time after sending the RRCRestablishment message so that the UE can send RRCRestablishmentComplete message. Otherwise the UE will initiate the RACH process because the PUCCH resource of the SR (schedule request) is not configured, thus delaying the transmission of the message.

If the UE context cannot be identified after receiving the RRCRestablishmentRequest message, the gNB sends an RRCSetup

Table 11.5 Abnormal flow in RRC resume procedure.

gNB response message	UE handling	Target RRC state
RRCSetup	UE releases the original context and starts the RRC connection establishment process	RRC_CONNECTED
RRCRelease	UE releases the original context	RRC_IDLE
RRCRelease (with SuspendConfig)	UE receives new incoming Configuration of RRC_INACTIVE state	RRC_INACTIVE
RRCReject	UE suspend SRB1	RRC_INACTIVE

Table 11.6 RRC reestablishment message comparison between NR and LTE.

RRC reestablishment procedure	LTE	NR
RRCReestablishmentRequest	Send via SRB0 and without any security protection.	Send via SRB0 and without any security protection.
RRCReestablishment	Send via SRB0 and without any security protection. To setup SRB1, configure the radio configuration and NCC.	Send via SRB with integrity protection but no encryption. Apply default configuration for SRB1. It is used to configure NCC.
RRCReestablishmentComplete	Send via SRB1 with integrity protection and encryption.	Send via SRB1 with integrity protection and encryption.

message to respond RRCReestablishmentRequest message. After receiving the RRCSetup message, the UE will clear the existing UE context and start the RRC establishment procedure. This exception handling is not available in LTE systems. Refer to figure 5.3.7.1-2 of Ref. [4] for the flowchart in Fig. 11.10.

When the Mr−DC is introduced in the NR system, the control plane is also improved. From the perspective of modeling, there are two places to generate RRC message content in the network, namely MCG and SCG [6]. The flowchart in Fig. 11.11 can be referred to figure 4.2.1-1 of Ref. [6], in which the master node is also referred to as MN and the secondary node is also referred to as SN.

In the Mr−DC architecture network a new SRB called SRB3 for SCG can also be configured. The initial configuration of SCG always needs to be sent to the UE through SRB1 of MCG. In this case SCG is only responsible for RRC content and ASN.1 coding while MCG is

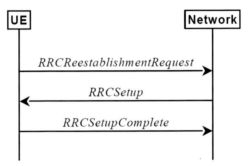

Figure 11.10 RRC fallback procedure.

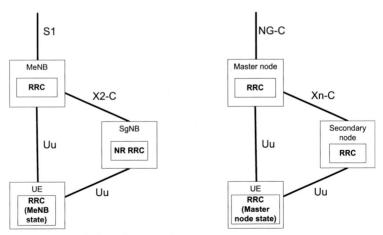

Figure 11.11 Control plane for Mr−DC.

responsible for PDCP security processing. After that the network can choose to send on SRB1 or SRB3. If SRB3 is used, the security processing is completed by the PDCP protocol layer of SCG. The security keys in SCG and MCG are different but the security algorithm may be the same. Although it is possible to use SRB1 or SRB3 to send RRC signaling, there is only one RRC state machine in the UE.

In the Mr−DC architecture the mechanism of MCG and SCG link maintenance and recovery are independent of each other. The causes of SCG radio-link failure may be:

- SCG RLF on PSCell
- SCG RACH failure on PSCell
- SCG configuration failure (only for messages on SRB3)
- SCG RRC integrity check failure (on SRB3)

After these events occur the UE temporarily suspends the transmission and reception of SRB3 and DRB and sends SCGFailureInformation to the network through SRB1 on MCG. This message contains the cause of SCG radio-link failure and some measurement results of serving cells and neighbor cells. After receiving this message, the network will take appropriate processing methods.

In R16, the NR system also introduces the procedure of MCG link failure and recovery, provided that the SCG link is still working normally. After a radio-link failure occurs in MCG, the UE will suspend the transmission and reception of SRBs and data radio bearers (DRBs) on MCG and send MCGFailureInformation to the network through SRB1 (assuming split SRB1 is configured) or SRB3 on SCG. Similarly in this message not only the cause of MCG link failure is included but also some measurement results of serving cells and neighbor cells are reported. After receiving this message, the network will adopt appropriate processing methods. The time of the UE waiting for network processing is limited by a timer t316 configured by network. If the t316 expires, the RRC reestablishment procedure will be triggered.

11.4 RRM measurement and mobility management

11.4.1 RRM measurement

11.4.1.1 RRM measurement model

The concept of beam-forming is introduced in the NR system so that the UE performs measurement on each beam separately. Beam measurement has some effects on the measurement model, mainly reflected in the fact that the UE needs to derive cell measurement results from beam measurement results.

As shown in Fig. 11.12, the UE's physical layer obtains multiple beam measurements results after performing RRM measurement and delivers them to the RRC layer. In RRC layer filtering is needed for beam measurement results before cell measurement results are derived. In terms of the filtering conditions, the following schemes are mainly involved in the discussion of 3GPP [7]:

- Scheme 1: select N best beams.
- Scheme 2: select N best beams with beam measurement result above certain threshold.
- Scheme 3: select the best beam and N-1 better beams whose beam measurements are within certain range of the best beam's measurement.

The intention of all three schemes is to select the best beams to derive the cell measurement result. Scheme 1 does not require the quality of the beam measurement results so it is not good for the network to accurately use the derived cell measurement results. The difference between Schemes 2 and 3 is not that much except that Scheme 3 does not strictly control the absolute value of the beam measurement results compared with Scheme 2. Eventually in the 3GPP RAN2 NR Adhoc June meeting Scheme 2 was agreed on. In the realization of Scheme 2 the UE obtains the threshold value and maximum beam number N through dedicated RRC signaling and takes the linear average of maximum N beam

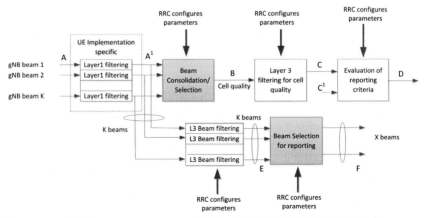

Figure 11.12 RRM measurement model. *From 3GPP TS 38.300 V16.2.0. Study on scenarios and requirements for next generation access technologies.*

measurement results that meet the threshold to derive the cell measurement result. When the threshold and N are not configured, the UE will use the best beam measurement result as the cell measurement result. In order to reduce the random interference in the measurement process, the derived cell measurement result needs L3 filtering before the measurement report can be triggered. That is the process for the UE reporting of the cell measurement result. In some cases, the network may require the UE to report beam measurement results directly. In this case, beam measurement results delivered by the physical layer can be reported after L3 filtering.

For a UE in RRC_CONNECTED state, measurement configuration of 5G system mainly inherits that of the LTE system, including the following parts.

The measurement object identifies the frequency information where the UE performs the measurement, which is the same as the LTE system. The difference is that the NR system supports the measurement of SSB and CSI−RS reference signals (see Section 6.4.1 for details). For SSB measurement, the frequency information is the SSB frequency associated with the measurement object. Since the 5G system supports the transmission of multiple different subcarrier spaces, the measurement object needs to indicate the measurement-related SSB subcarrier spaces. For the measurement configuration of SSB reference signal, the measuring object should also indicate the time window information of SSB measurement, namely SMTC information. The network can further indicate which SSBs to be measured by the UE within SMTC. For the measurement configuration of CSI−RS reference signal, the measurement object contains the CSI−RS resource configuration. In order to enable the UE to derive the cell measurement result from beam measurement results, the measurement object is also configured with filtering threshold values based on the beam measurement results of SSB and CSI−RS and linear average calculation of the allowed maximum beam number. For the L3 filtering of beam measurement and cell measurement results, detailed filtering coefficients are also indicated in the measurement objects according to different measurement reference signals.

Reporting configuration mainly includes reporting criteria, reference signal type, reporting type, etc. Like the LTE system, the NR supports periodic reporting, event-triggered reporting, CGI reporting for ANR purpose, and SFTD reporting for time difference measurement. For periodic reporting and event-triggered reporting, the reporting configuration indicates the type of reference signal (SSB or CSI—RS), the quantity of measurement report (any combination of RSRP, RSRQ, and SINR), and whether to report the beam measurement result and the maximum beam number. For event-triggered reporting, the reporting configuration specifies for each event a trigger quantity (e.g., RSRP, RSRQ, or SINR). Thus far, the 5G system inherits six intra-RAT measurement events of the LTE system (i.e., events A1 to event A6) and two inter-RAT measurement events (i.e., events B1 and B2).

Just like the LTE system, the NR system associates measurement identities with measurement objects and reports configurations as shown in Fig. 11.13. This association mode is flexible and can realize any combination of measurement objects and reporting configurations. One measurement object can associate multiple reporting configurations and vice versa. The measurement identities will be carried in the measurement report for reference on the network side.

Measurement configuration defines a set of measurement filtering configuration information for measurement event evaluation and reporting as well as periodic reporting. Each measurement quantity configuration in the measurement configuration contains beam measurement quantity configuration and cell measurement quantity configuration. Two sets of filtering configuration information are defined for SSB and CSI—RS, respectively, and three sets of filtering coefficients are defined for each filtering configuration information for RSRP, RSRQ, and SINR, respectively. The L3 filtering coefficient used in

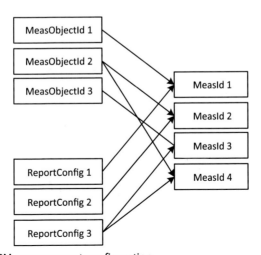

Figure 11.13 RRM measurement configuration.

the measurement object corresponds to a measurement quantity configuration is shown in Fig. 11.14 indicated by the index in the measurement quantity configuration sequence.

Similar to the LTE system, for the NR UE in RRC_CONNECTED state to perform interfrequency or inter-RAT measurements, the measurement gap needs to be configured by the network (see Section 6.4.2 for details). During the measurement gap the UE stops all ongoing traffic and serving cell measurement. For intrafrequency measurement the UE may also need a measurement gap; for example, when the current active BWP does not cover the SSB frequency to be measured. In terms of configuration NR supports per-UE and per-FR measurement gaps, take EN−DC as an example, when per-UE measurement gap is configured the frequency information of FR1 and FR2 to be measured by the SN is notified to the primary node (MN), which determines the final measurement gap and informs SN of the measurement gap configuration information. When the measurement gap is

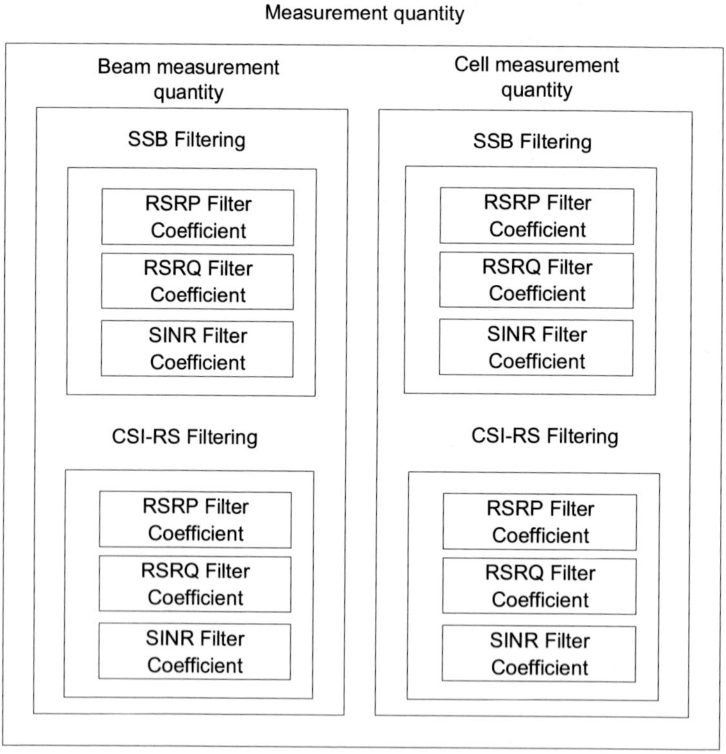

Measurement quantity

Figure 11.14 Measurement quantity configuration.

configured according to the frequency range SN will inform MN of the FR1 frequency information to be measured while MN will inform SN of the FR2 frequency information to be measured. Consequently MN will configure the measurement gap of FR1 while SN will configure the measurement gap of FR2. For the measurement gaps configured by MN, the UE refers to the radio frame number and subframe number of PCell when performing the measurement. Accordingly for SN-configured measurement gaps, the UE calculates measurement gaps based on PSCell's radio frame and subframe numbers.

In order to save power consumption, the network can include S-measure (RSRP value) parameters in the measurement configuration. The UE compares the RSRP measurement result of PCell with s-measure to control whether the UE performs the measurement of nonserving cells, which is the same as LTE system. Since NR supports SSB and CSI—RS measurements, different schemes from LTE system for s-measure configuration were brought up in the 3GPP discussion. It can be summarized into two types of schemes:

- Scheme 1: Two s-measure parameters are configured, one for SSB—RSRP value and the other for CSI—RSRP value. The two s-measure parameters control the start of neighbor cell SSB measurement and CSI—RS measurement, respectively.
- Scheme 2: Only one S-measure parameter is configured. The network indicates whether the threshold value is for SSB—RSRP or CSI—RSRP, and an s-measure parameter controls the start of all neighbor cell measurements (including SSB and CSI—RS measurements).

Although Scheme 1 is more flexible, Scheme 2 is simpler and sufficient to meet the demand of power saving of the terminal. Thus finally RAN2#100 meeting concluded on Scheme 2 (i.e., UE starts or stops SSB and CSI—RS measurement of all neighbor cells according to the only configured s-measure value).

The measurement reporting process of the NR system is basically the same as that of the LTE system. The difference is that SINR measurement results and beam measurement results are added to the NR system. The UE reports the beam index for identification when reporting the beam measurement results.

11.4.1.2 Measurement optimization

The first RRC reconfiguration message of the NR system generally cannot configure appropriate CA or Mr—DC functions for the UE because the measurement results of the UE have not been obtained on the network side at that time. The network can configure the appropriate CA or Mr−DC capabilities based on the reported results of the measurements via first RRC reconfiguration message. In a real implementation, this process has a significant delay because it takes a while from initial access to the measurement results to be reported. In order to configure SCell or SCG quickly, the network can ask the UE to perform early measurement in RRC_IDLE state or RRC_INACTIVE state, and report to the network when entering RRC_CONNECTED state. In this way the network can quickly configure SCell or SCG according to the results of early measurement. Network-configured target measurement frequency can include NR frequency list and E−UTRAN frequency list. Among them, the NR frequency list only supports SSB measurement but not CSI−RS measurement. SSB frequency includes synchronous SSB and asynchronous SSB.

Early measurement is configured using dedicated signaling (RRCRelease message) or system information (existing SIB4 and newly introduced SIB11). Among them, the measurement configuration in the system information broadcast is common for UEs in RRC_IDLE state and INACTIVE state. If early measurement configuration is received via RRCRelease, its content overrides the measurement configuration obtained from the system information. If the NR frequency configured by RRCRelease does not contain SSB configuration information, then SSB configuration information in SIB11 or SIB4 will be adopted.

Only when the cell system information indicates that the current cell supports early measurement reporting (by IE idleModeMeasurements), the UE will indicate to the network whether the UE has any measurement results that can be reported earlier through a parameter (idleMeasureAvailable) in the RRCResumeComplete or RRCResume message. Then the network side requires the UE to report the early measurement results in the UEInformationResponse message through the UEInformationRequest message. For the UE in the RRC _ INACTIVE state, the request and reporting of the measurement results in advance can also be done through the RRCResume message and the RRCResumeComplete message.

The UE performs early measurements within a specified time, which is controlled by the T331 configured in the RRCRelease message. The UE activates this timing after obtaining the measurement configuration for early measurement. The UE does not need to stop the timer when switching between RRC_INACTIVE and RRC_IDLE states.

Early measurements may also need to be performed within the validityAreaList. Validity areas can be configured in dedicated signaling or in SIB11. The validity area consists of a frequency and a list of cells of the frequency layer. If the network side does not configure a validity area that means there is no measurement area limit.

11.4.2 Mobility management

The UE mobility management in the NR system mainly includes cell selection and reselection process in RRC_IDLE or RRC_INACTIVE state as well as the handover process in the RRC_CONNECTED state.

11.4.2.1 RRC_IDLE/RRC_INACTIVE state mobility management

In the RRC_IDLE or RRC_INACTIVE state the premise for camping on a cell is that the cell's radio quality (including RSRP and RSRQ measurements) meets the cell selection S criterion in the same way as the LTE system does. The UE will continue the evaluation of cell reselection after selecting a suitable cell. The measurement to be performed in the evaluation of cell reselection is carried out according to the reselection priority of each frequency. Specifically:

- Neighbor cell measurements are always performed for higher priority frequency.
- For the intrafrequency frequency measurement, when the RSRP and RSRQ values of the serving cell are both higher than the intrafrequency measurement threshold configured by the network the UE can stop the intrafrequency neighbor cell measurement. Otherwise, the UE will measure.
- For the same priority and lower priority frequency measurement, when the RSRP and RSRQ values of the serving cell are both higher than the interfrequency measurement threshold configured by the network the UE can stop the neighbor cell measurement of the same priority and lower priority frequency. Otherwise the UE will measure.

After obtaining multiple candidate cells' measurement results, the process of determining the target candidates for cell reselection is basically the same as that of the LTE system. The principle of prioritizing those cells at higher priority frequency is adopted. Specifically:

- For cell reselection to higher priority frequency, it is required that the signal quality should be above a certain threshold over a specified length of time and the UE stays in the source cell for no less than 1 second.

- For cell reselection of intrafrequency or of the same priority frequency, the R criterion (ranking according to RSRP) should be met. The signal quality of the new cell should be better than the current cell over the specified time length and the UE should stay in the source cell for no less than 1 second.
- For cell reselection to lower priority frequency it is required that no cells at higher priority and the same priority frequency meet the above requirements. The signal quality of the source cell should be lower than a certain threshold and the signal quality of the neighbor cell at lower priority frequency should be higher than a certain threshold over a specified length of time while the UE camps on the source cell for no less than 1 second.

In the cell reselection process of intrafrequency or the same priority frequency, when multiple candidate cells meet the requirements the LTE system will select the best cell as the target cell by ranking RSRP. In the NR system in order to increase the probability of successful access through good beam, both cell signal quality and number of good beams should be taken into account when determining the target cell. In order to achieve this goal, the 3GPP discussed the following types of schemes:

- Scheme 1: the number of good beams is introduced into the ranking value. For example the number of good beams multiplied by a factor is appended to the cell measurement results and the UE selects the cell with the highest ranked cell as the target cell.
- Scheme 2: no change to the calculation of ranking value (i.e., to still use the cell measurement results for ranking). Before selecting the target cell, the UE selects the best few cells with similar signal quality first and then choose the cell with the largest good beam number as the target cell.

Eventually, the Scheme 2 was adopted through voting at the RAN2#102 meeting.

11.4.2.2 RRC_CONNECTED state mobility management

Mobility management in RRC_CONNECTED state is mainly realized through the network-controlled handover procedure. The NR system inherits the handover procedure of the LTE system, which mainly includes three stages: handover preparation, handover execution, and handover completion.

In the handover preparation phase, the source base station will make a handover decision and transfer a handover request to the target base station after receiving the measurement report from the UE. If the target cell accepts the request, a handover acknowledge message will be sent to the source base station through the interbase station interfaces (i.e., Xn interface). The message contains the configuration information of the target cell, namely, the handover command.

In the handover execution phase, the source base station sends the handover command to the UE. The UE disconnects with the source cell immediately after receiving the handover command and starts to establish downlink synchronization with the target cell. The UE then initiates a random access process to the target cell using the random access resources configured in the handover command and reports the handover completion message when the random access is completed. During the process of the UE accessing the target cell, the source base station forwards the data packets from UPF to the target base station and sends the status information of the uplink and downlink data packets in the source cell before forwarding to the target base station.

In the handover completion phase, the target base station sends a path switch request to the AMF to switch the data packet transmission path to the target base station. Once the AMF responds to the request that the path switch is successful the target base station can indicate the source base station to release the UE context information. At this point, the connection of the whole UE is handed over to the target cell.

As mentioned earlier, in the Rel-15 version of NR the handover process has not been changed and enhanced much compared with the LTE system. The difference from the LTE system is that handover in the NR system does not necessarily mean that the security keys will be updated. Such handover occurs between CUs. If the handover process occurs between different DUs under the same CU, the network can instruct the UE not to change the security keys during the handover process. Using the same security keys before and after the handover will not cause security risks. The PDCP entity will not be reestablished if security keys are not changed. Therefore PDCP reestablishment during handover in the NR system is also controlled by the network.

11.4.2.3 RRC_CONNECTED state mobility optimization
11.4.2.3.1 Shortening handover interruption time
The interruption time of mobility refers to the shortest time that the UE cannot interact with any base station for delivery of user plane data packets. In the existing NR handover procedure when the terminal receives

the handover command the UE will disconnect from the source cell and initiate a random access procedure to the target cell. During this period, the data interruption time of the UE is at least as long as 5 ms. In order to shorten the interruption time, the NR system introduced a new handover enhancement process (i.e., handover based on dual active protocol stack, called DAPS handover in this book).

The core idea of DAPS handover is that when the UE receives the handover command, it initiates random access to the target cell while maintaining data transmission with the source cell to achieve a close to 0 ms data interruption time in downlink. At the beginning of the standardization discussion, apart from DAPS there is another candidate scheme called handover based on DC (called DC-based handover in this book) [8,9].

Fig. 11.15 shows schematic diagrams of the protocol stacks between the UE and network side in the existing handover procedure where the UE will only keep connection with one cell and its corresponding protocol stack. The core idea of DC-based handover (Fig. 11.16) is to first add the target cell as a primary secondary cell (PSCell), then swap the role of PSCell and PCell. Finally the source cell changed into a PSCell is released. In DC-based handover interruption time could be close to 0 ms

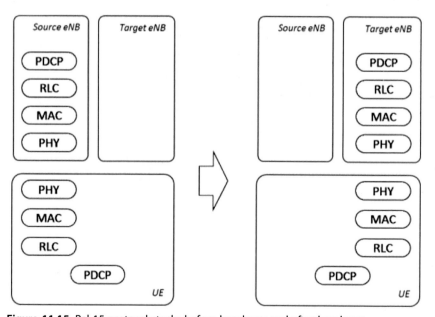

Figure 11.15 Rel-15 protocol stacks before handover and after handover.

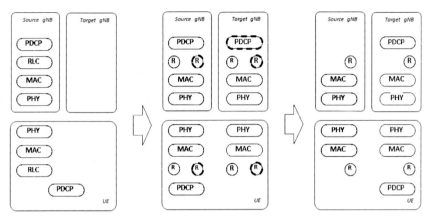

Figure 11.16 Protocol stacks of DC-based handover, before handover, during handover, and after handover.

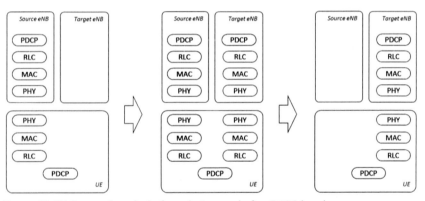

Figure 11.17 Protocol stacks before, during, and after DAPS handover.

and the reliability of handover is improved by maintaining the connection between two cells during handover. However, a new role change procedure needs be introduced as it was not adopted by the standard in the end.

As shown in Fig. 11.17, the protocol stack architecture of DAPS handover is relatively simple, which mainly includes establishing the protocol stack of the target side, keeping the protocol stack of the source cell during accessing the target cell, and releasing the protocol stack of the source cell when the handover is completed. The flowchart of DAPS handover is similar to that of legacy handover. DAPS handover can be configured per DRB (i.e., the network can configure some DRBs with high service interruption time requirements to perform DAPS handover). For DRB not configured with DAPS handover, the process of executing handover is the same as the legacy handover.

Considering the UE capability limitation, Rel-16 DAPS handover does not support DC and carrier aggregation (CA) at the same time. It means during DAPS handover, the UE only maintains the connection with the PCell of the source cell and the PCell of the target cell. Thus the source cell needs to release SCG and all SCells before sending the handover request.

The target base station determines the target cell configuration after receiving handover request and generates a handover command based on the received source cell configuration and UE capability. Then it sends the DAPS handover command to the source cell, which is transmitted by the source cell transparently to the UE.

The UE starts to execute DAPS handover after receiving the handover command. For DRB configured with DAPS, the UE will establish a protocol stack corresponding to the target cell which specifically includes the following points:

- Reconfigure the normal PDCP entity on the source side to the DAPS PDCP entity based on the configuration in the handover command (see details in Section 10.3).
- Establish the RLC entity on the target side and the corresponding logical channels.
- Create the MAC entity on the target side.

The handling of SRB is different from DRB. After receiving the handover command, the UE will establish the SRB's protocol stack on the target side based on the configuration information. Since the UE has only one RRC state, the UE will suspend the SRB of the source cell and switch the RRC signaling processing to the target cell. For DRB without DAPS handling of protocol stack is the same as legacy handover.

After completing the above steps, the UE starts to initiate a random access procedure to the target cell to obtain uplink synchronization with the target cell. As mentioned earlier, the basic idea of DAPS handover is to maintain the protocol stacks of source cell and target cell simultaneously. The UE maintains the connection with source cell while initiating random access procedure to the target cell. The data transmission between the UE and source cell is also maintained during that period.

If DAPS handover fails and the radio-link failure does not happen in the source cell, the UE can fall back to the connection with the source cell to avoid the RRC connection reestablishment procedure due to handover failure. At that point in time the protocol stack processing includes the following parts:

- For SRB, the UE will resume the suspended SRB of the source cell, report the DAPS handover failure to the network side, and release the PDCP entity, RLC entity, and the corresponding logical channel corresponding to the SRB of the target side.
- For DRBs configured with DAPS, the UE will reconfigure the DAPS PDCP entity to the normal PDCP entity and release the RLC entity on the target side and the corresponding logical channel, etc.
- For DRBs not configured with DAPS, the UE will fall back to the source cell configuration prior to receiving handover command, including SDAP configuration, PDCP, and RLC state variables, security configuration, and data stored in the buffers in the PDCP and RLC layer.
- At the same time, the UE releases all target-side configurations.

When the UE successfully accesses the target cell, the UE will switch the uplink data transmission from the source cell side to the target cell side. During standard discussions, there has been long debate about whether the UE supports single uplink data transmission or simultaneous uplink data transmission with the source cell and the target cell. Since the uplink anchor point on the network side is on the source cell side at that time, forwarding the received data by the target cell to the source cell will bring additional X2 interface transmission delay for simultaneous uplink transmission approach. Finally, the scheme of single uplink data transmission is agreed on.

On the other hand, "limited uplink transmission to source cell" is maintained. After the UE successfully completes the random access process, it will immediately switch to target to transmit data packet including unacknowledged PDCP SDUs while the UE will continue uplink HARQ and ARQ retransmission on the source side. If the source cell maintains downlink data transmission with the UE, HARQ feedback, CSI feedback, ARQ feedback, and ROHC feedback corresponding to these downlink data will also continue to be reported to the source cell.

After the UE successfully accesses the target cell and before releasing the source cell, the UE keeps the connection with the source cell and the target cell at the same time. Radio-link monitoring in both source and target side will be maintained. If the radio link fails in the target cell at this time, the UE will trigger the RRC connection reestablishment procedure. On the contrary, if the radio link fails in the source cell, the UE will not trigger the RRC connection reestablishment procedure but will suspend all DRBs on the source side and release the connection with the source cell.

When the target cell instructs the UE to release the source cell, the UE will release the connection with the source cell and stop the uplink data transmission and downlink data reception with the source cell, including resetting the MAC entity and releasing the MAC configuration, physical channel configuration, and security key configuration. For SRB, the UE will release its corresponding PDCP entity, RLC entity, and corresponding logical channel configuration. For DRB configured with DAPS, the UE will release RLC entity and corresponding logical channel on source side and reconfigure DAPS PDCP entity into normal PDCP entity.

11.4.2.3.2 Handover robustness optimization
Handover robustness optimization are mainly for high-speed mobility scenarios, such as high-speed train scenarios covered by cellular networks. The channel quality of the source cell monitored by the UE will drop sharply, which may lead to too late handover and higher handover failure rate. Specifically, it is reflected in the following two aspects:
- If measurement event parameters are set inappropriate, such as high measurement reporting threshold, the content of measurement reporting cannot be correctly received due to the sharp deterioration of link quality in the source cell.
- High-speed mobility brings new challenges to the handover preparation process. After the target cell responds with the handover command, the UE may not be able to correctly receive the handover command forwarded by the source cell due to the sharp deterioration of the link quality of the source cell.

CHO (Conditional Handover) is recognized as a promising technology that can improve handover robustness. Different from the traditional handover procedure triggered by the base station, the core idea of CHO is to configure the handover command of the target cell(s) to the UE in advance when the link quality of the source cell is good together with a handover execution condition associated with the target cell. When the configured execution condition is met, the UE can immediately initiate handover access to the corresponding target cell. Since the UE no longer triggers the measurement report when the execution condition is met and the UE has already acquired the configuration in the handover command in advance, the aforementioned problem is solved.

There are also three phases in the CHO procedure.

In the handover preparation phase, the source base station decides to initiate CHO preparation after receiving the measurement report sent by the UE (usually, the threshold configured for CHO measurement report will be lower than the reporting threshold configured for normal handover process). Then it sends a handover request message to the target base station. Once the target base station accepts the handover request, it will respond with a set of target cell configuration to the source base station. When forwarding the target cell configuration, the source base station will also configure a set of handover execution conditions to the UE.

In more detail, CHO configuration includes two parts: target cell configuration and handover execution condition configuration. Among them, the target cell configuration is the handover command generated by the target base station which must be forwarded completely and transparently to the UE without any modification. This principle is consistent with the traditional handover. When delta configuration for handover command is adopted, the latest source cell configuration will be used as the reference configuration.

Another part of CHO configuration is handover execution condition, which is mainly used by the UE to evaluate when to trigger handover. In the discussion of CHO configuration, the 3GPP adopted the principle of maximizing reuse of RRM measurement configuration and decided to turn measurement reporting events widely used in traditional handover into handover execution condition configuration: events A3 and A5. The difference is that when the event A3 or A5 as handover execution condition is triggered, the UE will not report the measurement result but execute the handover access procedure. During the discussion, some network vendors said that in order not to deviate from the network implementation to the greatest extent, many factors should be considered for the handover execution conditions such as multiple reference signals (SSB or CSI−RS); multiple measurement quantities (RSRP, RSRQ, and SINR); and multiple measurement events. On the contrary, terminal vendors hope that configuration of handover execution conditions can be as simple as possible for easy UE implementation. Finally, as a compromise, the following restrictions and flexibility are achieved for the configuration of handover execution conditions:

- Maximum two measurement identities
- At most one reference signal (SSB or CSI−RS)

- Maximum two measurement quantities (i.e., any two of RSRP, RSRQ, and SINR)
- Maximum two measurement events

Since the CHO configuration is sent to the UE in advance and the UE's movement direction has certain unpredictability, the source base station cannot accurately know which candidate cell the UE will eventually initiate handover access to. Therefore in the actual network deployment, the source base station will initiate handover requests to a multiple target base stations and configure corresponding handover execution conditions in the same message. In other words, the UE usually receives the CHO configuration for a group of candidate cells.

In the handover execution phase, the UE will continuously evaluate whether the measurement results of the candidate cells meet the handover execution conditions. Once the conditions are met, the UE immediately executes the handover procedure via a random access procedure and reports a handover completion message to the target base station when the random access is completed.

As mentioned earlier, since the moving direction of the UE is unpredictable the network usually configures the terminal with a set of CHO candidate cells. This actually provides the UE with great flexibility in selecting the target cell for handover execution. At the beginning of the standard discussion, there were mainly the following two types of schemes:

- Scheme 1: The UE can select multiple target cells, and can continue to evaluate whether other target cells meet the handover execution conditions after the first failed CHO attempt.
- Scheme 2: The UE is only allowed to select the target cell once. If the UE fails to access the target cell connection reestablishment procedure will be triggered.

The advantage of Scheme 1 is that it can maximize the use of the configured CHO candidate cells. The disadvantage is that the delay of the whole handover procedure is difficult to control due to multiple attempts with multiple configured timers. Scheme 2 has the advantage of simplicity and is beneficial for the UE's implementation. The disadvantage is that only one target cell can be selected for access which will cause waste of network resources. In the end, the 3GPP adopted Scheme 2. To overcome its shortcomings the 3GPP introduced an enhancement of connection reestablishment. Specifically if the selected cell for connection reestablishment is a CHO candidate cell, the UE can directly perform handover access based on the CHO configuration of the cell. Otherwise the UE performs the traditional connection reestablishment procedure. This kind of enhancement actually leverages some of the benefits of Scheme 1 (i.e., by taking advantage of multiple CHO candidate cells to some extent).

Although Scheme 2 is simple, there are still some problems to be solved. For example, how does the UE select one from multiple candidate cells that meet the conditions? Someone believes that the UE should select the cell with the best channel quality. Others believe that the UE's behavior in cell reselection process should be reused here (i.e., the UE should give priority to the cell with the largest number of good beams). There are also some views that the network should configure priorities for multiple candidate cells which can reflect the frequency priority and load situation of the candidate cells. As there are too many schemes to converge, the 3GPP finally decided not to standardize the UE's behavior (i.e., how to select the target cell among multiple candidates is up to the UE's implementation).

The handover completion phase is similar to the traditional handover procedure. It is worth mentioning that since the source base station cannot accurately predict when the UE meets the handover execution conditions, it is a problem for the source base station to decide when it should perform data forwarding. Basically, there are two options:

- Option 1: Early forwarding (i.e., to start the data forwarding process after sending the UE the CHO configuration so that the target base station can carry out data transmission immediately after the UE is connected to the target cell).
- Option 2: Later forwarding (i.e., when the UE selects and accesses to the target cell, the target base station notifies the source base station to forward the data).

Option 1 has the advantages of short data interruption time and good service continuity during handover while the disadvantage is that the source base station needs to forward data to multiple candidate base stations resulting in large network overhead. Option 2, on the contrary, has the advantage of forwarding data to only one target base station. The disadvantage is that the target base station cannot transmit data to the UE immediately after the UE successfully accesses the target cell until the source base station forwards the data. Each option has its own advantages and disadvantages. Finally, 3GPP RAN3 decided to support both types of data forwarding schemes.

11.5 Summary

The biggest difference between the NR system and LTE system in terms of system information broadcasting mechanism is the introduction of on-demand request. It is used to reduce unnecessary system information broadcasting, neighbor intrafrequency interference, and energy consumption. The NR not only introduces new paging initiation but also optimizes its transmission mechanism to adapt to beam management, which is more suitable for high-frequency bands. The introduction of the RRC_INACTIVE state makes a compromise between power saving and control plane delay, which results in a new RRC connection resume process. The LTE framework is followed basically for RRM measurement except for the introduction of a new reference signal, namely CSI—RS-based measurement. The biggest enhancement of mobility management in the RRC_CONNECTED state is the introduction of handover based on dual active protocol stack, which makes the user plane interruption time close to 0 ms while the CHO method greatly improves the robustness of handover.

References

[1] 3GPP TS 38.300 V16.2.0. Study on scenarios and requirements for next generation access technologies.
[2] 3GPP TS 38.304 V16.3.0. User Equipment (UE) procedures in idle mode and in RRC Inactive state.
[3] 3GPP TS 24.501 V16.6.0. Non-access stratum (NAS) protocol for 5G System (5GS).
[4] 3GPP TS 38.331 V16.3.1. Radio resource control (RRC) protocol specification.
[5] 3GPP TS 38.423 V16.4.0. Xn application protocol (XnAP).
[6] 3GPP TS 37.340 V16.1.0. Evolved universal terrestrial radio access (E-UTRA) and NR; multi-connectivity.
[7] R2—1704832. RRM measurements open issues. Sony.
[8] R2—1910384. Non DC based solution for 0ms interruption time, Intel Corporation, Mediatek Inc, OPPO, Google Inc., vivo, ETRI, CATT, China Telecom, Xiaomi, Charter Communications, ASUSTeK, LG Electronics, NEC, Ericsson, Apple, ITRI.
[9] R2—1909580. Comparison of DC based vs. MBB based approaches. Futurewei.

CHAPTER 12

5G network slicing

Haorui Yang, Tricci So and Yang Xu

12.1 General descriptions

This section mainly introduces the basic concepts of network slicing to help to establish a preliminary understanding of network slicing. At the same time, this section also hopes to lay a foundation for further understanding of network slicing.

12.1.1 Background

When assessing the most immediate requirements for the 5G system by various standards bodies and technology consortiums, such as the 3GPP, International Telecommunication Union, Next Generation Mobile Networks, etc., the following three main classes were identified:

1. Enhanced Mobile Broadband (eMBB)

   Mainly data-driven use cases for Mobile Broadband requiring extreme high data rates, up to 20 Mbps, across a wide coverage area to deliver a much smarter and better user experience.

2. Ultra-Reliable and Low-Latency Communications (URLLC)

   Services impose strict requirements on latency and reliability for mission critical communications, such as remote surgery, autonomous vehicles, or the tactile internet.

3. Massive Machine Type Communications (mMTC)

   Efficient communication system that needs to support a very high volume of devices in a small area, which may only send data sporadically, such as Internet of Things (IoT) use cases.

In addition to the main key performance indicators relevant to the service experience such as user data rate, latency, and density of devices as described above for the three main use cases, the following aspects related to the deployment and network operational perspective have also been identified:

- Service diversity: The ability to deliver different service classes (eMBB, URLLC, mMTC) over the same infrastructure independently and simultaneously.

5G NR and Enhancements
DOI: https://doi.org/10.1016/B978-0-323-91060-6.00012-X
621

- Performance guarantee: Guarantee for the performance KPIs in terms of Quality of Service (QoS) and Quality of Experience, such as user data rate or reliable service.
- Time-to-market: The ability to adapt to an existing or new business model and to provision network services deployment for customers in timely manner (e.g., within a day).
- Resource isolation: Guarantee that each network service tailored for a customer is secured and will not be impacted by the performance degradation from the other services.
- Agile networking: Increase the agility of network deployment and ease of adaptation of the network performance to unforeseen network conditions.
- Access technology convergence: Treat all access technologies (e.g., fixed and mobile access technologies) equally and deliver the same user experience regardless of the network access technology type.

As the 4G system could not keep up with the above use cases to meet the operational and performance objectives, the concept of network slicing emerged to meet these challenges [1]. Network slicing configuration is based on customized connectivity designed to address and to balance the specific operator and user requirements by offering a flexible way to segment the network to support particular services or business segments. Network slices are optimized by myriad characteristics including latency or bandwidth requirements. Since the network slices are isolated from each other in the control and user planes, the user experience of the network slice will be the same as if it was a physically separate network. Such network solution techniques are referred as as-a-service basis, which enhances operational efficiency while reducing time-to-market for new services.

During the early stage of the network slicing architecture design, there were discussions on how to support multiple network slices for a given UE when network resource and operation isolation need to be considered. Different services require different network slices, which means that an operator needs to deploy multiple network slices to satisfy the different services and the UE must be able to access multiple network slices simultaneously. Three levels of network resource isolation have been considered [2], as shown in Fig. 12.1:

- Option 1: all the core network functions are isolated per network slice.
- Option 2: only some of the core network functions are isolated per network slice.

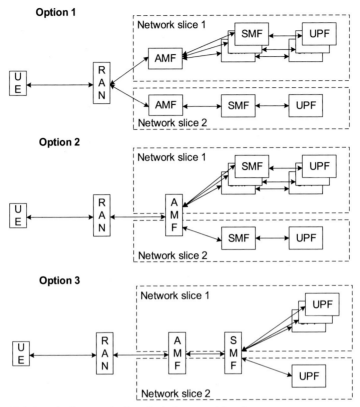

Figure 12.1 Level of network slices isolation considerations.

- Option 3: the control plane core network functions are shared by several network slices while the user plane core network functions are isolated per network slice.

Even though Option 2 was chosen as the base architecture to develop the standard protocols, in practice, all three options can be supported with some support of the preconfigured and local policy. Option 2 is considered as the most basic configuration because when a UE is served by multiple network slices, they all need to be coordinated and controlled by a common access and mobility management entity (MME). The UE's mobility management and service packet transmission management, however, can be operated independently. Hence, multiple network slices that serve the UE simultaneously could share the same Access and Mobility Management Function (AMF). Meanwhile, the network functions responsible for the specific service transmission management [i.e., the

Session Management Function (SMF) and the User Plane Function (UPF)] could be isolated to achieve the data session management and transmission resource isolation. The isolation here can be implemented at the software level through virtualization technology, or by the physical isolation with deploying the different network function equipment.

12.1.2 Network slicing terminologies and principles
12.1.2.1 Network slice identification
In 3GPP standard specification, the term of Single Network Slice Selection Assistance Information (S−NSSAI) is used to identify a network slice. A network slice is defined as a logical network that provides specific network capabilities and network characteristics.

An S−NSSAI is comprised of a slice/service type (SST) and Slice Differentiator (SD), as shown in Fig. 12.2 [3]. The SST field is used to distinguish the specific network service (e.g., eMBB, URLLC, mMTC, V2X, etc.) that the S−NSSAI is applied to. It occupies the upper eight bits of the S−NSSAI format. The SD field provides another level of distinction for the S−NSSAI within the same SST. The SD field is indicated by the lower 24 bits of the S−NSSAI. For example, if the SST is V2X, different car vendors are distinguished by the different SDs. An S−NSSAI can be the standardized values [i.e., such S−NSSAI is only comprised of an SST with a standardized SST value (i.e., eMBB, URLLC, MIOT, and V2X) and no SD] or nonstandard values (i.e., such S−NSSAI is comprised of both an SST and an SD or only an SST without a standardized SST value and no SD).

Considering the customized characteristics of network slices, different operators could deploy different network slices for their service offerings. However, there are some worldwide services that would be supported by most operators. For example, the mobile broadband data service, the voice service, etc. For the benefits of roaming interoperability, the standardized SST value would be very useful among 5G operators. According

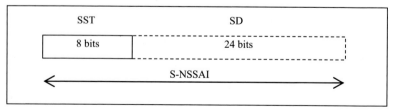

Figure 12.2 S−NSSAI format (from clause 28.4.2 in TS 23.122 [3]).

to 3GPP Rel-16, there are four standardized SST values: eMBB, URLLC, MIoT, and V2X.

An S–NSSAI with a nonstandard value identifies a single network slice within the PLMN with which it is associated. An S–NSSAI with a nonstandard value will not be used by the UE in access stratum procedures in any PLMN other than the one to which the S–NSSAI is associated.

12.1.2.2 Network slicing terminology and usage

When a network slice is deployed, it is referred as a network slice instance, which is comprised of a set of network function instances and the required resources (e.g., compute, storage, networking resources). As a network slice is end-to-end, the term NSI ID is defined as an identifier for identifying the core network part of a network slice instance when multiple network slice instances of the same network slice are deployed, and there is a need to differentiate between them in the 5G-Core.

A network slice instance will include the Core Network Control Plane and User Plane Network Functions and the serving PLMN will also include 5G–AN.

A UE may have subscriptions with multiple network slices corresponding to multiple S–NSSAIs. In general, NSSAI is a collection of S–NSSAIs. In order to describe how NSSAI is used at different stages of the UE operation, 3GPP TS 23.501 [4] has defined different categories of NSSAI, which could be the full set or a subset of the UE's subscribed S–NSSAIs. Further details on the different categories of NSSAI that are used for the various operations will be described in a later section of this chapter.

The various categories of NSSAI are:

- *Requested NSSAI:* NSSAI provided by the UE to the serving operator (i.e., serving PLMN) during registration.
- *Allowed NSSAI:* NSSAI provided by the Serving PLMN during a Registration procedure, for example, indicating the S–NSSAI values the UE could use in the Serving PLMN for the current Registration Area (RA).
- *Subscribed S–NSSAI:* S–NSSAI based on subscriber information, which a UE is subscribed to use in a PLMN.
- *Default Configured NSSAI:* If it is configured in the UE, it is used by the UE in a Serving PLMN only if the UE has no Configured NSSAI for the Serving PLMN. The values used in the Default Configured

NSSAI are expected to be commonly decided by all roaming partners (e.g., by the use of SST values standardized by 3GPP or other bodies).

- *Configured NSSAI:* NSSAI provisioned in the UE applicable to one or more PLMNs. A Configured NSSAI may either be configured by a Serving PLMN and apply to the Serving PLMN, or may be a Default Configured NSSAI configured by the HPLMN and that applies to any PLMNs for which no specific Configured NSSAI has been provided by the Serving PLMN. There is at most one Configured NSSAI per PLMN.

- *Pending NSSAI:* NSSAI provided by the Serving PLMN during a Registration procedure, indicating the S−NSSAI(s) for which the network slice-specific authentication and authorization procedure is pending.

- *Rejected S−NSSAI:* One or more rejected S−NSSAIs with cause and validity of rejection from the AMF for the given UE's operation. An S−NSSAI may be rejected for the entire PLMN, for the current RA, or failed or revoked Network Slice Specific Authentication & Authorization (NSSAA).

Standardized SST values can be used by any category of the S−NSSAI above. Note that an S−NSSAI can belong to more than one category. For example, an S−NSSAI that is part of the Subscribed NSSAI could be included in the Requested NSSAI of the UE's Registration Request; the same S−NSSAI could also be dedicated as the Default S−NSSAI in the UE's subscription. Furthermore, the same S−NSSAI may become part of the Allowed NSSAI as well as Configured NSSAI, which are included in the Registration Accept message to the UE.

Based on operator's policy, one or more Subscribed S−NSSAIs can be marked as a Default S−NSSAI. The network is expected to serve the UE with a related applicable network slice instance corresponding to the Default S−NSSAI only when the UE does not provide any valid S−NSSAI in the Registration Request message as part of the Requested NSSAI. The network verifies the Requested NSSAI, if any, provided by the UE in the Registration Request against the Subscribed S−NSSAI(s) before identifying the possible Allowed NSSAI. It is the UE's serving AMF or NSSF, based on the operator's configuration, that is used to determine the Allowed NSSAI at the UE's RA, which is then included in the Registration Accept message to the UE.

The S−NSSAIs in the Configured NSSAI, the Allowed NSSAI, the Requested NSSAI, and the Rejected S−NSSAIs except for the

S—NSSAIs rejected due to the failed or revoked NSSAA contain only values from the UE's Serving PLMN. The Serving PLMN can be the HPLMN or a VPLMN.

Once the UE has registered with the Serving PLMN and provisioned with Allowed NSSAI, the UE can request a PDU session establishment to enable data transmission for its application, which corresponds to one of its allowed S—NSSAIs. A Data Network (DN) associated with the S—NSSAI would have also been included in UE's subscription and provided to the UE. As a UE could have been associated with multiple network slices [i.e., S—NSSAI(s)], the operator may provision the UE with a Network Slice Selection Policy (NSSP), which is part of the UE Route Selection Policy (URSP). The NSSP consists of a set of rules (at least one rule), where each rule attributes an application with a certain S—NSSAI. A default rule may exist that matches all applications to an S—NSSAI.

Only the S—NSSAI from the Allowed NSSAI can be used by the UE to be included in the PDU Session Establishment request.

In order to support roaming, the S—NSSAI in the PDU Session Establishment Request contains one Serving PLMN S—NSSAI value and in addition may contain a corresponding HPLMN S—NSSAI value to which this first value is mapped.

The optional mapping of Serving PLMN S—NSSAIs to HPLMN S—NSSAIs contains Serving PLMN S—NSSAI values and corresponding mapped HPLMN S—NSSAI values. 3GPP TS 23.501 [4] provides further detailed instructions on how the different categories of S—NSSAI/NSSAI are used during various UE and network operations and how the mapping is performed.

There can be at most eight S—NSSAIs in the Requested and Allowed NSSAIs sent in signaling messages (e.g., Registration Request and Registration Accept, respectively) between the UE and the network. The Requested NSSAI signaled by the UE to the network allows the network to select the Serving AMF, Network Slice(s), and Network Slice instance (s) for this UE.

Based on the operator's operational or deployment needs, a network slice instance can be associated with one or more S—NSSAIs, and an S—NSSAI can be associated with one or more network slice instances. Multiple network slice instances associated with the same S—NSSAI may be deployed in the same or in different Tracking Areas.

Based on the UE-provided Requested NSSAI and the UE's Subscription Information, the 5G-Core is responsible for selection of a network slice instance(s) to serve a UE including the 5G-Core Control Plane and User Plane Network Functions corresponding to this network slice instance(s).

The (R)A may use Requested NSSAI in access stratum signaling to handle the UE Control Plane connection before the 5G-Core informs the (R)AN of the Allowed NSSAI. The Requested NSSAI is used by the RAN for AMF selection. The UE will not include the Requested NSSAI in the RRC Resume when the UE asks to resume the RRC connection and is CM-CONNECTED with RRC Inactive state.

When a UE is successfully registered over an Access Type, the 5G-Core informs the (R)A by providing the Allowed NSSAI for the corresponding Access Type.

Details of how the RAN uses NSSAI information are not included in this chapter and readers should refer to 3GPP TS 38.300 [5] for more information.

12.1.2.3 Storage of S—NSSAI/NSSAI in 5G UE

In order for the UE to be served properly for its subscribed service from the network, the UE is required to ensure the valid S—NSSAI/NSSAI included in its request to the network. The UE is provisioned or provided with different categories of S—NSSAI(s)/NSSAI(s) and the associated mappings between the UE's serving PLMN and HPLMN during the UE Registration and the UE Configuration Update (UCU) procedures. The UE is responsible for storing them, maintaining their latest status, and ensuring their validity for use in its future network request (e.g., Registration, Service Request, PDU Session Establishment, Slice Specific Authentication, Authorization, etc.), even though the UE may have moved to a different location, registered with a different PLMN, switched between different power-saving modes and powered on/off, etc. 3GPP TS 23.501 [4] specifies more in-depth procedures on how the UE should manage and store its provisioned S—NSSAI(s)/NSSAI(s) and their associated mappings.

12.2 Network slicing as a service in the 5G system

The following sections describe in detail how the 5G UE leverages the Network Slicing feature from its 5G operator to establish service for its application. Further in-depth details can be found in 3GPP TS 23.501 [4] for an overview, TS 23.502 [6] for detailed control flows, and TS 23.503 [7] for policy control support.

12.2.1 Network slicing service registration

Network Slicing is a mandatory feature in a 5G system. Once the UE is registered, it will be associated with one or more S—NSSAIs corresponding to the UE's subscription. Prior to the UE's initial registration with the 5G system, the UE may have been preconfigured with zero, Default Configured S—NSSAIs, or a set of Configured NSSAIs corresponding to a particular PLMN(s). If the UE has registered with a 5G system previously, the UE may have obtained the Allowed NSSAI as well as Configured NSSAI corresponding to particular PLMN(s). Furthermore, the UE may have been provisioned with a NSSP that consists of a set of rules (at least one rule), where each rule attributes an application with the associated S—NSSAI. A default NSSP rule may be provisioned instead in order to match all applications to an S—NSSAI. The high-level descriptions for the UE Network Slicing service registration procedure are as follows and the simplified control flow is shown in Fig. 12.3:

1. The UE may determine the Requested NSSAI based on the Default Configured NSSAI, Configured NSSAI, or Allowed NSSAI associated with the selected PLMN for the given access. Note that UE Network Slicing registration is per access type.

   The Requested NSSAI is comprised of one or more S—NSSAIs corresponding to the UE applications' services the UE may initiate. The UE may have been provisioned with a NSSP that consists of a set of rules (at least one rule), where each rule attributes an application with the associated S—NSSAI. A default NSSP rule may be provisioned instead in order to match all applications to an S—NSSAI.

   The UE includes the Requested NSSAI in the Registration Request message and sends it to the source AMF. It is possible the UE may not include the Requested NSSAI if there is no valid S—NSSAI for the UE (e.g., IoT device or first initial UE registration).

2. When a UE registers with a PLMN, if for this PLMN the UE has not included a Requested NSSAI nor a GUAI while establishing the connection to the (R)AN, the (R)A will route all NAS signaling from/to this UE to/from a default AMF. When receiving from the UE a Requested NSSAI and a 5G-S—TMSI or a GUAMI in RRC Connection Establishment or in the establishment of a connection to N3IWF/TNGF, if the 5G—A can reach an AMF corresponding to the 5G-S—TMSI or GUAMI, then 5G—AN forwards the request to this AMF. Otherwise, the 5G—AN selects a suitable AMF based on the

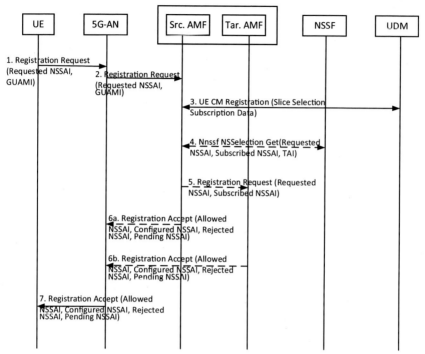

Figure 12.3 Network Slice Service registration.

Requested NSSAI provided by the UE and forwards the request to the selected AMF. If the 5G—AN is not able to select an AMF based on the Requested NSSAI, then the request is sent to a default AMF.

3. After the source AMF receives the Requested NSSAI, the source AMF needs to determine which S—NSSAI(s) from the Requested NSSAI are allowed for the UE (i.e., the Allowed NSSAI). If the UE is authenticated by the source AMF successfully, the source AMF obtains the slice selection subscription data of the UE from the Unified Data Management Function. The slice selection subscription data includes one or more S—NSSAIs subscribed by the UE [i.e. Subscribed S—NSSAI(s)], in which all or a subset of the Subscribed S—NSSAI(s) may be marked as default [i.e., Default S—NSSAI (s)]. Certain Subscribed S—NSSAI(s) may also require NSSAA (see further details on NSSAA in a later section).

4. Based on local operator's policy, the source AMF may consult with Network Slice Selection Function (NSSF) to determine the potential Allowed NSSAI. The source AMF forwards the Subscribed NSSAI and the Requested NSSAI as well as the UE's RA info to the NSSF.

The NSSF determines the potential Allowed NSSAI and the target AMF that can serve the Allowed NSSAI.

5. If the source AMF cannot support the potential Allowed NSSAI, it will trigger AMF redirection to the target AMF that can support the Allowed NSSAI. The source AMF will then transfer the Registration Request message and the potential allowed NSSAI to the target AMF.

6. 6a. is for the case where the source AMF is also the target AMF when no AMF redirection was triggered, whereas 6b. is for the case where source new target AMF is selected due to the AMF redirection.

The target AMF includes the authorized S—NSSAI(s) that serves the UE's RA, which is from the UE's Requested NSSAI as well as part of the UE's Subscribed S—NSSAI(s) in the Allowed NSSAI. The target AMF may also include the Configured NSSAI, Rejected NSSAI as well as Pending S—NSSAI(s) in a Registration Accept message according to the UE's RA and other UE mobility-related policy. The target AMF sends the Registration Accept message to the UE. For the Rejected NSSAI, the AMF will also inform the UE with the unallowed area per the rejected S—NSSAI (e.g., the entire PLMN or the current RA of the UE).

For the Pending S—NSSAI, once the NSSAA procedure is successful, the Pending S—NSSAI will then be included in the Allowed NSSAI. Otherwise, the Pending S—NSSAI is rejected. If the NSSAA ends after registration, the AMF updates the Allowed NSSAI and Rejected NSSAI with the UE, accordingly via the UCU command procedure. Further details on NSSAA will be described in a later section.

12.2.2 Traffic routing in Network Slicing

After the UE has successfully registered to the Allowed NSSAI, it needs to initiate the PDU Session Establishment to associate the target application with the target S—NSSAI(s). Only after the corresponding PDU session is established and activated can the UE's application become active to transmit and/or receive the service packets.

12.2.2.1 UE Route Selection Policy support for Network Slicing traffic routing

URSP is used to bind service data streams to different PDU sessions for traffic routing as needed. Since a given PDU session is to serve a particular S—NSSAI, URSP rules can be used to achieve the purpose of routing the specific business data streams over different network slices, as shown in Fig. 12.4. A PDU session can be used to transport application traffic to

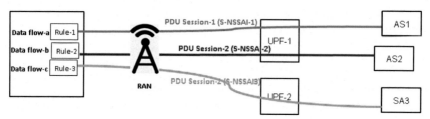

Figure 12.4 Binding application data to PDU session based on URSP rules.

the target Application Server associated with a particular S—NSSAI out-side the network.

URSP policy can support multiple URSP rules and each URSP rule specifies a Traffic Descriptor (TD) to match a specific service. Based on the 3GPP definition, the TD can use the following parameters to match the different data streams:

- *Application Identifier:* including Operation System (OS) ID and Application ID. OS ID is used to distinguish the different OS of the device vendors and Application ID is used to identify an application under one OS.
- *IP descriptor:* including destination address, destination port, protocol version.
- *Domain descriptor:* Fully Qualified Domain Name.
- *Non-IP descriptor:* defining the Non-IP descriptor; for example, Virtual Local Area Network ID, Media Access Control address.
- *DN Name (DNN):* DNN is a parameter defined by the operator that corresponds to the exit of the core network and can access a specific external DN, such as IP Multimedia System (IMS)—DNN, Internet—DNN.
- *Connection capability:* a parameter defined by Android system and used by application to indicate the purpose of the connection establishment, such as "ims," "mms," and "internet."

In order to realize URSP, the 3GPP specification requests the UE to implement the URSP handling layer. As shown in Fig. 12.5, when a new application initiates data transmission, the OS/APP initiates a connection request message to the URSP handling layer. The connection request message will contain the characteristic information of the application data streams. The URSP handling layer matches the TD of each URSP rule according to the parameters provided by the upper layer. If one URSP rule is matched, the application data stream should be transported by one

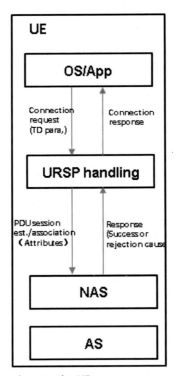

Figure 12.5 URSP handling layer in the UE.

PDU session that has the attributes in the matched URSP rule. If the PDU session already exists, then the data is transported over this existing PDU session; if the PDU session does not exist, the UE will first establish the PDU session for the application data. The application will then be bound with the PDU session. After the binding is established, the new PDU session will then be used to transport the subsequent application data streams according to the new binding.

Further details of the URSP support for network slicing can be found in TS 23.503 [7].

12.2.2.2 Service data path establishment

When the UE initiates its application service activation, as described in the section above, it determines the attributes of the PDU session corresponding to the service according to the URSP, such as DNN, S—NSSAI, and Session and Service Continuity mode. In addition, the URSP rule may also indicate the type of access technology that needs to be preferentially used for the PDU session (e.g., 3GPP access or Non-3GPP access). 3GPP access includes LTE and NR while Non-3GPP access includes the Wireless Local Area Network. The 5G system supports Network Slicing over both 3GPP access and Non-3GPP access. The UE includes the PDU session attributes when initiating the PDU session establishment procedure. The simplified procedure is as follows and can be seen in Fig. 12.6, and the detailed procedure can be found in TS 23.502 [6].

1. When initiating the establishment of a PDU session, the UE includes the DNN, the S—NSSAI, other PDU session parameters, and the PDU Session Establishment Request in the UL NAS Transport message. The UE will then send the UL NAS Transport message to the AMF.

2. Based on the received DNN, S—NSSAI, and other PDU session parameters, the AMF selects an SMF belonging to the network slice identified by the S—NSSAI. After the successful SMF selection, the AMF forwards the DNN, S—NSSAI, PDU Session Establishment Request message, and other parameters to the selected SMF. Then the SMF selects a UPF to serve the PDU session.

3 and 4. If the SMF accepts the PDU session establishment, it sends the PDU Session Establishment Accept message to the UE. In the message, the allocated PDU address and QoS flow parameters are included so the UE can use the PDU address to satisfy the data transmission protocol.

When the UE subscription or the network deployment changes, the network can update the new Allowed NSSAI to the UE. After receiving the Allowed NSSAI, the UE will compare it with the S—NSSAIs of the

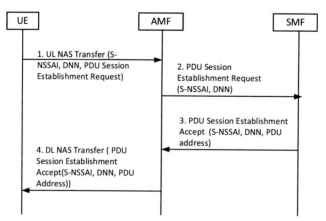

Figure 12.6 PDU Session Establishment in a network slice.

established PDU sessions. If one S—NSSAI of the established PDU session is no longer in the Allowed NSSAI, the UE will locally release this PDU session.

12.2.2.3 Support interworking between EPC and 5GC for Network Slicing

When the UE moves between EPC and 5GC, in order to ensure uninterrupted data transmission (i.e., service continuity), the Packet DN (PDN) connection over the EPC needs to be switched over to 5GC in order to maintain the same PDN address of the PDN connection (e.g., IPv4 address). To achieve this, the control plane of the PDN Gateway (PGW) in the EPC and the SMF in the 5GC are integrated to become one single network entity, PGW—C + SMF, for the interworking support.

The UE initiates the registration procedure when moving to 5GC from EPC, which includes the Requested NSSAI to request for the Allowed NSSAI. The UE in idle mode initiates the registration procedure right away, while the UE in the connected mode initiates the registration procedure only after the handover procedure. When considering to the procedure as described in Section 12.2.1 above, if the UE does not include the target S—NSSAI corresponding to the existing PDN connection in the Requested NSSAI, the target S—NSSAI may not be included in the Allowed NSSAI, which could cause the UE to locally release the PDN connection, and therefore the service continuity will not be maintained for the given PDU session. In order to ensure that the PDU session corresponding to the PDN connection is not released, when the UE moves to 5GC from EPC, the UE needs to include the target S—NSSAI associated with the PDN connection as the Requested NSSAI in the PDU Session Establishment Request. Therefore the UE is required to obtain the target S—NSSAI corresponding to the PDN connection before initiating the Registration Request with the 5GC. In order to obtain the target S—NSSAI, during the PDN connection establishment in the EPC, the PGW—C + SMF will provide the UE of the target S—NSSAI associated with the PDN connection according to the network configuration and the serving APN.

If the UE is in the connected mode, the eNB of the 4G system will trigger the handover procedure. During the handover process, the MME will select the target AMF based on the TAI and other information, and send the stored mobility management context of the UE to the AMF, which contains the UE's IP address or ID of the PGW—C + SMF. Then

the AMF locates the PGW−C + SMF and obtains the PDU session parameters from PGW−C + SMF, such as PDU session ID, S−NSSAI, etc.

Unfortunately, the target AMF selected by the MME may not be able
to support all the network slices associated with the established PDN connections, which will cause an unsuccessful PDN connection transfer from
EPC to 5GC and service interruption. During the R15 discussions, most
of the companies believed that only a few network slices should be
deployed during the initial phase of the 5G system. Hence, the issue as
described above will be unlikely for Rel-15 deployment and therefore it
was addressed.

It was resolved in Rel-16 and the solution is captured in TS 23.502
[6]. The resolution for this issue is that, after the initial target AMF obtains
the S−NSSAIs from PGW−C + SMF, and if the initial target AMF cannot support all the requested S−NSSAIs, the initial target AMF will select
a more suitable new target AMF, which is capable of serving all or more
requested S−NSSAIs and trigger the AMF relocation procedure as
described in Section 12.2.1. In other words, the AMF relocation procedure is embedded into the handover procedure. As a result, the PDU sessions can be reserved as much as possible and provide a better service
experience during the interworking between EPC and 5GC.

12.3 Network slice congestion control

It is possible that the upper bound of network resources allocated for a
network slice is reached. When a high volume of signaling is generated at
the same time, the network slice could be congested. For these congested
network slices, if the AMF or SMF receives a session management signaling message sent by the UE, AMF, or SMF it can reject it.

Once the signaling message is rejected, if the UE continues to send
the same signaling message repeatedly, this will further deteriorate the
load of the network slice. In order to prevent this overload situation, the
AMF or SMF can provide the UE with a back-off timer while rejecting
the signaling. In Rel-15, the UE associates the back-off timer with the
S−NSSAI included in the PDU session establishment request message. If
the UE does not provide S−NSSAI in the PDU session establishment
request message or the PDU session is transferred from EPC to 5GC, the
UE associates the back-off timer with "no S−NSSAI." Before the back-
off timer expires, the UE cannot initiate session establishment and a

session modification request for the congested S−NSSAI or "no S−NSSAI."

Taking into account that the UE may move across several PLMNs, in addition to setting the above timer, the network will also indicate to the UE whether the back-off timer is applicable to all PLMNs.

Note that the network slice congestion control is not applicable to high-priority UEs, emergency services, and UE signaling to update the PS data off state.

12.4 Network slice in roaming case

When the UE is roaming, because some services need to be routed from the VPLMN to the HPLMN, the PDU session will be established over both the VPLMN slice and the HPLMN slice. In the case of roaming, the following additional information will be included in the Registration and PDU Session Establishment procedures as described in Section 12.2.2 above accordingly:

- S−NSSAI includes VPLMN SST, VPLMN SD, and their respective mapping to HPLMN SST and mapped HPLMN SD. The mapping to HPLMN SST and SD are optional.
- During the registration procedure, if the UE has a Configured NSSAI associated with the VPLMN and the mapping to HPLMN S−NSSAIs, the UE needs to provide the VPLMN S−NSSAI and the respective mapping HPLMN S−NSSAI in the Requested NSSAI. If the UE does not have the Configured NSSAI associated with the VPLMN but there is the HPLMN S−NSSAIs of the established PDU sessions, the UE only provides the HPLMN S−NSSAIs as the Requested Mapped NSSAI.
- The AMF sends to the UE the mapping of VPLMN S−NSSAI and the HPLMN S−NSSAI in the Allowed NSSAI and Pending NSSAI.
- Rejected NSSAI is applicable to the current VPLMN. If the S−NSSAI is rejected due to NSSAA procedure failure, the Rejected NSSAI includes the HPLMN S−NSSAIs; otherwise the Rejected NSSAI includes the VPLMN S−NSSAIs.
- When the UE requests PDU session establishment, it provides both VPLMN S−NSSAI and HPLMN S−NSSAI.
- If the back-off timer is applicable to all the PLMNs, the timer is associated with the HPLMN S−NSSAI.

12.5 Network slice specific authentication and authorization

The NSSAA procedure is triggered for an S—NSSAI requiring Network Slice-Specific Authentication and Authorization with an AAA Server (AAA—S), which may be hosted by the H—PLMN operator or by a third party which has a business relationship with the H—PLMN, using the EAP framework as described in 3GPP TS 33.501 [8]. An AAA Proxy (AAA—P) in the HPLMN may be involved (e.g., if the AAA—S belongs to a third party).

The NSSAA procedure is triggered by the AMF during a registration procedure:

- When some network slices that the UE has requested and subscribed require Slice-Specific Authentication and Authorization;
- When the AMF determines that Network Slice-Specific Authentication and Authorization is required for an S—NSSAI in the current Allowed NSSAI (e.g., subscription change); or
- When the AAA—S that authenticated the network slice triggers a reauthentication.

The AMF performs the role of the EAP Authenticator and communicates with the AAA—S via the Network Slice Specific Authentication and Authorization Function (NSSAAF). The NSSAAF undertakes any AAA protocol interworking with the AAA protocol supported by the AAA—S.

The NSSAA procedure requires the use of a GPSI. In other words, a subscription that contains S—NSSAIs subject to NSSAA will include at least one GPSI.

If the UE does not support the NSSAA feature, the AMF will not trigger the NSSAA procedure for the UE and the UE's requested S—NSSAIs are rejected for the PLMN. If the UE supports the NSSAA feature and if the UE requests any of the S—NSSAIs in the Requested NSSAIs that are subject to NSSAA, they are included in the list of Pending NSSAIs for the PLMN if the UE's registration is completed before the NSSAA procedure.

Both the UE and the AMF need to keep track of the NSSAA status of the UE's requested S—NSSAI.

The NSSAA requires that the UE Primary Authentication and Authorization of the SUPI has been successfully completed. If the SUPI authorization is revoked, then the Network Slice-Specific authorization is also revoked.

12.6 Summary

This chapter introduced background knowledge of Network Slicing, registration to network slice, traffic routing in network slice, network slice congestion control, network slice-specific authentication and authorization, etc., and also discussed network slice-related adaptation for roaming scenarios. The intent of this chapter was to provide a general overview of the Network Slicing feature in the 5G system by describing the basic concepts, architecture principles, terminology, and key aspects of the network operations of Network Slicing.

References

[1] 3GPP TR 22.891 V1.0.0. *Feasibility study on new services and markets technology enablers (release 14)*; 2015-09.
[2] 3GPP TS 23.799 V14.0.0. *Study on architecture for next generation system (release 14)*; 2016-12.
[3] 3GPP TS 23.003 V16.1.0. *Numbering, addressing and identification (release 16)*; 2019-12.
[4] 3GPP TS 23.501 V16.4.0. *System architecture for the 5G System (5GS) (release 16)*; 2020-3.
[5] 3GPP TS 38.300 V16.4.0. *NR; NR and NG-RAN overall description (release 16)*; 2020-12.
[6] 3GPP TS 23.502 V16.3.0. *Procedures for the 5G system (5GS) (release 16)*; 2019-12.
[7] 3GPP TS 23.503 V16.4.0. *Policy and charging control framework for the 5G system (5GS) (release 16)*; 2020-3.
[8] 3GPP TS 33.501 V16.7.0. *Security architecture and procedures for 5G System (release 16)*; 2020-6.

CHAPTER 13

Quality of service control

Yali Guo and Tricci So

13.1 5G quality of service model

QoS means quality of service. The 5G network provides the data transmission service between the user equipment (UE) and external data network through a packet data unit (PDU) session, and can provide differentiated QoS guarantee for data flows transmitted in the same PDU session according to the service requirements. In order to understand the 5G network QoS model, we first review the QoS model of the 4G network.

In the 4G network, QoS control falls under the concept of evolved packed system (EPS) bearer. The EPS bearer is the finest granularity for QoS forwarding treatment in the 5G system, and provides the same QoS treatment for all service data flows (SDFs) transmitted on the same EPS bearer. For the SDFs with different QoS requirements, different EPS bearers need to be established to provide differentiated QoS treatments. In 3GPP (The 3rd Generation Partnership Project) specification, QoS control of the 4G network is defined in clause 4.7 of TS 23.401 [1].

As shown in Fig. 13.1, EPS bearer between the UE and PDN gateway (PGW) consists of a radio bearer between the UE and 5G—AN (5G Access Network), a S1 bearer between the 5G—AN and SGW (Serving Gateway), and S5/S8 bearer between the SGW and PGW (PDN Gateway). There is a one-to-one mapping relationship among radio bearer, S1 bearer, and

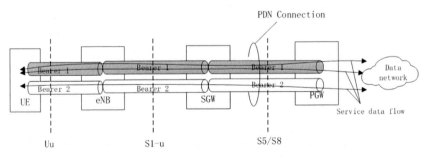

Figure 13.1 Evolved packed system bearers in the 4G network.

5G NR and Enhancements
DOI: https://doi.org/10.1016/B978-0-323-91060-6.00013-1

S5/S8 bearer. Each S1 bearer and S5/S8 bearer uses a separate GTP (GPRS Tunnelling Protocol) tunnel. The PGW is the decision point of QoS control, and is responsible for the control of establishment, modification, and release of each EPS bearer, the setting of QoS parameters, and the determination of the EPS bearer used for SDF transmission. The operation for radio bearers (e.g., establishment, modification and release) and QoS parameter setting at the 5G—AN completely follow the instructions from the core network. For the bearer management request from the core network, the 5G—AN can only accept or reject, it can not set or modify the QoS parameters of a radio bearer by itself.

In the 4G network, a single UE can support up to eight radio bearers over its radio interface; the QoS differentiation corresponding to the supporting of eight EPS bearers cannot meet the finer QoS control requirements. In the process of bearer management, the operation of each EPS bearer needs to be implemented by the establishment or release of a separate GTP tunnel. The signaling overhead is large and the processing is slow, and the adaptation to different applications' QoS requirements is not flexible enough. There are only a few values of standardized QoS class identifier (QCI) defined in the 4G network. If the QoS requirement is different from the preconfigured or standardized QCI of the current operator deployed network, the 4G network cannot provide accurate QoS guarantee for a service. The internet has rapidly evolved from supporting regular internet surfing and emails to high-quality voice, 4k live video, and real-time conferencing. Furthermore, the emergence of various non-public networks, industrial internet, internet of vehicles, machine communication, and other emerging services demand adaptive mobile broadband QoS support. The more stringent QoS requirements for the 5G system are far beyond the QoS control capabilities that can be provided by the 4G system. In order to provide better differentiated QoS treatments for a wide variety of services, 3GPP standardization body has revised its QoS model for the 5G system.

The QoS model of the 5G system is defined in section 5.7 of 3GPP specification TS 23.501 [2], as shown in Fig. 13.2.

The 5G system abandons the concept of bearer in the core network side and replaces it with QoS Flow. A PDU session can have up to 64 QoS Flows, and there is no longer any bearer between 5GC (5G Core) and radio access network (RAN). The GTP tunnel between 5GC and RAN is at PDU session level. The QoS flow identifier (QFI) is carried in the header of each packet transmitted in the GTP tunnel. The 5G—AN

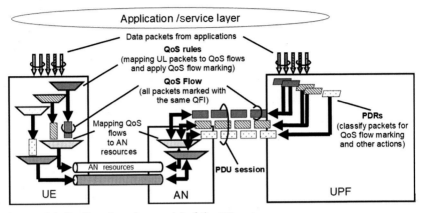

Figure 13.2 Quality of service model of the 5G system.

identifies different QoS Flows according to the QFI in the packet header. Therefore there is no need to modify the GTP tunnel every time the QoS Flow is established or released, which reduces the signaling overhead caused by session management procedures and improves the processing speed of session management procedures.

The number of radio bearers over the radio interface in the 5G network is also expanded to a maximum of 16. Each radio bearer can only belong to one PDU session, and each PDU session can include multiple radio bearers. In this way, a 5G UE can have up to 16 PDU sessions, thus supporting up to $16 \times 64 = 1024$ QoS Flows. Compared with the maximum 8 EPS bearers in the 4G network, the QoS differentiation can be finer. The mapping of QoS Flow to radio bearer is many-to-one mapping in the 5G network. The mapping is determined by the 5G−AN itself. To achieve this, in the 5G system the 5G−AN adds a special Service Data Adaptation Protocol (SDAP) layer to decide the mapping between QoS Flow and radio bearer. For the specific technology of the SDAP layer, Section 10.2 of this book should be referenced. According to the mapping relationship, the 5G−AN can establish, modify, and delete the radio bearer and set QoS parameters by itself, so as to make more flexible use of wireless resources.

The 5G system not only supports standardized 5G QoS Identifier (5QI) and operator preconfigured 5QI, but also adds the support of dynamic assigned 5QI, Delay Critical GBR (Guarantee Bit Rate) resource type, reflective QoS, QoS notification control, alternative QoS profile, and other features in order to provide better differentiated QoS treatments for a wide variety of services.

13.2 End-to-end quality of service control

13.2.1 General introduction

This section presents a general overview of the end-to-end QoS control for data transmission through the 5G network. However, the descriptions herein are based on how the UE interacts with its application server, which is referred to as an Application Function (AF) in 5G architecture, to establish communications that meet the application layer service requirements. The principles of 5G QoS as presented below also apply to other 5G service scenarios in general.

Before the application session is started for the service data transmission, the AF provides the Policy Control Function (PCF) with the application layer service requirements. If it is a trusted AF, it can provide information to the PCF directly. If it is a third-party AF that is not trusted by the operator, it can provide information to the PCF through the Network Exposure Function (NEF). The application layer service requirements include, for example, flow description information used for SDF detection. For IP (Internet Protocol) type packets, it is IP 5-tuple information comprised of source address, destination address, source port number, destination port number, and the protocol type above IP layer. Application layer service requirements also include QoS-related requirements, such as bandwidth requirements, service types, etc.

The PCF produces the PCC (Policy and Charging Control) rules according to the information collected from the SMF, AMF (Access and Mobility Management Function), CHF (Charging Function), NWDAF (Network Data Analytics Function), UDR (Unified Data Repository), AF (Application Function), and the preconfigured information in the PCF (Policy Control Function). The PCC rules are sent to the SMF (Session Management Function). The PCC rules are at SDF (Service Data Flow) level.

According to the received PCC rules, the configuration information of the SMF itself and the UE subscription information obtained from the UDM (Unified Data Management), SMF will map the received PCC rules to the appropriate QoS Flow, which is used to transmit the SDF corresponding to the PCC rule. Multiple PCC rules can be bound to the same QoS Flow, which implies a QoS Flow can be used to transmit multiple SDFs, which is a collection of SDFs with the same QoS requirements. The QoS Flow is the finest QoS granularity in the 5G network. Each QoS Flow is identified by the QFI and belongs to a PDU session.

All data in each QoS Flow have the same resource scheduling and guarantee within the radio interface. The SMF will determine the following information for each QoS Flow:

1. A QoS profile including the QoS parameters for this QoS Flow: 5QI, Allocation and Retention Priority (ARP), bitrate requirements, etc. The QoS profile is sent to the 5G−AN at the QoS Flow level.

2. One or multiple QoS rules. A QoS rule mainly includes the flow description information used for SDF detection and the QFI of QoS Flow used to transmit the SDF. The QoS rules are sent to the UE, mainly used by the UE for the detection of uplink data. The QoS rule is at the SDF level.

3. One or more packet detection rules (PDR) and corresponding QoS enforcement rules (QER), mainly including the flow description information for SDF detection and the QFI of QoS Flow used to transmit the SDF, which are sent to the UPF, mainly used by the UPF for the detection of downlink data, but can also be used for further verification of uplink data at the network side. The packet detection rule and QoS enforcement rule are at SDF level.

The SMF sends the QoS profile to the 5G−AN, sends QoS rules to the UE, and sends PDR and corresponding QER to the UPF. The SMF also provides bitrate-related QoS requirements of SDF level and QoS Flow level to the UE and UPF.

After receiving the QoS profile, the 5G−AN maps the QoS Flow to the appropriate radio bearer according to the QoS parameters of the QoS Flow, and allocates the corresponding radio resources. The 5G−AN also sends the mapping relationship between QoS Flow and radio bearer to the UE.

After the QoS Flow and radio resources are ready, the transmission of the service data will then be started.

For DL data:

1. The UPF uses the PDR received from the SMF to match the downlink data, marks the QFI to the matched packet header according to the corresponding QER, and then sends the packets to the 5G−AN through the PDU session level GTP tunnel between the UPF and the 5G−AN. The UPF discards the downlink packets that cannot match the PDR; in addition, the UPF also performs the rate control for the DL data.

2. The 5G−AN receives packets from the GTP tunnel, distinguishes different QoS Flows according to the QFI carried by the packet header,

and then forwards the packets to the UE through the corresponding radio bearer.

For UL data:

1. The UE matches the UL data with the flow description information in the QoS rule in order to determine the QoS Flow used to transmit the UL data. The UE then marks the QFI to the matched packet header. The access layer of the UE determines the corresponding radio bearer according to the mapping relationship between the QoS Flow and the radio bearer, and sends the uplink packets to the 5G–AN through the corresponding radio bearer. The UE will discard the uplink packets that cannot match the QoS rules. In addition, the UE controls the rate of uplink packets (Fig. 13.3).

2. The 5G–AN sends the received uplink data to the UPF through the GTP tunnel of the PDU session, and the QFI is carried in each of the uplink packet header.

3. The UPF uses the PDR received from the SMF to verify whether the uplink data carries the correct QFI. The UPF also controls the rate of uplink packets.

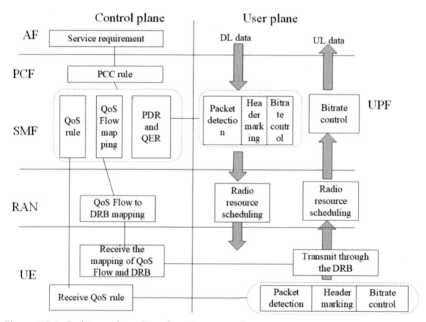

Figure 13.3 End to end quality of service control.

The above end-to-end QoS control is illustrated when the access network is a 3GPP 5G—AN. It can also be used for the UE to access the network through a non-3GPP access point.

The QFI of the QoS Flow can be dynamically allocated by the SMF, or the SMF can directly use the value of 5QI as the value of the QFI. In both cases, the SMF will send the QFI and the corresponding QoS profile to the 5G—AN through N2 signaling, which involves PDU session establishment, PDU session modification process. Every time when the User Plane of the PDU Session is activated, the SMF sends QoS profile to the 5G—AN again through N2 signaling.

However, when the UE accesses the network through non-3GPP access points, some non-3GPP access points may not support the QFI and corresponding QoS profiles from the SMF through control plane signaling. Therefore another QoS control method is designed in the 5G system. the SMF directly uses the value of 5QI as the value of the QFI, the SMF does not need to send the QFI and corresponding QoS profile to the 5G—AN through N2 signaling. The value of ARP is preconfigured on the 5G—AN. The 5G—AN gets the corresponding 5QI value according to the QFI carried in the downlink packet header, and then the corresponding radio resource can be allocated according to the derived 5QI and the preconfigured ARP. This QoS control method is only suitable for non-GBR QoS Flow, and mainly designed for non-3GPP access.

13.2.2 PCC rule

In the previous section, we discussed the end-to-end QoS control for data transmission in the 5G system. The PCF is responsible for generating PCC rules and sending these rules to the Session Management Function (SMF), which affects the establishment, modification, deletion of QoS Flow, and the setting of QoS parameters by the SMF. In this section, we will discuss the specific contents of PCC rules and how are they generated by the PCF.

The information contained in the PCC rules is mainly used for SDF detection and policy control, as well as the charging of SDF. According to the SDF template in the PCC rules, the data packets that match to the same packet filter from the SDF template detection constitute the SDF. In the 3GPP specification, the PCC rules of the 5G network are defined in clause 6.3 of TS 23.503 [3]. In Table 13.1, only the names and descriptions needed for QoS control are listed.

Table 13.1 Quality of service control-related parameters in the PCC rules.

Information name	Description
Rule identifier	Uniquely identifies the PCC rule, within a PDU Session. It is used between the PCF and the SMF for referencing the PCC rules.
Service data flow detection	*This part defines the method for detecting packets belonging to a service data flow.*
Precedence	Determines the order in which the service data flow templates are applied at service data flow detection, enforcement and charging.
Service data flow template	For IP PDU traffic: Either a list of service data flow filters or an application identifier that references the corresponding application detection filter for the detection of the service data flow. For Ethernet PDU traffic: Combination of traffic patterns of the Ethernet PDU traffic.
Policy control	*This part defines how to apply policy control for the service data flow.*
5G QoS identifier (5QI)	The 5QI authorized for the service data flow.
QoS notification control	Indicates whether notifications are requested from the 3GPP RAN when the GFBR can no longer (or can again) be guaranteed for a QoS Flow during the lifetime of the QoS Flow.
Reflective QoS control	Indicates to apply reflective QoS for the SDF.
UL-maximum bitrate	The uplink maximum bitrate authorized for the service data flow.
DL-maximum bitrate	The downlink maximum bitrate authorized for the service data flow.
UL-guaranteed bitrate	The uplink guaranteed bitrate authorized for the service data flow.
DL-guaranteed bitrate	The downlink guaranteed bitrate authorized for the service data flow.
Allocation and retention priority	The Allocation and Retention Priority for the service data flow consisting of the priority level, the preemption capability, and the preemption vulnerability.
Alternative QoS parameter sets	*This part defines Alternative QoS Parameter Sets for the service data flow.*

PCF, Policy Control Function; PDU, packet data unit; SMF, Session Management Function, QoS, quality of service; RAN, radio access network.

The PCF generates PCC rules according to the information collected from various network functions, such as the SMF, AMF, CHF, NWDAF, UDR, AF, and preconfiguration information on the PCF. The following is an example of the inputs for the PCF to generate PCC rules. It should be noted that the PCF does not necessarily need all the information for PCC rule generation. The following examples only list some information that the PCF may use.

The information provided by AMF includes, for example, SUPI, PEI of the UE, location information, RAT (Radio Access Technology) type, service area restriction information, PLMN ID (Public Land Mobile Network Identifier), and Slice ID.

The information provided by the SMF includes, for example, SUPI (Subscription Permanent Identifier), PEI (Permanent Equipment Identifier) of the UE, IP address of the UE, default 5QI, default ARP, PDU session type, and S−NSSAI, DNN.

The information provided by AF includes, for example, user identification, the UE IP address, media type, bandwidth requirement, SDF description information, and Application Service Provider information. The SDF description information includes source address, destination address, source port number, target port number, protocol type, and so on.

Information provided by UDR, such as subscription information of specific DNN or S−NSSAI.

Information provided by NWDAF, such as statistical analysis or prediction information of some network functions or services.

The PCF can activate, modify, and delete the PCC rules at any time.

13.2.3 Quality of service flow

The 5G QoS model supports two types of QoS Flow: The GBR type can guarantee the flow bitrate while the non-GBR type cannot guarantee the flow bitrate. Each QoS Flow has its corresponding QoS profile, which includes at least 5QI and ARP. For GBR type QoS Flows, GFBR and MFBR are also included in the QoS profile. The SMF provides the QFI and the corresponding QoS profile to the 5G−AN for radio resource scheduling.

For each PDU session, there is a default QoS Flow with the same lifetime as the PDU session, and the default QoS Flow is a non-GBR type QoS Flow.

The SMF determines the mapping between SDF and QoS Flow according to the PCC rules received from the PCF. The basic parameters used for binding the PCC rules to the QoS Flow is the combination of 5QI and ARP.

For a PCC rule, the SMF checks whether a QoS Flow has the same binding parameters as the PCC rule (i.e., the same 5QI and ARP). If there is such a QoS Flow, the SMF binds the PCC rule to the QoS Flow and may modify the QoS Flow. For example, increase the existing GFBR of the QoS Flow to support the GBR of the newly bound PCC rule. The GFBR of the QoS Flow will be set to the sum of the GBRs of the PCC rules bound to the QoS Flow. If there is no such QoS Flow, the SMF creates a new QoS Flow, allocates the QFI for the new QoS Flow, sets the QoS parameters of this QoS Flow according to the parameters in PCC rules, and binds PCC rules to this QoS Flow.

The SMF can also bind a PCC rule to the default QoS Flow according to the instructions from the PCF.

The QoS Flow binding can also be based on some other optional parameters, such as QoS notification control indication, dynamically assigned QoS characteristics, etc., which will not be described in detail here.

After the QoS Flow binding, the SMF provides the QoS profile to the 5G—AN, sends the QoS rules to the UE, and sends PDR and corresponding QER to the UPF. The SMF also provides SDF level and QoS Flow level bitrate requirements to the UE and UPF. As a result, the DL data can be mapped to the correct QoS Flow at the UPF side, and the UL data can be mapped to the correct QoS Flow at the UE side.

When the binding parameters in PCC rules change, the SMF needs to reevaluate PCC rules to determine the new QoS Flow binding relationship.

If a PCC rule is deleted by the PCF, the binding between PCC rule and QoS Flow will be deleted by the SMF. When the last PCC rule bound to a QoS Flow is deleted, the SMF will delete the QoS Flow accordingly.

When a QoS Flow is deleted, for example, when the radio resource cannot be guaranteed, based on the instructions from the 5G—AN, the SMF also needs to delete all PCC rules bound to the QoS Flow, and report these PCC rules to the PCF.

13.2.4 Quality of service rule

The UE performs packet detection for the uplink data based on the QoS rules received from the SMF so that it can map the uplink data to the correct QoS Flow. A QoS rule includes the following information:

1. QoS rule identifier assigned by the SMF, which is unique in a PDU session.

2. QFI of the QoS Flow used to transmit the SDF according to the QoS Flow binding result.

3. The flow description information used for SDF detection is generated according to the SDF template in the corresponding PCC rule, mainly including the UL SDF template; may also including the DL SDF template.

4. The priority of a QoS rule is generated according to the priority of the corresponding PCC rule.

QoS rules are at SDF level, and there can be more than one QoS rule associated with the same QoS Flow.

For each PDU session, there is a default QoS rule. For IP and Ethernet type PDU sessions, the default QoS rule is the only QoS rule of a PDU Session that may allow all UL packets, and it has the lowest priority. In other words, when a packet cannot match other QoS rules, it can match with the default QoS rule and transmit by the corresponding QoS Flow.

For an unstructured type PDU session, the packet header of data for this PDU session does not have a fixed format. This kind of PDU session only supports the default QoS rule, and the default QoS rules do not include the flow description information, so that all the UL data can be matched with the default QoS rule. The unstructured PDU session supports only one QoS Flow; all data between the UE and UPF have the same QoS treatment.

13.3 Quality of service parameters

13.3.1 5G quality of service identifier and the quality of service characteristics

The 5QI can be understood as a scalar that points to a set of QoS characteristics. These QoS characteristics are used to control the QoS-related operations in the access network, such as setting the scheduling weight, access threshold, queue management, link layer configuration, etc. The 5QI includes standardized 5QI, preconfigured 5QI, and dynamically assigned 5QI. For the dynamically assigned 5QI, not only the 5QI but also the complete set of QoS characteristics are included in the QoS profile. For the standardized and preconfigured 5QI, the QoS characteristics are not included in the QoS profile; the 5G—AN can resolve the set of QoS characteristics by itself. In addition, for a standardized or preconfigured 5QI, the core network can modify the corresponding QoS characteristics by providing a different value of the QoS characteristic. The standardized 5QI is mainly used for the more general and frequently used services. Providing the 5QI instead of the set of multiple QoS characteristic can optimize the signaling transmission. The dynamic allocation of 5QIs is mainly used for the infrequently used services that cannot be satisfied by standardized 5QIs.

The QoS characteristics are as follows:

1. Resource type: including GBR type, Delay critical GBR type, and Non-GBR type
2. Priority Level
3. Packet delay budget (PDB): represents the packet transmission delay from the UE to the UPF
4. Packet error rate (PER)
5. Average window: for GBR type and delay critical GBR type only
6. Maximum data burst volume (MDBV): for delay critical GBR type only

The resource type is used to determine whether the network will allocate dedicated network resources to a QoS Flow. The GBR type QoS Flow and delay critical GBR type QoS Flow need dedicated network resources to guarantee their GFBR. Non-GBR type QoS Flows do not require dedicated network resources. For a 5G system, the delay critical GBR type is added in to support services with high reliability and low latency requirements. Such services, such as automatic industrial control, remote driving, intelligent transportation system, energy distribution of power system, etc., have high requirements on transmission delay and transmission reliability. As shown in Table 13.2, when comparing the 5QIs corresponding to GBR types and non-GBR types, the PDB for the delay critical GBR type is significantly lower, and the PER is also relatively low. In the 4G system, for the bitrate control of GBR service, a second level time window is preconfigured at the 4G RAN for the calculating of GFBR. Based on this time window, the average bitrate to meet the GFBR requirements is calculated. However, because the time window is too long, there may be the case that the GFBR is guaranteed from the second level time window, but cannot be guaranteed from the millisecond level time window. In order to better support the high-reliability and low-latency services, the 5G network also adds MDBV for the delay critical GBR type 5QI, which is used to represent the amount of data that the service needs to transmit in a millisecond level time window, and it provides better guarantee for the support of high-reliability and low-latency services.

Table 13.2 Standardized 5QI to quality of service characteristics mapping.

5QI	Resource type	Priority level	Packet delay budget (ms)	Packet error rate	Maximum data burst volume	Average window (ms)
1	GBR	20	100	10^{-2}	N/A	2000
2		40	150	10^{-3}	N/A	2000
3		30	50	10^{-3}	N/A	2000
4		50	300	10^{-6}	N/A	2000
65		7	75	10^{-2}	N/A	2000
66		20	100	10^{-2}	N/A	2000
67		15	100	10^{-3}	N/A	2000
71		56	150	10^{-6}	N/A	2000
72		56	300	10^{-4}	N/A	2000
73		56	300	10^{-8}	N/A	2000
74		56	500	10^{-8}	N/A	2000
76		56	500	10^{-4}	N/A	2000
5	Non-GBR	10	100	10^{-6}	N/A	N/A
6		60	300	10^{-6}	N/A	N/A
7		70	100	10^{-3}	N/A	N/A
8		80	300	10^{-6}	N/A	N/A
9		90				
69		5	60	10^{-6}	N/A	N/A
70		55	200	10^{-6}	N/A	N/A
79		65	50	10^{-2}	N/A	N/A
80		68	10	10^{-6}	N/A	N/A
82	Delay critical	19	10	10^{-4}	255 bytes	2000
83	GBR	22	10	10^{-4}	1354 bytes	2000
84		24	30	10^{-5}	1354 bytes	2000
85		21	5	10^{-5}	255 bytes	2000
86		18	5	10^{-4}	1354 bytes	2000

Priority level is used for the radio resource allocation among the QoS Flows. It can be used among the QoS Flows of the same UE, or can be used among the QoS Flows of different UEs. When congestion occurs, the 5G–AN cannot guarantee the QoS requirements of all QoS Flows, so the priority level is used to guarantee the high-priority QoS Flows. When there is no congestion, the priority level can also be used to allocate resources between different QoS Flows, but it is not the only factor determining resource allocation.

PDB defines the maximum transmission latency between the UE and UPF. For a 5QI, the PDB of uplink and downlink data is the same.

When the 5G—AN calculates the PDB between the UE and 5G—AN, it subtracts the PDB of core network side from the PDB corresponding to the 5QI, in other words, the PDB between the 5G—AN and UPF is subtracted. The core network-side PDB can be statically configured on the 5G—AN, or it can be dynamically determined by the 5G—AN according to the connection with the UPF, or it can be indicated to the 5G—AN by the SMF. For a GBR type QoS Flow, if the transmitted data does not exceed the GFBR, the network needs to ensure that the transmission latency of 98% of the packets does not exceed the PDB corresponding to the 5QI. However, for delay critical QoS Flows, if the transmitted data does not exceed GFBR and the burst data does not exceed MDBV, each packet exceeding the PDB corresponding to the 5QI will be considered as packet loss and included in PER. For the non-GBR type QoS Flow, the transmission latency beyond the PDB and packet dropping caused by congestion are allowed.

The PER defines an upper bound for the rate of PDUs that have been processed by the sender of a link layer protocol but that are not successfully delivered by the corresponding receiver to the upper layer. It is used to describe packet loss in noncongestion situations. PER is used to affect link layer configuration, such as RLC and HARQ configuration. For a 5QI, the PER of uplink and downlink data is the same. For non-GBR and GBR type QoS Flows, packets exceeding the PDB will not be included in PER, but for delay critical QoS Flows, if the transmitted data does not exceed GFBR and the data burst does not exceed MDBV, packets exceeding the PDB will be considered as packet loss and included in the PER.

The average window is only used for GBR and delay critical GBR type QoS Flows, and is used as the time duration for calculating GFBR and MFBR. This parameter also exists in the 4G system, but it is not mentioned in the 4G specification; it is a time window preconfigured on the 4G RAN. The average window is added in the 5G specification, which enables the core network to change the value of average window according to the service requirements, so as to better adapt to different services. For the standardized and preconfigured 5QI, although the average window corresponding to the 5QI has been standardized or configured, the core network can also change it to a different value.

The MDBV is only used for delay critical GBR type QoS Flows, which represents the maximum amount of data to be processed by the 5G—AN for the QoS Flow during the period of PDB. For the standardized and preconfigured 5QI of delay critical GBR type, although the MDBV corresponding to the 5QI has been standardized or preconfigured at the 5G—AN, the core network may provide different value to modify the MDBV.

The UDM saves the default 5QI value for each DNN. The default 5QI is a standardized 5QI of the non-GBR type. After the SMF obtains the default value of 5QI from the UDM, it is used to configure the parameters of the default QoS Flow. The SMF can modify the value of the default 5QI according to the interaction with the PCF or according to the local configuration at the SMF.

The mapping of the standardized 5QI to the QoS characteristics is defined in clause 5.7 of the 3GPP specification TS 23.501 [2], as shown in Table 13.2. It should be noted that the values in this table may be changed slightly in different versions of TS 23.501, so this is only an example to understand the mapping of standardized 5QI and the QoS characteristics.

13.3.2 Allocation and retention priority

The ARP includes three types of information: priority level, preemption capability, and preemption vulnerability. It is used to determine whether to allow the establishment, modification, or handover of QoS Flow when resources are limited. It is generally used for admission control of GBR type QoS Flows. The ARP is also used to preempt the resources of the existing QoS Flow when the resource is limited, such as releasing the existing QoS Flow to accept and establish a new QoS Flow.

The priority level of the ARP is used to indicate the importance of QoS Flow. The value is 1—15, and 1 represents the highest priority. Generally speaking, 1—8 can be assigned to the services authorized by the serving network. 9—15 is assigned to the services authorized by the home network, so it can be used for roaming. The value can also be assigned according to the roaming protocol.

Preemption capability indicates whether a QoS Flow may get resources that were already assigned to another QoS Flow with a lower ARP priority level.

Preemption vulnerability defines whether a QoS Flow may lose the resources assigned to it in order to admit a QoS Flow with higher ARP priority level. It can be set to allow or disable.

The UDM stores the default ARP value for each DNN. After the SMF obtains the default ARP value from the UDM, it is used for the parameter configuration of the default QoS Flow. The SMF can modify the default ARP value according to the interaction with the PCF or according to the local configuration of the SMF.

For QoS Flows other than the default QoS Flow, the SMF sets the ARP priority level, preemption capability, and preemption vulnerability in the PCC rules bound to the QoS Flow as ARP parameters of the QoS Flow. If the PCF is not deployed in the network, the ARP can also be set according to the local configuration in the SMF.

13.3.3 Bitrate-related parameters

The bitrate control-related parameters include GBR, MBR (Maximum Bit Rate), GFBR (Guarantee Flow Bit Rate), MFBR (Maximum Flow Bit Rate), UE−AMBR (UE Aggregated Maximum Bit Rate), and Session-AMBR (Session Aggregated Maximum Bit Rate).

GBR and MBR are SDF level bitrate control parameters that are used for bitrate control of GBR type SDF. We introduced them in the PCC rule parameters in Section 13.2.2 earlier. MBR is necessary for GBR type SDF and optional for non-GBR type SDF. The UPF performs SDF level control of the MBR.

GFBR and MFBR are QoS Flow level bitrate control parameters that are used for bitrate control of GBR type QoS Flow. The GFBR instructs the 5G−AN to allocate enough resources to guarantee the bitrate of a QoS Flow within the average window. The MFBR is the maximum allowed bitrate of the QoS Flow; any data exceeding MFBR may be discarded. Bitrates above the GFBR value and up to the MFBR value may be provided with relative priority determined by the Priority Level of the QoS Flows. The MFBR of DL data of QoS Flow is controlled at the UPF; the 5G−AN also controls the MFBR of both UL and DL data. the UE can perform MFBR control of UL data.

Both the UE−AMBR and Session-AMBR are used for non-GBR type QoS Flow.

The Session-AMBR limits the aggregate bitrate that can be expected to be provided across all non-GBR QoS Flows for a specific PDU Session. When each PDU session is established, the SMF obtains the subscribed session-AMBR. The SMF can modify the value of Session-AMBR according to the interaction with the PCF or according to the local configuration of the SMF. The UE can control Session-AMBR of uplink data, and the UPF can also control session-AMBR of both uplink and downlink data.

UE−AMBR limits the aggregate bitrate that can be expected to be provided across all non-GBR QoS Flows of a UE. The AMF can obtain the subscribed the UE−AMBR from UDM and modify it according to the PCF's instructions. AMF provides the UE−AMBR to the 5G−AN, and the 5G−AN will recalculate the UE−AMBR. The calculation method is setting its the UE−AMBR to the sum of the Session-AMBR of all PDU Sessions with active user plane to this 5G−AN up to the value of the received the UE−AMBR from AMF. The 5G−AN is responsible for the uplink and downlink rate control of the UE−AMBR.

13.4 Reflective quality of service

13.4.1 Usage of reflective quality of service in the 5G system

Reflective QoS was originally introduced in 4G system, which was used for QoS control of the UE accessing 3GPP network through fixed broadband, such as WLAN. In fixed broadband network, QoS can be differentiated by DSCP of packet header. Reflective QoS mechanism of 4G system is defined in clause 6.3 of 3GPP specification TS 23.139 [4], as shown in Fig. 13.4. After the UE is connected to 4G core network through fixed broadband, for downlink data, PGW obtains SDF template and QoS control information based on the policy received from the PCF, it then marks corresponding DSCP on the matched packet header and sends it to fixed broadband network. Fixed broadband network performs QoS control based on DSCP and sends the packets to the UE. For the uplink data, the UE reflects the packet header of the downlink data to generate the corresponding packet filter and DSCP information for the

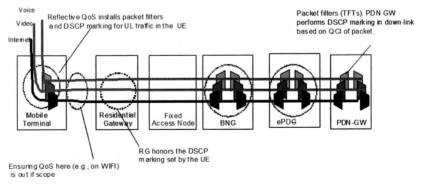

Figure 13.4 Reflective quality of service in 4G network for fixed broadband accessing.

uplink data, thus marking the corresponding DSCP on the matched uplink packet header and sending the packets to the fixed broadband network. the UE generates packet filter for uplink data according to the packet header of downlink data. For example, the source IP address and source port number of downlink packet header are taken as the destination IP address and destination port number of uplink packet filter, and the destination IP address and destination port number of downlink packet header are used as the source IP address and source port number of uplink packet filter. In other word, the packet header of downlink data is used to derive the packet header of uplink data.

Reflective QoS mechanism controls QoS through the user plane header, which avoids amount of signaling interactions between core network and access network, and between core network and the UE, as a result, it improves the speed of QoS control. Considering the support of multiple new services in the 5G network, many Internet services will use a lot of discontinuous address/port information or change address/port information frequently. Therefore the number of flow description information (i.e. packet filter) in QoS rules configured by network to the UE may be very large, which greatly increases the number NAS messages between network and the UE. When designing the QoS solution for 5G system, the Reflective QoS mechanism becomes one of the viable alternative.

13.4.2 The mechanism of reflective quality of service in the 5G system

Reflective QoS mechanism can be used for IP type or Ethernet type PDU sessions. The mapping of uplink data to QoS Flow can be realized at the UE side without QoS rules provided by the SMF. the UE generates QoS rules according to the received downlink packets. Reflective QoS mechanism is at SDF level. In the same PDU session, even within the same QoS Flow, there can be packets controlled by Reflective QoS and packets controlled without Reflective QoS.

If the UE supports the Reflective QoS function, the UE needs to indicate to the network that it supports Reflective QoS in the PDU session establishment procedure. In general, the indication that the UE supports Reflective QoS will not change during the lifetime of a PDU session. However, in some special cases, the UE can also revoke the indication of supporting Reflective QoS. In this case, the UE needs to delete all the QoS rules generated by the UE itself belonging to this PDU session. The network can also provide the UE with new QoS rules through signaling for the SDF previously controlled by the Reflective QoS. Because this scenario is quite special, the UE is not allowed to indicate to the network again that it supports Reflective QoS during the lifetime of this PDU session.

A QoS rule generated by the UE includes:

1. An uplink packet filter
2. QFI
3. Priority of QoS rule

For IP type PDU sessions, the uplink packet filters are generated based on the received downlink packets:

1. When the protocol is TCP or UDP, the source IP address, destination IP address, source port number, destination port number and protocol ID are used to generate uplink packet filters. For example, take the source IP address and source port number of the downlink packet header as the destination IP address and destination port number of the uplink packet filter, and take the destination IP address and destination port number of the downlink packet header as the source IP address and source port number of the uplink packet filter.
2. When the protocol ID is ESP, the source IP address, destination IP address, security parameter index, and protocol ID are used to generate uplink packet filters. It can also include IPSec-related information.

For the Ethernet type PDU session, the uplink packet filter is generated based on the source MAC address and destination MAC address of the received downlink packet, as well as other packet header field information of the Ethernet type packet.

The QFI in the QoS rule generated by the UE is set to the QFI carried by the downlink packet header. The priority of all the UE generated QoS rules is set to a standardized value.

When the network determines to activate Reflective QoS for a SDF, the SMF indicates to the UPF through the N4 interface signaling that Reflective QoS control is enabled for the SDF. After receiving the instruction, the UPF adds the QFI and Reflective QoS indication (RQI) to the packet header of each downlink packets of this SDF. The packets with the QFI and RQI in the packet header are sent to the UE through the 5G–AN.

A Reflective QoS timer will be set on the UE. The default value can be used, or the network can set the value for each PDU session.

After the UE receives a downlink packet with RQI:

1. if a QoS rule generated by the UE with a Packet Filter corresponding to the DL packet does not already exist,

    a. the UE will create a QoS rule with a Packet Filter corresponding to the DL packet; and

    b. the UE will start, for this QoS rule, a timer set to the Reflective QoS Timer value.

2. otherwise,

    a. the UE will restart the timer associated to this QoS rule generated by the UE; and

    b. if the QFI associated with the downlink packet is different from the QFI associated with the QoS rule generated by the UE, the UE will update the QoS rule generated by the UE with the new QFI.

After the Reflective QoS timer corresponding to a UE-generated QoS rule has expired, the UE will delete the corresponding QoS rule.

When the network decides not to use the Reflective QoS for a SDF, the SMF removes the RQI from the UPF by the N4 interface signaling. After receiving the deletion instruction, the UPF does not add the RQI to the downlink packets of this SDF.

The SDF using Reflective QoS can be bound to the same QoS Flow with other SDFs without Reflective QoS. As long as a QoS Flow includes at least one SDF using the Reflective QoS, the SMF will add the Reflective QoS attribute (RQA) in the QoS profile provided to the 5G−AN through N2 signaling. According to the RQA received, the 5G−AN can turn on the control mechanism of Reflective QoS on the 5G−AN side, such as the DRB mapping configuration at the SDAP layer.

13.5 Quality of service notification control

13.5.1 General description of quality of service notification control

In the LTE network, for a GBR type bearer, when the resources on the 4G RAN side cannot guarantee the required GBR, the 4G RAN will directly release the bearer. If the service can accept the downgrade to a lower GBR, it can reinitiate the establishment of a new bearer with a lower GBR value, but this will lead to service interruption and poor user experience.

In order to further improve the service experience in the 5G system and to consider the support for new services, e.g. V2X (Vehicle-to-Everything), which is sensitive to service interruption, the 5G network introduces the QoS Notification Control mechanism for GBR type QoS flows.

The QoS notification control mechanism requires the 5G−AN to maintain QoS Flow while informing the core network of the situation that the QoS cannot be guaranteed (i.e., 5G−AN will no longer to guarantee the QoS requirements of a QoS Flow). It is designed for GBR services that can adjust QoS requirements according to the network conditions, such as the bitrate adjustment. For this kind of service, the PCF sets the QoS notification control indication in the PCC rule. The SMF considers QoS notification control indication for QoS Flow binding, in addition to the binding parameters as described in earlier Section 13.2.3; the SMF sets the QoS notification control indication in the QoS profile and provides the QoS profile to 5G−AN.

For a GBR type QoS Flow, if the QoS notification control is enabled, when the 5G−AN cannot guarantee the GFBR, PDB, or PER of the QoS Flow, the 5G−AN will continue to maintain the QoS Flow and continue to try to allocate resources for the QoS Flow to meet the parameters in the QoS profile. The 5G−AN also sends a notification message to the SMF, indicating that the QoS cannot be guaranteed. Once the 5G−AN is able to determine that GFBR, PDB, and PER can be guaranteed again, it sends another notification message to the SMF to indicate that QoS can be guaranteed.

13.5.2 Alternative quality of service profile

The QoS notification control mechanism can solve the problem of service interruption due to the bearer release when the QoS cannot be guaranteed. The 5G−AN sends the notification to the core network that the QoS cannot be guaranteed. But the interaction between the core network and the application is slow. Hence, it is uncertain whether or not the application will adjust the QoS requirements. The 5G−AN can only allocate resources to meet the original QoS requirements while waiting for the adjustment instructions from the core network. At this point, the allocation of resources by the 5G−AN could become unsustainable. Due to such consideration, the alternative QoS profile is introduced as part of the QoS notification control mechanism in the 5G system. Such QoS profile is used to make the 5G−AN aware of the alternative QoS profiles when the original QoS cannot be guaranteed, so as to further improve the efficiency of resource allocation.

An alternative QoS profile is an optional optimization that depends on the QoS notification control mechanism and is only used for GBR QoS

Flows with QoS notification control enabled. The PCF provides one or more alternative QoS parameter sets in the PCC rules by interacting with the AF. Then the SMF generates one or more QoS profiles for the corresponding QoS Flow, and sends them to the 5G–AN.

When the QoS cannot be guaranteed, the 5G–AN evaluates whether the resources can be sustained by one of the alternative QoS profiles. If so, it will indicate the alternative QoS profile that can be guaranteed when sending notification to the SMF. The operation at the 5G–AN is as follows:

1. When the GFBR, PDB, or PER of a QoS Flow cannot be guaranteed, the 5G–AN evaluates whether the radio resources can guarantee an alternative QoS profile one by one according to the priority of the alternative QoS profiles. If an alternative QoS profile can be guaranteed, the 5G–AN indicates to the SMF the first matched alternative QoS profile. If none of the alternative QoS profiles can be guaranteed, the 5G–AN indicates to the SMF that QoS requirements cannot be guaranteed and there is no viable alternative QoS profile.

2. After that, when the 5G–AN determines that the currently fulfilled GFBR, PDB, or PER are different (better or worse) from the situation indicated in the last notification, the 5G–AN indicates the latest QoS status to the SMF again.

3. The 5G–AN always tries to fulfill the QoS profile with any alternative QoS profile that has the highest priority to meet the current situation.

4. The SMF needs to notify the PCF after receiving the notification from the 5G–AN.

5. The SMF may also inform the UE about changes in the QoS parameters.

13.6 Summary

This chapter introduced the QoS control mechanism of the 5G system. By reviewing and comparing with the QoS control mechanism of the 4G system, this chapter presented the architecture considerations of the 5G system QoS model. This chapter also provided a high-level overview of the new QoS control methods, QoS parameters, Reflective QoS, QoS notification control, and the alternative QoS profile introduced in 5G systems.

References

[1] 3GPP TS 23.401 V16.5.0 (2019—12). General packet radio service (GPRS) enhancements for evolved universal terrestrial radio access network (E-UTRAN) access (release 16).
[2] 3GPP TS 23.501 V16.4.0 (2020—03). System architecture for the 5G system (5GS) stage 2 (release 16).
[3] 3GPP TS 23.503 V16.4.1 (2020—04). Policy and charging control framework for the 5G System (5GS); stage 2 (release 16).
[4] 3GPP TS 23.139 V15.0.0 (2018—06). 3GPP system—fixed broadband access network interworking stage 2 (release 15).

Further reading

3GPP TS 23.502 V16.4.0 (2020—03). Procedures for the 5G system (5GS) stage 2 (release 16).
3GPP TS 23.799 V14.0.0 (2016—12). Study on architecture for next generation system (release 14).

CHAPTER 14

5G voice

Yang Xu, Jianhua Liu and Tricci So

When reviewing the 4G voice solutions, the mainstream solutions are listed in Table 14.1.

In the 5G era, due to the persistent evolution of pure IP-based mobile networking, legacy Circuit Switch (CS) domain voice is gradually becoming obsolete and will be replaced by 4G VoLTE and 5G VoNR voice based on IMS networks. The CS and legacy short message services will gradually be obsoleted in the 5G era, replaced by VoNR/VoLTE and RCS services. This evolutionary trend helps operators reduce network maintenance costs while further improving the user experience of related services.

With the advent of 5G, although the importance of voice is lower than that of other services compared to 4G, it is still an important source of revenue and service content for operators. Compared with OTT voice

Table 14.1 4G mainstream voice solution.

Voice solution	Description
VoLTE/ eSRVCC	The voice service is provided based on IMS and supports voice continuity from LTE to 2G/3G networks. Both the UE and the network need to support the IMS protocol stack, and need to support the SRVCC function, with major changes.
CSFB	The UE is single standby. When the voice service is required, the UE actively requests to fall back to 2G/3G and establish a voice connection with the circuit. Both the UE and the network need to support the fallback function of redirection or handover, and reuse the voice function of the 2G/3G network. The change is moderate, but the call establishment time is longer.
SvLTE	The UE is dual standby, and voice services are provided by traditional 2G/3G networks. The network changes are small, but the UE needs to support dual standby, which has higher requirements for mobile phone chips and consumes high power.

5G NR and Enhancements
DOI: https://doi.org/10.1016/B978-0-323-91060-6.00014-3

functions (such as WeChat voice), 5G voice has the following irreplaceable advantages:

- Embedded native service, no need to install third-party APP.
- You can make a call by using the phone number, without OTT software for friend authentication.
- It can communicate with 2G/3G/4G users and fixed-line users.
- With dedicated bearer and leading coding capabilities to ensure the voice quality of users. Compared with the simultaneous transmission of OTT voice data and ordinary Internet traffic, 5G voice data will be transmitted preferentially in the operator's network, and the coding efficiency has advantages.
- Emergency call function. The emergency call function allows users to dial an emergency call number (such as 110 in China, 911 in the United States) without a SIM card or under any carrier network to ensure personal safety in emergency situations.

Compared with the 4G era, 5G voice services have the following development trends:

- Better voice quality, using EVS high-definition voice coding, effectively improving the MOS value. Voice billing is combined with traffic billing (i.e., billing is done through voice IP data instead of duration billing).
- Business continuity is further improved, and 5G users in VoNR calls can also be seamlessly switched to EPS.

14.1 IP multimedia subsystem (IMS)

VoLTE and VoNR are more viable and supreme solutions for voice for the 5G system. Both VoLTE and VoNR are both based on the IMS protocol. The main difference is that VoLTE is the transmission of IMS data packets through the 4G LTE, while VoNR uses 5G NR for transmission.

In the 5GS the overall architecture of the IMS network is similar to the EPS. The main difference between VoNR and VoLTE is that VoNR supports two additional ultrahigh-definition voice codec technologies, EVS WB and SWB, in order to further improve voice quality.

This chapter will introduce the basic functions and procedures of IMS, and focus on the capabilities required by UEs to support IMS from the perspective of the UE.

14.1.1 IMS registration

In order to implement IMS services, the IMS protocol stack needs to support the following two important IMS layer user identifications:

* PVI (Private User Id):

Also known as the IP Multimedia Private Identity (IMPI), it is an identifier of the network layer and is globally unique. It is stored on the SIM card and used to indicate the subscription relationship between the user and the network. Generally, it can also uniquely identify a UE. The network can use PVI for authentication to identify whether the user can use the IMS network. PVI is not used for call addressing and routing. PVI can be represented by "username@home network domain name" or "user number@home network domain name." It is customary to use the "user number@home network domain name" representation.

* PUI (Public User Id):

It is also called the IP Multimedia Public Identity (IMPU), which is a service layer identifier assigned by the IMS layer to identify service contract relationships, charging, etc. It also represents user identity and is used for IMS message routing. The PUI does not require authentication. A single device can be associated with more than one IMPU, and multiple devices can share the same IMPU. The PUI can adopt the format of SIP URI or Tel URI. The format of SIP URI follows RFC3261 and RFC2396, such as: 1234567@domain, Alex@domain. The format of Tel URI follows RFC3966, such as: tel: +1-201-555-0123, tel: 7042; phone-context = example.com.

In order to perform IMS voice services, the UE needs to perform IMS registration. The purpose of the registration process is to establish a logical path between the UE and the IMS network, and the path is used to transfer subsequent IMS data packets. After IMS registration is completed:

* The UE can use IMPU to communicate.
* The association(s) can be established between IMPU(s) and user IP address(es).
* The UE can obtain current location information and service capabilities.

The registration process is shown in Fig. 14.1. Before using the IMS service, the UE should perform IMS registration, and the IMS network maintains the user registration status. The IMS registration is initiated by the UE, and the S-CSCF performs authentication and authorization, and maintains the user status. The UE can maintain the registration

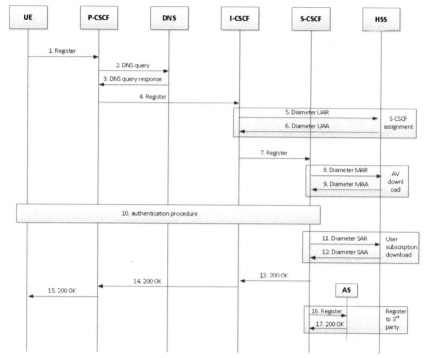

Figure 14.1 IMS registration.

information through periodic registration updates. If the status of the IMS registration is not maintained over time, the network side will drop the user.

After the registration is completed, the UE, the core network element, and the IMS network element will obtain the information in Table 14.2. In summary, after the UE is registered, all communication paths are opened. The IMS network elements will locate the UE and obtain the UE's information. The UE will also have all the necessary parameters for performing voice services.

After IMS registration is completed, the following subsequent registration could happen:
- Periodic registration;
- Registration triggered by capability change or new business request;
- Reregistration (reregistration uses the same "registration request" message).

For more detailed descriptions of the IMS registration process, refer to Chapter 5, 5G Flexible Scheduling, of TS23.228 [1].

Table 14.2 Data acquired by UE and NF before and after registration.

Network element	Before register	Registering	After register
UE	IMPU, IMPI, domain name, P-CSCF name or address, authentication password	IMPU, IMPI, domain name, P-CSCF name or address, authentication password	IMPU, IMPI, domain name, P-CSCF name or address, authentication password
P-CSCF	DNS address	I-CSCF address, UE IP address, IMPU, IMPI	I-CSCF address, UE IP address, IMPU, IMPI
I-CSCF	HSS/UDM address	S-CSCF address	N/A
S-CSCF	HSS/UDM address	HSS/UDM address, User subscription service information, P-CSCF address, P-CSCF network identification, UE IP address, IMPU, IMPI	HSS/UDM address, User subscription service information, P-CSCF address, P-CSCF network identification, UE IP address, IMPU, IMPI
HSS/ UDM	User-contracted service information and authentication data	P-CSCF address	S-CSCF address

14.1.2 IMS call setup

Once the IMS registration process has completed, each IMS network element on the calling side (MO process) and called side (MT process) can perform next-hop routing according to the called PUI identifier in the SIP INVITE message.

The routing between the calling domain and the called domain needs to be implemented through ENUM network elements. Specifically, the calling I-CSCF sends the called PUI to the DNS to query the called

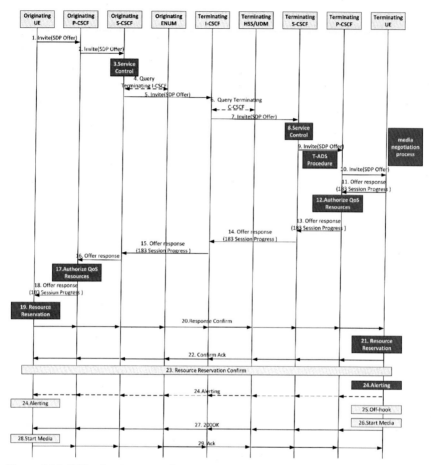

Figure 14.2 IMS call setup procedure.

I-CSCF address, and the called I-CSCF sends query information to the called HSS to obtain the called S-CSCF address.

A typical IMS call establishment process is shown in Fig. 14.2. The call establishment process mainly includes the following important contents:

- Media negotiation

    The media type and coding scheme supported by the calling UE are negotiated through the SDP protocol between the called and the calling UEs (usually through the SDP request-response mechanism in SIP Invite and Response messages) and the media types negotiated by both parties include audio, video, text, etc. Each media type can include multiple encoding formats.

- Service control

  Both the calling and called S-CSCFs can determine whether the service requested in the SIP Invite message is allowed to be executed according to the maintenance contract information obtained during the IMS registration process.

- QoS resource authorization

  In order to support dedicated QoS flow/bearer establishment for IMS call, the S-CSCF triggers the UE's dedicated carrier/data flow establishment process, and triggers the core network session modification process after the first SDP request response.

- T-ADS (Terminating Access Domain Selection)

  In order to determine which domain the called user resides recently (such as PS domain or CS domain), S-CSCF triggers AS to query the HSS in which domain that the called UE currently resides (e.g., PS or CS). Then the S-CSCF sends a message to the specified domain that the called UE resides.

- Resource reservation

  To ensure that the negotiated media plane can be successfully established, resources need to be reserved on the calling and called sides. Resource reservation usually occurs after the media negotiation process is completed and the peer's confirmation is obtained.

- Alerting

  Occurs after the resource reservation is successful.

Refer to clause 5 of 3GPP TS23.228 [1] for detailed IMS call establishment process and description.

14.1.3 Abnormal case handling

As the mobile network resources are limited, some abnormal scenarios could happen during the 5G voice session while the UE is moving. Common exception scenarios and the handling methods on the UE side are as follows:

1. PDU session is dropped

  If the PDU session between the UE and the network is dropped, the network side must terminate the ongoing SIP session. For this reason, the UE should try to reestablish a PDU session. In this case, the network side will reestablish a new QoS data stream for the UE.

In order to support the IMS voice service, the IP address of the UE cannot be changed. Although the SSC Mode-2 and -3 (i.e., the service may still continue when the IP address changes) are introduced in the 5G standard, these modes are not suitable for 5G voice service.

2. Voice QoS data flow is lost

Generally, the QoS data stream used for voice is established by the QoS parameter corresponding to $5QI = 1$. In order to achieve a QoS data stream with $5QI = 1$, the 5G-AN uses more resources than the QoS data stream used to transmit ordinary Internet data. In the case of network congestion or weak coverage, the 5G-AN resources may not be able to guarantee a QoS data stream with $5QI = 1$ and may release the QoS data stream. In this case, the UE's voice service data will be mapped to the "match-all" QoS data stream according to the QoS rule. Although the QoS data stream is generally not guaranteed by GFBR, it can guarantee the transmission of 5G voice data as much as possible through the "best effort" method to minimize the service interruption for the voice service.

3. The network does not support the indication of IMS voice

In the deployment of the operator's network, there may be some areas that do not support IMS voice. This situation is generally seen when the density of 5G-AN(s) and the capacity of air interface of the operator's network are limited. The operator network may be able to sustain the transmission of general Internet data (ordinary Internet data has lower requirements for QoS), but it would not be able to guarantee that a high volume of users can initiate voice services with $5QI = 1$ QoS data stream at the same time in the same area. In this case, the network-side AMF will indicate the UE "IMS Voice over PS Session is not supported" in the NAS message so that the UE will not initiate voice services over the 5GS in this area.

However, when the UE with the existing PS IMS voice session moves into the area that is overloaded with voice services, even though the UE will still receive the "IMS Voice over PS Session is not supported" indication, the ongoing voice call from such UE will not be impacted. The UE does not need to release the ongoing IMS voice session. Once the current IMS voice session ends, the network will not accept any new IMS voice calls in this overloaded area.

4. The UE fails to perform voice services after EPS fallback

3GPP Rel-16 introduced a new parameter "EPS Fallback indication" in the RRC Release or Handover Command message. EPS fallback is an additional mobility trigger for improving voice KPIs. EPS Fallback enables phones to use the 5GC with NR, but RAN may trigger moving the phone LTE connected to EPC during call establishment. In this case, even if the UE fails to establish a connection with the target 5G NR cell, it may still be able to maintain the voice service as much as possible. For further details, TS38.306 [2] clause 5 can be referenced.

Further details of the EPS Fallback procedure are also provided in Section 14.2.2 of this chapter.

14.2 5G voice solutions and usage scenarios

In addition to the functionalities described in Section 14.1 to support 5G voice, support for 5G voice is also affected by core network capabilities and coverage issues.

The core network capabilities mainly consider the following aspects:
- Lawful interception;
- Charging;
- Roaming;
- Upgrade and transformation of existing networks;
- Compatibility issues between 5G network and 4G network.

The service requirements above necessitate support from the core network, but the above requirements do not apply to ordinary service data.

In the early stages of 5G deployment, the coverage will not be widespread. Therefore one should expect the frequent 4G/5G cross-system handover, which requires careful considerations when handling this scenario.

As described earlier, the 5G system supports two types of voice solutions: VoNR and EPS/RAT Fallback. In the former the 5G UE completes the call establishment of the IMS call on the 5GS (including MO Call and MT Call), and in the latter the UE falls back to the EPS to complete the IMS call establishment. Whether it is VoNR or EPS/RAT Fallback, the UE performs core network registration and IMS registration on the 5GS, and performs data services on the 5GS. The behavior is different only when a voice call occurs. EPS/RAT Fallback can better meet the voice needs of operators in the initial stage of 5G deployment, but VoNR is the ultimate goal. VoNR and EPS/RAT Fallback will be described in more detail in subsequent sections.

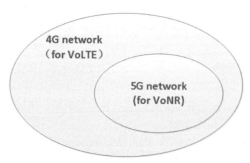

Figure 14.3 5G network supports VoNR.

Note that the 4G voice CSFB scheme is no longer supported by the 5G system, which means that the UE cannot send a Service Request message to request the voice call to fall back to the 2/3G CS domain to perform call establishment.

14.2.1 VoNR

VoNR refers to when the voice call is handled over the 5GC connected to the NR. This section provides further details on this approach. The EPS/RAT Fallback refers to the voice call handling over the EPC or on the 5GC connected to E-UTRA. Further details on this approach is described in subsequent sections. As shown in Fig. 14.3, the premise of VoNR is that the 5G network supports IMS voice services, and the UE can complete the call directly on the 5G system when establishing a call.

When establishing a VoNR call, how each network layer supports such operation as described in Section 14.1 is further explained below (Fig. 14.4).

1. RRC connection establishment

   The UE completes the RRC connection establishment with the base station through random access. After the UE has established the RRC connection, the base station can request the UE to report voice-related capability parameters (IMS parameters), including: VoNR support, VoLTE support, and the voice capability-related parameters reported by the UE at the AS layer, which can be used in the core network to decide whether to support voice services. The core network will also store relevant capability parameters for subsequent use.

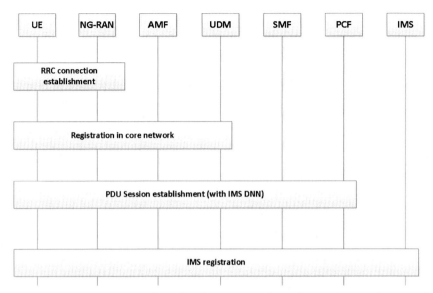

Figure 14.4 Relevant procedures for the UE to perform for VoNR at each network layer.

2. Core network registration procedure

After completing the RRC establishment, the UE sends a Registration Request message to the AMF. The registration request contains parameters related to voice capabilities: attachment with the "handover" flag, Dual Connectivity with NR, SRVCC capability, and class mark-3.

After the AMF receives the Registration Request, it obtains the subscription information from the UDM. The voice-related parameters in the subscription information include: STN-SR (session transfer number), C-MSISDN (associated mobile station international user identification code), used for execution 5G-SRVCC (5G single wireless voice continuity) are used.

In addition, the AMF can also initiate the "UE capability match" process to obtain the UE's voice support parameters on the base station, including the voice-related parameters sent by the UE to the base station during the RRC establishment process. 3gpp TS23.502 clause 4.2.8a should be referenced for further details.

According to the registration request, the subscription information, the wireless capability parameters on the 5G-AN, and the local configuration, the AMF determines whether to send instructions to the UE to support voice services. For further details, clause 5.16.3.1 of TS23.501 should be referenced. It should be noted that regardless of whether the AMF decides to perform VoNR or EPS/RAT Fallback (EPS/RAT Fallback) for the UE, it will send the same indication to the UE (i.e., the NAS message "Registration Reply" message carries "IMS voice over PS session is supported (IMS voice supported)" indication). The UE should not need to determine whether to implement the voice service through the VoNR or EPS/RAT Fallback process. The decision making for determining which voice service type to apply to the UE is the responsibility of the network. For the UE side, as long as the "IMS voice over PS session is supported" indication is received from the AMF in the NAS message response, the UE expects that the network supports IMS voice in its registration area and will proceed with the PDU session establishment procedure.

For a detailed description of the entire registration process, clause 4.2.2 of 3GPP TS23.502 [3] should be referenced.

3. PDU session establishment

After a UE completes the registration process in the core network, multiple PDU sessions can be established. Among them, the UE needs to specifically initiate a PDU session establishment for voice services (i.e., the UE sends a NAS message "PDU Session Establishment Request" to the SMF) with the following voice-related parameters:

a. Ms DNN (IMS data network name) specifically used for DNN parameters of IMS services.

b. SSC Mode (Conversational and business continuity mode) for IMS voice services; currently, only SSC Mode = 1 can be selected to ensure the service continuity is maintained during a call, and its core network user plane gateway (UPF) cannot be changed.

c. PDU Session Type (PDU Session Type) for IMS service PDU sessions; the PDU session types can only be IPv4, IPv6, or IPv4v6. Ethernet (Ethernet) and Unstructure (unstructured) types are not supported. In addition, due to the exhaustion of the number of IPv4 addresses, the 5G system gives priority to using IPv6 types.

After the SMF receives the "PDU Session Establishment Request" message, it can obtain session management-related subscription information from the UDM, and also the PCC policy from the PCF. The PCC policy contains the PCC rules used by the SMF to determine the QoS data stream to transmit IMS signaling. The rules are provided to 5G-AN and UPF to establish a QoS data stream with 5QI = 5, which is used to

Table 14.3 VoNR-relevant QoS flow.

5QIvalue	Resource type	Default priority	Packet delay budget	Packet error rate	Default maximum data burst	Default averaging window	Example services
1	GBR	20	100 ms	10^{-2}	N/A	2000 ms	Conversational voice
5	Non-GBR	10	100 ms	10^{-6}	N/A	N/A	IMS signaling

carry IMS signaling. It should be noted here that the QoS data stream with 5QI = 5 is completed during the establishment of the PDU session, while the QoS data stream with 5QI = 1 is used to carry IMS voice data, which is completed during the establishment of the voice call and will be released after the voice call ends. The key parameters for the two important 5G voice-related data streams 5Q1 = 1 and 5QI = 5 are shown in Table 14.3.

For a detailed description of QoS data flow, Chapter 13, Quality of Service Control of this book should be referenced.

In addition to the establishment of voice-related QoS data streams, another important thing the SMF needs to do is to discover the P-CSCF for the UE. As described in Section 14.1.1, the P-CSCF is the access point of the IMS network. Both the IMS signaling and voice data of the UE need to be sent to other network elements of the IMS through the P-CSCF for routing and processing.

After the SMF executes the PDU session establishment process, it will carry the following voice-related parameters in the NAS message "PDU Session Establishment Reply":

- The IP address or domain name information of the P-CSCF is used for the UE to communicate with the correct P-CSCF.
- UE IP address.

For a detailed description of the complete PDU session establishment, clause 4.3.2.2 of 3GPP TS23.502 [3] should be referenced.

4. IMS registration

After the PDU session for the IMS service is established, the UE can initiate an IMS registration request process to the P-CSCF through a QoS data flow of 5QI = 5. The specific IMS registration process has been elaborated on Section 14.1.1.

14.2.2 EPS fallback/RAT fallback

1. EPS Fallback

As mentioned at the beginning of Section 14.2, due to the requirements of the core network and the 5G radio coverage, it is difficult to implement VoNR in the early stages of 5G deployment. Since the VoLTE infrastructure on the current network is quite mature, many operators are considering using VoLTE in the early stages of 5G deployment instead. As a result, 5G UEs will use the 5GS for non-voice services. When voice services are required, they will fall back to

Table 14.4 Comparison between EPS fallback and CS fallback.

	EPS fallback	**CS fallback**
Support IMS stack	Required. It is an IP call.	Not required. It is a CS domain voice.
HD voice coding HD	Supports EVS AMR-WB (NOTE 1).	Supports AMR-WB at most.
Perform registration before the voice call is established	The UE must register with the IMS after registering with the core network to perform the voice establishment process.	The UE only performs joint attachment in the core network.
The first voice call establishment message	The UE does not need to wait for the fallback to complete, and can send SIP Invite/183 Ack messages on the 5G network.	The UE needs to wait for the fallback to complete and send the Call setup message in the CS.
UE initiates a fallback request	Not required; the network side triggers the fallback to the 4G network through handover or redirection during the IMS voice establishment process.	The UE is required to initiate an Extended Service Request message to trigger the network to perform handover or redirection to the 2/3G network.

NOTE 1: For a detailed description of HD speech coding, refer to section 5.2.1 of 3GPP TS26.114 [7].

the EPS to perform VoLTE voice calls through the EPS Fallback process.

Many people may see the word "EPS Fallback" and think of the "CS Fallback (CSFB)" scheme used in 4G voice. Indeed, the basic idea of both is to bring the UE back from the "n" G network to" n-1" G network to perform voice services, but there are some essential differences between the two. Table 14.4 illustrates the main differences between the two mechanisms from the UE perspective.

The EPS Fallback process is shown in Fig. 14.5. The UE can fall back by triggering redirection or handover. No matter which fallback method is used, its UPF (PGW + UPF) cannot be changed (i.e., the IP address will not change). After the UE falls back to the EPS, the core network

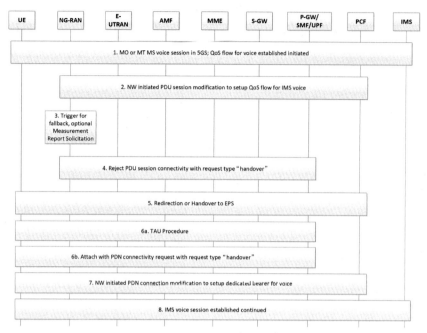

Figure 14.5 EPS fallback.

will again trigger the establishment of a dedicated voice bearer for voice services. At the same time, it should noted that the entire EPS Fallback process does not affect the signaling transmission of the IMS layer. The IMS call establishment process as described in Section 14.1.2 performed by the IMS layer will not be interrupted due to the EPS Fallback process.

For a detailed description of the EPS Fallback process, clause 4.13.6.1 of TS23.502 [3] should be referenced.

2. RAT Fallback

In addition to EPS Fallback, 3GPP also introduced the RAT Fallback scenario, which is very similar to the EPS Fallback process. The only difference is that the core network 5GC remains unchanged (i.e., the UE falls back from the 5G NR connected to 5GC to the E-UTRA, which is also connected to 5GC).

The scenario of RAT Fallback mainly considers that the coverage of 5G NR in the early stage of 5G deployment is small, and the use of E-UTRA will be able to avoid frequent Inter-RAT handovers, thereby reducing the network burden and ensuring business continuity. However, due to the early stages of 5G deployment, the 5GC is also a newly deployed network. As mentioned at the beginning of Section 14.2 of this book, even if only 5GC supports voice services, it is not an easy task. Having the new 5GC to support RAT Fallback could still be challenging. It is foreseeable that in the first few years of 5G deployment, EPS Fallback will be the fallback method used by many operators, and then gradually be transitioned to VoNR.

For a detailed description of RAT Fallback, clause 4.13.6.2 of TS23.502 [3] should be referenced.

14.2.3 Fast return

Fast Return means that the UE can quickly return to the 5G network after executing EPS Fallback or RAT Fallback and completing the voice service.

When using the Fast Return function, the UE can receive the LTE neighbor cell information sent by the E-UTRA, and return to the NR cell according to the redirection or handover procedure performed by the base station. This eliminates the need for the UE to measure the signal strength of the neighbor cell and read the Fetch SIB message as well as wait time. Therefore the return speed is faster and the accuracy is higher. This approach can prevent the UE from being affected by the complex environment of the existing network.

In contrast, if there is no fast return support after the UE finishes the voice service under the E-UTRA base station, the UE can only wait until there is no other data service transmission and return to the idle state, and then perform cell reselection according to the "frequency priority" back to the NR cell. In the field environment, the UE is easily affected by the complex environment of the existing network, and the access request is highly likely to be rejected. At this time, the UE needs to reselect other cells to access, which takes significantly longer than the Fast Return.

The key point of the Fast Return function is that the E-UTRA needs to be able to accurately determine that the UE can be redirected or switched to the target cell, which is caused by the EPS Fallback and not due to other reasons (such as UE mobility, poor signal quality in NR cells, etc.).

For this reason, during the EPS Fallback or RAT Fallback process, the E-UTRA will receive the Handover Restriction List and RFSP Index from the core network side. The HRL contains the last used PLMN ID (the last used PLMN ID) and the RFSP Index (connected to the standard and frequency selection strategy index) is used to indicate whether the

UE can access the NR frequency point. The E-UTRA combines the two parameters and the local policy configuration to determine whether the UE has performed EPS Fallback or RAT Fallback. After the execution of the voice service (the QoS data stream with QCI = 1 will be deleted after the voice service is completed), the E-UTRA node actively initiates a redirection or handover process to make the UE return to the 5G network.

Although the Fast Return function is simple, it can effectively improve the user experience, allowing users to stay on the 5G network as much as possible and experience better service quality. Therefore in 4G CSFB, most operators deploy the Fast Return mechanism, which is believed to be widely used in 5G EPS Fallback and RAT Fallback.

For a detailed description, clause 5.16.3.10 of TS23.501 [4] should be referenced.

14.2.4 Voice continuity

Voice service requires mobility support. For IMS voice, the voice service during a call needs to ensure no change to IP address. Hence, the anchored UPF cannot be changed during the call. Although the SSC Mode-3 mode in 5G supports the service to be seamlessly switched from UPF-1 to UPF-2 by means of make before break switching, this mode requires changes at the application layer. At present, the IMS voice service still only supports SSC Mode-1 mode (i.e., the anchor point during the call cannot be changed to avoid service changes).

In addition to the UPF anchor point remaining unchanged, the UE also needs to support the following functions to ensure the continuity of voice services during the movement:

- Handover within 5G system: The UE can perform operations according to the process of cross-base station handover in the same system; clause 4.9 of 3GPP TS23.502 [3] should be referenced for further details.
- Handover between 5G and 4G systems: The UE switches between 5GS and EPS systems; clause 4.11 of 3GPP TS23.502 [3] should be referenced for further details.
- In addition to the above two mandatory functions, 3GPP R16 also defines the optional function 5G-SRVCC (Single Wireless Voice Continuity). This is to support the UE with voice services that move out of the 5G coverage area, and where there is also no 4G coverage.

If the coverage does not support VoLTE, the 5G-AN triggers the SRVCC process from NG-RAN to UTRAN, and switches to the 3G CS domain to perform voice services. Since SRVCC involves two processes, Inter-RAT handover (cross-RAT handover) and Session Transfer (session transfer), both air interface and network-side deployment requirements are very high. Therefore as of now no operators have considered the 5G-SRVCC function. For detailed description and process, the relevant causes of 5G-SRVCC in 3GP P TS23.216 [5] should be referenced.

14.3 Emergency call

Emergency call means that when dialing emergency rescue numbers (such as 911 in the United States, 119 and 110 in China), the UE can always pass through even if there is no SIM card, the tariff is insufficient, or it is out of the operator's network coverage of the SIM card. Any operator's network that supports the emergency call service will be able to make an emergency call.

Emergency calls are very important functions in Europe and the United States, and are increasingly being valued in China. Although broadly speaking, emergency calls are also a type of voice service, the details are different from ordinary voice call services as shown in Table 14.5.

For 5G emergency calls, clause 5.16.4 of 3GPP TS23.167 [6] and 3GPP TS23.501 [4] should be referenced for further detailed descriptions.

14.4 Summary

This chapter presented a detailed introduction to 5G voice, revealed the development of 4G voice, disclosed the development trend of future voice services, and identified the relevant important requirements of successful 5G voice solutions from the perspective of the UE and the reality of 5G deployment, so that readers can recognize the future of 5G voice and the following points:

- The evolution trend of voice services and the advantages of 5G voice services.
- Lead 5G voice solutions from the perspective of 5G network deployment: VoNR, EPS/RAT Fallback.

Table 14.5 The differences between emergency call and normal call.

	Emergency call	Normal voice call
Authentication during registration	Not needed.	Needed.
Call restriction	The service has the highest preemption priority, and general access and mobility are not suitable for emergency call services.	Subject to access and mobility restrictions.
Mobile Originated (MO) and Mobile Terminated (MT)	Generally, only the MO function is supported (so there is no need to perform IMS registration in advance), but some countries have a "call back" function (i.e., after an emergency call is made, the called party can call the user).	Both MO and MT are mandatory, and IMS registration is required in advance.
IMS network to perform the call	In most cases, the emergency call number of the UE in the roaming area will be served by the IMS server (proprietary CSCF) in the roaming network.	In most cases, the call flow is executed by the home network.
PDU session establishment	The PDU session is established during the call.	Under normal circumstances, establish well in advance.
Domain selection	The UE can autonomously leave and select other networks when performing an emergency call according to the support of the emergency call on the network side.	The UE cannot choose other networks independently, and must follow the commands of the network side to execute.
Initiating fallback	If allowed by the network, the UE can actively initiate the EPS/RAT Fallback process.	Whether to perform Fallback is completely determined by the network side, and the UE cannot actively request or need to perceive.

- The progress of the important features of 5G voice from the AS layer, NAS layer, and IMS layer.
- The main features of the emergency call service.

References

[1] 3GPP TS23.228 V16.4.0 (2020−03). IP multimedia subsystem (IMS); stage 2.
[2] 3GPP TS38.306 V16.0.0 (2020−04). NR; user equipment (UE) radio access capabilities.
[3] 3GPP TS23.502 V16.4.0 (2020−03). Procedures for the 5G system (5GS).
[4] 3GPP TS23.501 V16.4.0 (2020−03). System architecture for the 5G system (5GS).
[5] 3GPP TS23.216 v16.3.0 (2019−12). Single radio voice call continuity (SRVCC); stage 2.
[6] 3GPP TS23.167 V16.1.0 (2019−12). IP multimedia subsystem (IMS) emergency sessions.
[7] 3GPP TS26.114 v16.5.2 (2020−03). IP multimedia subsystem (IMS); multimedia telephony; media handling and interaction.

CHAPTER 15

5G Ultra-reliable and low-latency communication: PHY layer

Jing Xu, Yanan Lin, Bin Liang and Jia Shen

15.1 Physical downlink control channel enhancement

15.1.1 Introduction to compact downlink control information

A typical physical layer transmission process includes both control information transmission and data transmission. Take downlink transmission as an example. A packet transmission on the downlink includes downlink control information transmission and downlink data transmission. Therefore the reliability of downlink physical layer transmission depends on the reliabilities of both the downlink control channel and downlink data-sharing channel. To be more specific, $P = P_{\text{PDCCH}} \times P_{\text{PDSCH}}$, where P is the reliability of physical layer transmission, P_{PDCCH} is the reliability of the downlink control channel, and P_{PDSCH} is the reliability of the downlink data-sharing channel. If retransmission is considered, the reliability of Hybrid Automatic Repeat reQuest-Acknowledge (HARQ-ACK) feedback and reliability of the successive transmissions should also be considered. For the reliability requirements of URLLC, such as 99.999% in R15 and 99.9999% for some cases in R16, the reliability of the downlink control channel should at least reach 99.999% or 99.9999%. Therefore reliability enhancement for the downlink control channel has to be considered. Chapter 5, 5G Flexible Scheduling already described in detail the higher aggregation level ($= 16$) and distributed control channel element (CCE) mapping that are adopted in R15. This chapter mainly discusses the physical downlink control channel (PDCCH) enhancement scheme for URLLC in R16, which was selected from the following two candidate schemes to improve PDCCH reliability.

- Candidate Scheme 1: Reducing the size of Downlink Control Information (DCI)

    A smaller number of DCI bits can improve energy per single information bit, and therefore improve PDCCH reliability. Reducing the size of DCI is usually achieved by compressing DCI fields or defining the default indication.

- Candidate Scheme 2: Increasing PDCCH resource

More time-frequency resources are consumed for a PDCCH transmission; for example, to increase the aggregation level of PDCCH or to apply repetition transmissions.

Scheme 1 can not only improve reliability of the downlink control channel, but also alleviate congestion of PDCCH. Even though the compact DCI inevitably introduces the scheduling constraints, such constraints can be ignored for URLLC transmission due to the service requirements and transmission characteristics of URLLC traffic, such as a small amount of data and large bandwidth transmission. For Scheme 2, the evaluation [1] showed that the aggregation level 16, which was introduced in R15 and already higher than that in LTE, can provide the reliability quite close to or even above the reliability requirement. In addition, a higher aggregation level will inevitably increase time-frequency resource overhead, which is not suitable for small bandwidth transmission. Therefore further increase of aggregation level was not considered in R16. PDCCH repetition transmission was not adopted either because repetition transmission scheme may lead to latency and complexity issues. Consequently, Scheme 1 was accepted in 3GPP RAN1.

15.1.2 Compact downlink control information

With performance evaluation, 3GPP RAN1 determined the following design goals of compact DCI.
- Support configurable DCI format, and the range of DCI size is:
  - The maximum number of DCI bits can be larger than the size of DCI format 0_0/1_0.
  - The minimum number of DCI bits is 10−16 bits less than the size of DCI format 0_0/1_0.

With the requirements and characteristics of URLLC transmission, the design of compact DCI mainly considers the following three aspects:
- Removal of information fields relating to large data transmission, including:
- MCS for the 2nd codeword
- New Data Indicator (NDI) for the 2nd codeword
- Redundancy Version (RV) for the 2nd codeword
- Codebook Group (CBG) transmission information
- CBG flushing out information
- Optimization for information fields relating to the characteristics of URLLC data transmission, including:
- Frequency-domain resource assignment

- Time-domain resource assignment
- RV indication for the 1st codeword
- Antenna ports indication
- Most information fields remain to have the same size as in DCI format 0_1/1_1, including carrier indication, PRB binding size indication, rate-matching indication, etc.

The compact DCI with the above design aspects is categorized as DCI format 0_2/1_2. The DCI size alignment procedure relating to DCI format 0_2/1_2 is introduced to guarantee the same blind decoding complexity as in R15.

15.1.3 Physical downlink control channel monitoring capability per monitoring span

In theory, enhancement of PDCCH monitoring capability can improve the URLLC scheduling flexibility and meanwhile reduce PDCCH congestion and transmission latency. However, upgrading PDCCH monitoring capability usually means to increase terminal complexity. In order to avoid increasing implementation complexity, one way is to restrict configuration relating to PDCCH (e.g., by limiting the number of carriers supporting enhanced PDCCH monitoring capability to keep the total PDCCH monitoring capability unchanged on all involved carriers); or to reduce PDSCH/PUSCH processing time (e.g.., restricting the PDSCH/PUSCH transmission to use preconfigured parameters), including the number of PRBs, the number of transmission layers, and the size of transmission blocks so that the processing time saved by simplifying PDSCH/PUSCH transmissions can be used for PDCCH monitoring while the total processing time is kept unchanged. Another way is to restrict PDCCH monitoring span and to avoid PDCCH accumulation. Specifically, R15 adopts PDCCH monitoring capability definition based on slot, and R16 introduces PDCCH monitoring capability definition based on monitoring span, which is a shorter time range than a slot.

Once the PDCCH monitoring capability is determined to be based on monitoring span, the value representing the PDCCH monitoring capability is considered mainly from two aspects: URLLC scheduling flexibility and terminal complexity. 3GPP RAN1 decided that the cumulative PDCCH monitoring capability based on monitoring span in a slot is about twice as much as that of PDCCH based on slot.

The PDCCH monitoring capability based on the monitoring span includes the maximum number of nonoverlapping CCEs (C) and the maximum number of PDCCH candidates (M) the UE can monitor per span.

15.1.4 Physical downlink control channel monitoring for CA

In the CA scenario, when the number of configured carriers is greater than the number of carriers corresponding to the PDCCH monitoring capability reported by the terminal, the total number of monitored PDCCH candidates across multiple configured carriers should be less than or equal to the product of the number of carriers corresponding to the PDCCH monitoring capability reported by the terminal and the PDCCH monitoring capability for a single carrier. Similar to the PDCCH monitoring capability based on slot, the PDCCH monitoring capability based on monitoring span can be also impacted by multicarrier scaling. However, the difference is that the PDCCH monitoring capability based on the monitoring span may have problem with span misalignment over different carriers, as shown in Fig. 15.1. For CA, scaling of PDCCH monitoring capability based on the monitoring span is restricted to aligned carriers.

In addition, the numbers of carriers are independently configured for PDCCH monitoring capability based on slot and PDCCH monitoring capability based on monitoring span. The specification also supports the combined reporting of the number of carriers for both PDCCH monitoring capability based on monitoring span and PDCCH monitoring capability based on slot, with the following constraints

- The minimum values for pdcch-BlindDetectionCA-R15 and pdcch-BlindDetectionCA-R16 are both 1.
- The range of pdcch-BlindDetectionCA-R15 + pdcch-BlindDetectionCA-R16 is 3−16.

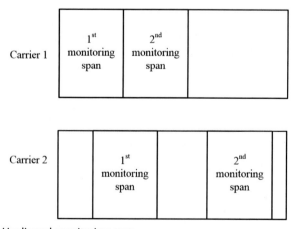

Figure 15.1 Unaligned monitoring span.

- The range of pdcch-BlindDetectionCA-R15 is 1—15.
- The range of pdcch-BlindDetectionCA-R16 is 1—15.

15.2 UCI enhancements

As described in Chapter 5, 5G Flexible Scheduling, a major innovation of R15 5G NR compared with LTE is the flexible "symbol-level" time-domain scheduling, which can achieve the lower transmission delay. In particular, the introduction of symbol-level "short PUCCH" can facilitate the fast UCI transmission, which enables the low-delay HARQ—ACK feedback and scheduling request (SR) transmission. However, compared to the requirements of low-delay and high-reliable transmission of URLLC traffics, R15 NR PUCCH still needs some improvements in the following aspects:

- Although the R15 NR PUCCH has been shortened to symbol-level, the transmission opportunity of UCI is still on "slot-level." Only one PUCCH-carrying HARQ—ACK can be transmitted in a slot, which cannot realize "to feedback HARQ—ACK at any time."
- In R15 NR, an URLLC UCI and an eMBB UCI cannot be generated independently. When an URLLC HARQ—ACK and an eMBB HARQ—ACK have to be built in one codebook, the delay and reliability of the URLLC HARQ—ACK may be affected by the eMBB HARQ—ACK, because an eMBB HARQ—ACK usually contains a large number of bits and results in a large-size HARQ—ACK codebook, which makes a high-reliability transmission difficult.
- Regardless whether the UCI is separately generated for URLLC and eMBB services, the R15 physical layer cannot distinguish the two UCIs with their "priorities" (e.g., high-priority vs low-priority); therefore the resource cannot be dedicatedly mapped to URLLC UCI or eMBB UCI to "guarantee" the transmission performance for the UCI with "high priority."
- Finally, even if the priority of UCI were defined and the "dedicated" resources were respectively allocated to the "high-priority" and "low-priority" UCIs, it could be still difficult to avoid the resource conflicts due to the scheduling order of the gNB. When the resource scheduled for a high-priority UCI (e.g., for URLLC service) overlaps with a previously-scheduled resource for a low-priority UCI (e.g., for eMBB service), R15 faces a lack of a resource conflict resolution mechanism to ensure the performance of the high-priority UCI.

These problems in R15 NR were addressed in the R16 URLLC work item. The corresponding R16 enhancements are described in this section [2–7].

15.2.1 Multiple HARQ–ACK feedbacks in a slot- and subslot-based PUCCH

Although R15 NR introduced "short PUCCH" with only a few symbols, which provides a good design basis for the fast UCI feedback; a UE can only transmit one PUCCH-carrying HARQ–ACK in a slot. This restriction cannot fulfill the real-time feedback if a second HARQ–ACK is generated in the same slot. As shown in Fig. 15.2A, the UE receives PDSCH 1 and feeds back HARQ–ACK in uplink Slot 1. At a later time, the UE receives PDSCH 2 requiring lower latency and needs to feed back the HARQ–ACK as soon as possible. If the HARQ–ACK corresponding to PDSCH 2 is also sent in uplink Slot 1, in order to ensure the UE to have enough processing time, only a PUCCH resource in the rear part of Slot 1 can be allocated to the HARQ–ACK. According to R15 NR mechanism, the HARQ–ACK for PDSCH 1 has to be also

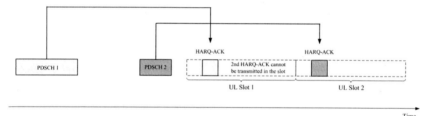

(A) HARQ-ACK feedback is delayed because multiple HARQ-ACKs cannot be transmitted in a slot

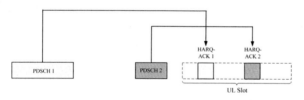

(B) Guarantee HARQ-ACK feedback at any time by supporting multiple HARQ-ACKs in a slot

Figure 15.2 Motivation to support multiple HARQ–ACKs in a slot. (A) HARQ–ACK feedback is delayed because multiple HARQ–ACKs cannot be transmitted in a slot. (B) Guarantee HARQ–ACK feedback at any time by supporting multiple HARQ–ACKs in a slot.

transmitted in the later PUCCH resource, which delays the HARQ−ACK feedback corresponding to PDSCH 1. The alternative is to transmit the second HARQ−ACK in the next uplink slot, which delays the HARQ−ACK feedback of PDSCH 2. In either case, R15 UE can only transmit one PUCCH-carrying HARQ−ACK in a slot. This inevitably increases the feedback delay of downlink services.

To solve this problem, it is necessary to provide multiple HARQ−ACK feedback opportunities in one uplink slot. As shown in Fig. 15.2B, if the HARQ−ACK for PDSCH 2 can be transmitted in the same slot, it is helpful to enable the HARQ−ACK feedback at any time and guarantee the delay requirement of the downlink URLLC service.

15.2.1.1 Subslot-based HARQ−ACK

As introduced in Section 15.5.5, the R15 NR PUCCH resource allocation is still slot-based. Although from the perspective of gNB, the "short PUCCH" from multiple UEs can be multiplexed in a slot, one UE can only have one PUCCH resource carrying HARQ−ACK in a slot. In order to support multiple HARQ−ACK feedbacks in a slot for a single UE, some aspects of NR PUCCH resource allocation need to be redesigned, including:

- How to separate multiple HARQ−ACKs in one time slot
- How to associate the PDSCHs with the multiple HARQ−ACKs
- How to allocate time-frequency resources for the multiple HARQ−ACKs
- How to indicate the time-domain offset (i.e., K_1) from the PDSCH to the HARQ−ACK

To address these design aspects, several candidate schemes were proposed in R16 NR standardization, including the "subslot scheme" and "PDSCH-grouping scheme."

The subslot scheme is used to further divide an uplink slot into shorter subslots. A subslot only contains several symbols. The R15 HARQ−ACK codebook construction, PUCCH resource allocation, and UCI multiplexing can be reused as much as possible, but just in unit of subslot. The details of the schemes are described as follows:

- K_1 is defined in unit of subslot.
- All HARQ−ACKs transmitted in the same subslot are multiplexed into one HARQ−ACK codebook.
- A virtual "subslot" grid is used in the downlink slots to determine the association between PDSCH and HARQ−ACK (i.e., assuming the virtual subslot containing the end of PDSCH as the reference point

for indication of K_1). Based on the reference point and K_1 indication, the subslot containing HARQ−ACK feedback can be determined.

- The start of the subslot containing HARQ−ACK is used as the reference point to indicate the starting symbol of the PUCCH conveying the HARQ−ACK. Subslot-based PUCCH resource sets are defined accordingly.

Take Fig. 15.3 as an example. Suppose that an uplink slot is divided into seven subslots. Each subslot is two symbols long. Based on the "virtual subslots" containing the ends of PDSCH 1 and PDSCH 2 as well as the corresponding K_1 values for the two PDSCH, it can be determined that their corresponding HARQ−ACKs fall in Subslot 1. The starting symbols for the two HARQ−ACKs are Symbol 0 and Symbol 1, respectively. Therefore the two HARQ−ACKs are multiplexed and transmitted in Subslot 1. Similarly, based on the "virtual subslots" containing the ends of PDSCH 3 and PDSCH 4 as well as the K_1 values, the two HARQ−ACKs for these two PDSCH are multiplexed and transmitted in Subslot 4, with the starting symbols being Symbol 0 and Symbol 1, respectively. It should be noted that the actual subslot structure does not indeed exist in PDSCH construction. The "virtual subslot grid" is only used to locate the starting reference time for the offset K_1.

The PDSCH-grouping scheme does not rely on concept of subslot to determine which HARQ−ACKs of PDSCH are multiplexed based on the time-domain position of the PUCCHs. Instead, it can determine the association between PDSCH and PUCCH in a more "explicit" way. The details of the scheme are given as follows:

- K_1 is still indicated in unit of slot.
- The HARQ−ACKs that are multiplexed into a HARQ−ACK codebook correspond to the PDSCHs in the same PDSCH group.

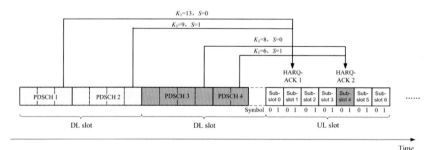

Figure 15.3 Subslot-based HARQ−ACK feedback mechanism.

An indicator in the DCI scheduling the PDSCH (e.g., PRI (PUCCH resource indicator) or a newly added "PDSCH grouping indicator") is used to indicate which PDSCH group the PDSCH belongs to.

- The R15 definition of PUCCH resource sets is reused (e.g., the "slot-based" PUCCH resource and its position relative to the slot boundary).

Fig. 15.4 shows an example for PDSCH-grouping scheme. The DCI scheduling the PDSCH includes a 1-bit PDSCH grouping indicator to indicate which HARQ—ACK codebook the PDSCH is associated with. Both PDSCH 1 and PDSCH 2 are indicated to be associated with HARQ—ACK 1, and both PDSCH 3 and PDSCH 4 are indicated to be associated with HARQ—ACK 2. Then HARQ—ACK can be fed back multiple times in a slot without relying on a "subslot" structure.

The above two schemes can both support multiple HARQ—ACK feedbacks in a slot. The advantage of subslot scheme is that the R15 NR HARQ—ACK codebook generation method can be reused, as long as the subslot is used to replace slot as the unit for the operation. However, the definitions of K_1 and PUCCH resources need to be changed to "subslot level." The advantage of PDSCH grouping scheme is that the PUCCH resource configuration and the K_1 definition in R15 can be reused. The slot-based K_1 becomes more efficient than the subslot-based K_1 when indicating a HARQ—ACK that is far away from the corresponding PDSCH, but URLLC usually uses fast HARQ—ACK feedback which is very close to PDSCH. The slot-based scheme needs to additionally design a PDSCH grouping mechanism, such as adding a new indication field in DCI. Even though the PDSCH grouping scheme seems more flexible because the position of HARQ—ACK is not limited by the subslot grid, this scheme may lead to a PUCCH resource overlapping problem, which may require a complicated resource conflict resolution.

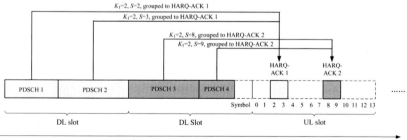

Figure 15.4 PDSCH-grouping-based HARQ—ACK feedback mechanism.

After studies and discussions, 3GPP RAN1 decided to adopt the subslot-based scheme in R16 to support multiple HARQ—ACK feedbacks in a slot.

15.2.1.2 Whether to support subslot PUCCH resource across subslot boundary?

A subslot-based PUCCH starts within a subslot, and the starting symbol of the PUCCH resource is defined relative to the boundary of the subslot. Take the example shown in Fig. 15.3. The starting symbol S of PUCCH has only two choices: Symbol 0 and Symbol 1. An open issue is whether the length of the PUCCH resources should be restricted within subslot as well; or equivalently whether a subslot PUCCH can cross the subslot boundary and enter the next subslot. The two following options were discussed in 3GPP RAN1.

- Option 1: Do not support subslot PUCCH resource across subslot boundary

   If a subslot PUCCH cannot cross the subslot boundary, it actually means that the long PUCCH cannot be used. The two–symbol subslot structure can only hold one-symbol PUCCH or two–symbol PUCCH, which can only be used in the scenario where the uplink coverage is good and a relatively small UCI capacity is sufficient.

- Option 2: Support subslot PUCCH resource across subslot boundary

   If PUCCH resources are allowed to cross the subslot boundary, the R15 NR standard can flexibly support long PUCCH and short PUCCH with various lengths (as described in Section 15.5.5). Even for a very short subslot, the long PUCCH can always be used to support cell edge coverage and to convey large-capacity UCI.

   The questions raised against Option 2 in RAN1 discussion include: Does the URLLC service really need a long PUCCH to transmit HARQ—ACK? Can a long PUCCH achieve the low latency? Fairly speaking, even when a long PUCCH is used, Option 2 can still reduce UCI latency by providing more opportunities in a slot for "starting a PUCCH." Using a short PUCCH only is not the necessary condition for latency reduction.

   However, Option 1 is a relatively simple but sufficient scheme. It assumes that the coverage for a UE is relatively "stable" and changes "slowly." If a gNB configures two-symbol subslots for a UE, it means that the current coverage is good enough to just use short PUCCHs and long PUCCHs are not necessary. It is not a common case for the coverage to suddenly become worse. Moreover, subslot PUCCH is mainly used to transmit HARQ−ACK for URLLC service. It is not considered a common case for URLLC that a large number of HARQ−ACKs are multiplexed into a large HARQ−ACK codebook (such as in a TDD carrier aggregation system, HARQ−ACKs for many DL slots and carriers are multiplexed in a slot). The large UCI capacity provided by Option 2 is less beneficial to URLLC. Meanwhile, Option 2 does not confine a PUCCH into a subslot. The "crossing of subslot boundary" may lead to complicated PUCCH overlapping events, which require a complicated conflict resolution mechanism to be designed.

3GPP RAN1 finally decided that a subslot PUCCH is not allowed to cross the subslot boundary (e.g., $S+1 \leq$ subslot length). This restriction is similar to the one in the R15 NR design (i.e., $S+1 \leq$ slot length), except that the slot is replaced by the subslot. It should be noted that both slot-based PUCCH and subslot-based PUCCH are not allowed to cross the slot boundary.

The motivation for introducing the subslot PUCCH structure is to support multiple HARQ−ACK feedbacks in a slot. But questions arise: Can other types of UCI (such as SR, CSI (channel state information) reports) be carried by subslot-based PUCCH? If yes, is the restriction of "no subslot boundary crossing" also applicable to the subslot-based PUCCH? One opinion in RAN1 was that the length of SR and CSI should not be confined in a subslot. This is because, different from HARQ−ACK, SR can be transmitted multiple times in a slot since R15 NR, and CSI feedback may have large payload that is hardly accommodated into a subslot. On the other hand, the opposite opinion was that the purpose of confining a HARQ−ACK in a subslot is to avoid complicated PUCCH conflict events caused by subslot boundary crossing, which can happen not only for HARQ−ACK, but also for other types of UCI. Only when all UCIs are confined in subslots can the R15 NR conflict resolution mechanism be reused by simply changing the unit from slot to subslot. After discussions, the RAN1 decided that in a subslot-based *PUCCH-config*, all PUCCH resources including the ones carrying SR and CSI cannot cross the subslot boundary.

15.2.1.3 Subslot length and PUCCH resource set configuration

The next relevant question in RAN1 discussion was: Can different PUCCH resource sets be configured for different subslots? Or equivalently should all subslots share the same PUCCH resource set? The answer to this question is related to two other issues: Are PUCCH resources across subslot boundary allowed? Is there a scenario where subslots of different lengths are used at the same time?

Take what Fig. 15.5 shows as an example, where one slot contains two subslots (subslot length = 7 symbols). If PUCCH resources are allowed to cross the subslot boundary but still not allowed to cross–slot boundary, PUCCH resources of more than 7 symbols (such as PUCCH resources 3 and 4 of subslot in the figure) can be used for Subslot 0, while only PUCCH resources of no more than 7 symbols can be used for Subslot 1. In this case, if the two subslots share the same PUCCH resource set, only PUCCH resources with $S + L \leq 7$ can be configured for both subslots, where S and L are the starting symbol index and the total number of symbols for the corresponding PUCCH resource, respectively. This restriction substantially offsets the benefit in allowing PUCCH resources crossing subslot boundary. On the other hand, if the two subslots do not share the same PUCCH resource set, PUCCH

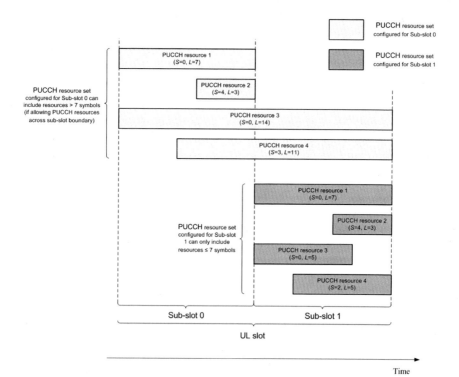

Figure 15.5 Example of different PUCCH resource sets across different subslots.

resources with $S + L > 7$ can be configured in the resource set, but can only be used by PUCCH starting from Subslot 0. The PUCCH resources with $S + L > 7$ cannot start from Subslot 1, which reduces the number of available PUCCH resources for Subslot 1. Another approach is to "truncate" the configured long PUCCH resources for Sub slot 1, which brings in extra complexity. Therefore if PUCCH resources are allowed to cross the subslot boundary, it should also be allowed to configure different PUCCH resource sets for different subslots (for example, the subslot at the beginning of the slot and the subslot at the end of the slot can be configured with different resource sets). However, as mentioned above, it has been decided that R16 NR PUCCH resources are not allowed to cross subslot boundaries. Hence the main reason for configuring the subslot-specific PUCCH resources does not exist.

Another related question is about the length of subslot. A slot of 14 symbols can be equally divided into seven 2-symbol subslots or two 7-symbol subslots. Other subslot divisions will lead to unequal-length subslot structures. As shown in Fig. 15.6, if a slot contains 4 subslots, the slot can be divided into two 3-symbol subslots (Subslot 0 and Subslot 3 in the figure) and two 4-symbol subslots (Subslot 1 and Subslot 2 in the figure). The PUCCH resource with $S + L = 4$ can only be used for Subslot 1 and Subslot 2, but not for Subslot 0 and Subslot 3. Therefore if it is desirable to fully use the resources in the PUCCH resource set, the different PUCCH resource sets should be separately configured for 3-symbol subslots and 4-symbol subslots.

Due to the complexity caused by the dependency between resource set configurations and subslot lengths, RAN1 finally decided to adopt a relatively simple design: Only 2-symbol subslot and 7-symbol subslot are supported, while the unequal-length subslot structure in a slot is not supported. For a given subslot structure, the same PUCCH resource sets are configured for all subslots. In other words, the subslot PUCCH resources are "subslot-common."

15.2.2 Multiple HARQ−ACK codebooks

The aim of constructing the HARQ−ACK codebook is to improve the HARQ−ACK feedback efficiency by multiplexing the HARQ−ACKs of multiple PDSCHs. Compared to the eMBB service, the URLLC service requires higher reliability and lower latency, but does not pursue higher spectrum efficiency and user capacity. To be more specific, URLLC

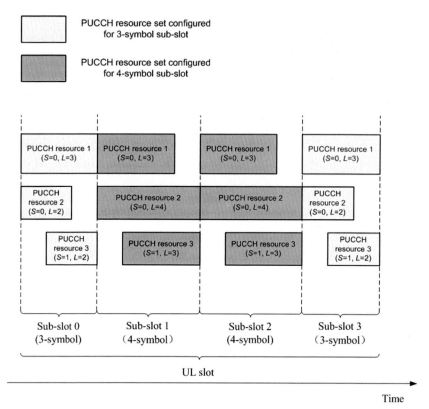

Figure 15.6 Different PUCCH resource sets should be configured for subslots with different lengths.

PDSCH requires fast feedback and multiple feedback opportunities in a slot, while it is normally good enough for eMBB HARQ—ACK to feedback only once in a slot. If the two kinds of HARQ—ACKs are multiplexed into one codebook, even if the URLLC HARQ—ACK is subslot-based, it has to wait to be multiplexed with eMBB HARQ—ACK before transmission. Thus the latency performance cannot be guaranteed. Regarding the HARQ—ACK reliability, the URLLC HARQ—ACK codebook in a codebook construction window can usually maintain a relatively small size to achieve the high transmission reliability. If URLLC HARQ—ACK must be multiplexed with eMBB HARQ—ACK in the same codebook, the large payload of the multiplexed HARQ—ACK codebook will impact the transmission reliability. Therefore R16 did not pursue to multiplex HARQ—ACKs of both URLLC PDSCH and eMBB PDSCH into one HARQ—ACK codebook.

Consequently, another enhancement for R16 URLLC PUCCH is to support the parallel constructions of multiple HARQ−ACK codebooks for services with different priorities (refer to Section 15.2.3 for the physical (PHY) priority indication). To simplify the design, R16 NR supports construction of two HARQ−ACK codebooks for the two service priorities: high priority for URLLC and low priority for eMBB. Then two problems need to be solved: The first is how to associate a PDSCH and its corresponding HARQ−ACK codebook with one of the two priorities; the second is which resources and configurations are used to construct the two codebooks.

The first problem is similar to the PDSCH grouping described in Section 15.2.1, which is eventually solved through the priority indication in DCI (refer to Section 15.2.3 for details). The second problem is related to the parameters configured for different types of PUCCHs. In R15 NR, one uplink bandwidth part (UL BWP) is configured with only one set of PUCCH parameters (as in *PUCCH-config*). When it comes to URLLC in R16, in order to enable the separate optimizations for different service types, some parameters in *PUCCH-config* should be separately configured for eMBB and URLLC, including but not limited to:

- Subslot configuration: a selection from 2-symbol subslot, 7-symbol subslot, and 14-symbol slot.
- K_1 set: the candidate timing intervals between the PDSCH and the corresponding HARQ−ACK.
- PUCCH resource set: time-frequency resources used for PUCCH conveying HARQ−ACK feedback in a slot or a subslot.

Because the subslot structure is introduced for URLLC service, it is natural to allow that eMBB HARQ−ACK codebook and URLLC HARQ−ACK codebook adopt different subslot configurations. With different lengths of subslots and different indication units of K_1 (e.g., slot or subslot), it is unreasonable to share the same K_1 sets between URLLC and eMBB. Separate configurations for K_1 sets should be supported. Similarly, due to different subslot configurations, PUCCH resources should also be configured separately. As described in Section 15.2.1, given PUCCH resources are confined in subslot, different PUCCH resource sets should even be configured for two subslots with different lengths, not to mention that different PUCCH resource sets should be configured for subslot-based PUCCH and slot-based PUCCH. The starting symbol and duration of PUCCH should be configured separately for the two service kinds to realize the separate optimization for URLLC and eMBB PUCCH resource allocation. Other parameters, such as spatial transmission parameter and power control parameter, are also suitable to be separately optimized. Therefore it was finally decided in RAN1 that all HARQ−ACK-related parameters in *PUCCH-Config* can be separately configured for URLLC and eMBB. It should be noted that, although the two sets of *PUCCH-Config* parameters are motivated for eMBB and URLLC, respectively, the specification does not explicitly stamp a set of *PUCCH-Config* parameters with a title of either URLLC or eMBB.

In addition, to maintain flexibility of the standard, the following combinations in *PUCCH-Config* configurations are all supported:

- Among the two *PUCCH-Configs*, one is slot-based and another is subslot-based. This configuration is suitable for multifunctional UEs (e.g., a UE running eMBB service as well as a URLLC service).
- The two *PUCCH-Configs* are both subslot-based. This configuration is suitable for a UE running different types of URLLC services.
- The two *PUCCH-Configs* are both slot-based. This configuration is suitable for eMBB UE running different types of eMBB services.

It should be noted that not only URLLC services, but also some eMBB services [such as virtual reality (VR) and augmented reality (AR) applications], require low transmission latency. In this sense, subslot-based HARQ−ACK feedback can also be used for eMBB services.

15.2.3 Priority indication

Taking into account the reliability and latency requirement of URLLC service and system efficiency, 3GPP RAN1 determines that upto two HARQ−ACK codebooks can be simultaneously constructed for supporting different service types for a UE, and a HARQ−ACK codebook can be identified based on PHY indication or PHY property. The following physical layer methods were proposed for identifying the HARQ−ACK codebook used to transmit the ACK/NACK corresponding to a PDSCH scheduled by a DCI indicating the dynamic scheduling or a DCI indicating Semipersistent scheduling (SPS) release:

- Option 1: A HARQ−ACK codebook is identified based on the DCI format.

  For R16 URLLC, compact DCI format is supported to improve both the reliability and efficiency of PDCCH. Therefore the DCI formats can be used to implicitly distinguish the HARQ−ACK codebook: DCI format 0_1/1_1 corresponds to one HARQ−ACK codebook, and the new DCI formats correspond to another HARQ—ACK codebook.

The main disadvantage of this option is that it has great restrictions on scheduling. R15 DCI format 0_0/0_1/1_0/1_1 can only be used to schedule eMBB service, and the new DCI formats can only be used to schedule URLLC service. In addition, if different services are mandated to use different DCI formats for scheduling, the UE needs to blindly decode more DCI formats and the maximum number of blind decoding will increase.

- Option 2: A HARQ—ACK codebook is identified based on the RNTI used for scrambling the DCI CRC.

In R15, two MCS tables are supported for the data transmission. One of the two tables is designed to include some lower coding rates and to ensure the reliability of URLLC service. If this MCS table is used for transmission, a specific MCS—C—RNTI is used for scrambling the DCI CRC, as described in Section 15.4.1. Because the introduction of MCS—C—RNTI is to satisfy the higher reliability requirement of URLLC service, it is a straightforward solution to use the MCS—C—RNTI to identify the HARQ—ACK codebook for URLLC service.

The disadvantage of this option is the restriction on scheduling. The PDSCH corresponding to the given HARQ—ACK codebook has to use the MCS table with lower coding rates, but not be able to use another MCS table.

- Option 3: A HARQ—ACK codebook is identified based on an explicit indication in the DCI.

The option can be further categorized into using the existing information field or new information field to indicate the HARQ—ACK codebook:

- Option 3—1: PDSCH duration or mapping type is used to identify the HARQ—ACK codebook. For example, if the duration of PDSCH for URLLC transmission is less than a given length or PDSCH mapping type is indicated as Type B, the corresponding HARQ—ACK codebook is identified for URLLC service.
- Option 3—2: HARQ process ID is used to identify the HARQ—ACK codebook. This requires separate sets of HARQ process ID's be used for eMBB and URLLC services.
- Option 3—3: The K_1 entries in the set of PDSCH-to-HARQ feedback timing are used to identify a HARQ—ACK codebook. In

particular, the URLLC service uses the large K_1 entries in the set and eMBB service uses the small K_1 entries in the set.

- Option 3−4: An explicit field in DCI is used to identify the HARQ−ACK codebook.

In essence, Option 3−1 ∼ Option 3−3 introduce scheduling restrictions to eMBB transmission, which will lead to the performance loss of eMBB transmission. In the RAN1 discussion, many companies did not agree with leaving restrictions to eMBB transmission. The disadvantage of Option 3−4 is the increase of DCI overhead. However, given only two HARQ−ACK codebooks are supported, only one bit is added into the DCI. The increased overhead is fairly acceptable.

- Option 4: A HARQ−ACK codebook is identified based on the CORESET or search space used to transmit the DCI.

If the CORESET or search space is independently configured for eMBB and URLLC, the HARQ−ACK codebook can be identified based on the CORESET or search space. However, using independent CORESET or search space to schedule different services increases the probability of PDCCH blocking and the number of PDCCH blind decoding. Furthermore, it also introduces a search space collision issue, which increases complexity of detection.

According to the above analysis, all options except Option 3−4 introduce scheduling restrictions, and lead to either the performance loss of eMBB transmission or the complexity increase of PDCCH blind decoding. In comparison, Option 3−4 does not introduce any scheduling restrictions and the corresponding DCI overhead increases by only one bit, which has little impact on the detection performance. Therefore it was determined that a 1-bit information field can be configured in the DCI to indicate the priority information of the corresponding HARQ−ACK codebook. In addition, the priority information is also used to solve the collision between ACK/NACK feedback and other uplink transmissions. Accordingly, other uplink transmissions should also support priority indication, including:

- For dynamic PUSCH, the priority is indicated by an explicit field in the DCI.
- For the PUSCH configured by high layer signaling or SR, or HARQ−ACK codebook for SPS PDSCH, the priority is indicated by high layer signaling.
- The priority of CSI on PUCCH is low.
- The priority of CSI on PUSCH depends on the priority of the PUSCH.

15.2.4 Intra-UE collision of uplink channels

In NR R15, when multiple uplink channels overlap in time, all uplink information is multiplexed in one uplink channel. The two main factors affecting the multiplexing transmission are the formats of overlapping PUCCHs and the processing timeline for multiplexing.

At the early stage of NR R16 discussion, in order to improve the reliability and reduce the latency of URLLC transmission, potential solutions, shown in Table 15.1, were provided for different transmission overlapping cases. This section focuses on physical layer solutions for handling intra-UE collision between uplink transmissions for different services, while the high layer solutions are described in Chapter 16, URLLC in high layer.

In order to support multiple HARQ–ACK codebooks (as described in Section 15.2.1), RAN1 agreed to support two-level priorities, high priority and low priority, in the physical layer. Furthermore, for intra-UE collision handling in the physical layer, in case a high-priority UL transmission overlaps with a low-priority UL transmission, the low-priority UL transmission is dropped with certain constraint. The constraint is mainly considered for the processing time needed to cancel low-priority uplink transmission and to prepare high-priority uplink transmission. Because the UE needs to determine the cancellation before the preparation, additional time compared to R15 PUSCH preparation time should be added. Specifically, the gap between the ending of PDCCH with high priority and the starting of corresponding PUSCH is no less than $T_{proc,2} + d1$, where $T_{proc,2}$ is R15 PUSCH preparation procedure time, and the value of $d1$, reported as a UE capability, can be 0, 1 or 2. When the above constraint is satisfied, the UE cancels the low-priority transmission before the first overlapping symbol. In addition, in order to reduce the complexity of UE implementation, it was agreed that when a high-priority UL transmission overlaps with a low-priority UL transmission in a slot, the UE is not expected to transmit the low-priority transmission in the nonoverlapping symbols.

When the above conclusion is extended to resolve the collision of multiple uplink channels with different priorities, in order to not introduce additional processing procedure to the UE, the following steps are used to determine the uplink channel to be transmitted:

- Step 1: Resolve collision between UL transmissions with same priority based on R15 mechanism.
- Step 2: Resolve collision between UL transmissions with different priorities.

Table 15.1 Solutions for the overlapping uplink transmissions.

	URLLC SR	URLLC HARQ–ACK	URLLC PUSCH
URLLC HARQ–ACK	Reuse R15 mechanism	—	—
URLLC PUSCH	SR (non–BSR) is directly multiplexed in PUSCH.	Reuse R15 mechanism	—
CSI	Option 1: Drop CSI. Option 2: If certain conditions are satisfied, CSI is multiplexed with URLLC UCI or into URLLC PUSCH; otherwise, drop CSI. The conditions include timing, latency, reliability, the priority of CSI resource, etc.		
eMBB SR	Option 1: Drop SR with low priority. Option 2: It is upto UE implementation.	Option 1: Drop eMBB SR. Option 2: If certain conditions are satisfied, UCIs are multiplexed; otherwise, drop eMBB SR. The conditions include: timing, latency, reliability, PUCCH format, etc.	Option 1: Drop eMBB SR. Option 2: If certain conditions are satisfied, eMBB SR is multiplexed into PUSCH; otherwise, drop eMBB SR. The conditions include timing, and whether UL–SCH is included in the PUSCH.
eMBB HARQ–ACK	Option 1: Drop eMBB HARQ–ACK. Option 2: If certain conditions are satisfied, UCIs are multiplexed; otherwise, drop eMBB HARQ–ACK. The conditions include: timing, latency, reliability, PUCCH format, dynamic indication, etc.		Option 1: Expand the range of beta values (i.e., the value of beta can be less than 1). When the value of beta is 0, eMBB HARQ–ACK is not multiplexed in the PUSCH. Option 2: Drop eMBB HARQ–ACK.
eMBB PUSCH	Option 1: When URLLC SR is positive, eMBB PUSCH is dropped. Option 2: If the requirements of latency and reliability are satisfied, SR is multiplexed in eMBB PUSCH; otherwise, drop eMBB PUSCH.	Option 1: Drop eMBB PUSCH. Option 2: If the requirements of latency and reliability are satisfied, HARQ–ACK is multiplexed in eMBB PUSCH; otherwise, drop eMBB PUSCH.	Option1: Drop eMBB PUSCH. Option 2: Leave the prioritization to higher layer protocol, which is in 3GPP RAN2 study scope.

15.3 UE processing capability enhancements

15.3.1 Introduce of processing capacity

In order to support multiple services and scenarios, the NR system introduces flexible scheduling mechanism, such as flexible time-domain resource allocation and flexible scheduling schedule. At the same time, the corresponding processing capacity is defined in the standard to ensure that the scheduling timing indicated by the base station can give the terminal enough time to process data. The R15 standard defines two processing capabilities for ultralow delay requirements and conventional delay requirements. The terminal supports UE capability 1 by default to meet conventional delay requirements, while UE capability 2 needs to be reported to meet ultralow delay requirements. The terminal report mode is adopted for the fast processing capability, because the fast processing capability needs more advanced chip technology, and the cost and complexity of the chip are also increased correspondingly. Although the two processing capabilities are introduced by taking into account the requirements of multiple services, such as URLLC and eMBB, the corresponding relationship between processing capabilities and service requirements is not explicitly defined by the specification. Therefore the content of this section is not limited to URLLC.

At present, the processing capacity defined in the standard includes [8]:

- The PDSCH processing time N_1 is used to define the minimum processing time requirement between the end of the last symbol of PDSCH and the starting of the first symbol of PUCCH carrying the corresponding HARQ−ACK.
- PUSCH processing time N_2, which is used to define the minimum processing time requirement between the end of the last symbol of PDCCH and the starting of the first symbol of PUSCH that is granted by the PDCCH.

15.3.2 Processing time determination

Processing time is determined based on typical processing framework analysis. Fig. 15.7 is a typical downlink data processing model [9]. In order to reduce processing time, some functional modules can be processed in parallel, which is greatly affected by physical layer signal design. To be more specific, the left figure shows downlink data processing based on front-load pilot only, and the right figure shows downlink data processing based on both front-load pilot and additional pilot. For the left figure,

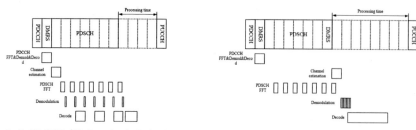

Left: PDSCH with front-load pilot only   Right: PDSCH with front-load and additional pilot

Figure 15.7 Downlink data processing. *Left*: PDSCH with front-load pilot only. *Right*: PDSCH with front-load and additional pilot.

data demodulation–decoding and data receiving are in parallel. Therefore only a small amount of data processing time is additionally needed after data receiving is completed. For the right figure, due to the additional pilot, data demodulation–decoding starts after data reception is completed, so a longer data processing time is needed. It can be seen that processing time is greatly affected by pilot position. Similarly, data mapping also affects data processing time. For frequency-first mapping mode, the terminal can carry out code block (CB)-level detection after receiving a small number of symbols. For interleaved mapping mode, the terminal reconstructs a complete CB and decodes it after all symbols are received.

There are still many other influencing factors (e.g., data bandwidth, data volume, PDCCH blind detection, SCS configuration, etc.). They are put into an evaluation hypothesis, as shown in Table 15.2 [10] where the candidate factors listed in brackets can be optionally considered for (N_1, N_2). With this evaluation hypothesis being agreed on in RAN1, each company reported its own experience values for N_1 and N_2.

The processing capacity defined in specification is based on above transmission assumptions. For cases beyond above conditions, special procedure is required to reduce UE complexity, especially when processing capacity 2 is reported. When the following two conditions are met, the terminal falls back to processing capacity 1:

- For UE processing capability 2 with scheduling limitation when $\mu_{PDSCH} = 1$, if the scheduled RB allocation exceeds 136 RBs, the UE defaults to capability 1 processing time. The UE may skip decoding a number of PDSCHs with the last symbol within 10 symbols before the start of a PDSCH that is scheduled to follow Capability 2, if any of those PDSCHs are scheduled with more than 136 RBs with 30 kHz SCS and following Capability 1 processing time.

Table 15.2 Candidate factors for UE processing time (N_1, N_2)

	N_1	N_2
Nominal assumptions	Single carrier/Single BWP/Single TRP • Full range of MCS and multilayer support upto the four-layer MIMO and 256–QAM • Upto 3300 active subcarriers PDCCH • Same numerology/BWP as PDSCH • Single grant monitored for PDSCH • 44 blind decodes, single symbol CORESET PDSCH • PDSCH does not precede PDCCH • 14-symbol slot-based scheduling • Frequency-first RE-mapping, no time-interleaving of CBs across TB PUCCH • Short formats for HARQ–ACK	Single carrier/Single BWP/Single TRP • Full range of MCS and multilayer support upto the two-layer MIMO and 64–QAM • Upto 3300 active subcarriers PDCCH • Same numerology/BWP as PUSCH • Single grant monitored for PUSCH • 44 blind decoding, single symbol CORESET PUSCH • 14-symbol slot-based scheduling • No time-interleaving of CBs across TB • DFTs OFDM or OFDM • Front-loaded DMRS for low latency • No UCI multiplexing
Candidate factors	• SCS • DMRS configuration • [Percentage of peak rate] • [RE-mapping]	• SCS • RE-mapping (depending on specification) • [Percentage of peak rate]

- Additional demodulation reference signal (DMRs additional position ≠ pos0) is configured.

15.3.3 Definition of processing time

As described in Chapter 5, 5G Flexible Scheduling, the NR supports flexible scheduling configurations. In order to ensure enough processing time for terminal, the flexibility of scheduling configuration needs to be subject to certain timing conditions. The timing conditions for PDCCH-to-PUSCH and PDSCH-to-PUCCH are defined in the standard. This section describes the corresponding rules in detail.

As shown in Fig. 15.8, the terminal receives in slot n a PDCCH that grants PUSCH transmission in slot $n + K_2$, where K_2 is either indicated by DCI or configured by high layer parameters. Because the terminal needs to demodulate PDCCH and then to prepare PUSCH, the start time of PUSCH determined based on K_2 should not be earlier than the first uplink symbol that is $T_{proc,2} = \max\left((N_2 + d_{2,1})(2048 + 144) \times \kappa 2^{-\mu} \cdot T_C, d_{2,2}\right)$ after the end of corresponding PDCCH. In the equation of $T_{proc,2}$, N_2 is determined according to terminal capability and subcarrier spacing. If the first OFDM symbol in PUSCH only sends DMRS, then $d_{2,1} = 0$; otherwise, $d_{2,1} = 1$. If the scheduling DCI triggers the BWP handover, $d_{2,2}$ is equal to the handover time; otherwise, $d_{2,2}$ is 0.

After receiving a PDSCH, the terminal needs to send feedback information to gNB to inform gNB whether the PDSCH is received correctly. As shown in Fig. 15.9, once the terminal receives PDSCH in slot n, it sends feedback information in slot $n + k$, where k is PDSCH-to-HARQ feedback timing indicated in DCI or configured by high layer parameters. With the processing time for PDCCH and PDSCH receptions being taken into account, the value of k should be determined in such a way that the starting symbol of PUCCH-carrying HARQ−ACK is no earlier than the first uplink symbol that is $T_{proc,1} = \left(N_1 + d_{1,1}\right)(2048 + 144) \times \kappa 2^{-\mu} \times T_C$ after

Figure 15.8 Processing timing for PUSCH.

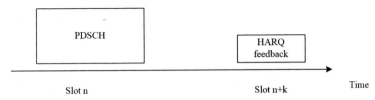

Figure 15.9 Processing timing for PDSCH.

the end of PDSCH, where N_1 is determined according to terminal capability and the subcarrier spacing, and $d_{1,1}$ is related to the capability level of the terminal and the resources of PDSCH.

15.3.4 Out-of-order scheduling/HARQ

NR R15 supports flexible scheduling, as described in Chapter 5, 5G Flexible Scheduling. In order to reduce the complexity of UE implementation, strict rules on scheduling/HARQ order are defined for data processing within a carrier, including:
- Scheduling/HARQ order for PDSCH
  - The UE is not expected to receive two overlapped PDSCHs.
  - The UE is not expected to receive another PDSCH for a given HARQ process until after the end of the expected transmission of HARQ–ACK for that HARQ process. As shown in Fig. 15.10, the UE does not expect to receive PDSCH #2, which belongs to the same HARQ process as PDSCH #1, since the transmission of ACK/NACK corresponding to PDSCH #1 has not yet finished.
  - The UE is not expected to receive a first PDSCH in slot i, with the corresponding HARQ–ACK assigned to be transmitted in slot j, and a second PDSCH starting later than the first PDSCH with its corresponding HARQ–ACK assigned to be transmitted in a slot earlier than slot j, as shown in Fig. 15.11.
  - If the UE is scheduled to receive a first PDSCH by a PDCCH ending in symbol i, the UE is not expected to be scheduled to receive a PDSCH starting earlier than the end of the first PDSCH by a PDCCH that ends later than symbol i, as shown in Fig. 15.12.
- Scheduling order for PUSCH
  - The UE is not expected to transmit two overlapped PUSCHs.
  - If the UE is granted to start a first PUSCH transmission by a PDCCH ending in symbol i, the UE is not expected to be granted to transmit a PUSCH starting earlier than the end of the first

PUSCH by a PDCCH that ends later than symbol i, as shown in Fig. 15.12.

At the beginning of NR R16 discussion, in order to better satisfy the latency requirement of URLLC transmission, out-of-order scheduling/ HARQ has been proposed to allow the URLLC data that is scheduled/ granted later being transmitted or acknowledged earlier, due to common sense being that the channel emerging later but with higher priority should be transmitted without waiting for the completeness of channel transmission

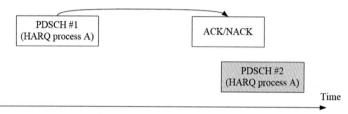

Figure 15.10 Scheduling/HARQ order for PDSCH.

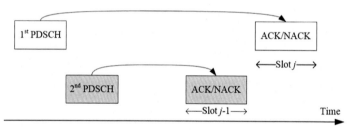

Figure 15.11 HARQ order for PDSCH.

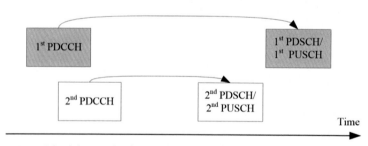

Figure 15.12 Scheduling order for PDSCH/PUSCH

that emerges earlier but with lower priority. But the main controversial debate was how to deal with the earlier channel with low priority:

- Option 1: The UE always processes the later channels with high priority. The UE may or may not drop the processing of the earlier channel with low priority.
- Option 2: The UE processes both the earlier channel and later channel as a UE capability with no exception.
- Option 3: The UE processes both the earlier channel and later channel under some conditions (e.g., using the CA capability). If the conditions are not satisfied, the UE behavior is not defined.
- Option 4: A UE drops the processing of the earlier channel with low priority.
- Option 4−1: The UE always drops the earlier channel with low priority.
- Option 4−2: Some scheduling conditions should be defined (e.g., based on the number of RBs, TBS, number of layers, the gap between the earlier and later channels, etc.). The UE drops the processing of the earlier channel based on the testing of these conditions.

The implementations corresponding to the above options can be roughly divided into two categories:

1. The UE uses one pipeline to process the two channels with different priorities. Thus, the UE needs to have the capability to cancel the processing of the low-priority channel and switch the processing to the high-priority channel.
2. The UE uses two pipelines to process the two channels with different priorities.

Different implementations have different impacts on product research and development. Because no consensus can be reached by the companies in RAN1, out-of-order scheduling/HARQ is not supported in R16. Therefore R15 rules on scheduling/HARQ order are maintained.

15.4 Data transmission enhancements

15.4.1 CQI and MCS

Usually, the appropriate modulation and coding scheme is configured to adapt to channel variation and to meet reliability requirement. A normal MCS table designed for eMBB cannot satisfy the reliability requirement for URLLC transmission. In a later stage of R15, MCS and CQI mechanisms were optimized for the URLLC scenario.

15.4.1.1 CQI and MCS table design

A CQI table with ultralow code rate is designed to meet the high-reliability requirements of URLLC. The reliability requirement of URLLC is 99.999%, and the corresponding target BLER is 10^{-5}.

In addition to adding the low code rate, CQI table design also needs to consider CQI indication accuracy and overhead to reach tradeoff between system resource efficiency and CQI signaling overhead. Due to specification workload, CQI table design for URLLC intended to reuse existing CQI value with small modifications, including adding some new CQI values with lower code rate and removing some existing CQI values with higher code rate.

MCS table design is similar to CQI table design, which is mainly consistent with the spectrum efficiency of the CQI table. In the existing MCS table, MCS entries corresponding to low spectral efficiency are supplemented, and MCS entries corresponding to high spectral efficiency are removed.

15.4.1.2 Configuration for CQI and MCS tables

The NR system contains multiple CQI and MCS tables. The CQI table configuration is determined by cqi-Table in *CSI-ReportConfig* and target BLER. For the MCS table, due to dynamic arrival of URLLC and eMBB traffic, some dynamic indication methods are proposed for the MCS table. However, because a MCS table covers a large range of spectrum efficiency, and channel environment changes continuously, some companies believe that the MCS table should be configured semistatically.

For semipersistent transmission modes, such as configured grant and SPS transmission, since transmission parameters are configured through RRC signaling, it is natural to configure MCS tables by RRC signaling as well.

For dynamic transmission, most transmission parameters can be indicated dynamically. Various dynamic MCS table indication schemes were fully discussed and converged to the following two schemes:
- Scheme 1: MCS table is indicated by searching space and/or DCI format
  - Scheme 1−1: if the new MCS table is configured, DCI format 0_0/1_0 in the common search space refers to the legacy MCS table and DCI format 0_0/0_1/1_0/1_1 in UE-specific search space refers to the new MCS table; otherwise, the legacy MCS scheme is always adopted.

- **Scheme 1−2**: if the new MCS table is configured, DCI format 0_0/ 1_0 in the common search space refers to legacy MCS table, DCI format 0_0/1_0 in UE-specific search space refers to a first MCS table and DCI format 0_1/1_1 in UE-specific search space refers to a second MCS table, where the first MCS table and the second MCS table are configured by high layer signaling and can be individually configured as either new MCS table or legacy MCS table; otherwise, the legacy MCS scheme is always adopted.
- Scheme 2: MCS is indicated by RNTI
  - If a new MCS-related RNTI is configured, the MCS table is indicated by RNTI; otherwise, the legacy MCS scheme is always adopted.

Scheme 1−1 has the advantage of no impact on false-alarm probability and RNTI space, but suffers the restrictions on scheduling flexibility. In addition, because the MCS table is configured with RRC signaling and RRC (re)configuration may depend on using DCI format 0_0/1_0, there could be an ambiguity issue during RRC reconfiguration. Scheme 1−2 further supports indication of MCS table configuration based on Scheme 1−1. When two DCI formats are configured, MCS table configuration can be switched dynamically, but the PDCCH blind detection budget may be affected due to an increase of monitored DCI formats. When only one DCI format is configured, only one MCS table configuration can be supported, which still causes the restriction on scheduling flexibility. Scheme 2 supports MCS table indication based on RNTI, and can quickly switch MCS table configuration. But this method consumes RNTI values and increases false-alarm probability in PDCCH detection. However, the influence of false-alarm probability can be controlled by appropriate search space configuration. The above schemes have their own advantages and disadvantages. Finally, 3GPP RAN1 integrated the two schemes and agreed on the following scheme:

- If the new MCS-related RNTI is not configured, semistatic configuration of the new MCS table is supported by extending RRC parameter for MCS table. When the new MCS table is configured, the DCI format 0_0/1_0 in the common search space refers to the legacy MCS table, and DCI format 0_0/0_1/1_0/1_1 in the UE-specific search space refers to the new MCS table.
- If the new MCS-related RNTI is configured, the MCS table is indicated by RNTI. Specifically, if DCI is scrambled by the new RNTI, the new MCS table is used; otherwise, the legacy MCS table is used. The above scheme is also applicable to DCI format 0_2/1_2.

15.4.2 PUSCH enhancement

In R15, uplink transmission delay is reduced by optimizing transmission procedure (e.g., introducing a configured grant transmission). However, there are still some limitations in uplink transmission, which affect the scheduling delay:

- One uplink transmission cannot across different slots.
- Data transmission of the same HARQ process needs to meet certain scheduling timing requirements.
- The transmission repetition is based on slot unit.

In order to reduce the delay caused by these limitations, R16 further enhanced uplink transmission by introducing the back-to-back repetition mechanism, which has the following characteristics:

- The adjacent repetition transmission resources are back-to-back in the time domain.
- The resources for one grant can cross slots in the time domain. In this way, it can ensure that the services reached at the back of the slot can be scheduled with enough resources in real time.
- The number of repetitions is indicated dynamically to adapt to dynamic changes of traffic and channel environment.

Three kinds of back-to-back transmission schemes were discussed.

- Scheme 1: Mini-slot repetition scheme

The time-domain resource allocation indicates time-domain resource for the first repetition transmission, and the time-domain resources of the remaining repetition transmissions are determined according to the time-domain resources of the first repetition transmission and uplink/downlink allocation configuration. Each repetition transmission contains contiguous symbols (Fig. 15.13).

- Scheme 2: Auto-splitted scheme

The time-domain resource allocation indicates total resources of all repetition transmissions. Based on uplink/downlink allocation configuration, slot boundary, and other information, the above resources are split automatically. Each repetition transmission contains contiguous symbols (Fig. 15.14).

- Scheme 3: Multiple grant scheme

Multiple UL grants schedule multiple uplink transmissions, respectively. Multiple uplink transmissions are continuous in the time domain. UE is allowed to transmit the i-th granted uplink transmission before the end of the (i-1)-th granted uplink transmission (Fig. 15.15).

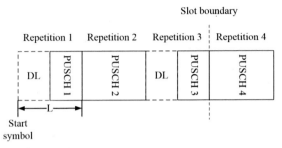

Figure 15.13 Mini-slot repetition.

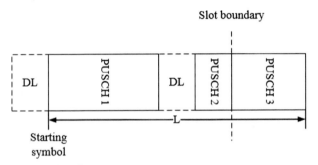

Figure 15.14 Auto-split scheme.

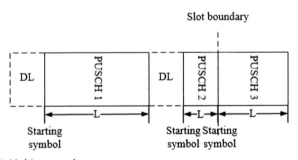

Figure 15.15 Multigrant scheme.

Scheme 1 has the simple resource indication; but in order to adapt to the slot boundary, the length of each repeat transmission needs to be carefully trimmed. The increasing number of repetition can lead to the increase of reference signal overhead. In Scheme 2, the resource indication is also simple, but the length differences among the repetition transmissions can be large, which increases the complexity in PUSCH demodulation and decoding. In Scheme 3, resource indication is flexible and accurate, but control signaling overhead is the largest among three schemes. 3GPP RAN1 firstly excluded Scheme 3 due to its high overhead of control signaling, and then combined Scheme 1 and Scheme 2 to reach a final solution: the time-domain resource is determined according to TDRA; when the slot boundary or unavailable symbols are encountered, the time-domain resource of PUSCH is split.

The enhanced time-domain resource indication of uplink transmission follows R15 time-domain resource indication framework (i.e., the high layer signaling configures multiple time-domain resource locations, and the physical layer signaling indicates which one is used). Further, the SLIV indication was enhanced with an additional field indicating the number of repetitions.

The above uplink repetition transmission is called type B PUSCH repetition. R16 not only introduces type B PUSCH repetition, but also enhances the slot-level repetition transmission of R15, which is called type A PUSCH repetition (i.e., the number of repetition transmissions can be dynamically indicated). The type of PUSCH repetition is determined by the high layer configuration.

15.4.3 Time-domain resource determination

The time-domain resource indication determines the total resource range for repetition transmission, but there can be some unavailable symbols within the range (e.g., the downlink symbols, the symbols used for periodic uplink detection signal transmission, etc.). Therefore the available uplink transmission resources need to be further clarified. In R16 uplink transmission enhancement, there are two time-domain resource definitions:

- Nominal PUSCH repetition: this is indicated by TDRA directly. The nominal length of each PUSCH repetition is the same. Nominal PUSCH repetition is used to determine TBS, uplink power control, and UCI multiplexing resources.
- Actual PUSCH repetition: this is what remains available after removal of the unavailable symbols from the nominal PUSCH repetition. The lengths of different actual PUSCH repetitions are not always the same. The actual PUSCH repetition is used to determine DMRS symbol, actual transmission rate, RV, and UCI multiplexing resources.

15.4.4 Frequency hopping

In general, uplink transmission obtains frequency diversity gain by frequency hopping. In R15, uplink repetition transmission is based on slot repetition, so interslot frequency hopping and intraslot frequency hopping are used to obtain frequency diversity gain. For back-to-back repetition transmission in R16, the repetition granularity becomes finer, and four frequency hopping schemes were discussed:

- Scheme 1: interslot frequency hopping.
- Scheme 2: intraslot frequency hopping.
- Scheme 3: frequency hopping between PUSCH repetitions, transmission repetition can be nominal PUSCH repetition or actual PUSCH repetition.
- Scheme 4: frequency hopping within each PUSCH repetition, transmission repetition can be nominal PUSCH repetition or actual PUSCH repetition.

The frequency hopping scheme is designed to mainly pursue the equal frequency diversity gain for the uplink transmission resources, which means the lengths of time-domain resources corresponding to the two frequency-domain resources should be as close to each other as possible. In addition, the design should avoid too much pilot overhead due to frequent hopping. At last, 3GPP RAN1 adopted interslot frequency hopping and nominal PUSCH repetition-based frequency hopping.

15.4.5 UCI multiplexing

UCI multiplexed in PUSCH is enhanced for determination of the multiplexing timeline and the PUSCH repetition used for UCI multiplexing.

15.4.5.1 Multiplexing timing

Multiplexing timing is defined to ensure the terminal has enough time to multiplex UCI in PUSCH. There are two main schemes for multiplexing timing determination (Figs. 15.16 and 15.17):

- Scheme 1: Multiplexing timeline ends at the starting symbol of the first actual PUSCH repetition overlapping with the uplink control channel.
- Scheme 2: Multiplexing timeline ends at the starting symbol of the i-th actual PUSCH repetition overlapping with the uplink control channel, where the interval between the starting symbol of the i-th actual PUSCH repetition and the last corresponding DCI is not less

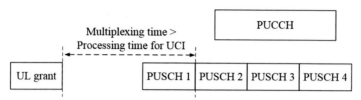

Figure 15.16 UCI multiplexing Scheme 1.

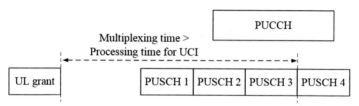

Figure 15.17 UCI multiplexing Scheme 2.

than the processing time of the uplink control information multiplexing, and i is less than or equal to the maximum number of the actual PUSCH repetitions.

Scheme 1 follows the design logic of R15, and Scheme 2 allows the uplink control information to be multiplexed to the first overlapped PUSCH repetition satisfying the processing time requirement. Scheme 2 offers more flexibility and smaller delay for uplink control information transmission and uplink data transmission with repetition. However, in Scheme 2, a specific PUSCH repetition needs to be selected, which increases the terminal complexity, and TA estimation error may lead to misunderstanding on the selected PUSCH repetition between base station and the terminal. Therefore Scheme 1 was adopted.

15.4.5.2 PUSCH repetition used for UCI multiplexing

- Scheme 1: The UCI multiplexing occurs in the first actual PUSCH repetition overlapping with uplink control channel. As shown in Fig. 15.18, the second actual PUSCH repetition is the first actual PUSCH repetition overlapping with PUCCH, and therefore is the PUSCH repetition on which the uplink control information is multiplexed. However, there could be a chance for the resources of the second actual PUSCH repetition to be too small to carry all of the uplink control information.

- Scheme 2: The UCI multiplexing occurs in the earliest PUSCH repetition having enough multiplexing resource and meanwhile overlapping with uplink control channel. As shown in Fig. 15.19, the third actual PUSCH repetition is the first one overlapping with PUCCH and meanwhile being able to carry all uplink control information. The uplink control information is then multiplexed on the third actual PUSCH repetition.
- Scheme 3: The UCI multiplexing occurs in the earliest actual PUSCH repetition having the largest number of symbols and meanwhile overlapping with the uplink control channel. As shown in Fig. 15.20, the fourth actual PUSCH repetition has the largest number of symbols among all four repetitions and meanwhile overlaps with PUCCH. The uplink control information is then multiplexed on the fourth PUSCH repetition.

Scheme 1 follows R15 design logic, which is simple for implementation and has low transmission delay for uplink control information, but

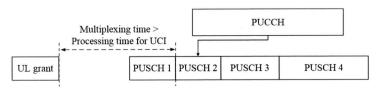

Figure 15.18 UCI multiplexed in PUSCH repetition (Scheme 1).

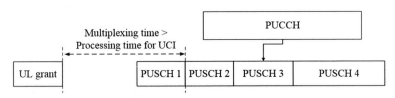

Figure 15.19 UCI multiplexed in PUSCH repetition (Scheme 2).

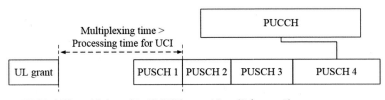

Figure 15.20 UCI multiplexed in PUSCH repetition (Scheme 3).

may suffer loss of uplink control information. Scheme 2 reduces the transmission delay of uplink control information as much as possible on the premise of ensuring the uplink control information is not lost. In Scheme 3, the resource selection strategy is simple, and the uplink control information transmission resource is maximized to avoid the loss of uplink control information, but the uplink control information transmission delay is uncontrollable. Eventually, Scheme 1 was adopted in the standard.

15.5 Configured grant transmission

The uplink service is normally initiated by the terminal, and the base station initiates a reasonable grant after knowing the uplink service requirement. Therefore the traditional uplink transmission process is relatively complex, and mainly includles (1) SR by terminal, (2) UL scheduling by base station to obtain buffer status, (3) buffer status report by terminal, (4) UL scheduling by base station for data transmission, and (5) UL transmission by terminal. Scheduling delay is inevitable in complex uplink data transmission processes. For URLLC, the scheduling delay reduces effective transmission time. Therefore in the NR phase, configured grant transmission technology is introduced to meet low-latency requirement of URLLC. The basic idea of configured grant transmission technology is that base station allocates uplink transmission resources for the terminal in advance, and the terminal can directly initiate uplink transmission on pre-allocated resources according to service requirements. There are some similarities between configured grant transmission technology and the SPS technology of LTE system (i.e., resources are all preconfigured). However, to meet the low-latency requirement of URLLC, configured grant transmission technology further optimizes physical layer design with flexible initial transmission occasion, resource allocation, and multiple sets of configured grant resource mechanism. See Chapter 16, URLLC in High Layer for details of high layer technology of configured grant.

15.5.1 Flexible initial transmission occasion

The transmission resources of configured grant transmission are preallocated, which although simplifies uplink transmission process, may result in low resource utilization. In order to avoid the collision and interference between uplink transmissions from different users, configured grant resources are not dynamically reclaimed for different users. However, for uncertain services, some preallocated resources are not necessarily used by the target terminal, which leads to resource waste. Generally, configured grant resources are determined by traffic arrival characteristics (e.g., periodicity and jitter) of uplink services and transmission delay requirements. Take periodicity as an example. For URLLC, long periodicity may lead to long waiting time for uncertain service or jitter cases. As shown in Fig. 15.21, for a service with a periodicity of 8 Ms and jitter range from − 1 to 1 Ms, the base station configures a configured grant resource with a period of 8 Ms (i.e., the transmission resource appears once every 8 Ms). Due to jitter of traffic arrival, the data can hardly arrive right before 2 Ms. As shown in Fig. 15.21, when the service arrives at 2.5 Ms, it needs to wait 7.5 Ms for the next resource. On the other hand, if periodicity is set short, system transmission resources will be wasted significantly. As shown in Fig. 15.22, for the same incoming traffic, in order to make sure the delay of waiting resources is within 1 Ms, the base station configures a configured grant resource with 1 Ms period. For a service with period of 8 Ms, configured grant resources reaches eight times the requirement, resulting in a lot of resource being wasted.

In order to balance transmission delay and system resource efficiency, the NR system introduces flexible initial transmission occasion on the basis of repetition transmission, which not only avoids excessive resource allocation, but also overcomes the long waiting time caused by service jitter. Specifically, the base station can configure configured grant resource periodicity according to service arrival periodicity. According to the demand of service jitter and delay, the base station can configure resources with a certain number of repetitions within a periodicity range, and allow a terminal to flexibly initiate transmission on these repetition resources.

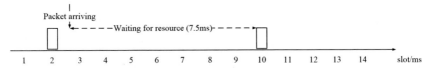

Figure 15.21 Traditional resource allocation 1.

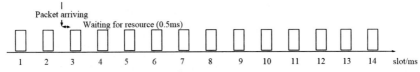

Figure 15.22 Traditional resource allocation 2.

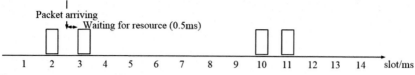

Figure 15.23 Resource allocation with flexible initial transmission occasion.

Even if traffic arrives with jitter, the terminal can initiate transmission in a small time range without waiting for resource in the next periodicity. In the same use case as the above example, as shown in Fig. 15.23, the base station configures a configured grant resource with a periodicity of 8 Ms and a repetition number of 2. When a service arrives at 2.5 Ms, it can be transmitted on the second repetition resource in the first periodicity. The resource waiting delay is 0.5 Ms, but the reserved resource is only two times the required resource. Compared with the traditional resource allocation method 2, the resource cost is reduced to 25%.

In addition, in order to avoid the problem where the base station cannot distinguish uplink data in multiple periodicity due to resource overlapping, repetition for one HARQ process is restricted to one SPS periodicity.

The flexible initial transmission occasion greatly reduces resource waiting delay. However, the flexible initial transmission occasion based on traditional RV sequence {0,2,3,1} may lead to incomplete systematic coded bit information due to possible missing the RV0 transmission.

In order to ensure the uplink transmission with flexible initial transmission occasion contains the complete systematic coded information, the NR system restricts the RV of the initial transmission to adopt rv0 only. At the same time, in order to support a flexible initial transmission occasion, two RV sequences, {0,0,0,0} and {0,3,0,3}, are introduced. In the case of RV sequence {0,0,0,0}, the terminal can send the initial transmission at any occasion within the K-repetition, as shown in Fig. 15.24A, except for the last transmission in case of $K \geq 8$ repetitions. For RV sequence {0,3,0,3}, the terminal can perform the initial transmission at any occasion with RV $= 0$, as shown in Fig. 15.24B. For RV sequence {0,2,3,1}, the terminal can only perform the initial transmission at the first transmission occasion within the K-repetition, as shown in Fig. 15.24C.

In R15, only one set of configured grant transmissions can be active in one BWP. In R16, multiple active sets of configured grant transmission are supported. For service traffic arriving at different times, the

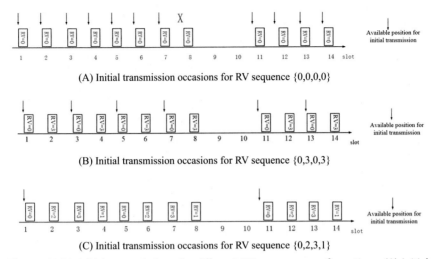

Figure 15.24 Initial transmissions for different RV sequence configurations. (A) Initial transmission occasions for RV sequence {0,0,0,0}, (B) Initial transmission occasions for RV sequence {0,3,0,3}, (C) Initial transmission occasions for RV sequence {0,2,3,1}.

corresponding set of configured grant transmission can be picked to match the service traffic.

15.5.2 Resource allocation configuration

There are two kinds of configured grant resource allocation: type 1 and type 2.

- For type 1 configured grant, all of the resource configuration parameters are configured by high layer signaling. Once the corresponding high layer configuration completes, the configured grant resources are activated.
- For type 2 configured grant, part of the resource parameters are configured by high layer signaling, and the other part of the parameters are indicated by downlink control information, which also activates the configured grant resources.

Compared with type 1 configured grant, type 2 configured grant provides flexibility of dynamic resource activation/deactivation and reconfiguration of some resource parameters. However, because it is activated by downlink control information, activation delay is introduced. Both configured grant types have their own advantages and disadvantages, which can fit to different service types. For example, type 1 can be used for delay-sensitive URLLC services, and type 2 can be used for voice over Internet Protocol services.

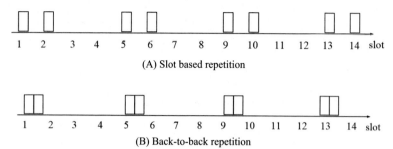

Figure 15.25 Repetition schemes. (A) Slot-based repetition, (B) Back-to-back repetition.

Repetition is a method for URLLC to ensure high reliability. There is slot-based repetition transmission and back-to-back repetition transmission, as shown in Fig. 15.25.

- The transmission resources of slot-based repetition transmission are repeated based on slot, and the position of repetition resources in each slot is the same. When the time-domain resources are less than 14 symbols in one repetition transmission, there is a gap between successive repetitions. This method is simple in the resource determination, but the potential gap between repetition transmissions leads to certain delay.

- The time-domain resources of back-to-back repetition transmission are contiguous, leaving no gap between repetition transmissions, which fit low-latency service. However, the determination of back-to-back repeat transmission resources is complex, including issues due to cross-slot, SFI interoperability, etc.

The R15 standard adopted the slot-based repetition transmission scheme due to lack of time for a more advanced scheme before specification being frozen. In R16, the back-to-back repetition transmission was introduced to meet the low-latency requirement of URLLC. See Sections 15.4.2–15.4.5 for details. As of this writing, the NR system supports two transmission schemes: slot-based repetition transmission and back-to-back repetition transmission.

15.5.3 Multiple configured grant transmission

In R15, due to the complexity of system design and diversity of application requirements, one configured grant resource can be active in one BWP. However, in R16, the following two issues motivated the creation of a mechanism allowing multiple configured grant transmissions for a terminal:

- There are many kinds of service types with different requirements, such as service periodicity and size of service package. In order to adapt to various services, it is necessary to introduce multiple configured grant resources. As shown in Fig. 15.26, configured grant resource 1 is used for packet services with short periodicity and sensitive delay. Configured grant resource 2 is used for services with long periodicity and large packet size. The variation in resource allocation method can not only adaptively meet service transmission requirements, but also optimize the system resource utilization.

- The flexible starting occasion helps to meet low-latency requirement, but due to nonflexible ending position, the available resources can be limited and service reliability cannot be guaranteed. In order to meet both latency and reliability requirements at the same time, multiple configured grant resources are introduced to adapt to different arriving times and to ensure sufficient repetitions. As shown in Fig. 15.27, when services arrive in slots 1, 5, 9, 13, configured grant resource 1 is used. When the service arrives in slots 2, 6, 10, 14, configured grant resource 2 is the better choice to use.

In R15, for a type 2 configured grant transmission, a downlink control signaling used for activation and deactivation is required. When it comes to R16, where multiple configured grant transmissions are allowed for a terminal, if R15 activation/deactivation signaling is reused, the downlink control signaling overhead would increase. In order to reduce downlink control signaling overhead, both independent scheme and joint scheme were studied for activation/deactivation signaling.

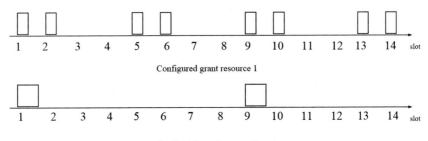

Figure 15.26 Multiple configured grants.

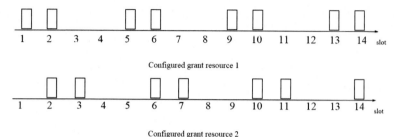

Configured grant resource 1

Configured grant resource 2

Figure 15.27 Multiple configured grants.

For the deactivation signaling, because it has no resource allocation function, joint deactivation does not need too much additional downlink control signaling design work, and the benefits of downlink control signaling overhead reduction are obvious. Therefore the joint deactivation mechanism is adopted. In order to support joint deactivation, it is necessary to configure in advance the configured grant transmission resource mapping tables for the purpose of joint deactivation, so that the deactivating downlink control signaling can indicate the joint configured grant transmissions in the table.

15.5.4 Nonorthogonal multiple access

As an advanced multiple access technology, nonorthogonal multiple access (NOMA) was an independent study item in R16. It has very wide application scenarios [11] across URLLC, eMBB, and mMTC, as shown in Table 15.3; it also has superior performance over conventional OFDMA in terms of total capacity, energy efficiency, and spectral efficiency. On the other hand, there had been concerns on implementation challenges; moreover, as mentioned later, various transmission processing principles were proposed as competing NOMA solutions, from which the 3GPP was not able to reach a consensus after a long time of research and discussion. NOMA technology was finally not adopted by the 3GPP. Therefore NOMA is not introduced in this book as an independent chapter. Nevertheless, as a potential technique to improve capacity for configured grant, it is briefly described here in Section 15.5.

NOMA is fundamentally different from the legacy techniques of multiple access, which provide orthogonal access to the users in the domain of time, frequency, code, or space. In NOMA, users can operate on the same time-frequency resource where they are distinguished by different power levels or unique multiple access (MA) signatures so that the

Table 15.3 Scenarios of NOMA [12].

Scenarios	Motivation	Potential benefits

URLLC	• High reliability
	• Low latency
	• High capacity/
	   traffic loading
		• Higher reliability through diversity gain
		   achieved by spreading and coding
		• Higher capacity grant-free transmission
		• Ability to multiplex mixed traffic types

eMBB	• High spectrum
	   efficiency
	• High user
	   density
	• Uniform user
	   experience
		• Larger capacity region by nonorthogonal user
		   multiplexing
		• Robustness to fading and interference with
		   code-domain design
		• Efficient link adaptation with relaxed CSI
		   accuracy

mMTC	• Massive
	   connections
	• High power
	   efficiency
		• Higher connection density with high
		   overloading
		• Robust and high-capacity grant-free
		   transmission

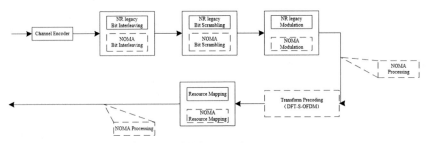

Figure 15.28 Processing chain in NOMA transmitter.

successive interference cancelation (SIC) can apply in the receiver to separate the signal for specific user.

The transmission processing chain for NOMA is implemented by replacing some existing processing steps in the legacy processing chain with NOMA processing steps, as shown in Fig. 15.28.

NOMA transmission processing takes MA signature into account. The implementation schemes of MA signatures and NOMA receiving algorithms are summarized as follows. In addition, power assignment can also be considered in the design of MA signatures.

15.5.4.1 Transmission schemes for multiple access signatures
• Scheme 1: Bit-level processing

   Bit-level processing achieves the user/branch separation by randomization, which can be based on either scrambling or interleaving.

- Scheme 2: Symbol-level processing

     The popular symbol-level processing schemes include spreading based on legacy modulation, spreading based on modified modulation, symbol-level scrambling, and interleaving with zero padding.
- Scheme 3: Sparse RE-mapping

     Sparse RE mapping is proposed as MA signature, including sparse code multiple access, pattern division multiple access, and interleave grid multiple access.
- Scheme 4: Staggered transmission pattern

Staggered starting time of transmission is also a part of MA signature in asynchronous code multiple access.

15.5.4.2 Receivers for nonorthogonal multiple access

- There are a few types of receivers, such as MMSE−IRC, MMSE−hard IC, MMSE−soft IC, ESE + SISO, EPA + hybrid IC, etc.

15.6 Semipersistent transmission

Semipersistent transmission is a scheduling-free transmission technology for downlink. It is mainly used for periodic transmissions of small packets, which can reduce the downlink control signaling overhead. In R15, LTE-alike scheme is basically used for periodic service traffic. In R16, due to the characteristic of URLLC, such as jitter and low latency, semipersistent transmission technology and corresponding HARQ−ACK feedback are enhanced [13]. This chapter only focuses on the physical layer enhancements, while the high layer enhancements are detailed in Chapter 16, URLLC in High Layer.

15.6.1 Semipersistent transmission enhancement

R16 enhancements for semipersistent transmission include multiple semipersistent transmissions and short period semipersistent transmissions. To handle service arrival jitter, R16 introduces the mechanism supporting multiple semipersistent transmissions for a terminal. Similar to configured grant, multiple semipersistent transmissions can be deactivated jointly.

     There are two differences between multiple semipersistent transmissions and multiple configured grant transmissions.

- Due to the constraint that downlink transmissions with the same priority correspond to the same HARQ−ACK codebook, the multiple semipersistent transmissions deactivated jointly should have the same priority.

- When time-domain resources for multiple semipersistent transmissions overlap directly, the semipersistent transmission with smaller SPS index is received and the corresponding HARQ−ACK is sent back. The reasonable configuration is supposed to keep the SPS index consistent with the priority (i.e., the smaller the SPS index, the higher the priority). In the case of indirect overlapping, the terminal can receive multiple semipersistent transmissions that are indirectly overlapping. The number of downlink transmissions that a terminal can receive in a slot is also determined by the UE capability.

In R15, the minimum periodicity of semipersistent transmission is 10 Ms, which is difficult to meet the demand of URLLC service. R16 reduces the minimum periodicity of semipersistent transmission to one slot.

15.6.2 Enhancements on HARQ−ACK feedback

In R15, upto one DL SPS configuration can be configured to a UE, and the minimum SPS periodicity is 10 Ms. If not multiplexing with ACK/NACK corresponding to the dynamic PDSCH, the ACK/NACK corresponding to SPS PDSCH is obtained in a separated PUCCH format 0 or PUCCH format 1 that is determined based on the semistatically configured PUCCH resource and K_1 indicated in the activation signaling, as shown in Fig. 15.29.

As multiple DL SPS configurations are supported in R16, the legacy one-to-one ACK/NACK feedback mechanism is no longer applicable. The following two issues were discussed.

15.6.2.1 How to determine the HARQ−ACK feedback timing corresponding to each semipersistent scheduling PDSCH

Take Fig. 15.30 as an example. Four DL SPS configurations with 2-Ms periodicity are configured to a UE. The following two options were proposed to determine the slot for transmission of ACK/NACK corresponding to each SPS PDSCH:

- Option 1: the ACK/NACK is deferred to the first available uplink slot.

  For SPS configuration #1 in Fig. 15.30, if the value of K_1 indicated by the gNB is 2, the ACK/NACK information corresponding to SPS configuration #1 in slot 1 needs to be deferred from slot 3 to slot 5, because the slot 3 derived from K_1 is a downlink slot. In comparison, the ACK/NACK information corresponding to SPS configuration #1 in slot 3 can be transmitted in slot 5 without deferring.

- Option 2: K_1 is separately configured for each SPS transmission within one SPS configuration.

In Fig. 15.30, K1 is configured separately for all SPS transmissions in slot 1 ~ slot 4. For example, the value of K_1 for the SPS transmission in slot 1 is 4, the value of K_1 for the SPS transmission in slot 2 is 3, the value of K_1 for the SPS transmission in slot 3 is 2, and the value of K_1 for the SPS configuration #3 in slot 4 is 1. The ACK/NACK corresponding to SPS configuration #4 in slot 4 cannot be transmitted on a PUCCH in slot 5, because the decoding cannot be finished. A different PUCCH in a later slot along with a different value of K_1 should be used.

The UE implementation of Option 1 is more complex, while the signaling overhead of Option 2 is higher. During the 3GPP discussion, neither Option 1 nor Option 2 was approved. The R15 mechanism was used to indicate a K_1 for each SPS configuration, and there is no enhancement on the determination of HARQ−ACK feedback timing to support multiple DL SPS configurations in R16. Take SPS configuration #1 in Fig. 15.30 as an example. If the value of K_1 indicated by the gNB is 2, the ACK/NACK information corresponding to SPS configuration #1 in slot 1 cannot be transmitted accordingly, since slot 3 is a downlink slot. The ACK/NACK corresponding to SPS configuration #1 in slot 3 is transmitted in slot 5.

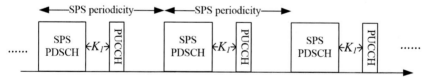

Figure 15.29 R15 ACK/NACK feedback for SPS PDSCH.

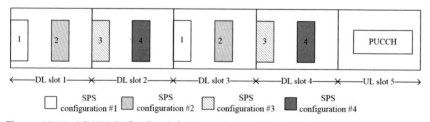

Figure 15.30 ACK/NACK feedback for multiple SPS configurations.

15.6.2.2 How to configure the PUCCH resources

When multiple SPS configurations are configured to a UE, the ACK/NACK information corresponding to multiple SPS PDSCHs may be transmitted in the same slot. Take Fig. 15.30 as an example. If the value of K_1 indicated by the gNB for SPS configuration #1 and #2 is 4, and the value of K_1 indicated by the gNB for SPS configuration #3 and #4 is 3, four ACK/NACK indications are transmitted on the PUCCH in slot 5. However, in R15 the SPS PDSCH transmission can be associated with PUCCH formats 0 or 1 only, which cannot carry more than two bits in a payload. Moreover, the number of ACK/NACK information bits corresponding to SPS PDSCH carried in different slots may be different. How to determine the actual PUCCH resource used in each slot is another question pending for discussion.

Option 1: Multiple common PUCCH resources are configured for multiple SPS configurations. One PUCCH is selected from the multiple common PUCCH resources based on the number of ACK/NACK bits.

Option 2: Multiple sets of common PUCCH resources are configured for multiple SPS configurations. One set is selected from the multiple sets based on the number of ACK/NACK bits. Further, one PUCCH from the set is indicated by the activation DCI.

To determine PUCCH resource based on the activation signaling limits scheduling flexibility, while the multiple configured PUCCH resource sets increase the overhead significantly. Therefore 3GPP RAN1 agreed to adopt Option 1 to determine the PUCCH resource for multiplexing the ACK/NACK information corresponding to multiple SPS PDSCHs.

15.7 Inter-UE multiplexing

In the NR system, URLLC and eMBB can be supported separately in independent deployments or jointly in the same deployment. In either case, the URLLC service requirement is different from the eMBB service requirement. The URLLC traffic transmission needs to be quickly scheduled and transmitted, with a latency as low as 1 Ms. The latency requirement of eMBB service is more relaxed than that of URLLC service. Due to different scheduling timing, there may be transmission collisions between the two services in the case of joint deployment. In particular, to ensure timely scheduling of URLLC, gNB needs to allocate URLLC transmission with the resources that are already scheduled to a different terminal for eMBB transmission as shown in Fig. 15.31. At this time, the

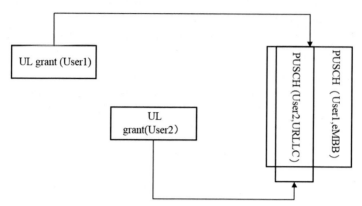

Figure 15.31 Collision between URLLC and eMBB.

data transmissions from the URLLC terminal and eMBB terminal interfere with each other, making it difficult to meet the reliability requirements of URLLC and eMBB. For the downlink transmission, a similar resource collision between terminals may occur as well. In order to solve this kind of inter-UE collision, 3GPP studied a preemption technique for the downlink, as well as transmission cancelation and power adjustment for the uplink [14–20].

15.7.1 Multiplexing solutions

During the RAN1 discussion, resource collision in both downlink transmission and uplink transmission were considered.

For the resource collision between eMBB PDSCH and URLLC PDSCH, the candidate schemes include:

- Scheme 1: Discard eMBB transmission

    gNB preempts the resources previously scheduled for eMBB transmission and sends signaling to inform the affected terminal of this preemption.

- Scheme 2: Delay eMBB transmission

    gNB preempts the resources previously scheduled for eMBB transmission and restores the eMBB transmission on the preempted resources after the end of the URLLC transmission. The resources used for the restored transmission can be indicated by another DCI or configured by high layer signaling.

    Because it is not always easy or even feasible to find suitable resources to restore the preempted eMBB transmission, 3GPP RAN1 finally adopted Scheme 1.

For the resource collision between eMBB PUSCH and URLLC PUSCH, the candidate schemes include:

- Scheme 1: Cancel the eMBB PUSCH transmission

  Uplink cancelation indication (UL CI) is used to indicate the colliding resources between eMBB PUSCH and URLLC PUSCH. The terminal determines whether and how to cancel transmission based on its own transmission information and the colliding resources indicated by UL CI.

- Scheme 2: Adjust power of URLLC PUSCH transmission

In the case of resource collision between eMBB PUSCH and URLLC PUSCH, the terminal to transmit URLLC PUSCH applies higher transmission power based on the open-loop power adjustment indication signaling.

Scheme 1 can eliminate the interference from eMBB PUSCH to the URLLC transmission. However, it increases the complexity of the eMBB terminal because of the requirement for fast processing of UL CI, while the gain of this fast processing is in the URLLC terminal. Moreover, given there are always some terminals not capable of supporting cancelation, the interference from these terminals cannot be eliminated. On the other hand, Scheme 2 overcomes the interference by enhancing the competition capability of the URLLC terminal against eMBB terminal. However, Scheme 2 does not eliminate interference completely, and therefore the URLLC transmission reliability cannot be guaranteed for terminals with limitations on transmission power. The two schemes have their own advantages and disadvantages, and are complementary to each other. Both of the schemes are adopted by the 3GPP.

In Scheme 1, once the resource indicated by UL CI overlaps with the transmission resource, the terminal cancels the whole overlapping transmission without recovery. In the case of transmission with repetition, cancelation for transmission is done on a per repetition basis, as shown in Fig. 15.32.

UL CI is delivered in group common signaling, so it can be received by not only eMBB UE but also URLLC UE. To avoid canceling the URLLC service, the cancelation is further proposed to be based on PUSCH priority. If an uplink transmission is indicated to have high priority, it cannot be canceled upon an UL CI; otherwise, it can be canceled. However, because priority indication does not indicate absolute priority level but relative priority level (i.e., the low-priority indication does not necessarily mean the lowest transmission priority), some companies did

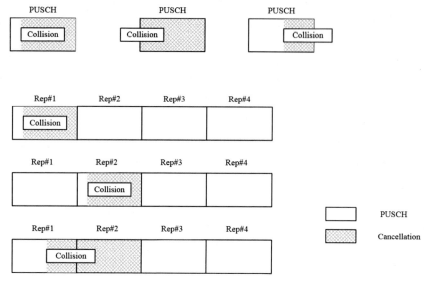

Figure 15.32 Cancellation for PUSCH.

not want to bundle priority indication into cancelation procedure. As a compromise, a new high layer parameter was introduced to tell whether the cancelation is based on the relative priority indication.

15.7.2 Signaling design

The signaling design for handling of inter-UE transmission collision contains the timing of DL Preemption Indication (PI) and UL CI as well as the corresponding DCI format to carry the signaling.

15.7.2.1 Timing of DL preemption indication

The following timing schemes for the DL PI signaling transmission were discussed. As shown in Fig. 15.33, for a resource conflict in slot n:

- Scheme 1: DL PI is sent in the slot n and before the conflicting resource.
- Scheme 2: DL PI is sent in the slot n and after the conflicting resource.
- Scheme 3: DL PI is sent in the next downlink slot after the slot n.
- Scheme 4: DL PI is sent k slots after the slot n.

The simulation comparisons among the above schemes show that Scheme 1 has no obvious impact on the performance, but the implementation complexity of Scheme 1 and Scheme 2 is relatively high. For the comparison between Scheme 3 and Scheme 4, the earlier the DL PI is to be delivered, the shorter the delay to be experienced in the retransmission of discarded PDSCH. Therefore the standard finally adopted Scheme 3 (i.e., the DL PI is sent in the first downlink slot after the preempting slot).

15.7.2.2 Downlink control information format for DL preemption indication

DCI format 2_1 is defined for DL PI. DCI format 2_1 carries multiple preemption information fields, each of which is a 14-bit bitmap for a given time–frequency resource in a given carrier. The minimum size of DCI format 2_1 is 14 bits and the maximum size is 126 bits. One specific search space is configured for the PDCCH-carrying DCI format 2_1.

15.7.2.3 Downlink control information format for uplink cancelation indication

UL CI signaling design follows design logic of DL PI with small modification, as shown in Table 15.4.

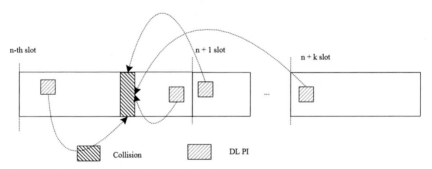

Figure 15.33 Transmission timing for DL PI.

Table 15.4 Comparisons between DL PI and UL CI.

	DL PI	UL CI
Signaling type	Group common DCI	Group common DCI
Timing	Later than collision;	Earlier than collision;
DCI size	14 bits as minimum, 126 bits as maximum.	14 bits as minimum, 126 bits as maximum.
Resource indication type	Indication granularity is configurable. Indicated resource range is fixed.	Indication granularity is configurable. Indicated resource range is configurable.

The main differences between UL CI and DL PI lie in timing and resource indication. For the timing, the DL PI can be sent after the colliding resources. The UL CI is used to cancel PUSCH transmission and therefore has to be sent before the colliding resource. Moreover, the timing of the UL CI transmission must reserve enough processing time for the UE to cancel the transmission. For the resource indication, both DL PI and UL CI use joint time-frequency resource indication. However, for DL PI, the indicated resource range is fixed (i.e., period of DL PI) and the indication granularity is selected as either the whole bandwidth or half bandwidth in the frequency domain; for UL CI, both the indicated resource range and the indication granularity can be configured flexibly.

15.7.3 Uplink power adjustment scheme

When there is a transmission resource conflict between eMBB PUSCH and URLLC PUSCH, in addition to canceling the transmission of eMBB PUSCH, the standard also intends to improve the URLLC receiving performance by adjusting the transmission power of URLLC PUSCH.

The power adjustment in this scheme has a different purpose from the traditional uplink power adjustment. It has larger adjustment range and its effect may not be necessarily maintained over time. Accordingly, some enhancements are imposed on the traditional power control procedure to improve the URLLC reliability in real-time, including:

- One of the open-loop power control parameters, P_0, can be configured with two or three candidate values in the higher layer configuration, where only one candidate value comes from the R15 configuration.
- In R16, the selection from the candidate values for P_0 is indicated by UE-specific DCI.

15.8 Summary

This chapter introduced physical-layer technologies optimized for URLLC in the NR system, mainly R15 techniques such as UE processing capability, MCS/CQI design, configured grant transmission, and downlink preemption, and R16 enhancements such as downlink control channel enhancement, uplink control information enhancement, uplink data-sharing channel enhancement, processing timing enhancement, configured grant transmission enhancement, semipersistent transmission enhancement, and inter-UE prioritization/multiplexing.

References

[1] R1−1903349. Summary of 7.2.6.1.1 potential enhancements to PDCCH, Huawei, 3GPP RAN1 #96, 25 February−1 March 2019, Athens, Greece.

[2] R1−1905020. UCI enhancements for eURLLC, Qualcomm Incorporated, 3GPP RAN1 #96b, 8th−12th April 2019, Xi'an, China.

[3] R1−1906752. On UCI enhancements for NR URLLC, Nokia, Nokia Shanghai Bell, 3GPP RAN1#97, Reno, 13th−17th May 2019, Nevada, United States.

[4] R1−1906448. UCI enhancements for URLLC, OPPO, 3GPP RAN1#97, Reno, 13th−17th May 2019, Nevada, United States.

[5] R1−1907754. Summary on UCI enhancements for URLLC, OPPO, 3GPP RAN1#97, Reno, 13th−17th May 2019, United States.

[6] R1−1909645. Offline summary on UCI enhancements for URLLC, OPPO, 3GPP RAN1#98, 26th−30th August 2019, Prague, CZ.

[7] R1−1912519. UCI enhancements for URLLC, OPPO, 3GPP RAN1#99, 18th−22nd November 2019, Reno, United States.

[8] 3GPP TS 38.214. NR; Physical layer procedures for data, V15.9.0 (2020−03).

[9] R1−1717075. HARQ timing, multiplexing, bundling, processing time and number of processes, Huawei, 3GPP RAN1 #90bis, 9th−13th October 2017, Prague, Czech Republic.

[10] R1−1716941 Final Report of 3GPP TSG RAN WG1 #90 v1.0.0, MCC Support, 3GPP RAN1 #90bis, 9th−13th October 2017, Prague, Czech Republic.

[11] 3GPP TR 38.812 V16.0.0 Study on non-orthogonal multiple access (NOMA) for NR.

[12] R1−1801397. Discussion on application scenarios for NoMA, Huawei, HiSilicon, 3GPP TSG RAN WG1 NR Ad Hoc Meeting #92, 26th February−2nd March 2018, Athens, Greece.

[13] R1−1911554. Summary#2 of 7.2.6.7 others, LG Electronics, 3GPP RAN1 #98bis, 14th−20th October, 2019, Chongqing, China.

[14] R1−1611700. eMBB data transmission to support dynamic resource-sharing between eMBB and URLLC, OPPO, 3GPP RAN1 #87, 14th−18th November 2016, Reno, USA.

[15] R1−1611222. DL URLLC multiplexing considerations, Huawei, HiSilicon,3GPP RAN1 #87, 14th−18th November 2016, Reno, USA.

[16] R1−1611895. eMBB and URLLC multiplexing for DL, Fujitsu, 3GPP RAN1 #87, 14th−18th November 2016, Reno, USA.

[17] R1−1712204. On pre-emption indication for DL multiplexing of URLLC and eMBB, Huawei, HiSilicon, 3GPP RAN1 #90, 21−25 August 2017, Prague, Czech Republic.

[18] R1−1713649. Indication of Preempted Resources in DL, Samsung, 3GPP RAN1 #90, 21−25 August 2017, Prague, Czech Republic.

[19] R1−1910623. Inter UE Tx prioritization and multiplexing, OPPO, RAN1 #98bis, 14th−20th October 2019, Chongqing, China.

[20] R1−1908671. Inter UE Tx prioritization and multiplexing, OPPO, RAN1 #98, 26th−30th August 2019, Prague, Czech.

CHAPTER 16

Ultra reliability and low latency communication in high layers

Zhe Fu, Yang Liu and Qianxi Lu

A time-sensitive network (TSN) is a common network scenario in industry. One key vision of the New Radio (NR) R16 is to support Industrial Internet service. Therefore the Industry Internet of Things (IIoT) project focuses on how TSN services can be supported in 5G system. The service requirements for TSN service on the 5G system is specified in TR 38.825 [1], and the details are described in Table 16.1.

Since TSN services are usually delay-sensitive, TSN use cases such as robotic arms on production lines have specific requirements for time synchronization. The 3GPP has also researched this area. Specifically, refer to TR 22.804 [2]. The time synchronization requirements of the TSN service are shown in Table 16.2.

To support ultrahigh-reliability and low-latency service transmission requirements, the IIoT project also studied two issues. One is how to prioritize one grant among multiple collided resources for a certain user, and another is to support more than two-leg duplication transmission. This chapter will address these issues individually.

Note that the solutions related to the physical layer are included in Chapter 15, 5G URLLC: PHY layer.

16.1 Timing synchronization for industrial ethernet

In a typical application scenario such as a smart factory environment, product assembly services first require the main controller to send information such as the relevant operating instructions for the action unit(s) and the expected assembly completion time to the UE. Upon reception of this information, the UE informs the targeted action unit(s) according to the given instructions at the proper time. It is expected that if strict clock synchronization is not performed between the UE, the action unit(s), and the main controller, the action unit(s) might perform the operation at an unexpected time point, which will lead to the result of unqualified products. As mentioned earlier,

Table 16.1 Use cases and performance requirements for TSN service [1].

Case	SUE 3	Communications service availability	Transmit period (ms)	AllowedE2E latency	Survival time	Packet size	Service area	Traffic periodicity	Use case
I	20	99,9999% to 99,999999%	0–5	≤ Transmit period 3	Transmit period	50 bytes	$15 \times 15 \times 3 \ m^3$	Periodic	Motion control and control-to-control use cases
II	50	99,9999% to 99,999999%	1	≤ Transmit periods	Transmit period	40 bytes	$10 \times 5 \times 3 \ m^3$	Periodic	Motion control and control-to-control use cases
III	100	99,9999% to 99,999999%	2	≤ Transmit periods	Transmit period	20 bytes	$100 \times 100 \times 30 \ m^3$	Periodic	Motion control and control-to-control use cases

Table 16.2 Time synchronization requirements [2].

Clock synchronicity accuracy level	Number of devices in one Communication group for clock synchronization	Synchronization clock synchronicity requirement	Service area	Scenario
1	Upto 300 UEs	<1 μs	≦ 100 × 100 m	Motion controls Control-to-control communication for industrial controller
2	Upto 10 LIES	<10 μs	2500 m^2	High data rate video streaming
3	Upto 100 UEs	<1 μs	<20 km^2	Smart Grid: synchronicity between PMUs

a survey of the demanding level of the industrial Ethernet synchronization has been presented in TR 22.804.

As Table 16.2 shows the most stringent synchronization accuracy performance requirement is to provide "< 1 Ms" clock synchronization service toward 300 UEs within the coverage area of one gNB. In 5G NR R15, the SystemInformationBlock9 broadcasts the 5G internal clock with a granularity of 10 Ms toward UEs, which cannot satisfy performance requirements of industrial Ethernet. As a result, enhancements are needed to meet the timing synchronization performance requirements for the industrial IOT in R16.

A solution for 5G NR to support industrial Ethernet clock synchronization has been provided by TR 38.825 [1]. Specifically, the 5G system acts as a bridge for TSN to undertake the communication task between the TSN network system and the TSN end stations. Among them, the TSN adapters at the edge of the 5G system (such as UE and UPF) need to support the function of the IEEE 802.1AS clock synchronization protocol. In contrast, 5G system components such as UE, gNB, and user plane functions (UPFs) are not impacted: they only need to be synchronized with the 5G clock. In this way, except for the introduction of the TSN adaptation layer on the edge of the 5G system, the impacts on the functions and standards of 5GS are minimized from the need for industrial Ethernet clock synchronization.

Next, the author will introduce the clock synchronization mechanism provided by the TSN adaptation layer for the TSN network and TSN end stations. In Fig. 16.1, first, the node in the right TSN network needs to send the clock synchronization message to the 5G edge network element UPF. After that, the TSN adapter on the UPF uses the 5G system internal clock to record the time moment TSi corresponding to receiving the gPTP clock message. Then, the UPF transmits this gPTP clock message again to the UE via gNB. After processing by the TSN adapter on the UE-side, the UE continues to send this message toward the TSN end station to complete the clock synchronization task. In the message, the UE-side adaptation layer will compute and add the 5G system internal message processing delay in the correction field, which is equal to TSe−TSi. Here, TSe denotes the 5G system internal time when the gPTP clock is sent to the TSN end station. Therefore the clock maintained at the TSN end station, when further receiving the clock information, should be tuned as follows:

$$T_{\text{end-station}} = T_{\text{TSN system}} + \text{TSe} - \text{TSi} + T_2 + T_1 \qquad (16.1)$$

In Eq. (16.1), T_1, T_2 denote the transmission delay of the gPTP message from TSN system to UPF and from the UE to TSN node,

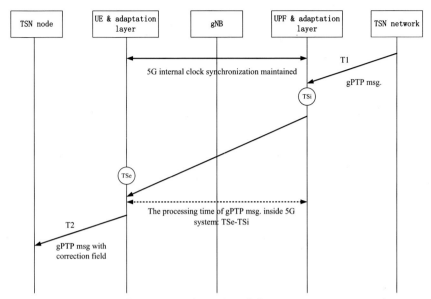

Figure 16.1 Diagram of 5G system being used for time-sensitive network timing synchronization.

respectively. They could be derived with the usage of the peer–delay algorithm, as given in Ref. [3]. In addition, it could be also found that if the clocks of the UEs in the 5G system and the UPF cannot be synchronized at a high level, inaccurate estimation of the transmission time of gPTP messages in the 5G system will resulted, which leads to a deteriorated TSN end-station clock.

From the perspective of the radio access network (RAN), the source of the synchronization errors consists of two parts: the air interface synchronization error from the gNB to the UE and the synchronization error from the TSN clock source to the gNB. 3GPP RAN1/2 and RAN3 completed the corresponding evaluation work on the synchronization performance on these two interfaces, respectively, which can be found in TR 38.825 [4]. In addition, TR 38.825 [4] also summarizes the synchronization performance analysis results given by the RAN1/2/3 group, and concludes that the overall synchronization accuracy error is 665 ns, which meets the most stringent synchronization performance requirements of industrial Ethernet, with the assumptions of the clock error deviation of 100 ns between the TSN clock source and the gNB, and the SCS of 15 kHz applied.

As mentioned earlier, in R15 NR, the time granularity defined by TS 38.331 [5] is 10 Ms (i.e., the value representing clock information is increased by 1 every 10 Ms) for system clock information in SIB9, which cannot meet the requirements of industrial Ethernet. To tackle the problem, in R16 NR, 3GPP RAN Group 2 decided to introduce a new system clock information IE—ReferencetimeInfo with a time granularity of 10 ns in SIB9. The mechanism of applying this system clock information is similar to R15—the clock time information derived from *ReferencetimeInfo* IE corresponds to the boundary of *ReferenceSFN* given in SIB9. It can be easily found that the potential source of errors in the synchronization mechanism is the difference between the time when the gNB sends the downlink reference frame and the time when the UE receives the downlink reference frame, which is relevant to the propagation delay between UE and gNB.

During the discussion of 3GPP RAN2, UEs, chips, and network equipment vendors had a fierce debate about whom, the gNB or the UE, should take the responsibility for propagation delay compensation for the timing synchronization between UE and gNB. Generally, there exist two solutions on the table as follows:

- Option 1: The UE obtains the timing advance (TA) from the gNB via either by RACH procedure or receiving TA command MAC CE (TA command Media Access Control Element) to correct the received time clock information.
- Option 2: According to the sounding reference signal or localization result, the network derives the information of the distance between the UE and the gNB. After that, clock information sent to the UE is precompensated at the gNB.

Network infrastructure vendors believed that if it is the gNB to take the task, the workload is significantly large considering the potentially large number of UEs are within the coverage. On the other hand, UEs can update the TA through legacy RACH procedure, etc. It is more appropriate to choose Option 1. The UE and chip vendors, however, insisted that new features, such as 5G high-precise positioning, would be introduced in NR Release-16, and the network naturally has more methods to derive the distance information more accurately. In addition, propagation delay precompensation to be done at gNB could avoid the TA quantization problem completely. In short, Option 2 should be selected. Finally, considering that R-16 NR industrial Ethernet is mainly to be deployed in small cells and the impact of propagation delay on the clock synchronization performance in the 5G system is limited, it was decided that the UE is free to decide whether or not to do propagation delay compensation.

Moreover, UEs are allowed to obtain clock information via broadcast or RRC-dedicated signaling in R16 NR. When in the RRC_IDLE/RRC_INACTIVE state, the UE obtains SIB9 by monitoring the system broadcast; if from the SIB1 information the UE finds that the system does not schedule SIB9 broadcast currently, the standard also allows the UE to retrieve the clock information via the on-demand SI mechanism, which can be found in Section 11.1 of this book. On the other hand, for the UE in the RRC_connected state, a new retrieving mechanism is introduced in R16, indicated as follows:

- When in the need of request for the ReferencetimeInfo IE, the UE sets the value of the reference information request related flag (referenceTimeInfoRequired) as "1" in the UEAssistanceInfomation message.
- In response, the network will unicast downlink RRC signaling carrying ReferenceTimeInfo IE toward the UE or broadcast the SIB9 to the UE.

16.1.1 Intra-UE prioritization

To support multiple ultra reliability and low latency communication (URLLC) services, and to meet the stringent delay requirements of URLLC services, NR R16 considers more resource conflict scenarios. For uplink resource conflicts within the same user, R16 mainly considers the following conflict scenarios:

- Conflict between data and data: Specifically, according to the type of resource, it can be divided into three subscenarios: the conflict between CG (Configured Grant) and CG, the conflict between CG and dynamic authorization (DG), and the conflict between DG and DG.
- Conflict between data and scheduling request (SR): Similarly, according to the type of resource, it can be divided into two subscenarios: the conflict between CG and SR, and the conflict between DG and SR. The same rule for the two subscenarios involving SR can be adopted.

The details for each scenario are elaborated on in the following.

16.1.2 The conflict between data and data

When considering the conflict between data and data, only the conflict between DG and CG is considered in R15, where DG is always prioritized. More complex resource conflict scenarios are considered in R16, including CG versus CG, CG versus DG, DG versus DG. In order to ensure the transmission requirements of URLLC services, some enhancement is required in R16 on intra-UE prioritization.

16.2 Dynamic authorization versus configured grant and configured grant versus configured grant

In order to support multiple URLLC services, and to meet the stringent delay requirements of URLLC services, there may be resource overlapping between CG and CG or CG and DG. Considering there may be no available data to be transmitted on CG, for the collision involving CG, several potential solutions can be considered in the following:

- Solution1: Intra-UE prioritization is only performed by physical layer.
- Solution2: Intra-UE prioritization is performed by physical layer and MAC layer together.

Some opinion is MAC layer can only resolve part of conflict scenarios, such as the conflict between PUSCH (Physical Uplink Shared Channel) and PUSCH, or the conflict between PUSCH and SR, but it cannot resolve other cases involving UCI (Uplink control information), such as the conflict between HARQ-ACK (Hybrid Automatic Repeat reQuest Acknowledge) and PUSCH. In addition, for the case of multiple resource conflicts, such as the conflict among CG PUSCH, HARQ-ACK and DG PUSCH, some companies believe that even if MAC layer selects to prioritize CG, the physical layer will finally choose another resource to transmit since the physical layer may think CG PUSCH is a deprioritized one. Thus, it is better to perform intra-UE prioritization at physical layer. If we follow this way (i.e., intra-UE prioritization is only performed by physical layer), the issue of interlayer interaction and delay will be introduced, since the physical layer needs to obtain data available information before it makes any decision.

In addition, unnecessary MAC PDU generation and data transmission delay can be avoided if the MAC layer is involved in the loop of intra-UE prioritization. Therefore Solution2 is selected (Fig. 16.2).

When the MAC layer performs prioritization processing, LCH-based prioritization is applied (i.e., logical channel priority is used in the MAC layer for selecting a prioritized resource).

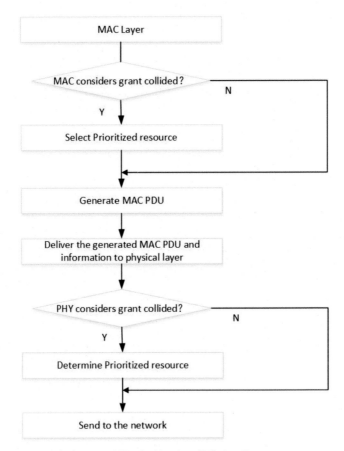

Figure 16.2 Example for intra-UE prioritization (Solution2).

Specifically, when LCH–based prioritization is configured to the MAC entity and a collision occurs, the MAC layer will prioritize the uplink grant with a high priority. The priority of an uplink grant is determined by the highest priority among priorities of the logical channels with data available that are multiplexed or can be multiplexed in the MAC PDU, according to the mapping restrictions. Besides legacy LCH mapping restrictions, additional mapping restrictions for CG and DG is introduced seperately in R16 based on reliability requirement, which are used to limit the logical channels that can be transmitted on CG or DG.

For the case of DG overlapping CG, DG is always prioritized when the logical channel priorities of the two resources are the same. For the case of CG overlapping CG, it is upto UE implementation to choose one of the two when logical channel priorities of the two resources are the same.

The scenarios of resource conflicts can be divided into the following situations:

- Case 1: If no MAC PDU for any collided resource has been generated when the UE is performing intra-UE prioritization, only one MAC PDU will be generated.
- Case 2: If one MAC PDU has been generated and the priority of associated uplink grant is high when the UE is performing intra-UE prioritization, the MAC layer will not generate another MAC PDU.
- Case 3: If one MAC PDU has been generated and the priority of the associated uplink grant is low when the UE is performing intra-UE prioritization, the MAC layer can generate another MAC PDU. Accordingly, the grant with a low priority is considered a deprioritized one.

When the low-priority resource is a CG resource, and the associated MAC PDU has been generated for this low-priority CG, this MAC PDU can be called a deprioritized MAC PDU corresponding to the CG. For such MAC PDU, the network may not schedule retransmission for this CG, since the network does not know whether the CG resource is not transmitted due to low priority or no available data. Once such MAC PDU is generated but the retransmission is not assured, it is inevitable for MAC PDU discarding and thus data loss. To resolve this issue, UE autonomous transmission for such MAC PDU is introduced, as a supplement to the method of the network retransmission scheduling.

Specifically, the network can instruct the UE whether to use the automatic transmission function by configuring autonomousTx. If autonomousTx is configured, the UE can use subsequent CG resources that are with the same HARQ process as the deprioritized CG and belong to the same CG configuration as the deprioritized CG to transmit such MAC PDUs. It depends on UE implementation on which subsequent CG resource is specifically selected to transmit the deprioritized MAC PDU. In addition, since the automatic transmission mechanism is a supplement to the network retransmission scheduling method, when the UE receives the retransmission resource for this deprioritized MAC PDU scheduled by the network, even if the automatic transmission function is configured, the UE will not transmit the deprioritized MAC PDU automatically again.

However, at the end of the R16, RAN2 downselected the collision scenarios supported due to RAN1 restriction and finally reached the following conclusions:

- For a scenario where DG and CG conflict, regardless of whether the physical layer priority is the same or not: MAC generates only one MAC PDU to the physical layer.
- The scenario where CG and CG conflict.
- If the physical layer priorities of the conflicting CG resources are the same, the MAC generates only one MAC PDU to the physical layer.
- If the physical layer priorities of the conflicting CG resources are different, the MAC can generate multiple MAC PDUs to the physical layer. It depends on UE implementation to assure that low-priority resources are canceled and high-priority resources are transmitted.

16.3 Dynamic authorization versus dynamic authorization

Generally speaking, this scenario can occur in the following situations: after the network schedules the DG resources for the eMBB services, the network finds that the URLLC service is available and its delay requirements are high, and the network has to schedule another DG resource for the URLLC service, which leads to two DG resources overlapping. However, this scenario is finally not supported since RAN1 believes that this conflict scenario will not occur.

16.3.1 The conflict between data and scheduling request

In R15, when the data and the SR conflict, the MAC will not instruct the physical layer to send the SR. In R16, in order to better support the transmission requirements of URLLC services, RAN2 discussed the possibility of the transmission of the conflicting SR, and whether to prioritize SR depends on the priority of LCH that triggers SR.

Specifically, when LCH-based prioritization is configured to the MAC entity and collision occurs between UL−SCH and SR, and the priority of the logical channel that triggers the SR is higher than the priority of the UL−SCH resource, the MAC layer will instruct the physical layer to perform SR transmission. If the UL−SCH resource conflicts with the resource for SR transmission, and the SR with a high priority is triggered before MAC PDU generation, the MAC will not generate the corresponding MAC PDU. Conversely, if the SR with a high priority is triggered after MAC PDU generation, the MAC PDU will be considered as a deprioritized MAC PDU. How to handle the deprioritized MAC PDU can be found in the description in Section 16.2.1.

However, at the end of the R16, RAN2 downselected the collision scenarios supported due to RAN1 restriction and finally reached the following conclusions:

- If the physical layer priorities of the conflicting resources are the same, the MAC will not instruct the physical layer to send SR.
- If the physical layer priorities of the conflicting resources are different, the MAC may instruct the physical layer to send SR.

In order to facilitate readers' understanding, we summarize and compare intra-UE prioritization schemes in R16 and R15 in Table 16.3.

16.4 Enhancements to the semipersistent scheduling

According to Table 16.1, it can be found that the Transmit period of TSN network data packets varies between 0.5 and 2 Ms. On the RAN side, if the uplink and downlink data is transmitted via the dynamic scheduling method, the signaling overhead (SR, BSR, DCI, etc.) is high. As a result, it is pursued to use semipersistent scheduling approach for scheduling of the TSN data packets (including instruction and feedback message) in 5G network. Accordingly, 3GPP RAN2 found that the semipersistent scheduling (SPS) needs to be enhanced in many aspects to guarantee the QoS for TSN data transmission, for which details are described in the following four subsections.

16.4.1 Support shorter period for semipersistent scheduling resource

The semistatic scheduling period defined in R15 is identified with the following two drawbacks to adapt to the TSN data transmission:

- the difference between the finest granularity of the uplink SPS period versus when the downlink SPS period is too large [the minimum period for downlink is 10 Ms, while the minimum period for uplink is two symbol periods (varying between 18 us and 143 us, depending on the SCS applied)].
- the values of the uplink/downlink SPS period could be chosen are limited.

Note that if the period of the SPS resource is set too large, compared with the IIOT instruction period, there will be a problem that the TSN

Table 16.3 Comparison on intra-UE prioritization.

Scenario		DG versus CG	CG versus CG	DG versus DG	CG versus SR	DG versus SR
	R15	√			√	√
	R16	√			√	√
Principle of intra-UE prioritization	R15	Prioritize DG	√		Prioritize UL-SCH	Prioritize UL-SCH
	R16	DG versus CG: only one MAC PDU is delivered to physical layer. CG versus CG: If the physical layer priorities of the conflicting CG are the same, the MAC generates only one MAC PDU to the physical layer, otherwise, the MAC can generate two MAC PDUs to the physical layer. Whether to generate another MAC PDU depends on LCH priority. Data versus SR: If the physical layer priorities are different, the MAC can deliver SR and data to the physical layer, otherwise, SR cannot be instructed to physical layer. Whether to deliver both SR and MAC PDU depends on LCH priority.				
Deprioritized MAC PDU	R15	Not				
	R16	Possible. UE can be configured with autonomousTX for CG.				

instruction needs to wait a long time before the SPS resource becomes available, as shown in Fig. 16.3:

To tackle the problem, the 3GPP RAN1 physical layer standard group decided that 5G NR R16 will support the configuration of downlink semipersistent scheduling resources with a minimum period of one time slot for all SCS options.

In addition, as described in TR 38.825 [1], the cyclic generation time of instruction/feedback data packets is in the range of 1 to 10 Ms, and the specific value of the period is not fixed (depending on specific application requirements). In this case, if the short-period SPS resource configuration is not supported, the resource period may not match the data packet generation period. As shown in Fig. 16.4, after the network activates the semipersistent scheduling resource, the time when the URLLC uplink data reaches the communication protocol stack will gradually be staggered

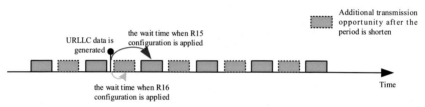

Figure 16.3 Wait time is decreased, when a shorter period of semipersistent scheduling resource is applied.

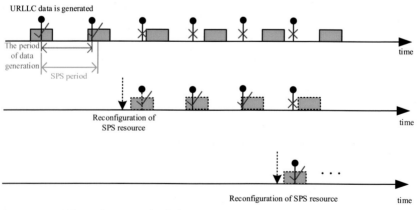

Figure 16.4 When the period of data generation and semipersistent scheduling (SPS) resource is mismatched, the network needs to reconfigure the SPS resource frequently.

from the time when the semipersistent scheduling resource appears, and therefore application data that cannot be sent out in time will become a common issue. To deal with this problem, the network is forced to reconfigure SPS resources frequently.

16.4.2 Configuration of multiple active semipersistent scheduling resource simultaneously

In order to support the requirements of URLLC service transmission, R16 supports the configuration of multiple sets of semistatic scheduling resources for the UE. For specific analysis, refer to Chapter 15, 5G URLLC: PHY layer.

In order to assist the UE to map the data of the logical channel with different communication performance requirements to the appropriate uplink semipersistent scheduling resource (CG), R16 NR decided to assign the simultaneously configured CGs with different labels (IDs). Firstly, when the network uses RRC signaling to configure a certain CG for the UE, such label information is also provided. Secondly, when configuring a certain logical channel for the UE, the UE is optionally configured with a list (allowedCG-List-r16) containing at least one CG ID, which implies that the data from this logical channel can be transmitted on these CG resources. In this way, when the transmission opportunity of a certain CG arrives, the UE MAC entity can put the data from the qualified logical channel on the CG, as shown in Fig. 16.5. NR R16 will inform the UE of the CG resources that need to be added/changed or released, when configuring the uplink BWP for the UE.

In addition, for type-2 CG, when the UE receives the CG activation/deactivation DCI from the network, it needs to send the corresponding confirmation information to the network. In R15 NR, the network

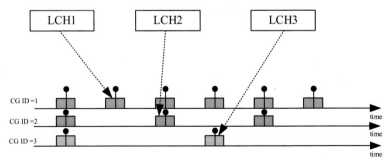

Figure 16.5 Different Configured Grant carries data coming from different logical channels.

can only configure at most one type-2 CG resource for the UE. Correspondingly, the structure of the CG confirmation MAC CE is very simple. It only contains the MAC PDU subheader (0 payload) with a dedicated logical channel ID. However, in R16 NR, as described above, since the network can configure multiple CG resources for the UE, when the UE responds to the network with a confirmation message, it also needs to tell the network for which CG the activation/deactivation instructions it has received. Note that, in order to support transmission of service data with different profiles, the 3GPP decided that each MAC entity of the R16 UE would support upto 32 activated CG resources. Correspondingly, the length of the multiple configuration uplink confirmation MAC CE is also 32 bits. The specific payload format can be found in TS 38.321 [6], wherein the x-th position 1 or 0 represents that the UE receives or does not receive the DCI indication information for the CG with ID = x.

In fact, in the process of standardization discussion, some companies have objections to the meaning of bit position 0 or 1 in the MAC CE payload. Suppose that the UE has received both the DCI activation and deactivation indication for the CG resource with ID = 2 within a short period of time. If the UE does not send a confirmation MAC CE to the network until the later DCI indication is received, the MAC CE cannot tell the network side whether it has received the first or second indication or both. However, in practice, the gNB generally does not send two instructions continuously in a short time, so the assumption of the problem raised by this objection is too extreme, which ultimately led to the objection not being widely accepted.

Another point that should be noted is that although the multiple configuration uplink confirmation MAC CE shown in TS 38.321 [6] can identify 32 CGs to indicate the reception status, the network not only can assign type-2 CG but also type-1 CG to the terminal. Note that type-1 CG is activated immediately after receiving the RRC configuration of the network, without waiting for the further DCI indication information. Correspondingly, after receiving the multivariate CG confirmation MAC CE, the network will ignore the value of the corresponding bits of all type-1 CG IDs on the MAC CE.

16.4.3 Enhancement to the semistatic scheduling resource time-domain position determination formula

In LTE and R15 NR, the uplink/downlink semipersistent resource period configured by the network for the UE can be divided by the duration of a system hyperframe (1024 frames = 10,240 Ms). However, in R16 NR, as mentioned above, because the network supports semipersistent scheduling resources with a period of any integer multiple of unit time slots configured for the UE, the uplink/downlink semipersistent scheduling resource period may not be divisible by the system hyperframe duration. Accordingly, when the hyperframe number is changed, the spacing between neighboring SPS positions becomes abnormal, as shown in Fig. 16.6.

According to TS 38.321 [6], the position of NR R15 type-1 uplink grant depends on three factors: the system frame number (SFN, System Frame Number), the number of time slots in each frame (slot number in the frame), and the number of symbols in each slot (symbol number in the slot). The UE calculates the position where the uplink grant appears, according to S, the time-domain offset value (timeDomainOffset) and the period (periodicity) by incrementing N from 0, starting from SFN = 0 of a new hyperframe. Among them, S is derived from SLIV (Start and Length Indicator) [7], which gives the specific first uplink grant OFDM start symbol position.

It could be seen from Fig. 16.6 that in two consecutive hyperframes, the position of the first uplink grant is the same relative to the two SFN = 0 boundaries (the distance from the SFN = 0 boundary is determined by Given the time-domain offset), the result is that the distance between the first uplink grant after the boundary of each hyperframe and the last uplink grant before the boundary of the hyperframe does not match the distance given by the periodicity parameter. The same problem is also raised for the type-2 CG, wherein the position of the first uplink UL is given by the DCI activation indication received by the UE. So how is this problem solved? In fact, the method is very simple. First, the

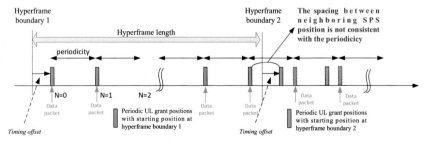

Figure 16.6 The time-domain interval between the semipersistent scheduling resource across the hyperframe boundary is abnormal.

following SPS time-domain position is anchored to the previous time-domain position with a certain period. Secondly, when crossing the hyperframe boundary, N is no longer reset to 0. Correspondingly, the 3GPP has decided to modify the formulas for determining the cyclic time domain of type-1 and type-2 CG. For details, refer to TS 38.321 [6].

Finally, another change that needs to be noticed: for type-1 CG, in TS 38.321 [6], the time-domain position calculation formula has added timeReferenceSFN related items. The main motivation is that RLC retransmission or air interface transmission delay uncertainty may cause excessive delay between the time when the RRC configuration signaling is sent by the network and the time when the UE successfully receives the signaling. Suppose that when sending the signaling, the network obtains the time-domain offset and other parameters according to the position of SFN = 0 in the current hyperframe and the expected position of the uplink grant period. If the message is delayed such that it arrives at the UE in the next hyperframe, it will refer the configured time-domain offset to the SFN = 0 position in the next hyperframe to obtain the time-domain position of the first uplink grant. In this case, the actual time-domain location of the uplink grant does not match the network requirements, which affects data transmission. So how to solve this problem? 3GPP decided to additionally configure the reference SFN value for the UE in the RRC signaling (timeReferenceSFN, the default value is 0). When the network estimates that the scenario that the time when the terminal receives the RRC signaling and the time when the network sends the RRC signaling are distributed on opposite side across the boundary might emerge, the network can set timeReferenceSFN to 512 in the signaling and set the time-domain offset according to the next hyperframe boundary.

16.4.4 Redefine hybrid automatic repeat request ID

In R15 NR, for the uplink/downlink grant of semipersistent scheduled transmission, the HARQ process ID is only related to the time-domain start position of the first symbol in the time-frequency resource.

As illustrated in Section 16.3.2, NR R16 supports the configuration of multiple active SPS resources for a UE. According to the HARQ process ID determination formula shown in TS 38.321 [6], for the uplink grants of multiple semipersistent scheduling resources configured by the network in a certain period of time, if the floor operation result of dividing the start positions of the symbols of the respective SPS UL grant in the time domain (CURRENT_symbol) by the transmission period is equal, their HARQ process IDs will also be the same. And therefore the MAC PDUs required to be carried by such uplink grants need to be stored in the same buffer corresponding to the same HARQ process. Even if the standard allows the buffer of the same HARQ process to store these MAC PDUs at the same time, once a transmission error occurs and a request of the retransmission (using the HARQ ID) occurs, the network cannot figure out for which MAC PDU the UE is requesting.

To solve the problem, the 3GPP RAN2 group decided to introduce a new IE named harq-procID-offset-r16. For each group of SPS resources, the HARQ process ID is not only related to the time-domain start position of the first symbol, but also related to the newly introduced IE configured by the network. From TS 38.331 [5], it could be found that harq-procID-offset-r16 IE is an integer, ranging from 0 to 15. In this way, for multiple uplink/downlink SPS resources with the same start position in the time domain, the MAC PDUs carried on them will be put into different HARQ entities and corresponding buffers.

16.5 Enhancement to packet data convergence protocol data packet duplication

16.5.1 R15 new radio packet data convergence protocol data duplication mechanism

In R15 NR, in order to meet the basic high-reliability requirements of URLLC data transmission, 3GPP RAN2 determined the packet data convergence protocol (PDCP) and data packet duplication mechanism during the standardization process. Specifically, in the carrier aggregation scenario, after the PDCP packet duplication transmission is enabled, the data packets configured to be conveyed on the signaling radio bearer (SRB)/data radio bearer (DRB) could be transmitted on the corresponding two logical channels (one corresponds to the primary RLC entity and another one corresponds to the secondary RLC entity). Note that if two RLC entities serve the same radio bearer, the srb-Identity or drb-Identity in their corresponding configuration RLC-BearerConfig will be set to the same value, and finally when the MAC layer builds the MAC PDU, it is mapped to the transmission resources of the corresponding different carriers, for which the architecture is shown in Fig. 16.7; in the dual-connection scenario, after the PDCP data packet duplication transmission

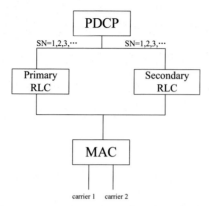

Figure 16.7 Packet data convergence protocol packet duplication for CA scenario (introduced in R15).

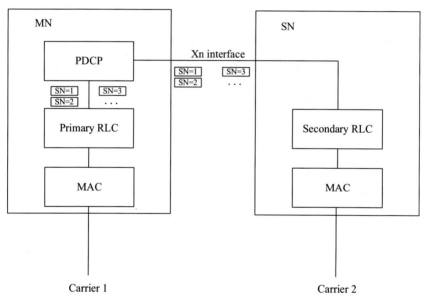

Figure 16.8 Packet data convergence protocol packet duplication for DC scenario (introduced in R15).

is enabled, the primary RLC entity and the secondary RLC entity at the UE will send the same data packet to the UE, as shown in Fig. 16.8.

In both scenarios, after reception of the PDCP packets, the PDCP layer of the UE needs to complete the task of identifying and discarding the redundant packet according to the PDCP SN number. In addition, if it is confirmed that the data packet is successfully transmitted on one of the communication links, the PDCP layer will also try to prevent the same data from being sent by the other link to save air interface transmission resources.

For SRB, the status of the duplication transmission is always active. For DRB, the state is determined by the network by means of RRC signaling or MAC CE. If RRC configuration signaling is used, the value of pdcp-Duplication IE in PDCP-Config can be set to "true" or "false," indicating whether the UE behavior of the PDCP duplication is enabled or not after reception of this RRC signaling. In addition, the network can also dynamically turn on/off the data packet duplication transmission of the bearer by sending the duplication activation/deactivation MAC CE (as shown in Fig. 16.9). Among them, the value of the i-th bit (0/1) indicates that the UE needs to deactivate/activate the i-th DRB (the DRB ID of the i-th DRB is the i-th one among of all configured DRB ID in the ascending order). After PDCP packet duplication transmission (through MAC CE or RRC signaling) is in the inactive state or deactivated, the primary RLC entity and logical channel will still be responsible for data packet transmission, while the secondary RLC entity and the corresponding logical channel will not be used.

16.5.2 Enhancement on duplication transmission in R16

In the R16 NR specification Study Item process, in order to meet the more stringent data transmission reliability requirements of industrial Ethernet, some European and American network operators proposed to allow the UE to use more than two RLC transmission links in the PDCP packet duplication transmission mode. After online discussion, 3GPP RAN2 reached the conclusion that it allows the network to configure upto four RLC transmission links for the PDCP duplication transmission.

In practice, the network will configure upto 4 RLC transmission links for each targeted DRB (DRB ID/SRB ID for more than two RLC entities are set as the same). Similar to the R15 duplication, it is implied that the network has configured duplication transmission for the UE, when

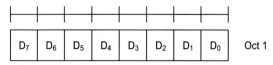

Figure 16.9 Duplication activation/deactivation MAC CE payload.

PDCP-duplication IE has been configured in the PDCP-config IE for the RB. For SRB, if the value of the PDCP-duplication IE is set as 1, all involved RLC entities are active for PDCP duplication. On the other hand, for DRB, when the PDCP-duplication IE is set as 1, a further check of the activation state of each involved RLC entity is needed. To achieve this goal, in the newly introduced moreThanTwoRLC-r16 IE, duplicationState IE is defined:

- This IE is formatted as a 3-bit bitmap, which implies the current activation state of each secondary RLC transmission link. If the bit value is set to 1, the corresponding secondary RLC transmission link is considered as in the active state.
- If the number of secondary RLC links used for PDCP duplication is 2, the value of the highest bit in the bitmap will be ignored.
- If the duplicationState IE is not included in the RRC configuration, all secondary RLC links is deactivated for PDCP duplication.
- In the RRC configuration, the configuration of PDCP-duplication IE and duplicationState IE need to be consistent to a certain extent, as shown in Table 16.4.

Similar to R15 NR, the network supports the option of falling back to the split bearer from PDCP duplication for the UE: a logical channel corresponding to the split bearer could be configured in the morethanTwoRLC-r16 IE. When falling back to a split bearer, except for the main transmission link, the UE could only perform data transmission on this logical channel.

After the completion of RRC configuration, the network can dynamically change the active transmission link for the UE based on the monitoring result if the channel condition by the network or the channel condition fed back by the UE. A new RLC activation/deactivation MAC CE is introduced in R16 and is used to dynamically change the currently

Table 16.4 The configuration relationship between PDCP-duplication and duplicationState.

IE	Configuration #1	Configuration #2	Configuration #3
PDCP-duplication	Not included	Set as 1	Set as 0
duplicationState	Not included	At least 1 bit is set as 1	Absent or all bits are set as 0

Figure 16.10 Payload of the RLC duplication activation/deactivation MAC CE.

activated RLC duplication transmission link. The payload format of the MAC CE is composed of DRB ID and related RLC activation status flags, as shown in Fig. 16.10.

The DRB ID in the duplicate RLC activation/deactivation MAC CE shown in Fig. 16.10 identifies the target bearer. The bit position 0/1 indicates that the UE needs to deactivate/activate the corresponding RLC transmission link (the RLC transmission link with index 0 to 2 corresponds to the secondary RLC transmission link in ascending order of logical channel ID). Using MAC CE, the network can promptly instruct the UE to use which of the configured secondary RLC transmission links to perform packet duplication transmission on a given bearer. When the bits corresponding to all secondary RLC transmission links are set to 0, for this bearer, there are two situations:

- The UE stops the packet duplication transmission, and only uses the primary RLC transmission link given in the RRC signaling to transmit PDCP PDUs.
- The UE suspends the data packet duplication transmission and returns to the split bearer state.

16.5.3 The concept of UE autonomous duplication transmission

As mentioned earlier, in the PDCP duplication enhancement mechanism based on network instructions, the UE first needs to report channel conditions and other information to the network. The network then makes a judgment whether the duplication transmission mechanism needs to be activated based on the information reported by the UE, and the network makes a series of decisions based on the information reported by the UE, such as whether the duplication transmission needs to be activated, how many RLC transmission links should be activated, whether a specific activated RLC transmission link needs to be changed, and so on. After that, the network will send the judgment result to the UE in the form of the MAC CE as shown in Fig. 16.10. Finally, the UE changes the activation state of the corresponding bearer according to the received MAC CE (if necessary).

Suppose that the UE is allowed to independently determine the duplication transmission activation status the first time if the UE finds out that the channel condition of the main transmission link, the HARQ feedback condition, the data packet transmission delay, or other reference information meets certain conditions. The duplication transmission will take effect in a more real-time fashion. However, some 3GPP members, such as network equipment vendors, are worried that the UE and network equipment may have to experience an awkward short-term transmission activation status mismatch period. During R16 meetings, a lot of discussions on this topic were carried on. For further details, refer to the related email discussion [8] and RAN2 107th chairman note. The final conclusion was to temporarily postpone the support of this feature.

Another variant is to let the UE implement the duplication transmission on specific data packets. Specifically, the survival time mechanism could be applied to some IIOT APPs. For instance, when the survival time is set as two cyclic time, if the first data packet is not successfully transmitted, the system has another opportunity for data transmission. If, unfortunately, the second-time data packet transmission fails again, the IIOT system will be in a failed state, which is considered as a serious problem. Obviously, a good idea is to duplicate the transmission of the second data packet to improve the reliability of the transmission, for which the UE autonomous transmission could assist a lot. In the future, it is foreseen that UE autonomous transmission will be discussed again at the proper time.

16.6 Ethernet header compression

Time-Sensitive Communication (TSC) services usually use the encapsulation format of Ethernet frames. Considering that TSC services to be transmitted on 5G system, the proportion of Ethernet frame header and load, and the promotion of the resource utilization of Ethernet frame transmission on the air interface, R16 introduced a Ethernet header compression (EHC) mechanism.

Since R15 NR does not support header compression for Ethernet frames, the first question is how to implement EHC. Considering that 5G already supports the Robust Header Compression (RoHC) mechanism for Internet Protocol (IP) headers, some people believe that the same implementation principles as ROHC can be adopted (i.e., the RoHC algorithm is organized by other organizations, and as a result, 5G networks only need to configure the corresponding RoHC parameters, and use the configured RoHC parameters and algorithms specified by other organizations to perform header compression and decompression processing). Others believe that if the same principles are adopted, other organizations need to be triggered to study and standardize Ethernet packet header compression. Accordingly, RAN2 work will be limited by the progress of other working groups. This will bring a lot of time delay and communication between groups, which is not conducive to the progress of 3GPP standardization. Therefore it was concluded that EHC work is completed independently by the 3GPP.

In the specific design, EHC adopts a design principle similar to RoHC (i.e., based on context information to save, identify, and restore the compressed header).

When designing context information, it is easily concluded that context identifier (CID) is the context information of EHC, but it is hard to reach a conclusion on whether to include the profile identifier. Some companies believe that profiles can be used to distinguish whether the Ethernet frame contains Q-tags and how many Q-tags are included. Other companies believe that profiles can be used to distinguish different high-level protocol types. The opposing view is that we can give a larger range of the CID, which is also used to distinguish various information. For simplification, RAN2 finally determined that there is no need to distinguish different high-level protocol types in R16, and only CID is used as the context information of EHC.

In other words, the decompressor recognizes and restores the compressed Ethernet frame based on the context identifier. Specifically, the compressor and the decompressor mark the original header information that needs to be compressed as a context. Each context is uniquely identified by a context identifier. In R16, two lengths of CID are supported and configured by RRC, which are 7 bit and 15 bit, respectively.

For the compressor, when the context is not established, the compressor will send a data packet containing the full header (FH) to the decompressor. After the context is established, the compressor will send a data packet containing the compressed header (CH) to the decompressor. Then, one issue is how to determine whether the context has been established or whether the state transition can be started? The following two options were considered in discussion:

- Solution1: feedback-based state transition
- Solution2: N-time FH packet-based state transition

If solution2 is adopted, 3GPP needs to consider how to standardize the number N. Generally speaking, if it depends on compressor implementation to set the value of N, it will introduce an abnormal situation where the decompressor has to decompress the compressed packet when without established context. But it is not easy work to determine an appropriate value of N, since it is related to various factors such as the decompressor processing capability and channel quality. In addition, since only a bidirectional link is supported in R16 for EHC, solution1 was finally adopted by the 3GPP.

Correspondingly, the EHC compression process is as follows: For an Ethernet frame packet stream, the EHC compressor first establishes an EHC context and associates it with a context identifier. Then, the EHC compressor sends data packets containing the full header and CID to the decompressor. After receiving the data packet containing the full header and CID, the decompressor establishes context information associated with the included CID. When the context information is established at the decompressor, the decompressor sends EHC feedback packet to the compressor, to indicate the context has been established successfully. When the compressor receives the EHC feedback packet, the compressor starts to send data packets containing the compressed header and CID to the decompressor. When the decompressor receives such a data packet, it will restore the original header of the compressed packet-based on CID and the stored original header information corresponding to the context identifier. The decompressor can determine whether the received data packet is the one with a full header or compressed header according to the packet type indication carried in the packet header (i.e., F/C). The packet type indication occupies one bit. The processing flow for EHC compression is shown in Fig. 16.11.

When EHC is configured, the following fields can be compressed for an Ethernet frame header: destination address, source address, 802.1Q-tag, and length/type. Since the preamble, Start-of-Frame Delimiter (SFD), and Frame Check Sequence (FSC) will not be transmitted through an air interface in 3GPP system, there is no need to consider how to compress such fields.

Similar to RoHC, the function of EHC is implemented at the PDCP layer. The RRC layer can configure EHC parameters for the uplink and downlink separately for the PDCP entity associated with the DRB. If EHC is configured, the compressor will perform Ethernet header compression on the data packets carried through the DRB. It should be noted

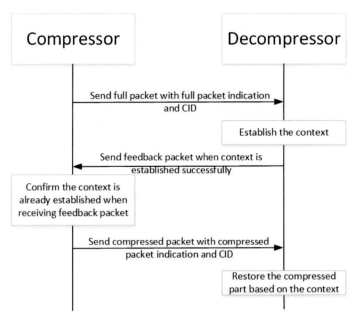

Figure 16.11 The processing flow for Ethernet header compression.

that EHC is not applied to the Service Data Adaptation Protocol (SDAP) packet header and SDAP control PDU.

Both EHC and ROHC can be supported in R16 NR, where RoHC is used for IP header compression, and EHC is used for Ethernet frame header compression. For a DRB, EHC and RoHC are configured independently. When both RoHC and EHC are configured for a DRB, the RoHC header is located behind the EHC header. If a PDCP SDU including non-IP Ethernet packet is received from the upper layers, the EHC compressor will bypass the ROHC compressor and submit the EHC compressed non-IP Ethernet packet to the lower layers. If a PDCP Data PDU including non-IP Ethernet packet is received from the lower layers, the EHC decompressor will bypass the ROHC decompressor and deliver the EHC decompressed non-IP Ethernet packet to the upper layers.

16.7 Summary

This chapter introduced the relevant content and conclusions of IIoT technology. It mainly involves time synchronization, scheduling enhancement, Ethernet header compression, intra-UE prioritization for uplink and PDCP duplication. These technologies can enable the 5G system to support/guarantee for URLLC/TSC services and meet the requirements of ultrahigh-reliability and low-latency for such services.

References

[1] 3GPP TR 38.825: study on NR Industrial Internet of Things (IOT), V16.0.0, 2019−03.
[2] 3GPP TR 22.804: Study on Communication for Automation in Vertical Domains, V2.0.0, 2018−05.
[3] Lee KB, Eldson J. Standard for a precision clock synchronization protocol for networked measurement and control systems. In: 2004 Conference on IEEE 1588, Standard for a Precision Clock Synchronization Protocol for Networked Measurement and Control Systems; 2004.
[4] 3GPP TR 23.734: Study on enhancement of 5G System (5GS) for vertical and Local Area Network (LAN) services, V16.2.0, 2019−06.
[5] 3GPP TS 38.331: Radio Resource Control (RRC) protocol specification, V16.0.0, 2020−03.
[6] 3GPP TS 38.321: Medium Access Control (MAC) protocol specification, V16.0.0, 2020−03.
[7] 3GPP TS 38.214: NR: Physical layer procedures for data, V16.1.0, 2020−03.
[8] R2−1909444 Summary of e-mail discussion: [106#54] [IIoT] Need for and details of UE-based mechanisms for PDCP duplication, CMCC.

CHAPTER 17

5G V2X

Zhenshan Zhao, Shichang Zhang, Qianxi Lu, Yi Ding and Kevin Lin

17.1 NR−V2X slot structure and physical channel

17.1.1 Basic parameters

New Radio - Vehicle-to-Everything (NR−V2X) developed by 3GPP in Release 16 based on the latest 5G technology is designed to operate in dedicated Intelligent Transportation System (ITS) spectrums. At the same time, in order to expand the application of NR−V2X, it is also able to coexist with NR Uu in licensed frequency bands. Furthermore, NR−V2X is capable of operating in FR1 and FR2 spectrum bands. However, besides its support for Phase-Tracking Reference Signal (PT−RS), the Release 16 version of NR−V2X does not have other optimizations for operating in FR2. Therefore NR−V2X does not support enhancement for FR2 such as beam management in Release 16 [1].

The subcarrier spacing (SCS) and corresponding Cyclic Prefix (CP) length supported by NR−V2X in FR1 and FR2 are the same as in NR Uu, as shown in Table 17.1. In NR Uu, the network can configure an individual and different active BWP for each terminal in a cell and each Bandwidth Part (BWP) with its own SCS. From the perspective of the system, the system can support multiple SCSs at the same time. However, for NR−V2X if different terminals are configured with different sidelink BWP with different SCS, it is necessary for the receiving terminal to support multiple SCSs at the same time in order to receive all the data sent by other terminals. Therefore in order to reduce the complexity of UE implementation, only one CP length and one SCS can be configured on a sidelink carrier.

Two waveforms are supported in the uplink of NR Uu, namely CP−OFDM and DFT−s−OFDM, and they were both considered for NR sidelink as well since they are to coexist in the same carrier when operating in a licensed band. In the RAN1#94 and RAN1#95 meetings, RAN1 discussed the waveforms to be supported by NR−V2X. Some companies suggested that NR−V2X can follow the NR uplink design and support the above two waveforms, whereas most companies suggested that NR−V2X only needs to support CP−OFDM. One main reason to

5G NR and Enhancements
DOI: https://doi.org/10.1016/B978-0-323-91060-6.00017-9

Table 17.1 Subcarrier spacing and CP length supported by NR–V2X in different frequency ranges.

	FR1			FR2	
Subcarrier spacing	15 kHz	30 kHz	60 kHz	60 kHz	120 kHz
Cyclic Prefix (CP) length	Normal	Normal	Normal and extended	Normal and extended	Normal

support DFT–s-OFDM was due to the fact that Peak to Average Power Ratio (PAPR) of this waveform is lower than that of CP–OFDM, and thus it would increase the coverage of sidelink transmission, especially for Sidelink Synchronization Signal (S–SS), Physical Sidelink Control Channel (PSCCH), and Physical Sidelink Feedback Channel (PSFCH). Due to the increased coverage of S–SS and Physical Sidelink Broadcast Channel (PSBCH), it can help to minimize multiple groups of the sidelink UE using different transmission/synchronization timings when they are not operating in the coverage of a cellular network and GNSS signal. Increasing the coverage of PSCCH is beneficial to improve the performance of sensing-based resource selection to avoid transmission collisions of using same sidelink resources. In addition, since the PSFCH only occupies one OFDM symbol in a slot without repetition, in some extreme communication range cases the increased coverage of PSFCH would help to improve Hybrid Automatic Repeat Request (HARQ) reception performance even when the maximum transmission power is used. On the other hand, if both waveforms are supported, sidelink-capable UE will need to support transmission and reception of DFT–s-OFDM at the same time, whereas UEs in NR Uu only need to support the transmission of DFT–s-OFDM but not the reception, so the complexity of sidelink UE implementation will be significantly increased. In addition, in NR–V2X, the PSCCH and Physical Sidelink Shared Channel (PSSCH) are multiplexed on different Physical Resource Blocks (PRB) in some OFDM symbols, meaning the UE needs to send PSCCH and PSSCH simultaneously. In this case, the low PAPR advantage of DFT–s-OFDM will no longer exist since a separate DFT spread operation is performed for the two channels. Furthermore, the Zadoff-Chu (ZC) sequence is adopted for S–SS and PSFCH, and DTF–s–OFDM will not bring additional gain in terms of PAPR. In comparison, the disadvantages of DTF–s–OFDM are far greater than the benefits. As such, at the

RAN1#96 meetings RAN1 decided that only the CP−OFDM wave-form is supported for NR−V2X.

Similar to the NR Uu interface, Sidelink (SL) BWP configuration is supported on NR−V2X carriers. Due to the presence of broadcast and groupcast services in sidelink communication, the UE needs to send side-link signals to multiple receiving UEs, and UEs may also need to receive sidelink signals transmitted by multiple UEs at the same time. To avoid transmitting or receiving on multiple BWPs at the same time, at most one SL BWP can be configured on a carrier, and the configured SL BWP is to be used for both transmitting and receiving SL signals and channels. In a licensed frequency band, if the UE is configured with SL BWP and UL BWP at the same time, the SCS needs to be the same. Such restriction will prevent UEs from using two different SCSs simultaneously.

In NR−V2X, the UE can be also configured with one or multiple sidelink Resource Pools (RPs), which limits the range of time and fre-quency resources for sidelink communication. The minimum time-domain granularity of a sidelink RPs is one slot, and a RP can contain noncontiguous slots in the time domain. In the frequency domain, the minimum granularity is one subchannel, which comprises multiple conse-cutive PRBs. In NR−V2X, a subchannel can contain 10, 12, 15, 20, 25, 50, 75, or 100 PRBs. Since only the CP−OFDM waveform is supported, subchannels of a RP must be continuous in the frequency domain in order to reduce the PAPR of sidelink transmission. In addition, frequency resources of SL RPs should be within the bandwidth of an SL BWP, as shown in Fig. 17.1.

17.1.2 Sidelink slot structure

There are two different slot structures in NR−V2X. The first type of slot structure contains the PSCCH, PSSCH, and possibly PSFCH, which is

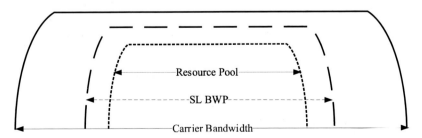

Figure 17.1 The relationship between carrier, SL BWP, and resource pool.

referred to as the normal slot structure. The second type of slot structure contains S—SS and PSBCH, which is referred to as the S—SSB slot structure.

Fig. 17.2 shows the first type of slot structure in NR—V2X. It can be seen that the first OFDM symbol is allocated for Automatic Gain Control (AGC) at the sidelink reception UE. On this AGC symbol, the UE copies the signal generated for the second OFDM symbol and transmits it in the first OFDM symbol as a repetition. And there is a symbol at the end of the slot for receiving and sending conversions, which is used for UEs to switch from the sending (or receiving) state to the receiving (or sending) state. For other OFDM symbols in a sidelink slot, the PSCCH occupies two or three OFDM symbols starting from the second OFDM symbol. In the frequency domain, the number of PRBs occupied by the PSCCH is within the range of a PSSCH subchannel. If the number of PRBs occupied by the PSCCH is less than the size of a subchannel of the PSSCH, or the frequency-domain resource of the PSSCH includes multiple subchannels, the PSCCH can be frequency-division multiplexed with the PSSCH on the OFDM symbols of the PSCCH.

In the RAN1#94 meeting, RAN1 discussed the multiplexing of PSCCH and PSSCH. Four different options of multiplexing PSCCH and PSSCH in the time and frequency domains were considered, as shown in Fig. 17.3.

17.1.2.1 Option 1

The PSCCH and PSSCH occupy nonoverlapping OFDM symbols in the time domain and occupy the same PRBs in the frequency domain, meaning they are purely time-division multiplexed only. This multiplexing

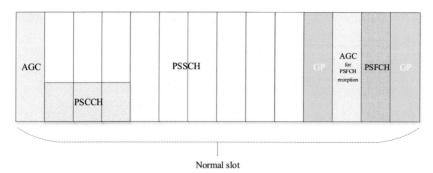

Normal slot

Figure 17.2 NR—V2X slot structure for 14 OFDM symbols.

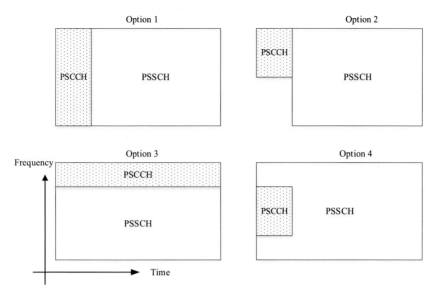

Figure 17.3 Multiplexing of PSCCH and PSSCH.

option helps to reduce the decoding delay of the PSSCH, because the decoding of PSCCH can start immediately after the last PSCCH symbol. However, since PSCCH and PSSCH occupy the same PRBs in the frequency domain, the number of PRBs occupied by the PSCCH in the frequency domain will vary with the number of PRBs used by the PSSCH. In NR−V2X, the traffic load and coding rate may vary significantly, resulting in a large dynamic range of the number of PRBs for the transmitting PSSCH. Furthermore, the PSSCH can start from any subchannel. Therefore the receiving UE needs to blindly detect the PSCCH for all possible PSCCH frequency-domain sizes at the beginning of each subchannel.

17.1.2.2 Option 2

Same as Option 1, the PSCCH and PSSCH also occupy nonoverlapping OFDM symbols. As such, Option 2 and Option 1 will have the same decoding performance in terms of latency. However, different from Option 1, the number of PRBs occupied by the PSCCH in Option 2 does not change with the size of PSSCH, so the receiving UE can avoid blind detection of the PSCCH according to different hypotheses of PSCCH size. However, resources on the OFDM symbols occupied by the PSCCH will be wasted since the number of PRBs occupied by the PSSCH is usually more than that of the PSCCH in this case.

17.1.2.3 Option 3

Option 3 of multiplexing PSCCH and PSSCH is arranged in the same manner as adopted in Long Term Evolution—Vehicle-to-Everything (LTE—V2X). This means the PSCCH and PSSCH occupy nonoverlapping frequency-domain resources, but coexist in the same OFDM symbols. In this way, the PSCCH occupies all OFDM symbols in a slot and it is possible to boost the power spectral density of PSCCH by 3 dB to improve the reliability of the PSCCH as in LTE—V2X. However, the receiving UE can only start to decode the PSCCH at the end of the slot. As such, it will ultimately lead to a higher decoding delay for the PSSCH than Option 1 and Option 2.

17.1.2.4 Option 4

In this option, the PSCCH and a part of the PSSCH are multiplexed on the same OFDM symbols in nonoverlapping frequency-domain resources, while other parts of PSSCH are transmitted on the remaining/nonoverlapping OFDM symbols. Option 4 has the same advantage of low latency as Option 1 and Option 2. However, blind detection can be avoided because the frequency-domain size of the PSCCH is kept constant. In addition, on the OFDM symbols occupied by the PSCCH, if the number of PRBs occupied by PSCCH is less than that of the corresponding PSSCH, the remaining PRBs can still be used for the PSSCH transmission, so the problem of resource waste in Option 2 can be avoided. Since Option 4 has the same advantages as in other options, it was eventually adopted for PSCCH and PSSCH multiplexing in NR—V2X.

In NR—V2X, PSFCH resources are configured periodically, which can be {0, 1, 2, 4} slots within a sidelink RP. If the period is set to 0, it means that there is no PSFCH resource configured in the RP. When the period is set to two or four slots, it reduces the resources allocated for the PSFCH in the system compared to when the period equals to one slot. For a slot that contains PSFCH, PSFCH resources are only allocated in the second last OFDM symbol in the slot. Since the received power may change on the OFDM symbol where the PSFCH is located compared to other OFDM symbols containing the PSSCH, the preceding OFDM symbol (the third symbol last in the slot) is also used for PSFCH transmission to assist receiving UEs with AGC adjustment. The signal to be transmitted on this OFDM symbol (the third last symbol) is a repetition of the signal for the PSFCH transmission on the second last symbol. In addition, since the UE sending the PSSCH may be different than the UE sending the PSFCH in the same slot, it is also necessary to have a gap symbol before the two PSFCH symbols for switching between transmission and reception, as shown in Fig. 17.2.

To support synchronization among UEs operating in the out-of-coverage of cellular network and GNSS, NR−V2X UEs need to send the S−SS and PSBCH to align sidelink transmission and reception timings. The transmission of S−SS and PSBCH occupies a slot, namely the S−SSB slot, as shown in Fig. 17.4. The S−SS consists of a Sidelink Primary Synchronization Signal (S−PSS) and a Sidelink Secondary Synchronization Signal (S−SSS). The S−PSS occupies the second and third OFDM symbols, and the S−SSS occupies the fourth and fifth OFDM symbols in the slot. The last symbol is used as a GP symbol and the remaining symbols in the slot are used for PSBCH. The two S−PSS and S−SSS are continuous in the time domain, so that channel estimation results obtained from S−PSS can be applied for S−SSS as well to improve detection performance of S−SSS.

17.1.3 Physical sidelink channel and sidelink signal

17.1.3.1 Physical sidelink control channel

In NR−V2X, the PSCCH is used to carry SL control information related to sidelink resource allocation and sensing (as elaborated on in Section 17.2.4) and decoding of the PSSCH. The PSCCH occupies two or three OFDM symbols in the time domain and {10, 12, 15, 20, 25} PRBS in the frequency domain. The number of OFDM symbols and the number of PRB occupied by the PSCCH in a RP are (pre)configured by the network. To avoid any restrictions on PSSCH resource selection or allocation, the number of PRBs

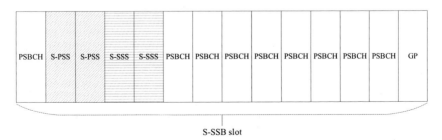

S-SSB slot

Figure 17.4 S−SS/PSBCH slot structure.

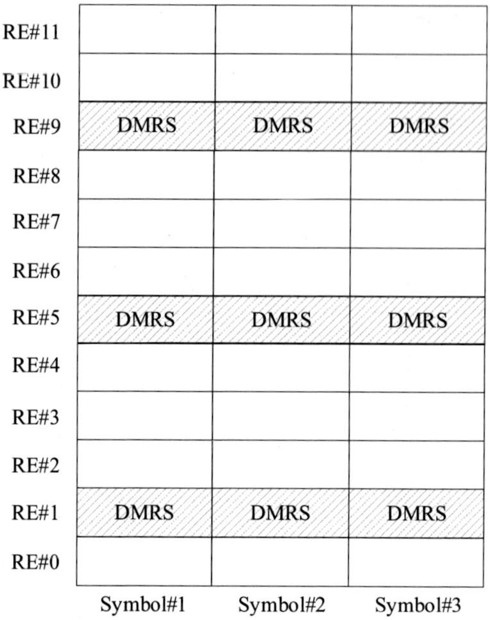

	Symbol#1	Symbol#2	Symbol#3
RE#11			
RE#10			
RE#9	DMRS	DMRS	DMRS
RE#8			
RE#7			
RE#6			
RE#5	DMRS	DMRS	DMRS
RE#4			
RE#3			
RE#2			
RE#1	DMRS	DMRS	DMRS
RE#0			

Figure 17.5 PSCCH DMRS pattern.

occupied by the PSCCH should be less than or equal to the number of PRBs allocated for a subchannel in the RP.

Blind detection of the sidelink control channel has a significant impact on the complexity of a receiver UE. In order to reduce the number of PSCCH blind detections, the number of PSCCH symbols and frequency PRBs are (pre)configured in a RP, that is to say, PSCCH has only one aggregation level. In addition, the PSCCH adopts QPSK modulation, which is the same as the downlink control channel in the Uu interface, and always uses Polar coding. Moreover, for broadcast, groupcast, and unicast, the number of bits carried in the PSCCH is always the same.

The DMRS pattern for the PSCCH is the same as that of the PDCCH. The DMRS for the PSCCH is transmitted in all OFDM symbols allocated for the PSCCH, and it is located in {# 1, # 5, # 9} RE of a PRB in the frequency domain, as shown in Fig. 17.5. The DMRS sequence for thePSCCH is generated by the following formula:

$$r_l(m) = \frac{1}{\sqrt{2}}(1 - 2c(m)) + j\frac{1}{\sqrt{2}}(1 - 2c(m + 1)) \tag{17.1}$$

where $c(m)$ is initialized by $c_{init} = (2^{17}(N_{symb}^{slot}n_{s,f}^{\mu} + l + 1)(2N_{ID} + 1) + 2N_{ID}) \mod 2^{31}$, l is the index of OFDM symbol for DMRS in a slot, $n_{s,f}^{\mu}$ is the index of the slot where the DMRS is located within the system frame, $N_{ID} \in \{0, 1, \cdots, 65535\}$, and N_{ID} is (pre)configured by the network in a RP.

In sidelink communication, the UE selects transmission resources autonomously or determines the transmission resources according to gNB scheduling. When each UE autonomously selects its own transmission resources, it is possible that different UEs may select and send the PSCCH in the same time-frequency resource. In order to ensure that a receiver UE can detect at least one PSCCH in the case of PSCCH resource collision, PSCCH DMRS randomization is used in LTE−V2X. Specifically, when the UE sends the PSCCH, it randomly selects a value from $\{0, 3, 6, 9\}$ as the cyclic shift of DMRS. When multiple UEs transmit PSCCH DMRS on the same time-frequency resource with different cyclic shift, the receiver UE should still be able to detect and decode at least one PSCCH thanks to the orthogonality property of the DMRS sequence. For the same purpose, three PSCCH DMRS frequency-domain OCCs are introduced in NR−V2X for random selection by the transmitter UE (TX UE), thus different UEs transmitting the PSCCH on the same time/frequency resource can be differentiated. Finally, the DMRS symbol on each RE in a PRB can be represented as:

$$a_{k,l}^{(p,\mu)} = \beta_{DMRS}^{PSCCH} w_{f,i}(k')r_l(3n + k')$$
$$k = nN_{sc}^{RB} + 4k' + 1$$
$$k' = 0, 1, 2 \tag{17.2}$$
$$n = 0, 1, \ldots$$

where β_{DMRS}^{PSCCH} is the amplitude scaling factor. The quantity $w_{f,i}(k')$ is given by Table 17.2 and $i \in \{0,1,2\}$ will be randomly selected by the UE.

Table 17.2 The quantity $w_{f,i}(k')$.

k'	$w_{f,i}(k')$		
	$i = 0$	$i = 1$	$i = 2$
0	1	1	1
1	1	$e^{j2/3\pi}$	$e^{-j2/3\pi}$
2	1	$e^{-j2/3\pi}$	$e^{j2/3\pi}$

Sidelink Control Information (SCI) carried on the PSCCH in Release 16 is the SCI format 1—A, which is a 1nd-stage SCI. The following information fields are contained in the SCI format 1—A:

- Priority—3 bits
- Frequency-resource assignment
  - When the SCI is used to indicate the current transmission resource and one retransmission resource— $\left\lceil \log_2 \left(\frac{N_{Subchannel}^{SL}(N_{Subchannel}^{SL}+1)}{2} \right) \right\rceil$ bits
  - when the SCI is used to indicate the current transmission resource and two retransmission resources— $\left\lceil \log_2 \left(\frac{N_{Subchannel}^{SL}(N_{Subchannel}^{SL}+1)(2N_{Subchannel}^{SL}+1)}{6} \right) \right\rceil$ bits
- Time resource assignment
  - when the SCI is used to indicate the current transmission resource and one retransmission resource—5 bits
  - when the SCI is used to indicate the current transmission resource and two retransmission resources—9 bits
- PSSCH DMRS pattern—$\log_2 N_{pattern}$ bits, where $N_{pattern}$ is the number of DMRS pattern allowed in the RP
- 2nd-stage SCI format—2 bits
  - "00" indicates SCI format 2—A, "01" indicates SCI format 2—B, "10" and "11" are reserved
- Beta_offset indicator—2 bits
- Number of PSSCH DMRS port—1 bit
- Modulation and coding scheme (MCS)—5 bits
- MCS table indicator—0—2 bits, depends on the number of MCS tables (pre)configured in the RP
- PSFCH overhead indication
  - if PSFCH period is 2 or 4—1 bit
  - otherwise—0 bit
- Resource reservation period—4 bits
- Reserved—2—4 bits, is configured or preconfigured by gNB, with the value set to zero

17.1.3.2 PSSCH

The PSSCH is used to carry 2nd-stage SCI and data information. Firstly, it is necessary to introduce the design of the two stages of SCI adopted for NR—V2X.

Since NR—V2X supports three transmission types—broadcast, groupcast, and unicast—different transmission types will require different SCI formats to support PSSCH transmission. Table 17.3 summarizes the SCI fields that may be required under different transmission types. It can be seen that the SCI fields required by different transmission types have some commonalities, but compared to broadcast transmission, groupcast and unicast transmissions will require more fields. If the same SCI size is adopted for all transmission types, many redundant bits will be included in the SCI for broadcast transmissions, thus affecting the efficiency of resource utilization. On the other hand, if different SCI sizes are used, receiver UEs will need to blindly detect different SCI formats.

In addition, for the unicast transmissions, a different code rate for the SCI could be adapted according to different propagation channel conditions. If a fixed SCI code rate is used assuming the worst propagation channel condition, sidelink resource utilization efficiency would degrade significantly. On the other hand, if dynamically adjusting the SCI code rate according to the propagation channel condition, receiver UEs will also need to blindly detect different SCI formats or resource aggregation levels.

After fierce competition in several meetings, 2nd-stage SCI finally won the support of most companies. At the RAN#98 meeting in August 2019, it was finally decided that NR−V2X would support 2nd-stage SCI. The principle of the 2nd-stage SCI is to minimize the number of bits of the 1nd-stage SCI and to ensure that the number of bits of the 1nd-stage SCI does not change with the transmission type, propagation channel condition, and/or other factors. As such, NR−V2X does not need to adjust the aggregation level of 1nd-stage SCI based on different operating scenarios or transmission type. Based on this principle, 1nd-stage SCI is designed to carry information

Table 17.3 Sidelink Control Information (SCI) fields of different transmission types.

SCI field	Broadcast	Groupcast	Unicast
Time-frequency resource assignment	✓	✓	✓
Physical Sidelink Shared Channel (PSSCH) priority	✓	✓	✓
Modulation and coding scheme	✓	✓	✓
Hybrid Automatic Repeat Request (HARQ) process	✓	✓	✓
Source ID	✓	✓	✓
Destination ID		✓	✓
New Data Indicator (NDI)	✓	✓	✓
HARQ feedback indicator		✓	✓
Zone ID		✓	
Communication range requirement		✓	
Channel State Information (CSI) request			✓

only related to the allocation/reservation of time-frequency resources of the scheduled PSSCH (for data reception and resource sensing), and to indicate the code rate and format of the associated 2nd-stage SCI. In contrast, the 2nd-stage SCI provides other information required for PSSCH decoding. Since the 1nd-stage SCI provides information about the associated 2nd-stage SCI, the 2nd-stage SCI is allowed to use different formats and code rates, so that the receiver UE does not need to perform blind detection of the 2nd-stage SCI. Therefore two stages of SCI can effectively reduce the number of bits required for the 1nd-stage SCI and thus improving its decoding performance and the accuracy of resource sensing. Furthermore, when the number of the 1nd-stage SCI bits remains constant, it also allows coexistence of different transmission types (broadcast, groupcast, and unicast) in the same RP without affecting the reception performance of the PSCCH.

Since the 2nd-stage SCI uses polar coding, a fixed QPSK modulation and the same transmission port as sidelink data in PSSCH, the DMRS of PSSCH can be used for demodulation of the 2nd-stage SCI. Additionally, similar to the transmission of sidelink data in PSSCH, when dual-layer transmission is used to transmit the sidelink data in PSSCH, the modulated symbols of the 2nd-stage SCI are transmitted in the exactly the same manner over two layers. This design can help to guarantee reception performance of the 2nd-stage SCI in a highly correlated channel. Furthermore, the code rate of the 2nd-stage SCI can be dynamically adjusted within a certain range. The specific code rate is indicated by the "Beta_offset indicator" field in the 1nd-stage SCI, so a receiver UE does not need to perform blind detection of the 2nd-stage SCI even when the code rate is changed. The modulated symbols of the 2nd-stage SCI are mapped from the first symbol of the PSSCH DMRS in the frequency domain and then the time domain, and they are multiplexed with the REs of the DMRS by interleaving, as shown in Fig. 17.6.

There are two 2nd-stage SCI formats defined in 3GPP Release 16, namely SCI format 2−A and SCI format 2−B. SCI format 2−B is suitable for groupcast communication with sidelink HARQ feedback based on geographical location and communication range; SCI format 2−A is suitable for other scenarios, such as transmissions (including unicast, groupcast, and broadcast) that do not require sidelink HARQ feedback, unicast communication that requires sidelink HARQ feedback, and groupcast communication that requires ACK or NACK feedback.

The following information is carried in the SCI format 2−A:
- HARQ process—$\log_2 N_{process}$ bits, where $N_{process}$ is the number of HARQ process

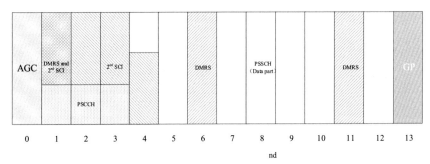

Figure 17.6 Mapping method of 2nd-stage SCI.

- New data indicator—1 bit
- Redundancy version—2 bits
- Source ID—8 bits
- Destination ID—16 bits
- HARQ feedback enabled/disabled indicator—1 bit
- Cast type indicator—2 bits
    - ■ "00" indicates broadcast, "01" indicates groupcast that requires ACK or NACK feedback, "10" indicates unicast, "11" indicates groupcast that requires NACK-only feedback
- Channel State Information (CSI) request—1 bit

SCI format 2−B is designed only to support sidelink groupcast communication. In comparison to SCI format 2−A, SCI format 2−B does not include the "Cast type indicator" and "CSI request" fields, but additionally carries the following two fields:

- Zone ID—12 bits
- Communication range requirement—4 bits

The "Zone ID" field is used to indicate the zone corresponding to the geographic location of the transmitting UE, and the "Communication distance requirement" field is used to indicate the target communication range of the current transmission. In this groupcast communication mode, if a receiving UE is located within the indicated communication range from the transmitting UE but fails to decode the PSSCH, it should feedback NACK. Otherwise, it should not feedback any HARQ information. For details, refer to Section 17.3.1.

The sidelink data part of the PSSCH uses LDPC coding, supports up to 256QAM modulation, and two layer transmission. Multiple MCS tables can be (pre)configured in a RP for PSSCH transmission, including a regular 64QAM MCS table, a 256QAM MCS table, and a low spectrum efficiency 64QAM MCS table [2]. The MCS table used in one transmission is indicated by "MCS table indicator" field in the 1nd-stage SCI. In order to minimize PAPR, the PSSCH is be transmitted using continuous PRBs. Since a subchannel is the minimum frequency-domain granularity for the transmitting PSSCH, it is required that PSSCH transmission occupies continuous subchannels.

Similar to the NR Uu interface, the PSSCH supports multiple time-domain DMRS patterns. In a RP, if the number of useable PSSCH symbols in a slot is larger than or equal to 10, upto three time-domain DMRS patterns can be (pre)configured (i.e., two, three, or four symbol time-domain DMRS patterns). If the number of useable PSSCH symbols in a slot is 8 or 9, then two or three symbols time-domain DMRS patterns can be (pre)configured. For a shorter PSSCH length, only the two-symbol time-domain DMRS pattern can be (pre)configured. It should be noted that the number of PSSCH symbols mentioned above does not include the first OFDM symbol for AGC, the last OFDM symbol used as a GP, the PSFCH symbol, and the AGC and GP symbols before the PSFCH symbol. If multiple time-domain DMRS patterns are (pre)configured in the RP, a specific time-domain DMRS pattern is selected by the transmitting UE and indicated in the 1nd-staeg SCI. Fig. 17.7 shows time-domain DMRS patterns that can be used for the 12-symbol-length PSSCH. Such a design allows high-speed moving UEs to select high-density DMRS patterns, thereby ensuring the

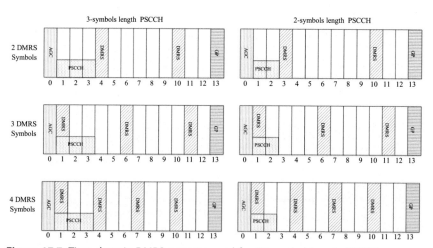

Figure 17.7 Time-domain DMRS patterns used for 12-symbol-length PSSCH.

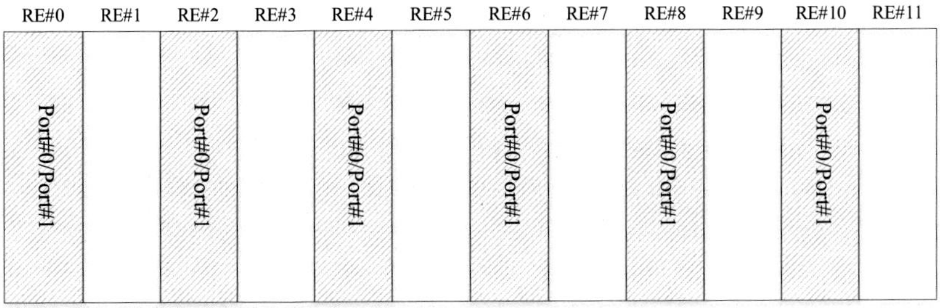

RE#0	RE#1	RE#2	RE#3	RE#4	RE#5	RE#6	RE#7	RE#8	RE#9	RE#10	RE#11

Port#0/Port#1 (RE#0) · Port#0/Port#1 (RE#2) · Port#0/Port#1 (RE#4) · Port#0/Port#1 (RE#6) · Port#0/Port#1 (RE#8) · Port#0/Port#1 (RE#10)

Figure 17.8 Single-symbol Configuration type 1 DMRS pattern.

accuracy of channel estimation. For low-speed moving UEs, low-density DMRS patterns should be used to improve spectrum efficiency.

The NR Uu interface supports two frequency-domain DMRS patterns, namely DMRS Configuration type 1 and DMRS Configuration type 2. For each type, there are single-symbol DMRS patterns and dual-symbols DMRS patterns. The single-symbol Configuration type 1 DMRS pattern supports four DMRS ports, and the single-symbol Configuration type 2 DMRS pattern can support six DMRS ports. In the case of dual symbols, the number of supported ports is doubled. However, in NR−V2X, since only two DMRS ports need to be supported at most in Release 16, only the single-symbol Configuration type 1 DMRS pattern is supported, as shown in Fig. 17.8.

17.1.3.3 PSFCH

In Release 16 version of NR−V2X, only sequence based PSFCH is supported, called PSFCH format 0. The transmission of the PSFCH occupies one PRB in the frequency domain and one OFDM symbol in the time domain. The sequence design is the same as PUCCH format 0. In a RP, the PSFCH resources are periodically configured with one, two, or four slots. In a slot configured with PSFCH, the PSFCH resources are located on the last OFDM symbol that can be used for sidelink transmission. However, in order for a UE to switch between PSSCH/PSFCH transmission or reception, and to perform AGC adjustment, as shown in Fig. 17.2, two OFDM symbols preceding the PSFCH symbol are allocated for the transceiver switching and AGC adjustment, respectively. Consequently, no PSCCH and PSSCH are transmitted on these three OFDM symbols. In Release 16 NR−V2X, the PSFCH is only used to carry HARQ feedback information of just one bit.

It can be seen that among the three OFDM symbols allocated to the PSFCH transmission, only one OFDM symbol is used for the actual transmission of feedback information, and the other two OFDM symbols are only used for transceiver switching or AGC adjustment. Hence, the resource utilization efficiency is relatively low in PSFCH slots. Therefore in the development of NR−V2X specifications, a long PSFCH structure was once considered with an aim to improve resource utilization efficiency. As shown in Fig. 17.9, the long PSFCH structure occupies one PRB in the frequency domain and 12 OFDM symbols in the time domain (all OFDM symbols except AGC symbol and GP symbol in a slot). With this structure, assuming that a total number of PFSCH resources configured in a RP is N, then the total number of REs used for PSFCH should be $N \times 12 \times 12$. However, with a short PSFCH structure, assuming that the number of PRBs contained in the RP is B, then the total number of REs occupied by the three OFDM symbols related to PFSCH in the RP is $B \times 12 \times 3$. Comparing the number of REs occupied by the two structures, it can be found that when the number of sidelink PRBs allocated in the RP is large and the number of required PSFCH resources is small, the long PSFCH structure can effectively reduce resources used for the PSFCH. For example, when $N = 10$ and $B = 100$, the number of resources required by the long PSFCH structure is only 40% of the short PSFCH structure. On the other hand, for sidelink groupcast communication, each PSSCH transmission may require multiple PSFCH feedback resources (related to the number of UEs within a group). With this increased require number of PSFCH resources in the system for groupcast, the advantage of using the long PSFCH structure in terms of resource efficiency become less obvious. In addition, the delay of the long PSFCH structure is greater than that of the short PSFCH structure. Therefore the long PSFCH structure was not adopted in the end.

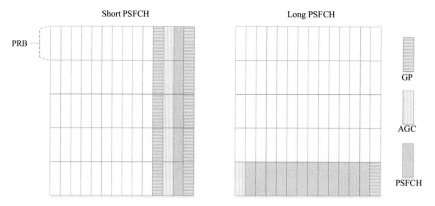

Figure 17.9 Short PSFCH structure and long PSFCH structure.

17.1.3.4 *Physical sidelink broad channel*

For NR−V2X, multiple types of synchronization sources are supported and these include {GNSS, gNB, eNB, UE}. An NR−V2X UE obtains synchronization information from one of the synchronization sources types and transmits S−SS and PSBCH over sidelink to assist synchronization of other UEs. If a UE cannot obtain synchronization information or timing from GNSS and gNB/eNB, it will search for S−SS sent by other UEs on the sidelink to obtain synchronization timing and system information carried in PSBCH. If a UE is able to detect S−SSs sent by another UE and uses it as a synchronization source, the UE forwards synchronization information contents in the PSBCH based on the synchronization information that it receives from the selected synchronization source.

The following information is carried on the PSBCH in NR−V2X:

- sl−TDD−Config: sidelink TDD configuration, this field is to provide information about uplink slots that can be used for sidelink communication, and the field is determined according to a higher layer parameter "TDD−UL−DL−ConfigCommon." This field has a size of 12 bits, and specifically includes one bit to indicate the number of patterns, four bits to indicate the period of the pattern, and seven bits to indicate the number of uplink slots in each pattern.
    - If the number and position of uplink symbols in a slot determined by higher layer parameter "TDD−UL−DL−ConfigCommon" meets the minimum requirement for sidelink transmission (i.e., the symbols in a slot $\{Y, Y + 1, Y + 2, ..., Y + X − 1\}$) are uplink symbols, then the slot can be used for sidelink transmission. Y represents the position of the start symbol used for side-line transmission and X represents the number of symbols used for sidelink transmission. Fig. 17.10 shows a specific example.
- inCoverage: this field is used to indicate whether the UE sending the PSBCH is within the network coverage.
- directFrameNumber: this field is used to indicate the DFN index of the S−SSB.
- slotNumber: this field is used to indicate the slot index of the S−SSB in DFN.

For a UE operating outside of the network coverage, the contents of the PSBCH are determined according to the preconfiguration information. For a UE operating inside of a network cell coverage, the contents of the PSBCH are determined according to the network configuration information. In a NR system, the network semistatically configures TDD configuration for the

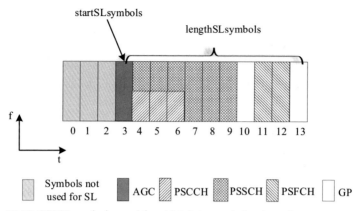

Figure 17.10 OFDM symbols used for sideink transmission in a slot.

cell through the "TDD−UL−DL−ConfigCommon" field. Information on this configuration parameter can be found in Section 5.6.2. Besides the TDD configuration, the configuration information also includes a reference SCS, which can be used to determine the timing boundary of the TDD pattern indicated in "TDD−UL−DL−ConfigCommon" field. However, when the UE transmits PSBCH, it is sent according to the SCS configured for the sidelink. The slot information indicated by the "sl−TDD−Config" field is also determined according to the SCS configured for the sidelink. Therefore in this case the UE needs to convert the number of uplink slots or uplink symbols indicated by "TDD−UL−DL−ConfigCommon" (and determined according to the reference subcarrier space) to the number of the uplink slots and uplink symbols corresponding to the SCS configured for the sidelink (Table 17.4).

When the "TDD−UL−DL−ConfigCommon" field configures only one pattern:

- 1 bit—pattern indicator sets to 0.
- 4 bits—period indicator as shown in Table 17.4.
- 7 bits—uplink resource indicator: the maximum period is 10 Ms for a single pattern. The maximum SCS—120 kHz—is used in the sidelink. The sidelink includes 80 uplink slots at most, which can be fully indicated by the seven bits.

When the "TDD−UL−DL−ConfigCommon" field configures two patterns:

- 1 bit—pattern indicator sets to 1.
- 4 bits—period indicator: when the network is configured with two patterns, the total period of the two patterns $P + P2$ (where P

Table 17.4 PSBCH period indicator when only one pattern is configured.

Physical Sidelink Broadcast Channel (PSBCH) period indicator index	Period (ms)
0	0.5
1	0.625
2	1
3	1.25
4	2
5	2.5
6	4
7	5
8	10
9–15	Reserved

Table 17.5 PSBCH period indicator when two patterns are configured.

Physical Sidelink Broadcast Channel (PSBCH) period indicator index	Total period (P + P2) (ms)	Period of each pattern	
		P(ms)	P2(ms)
0	1	0.5	0.5
1	1.25	0.625	0.625
2	2	1	1
3	2.5	0.5	2
4	2.5	1.25	1.25
5	2.5	2	0.5
6	4	1	3
7	4	2	2
8	4	3	1
9	5	1	4
10	5	2	3
11	5	2.5	2.5
12	5	3	2
13	5	4	1
14	10	5	5
15	20	10	10

represents the period of the first pattern, P2 represents the period of the second pattern) can be divided by 20 Ms. Then the possible combinations of periods are as shown in Table 17.5.

Table 17.6 PSBCH slot indicator granularity when two patterns are configured.

Physical Sidelink Broadcast Channel (PSBCH) period indicator index	Total period (P + P2) (ms)	Period of each pattern		Granularity of different SL SCS			
		P	P2	15 kHz	30 kHz	60 kHz	120 kHz
0	1	0.5	0.5	1			
1	1.25	0.625	0.625				
2	2	1	1				
3	2.5	0.5	2				
4	2.5	1.25	1.25				
5	2.5	2	0.5	1			2
6	4	1	3				
7	4	2	2				
8	4	3	1				
9	5	1	4				
10	5	2	3				
11	5	2.5	2.5				
12	5	3	2				
13	5	4	1				
14	10	5	5	1		2	4
15	20	10	10	1	2	4	8

- 7 bits—uplink resource indicator: the maximum period is 20 Ms for two patterns. In the case of different sidelink SCS, seven bits cannot fully indicate all possible combinations of the number of uplink slots for the two patterns. Therefore the indication information needs to be interpreted with coarser-granularity, specifically, different granularities are used to indicate uplink slots and uplink symbols in different combinations of SCS and different periods as shown in Table 17.6.

17.1.3.5 Sidelink synchronization signal

S—SS in NR—V2X consists of the S—PSS and S—SSS. The S—PSS is generated by a M-sequence with a sequence length of 127 points. The S—SSS is generated by a Gold sequence with a sequence length of also 127 points. The sequence generation formula for S—PSS and S—SSS can be found in Ref. [3]. Once the sequence for the S—PSS is generated, the same sequence is mapped to the second and third OFDM symbols in a synchronization slot. A similar operation is done for S—SSS, where once the sequence for the S—SSS is generated, the same sequence is mapped to the fourth and fifth OFDM symbols.

In NR—V2X, 672 Sidelink Synchronization Signal Identity (SL SSID) are supported, where 2 and 336 identities are allocated for the S—PSS and S—SSS, respectively, as:

$$N_{ID}^{SL} = N_{ID,1}^{SL} + 336 \times N_{ID,2}^{SL} \qquad (17.3)$$

where $N_{ID,1}^{SL} \in \{0, 1, \cdots, 335\}$, $N_{ID,2}^{SL} \in \{0, 1\}$.

17.1.3.6 SL PT—RS

In Release 16 NR—V2X, the SL PT—RS is supported in FR2 only. The PT—RS pattern and sequence generation for the sidelink are the same as those for the NR uplink CP—OFDM. For the physical resource mapping of SL PT—RS, the RB offset is determined by the 16 least significant bits of the 1nd-stage SCI CRC, and the SL PT—RS cannot be mapped to the REs that are occupied by PSCCH, 2nd-stage SCI, SL CSI—RS, and PSSCH DMRS.

17.1.3.7 SL CSI—RS

In order to enhance the performance of sidelink unicast communication, NR—V2X supports SL transmission of CSI—RS for RI and CQI reporting from a receiver UE. SL CSI—RS is sent only when the following three conditions are met:

- UE also transmits a PSSCH,
- Higher layer signaling activates SL CSI reporting, and

- When higher layer signaling activates SL CSI reporting, and the corresponding bit in the 2nd-stage SCI triggers SL CSI reporting.

That is, the UE cannot send the SL CSI−RS alone without any corresponding PSSCH data. The time-frequency position of the SL CSI−RS is determined by the transmitting UE and notified to the receiving UE through PC5−RRC. In order to avoid impact to the resource mapping of the PSCCH and 2nd-stage SCI, the SL CSI−RS should not be mapped to the same time-frequency resources occupied by the PSCCH and transmitted on the same OFDM symbol as the 2nd-stage SCI. Since channel estimation accuracy in the OFDM symbols occupied by PSSCH DMRS is expected to be high and SL CSI−RS of two -antenna ports will occupy two consecutive REs in the frequency domain, the SL CSI−RS should not be sent on the same OFDM symbol as the PSSCH DMRS. In addition, the SL CSI−RS should not collide with SL PT−RS.

17.2 Sidelink resource allocation

Different from the traditional cellular network where the UE and BS (gNB/eNB) transmit data information through the uplink and downlink, in V2X communication vehicle-mounted UEs directly exchange information with each other through a sidelink without data information routed through a central node or network. Since the V2X communication system is required to support deployment in both in network coverage and out-of-network coverage, NR−V2X defines two resource allocation modes, namely Mode 1 and Mode 2. The resource allocation Mode 1 is for sidelink resource allocation scheduled by gNB, and resource allocation Mode 2 is for UE autonomous resource selection. In resource allocation Mode 1, the gNB can provide resource allocation dynamically using DCI or semistatically using RRC configuration of sidelink configured grant (CG).

17.2.1 Resource allocation in time domain and frequency domain

17.2.1.1 Resource allocation in time domain

In NR−V2X, sidelink resource allocation granularity in the time domain is one slot and the UE needs to transmit the PSCCH and its associated PSSCH in the same slot. Refer to the multiplexing of the PSSCH and PSCCH in Section 17.1.2. PSCCH occupies two or three time-domain OFDM symbols in a slot. The network uses higher layer parameters *startSLsymbols* and *lengthSLsymbols* to configure the start and length of time-domain symbols that can be used for sidelink transmission in a slot. The first and last of these symbols are designated for AGC and GP, respectively. The PSSCH and PSCCH can only use the remaining symbols, as such. If the PSFCH resources are configured in a slot, the PSSCH and PSCCH cannot occupy the symbol used for PSFCH, as well as the AGC and GP symbols before the symbol for the PSFCH (see Fig. 17.2).

As shown in the Fig. 17.10, a network gNB configures with parameters *sl-StartSymbol* = 3, *sl-LengthSymbols* = 11, meaning 11 time-domain OFDM symbols in a slot starting from symbol index 3 can be used for sidelink transmission and reception. If there are also PSFCH resources configured in this slot, the PSFCH will occupy Symbol 11 and Symbol 12, where Symbol 11 is to be used as the AGC symbol for PSFCH reception and Symbol 10 and 13 are to be used as GP. Symbol 3 to Symbol 9 can be used for PSSCH transmission. The PSCCH occupies three symbols—4, 5, and 6—and Symbol 3 is used as the AGC symbol.

17.2.1.2 Resource in frequency domain

In NR−V2X, sidelink resource allocation granularity in the frequency domain is one subchannel, which has a size of N1 consecutive PRBs. Frequency-resource assignment information for a PSSCH transmission in a sidelink RP is determined by an initial subchannel index and a number of allocated subchannels. The PSSCH and the associated PSCCH are aligned in the frequency starting position. The PSCCH is transmitted only in the first subchannel of the PSSCH and it occupies N2 consecutive PRBs within the subchannel, where N1 and N2 are configurable parameters and N2 ≤ N1. The value range for N1 is {10, 12, 15, 20, 25, 50, 75, 100} and the value range for N2 is {10, 12 15, 20, 25}.

17.2.2 Sidelink dynamic resource allocation in resource allocation Mode 1

For sidelink dynamic resource allocation, a network gNB allocates/schedules transmission resources within a sidelink RP to a NR−V2X UE by using Downlink Control Information (DCI). In NR−V2X, there are two types of data message traffic that can be generated by a sidelink TX UE: namely periodic traffic and aperiodic traffic. For the periodic traffic, as the name suggests, the UE periodically generates sidelink messages with a relatively fixed/predictable interval and transmits these periodic messages using semipersistent scheduled resources provided by the network gNB.

For the aperiodic traffic, the message generation on the other hand can be random at any time and unpredictable and the message packet size is also variable. As such, it is difficult to make use of semipersistent resources provided by the network, and the UE usually adopts gNB dynamically allocated/scheduled resources for aperiodic transmissions. For sidelink dynamic resource allocation, the UE sends the Scheduling Request and Buffer Status Report to the network gNB, and the network allocates sidelink transmission resources for the UE according to the buffer information provided by the UE.

In LTE−V2X, basic safety messages are the main sidelink traffic used for assisted driving. For NR−V2X, advanced use cases like automatic driving, vehicle platooning, and extended sensor data sharing are expected to be supported. Therefore it is required that NR−V2X provide a sidelink communication with higher transmission reliability and shorter latency compared with LTE−V2X. In order to improve the reliability of sidelink transmission, a PSFCH is introduced in NR−V2X for a data message receiver UE to provide HARQ feedback information in the PSFCH to the message TX UE in accordance with a PSCCH/PSSCH decoding result. In sidelink resource allocation Mode 1, since all sidelink transmission resources are to be allocated/scheduled by the network gNB, the TX UE reports the sidelink HARQ feedback information received from the receiver UE to the network, so that the network can determine whether to allocate any further resources for retransmissions. For the sidelink HARQ feedback reporting to the network in Mode 1, the network provides the PUCCH resources in the uplink to the TX UE (Fig. 17.11).

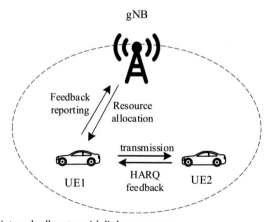

Figure 17.11 Network allocates sidelink resources.

In order for a network gNB to provide sidelink resource allocation in Mode 1, a new DCI format 3_0 is introduced for NR−V2X. When DCI format 3_0 is scrambled by a SL−RNTI, dynamic sidelink resource allocation is scheduled by the network gNB. Otherwise, when DCI format 3_0 is scrambled by a SL_CS_RNTI, a semipersistent scheduled Type 2 sidelink CG is activated/deactivated by the network gNB.

Specifically, the following Mode 1 scheduling information is included in DCI format 3_0:

- RP index: When multiple Mode 1 sidelink RPs are configured, the index of the intended RP needs to be indicated when the network gNB schedules sidelink resources using DCI.
- Sidelink resource allocation information: A network gNB can allocate upto N sidelink resources to a UE for PSCCH and PSSCH transmissions, where $1 \leq N \leq Nmax$ and Nmax can be configured to be 2 or 3. The network gNB indicates N sidelink time-frequency resources, according to the following information:
- Time gap: indicates the time gap in the number of slots between the first sidelink transmission resource and the slot in which this sidelink scheduling DCI is provided.
- Lowest index of the subchannel allocated for the initial transmission.
- Frequency-resource assignment: This field indicates the frequency size of the allocated sidelink transmission resource(s) (i.e., number of subchannels), and the frequency starting position of other N-1 sidelink transmission resources except the first resource. This field should be the same as the "frequency resource assignment" in SCI format 1−A.
- Time resource assignment: This field indicates the slot gaps of N-1 sidelink transmission resources, other than the first resource, relative to the first sidelink transmission resource. This field should be the same as the "time resource assignment" in SCI format 1−A.
- PUCCH resource allocation information: Provides information related to a PUCCH transmission resource to be used for UE reporting of sidelink HARQ feedback information to the network gNB. DCI format 3−0 indicates the PUCCH resource according to the following two parameter fields:
  - PUCCH resource indicator: Generally, a network gNB configures a PUCCH resource set by RRC configuration signaling, and indicates a specific PUCCH resource for HARQ reporting using this field in DCI. A detailed method of PUCCH resource indication can be found in Section 5.5.4.

- • PSFCH-to-HARQ feedback timing indicator: This parameter indicates the time gap between a SL HARQ report received in PSFCH and the PUCCH resource to be used for relaying the SL HARQ report to the network gNB. It is expressed in number of slots corresponding to the PUCCH SCS. If the sidelink transmission resources allocated by the network gNB correspond to more than one PSFCH resource, this field represents the slot gap from the last PSFCH resource and the PUCCH resource that should be used for SL HARQ feedback.
- HARQ process number: This is a HARQ process number associated with the sidelink transmission resource(s) allocated by the network gNB. When the UE uses the sidelink resource(s) allocated by the DCI for sidelink transmission, the sidelink HARQ process number indicated in the 2nd-stage SCI (denoted as the first HARQ process number) may be different from the HARQ process number assigned by the network in the DCI (denoted as the second HARQ process number). It is upto UE implementation to decide the sidelink HARQ process number to be signaled in the 2nd-stage SCI, but the UE needs to define an association/relationship between the first and the second HARQ process numbers. When the network schedules retransmission resources, the same second HARQ process number will be indicated again in the new scheduling DCI, but the NDI field will not be inverted. Hence, the UE will be able to determine that the scheduling DCI is used for allocating sidelink resources for the purpose of retransmission, and associate the new sidelink transmission resources allocated by the new scheduling DCI to be used for retransmissions for the first HARQ process number.
- New Data Indicator (NDI): Indicates whether the current transmission is the initial transmission of a TB.
- Configuration index: When a UE is configured with SL−CS−RNTI, DCI format 3_0 can be used to activate or deactivate a sidelink Type-2 CG. When the UE is configured with multiple Type-2 CGs by the network, this index is used to indicate the exact Type-2 CG to be activated or deactivated by the DCI. When the UE is not configured with SL−CS−RNTI, the length of this field is 0 bit.
- Counter sidelink assignment index: Indicates the accumulated number of DCI used to schedule sidelink transmission resources. The UE uses this information to determine the right number of information bits when generating the HARQ−ACK codebook to report to the network gNB.

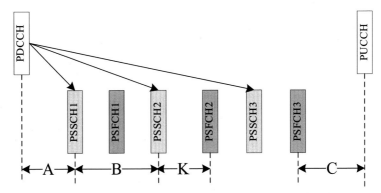

Figure 17.12 Time-sequence diagram of network allocation of side transmission resources.

Fig. 17.12 shows a timing relationship between different downlink (PDCCH), sidelink (PSSCH), and uplink (PUCCH) transmission resources. In this example, assuming a DCI in PDCCH allocates three sidelink resources for PSSCH transmissions and indicates a PUCCH transmission resource for SL HARQ feedback to the network gNB.

- A represents the time interval between the PDCCH-carrying sidelink scheduling DCI format 3_0 and the first scheduled sidelink transmission resource (PSSCH1), determined by the "Time gap" field in DCI format 3_0.
- B represents the slot interval between the second allocated sidelink transmission resource (PSSCH2) and the first sidelink transmission resource (PSSCH1), determined according to the "Time resource assignment" field in DCI format 3_0.
- C represents the slot interval between the last PSFCH resource (PSFCH3) and the indicated PUCCH resource, determined according to the "PSFCH-to-HARQ feedback timing indicator" field in DCI format 3_0. If there are multiple PSFCHs corresponding to the scheduled PSSCHs, the slot position of the last PSFCH is used.
- K represents the time interval between a scheduled PSSCH slot and its corresponding PSFCH slot. This parameter is determined according to RP configuration.

17.2.3 Sidelink configured grant in resource allocation Mode 1

The network usually allocates semipersistent scheduling (SPS) resources to a UE for transmitting periodic sidelink traffic. In LTE−V2X, it was done this way so that the network semistatically configures SPS sidelink resources to a UE and activates/deactivates the SPS resources to commence or terminate periodic sidelink transmissions based on UE traffic demand. NR−V2X adopts the concept of CG from NR Uu uplink (UL CG), which is the same or similar to LTE SPS resources, and introduces it to a sidelink as Sidelink CG (SL CG). So that if a UE is configured with SL CG, the UE can use the SL CG to transmit sidelink data without requesting transmission resources from the network when sidelink data arrives. Therefore SL CG can reduce sidelink transmission latency. SL CG has periodically occurring resources, so the UE can use it to transmit periodic sidelink traffic. When the UE has urgent aperiodic traffic, it can also use SL CG resources to transmit without waiting for network gNB to provide dynamic sidelink scheduling resources.

SL CG supports Type-1 SL CG and Type-2 SL CG:

- Type-1 SL CG: Similar to Type-1 UL CG, the network configures SL CG resources and transmission parameters for a UE by RRC signaling. The UE can freely uses Type-1 SL CG resources to transmit its sidelink data.

- Type-2 SL CG: Similar to Type-2 UL CG, the network configures a portion of sidelink transmission parameters for a UE by RRC signaling, and activates the SL CG by using DCI format 3_0 signaling in the PDCCH. The remaining sidelink transmission parameters such as assignment of sidelink resources are provided in the DCI. If the network requires the UE to report sidelink feedback information, DCI format 3_0 is also used to indicate the PUCCH resources for HARQ feedback reporting.

When a network configures a SL CG for a UE, the network allocates N sidelink transmission resources within each SL CG period, where $1 \leq N \leq Nmax$, and Nmax is a network configured parameter and can be 2 or 3. If the network requires the UE to report sidelink HARQ feedback information, it further allocates the PUCCH resource in each SL CG period. The PUCCH resource for SL HARQ reporting should be allocated after the PSFCH slot corresponding to the last PSSCH transmission resource in a SL CG period, so that the UE can determine the sidelink HARQ feedback information that needs to be reported to the network based on decoding results of all sidelink transmissions within the SL CG period.

NR−V2X supports multiple and parallel SL CG processes for a UE and each SL CG process can correspond to multiple SL HARQ processes.

The transmission of a sidelink transport block can only be carried out in one SL CG. And within a SL CG period, SL CG resources can only be used to transmit a new sidelink transport block. If a sidelink transport block needs extra retransmissions beyond what is available within a SL CG period, the network allocates additional retransmission resources by dynamic scheduling. The range in the time domain of the retransmission resources dynamically allocated by the network can exceed the SL CG period for the sidelink transport block.

As shown in Fig. 17.13 three PSSCH transmission resources (PSSCH1, PSSCH2, PSSCH3) and one PUCCH resource are configured within each SL CG period. If a UE transmits a transport block using the three configured PSSCH transmission resources in the first SL CG and the transmissions are not successful, the UE should report a NACK to the network using the configured PUCCH in the first SL CG. Then the network can schedule retransmission resources to the UE by DCI. In this case, the DCI schedules three PSSCH resources (PSSCH4, PSSCH5, PSSCH6) for the retransmission. These three retransmission resources can be extended to the next SL CG period if needed.

For type-2 SL CG, the network activates or deactivates the SL CG by DCI format 3_0 scrambled by SL−CS−RNTI. The "NDI" field should be set to 0, and

- If the "HARQ process number" field is set to all 0: it is used to activate a SL CG.
- If the "HARQ process number" field is set to all 1 and the "Frequency resource assignment" field is also set to all 1: it is used to deactivate a SL CG.

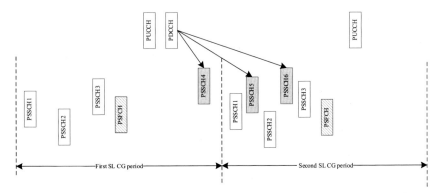

Figure 17.13 Retransmission scheduling for SL CG.

17.2.4 Sidelink resource allocation Mode 2

Sidelink resource allocation in Mode 1, where a cellular radio access network gNB manages all sidelink resources and allocates/schedules them to individual UE for direct their communication exchange was introduced in the previous section. In this section, sidelink resource allocation Mode 2 will be explained in detail. In direct contrast to a central node gNB controls all sidelink resources for every UE in a cell, in Mode 2 a UE fully on its own selects time-frequency resources from a sidelink RP, which can be preconfigured or configured by the network to the UE, based on resource sensing or random selection to send sidelink data messages. Therefore a more accurate description of resource allocation Mode 2 should be UE autonomous resource selection.

During the technology study phase of NR−V2X in 3GPP, four variant resource allocation schemes of Mode 2 were considered, namely Mode 2(a), Mode 2(b), Mode 2(c), and Mode 2(d).

In Mode 2(a), a UE autonomously selects resources from a sidelink RP for its own transmissions based on SCI decoding and measuring of received sidelink power from other UEs. Similar to one of resource allocation schemes (Mode 4) in LTE−V2X, the UE selects sidelink resources that are not already reserved by others or resources that have low received power measured, in order to avoid or minimize the impact of resource collision to others and thus ensuring reliability of sidelink communication. Since this principal mechanism of UE autonomous resource selection scheme has proven to work well in LTE−V2X, Mode 2(a) was chosen to be the main and only scheme for sidelink resource allocation Mode 2 during the technology development phase of NR−V2X in 3GPP Release 16. Hereafter, sidelink resource allocation Mode 2 refers to this mode 2(a).

In Mode 2(b), a coordination-based resource selection scheme was studied, where a UE selects sidelink resources for transmission based on assistance information provided by other UEs. The assistance information may be the results of resource sensing or resources recommended from another UE. For example, in Fig. 17.14 Mode 2(b), UE C sends assistance information to UE A and B. UE A and B use the received assistance information and combine it with results from their own resource sensing to select resources for sidelink transmission. Alternatively, UE A and B directly select sidelink resources based on the assistance information provided by UE C without performing any sensing [4]. This type of operation is also a form of Mode 2(b). However, since Mode 2(b) requires

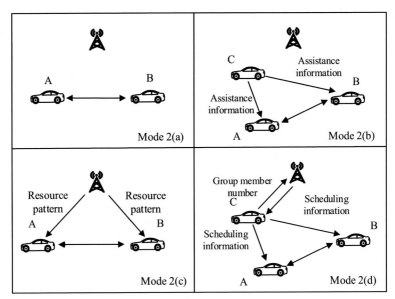

Figure 17.14 Mode 2 (a), 2 (b), 2 (c), 2(d).

sharing of assistance information between sidelink UEs, this scheme is more suitable for use in a group scenario where there is a stable connection between group member UEs. Additionally, it was found there is no obvious difference in resource sensing compared to Mode 2(a), this mode of sidelink resource selection was eventually not adopted for the development of NR−V2X in 3GPP Release 16. However, this mode of UE autonomous resource selection based on coordination between UEs is being reconsidered for NR sidelink in Release 17.

In Mode 2(c), one or more sidelink resource patterns are network configured or preconfigured for a UE, and the UE selects one of the patterns for sidelink initial transmission and retransmissions of a TB to achieve low transmission latency. Furthermore, the pattern selection could be also based on resource sensing or geographic location information. In addition, the negative impact caused by the half-duplex problem can be significantly reduced by ensuring that any two resource patterns do not overlap in the time domain [5,6]. However, Mode 2(c) is similar to SL CG in Mode 1 where it has a major disadvantage of being inflexible, such as releasing any unused resources in a selected pattern and adapting to aperiodic traffic transmissions. And therefore it was not further considered during the development phase of NR−V2X.

Mode 2(d) is very similar to Mode 2(b) for which one sidelink UE (not necessarily a sidelink data TX UE) also provides resource allocation related information to another UE. The difference between these two submodes is that in Mode 2(d) one UE directly provides scheduling of precise sidelink time-frequency resources to another UE. Therefore the Mode 2(d) scheme is also suitable for group communication scenarios when there is a stable connection between the member UEs. However, there are also a few issues related to the Mode 2(d) resource allocation scheme, such as determination of a suitable scheduling UE, design of physical layer signaling or higher layer signaling for the scheduling, and define proper behaviors for other UEs when the scheduling UE stops providing scheduling information (e.g., loss of connection), high processing complexity to implement the scheduling function within a UE, etc. This Mode 2(d) scheme was eventually simplified to a version as shown in Mode 2(d) of Fig. 17.14, where a UE C reports number of members in the group to the gNB the gNB provides sidelink scheduling decision/information to UE C, then UE C forwards the scheduling information to other UEs in the group without modifying the scheduling content from the gNB. And all of these information exchange among the gNB and UEs are to be carried by higher layer signaling, such as RRC signaling. However, the simplified Mode 2(d) sidelink resource allocation scheme was still considered to have high complexity and it will incur additional PC5-RRC design for sidelink groupcast communication. As such, it was also not further considered during the development phase of NR−V2X.

One of the key prerequisites/ingredients for sidelink resource allocation Mode 2 to work is UE reservation of sidelink resources for future transmissions, for which a UE sends SCI to reserve/announce time-frequency resources that it intends to use for sidelink transmissions in the future. In NR−V2X, the reservation of future sidelink resources can be performed for the same TB, a different TB, or both.

Specifically, the "Time resource assignment" and "Frequency resource assignment" fields in SCI format 1−A (i.e., 1nd-stage SCI) provide information for reserving resources for the same TB. When a UE sends SCI format 1−A in PSCCH, these two fields can indicate up to N time-frequency resources (including the resource for the current transmission) for transmitting the current TB, where. $1 \leq N \leq Nmax$ and Nmax equals 2 or 3 depending on sidelink RP configuration. As such, the UE can reserve up to 1 or 2 future sidelink resources for retransmission of the same TB. Following this reservation principal, the design should aim to reserve as many sidelink resources as possible (e.g., not just 1 or 2 future resources) to ensure collision-free for all retransmissions of the same TB. However, to control the number of information bits required to indicate future resources in SCI format 1−A, the number of time-frequency resources that can be indicated and the time window for indication should be limited. As such, the above Nmax was decided and the time window (W) is equal to 32 in NR−V2X. For example, as can be seen for TB1 in Fig. 17.15, a UE uses the above described two SCI format 1−A fields in PSCCH to indicate N = 3 time-frequency resources for the current/initial transmission, retransmission 1 (re-Tx 1) and retransmission 2 (re-Tx 2) when performing the initial transmission of TB1. By doing so, time-frequency resources for future retransmission 1 and retransmission 2 are reserved, and all three sidelink transmission resources (i.e., for the initial, re-Tx 1, and re-Tx 2), are distributed within 32 slots of the RP.

Furthermore, when the UE decides the number of sidelink resources (N) to indicate by the two fields (time- and frequency-resource assignment) in SCI format 1−A, it should be further restricted by N = min (Nselect, Nmax), where Nselect is the number of time-frequency sidelink resources selected by the UE within a 32-slot limit (including the resource of current transmission). For the TB1 example in Fig. 17.15, if the UE has not been able to select any sidelink resource within 32 slots from the initial transmission resource, N_{select} is set to 1 in this case and the UE will only indicate the current time-frequency resource in the SCI when performing the initial transmission. In another case, if the re-Tx 1 resource is selected within 32 slots from the initial transmission, N_{select} is then equal to 2 and the UE will indicate time-frequency resources for the current transmission and re-Tx 1 in the SCI when performing the initial transmission.

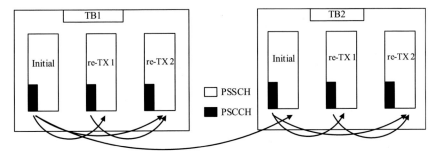

Figure 17.15 Illustration of sidelink resource reservation for the same TB and sidelink resource reservation for a different TB.

In SCI format 1−A, a "Resource reservation period" field can be additionally indicated by the UE to reserve time-frequency sidelink resources in the next time period for transmission of another TB. For the other example illustrated in Fig. 17.15, when the UE sends SCI for TB1, the "Resource reservation period" field can be used to indicate time-frequency resources in the next time period, and the indicated resources in the next period are for transmission of TB2. Let us first denote the three sidelink transmission resources for the initial, re-Tx 1 and re-Tx 2 of TB1 indicated by the SCI as $\{(n1, k1), (n2, k2), (n3, k3)\}$, where n1/n2/n3 and k1/k2/k3 are the slot timing positions and frequency RB positions of the three transmission resources for TB1, respectively. When the "resource reservation period" field in the SCI is set to 100, the SCI will reserve three more sidelink transmission resources in $\{(n1 + 100, k1),$ $(n2 + 100, k2), (n3 + 100, k3)\}$, and they can be used to transmit TB2. In NR−V2X, the set of possible values for resource reservation periodicity are 0, 1−99, 100, 200,..., 1000 Ms, which is more flexible than LTE−V2X, and no more than 16 reservation periods can be (pre)configured in a RP. The TX UE chooses one of the (pre)configured values for periodic resource reservation.

In sidelink resource allocation Mode 2, a TX UE should perform resource sensing to obtain resource reservation information by decoding SCI format 1−A transmitted from other UEs and exclude resources that are already reserved by others to avoid transmission collisions and thus to improve the reliability of sidelink communication. However, the "Resource reservation period" field may not be always present in SCI format 1−A according to the network configuration or preconfiguration in NR−V2X. In this case, transmission of a TB cannot reserve resources for transmitting another TB SCI. As shown in Fig. 17.15, when the SCI does not contain the "resource reservation period" field, the initial transmission resource of TB2 is not indicated by any SCI beforehand. As such, another UE performing resource sensing cannot obtain any information relating to the resource location for the initial transmission of TB 2 in advance. Consequently, this transmission resource for TB2 cannot be excluded by other UEs and resource collision may happen. In order to solve this problem, the concept of Standalone PSCCH was put forward for consideration during NR−V2X study in Release 16 [7,8].

In short, the Standalone PSCCH refers to a concept that an independent PSCCH (without an associated PSSCH transmission in a same slot) is sent prior to the initial transmission of the PSSCH. For example, in subfigure 1 of Fig. 17.16, the UE sends a SCI carried in the Standalone PSCCH to indicate/reserve time-frequency resources for its scheduled data transmissions to other UEs in advance, so that when another UE performing resource sensing will be able to detect and exclude these reserved resources during the resource selection process. However, Standalone PSCCH transmission should be multiplexed with PSSCH in frame structure Option 1 or Option 2 (see Section 17.1.2 for details). But since frame structure Option 4 was adopted, the Standalone PSCCH concept was not pursued any further in Release 16. In order to support a resource reservation scheme similar to Standalone PSCCH in Option 4, another solution was considered as shown in subfigure 2 of Fig. 17.16. In this solution, a sidelink TX UE sends PSCCH + PSSCH together and occupies only a single-frequency subchannel in lieu of the Standalone PSCCH. However, this solution would lead to variable number of frequency subchannels used for multiple transmissions of the same TB. Consequently, this would greatly increase the signaling design complexity and overhead of SCI. In the end, this solution was not also adopted.

In the following, the steps of sidelink resource allocation Mode 2 in PHY layer are elaborated on in detail.

In Fig. 17.17, a scenario where a UE is triggered to perform sidelink resource selection after arrival of a data packet TB from higher layer in slot n is depicted. At this point, the UE first determines a resource selection window which is defined to start from $n + T_1$ to $n + T_2$, where UE selection of T_1 and T_2 should fulfill the condition of $0 \leq T_1 \leq T_{proc,1}$ and $T_{2min} \leq T_2 \leq$ remaining Packet Delay Budget (PDB). When the SCS of the sidelink BWP is 15, 30, 60, or 120 kHz, the length of $T_{proc,1}$ is 3, 5, 9, or 17 slots, respectively. The value for T_{2min} is $\{1,5,10,20\} \times 2^{\mu}$, where

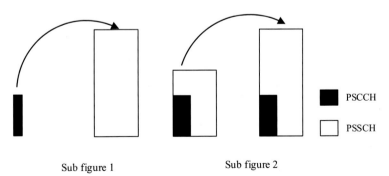

Sub figure 1   Sub figure 2

Figure 17.16 Standalone PSCCH.

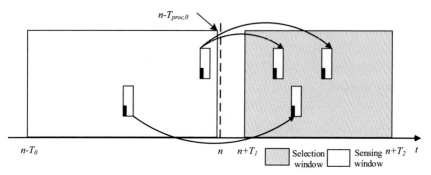

Figure 17.17 Resource allocation Mode 2.

$\mu = 0$, 1, 2, or 3 that correspond to the SCS of 15, 30, 60, or 120 kHz, respectively. The UE determines T_{2min} from this value set according to the priority of the packet TB to be sent. When T_{2min} is greater than remaining PDB, T_2 is set to the remaining PDB to ensure that the UE transmits the packet within the required latency.

The UE also performs resource sensing within a sensing window from $n - T_0$ to $n - T_{proc,0}$, where the value of T_0 is (pre)configured to be 100 or 1100 Ms, depending on whether resource reservation for a different TB is enabled/disable for the selected sidelink RP for transmission. When the SCS of the sidelink BWP is 15, 30, 60, or 120 kHz, the length of $T_{proc,0}$ is 1, 1, 2, or 4 slots, respectively.

17.2.4.1 Step 1: UE determines a set of candidate resources for selection

Based on the resource selection window determined above, the UE constructs a set of candidate resources (S_A) taking all sidelink resources within the resource selection window as an initial candidate set. Then the UE performs exclusion of resources from the initial candidate set S_A, based on the following two processes:

1. The first process to exclude any resource that falls within the resource selection window due to slots that were not monitored by the UE during the sensing operation. For example, if the UE performed SL or UL transmission in a slot within the sensing window, the UE would not have monitored/performed blind detection of any SCI sent from other UEs in this slot due to the half-duplex constraint (not receiving while transmitting). The UE should therefore assume a hypothetical SCI format 1−A was transmitted in this slot and exclude resources

from the candidate set S_A that correspond to periodic resource reservation from the hypothetical SCI to avoid any potential transmission collision with other UEs. Specifically, when the UE determines resources in selection slots that correspond to an nonmonitored slot for exclusion, the UE considers all possible resource reservation periodicities allowed in the RP to determine the selection slots and excludes all candidate resources within these selection slots.

2. The second process is to exclude candidate resources already reserved by other UEs from the set S_A. During resource sensing, when the UE successfully detects a SCI format $1-A$ in a resource from blind decoding, the UE additionally measures a sidelink RSRP of the received PSCCH or sidelink RSRP of PSSCH scheduled by the PSCCH depending on a RP (pre)configuration. If the measured sidelink RSRP is greater than a corresponding SL$-$RSRP threshold and the indicated/reserved resource (s) are within the resource selection window according to the time and frequency assignment fields and/or periodic reservation information in the decoded SCI, the UE will exclude these resource(s) from the candidate set S_A. If the remaining candidate resources in set S_A are less than X% of the total candidate resources before the resource exclusion process 1 and 2 (i.e., the initial set S_A), the UE will increase the SL$-$RSRP threshold by 3 dB and repeat this entire Step 1 until the X% of remaining candidate resources is reached. A set of X% is (pre)configured for the sidelink RP and the possible values are {20, 35, 50}. The UE determines a X% from the value set according to the priority of sidelink data packet to be sent. Similarly, the SL$-$RSRP threshold(s) that should be used for resource exclusion is also related to the transmission priority of the sidelink data packet to be sent, as well as the reception priority indicated in the received SCI. The UE then takes the remaining resources in the set S_A after the resource exclusion process 1 and 2 as the candidate resource set for final selection in Step 2.

17.2.4.2 Step 2: The UE randomly selects one or more sidelink resources from the remaining candidate resource set for sidelink data packet (re)transmissions

Overall, the sidelink resource allocation Mode 2 in NR$-$V2X operates very similar to the resource allocation Mode 4 in LTE$-$V2X, but there are some differences:

1. Since aperiodic transmissions have become one of the main traffic types in NR$-$V2X, the procedure of sorting the remaining candidate

resources according to SL RSSI measurement after resource exclusion in LTE−V2X is no longer performed in Mode 2 of NR−V2X.

2. In NR−V2X Mode 2, sidelink RSRP measurement is based on the DMRS of the PSCCH or PSSCH depending on RP configuration. This was done intentionally due to potential small allocation of frequency PRBs and number of OFDM symbols for the PSCCH in a RP. In this case, better sidelink RSRP measurement accuracy can be obtained from using the DMRS of the PSSCH.

3. The length of the sensing window is 100 Ms or 1100 Ms in NR−V2X Mode 2 (depending on resource reservation for a different TB is enabled/disabled), while the length of the sensing window is always constant at 1000 Ms in LTE−V2X Mode 4. In addition, the upper bound of resource selection window is determined by the remaining PDB of the sidelink transmission packet in NR−V2X, while the upper bound of the resource selection window is fixed at 100 Ms in LTE−V2X due to the assumption that basic safety messages always have a latency requirement of 100 Ms.

4. In the above Step 2 of sidelink resource allocation Mode 2, there are two timing restrictions that should be met by the UE during the final selection of sidelink resources for transmission.

Firstly, the UE is required to select a transmission resource that can be indicated by a prior SCI (a.k.a. chain reservation) with a possibility of some exemption cases. That is, if selection of a resource is not intended for an initial transmission of a sidelink TB, the UE will select a resource that is within 32 slots, which is the maximum time gap that can be indicated by the "time resource assignment" field in SCI format 1−A, from the previous selected sidelink resource for the same TB. As such, the time-domain interval between any two consecutive resources should be less than 32 slots. For example, as shown in Fig. 17.15, when the UE selects the three resources for TB1, the resource for re-Tx 1 should be selected in a way that can be at least indicated by the SCI when performing the initial transmission, and the resource for re-Tx 2 should also at least be indicated by the SCI when transmitting re-Tx 1. The abovementioned exceptions include that there is no other available resources from the remaining candidate resource set that the UE could select to meet the required timing restriction, and resource dropping or reselection due to preemption checking, SL/UL prioritization, or congestion control to limit the number of sidelink transmissions.

Secondly, the UE is also required to guarantee a minimum time gap Z between any two selected and consecutive sidelink resources when sidelink

HARQ feedback is enabled for the transmitted TB. The minimum time gap Z includes the slot gap between PSSCH transmission from the UE and the corresponding PSFCH slot in which the UE expects to receive sidelink HARQ feedback, and the amount of time to prepare for the next retransmission. For example, in Fig. 17.15, when the UE selects the three transmission resources for TB1, if sidelink HARQ feedback is enabled for the initial transmission, the time-gap interval between the initial transmission and re-Tx 1 should be at least Z slots. When the resource selection cannot meet this time domain restriction, it is up to UE implementations to drop or disable sidelink HARQ feedback for some of sidelink transmissions.

Besides the above described process for sidelink resource allocation in Mode 2, NR−V2X also supports reevaluation and preemption checking mechanisms for preselected and reserved resources, respectively. After a UE performs a Mode 2 resource selection process, it is possible for one or more preselected but not yet indicated by a SCI resources to be reserved by another before the UE has a chance to announce it. If no additional collision avoidance solution is implemented, collisions in resource selection will occur and sidelink transmission performance will be degraded. To mitigate this problem in NR−V2X, a mechanism for resource reevaluation is introduced for a UE continue to perform resource sensing and checking if a preselected resource has been reserved by another just before the transmission [9] and [10]. If a preselected but not yet announced resource is reserved by another UE, the UE reselects another resource for its transmission.

In Fig. 17.18, resource w, x, y, z, and v are time-frequency resources selected by a UE for sidelink transmissions of a TB, and the slot timing for the selected resource x is located at slot m. Resource y and z are preselected

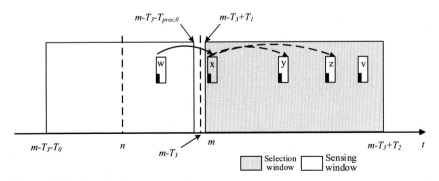

Figure 17.18 Reevaluation mechanism.

but not yet announced resources which will be first time indicated by a SCI in resource x (where resource x has been previously indicated by a SCI transmitted in resource w). To perform resource reevaluation for resource y and z, the UE carries out the above described Step 1 of sidelink resource allocation process for Mode 2 at least at slot $m - T_3$. If resource y and/or z is no longer part of the remaining candidate resource set S_A, meaning the resource has been excluded during Step 1, the UE performs Step 2 to reselect new time-frequency resource(s) to replace resource y and/or z. And it is upto UE implementation to reselect any other preselected but not yet reserved resources during the reselection process such that all selected resources can still be indicated by a prior SCI. Although it is found in some cases that performing resource reevaluation in every slot is beneficial for a UE to reselect new resources as early as possible to improve communication reliability [11,12], this operation behavior is expected to consume significant amount of processing power for the UE [13,14]. As such, it is also upto UE implementation to perform resource reevaluation for any preselected resource at any slot prior to the moment $m - T_3$, where T_3 is equal to $T_{proc,1}$. Note that the dashed arrows in Fig. 17.18 represent resources about to be indicated by a SCI for the first time, and the solid arrow indicates a resource that has been already indicated/reserved by a SCI.

As mentioned earlier, NR−V2X also supports preemption checking by a sidelink TX UE for already reserved sidelink resources. As such, this implies that preemption of sidelink resources from other UEs is allowed in NR−V2X. However, the UE behavior for the resource preemption mechanism is described only from the perspective of a preempted UE. For the resource preemption checking, similar to the resource reevaluation process, a UE continues to sense and perform checking (Step 1 of sidelink resource allocation process for Mode 2) to determine if any of its reserved resources has been booked/preempted by another UE at least in slot $m - T_3$ before sidelink transmission using the reserved resource. When all of the following three resource preemption conditions are met, the UE triggers the resource reselection procedure for the preempted resource(s).

1. A resource indicated by a decoded/preempting SCI during resource sensing overlaps fully or partially with one of the sidelink resources already reserved by the UE.
2. The measured sidelink RSRP based on the PSCCH DMRS or PSSCH DMRS of the decoded/preempting SCI being greater than the final sidelink RSRP threshold used during the resource exclusion procedure in Step 1.

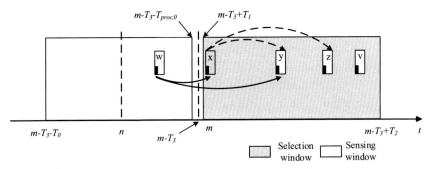

Figure 17.19 Resource preemption checking mechanism in NR—V2X.

3. The indicated priority in the decoded/preempting SCI is higher than the transmission priority of the sidelink data from the UE.

In Fig. 17.19, an example of resource preemption checking is illustrated. Similar to the resource reevaluation example in Fig. 17.18, resources w, x, y, z, and v are time-frequency resources that a UE has selected for transmission of a sidelink TB, and the slot timing for the selected resource x is located at slot m. As part of sidelink PSCCH transmission in resource w, resources x and y are indicated and reserved by the transmitted SCI. Before the planned sidelink transmissions in resources x, the UE performs resource preemption checking for resources x and y at least at slot $m - T_3$ before the resource x by carrying out the above described Step 1 of sidelink resource allocation process for Mode 2. If resource(s) x and/or y is no longer part of the candidate resource set S_A, meaning the resource(s) has been excluded during Step 1, the UE further determines whether the excluded resource(s) meets the above three conditions for preemption. If these conditions are met, the UE performs resource reselection Step 2 to find replacement resource(s). In addition, when the resource reselection is triggered for the excluded resource(s), it is upto UE implementation to reselect any preselected but not yet reserved resources to meeting the resource timing restriction of Step 2.

17.3 Sidelink physical layer procedure

17.3.1 Sidelink HARQ feedback

In LTE—V2X, only broadcast transmission is supported for direct sidelink communication among vehicle and pedestrian UEs. To ensure a basic reliability requirement for the sidelink communication is met, data messages are blindly retransmitted, even when an initial transmission is correctly decoded by a receiver UE. In sidelink blind retransmissions, the TX UE does not perform repeated transmission of the same data TB according to a HARQ feedback from the Receiver UE (RX UE), but autonomously retransmits the same sidelink data TB a certain number of times before it performs transmission for a new data TB.

For advanced V2X use cases, the reliability requirement for direct sidelink communication became significantly more stringent. If blind retransmission is still used as the main mechanism to fulfill the reliability target, the required number of blind retransmissions for each data TB would be large and such a scheme would not be very spectral efficient in scenarios where channel condition between the TX and RX UE is good, communication range is short, or there is only a small number of RX UEs. Furthermore, transmission latency would likely be prolonged as well due to unnecessary/excessive retransmissions. Therefore in order to meet the higher transmission reliability as well as improve the spectral utilization efficiency and transmission latency, two sidelink HARQ feedback schemes were developed and introduced for NR−V2X, namely ACK/NACK feedback and NACK-only feedback. Overall, in sidelink HARQ feedback, the RX UE decodes PSCCH/PSSCH sent by a TX UE, and then feeds back HARQ information using PSFCH to the TX UE when the sidelink HARQ indicator is set to enabled in SCI of the received PSCCH.

17.3.1.1 Sidelink HARQ feedback scheme

NR−V2X supports three transmission types: unicast, groupcast, and broadcast. Sidelink HARQ feedback is only applicable in unicast and groupcast transmissions, but not for broadcast. In the broadcast transmission mode, it is the same as LTE−V2X; a TX UE blindly performs retransmissions for multiple times (network configured or preconfigured for the RP) to ensure the target transmission reliability is achieved.

In the unicast transmission mode, after a TX UE and a RX UE have established a unicast communication link (i.e., a PC5-RRC connection) and the TX UE has sent a sidelink data transmission intended for the RX UE, the RX UE transmits PSFCH in a corresponding slot and resource containing HARQ information to the TX UE according to the decoding result as shown in Fig. 17.20. For sidelink unicast transmission, only the ACK/NACK feedback scheme is supported.

In groupcast transmission mode, both sidelink HARQ feedback schemes of ACK/NACK feedback and NACK-only feedback are

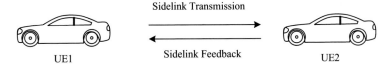

Figure 17.20 Sidelink feedback.

supported. The TX UE directly indicates the HARQ feedback scheme to be used by the receiver in the 2nd-stage SCI when transmitting the PSSCH to the RX UE.

- In the NACK-only feedback scheme, if a RX UE failed to decode a received PSSCH, it feeds back a NACK in PSFCH. If another RX UE successfully decodes the PSSCH, it does not send back any HARQ information. For all other RX UEs that needs to send a NACK due to decoding failure, they will send the same HARQ information (only NACK) using the same PSFCH resource corresponding to the sidelink resource in which the PSSCH was received. This HARQ feedback scheme was originally designed for connection-less groupcast transmissions, where no specific communication group is established. As such, this HARQ feedback scheme is very suitable for a type of V2X groupcast communication that is based on a communication distance requirement, and only the UEs within an indicated distance range from the TX UE need to feedback sidelink HARQ information to the TX UE. UEs that are outside the communication distance range need not send sidelink HARQ feedback. This NACK-only HARQ feedback scheme was later during the NR−V2X development phase extended to support also connection-based groupcast transmissions where the number of members in the group and group member ID are known to the TX UE. Since there may be a limited number of PSFCH resources that can be (pre)configured in a RP, there may not always be enough PSFCH resources available for sensing ACK/NACK reports from every UE in a group, especially when the group size is large (e.g., more than 12). As such, the NACK-only feedback scheme can be indicated in the 2nd-stage SCI from the TX UE when transmitting sidelink data in the PSSCH.

For the communication distance range groupcast transmissions, the concept of geographical zone is utilized to support the NACK-only sidelink HARQ feedback scheme. The basic mechanism behind the geographical zone concept is to divide the surface of the earth into different zones, where each zone is identified by a zone ID. For the NACK-only

HARQ scheme, TX UE's SCI carries a zone ID corresponding to the zone to which the TX UE belongs and a communication distance range. UEs that are in geographical zones within the distance range from the transmitter zone are required to provide NACK-only reports when there is failure to decode the PSSCH from the TX UE. Specifically, when a RX UE receives SCI format 1—B sent from a TX UE with communication distance range and zone ID indicated, it determines the distance to the TX UE according to the indicated zone ID and the zone where the RX UE is located which can be derived base on its own real-time geo-location (e.g., from receiving GNSS and/or cellular network signals). However, the RX UE only knows the zone ID of the TX UE and does not know the actual geographic location of the TX UE. Therefore the RX UE determines the distance to the TX UE according to its real-time geographic location and the zone center location of the TX UE. If the distance determined by the RX UE is less than or equal to the distance range indicated in the received SCI and the associated PSSCH is not successfully decoded, the RX UE is required to feedback a NACK in the PSFCH. If the associated PSSCH is successfully decoded, no HARQ information should be fed back. If the calculated distance separation between the TX and RX UEs is larger than the distance indicated by the communication distance requirement, no HARQ information should be fed back either.

As shown in Fig 17.21 below, a TX UE transmits sidelink control and data (PSCCH/PSSCH) in Zone 4 and indicates its zone ID and distance

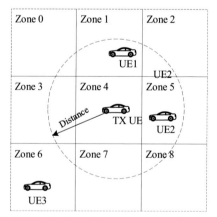

Figure 17.21 HARQ feedback based on zone and communication distance requirement.

range information in SCI format 1–B. When receiver UE 1 and UE 2 determined its distance to the TX UE is less than the indicated communication distance requirement, they will need to feed back a NACK if there is a failure to decode the transmitted PSSCH from the TX UE; otherwise, no sidelink HARQ information feedback is necessary from either UE. UE 3 determines its distance to the TX UE is larger than the communication distance range indicated and thus will transmit no sidelink HARQ feedback information to the TX UE nor it will attempt to decode the transmitted PSSCH from the TX UE (Fig. 17.21).

- In the ACK/NACK-based HARQ feedback scheme, if a UE successfully decodes a received PSSCH and the sidelink HARQ feedback indicator is set to enabled in the scheduling SCI, then the UE is required to feedback an ACK; otherwise it will feedback a NACK if decoded unsuccessfully. This sidelink HARQ feedback scheme is generally suitable for connection-based sidelink groupcast and unicast communications. In connection-based groupcast communication, a group of UEs constitute a communication group, and each UE in the group is given a corresponding member ID. For example, as shown in Fig. 17.22, a communication group consists of 4 UEs (i.e., the group size is 4) and each UE is given a member ID such as ID#0, ID#1, ID#2, and ID#3. Furthermore, each UE is also given information on the number of members within the group and the member IDs of all group member UEs. When a member UE transmits PSCCH/PSSCH (i.e., all other member UEs in the group

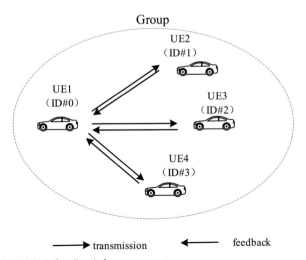

Figure 17.22 Sidelink feedback for groupcast.

are RX UEs), each RX UE will determine whether to feedback an ACK or a NACK to the TX UE according to its PSSCH decoding results. In this case, each RX UE uses a different PSFCH resource to feedback its sidelink HARQ information, and sidelink HARQ feedback information from all RX UEs are FDM'ed and/or CDM'ed in PSFCH.

17.3.1.2 Sidelink HARQ feedback resource configuration

In the PSSCH RP configuration information, sidelink HARQ feedback transmission resources (i.e., PSFCH resources) can be network configured or preconfigured in RRC. The configuration parameters for sidelink HARQ feedback transmission resources include the following four components:

- Sidelink HARQ feedback resource period: For a sidelink RP, PSFCH resources for sidelink HARQ feedback can be configured in every transmission slot. But since not every sidelink data transmission requires SL HARQ feedback from the receiver UE, an occurrence period (P) for PSFCH resources can be (pre)configured to reduce the overhead of sidelink HARQ feedback resources, where P can be 0, 1, 2, or 4, expressed in number of slots in the sidelink RP. P = 0 means no PSFCH feedback resource is (pre)configured and all sidelink HARQ feedback is disabled in the sidelink RP.
- Time interval: This is used to indicate a minimum time gap between the sidelink HARQ feedback resource and its corresponding PSSCH transmission resource, expressed in number of slots.
- Frequency-domain resource set of sidelink HARQ feedback resources: This is used to indicate the position and number of Resource Blocks (RBs) that can be used to transmit the PSFCH in a RP. This parameter is indicated in the form of a bitmap, and each bit in the bitmap corresponds to a RB in the frequency domain.
- Number of Cyclic Shift Pairs (CS pairs) in a RB: The sidelink HARQ feedback information is carried in a form of pseudo-random sequences. ACK and NACK information bits are represented by different sequences, which make up a CS pair. This parameter is used to indicate the number of CS pairs (i.e., the number of UEs that can be multiplexed by CDM in a RB).

17.3.1.3 Sidelink HARQ feedback resource determination

The PSFCH transmission resource is determined according to the time-frequency position of its corresponding PSSCH resources used for transmission. In NR−V2X, the following two PSFCH resource determination options are supported. The specific option for determining theh PSFCH resource is configured according to higher layer signaling.

- Option 1: PSFCH resource is determined by the first subchannel used for the corresponding PSSCH transmission.
- Option 2: PSFCH resource is determined by all the subchannels occupied by the corresponding PSSCH transmission.

For PSFCH resource determination Option 1, since PSFCH resources are determined only by the first subchannel occupied by the PSSCH (i.e., not dependent on the number of subchannels the PSSCH occupied), the total number of corresponding PSFCH feedback resources is fixed for each PSSCH transmission. For Option 2, the number of PSFCH transmission resources is determined according to the number of subchannels occupied by PSSCH. Therefore the more subchannels occupied by a PSSCH transmission, the more PSFCH transmission resources are available for sidelink HARQ feedback. As such, Option 2 is more suitable for scenarios where a large number of member UEs are involved in a sidelink groupcast communication that requires more sidelink HARQ feedback resources for the ACK/NACK feedback scheme.

The corresponding PSFCH transmission resource sets $R_{\text{PRB,CS}}^{\text{PSFCH}}$ that are available for multiplexing sidelink HARQ information can be determined according to the slot and subchannel(s) used for the PSSCH transmission. The index of PSFCH resources in the resource set is assigned first in the ascending order of the allocated PFSCH frequency RBs, then in the ascending order of configured CS pairs. To determine the exact resource for PSFCH transmission, the RX UE uses the following formula:

$$(P_{ID} + M_{ID}) \bmod R_{\text{PRB,CS}}^{\text{PSFCH}} \qquad (17.4)$$

where P_{ID} is the Source ID of the TX UE indicated in the received 2nd-stage SCI. For unicast and NACK-only groupcast sidelink HARQ feedback schemes, M_{ID} is set to zero. For the ACK/NACK groupcast sidelink HARQ feedback scheme, M_{ID} is the member ID of the RX UE provided by the UE higher layer.

17.3.2 Sidelink HARQ feedback reporting in Mode 1

In sidelink resource allocation Mode 1, all sidelink resources to be used by UE for initial and retransmissions of a TB are allocated by the network gNB. However, in order for the network gNB to decide whether to allocate resources for the retransmissions, the TX UE reports to the network gNB sidelink HARQ feedbacks it received from the RX UE to indicate whether or not the sidelink data PSSCH was received correctly. When the network receives a NACK report from the Tx UE, it allocates retransmission resource(s). If the network receives an ACK, it stops allocating resources for the same sidelink TB transmission.

Sidelink HARQ feedback reporting on the PUCCH and PUSCH is supported in NR−V2X. In order to reduce the complexity of the uplink feedback reporting process, NR−V2X does not support multiplexing of sidelink HARQ feedback and Uu interface uplink control information (Uu UCI) in the same PUCCH or PUSCH.

When a sidelink Tx UE is required to report sidelink HARQ feedback information to the network in Mode 1, a corresponding PUCCH transmission resource is indicated to the Tx UE during the scheduling of sidelink transmission resource(s) to the UE. The RX UE sends sidelink HARQ feedback after decoding of the PSCCH/PSSCH transmitted from the TX UE, and the TX UE reports sidelink HARQ feedback to the network via the indicated PUCCH. If the network gNB also schedules an uplink data transmission in the same slot of the sidelink HARQ reporting PUCCH, the sidelink TX UE multiplexes sidelink HARQ feedback on the PUSCH to the network.

For the dynamic sidelink resource allocation scheme, the network schedules sidelink resource(s) and a PUCCH reporting resource for a UE using DCI format 3_0. For resource allocation using SL CG, the network allocates one PUCCH transmission resource for the UE in each SL CG period. The time-slot position of the PUCCH transmission resource is located after the PSFCH resource corresponding to the last sidelink transmission resource scheduled by the DCI.

NR Uu supports uplink HARQ feedback based on HARQ codebook (i.e., HARQ feedback of multiple scheduled PDSCH slots is multiplexed onto the same PUCCH or PUSCH). NR−V2X follows this reporting mechanism and supports multiple sidelink HARQ−ACK bits to be multiplexed in the same PUCCH or PUSCH.

17.3.3 Sidelink measurement and feedback

17.3.3.1 CQI/RI

In NR−V2X, CSI reporting of CQI and RI is supported for unicast communication only, but PMI reporting is not supported. A RX UE is required to report CQI and RI at the same time in a single SL CSI report. Since the design of sidelink PSFCH is only intended to carry sidelink HARQ reports,

CQI/RI reporting is currently carried by MAC CE. In addition, if SL CSI reporting UE operates in resource allocation Mode2, it is not guaranteed that the UE can obtain periodic resources for SL CSI report. Therefore only aperiodic SL CSI reporting is supported in NR−V2X.

SL CSI reporting of CQI and RI can be triggered by the TX UE using the "CSI request" field in the 2nd-stage SCI. In order to ensure the validity of SL CSI reports, the RX UE is required to report SL CSI within a maximum delay after receiving the SL CSI trigger. The maximum delay is determined by the TX UE and signaled to the RX UE via PC5−RRC.

17.3.3.2 Channel busy ratio/channel occupancy ratio

Channel Busy Ratio (CBR) and Channel Occupancy Ratio (CR) are two basic measurements used in both LTE− and NR−V2X to support sidelink congestion control. CBR is defined as the ratio of subchannels with measured SL RSSI higher than a (pre)configured threshold to the total number of subchannels in the CBR measurement window $[n - c,$ $n - 1]$, where c is equal to 100 or 100 2^{μ} slots. CR is defined as the ratio of the number of subchannels that the UE has used for sidelink transmission in the range of $[n - a, n - 1]$ and the UE has reserved in the range of $[n, n + b]$ to the total number of subchannels belonging to the RP within the range of $[n - a, n + b]$. CR should be calculated separately for different priorities, and a is defined a positive integer, b is 0 or a positive integer, and the values of a and b are to be determined by the UE while fulfilling the following three conditions:

1. $a + b + 1 = 1000$ or $1000 \cdot 2^{\mu}$ slots
2. $b < (a + b + 1)/2$
3. $n + b$ does not exceed the last retransmission of the current transmission indicated by SCI.

For a UE in the RRC connected state, the CBR should be measured and reported according to network gNB configuration. Otherwise, the UE needs perform congestion control based on both measured CBR and CR. Specifically, in a RP, the congestion control process can limit the following PSCCH/PSSCH parameters:

- MCS range supported in the RP
- Number of subchannels that can be used
- The maximum number of retransmissions in Mode 2
- Maximum transmission power

$$\Sigma_{i \geq k} CR(i) \leq CR_{Limit}(k)$$

Table 17.7 Congestion control processing capability.

μ	Processing capability 1/slots	Processing capability 2/slots
0	2	2
1	2	4
2	4	8
3	8	16

where $CR(i)$ is the measurement CR of the priority i sidelink transmission in slot $n - N$, and $CR_{Limit}(k)$ is the CR limit for the priority k sidelink transmission and CBR measured in slot $n - N$ configured by the network. N indicates the time required for the UE to process congestion control and is related to sidelink SCS, μ. For NR−V2X in Release 16, two UE congestion control processing capabilities (processing capability 1 and processing capability 2) are defined as shown in Table 17.7.

17.3.3.3 SL−RSRP

In order to support sidelink pathloss-based power control (see Section17.3.3 for details), the TX UE needs to perform estimation of sidelink pathloss between itself and a RX UE. During the development phase of NR−V2X in 3GPP Release 16, two separate schemes of obtaining sidelink pathloss were considered. One of the schemes is for the TX UE to indicate its sidelink transmission power of a reference signal (e.g., in SCI or PC5−RRC) and the RX UE estimates a sidelink pathloss between the two UEs and reports the estimated pathloss to the TX UE. The other is that the RX UE performs a measurement of SL−RSRP, reports the measured SL−RSRP to the TX UE, and the TX UE estimates sidelink pathloss. Since the first scheme needs to indicate the transmission power of reference signal in SCI, which will significantly increase the number of SCI bits, the second scheme was adopted in the end.

To perform the measurement of SL−RSRP, the RX UE uses the sidelink DMRS for PSSCH and the measurement result should be L3 filtered before it is reported to the TX UE using PC5−RRC signaling.

17.3.4 Sidelink power control

For PSCCH and PSSCH transmissions, NR−V2X supports two different types of power control, namely downlink pathloss-based power control and sidelink pathloss-based power control.

Downlink pathloss-based power control is mainly used to reduce/limit the amount of interference from sidelink transmissions between UEs to uplink reception at BS gNB, as shown in Fig. 17.23. When sidelink communication resources are configured on the same carrier with NR Uu uplink, sidelink transmissions between UE#2 and UE#3 (if they are close to the BS) may cause interference to the uplink reception between the base station and UE#1. In the downlink pathloss-based power control, radio transmission power for the sidelink communication between UE#2 and UE#3 will decrease according to downlink pathloss estimation, so as to control the interference to the uplink.

The main purpose of sidelink pathloss-based power control is to reduce interference between different sidelink communication groups/pairs. Since sidelink pathloss-based power control relies on SL−RSRP reporting from the RX to the TX UE to calculate sidelink pathloss, sidelink pathloss-based power control is only available for the unicast communication in Release 16 version of NR−V2X.

The transmit power of PSSCH on OFDM symbols with only PSSCH (i.e., without PSCCH) can be determined using the following formula:

$$P_{PSSCH}(i) = \min\left(P_{CMAX}, P_{MAX,CBR}, \min\left(P_{PSSCH,D}(i), P_{PSSCH,SL}(i)\right)\right) \ [dBm]$$

$$(17.5)$$

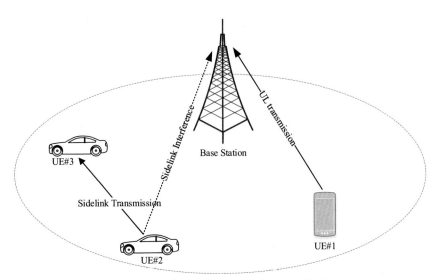

Figure 17.23 The interference of sidelink transmission to uplink reception.

where P_{CMAX} is the maximum transmit power allowed for the UE and $P_{MAX,CBR}$ represents the maximum transmit power allowed for the measured CBR level and the data transmission priority under congestion control. $P_{PSSCH,D}(i)$ and $P_{PSSCH,SL}(i)$ are respectively determined according to the following formulas:

$$P_{PSSCH,D}(i) = P_{0,D} + 10\log_{10}\left(2^{\mu} \times M_{RB}^{PSSCH}(i)\right) + \alpha_D \times PL_D \times [\text{dBm}]$$

(17.6)

$$P_{PSSCH,SL}(i) = P_{0,SL} + 10\log_{10}\left(2^{\mu} \times M_{RB}^{PSSCH}(i)\right) + \alpha_{SL} \times PL_{SL} \times [\text{dBm}]$$

(17.7)

where $P_{0,D}/P_{0,SL}$ is the basic operating point configured by higher layer signaling for downlink/sidelink pathloss-based power control, α_D/α_{SL} is a downlink/sidelink pathloss compensation factor configured by higher layer signaling, PL_D/PL_{SL} is the downlink/sidelink pathloss estimated by the UE, and $M_{RB}^{PSSCH}(i)$ represents the number of PRBs occupied by the PSSCH.

When an OFDM symbol contains both PSCCH and PSSCH, the UE allocates transmission power $P_{PSSCH(i)}$ to both PSCCH and PSSCH according to the ratio of the number of PSCCH PRBs and PSSCH PRBs. Specifically, the transmission power for PSSCH $- P_{PSSCH2(i)}$ is determined by:

$$P_{PSSCH2}(i) = 10\log_{10}\left(\frac{M_{RB}^{PSSCH}(i) - M_{RB}^{PSCCH}(i)}{M_{RB}^{PSSCH}(i)}\right) + P_{PSSCH}(i) \times [\text{dBm}]$$

(17.8)

The transmission power for PSCCH is determined by:

$$P_{PSCCH}(i) = 10\log_{10}\left(\frac{M_{RB}^{PSCCH}(i)}{M_{RB}^{PSSCH}(i)}\right) + P_{PSSCH}(i) \times [\text{dBm}]$$ (17.9)

where $M_{RB}^{PSCCH}(i)$ is the number of PRBs occupied by PSCCH.

For PSFCH transmission power control, since PSFCH format 0 does not contain DMRS and there may be several PSFCHs multiplexed in a CDM manner on a single PSFCH resource, the RX UE cannot estimate sidelink pathloss-based on PSFCH reception. Thus PSFCH format 0 only supports downlink pathloss-based power control in the same way as PSCCH and PSSCH. The S—SS and PSBCH are transmitted in sidelink broadcast mode without any SL—RSRP reporting, so only the downlink pathloss-based power control is supported for S–SS and PSBCH transmissions.

References

[1] RP–190766. New WID on 5G V2X with NR sidelink, LG Electronics, Huawei, RAN Meeting #83, Shenzhen, China, March 18–21, 2019.
[2] 3GPP TS 38. 214. Third generation partnership project; Technical specification group radio access network, NR. Physical layer procedures for data (release 16).
[3] 3GPP TS 38. 211. Third generation partnership project; Technical specification group radio access network,NR. Physical channels and modulation (release 16).
[4] R1-1904074, Discussion on mode 2 resource allocation mechanism, vivo, RAN-1 #96bis, Xi'an, China, April 8th – 12th, 2019.
[5] R1–1812409. Enhancements of configured frequency-time resource pattern in NR V2X transmission, Fujitsu, RAN-1 #95, Spokane, Washington, USA, November 12th–16th, 2018.
[6] R1–1812209. Sidelink resource allocation mode 2, Huawei, RAN-1 #95, Spokane, Washington, USA, November 12th–16th, 2018.
[7] R1–1906076. Discussion of resource allocation for sidelink-mode 2, Nokia, Nokia Shanghai Bell, RAN-1 #97, Reno, USA, May 13th–17th, 2019.
[8] R1–1906392. Mode 2 resource allocation mechanism for NR sidelink, NEC, RAN-1 #97, Reno, USA, May 13th–17th, 2019.
[9] R1–1910213. Discussion on mode 2 resource allocation mechanism, vivo, RAN-1 #98bis, Chongqing, China, October 14th–20th, 2019.
[10] R1–1910650. Resource allocation mode-2 for NR V2X sidelink communication, Intel Corporation, RAN-1 #98bis, Chongqing, China, October 14th–20th, 2019.
[11] R1–2002539. Sidelink resource allocation mechanism for NR V2X, Qualcomm Incorporated, RAN-1 #100bis-e,April 20th–30th, 2020.
[12] R1–2001994. Solutions to remaining opens of resource allocation mode-2 for NR V2X Sidelink Design, Intel Corporation, RAN-1 #100bis-e, April 20th–30th, 2020.
[13] R1–2001749. Discussion on remaining open issues for mode 2, OPPO, RAN-1 #100bis-e, April 20th–30th, 2020.
[14] R1–2002126. On mode 2 for NR sidelink, Samsung, RAN-1 #100bis-e, April 20th–30th, 2020.

CHAPTER 18

5G NR in the unlicensed spectrum

Hao Lin, Zuomin Wu, Chuanfeng He, Cong Shi and Kevin Lin

18.1 Introduction

The NR was first developed for 5G wireless communication systems and introduced in the 3GPP Release 15 standard, which is a radio access technology that uses the existing and new licensed spectrum. The NR radio system is designed to achieve seamless coverage, high spectral efficiency, high peak rate, and high reliability for 5G cellular networks. In the 4G long-term evolution technology (LTE) system, the unlicensed spectrum has already been used in cellular networks as a supplementary band to the licensed spectrum. Similarly, the NR system can also use the unlicensed spectrum as part of the 5G cellular network technology to provide services to users. In the 3GPP Release 16 standard, use of the NR system used in the unlicensed spectrum was discussed and developed, which is commonly known as NR−Unlicensed (NR−U). The technical framework for the NR−U system was mostly developed within one year during the RAN1#96−99 conferences in 2019. The NR−U system supports two networking modes: licensed spectrum-assisted access and unlicensed spectrum independent access. In the former one, UE access to the cellular network relies on the use of the licensed spectrum, and the unlicensed spectrum can be used as a secondary carrier. In the latter, the UE is allowed to access the network and operate independently using the unlicensed spectrum directly. The range of the unlicensed spectrum specified in the 3GPP for NR−U operation and introduced in Release 16 is mainly within the range of the 5GHz and 6GHz bands, such as 5925−7125 MHz in the United States or 5925−6425 MHz in Europe. In the Release 16 standard, band 46 (5150−5925 MHz) is newly defined as the unlicensed spectrum for NR−U operation.

A so-called unlicensed spectrum or band is a radio-frequency spectrum decided by countries and regions around the world that can be used for wireless communication (including both transmission and reception) without needing to issue an operating license to one or a group of operators.

5G NR and Enhancements
DOI: https://doi.org/10.1016/B978-0-323-91060-6.00018-0

It is generally regarded as a shared spectrum, in which communication devices can use the spectrum as long as they observe the regulatory requirements set by the country or region for use of the spectrum. A user also does not need to apply for a license or authorization from the exclusive spectrum management agency of the country or region to use the unlicensed spectrum. But the use of the unlicensed spectrum needs to meet the requirements of specific regulations of each country and region; for example, wireless communication equipment uses the unlicensed spectrum in accordance with the principle of "listen before talk" (LBT). Therefore Release 15 version of the 5G NR technology/radio system needs to be enhanced in order to meet the regulatory requirements of unlicensed frequency bands, while enabling efficient use of the unlicensed radio frequencies to provide services. In the 3GPP Release 16 standard, the standardization of NR−U technology primarily includes the following aspects: Channel access procedure, initial access procedure, control channel design, HARQ feedback and scheduling, configured grant (CG) transmission, and so on. These technical mechanisms are introduced in detail in this chapter.

18.2 Channel sensing

In order to enable various communication technologies that operate in the unlicensed spectrum for wireless communication to coexist friendly with each other, countries or regions around the world have in place specific regulatory requirements that must be met by these wireless technologies for the use of the unlicensed spectrum. For example, according to European regulations, when using the unlicensed spectrum for communication, communication equipment follows the principle of LBT. That is, the communication equipment needs to perform LBT or rather channel-sensing before using any channel on the unlicensed spectrum for transmission. Only after the channel is sensed and determined to be idle or rather the LBT is successful, can the communication equipment transmit signals using the channel; on the other hand, if the channel is determined to be busy after sensing or rather the LBT fails, then the communication equipment is not allowed to transmit any signal on the channel. In addition, in order to ensure fair usage of the shared spectrum, if the communication equipment succeeds in LBT on a channel of the unlicensed spectrum, the time duration that the communication equipment can use the channel for transmission cannot exceed a certain period. By limiting the maximum transmission time duration after a successful LBT, the mechanism allows different communication equipment to have their opportunities to access the shared channel, so that different communication technologies/systems can coexist friendly on the same spectrum.

Although channel-sensing is not a global regulation, it is a feature that should be supported by communication equipment when the NR system is operating on an unlicensed spectrum, considering that channel-sensing can bring benefits of interference avoidance for friendly coexistence of transmissions between intra-RAT and inter-RAT systems on the shared spectrum. From a network deployment standpoint, channel-sensing by communication equipment can be performed based on two main mechanisms. One of them is the load-based equipment (LBE) mechanism, which is also commonly known as dynamic channel access or channel-access procedure for dynamic channel occupancy (CO). The other one is the frame-based equipment (FBE) mechanism, which is often known as semistatic channel access or channel-access procedure for semistatic CO. This section will introduce these channel-access procedures for the gNB and UE in a NR−U system.

18.2.1 Overview of channel access procedure

Before describing the channel-access procedures in NR−U, this section will first introduce some basic concepts that are involved in radio transmission on the unlicensed spectrum:

- Channel Occupancy (CO): NR gNB or UE transmission on an unlicensed channel after successfully performing the associated channel-access procedures.
- Channel Occupancy Time (COT): Total transmission time on an unlicensed channel by the gNB and/or UE after performing the associated channel-access procedure to initiate a CO or to share the CO. Radio transmissions within the COT can be continuous or discontinuous.
- Maximum Channel Occupancy Time (MCOT): The maximum transmission time duration on an unlicensed channel by the gNB and/or UE after performing the associated channel-access procedure to initiate a CO or to share the CO. Different channel-access procedure may correspond to different channel-access priority classes (CAPC), and different CAPC correspond to different MCOT. Also, different countries or regions may also have different requirements on MCOT. For example, in the Europe region, the MCOT cannot exceed 10 ms;

while in Japan, the MCOT cannot exceed 4 ms. If no additional regional requirement, the MCOT currently defined for NR−U system is 10 ms.

- gNB-initiated COT: This is also known as the gNB's COT. It refers to the channel-access time obtained by the gNB after successfully performing the associated channel-access procedure on the channel. The gNB's COT can be used not only for downlink transmission, but also for uplink transmission from the UE under certain conditions.
- UE-initiated COT: This is also known as the UE's COT. It refers to the channel-access time obtained by the UE after successfully performing the associated channel-access procedure on the channel. The UE's COT can be used not only for uplink transmission, but also for downlink transmission from the gNB under certain conditions.
- DL transmission burst: A set of continuous downlink transmissions performed by a gNB, which may include one or multiple downlink transmissions without any gaps for a time period greater than 16 μs. If a gap between two downlink transmissions performed by the gNB is more than 16 μs, then the two downlink transmissions are considered to belong to two separate downlink transmission bursts.
- UL transmission burst: A set of continuous uplink transmissions performed by a UE, which may include one or multiple uplink transmissions without any gaps greater for a time period than 16 μs. If a gap between two uplink transmissions performed by the UE is more than 16 μs, then the two uplink transmissions are considered to belong to two separate uplink transmission bursts.

Fig. 18.1 shows an example of channel-access time obtained by communication equipment after successful completion of the LBT procedure for

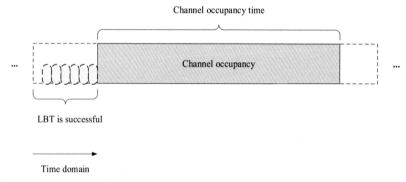

Figure 18.1 An example of channel-access time and channel occupancy.

gaining access to a channel in the unlicensed spectrum and using the resources of the channel within the channel-access time for radio transmission.

The channel-sensing procedure can be performed by energy detection. Usually the channel-sensing is performed without considering the effect of antenna array at the gNB or UE. The channel-sensing based on energy detection without considering the effect of antenna array can be called omnidirectional LBT. Omnidirectional LBT is easy to implement and widely used in Wireless Fidelity (Wi-Fi) and Licensed-Assisted Access (LAA)-based Long-Term Evolution (LTE) systems. Channel-sensing before transmission not only enables different competing equipment to coexist friendly and fairly on the unlicensed spectrum, but it also can be used for better channel multiplexing if the LBT procedure is enhanced. During the Study Item (SI) phase of NR−U, the factors that affected the design of LBT procedure for the NR−U system include: channel characteristics of the unlicensed spectrum, enhancements to the NR basic technology, and synchronization between different operators. For the second factor, since radio transmission on the unlicensed spectrum can be directional with beam-forming, it implies that the energy detected during the channel-sensing procedure may not match the actual interference to the receiver, resulting in "overprotection" to other competing equipment. For the third factor, since the NR technology supports flexible frame and slot structures, further enhancement can be based on network synchronization between different operator nodes. If the cellular networks between different operators are synchronized, different operators can use these characteristics to share the unlicensed spectrum more effectively, thus improving throughput, reliability, and quality of service.

Based on the above analysis, the enhancement topics considered during the study on the LBT procedures for the NR−U system include:

1. Enhancing the ability of spatial division multiplexing, such as using directional channel-sensing, multinode joint channel-sensing, etc., so that the probability of channel reuse can be improved but at the expense of certain degradation in reliability in some operating scenarios. As such, this can also be a tradeoff between reliability and channel reuse probability.

2. Improving the reliability of LBT results by using receiver-assisted channel-sensing in addition to energy detection-based channel-sensing to reduce the impact of hidden nodes.

3. Adopting a suitable LBT mechanism according to the operating environment, such as using different LBT mechanisms to comply with

regulations in different countries or regions, different operator network deployment scenarios, or different configurations. A better tradeoff between implementation complexity and system performance can thus be achieved as such, instead of supporting only one LBT mechanism.

18.2.1.1 Directional channel-sensing

Omnidirectional LBT is widely used in the Wi-Fi and LTE LAA systems. Since the use of narrow beam-forming can provide a higher link gain and also improves the ability of spatial division multiplexing, analog and digital beam-formings are supported for data transmission in NR system. As mentioned previously, the use of omnidirectional LBT may lead to an "overprotection" problem if the corresponding transmission is beam-forming based. For example, when a strong signal in one direction is detected using omnidirectional LBT, the signal cannot be transmitted in all directions. Since omnidirectional LBT may reduce the ability of apatial division multiplexing, energy detection-based channel-sensing that takes into consideration the effect of antenna array, also known as directional LBT [1,2], is also discussed during the SI phase.

As shown in Fig. 18.2, gNB 1 and gNB 2 provide data services to UE 1 and UE 2, respectively, and the directions of the two transmission links are different. If the directional LBT procedure that corresponds to beam-forming is used at each gNB before data transmission, it is possible to allow coexistence of multiple links at the same time. That is, gNB 1 may provide services to UE 1 and gNB 2 may provide services to UE 2 simultaneously using the same channel on the unlicensed spectrum, thus

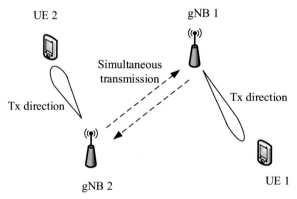

Figure 18.2 Beam-forming transmission based on directional listen before talk.

increasing apatial division multiplexing capacity and thereby increasing the overall throughput.

The most obvious advantage of directional LBT is that it improves the success rate of accessing the unlicensed channel, thus increasing spatial division multiplexing. However, during the discussion of directional LBT, some problems are also raised:

- The hidden−node problem may become more prominent since directional LBT only performs energy detection in limited directions.
- When a gNB performs the directional LBT procedure for data transmission corresponding to one transmit beam, the gNB can only provide data services to UEs corresponding to the same transmit beam. Therefore in order to provide services to UEs with different transmit beams, the gNB needs to perform multiple directional LBT procedure attempts to obtain the channel-access time that correspond to different directions. Compared with omnidirectional LBT, the time overhead of performing directional LBT would be increased. How to obtain the gain of spatial division multiplexing with less overhead is a problem that needs further study.
- Energy detection threshold setting for the directional LBT. The transmission power from the gNB, on one hand, can be reduced due to beam-forming gain and thus a higher energy detection threshold should be set according to the regulations. On the other hand, the energy detection threshold setting should also follow the principle of interference fairness (i.e., a higher energy detection threshold would lead to a higher channel access probability, but causing higher interference to other nodes). How to design and set a reasonable energy detection threshold considering various influencing factors needs further study.

Due to the above problems associated with adopting the directional LBT scheme, it is unclear whether directional LBT can improve the system spectral efficiency and how much gain it can achieve, which will require further studies. Although the conclusion of the SI phase is that directional LBT is beneficial to data transmission using beam-forming, due to time constraints and other reasons, there is no further study and standardization of directional LBT in the WI phase of NR−U.

18.2.1.2 Multinode joint channel-sensing

In the deployment of the NR−U system, neighboring nodes may belong to the same operator network, so multinode joint channel-sensing can be carried out among the multiple nodes through coordination [1,2]. Under the multinode joint channel-sensing mechanism, a group of gNB nodes exchange coordination information through backhaul signallings with each other to determine a common starting position of a downlink transmission burst. Each gNB node within the group performs own LBT procedure independently and may introduce a self-deferral after a successful LBT in order to send downlink transmission burst from the common starting position. By employing this channel-sensing mechanism, frequency reuse factor can be increased and the group of gNB nodes will be prevented from becoming hidden nodes or exposed nodes in the NR−U system. As such, a better coexistence with each node in the unlicensed spectrum is achieved. As shown in Fig. 18.3.

As multinode joint channel-sensing can increase frequency reuse for the system, multinode joint channel-sensing has been widely studied in the SI phase. Alignment of the starting time position for transmission, interaction, and coordination of LBT-related parameters between different gNBs or UEs, determining interference nodes, interference measurement and reporting from the UE side, and the adjustment of energy detection threshold are considered beneficial to support multinode joint channel-sensing. Some of the above techniques are standardized in the WI phase to improve frequency reuse in the NR−U system.

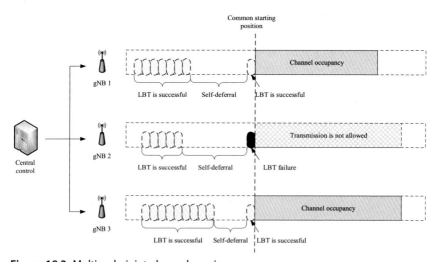

Figure 18.3 Multinode joint channel-sensing.

18.2.1.3 Receiver-assisted channel-sensing

Receiver-assisted channel-sensing for LBT is similar to the Request To Send/Clear To Send (RTS/CTS) mechanism in the Wi-Fi system. Before transmission in the downlink, the gNB first sends a signal similar to RTS to ask whether the UE is ready for data reception. After UE receives the RTS-like signal and if the LBT is successfully performed at the UE, the UE transmits a signal similar to CTS to the gNB containing information such as reporting its interference measurement results, informing the gNB that it is ready for data reception, and so on. The gNB then proceeds with downlink data transmission to the UE after receiving the CTS response. On the other hand, if the UE does not receive RTS or if the UE receives RTS but cannot send CTS due to LBT failure, it will not send the CTS response to the gNB. In this case, the gNB may abandon the downlink transmission to the UE if it does not receive the CTS response from the UE.

The advantage of receiver-assisted channel-sensing is that it can resolve the common hidden-node problem in a NR−U system. Based on the study during the SI phase, it is concluded that the handshake procedures between the transmitter and the receiver can effectively reduce or eliminate the effect of hidden nodes on the receiver UE. However, the hidden-node problem can also be mitigated by interference measurement and reporting from the UE side. As such, there is no continuing study and standardization effort spent on the receiver-assisted LBT in the WI phase of NR−U.

18.2.2 Dynamic channel-access procedure

The dynamic channel-access procedure can be considered as an LBE-based LBT mode. The principle of the dynamic channel-access procedure is that the communication equipment first perform LBT on an unlicensed channel after arrival of service data and transmits on the channel when the LBT is successful. The dynamic channel-access procedures includes Type-1 channel access and Type-2 channel access. Type-1 channel access is a multislot channel-sensing with random backoff scheme based on an adjusted contention window (CW) size, in which the corresponding CAPC p is selected according to the priority of the service data to be transmitted. Type-2 channel access is a channel-access procedure based on a deterministic duration of sensing slot. Type-2 channel access further includes Type-2A channel access, Type-2B channel access, and Type-2C channel access. Type-1 channel access is mainly used for communication equipment to initiate a CO, while Type-2 channel access is mainly used for communication equipment to share CO. A special case that needs to be mentioned is that when the gNB initiates a CO for SS/PBCH Block transmission without unicast data in a DRS window, if the duration of the DRS window does not exceed 1 ms and the duty cycle of the DRS window transmission does not exceed 1/20, then the gNB uses Type-2A channel access to initiate the CO for DRS transmission.

18.2.2.1 Default channel-access type for gNB: Type-1 channel access

Taking the gNB as an example, the channel-access parameters and values corresponding to various CAPC p to be used in the Type-1 channel-access procedure are shown in Table 18.1. In Table 18.1, m_p refers to the number of backoff slots, CW_p refers to the CW size, $CW_{min,p}$ refers to the minimum value of CW_p, $CW_{max,p}$ refers to the maximum value of CW_p, and $T_{mcot,p}$ refers to the maximum time duration for the CO.

The gNB selects a CAPC p according to the priority of the service data to be transmitted and performs the Type-1 channel-access procedure on the unlicensed channel according to the corresponding channel-access parameter values from Table 18.1, to initiate CO. That is, the COT is initiated by the gNB. Specifically, the Type-1 channel-access procedure includes the following steps:

1. Set the counter $N = N_{init}$, where N_{init} is a random number uniformly distributed between 0 and CW_p, and perform step (4).
2. If $N > 0$, the gNB subtracts 1 from the counter (i.e., $N = N - 1$).
3. Sensing the channel with the sensing slot duration T_{s1} (T_{s1} represents the LBT-sensing slot, the duration is 9 µs), if the sensing slot is idle, perform step (4); otherwise, perform step (5).

Table 18.1 Channel-access parameters and values for different channel-access priority classes p.

Channel access priority class (p)	m_p	$CW_{min,p}$	$CW_{max,p}$	$T_{mcot,p}$	Allowed CW_p values
1	1	3	7	2 ms	{3,7}
2	1	7	15	3 ms	{7,15}
3	3	15	63	8 or 10 ms	{15,31,63}
4	7	15	1023	8 or 10 ms	{15,31,63,127, 255,511,1023}

4. If $N = 0$, end the channel-access procedure; otherwise, perform step (2).

5. Sensing the channel with the sensing slot duration T_d (where $T_d = 16 + m_p \times 9$ μs), the sensing result of the sensing slot is that at least one sensing slot is occupied or all the sensing slots are idle.

6. If the sensing result of channel-sensing is that all the sensing slots are idle during T_d time, perform step (4); otherwise, perform step (5).

If the channel-access procedure on an unlicensed channel is completed successfully, the gNB performs DL transmission on that channel for the service data to be transmitted. The maximum time duration a gNB can occupy the channel for the DL transmission cannot exceed the $T_{mcot,p}$.

Before the gNB starts step 1 of the above Type-1 channel access, the gNB needs to maintain and adjust the CW size CW_p. For the initial transmission, the CW size CW_p is set by the gNB according to the minimum value $CW_{min,p}$; for the successive transmissions, the CW size CW_p can be adjusted by the gNB within the allowed CWp value range according to Acknowledgment (ACK) or Negative Acknowledgment (NACK) feedback information from the UE. If the CW size CW_p has been increased to the maximum value $CW_{max,p}$ and maintained for a certain number of times, the CW size CW_p can be reset to the minimum value $CW_{min,p}$.

18.2.2.2 Channel occupancy time sharing at gNB side

When the gNB initiates a COT, the resources in the COT not only can be used for downlink transmission, but also can be shared with UEs for uplink transmission. When the resources in the COT are shared with UEs for uplink transmission, the channel-access type the UE can use can be Type-2A channel access, Type-2B channel access, or Type-2C channel access, and these channel-access types are all based on a sensing slot with a fixed sensing duration.

- Type-2A channel access:

    This channel-access mode is a single-slot channel-sensing with a duration of 25 μs. Specifically, with Type-2A channel access, the UE can start to sense the channel 25 μs before the transmission starts, and perform transmission after the channel-sensing is successful.

- Type-2B channel access:

    This channel-access mode is a single-slot channel-sensing with a duration of 16 μs. Specifically, with Type-2B channel access, the UE can start to sense the channel 16 μs before the transmission starts, and perform transmission after the channel-sensing is successful. Note that the gap between the starting position of the transmission and the ending position of the previous transmission is 16 µs.

• Type-2C channel access:

In Type-2C channel access, a UE can perform transmission directly without channel-sensing after the end of a gap, wherein the gap between the starting position of the transmission and the ending position of the previous transmission should be less than or equal to 16 µs. The duration of the transmission cannot exceed 584 µs.

Different COT-sharing scenarios correspond to different channel-access types. For an uplink transmission burst occurring within a COT shared by the gNB, if the gap between the starting time position of the uplink transmission burst and the ending time position of the previous downlink transmission burst is less than or equal to 16 µs, the UE can be indicated to perform the Type-2C channel access before the uplink transmission. If the gap between the starting time position of the uplink transmission burst and the ending time position of the previous downlink transmission burst is equal to 16 µs, the UE can be told to perform the Type-2B channel access. If the gap between the starting time position of the uplink transmission burst and the ending time position of the previous downlink transmission burst is equal to or greater than 25 µs, then the UE performs the Type-2A channel access. In addition, the COT obtained by the gNB may include multiple uplink-to-downlink switching points. After the gNB shares its own COT with the UE for uplink transmission, the gNB may also use Type-2 channel access such as the Type-2A channel-access type for channel-sensing, and resume the downlink transmission within the COT after the channel-sensing is successful. Fig. 18.4 shows an example of COT-sharing at the gNB side.

When the gNB shares the acquired COT with the UE for uplink transmission, the principle of COT-sharing requires that the channel-access priority of the uplink service data sharing the COT transmission should not be lower than the channel-access priority used by the gNB when acquiring the COT, or rather the value of CAPC p of the uplink service data sharing the COT transmission should not be larger than the value of CAPC p used by the gNB when acquiring the COT. In addition, under certain channel-access types of COT-sharing at gNB, the gap-size between the starting time position of the uplink transmission burst and the ending time position of the previous downlink transmission burst needs to fulfill the requirements of 16 or 25 µs. The above COT-sharing principle and gap-size requirement should be guaranteed and indicated by

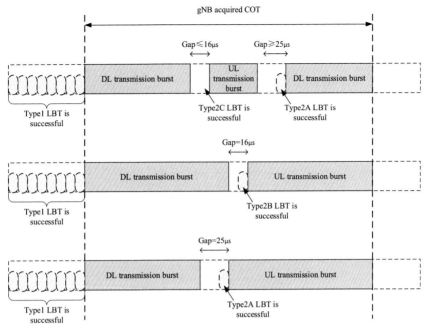

Figure 18.4 Channel occupancy time-sharing at the gNB side.

the gNB, and the gNB also indicates the channel-access type in the shared COT to the UE in an explicit or implicit manner. The methods of explicit and implicit indication are introduced in the next section.

18.2.2.3 Channel-access parameter indication

In the NR−U system, when a UE is scheduled for Physical Uplink Shared Channel (PUSCH) or Physical Uplink Control Channel (PUCCH) transmission, the gNB indicates the channel-access parameters for the PUSCH or PUCCH transmission via an UL grant or DL grant in a Downlink Control Information (DCI), respectively. Since the UE needs to fulfill the gap requirement of 16 or 25 μs for some certain channel-access types, the UE ensures the gap-size requirement between the two transmissions by transmitting a Cyclic Prefix Extension (CPE) of the first symbol from the uplink transmission. Accordingly, the gNB can indicate the CPE length of the first symbol for the uplink transmission to the UE.

Regarding the indication method, the gNB can explicitly indicate channel-access parameters such as CPE length, channel-access type, or CAPC p to the UE through joint coding. The following describes how

Table 18.2 The set of jointly indicating channel-access type and cyclic prefix extension length.

Indication	Channel access type	CPE length
0	Type-2C channel access	C2*symbol length–16μs–TA
1	Type-2A channel access	C3* symbol length –25μs–TA
2	Type-2A channel access	C1* symbol length –25μs
3	Type-1 channel access	0

to jointly indicate these channel-access parameters under different DCI formats.

- Fallback DCI for scheduling PUSCH transmission (DCI format 0_0):
  - A set of combined channel-access parameters is predefined in the specification as shown in Table 18.2, in which each row jointly indicates a channel-access type and a CPE length.
  - The fallback UL grant includes a 2-bit indication of channel-access parameters, which is used to indicate one row from the set shown in Table 18.2.
  - The indicated channel-access type and CPE length are used for the PUSCH transmission.
  - If the indicated channel-access type is Type-1 channel access, the UE selects CAPC p according to the priority of service data to be transmitted.
- Fallback DCI for scheduling Physical Downlink Shared Channel (PDSCH) transmission (DCI format 1_0):
  - The same set of channel-access parameter combinations in Table 18.2 is also used for this case, in which each row jointly indicates a channel-access type and a CPE length.
  - The fallback DL grant includes a 2-bit indication of channel-access parameters, which is used to indicate one row from the set shown in Table 18.2.
  - The indicated channel-access type and CPE length are used for PUCCH transmission, which carries ACK or NACK feedback information corresponding to the PDSCH scheduled by the fallback DL grant.
  - If the indicated channel-access type is Type-1 channel access, the UE determines the CAPC $p = 1$ for the PUCCH transmission.

In Table 18.2, the value of C1 is preset in the specification. When the subcarrier spacing is 15 or 30 KHz, C1 = 1; when the subcarrier spacing is

60 KHz, C1 = 2. The values of C2 and C3 are configured by higher-layer parameters. When the subcarrier spacing is 15 or 30 KHz, the value for C2 and C3 ranges from 1 to 28; when the subcarrier spacing is 60 KHz, the value for C2 and C3 ranges from 2 to 28.

- Nonfallback DCI for scheduling PUSCH transmission (DCI format 0_1)
  - A set of combined channel-access parameters is configured by higher-layer parameters, in which each row jointly indicates a channel-access type, a CPE length and a CAPC p.
  - The nonfallback UL grant includes an indication of channel-access parameters, which is used to indicate one row from the configured set.
  - The indicated channel-access type, CPE length, and CAPC p are used for PUSCH transmission.
  - If the indicated channel-access type is Type-2 channel access, then the CAPC p indicated at the same time is the CAPC p used by the gNB to obtain the COT.
  - The bit size of the indication is determined by the configured number of rows in the configured set. The maximum bit size for the indication is 6.
- Nonfallback DCI for scheduling PDSCH transmission (DCI format 1_1)
  - A set of combined channel-access parameters is configured by higher-layer parameters, in which each row jointly indicates a channel-access type and a CPE length.
  - The nonfallback DL grant includes an indication of channel-access parameters, which is used to indicate one row from the configured set.
  - The indicated channel-access type and CPE length are used for PUCCH transmission, which carries ACK or NACK feedback information corresponding to the PDSCH scheduled by the nonfallback DL grant.
  - If the indicated channel-access type is Type-1 channel access, the UE determines the CAPC $p = 1$ for the PUCCH transmission.
  - The bit size of the indication is determined by the configured number of rows in the configured set. The maximum bit size for the indication is 4.

In addition to the above explicit indication method, the gNB may implicitly indicate the channel-access type if the transmissions occur within the COT. When the UE receives an UL grant or DL grant sent by the gNB indicating that the channel-access mode for PUSCH or

PUCCH is the Type-1 channel access, if the UE determines that the PUSCH or PUCCH is within gNB's shared COT (e.g., UE receives a DCI format 2_0 sent by the gNB) and determines that the PUSCH or PUCCH is within gNB's shared COT according to a determined COT duration from the DCI format 2_0, then the UE can switch the channel-access mode from Type-1 channel access to Type-2A channel access for the PUSCH or PUCCH transmission.

18.2.2.4 Channel occupancy time sharing at UE side

When a UE uses Type-1 channel access to initiate a COT, not only the UE can use the resources in the COT for uplink transmissions, but the UE can also share the resources in the COT with the gNB for downlink transmissions. In the NR−U system, there are two cases where gNB may share the COT initiated by UE; one of them is gNB shares the COT of the scheduled PUSCH, and the other one is gNB shares the COT of a CG PUSCH.

In the case where the gNB shares the COT of the scheduled PUSCH, if the gNB has configured to the UE an energy detection threshold for COT-sharing, then the UE should use the configured energy detection threshold for channel-sensing. Since the channel-access parameters for PUSCH transmission are indicated by the gNB in this case, the gNB knows the MCOT value after the UE initiates the COT, therefore the COT-sharing from the UE for gNB transmissions can be supported transparently at the UE side.

For the case when the gNB shares the COT for CG−PUSCH, the CG−PUSCH transmission by the UE may carry a CG−UCI, which can include an indication of whether to share the COT acquired by the UE with the gNB. If the gNB configures an energy detection threshold for COT-sharing from the UE, then the UE should use the configured energy detection threshold for the channel-access procedure to acquire a COT if the UE intends to share the acquired COT with the gNB. Accordingly, the COT-sharing indication in the CG−UCI can indicate the starting time position and length of the shared UE COT with the gNB, and the CAPC p used by the UE when acquiring the COT. If the gNB does not configure to the UE an energy detection threshold for COT-sharing, then the COT-sharing indication in the CG−UCI includes 1 bit to indicate whether the gNB is allowed to share the COT acquired by the UE. Without an energy detection threshold configured for COT-sharing, the starting time position of the shared COT with the

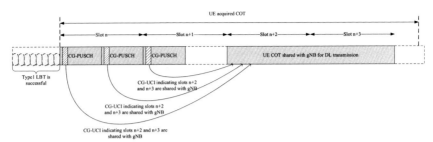

Figure 18.5 Channel occupancy time-sharing of configured grant—physical uplink shared channel.

gNB is determined according to higher-layer configured parameters and the maximum shared COT duration is predefined, and the gNB utilizes the shared COT only for transmitting common control information.

For the case of COT-sharing for CG—PUSCH transmission, the UE will ensure that the COT-sharing indications in multiple CG—UCIs transmitted over multiple CG—PUSCHs indicate the same shared COT with the same starting time position and length, as shown in Fig. 18.5.

18.2.3 Semistatic channel-access procedure

In the NR—U system, in addition to supporting the LBE-based channel-access mechanism, it also supports the FBE-based channel-access mechanism. The FBE-based channel-access mechanism offers benefits of frequency reuse, but it requires to operate in an interference-free environment and synchronization between network nodes when deploying NR—U networks on the unlicensed spectrum. In the FBE-based channel-access mechanism, or rather channel-access procedure for semistatic CO, the frame structure appears periodically. That is, the COT the gNB uses to provide services to UEs should appear periodically. The periodic frame structure has a Fixed Frame Period (FFP), and each FFP includes a COT and an Idle Period (IP). The duration of the FFP can be configured from a range of 1 to 10 ms, the duration of the COT in an FFP will not exceed 95% of the FFP duration, the duration of the IP is at least 5% of the FFP duration with the minimum value of the IP duration is 100 μs, and the IP is located at the end of the FFP.

The gNB performs LBT on the unlicensed channel for 9 μs duration in a sensing slot within the IP with the ending time position of the sensing slot aligned with the starting time position of the next FFP. If the LBT is successful, the COT in the next FFP can be used for transmission by the gNB; if the LBT fails, the COT in the next FFP cannot be used for transmission. In the NR−U system with FBE-based network deployment, currently the gNB is only allow to initiate a COT for semistatic CO. If a UE needs to perform uplink transmissions in this deployment scenario, it shares the COT from the gNB. For example, if UE detects downlink transmission bursts within a gNB's COT, the UE can perform uplink transmission in the same COT after the corresponding LBT is successful. It should be noted that allowing UE to initiate a COT for semistatic CO will be supported in the next release.

This semistatic channel-access mode can be configured by gNB via system information or higher-layer parameters. If a serving cell is configured by the gNB with semistatic channel-access mode, then the FFP duration of the serving cell is defined as T_x, the maximum COT duration included in the FFP of the serving cell is T_y, and the duration of the IP included in the FFP of the serving cell is T_z. Candidate values for the FFP duration T_x are configured by the gNB, and they include 1, 2, 2.5, 4, 5, or 10 ms. The UE determines the values of T_y and T_z based on the configured T_x. Specifically, starting from the even number of system frame, in every two consecutive system frames, the UE determines the starting position of each FFP according to $x \cdot T_x$, where $x \in \{0,1,\ldots, 20/T_x - 1\}$. The maximum COT duration in the FFP is $T_y = 0.95 \cdot T_x$, and the IP duration in the FFP is at least $T_z = \max(0.05 \cdot T_x, 100 \ \mu s)$.

Fig. 18.6 shows an example when the FFP duration is 4 ms. As shown in the figure, after receiving the configured FFP duration $T_x = 4$ ms from the gNB, the UE determines $x \in \{0,1,2,3,4\}$ according to $x \in \{0,1,\ldots, 20/T_x - 1\}$, and then the UE determines the starting time position of each FFP within every two consecutive system frames as 0, 4, 8, 12, 16 ms. Within each FFP, the maximum COT duration is $T_y = 3.8$ ms, and the minimum IP duration is $T_z = 0.2$ ms.

One disadvantage of the above semistatic channel-access mode is the channel-sensing slot for the gNB is always fixed at the end of an IP. If the

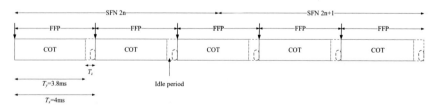

Figure 18.6 Semistatic channel occupancy.

channel-sensing result on the sensing slots is always busy, due to, for example, there being an interfering node using the LBE-based channel-access procedure such as the Wi-Fi system that keeps transmitting during the sensing slots on the same channel, then the gNB would be unable to perform any transmission in the following FFPs and to provide services to UEs in the cell for a long period of time. Therefore the FBE-based channel-access procedure is usually applied to a network deployment where there is no system deployed in the surrounding environment with the LBE-based channel-access procedure sharing the same unlicensed spectrum.

If there are multiple operators sharing the same unlicensed spectrum, since the frame structure is most likely not synchronized due to absence of a coordination mechanism among the operators, it is possible that the IP of a first operator and the COT of a second operator overlap in the time domain, resulting in the second operator's transmissions always occur during the IP in each FFP of the first operator. The first operator may not be able to access the channel for a very long period of time, thus being unable to provide services for its UEs. Therefore in network deployments with semistatic channel-access mode, synchronization of the frame structure among multiple operators is also required.

In addition, although the synchronization problem among different operators can be solved, the nodes of different operators may still interfere with each other under the network deployment. Therefore during the SI phase of the NR−U work, some companies proposed to enhance the semistatic channel-access mode [1]. As shown in Fig. 18.7, the IP duration may include multiple sensing slots. The gNB of different operators randomly selects one sensing slot from the multiple sensing slots for channel-sensing and starts transmission from the moment when the LBT is successful. This method can avoid the interference caused by transmissions from neighbor nodes belonging to different operators. However, due to time

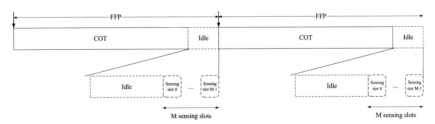

Figure 18.7 Enhancement for semistatic channel occupancy.

constraints and other reasons, this scheme did not progress further and was not standardized during the WI phase.

As mentioned above, if semistatic channel-access mode is configured for a NR−U system, the UE is not allowed to initiate a COT under the existing NR−U design. If the UE needs to perform uplink transmission, then the UE can only share the COT initiated by the gNB.

In the semistatic channel-access mode, if the gap between the starting time position of the uplink transmission burst and the ending time position of the previous downlink transmission burst in the same COT is up to 16 μs, the UE can perform the uplink transmission without sensing the channel. If the gap between the starting time position of the uplink transmission burst and the ending time position of the previous downlink transmission burst in the same COT is more than 16 μs, the UE will sense the channel in a sensing slot for 9 μs duration within a 25 μs interval ending immediately before the uplink transmission, and then it can perform the uplink transmission if the LBT is successful.

In the semistatic channel-access mode, the method of indicating channel-access parameters from the dynamic channel-access mode is reused. When the UE is scheduled with PUSCH or PUCCH transmission, the gNB can indicate the channel-access type and the CPE length of the first symbol of the PUSCH or PUCCH by an UL grant or DL grant. If the gNB indicates Type-1 channel access or Type-2A channel access, the UE should interpret the channel-access procedure as performing the LBT on the channel in a sensing slot for 9 μs duration within a 25 μs interval ending immediately before the uplink transmission.

18.2.4 Persistent uplink listen before talk detection and recovery mechanism

In the NR−U, a common problem is related to dealing with the impact of UE performing a series of LBT over multiple uplink transmissions. In particular, when the UE is experiencing persistent LBT failures attempting to transmit in the uplink, how do we deal with the problem where the UE enters a deadlock due to no access to the unlicensed radio channel? The UE will continue to make uplink transmission attempts, but these attempts will not succeed due to consistent LBT failures [3].

Below are a few examples illustrating that some Media Access Control (MAC) layer procedures will enter a deadlock when UE has consistent LBT failures. For example, as shown in Fig. 18.8, when a UE tries to transmit a random-access preamble or Scheduling Request (SR) in the

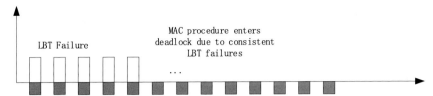

Figure 18.8 A schematic diagram of listen before talk failure during uplink transmission.

uplink, but due to persistent LBT failure the counters that correspond to the preamble transmission and SR transmission are not updated, the UE will keep attempting to transmit the preamble or SR by performing LBT, thus entering the deadlock in uplink transmission.

In order to solve the problem caused by persistent LBT failures, two solutions were discussed during the standardization of the NR−U:

The first is that the persistent LBT problem can be handled by their respective uplink transmission procedure. That is, when the UE is triggered to transmit a SR and encounters persistent LBT failures, the problem of not being able to transmit the SR should be handled by the SR procedure. Similarly, when an uplink transmission leads to the persistent LBT problem during the random-access process, the corresponding mechanism should be adopted to deal with this problem as part of the random-access procedure.

Another solution is that when the UE attempts to transmit any signal or channel in the uplink and experiences persistent LBT failures, it is necessary to design an independent mechanism to solve the persistent LBT failure problem. Table 18.3 summarizes the advantages and disadvantages of these two solutions:

After long discussions, it is finally decided to design an unified mechanism to deal with the persistent uplink LBT problem.

When discussing the use of an unified mechanism to deal with the persistent uplink LBT problem, the main consideration is how to detect persistent uplink LBT failures, and another problem is how to design a recovery mechanism when the UE declares a persistent failure in uplink LBT. Therefore the discussion mainly focused on the following aspects [4]:

The first aspect is whether to include all of the uplink transmissions or only those triggered by the MAC layer should be considered when designing a detection mechanism of persistent uplink failure. Transmissions that are triggered by the MAC layer include SR and random access related and PUSCH transmissions based on dynamic or semistatic scheduling. Besides

Table 18.3 Comparison of independent handling consistent listen before talk (LBT) failure and unified procedure handling consistent LBT failure.

	Advantages	Disadvantages
Handle the persistent uplink LBT failures in the respective procedure	• Simply make enhancements in existing procedure • Enhancements for different procedure can be different	• Significant impact on the MAC layer, and each procedure needs to be designed individually; • Uplink transmission triggered by various procedures may cause impact to each other and potentially lead to more problems.
Design an unified procedure to handle the persistent uplink LBT failure problem	• Individual enhancements to existing procedure are not required • AN unified mechanism to handle persistent uplink LBT failures caused by various uplink transmissions	• More discussions for the new mechanism will be triggered

these transmissions, some uplink transmissions are triggered by the physical layer, such as CSI, HARQ feedback, and SRS. After discussion, the 3GPP finally decided to take LBT failures caused by all types of uplink transmissions into account as part of the persistent uplink failure detection. As such, the physical layer is required to indicate the result of all corresponding uplink LBT failures to the MAC layer [5].

The second aspect is what mechanism should be used to detect a persistent failure in uplink LBT. During the discussion for the solution, most companies agreed that a timer was needed to decide whether to trigger/declare a persistent uplink LBT failure. In other words, "persistent" should be confined within a certain period of time, rather than simply accumulating number of LBT failures over any time period before triggering/declaring persistent LBT failures. Specifically, the network side is to configure an LBT detection timer and a threshold for detection counter. When the number of uplink LBT failures encountered by a UE reaches this configured detection counter threshold, a persistent uplink LBT failure is triggered/declared. Otherwise, when the timer expires, the UE resets the LBT failure counter. That is, if the LBT failure indication indicated by the physical layer is not received within a specified period of time, the LBT failure counter resets and the counting restarts. A specific example is shown as follows (Fig. 18.9):

The third aspect is, what granularity should the MAC layer use when counting the number of LBT failures? In other words, when all of the LBT failures indicated from the physical layer are combined during counting, whether an unified trigger/declaration of persistent uplink LBT failures is based on different uplink carriers, or only some uplink carriers are counted. First of all, the UE performs LBT independently for LBT subbands on each uplink Bandwidth Part (BWP). Considering that only one BWP can be activated on one carrier, it can be assumed that the UE executes LBT independently on each carrier. Therefore when counting the number of persistent uplink LBT failures, each carrier is treated independently. That is to say, a separate counter and timer should be maintained for each uplink carrier for counting LBT failures.

After designing the persistent uplink LBT failure detection mechanism, the remaining work is to consider how to perform a persistent uplink LBT failure recovery [6]. For the design of the recovery mechanism, considering that the persistent uplink LBT failures in different uplink carriers are triggered/declared independently, the recovery mechanism should then be different for different types of carrier.

- For the PCell (i.e., the primary cell), when the persistent uplink LBT failure is triggered/declared, the UE then directly triggers the radio-link failure procedure, and it finally recover from the persistent uplink LBT failure by performing a RRC reconfiguration procedure.
- For the PSCell in dual connectivity (i.e., the primary cell of the secondary cell group), when the persistent uplink LBT failure is triggered/declared, considering that the radio-link of the primary cell group is operating

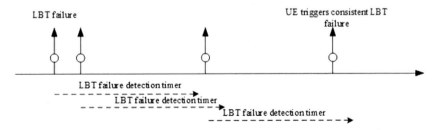

Figure 18.9 A schematic diagram of listen before talk detection timer mechanism.

normally, the UE reports the persistent uplink LBT failure event to the network via the PCell of the primary cell group. Generally, the network can perform RRC reconfiguration to solve the problem of persistent uplink LBT failures on the PSCell.

- For SCells (i.e., the secondary cells), when the persistent uplink LBT failure is triggered/declared, the UE reports the persistent uplink LBT failure by sending a MAC CE to the network.

18.3 Initial access procedure

This section introduces the design of the initial access in NR−U technology in the unlicensed spectrum. The initial access procedure of the NR−U system is similar to that of the NR system, and the related procedure can be found in Chapter 6, NR Initial Access. This section focuses on the differences between the initial access design for the NR−U system and the NR system.

18.3.1 SS/PBCH Block transmission

Similar to the normal NR operation in a licensed carrier, a NR−U UE also needs to acquire time and frequency synchronization with the network by searching SS/PBCH Blocks transmitted from the gNB as a part of the initial access procedure in the NR−U operation. Then the physical cell ID and Master information block (MIB) information carried in the PBCH can be obtained, and the search space set for PDCCH scheduling of the remaining minimum system message (RMSI) transmission in PDSCH is indicated in the MIB.

Before searching for SS/PBCH Blocks, the UE should determine the subcarrier spacing of the SS/PBCH Blocks. In the NR−U system, if the subcarrier spacing of the SS/PBCH Blocks is not provided by higher-layer signaling, the UE assumes the default subcarrier spacing of the SS/PBCH Blocks is 30 KHz. Through higher-layer signaling, the subcarrier spacing of the SS/PBCH Blocks for a secondary cell or cell group can be configured as 15 or 30 KHz. During the initial access procedure in NR−U, 30 KHz subcarrier spacing is also assumed at the UE for cell search, which is mainly due to the frequency band (5G−7 GHz) used by the NR−U technology introduced in the Release 16 standard [7].

The transmission scheme for SS/PBCH Blocks in the NR−U system was discussed during the RAN1#96-RAN1#99 meetings. According to the channel-access characteristics in the NR−U system, the transmission scheme for the SS/PBCH Blocks is enhanced, including the location of synchronization raster where the SS/PBCH Blocks are transmitted, the transmission pattern of a SS/PBCH Block in a slot, the transmission pattern of SS/PBCH Blocks in a half-frame, the Quasi-Co-Location (QCL) relationship between the positions of SS/PBCH Blocks in the half-frame, and so on.

In the NR−U system, the location of synchronization raster where the SS/PBCH Blocks are transmitted is redefined. The motivation and scheme of synchronization raster definition in the NR−U system have been discussed in detail [8−12]. First of all, in order to flexibly support a variety of channel bandwidths and deployment scenarios in the licensed spectrum, the number of synchronization raster defined in the NR system is relatively large. For the NR−U system, the channel bandwidth and location are relatively fixed for the SS/PBCH Blocks, and many synchronization rasters are unnecessary in a given channel range. The dense synchronization raster for the NR system is relaxed to reduce the complexity for the UE searching for SS/PBCH Blocks. Based on this consideration, only one synchronization raster is retained but the location of the synchronization raster allowed within the channel bandwidth in NR−U is another topic that needs to be decided. Two location scenarios were mainly discussed during the NR−U development work, that is, whether the synchronization raster should be placed roughly in the middle or at the edge of the channel bandwidth. The main factor of consideration is the location relationship between RMSI and SS/PBCH Blocks. If the synchronization raster is roughly in the middle of the channel bandwidth, then the number of RBs available for RMSI transmission would be limited, or it is necessary to perform rate matching of RMSI transmission around the time-frequency resources for the SS/PBCH Blocks. For the other location placement scenario, where the synchronization raster is roughly at the edge of the channel bandwidth, then rate matching is not necessary around the time-frequency resources of the SS/PBCH Blocks and the number of available RBs that can be used for RMSI transmission would be less restricted as well. In summary, in order to facilitate RMSI transmission and minimize the constraints for the NR−U system design, the NR−U system defines the synchronization raster location at the edge of the channel bandwidth.

In order to alleviate the impact of LBT failure on transmission of SS/PBCH Blocks, it is desirable for the network to send as many channels and signals as possible during channel occupation to reduce number of channel-access attempts from the gNB. A discovery reference signal (DRS) window similar to the one defined in LTE LAA was also introduced for the NR−U. In the DRS window, in addition to the transmission of SS/PBCH Blocks, it is also desirable to send channels related to RMSI transmission, including Type0-PDCCH and PDSCH. Time-domain multiplexing of SS/PBCH Blocks and Type0-PDCCH, similar to the SS/PBCH block and CORESET#0 multiplexing pattern 1 (see Section 6.2.2), is also applied in NR−U.

For the transmission pattern of SS/PBCH Blocks in a slot, it is necessary to consider how Type0-PDCCH, PDSCH and SS/PBCH Blocks are multiplexed and transmitted in the DRS window. At the RAN1#96 meeting, based on the consensus of most companies, the baseline for the transmission pattern of SS/PBCH Blocks in a slot is determined. That is, the symbol positions of the two SS/PBCH Blocks in a slot are (2,3,4,5) and (8,9,10,11), respectively. Considering the multiplexing of SS/PBCH Blocks and Type0-PDCCH, the Type0-PDCCH corresponding to the second SS/PBCH Block in the slot needs to support transmission on two consecutive symbols between the two SS/PBCH Blocks. For this reason, companies have proposed some schemes for SS/PBCH Blocks transmission pattern in a slot [13]. At the RAN1#96 meeting, the following two patterns for SS/PBCH Blocks transmission in a slot are concluded for further downselection, as shown in Fig. 18.10:

- Pattern 1: the symbol positions of the two SS/PBCH Blocks in a slot are (2,3,4,5) and (8,9,10,11), respectively.
- Pattern 2: the symbol positions of the two SS/PBCH Blocks in a slot are (2,3,4,5) and (9,10,11,12), respectively.

At the RAN1#96bis meeting, the above two patterns and the corresponding Type0-PDCCH CORESET symbol position in the slot for the two patterns were further discussed. At the RAN1#97 meeting, as the companies did not have consensus on the alternatives, it was decided to reuse the legacy SS/PBCH Block pattern Case A and Case C (Section 6.1.3) from the NR system, which is the above pattern 1. Although the standard eventually follows the original SS/PBCH Block pattern in the NR system, the discussion procedure of this issue can be helpful for us to understand the channel transmission in the DRS window in the NR−U system.

For the design of SS/PBCH Block transmission pattern in the DRS window, it was also considered how to reduce the impact of LBT failure on SS/PBCH Block transmissions. Design aspects considered included the

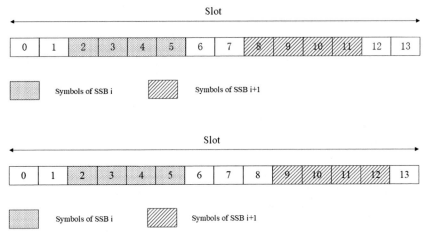

Figure 18.10 SS/PBCH Block transmission pattern in a slot.

length of the DRS window, transmission patterns for the SS/PBCH Blocks, and so on. In the NR—U system, the length of the DRS window is configurable with a maximum length of a half-frame and the configurable length includes{0.5,1,2,3,4,5}ms. During the initial access, before a UE receives the configuration information on the length of the DRS window, the UE assumes that the length of the DRS window is half of a frame. When the network transmits SS/PBCH Blocks, due to LBT, the starting time of obtaining channel occupation may not be the starting time of the DRS window. Based on this uncertainty in the starting time of channel occupation, how to design the transmission pattern of SS/PBCH Blocks in the DRS window has to be considered in the standard. For this reason, the concept of candidate SS/PBCH Block location within the DRS window is introduced. As mentioned earlier, there are two SS/PBCH Block transmission positions in each slot. According to the number of slots contained within the DRS window, then the transmission pattern of SS/PBCH Blocks in the DRS window can be obtained.

As an example the length of the DRS window could be 5 ms. For the SS/PBCH Blocks, considering the candidate subcarrier spacings are 30 and 15 KHz, the DRS window could contain 20 and 10 SS/PBCH Block locations, respectively. One of these SS/PBCH Block locations is called a candidate SS/PBCH Block location. Whether a SS/PBCH BLOCK is sent using the candidate SS/PBCH Block location depends on the result of the LBT. When the LBT is successful, the network transmits SS/PBCH Blocks using the candidate SS/PBCH Block locations in a continuous manner starting

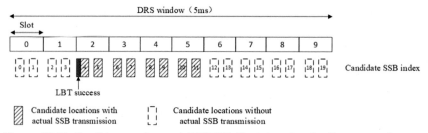

Figure 18.11 Candidate and actual SS/PBCH Block location in discovery reference signal window.

from the first candidate SS/PBCH Block location during the channel occupation. Possible candidate transmission locations and the actual transmission locations for the SS/PBCH Blocks within the DRS window are shown in Fig. 18.11, where each candidate SS/PBCH Block location corresponds to a candidate SS/PBCH Block index.

In order to acquire frame synchronization for NR−U UEs based on detected SS/PBCH Block(s), the locations of the SS/PBCH Block and corresponding indices in a frame are used. In the NR−U, the candidate SS/PBCH Block index is used to represent the index of the candidate SS/PBCH Block location within the DRS window. Assuming that the number of candidate locations within the DRS window is Y, and the range for the candidate SS/PBCH Block indices in the DRS window is 0,..., Y-1, the candidate SS/PBCH Block index is carried in the SS/PBCH Block. When the UE detects a SS/PBCH Block, it is then able to acquire the frame synchronization according to the candidate SS/PBCH Block index carried therein. The candidate SS/PBCH Block index is carried in the PBCH, and the indication method is described in Section 18.3.2.

As can be seen from the design of the candidate SS/PBCH Block locations in the DRS window, the network does not actually send the SS/PBCH Block in all of the candidate SS/PBCH Block locations. Among the periodic DRS windows, since the starting time from a successful channel access may be different, how to determine the QCL relationship between the SS/PBCH Block transmitted in different candidate SS/PBCH Block locations is an issue to be solved. The issue was discussed during the RAN1#96 and RAN1#97 meetings. Firstly, there is a QCL relationship between the SS/PBCH Block corresponding to the same candidate SS/PBCH Block index. In the Release 15 version of NR, the maximum number of SS/PBCH Blocks for the 3−6 GHz frequency band is 8, and the index range for the SS/PBCH Blocks is also 0–7, which is aligned with the maximum number and the QCL information of the SS/PBCH Blocks. Unlike the Release 15 version of NR, the number of candidate SS/PBCH Block locations in Release 16 NR–U is more than the maximum number of SS/PBCH Blocks that can be sent. As such, it is necessary to define the QCL relationship between the SS/PBCH Blocks sent at different candidate SS/PBCH Block locations. For this purpose, parameter Q was introduced in the RAN1#97 meeting to determine the QCL information corresponding to the candidate SS/PBCH Block index. When the two candidate SS/PBCH Block indexes have the same result after modulating Q, the SS/PBCH Block that corresponds to the two candidate SS/PBCH Block indexes has a QCL relationship. In the RAN1#99 meeting, two terminologies were adopted for the SS/PBCH Block indexes: candidate SS/PBCH Block indexes and SS/PBCH Block indexes, and the relationship between them is determined according to the following formula:

$$SS/PBCH \text{ block index} = modulo(candidate\ SS/PBCH \text{ block index},\ Q)$$

$$(18.1)$$

Fig. 18.12 shows a schematic diagram for determining QCL information of SS/PBCH block based on the candidate SS/PBCH block index and parameter Q.

That is, the result after the candidate SS/PBCH Block index modulates with the parameter Q is defined as the SS/PBCH Block index. Since SS/PBCH Blocks with different SS/PBCH Block indexes do not have a QCL relationship, then it can be seen that the parameter Q represents a maximum number of SS/PBCH Blocks without QCL relationship that can be sent by the network. In other words, the parameter Q represents

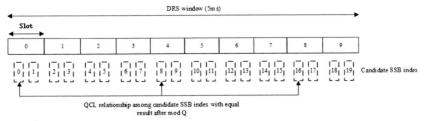

Figure 18.12 Determining quasi-co-location information of SS/PBCH Block based on the candidate SS/PBCH block index and parameter Q.

the maximum number of SS/PBCH Block beams. In the RAN1#98 meeting, the value range for the parameter Q was discussed. The main viewpoints included: all values in 1−8 or just some of them. The factors considered by companies mainly included flexibility of cell deployment, impact of channel access, and overhead of signaling for the parameter Q indication. Finally, a compromise way forward was adopted, and the value range for Q is {1, 2, 4, 8}.

In the RAN1#99 meeting, it was agreed that the number of SS/PBCH Blocks sent within a DRS window cannot exceed the limit of Q. In the RAN1#100bis-e meeting, it was concluded that SS/PBCH Blocks with the same SS/PBCH Block index can only be sent at most once within a DRS window. In order for the UE to obtain the QCL information of the detected SS/PBCH Block, the parameter Q is known to the UE. During the initial access, the parameter Q for the cell is indicated by PBCH, and the indication method is described in detail in Section 18.3.2. For neighboring cells, their Q parameters can be indicated using UE-specific RRC signaling or via SIB information, which is mainly used for UE's RRM measurement of neighboring cells during IDLE, INACTIVE, and CONNECTED states. The parameter Q corresponds to the parameters N_{SSB}^{QCL} defined in the standard.

After determining the subcarrier spacing, transmission pattern, and QCL information of the SS/PBCH Blocks, the UE finalizes the synchronization and MIB reception by using the detected SS/PBCH Block and further performing SIB reception and random access to complete the initial access procedure in the NR−U system.

18.3.2 Master information block

As described in the previous section, the candidate SS/PBCH Block index and the parameter used to determine the QCL information of the SS/PBCH Block are indicated in the PBCH. The transmission of PBCH includes the information carried by PBCH and the DMRS for decoding PBCH. The information carried by PBCH includes additional timing-related PBCH payload bits and MIB information from the higher layers. The additional timing-related bits are used to carry the SS/PBCH Block index and half-frame indication. This section introduces the differences in the information carried by PBCH in a NR−U system compared to a normal NR system.

Since the maximum length for the DRS window is 5 ms and the maximum number of candidate SS/PBCH Block locations within the

DRS window is 20 (for the case of subcarrier spacing it is 30 KHz), the range of candidate SS/PBCH Block indexes spans from 0,...,19 and five bits are needed in the PBCH to indicate the candidate SS/PBCH Block index as such. During the discussion in 3GPP RAN1, the first reached agreement was that the payload size for PBCH is not increased compared to Release 15, in order to minimize the impact on the standard and the design of the product. There are eight PBCH DMRS sequences defined in Release 15 and their indexes are used to indicate the lowest three bits of the SS/PBCH Block index. The NR−U also follows this approach and uses it to indicate the lowest three bits of the candidate SS/PBCH Block index. The remaining two bits are the same bits defined in Release 15 to indicate the 4th and 5th bits of the 6-bit indication of the maximum 64 SS/PBCH Block indexes in FR2.

In Release 16, the frequency bands supported by the NR−U system are within the range of FR1. Since the abovementioned two bits in the PBCH are reserved in FR1 from the Release 15 NR, the 4th and 5th bits are now redefined in the Release 16 NR−U system to indicate the candidate SS/PBCH Block index. In addition, the half-frame indication information in the PBCH is the same as that of Release 15. Overall, this is a scheme with minimal impact on the standards, and the relevant conclusions were reached in the RAN1#97−98 meeting. Fig. 18.13 shows a schematic diagram of the candidate SS/PBCH Block index indication.

In the NR−U system, the parameter Q for determining the QCL information of the SS/PBCH Block is new information that needs to be indicated to the UE. During the RAN1#98−99 meetings, various alternatives for indicating the parameter Q from the gNB were discussed, including the following:
- Scheme 1: Indicated by SIB1
- Scheme 2: Indicated by MIB

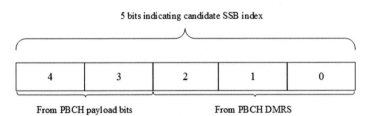

Figure 18.13 Candidate SS/PBCH Block index indication in PBCH.

- Scheme 3: Indicated by PBCH additional timing-related PBCH payload bits
- Scheme 4: Not indicated, fixed value

With the conclusion on the value range for the parameter $Q = \{1, 2, 4, 8\}$, Scheme 4 was excluded first. The remaining schemes mainly consist of two categories, indication by SIB1 or indication by PBCH. The focus of the choice among the schemes is whether or not the UE needs to know the parameter Q value before receiving the SIB1. Companies that supported SIB1-based indication believed that if parameter Q is indicated by PBCH, the UE cannot get parameter Q until it correctly receives PBCH, which is not helpful for PBCH decoding. For the reception of Type0-PDCCH, although the parameter Q is helpful to determine the monitoring occasions of Type0-PDCCH, in many scenarios such as RRM measurement, the parameter Q is indicated by SIB or using a dedicated RRC signaling. The scenario where the parameter Q is really needed is to determine the associated RACH resource from the QCL information of the SS/PBCH Block, which is after the SIB1 information is received, so the parameter Q can be indicated in the SIB1 [14].

Companies that supported PBCH-based indication of the parameter Q believed that a joint detection needs to be carried out between the SS/PBCH Block with the QCL relationship in different DRS windows, and also cell selection and beam selection should be carried out according to the detection results, for which the operation is similar to the function in Release 15. After long discussions, it was agreed on in the RAN1#99 meeting that two bits are used to indicate the parameter Q in MIB, which was preferred by the majority. The reuse of two bits in Release 15 MIB includes the following two further schemes, but finally Scheme 1 was agreed on in the RAN1#100 meeting. Due to the fact that the subcarrier spacing of Type0-PDCCH and SS/PBCH Blocks is always the same as defined in NR−U technology, the subcarrier spacing of Type0-PDCCH no longer needs to be indicated by SubcarrierSpacingCommon in MIB. At the same time, the subcarrier offset between the RB boundary of SS/PBCH Block and the common RB boundary is always even, which means that LSB of ssb-SubcarrierOffset is not needed anymore. Therefore the above two bits can be reused to indicate the value of parameter Q.

- Scheme 1:
  - SubcarrierSpacingCommon (1 bit)
  - LSB of ssb-SubcarrierOffset (1 bit)

- Scheme 2:
  - SubcarrierSpacingCommon (1 bit)
  - Spare bit in MIB (1 bit)

Fig. 18.14 shows a schematic diagram of the parameter Q indicated by MIB, and Table 18.4 shows the corresponding relationship between the value of parameter Q and two bits in MIB.

Since some of the bits in MIB are redefined in the NR−U system, the UEs in the NR and NR−U systems would have different interpretation of the MIB transmitted. However, as the definition and usage around the 6 GHz frequency spectrum may be different from one country/region to another, some bands are used as the licensed spectrum and some others are used as the unlicensed spectrum as shown in Fig. 18.15. In this case, the UE needs to identify whether the SS/PBCH Block detected in the 6 GHz frequency spectrum corresponds to a NR system or a NR−U system. In the RAN1#99 meeting, some companies proposed to introduce new MIB type indication into BCCH-BCH-Message to distinguish different MIB types [15]. But this kind of plan does not have the support of most companies. In the RAN1#100bis-e meeting, the following solutions to this problem were discussed [16]:

- UE tries to interpret two kinds of MIB

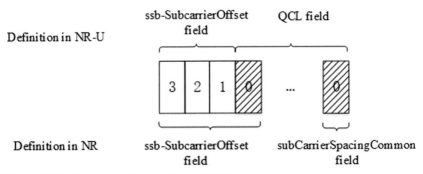

Figure 18.14 Parameter Q indication in PBCH.

Table 18.4 Parameter Q indication in PBCH.

subCarrierSpacingCommon	LSB of ssb-SubcarrierOffset	Q
scs15or60	0	1
scs15or60	1	2
scs30or120	0	4
scs30or120	1	8

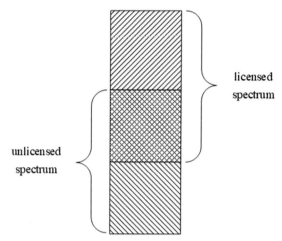

Figure 18.15 Overlapping of licensed and unlicensed spectrum.

- Different scrambling code for CRC of PBCH
- Separate synchronization raster for licensed and unlicensed spectrum
    As of the finalization of this book, there is still no agreed upon solution to this problem, which may be resolved in Release 16 or in a later standard release.

18.3.3 Remaining minimum system message monitoring

After the UE detection of SS/PBCH Block(s) and decoding of MIB information, the next step is for the UE to receive RMSI by monitoring Type0-PDCCH, whose CORESET and search space are indicated in the MIB. In Release 15 NR, there are three patterns for multiplexing SS/PBCH Block and Type0-PDCCH CORESET (see Section 6.2.2). In the Release 16 NR−U, the time-domain multiplexing of SS/PBCH Block and Type0-PDCCH CORESET is adopted, which is similar to the multiplexing pattern 1 in NR.

    In the RAN1#96 meeting, it was decided that the subcarrier spacing of Type0-PDCCH CORESET and SS/PBCH Block are always the same, and the frequency-domain resources for Type0-PDCCH CORESET are 48 RBs and 96 RBs in 30 and 15 KHz subcarrier spacing, respectively. In the RAN1#96bis meeting, it was agreed that Type0-PDCCH CORESET should occupy one or two symbols. In the RAN1#99 meeting, the number of offset RBs between the frequency-domain position of Type0-PDCCH CORESET and SS/PBCH block

Table 18.5 Type0-PDCCH CORESET configuration: {SS/PBCH block, PDCCH} SCS = {15, 15} KHz, with shared spectrum channel access.

Index	SS/PBCH block and CORESET multiplexing pattern	Number of RBs	Number of symbols	Offset (RBs)
0	1	96	1	10
1	1	96	1	12
2	1	96	1	14
3	1	96	1	16
4	1	96	2	10
5	1	96	2	12
6	1	96	2	14
7	1	96	2	16
8	Reserved			
9	Reserved			
10	Reserved			
11	Reserved			
12	Reserved			
13	Reserved			
14	Reserved			
15	Reserved			

was determined. For the case with 30 KHz subcarrier spacing, the RB offset is {0,1,2,3} RBs, and for the 15 KHz subcarrier spacing the RB offset is {10,12,14,16} RBs. Finally, the Type0-PDCCH CORESET mapping table for the NR−U was added to the Release 16 standard, as shown in Tables 18.5 and 18.6. The bit field indicating Type0-PDCCH CORESET in MIB is the same as in Release 15.

In the NR−U, the search space information for Type0-PDCCH is indicated using the same field and in the same manner as Release 15 in MIB (see Section 6.2.4). For SS/PBCH Block and Type0-PDCCH CORESET, with multiplexing pattern 1 as mentioned previously, the UE monitors PDCCH in the Type0-PDCCH CSS set over two consecutive slots starting from slot n_0. Each candidate SS/PBCH Block with an index $\bar{i}$ within a monitoring window contains the two consecutive PDCCH monitoring slots, where $0 \leq \bar{i} \leq \overline{L}_{\max} - 1$ and $\overline{L}_{\max}$ is the maximum number of candidate SS/PBCH Blocks. The index of slot n_0 is determined by the following formula:

$$n_0 = \left(O \cdot 2^{\mu} + \left\lfloor \bar{i} \cdot M \right\rfloor \right) \bmod N_{\text{slot}}^{\text{frame, } \mu} \tag{18.2}$$

Table 18.6 Type0-PDCCH CORESET configuration: {SS/PBCH block, PDCCH} SCS = {30, 30} KHz, with shared spectrum channel access.

Index	SS/PBCH block and CORESET multiplexing pattern	Number of RBs	Number of symbols	Offset (RBs)
0	1	48	1	0
1	1	48	1	1
2	1	48	1	2
3	1	48	1	3
4	1	48	2	0
5	1	48	2	1
6	1	48	2	2
7	1	48	2	3
8	Reserved			
9	Reserved			
10	Reserved			
11	Reserved			
12	Reserved			
13	Reserved			
14	Reserved			
15	Reserved			

After determining the index of slot n_0, the frame number SFN_C in which the monitoring window is located can be determined as follows:

- If $\left\lfloor \left(O \cdot 2^\mu + \lfloor \bar{i} \cdot M \rfloor \right) / N_{slot}^{frame,\mu} \right\rfloor \mod 2 = 0$, $SFN_C \mod 2 = 0$

- If $\left\lfloor \left(O \cdot 2^\mu + \lfloor \bar{i} \cdot M \rfloor \right) / N_{slot}^{frame,\mu} \right\rfloor \mod 2 = 1$, $SFN_C \mod 2 = 1$

Thus it can be seen that in the NR−U, the determination of Type0-PDCCH search space is similar to that of Release 15, except that each candidate SS/PBCH block index is associated with a set of Type0-PDCCH monitoring occasions in the NR−U.

In the NR−U system, there is another case where the UE may be required to receive RMSI in a secondary cell. This operation is required to support the automatic neighbor relations (ANR) function in the NR−U system to solve the problem of PCI conflicts that may occur when different operators deploy cells use the same frequency [17,18]. Due to the operation of ANR replies on the acquisition of RMSI, it is hence necessary to support the reception of RMSI in the secondary cell. After receiving the RMSI on the secondary cell, the UE reports the cell global identity (CGI) of the cell, which is used for the ANR function in the network. In order to receive RMSI, it is also necessary for the UE to receive SS/PBCH Block(s) on the secondary cell to obtain Type0-PDCCH information. Since the secondary cell is not the cell used for the initial access, the frequency location of the SS/PBCH Blocks used to carry Type0-PDCCH information is not located on the synchronization raster. In this case, it is necessary to determine how to obtain Type0-PDCCH information via the SS/PBCH Block received on an nonsynchronization raster in order to receive RMSI on the secondary cell. The procedure of receiving RMSI in a secondary cell was adopted during the RAN1#100-e meeting [19] and is as follows:

- Step 1: Detect ANR SS/PBCH block and decode PBCH for MIB acquisition.
- Step 2: Acquire subcarrier offset $\bar{k}_{SSB}$ from PBCH and determine k_{SSB} based on $\bar{k}_{SSB}$
    - If $\bar{k}_{SSB} \geq 24$, $k_{SSB} = \bar{k}_{SSB}$; else, $k_{SSB} = 2 \cdot \lfloor \bar{k}_{SSB}/2 \rfloor$.
- Step 3: Determine common RB boundary based on k_{SSB}.
- Step 4: Determine the first offset based on CORESET#0 information in MIB.
- Step 5: Determine the second offset based on the offset between the central frequency of the ANR SS/PBCH block and the GSCN defined in the LBT channel bandwidth.
- Step 6: Determine the frequency location of CORESET#0 based on the first and second offset.

This procedure is also illustrated in Fig. 18.16.

18.3.4 Random access

In the NR−U, enhancement for the RACH procedure mainly includes the following aspects:

- OCB of the PRACH
- PRACH preamble sequence

Figure 18.16 Procedure of receiving remaining minimum system message in the secondary cell for automatic neighbor relations.

- Channel access during random access
- Validity of PRACH resource
- Support of 2-step RACH.

In order to meet the OCB requirements of PRACH, in the RAN1#93 meeting, the comb structure (interlaced PRACH) for PRACH frequency resources was considered. In the RAN1#95 meeting, several interlace and noninterlace schemes were discussed. For the interlace scheme, the frequency resources of PRACH are discontinuously allocated with PRB- or RE-level interlace. For the noninterlace scheme, the frequency resources of PRACH are continuously allocated. The minimum OCB requirement is fulfilled by repeating the PRACH sequence in the frequency domain or introducing a long PRACH sequence. In the RAN1#96bis meeting, several companies provided simulation results that evaluated the coverage and capacity of the PRACH channel under the different schemes. The results showed that the scheme of PRACH sequence repetition or long PRACH sequence allocation in the frequency domain provided better MCL results, because PRACH takes up more bandwidth in the frequency domain. However, the disadvantage is that it will reduce the capacity of the PRACH channel. According to the simulation results and views of companies, it was finally agreed not to adopt the interlace scheme from the meeting, while considering the following candidates:

- Scheme 1: Repeat the legacy short PRACH sequence with a length of 139, mapped to continuous subcarriers
- Scheme 2: Longer sequence than 139, without repetition, and mapped to continuous subcarriers

The discussion on PRACH long sequence and the selection of PRACH sequence length continued for several RAN1 meetings after. Finally in the RAN1#99 meeting, it was concluded that besides the legacy PRACH short sequence with 139 length, a new long PRACH sequence is also supported:

- For 15 KHz subcarrier spacing, length of RA = 1151; for 30 KHz subcarrier spacing, length of RA = 571.
- Selection between legacy PRACH sequence with 139 length and the new long PRACH sequence is indicated via SIB1.

As shown in Table 18.7, additional long PRACH sequences are added to legacy preamble formats. For L_RA = 571, the PRACH includes 48 PRBs. For L_RA = 1151, the PRACH includes 96 PRBs.

Compared with the Release 15 NR, the transmission of messages in the RACH procedure needs to consider the impact of channel access in

Table 18.7 Preamble formats.

Format	L_{RA} $\mu \in \{0, 1, 2, 3\}$	$\mu = 0$	$\mu = 1$	Δf^{RA}	N_u	N_{CP}^{RA}
A1	139	1151	571	$15 \cdot 2^\mu$ kHz	$2 \cdot 2048\kappa \cdot 2^{-\mu}$	$288\kappa \cdot 2^{-\mu}$
A2	139	1151	571	$15 \cdot 2^\mu$ kHz	$4 \cdot 2048\kappa \cdot 2^{-\mu}$	$576\kappa \cdot 2^{-\mu}$
A3	139	1151	571	$15 \cdot 2^\mu$ kHz	$6 \cdot 2048\kappa \cdot 2^{-\mu}$	$864\kappa \cdot 2^{-\mu}$
B1	139	1151	571	$15 \cdot 2^\mu$ kHz	$2 \cdot 2048\kappa \cdot 2^{-\mu}$	$216\kappa \cdot 2^{-\mu}$
B2	139	1151	571	$15 \cdot 2^\mu$ kHz	$4 \cdot 2048\kappa \cdot 2^{-\mu}$	$360\kappa \cdot 2^{-\mu}$
B3	139	1151	571	$15 \cdot 2^\mu$ kHz	$6 \cdot 2048\kappa \cdot 2^{-\mu}$	$504\kappa \cdot 2^{-\mu}$
B4	139	1151	571	$15 \cdot 2^\mu$ kHz	$12 \cdot 2048\kappa \cdot 2^{-\mu}$	$936\kappa \cdot 2^{-\mu}$
C0	139	1151	571	$15 \cdot 2^\mu$ kHz	$2048\kappa \cdot 2^{-\mu}$	$1240\kappa \cdot 2^{-\mu}$
C2	139	1151	571	$15 \cdot 2^\mu$ kHz	$4 \cdot 2048\kappa \cdot 2^{-\mu}$	$2048\kappa \cdot 2^{-\mu}$

the NR−U. In order to minimize the delay in the RACH procedure caused by LBT failure, channel occupation sharing (COT-sharing, refer to Section 18.2.2) is also supported for the RACH procedure. When the base station transmits Msg2, the COT obtained by the base station can be shared with the UE for Msg3 transmission. Otherwise, the UE would need to perform its own LBT to acquire channel access for sending Msg3, and consequently this may lead to LBT failure and transmission delay for Msg3 delivery. When sending Msg2 from the base station, the base station can indicate the type of LBT in the Msg2 the UE needs to perform for sending Msg3 according to its own COT duration. On the other hand, when the base station sends Msg2, considering the potential LBT failure, it may not be able to acquire the channel access in time before the RAR reception window of the UE, which consequently leads to the inability to send Msg2. Therefore the maximum length of the RAR reception window is extended in the NR−U. That is, the maximum RAR reception window of 10 ms defined in Release 15 is extended to 40 ms in the NR−U, allowing more time for the base station to access the channel and to avoid cases where the RAR cannot be sent in time due to LBT failure.

Accordingly, since the periodicity of PRACH resources in Release 15 is at least 10 ms, the method of calculating RA-RNTI does not need to distinguish the SFN where the PRACH resource is located. After the maximum RAR reception window is extended to 40 ms, multiple UEs with overlapping RAR reception window due to PRACH transmissions from multiple ROs could cause more than one UE receiving RAR information scrambled by the same RA−RNTI in the extended RAR reception window. In order to help these UEs distinguish between different PRACH resources that correspond to the received RAR, two LSB bits of the SFN are indicated in DCI format 1_0, which is used to schedule the PDSCH carrying the RAR. The UE can then determine whether the received RAR corresponds to the SFN where the PRACH transmission from the UE is located [20−24].

The determination of PRACH resources follows not only the legacy definition from Release 15, but also an additional definition based on the channel-access procedure in the NR−U. The FBE channel-access type is introduced in Section 18.2.3. When the FBE channel-access type is configured, if the PRACH resource overlaps with a set of consecutive symbols before the channel occupation begins, the PRACH resource is considered invalid.

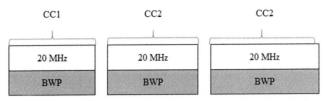

Figure 18.17 Using carrier aggregation concept to support wideband operation.

In addition, the two-step RACH procedure introduced in Release 16 is also supported in the NR−U. The enhancements to four-step RACH in NR−U are also applicable to the 2-step RACH, including the extension of the maximum RAR reception window, DCI format 1_0 indicating the two LSB bits of SFN, COT-sharing, and so on.

18.4 Wideband operation and physical channel enhancements

This section introduces the enhancements designed for supporting the NR−U with wideband operation in a standalone (SA) deployment. The enhancements were mainly motivated by addressing the issue caused by the LBT failure in the unlicensed spectrum. This section focuses on aspects related to control channel design and detection.

18.4.1 Wideband operation in NR−unlicensed

In the NR−U system, due to the nominal bandwidth requirement from some regional regulations, each transmission is based on a granularity of a 20 MHz bandwidth. The design of the NR in a previous release already supported wideband transmission. Therefore the transmission of the NR in the unlicensed spectrum does not need to be limited to a 20 MHz transmission bandwidth and larger bandwidth transmissions should be supported in the NR−U. Here the large bandwidth refers to the order of multiples of 20 MHz. At the beginning of the discussion for NR−U in 3GPP, two options were considered and both of them received an equal number of supporting companies.

The first option reused the carrier aggregation (CA) concept, as shown in Fig. 18.17, where each of the 20 MHz bandwidths is considered as a component carrier (CC).

The advantage of this option is that the CA feature has already been completed and standardized in the Release 15 version of NR. If the CA

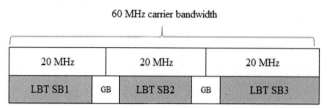

Figure 18.18 Wideband is composed of multiple listen before talk subbands with a guard band in between.

feature is reused/applied here to support NR−U wideband operation, there is no need for additional specification effort, which can greatly reduce Release 16 specification effort. But on the contrary, the shortcomings of this option are also identified: Firstly, the prerequisite for supporting wideband operation in the NR−U system is to support the CA feature, while the Release 15 version of NR does not have such implicit constraints. Note that in Release 15, a UE is able to support 100 MHz bandwidth in a single transmission without aggregating multiple carriers. But if this approach is adopted in the NR−U, it will mandate the UE to always use CA for any transmission that is greater than 20 MHz bandwidth. Secondly, a fixed BWP bandwidth of 20 MHz will always need to be configured and used for CA, which is contrary to the NR principle. Since the design and operating principle of the NR is to be able to configure the bandwidth for BWPs and switch between multiple BWPs flexibly, to a certain extent this CA-like option for NR−U widebank operation seems like a fallback design, abandoning the well-known flexibility of the NR as such.

The second option is shown in Fig. 18.18, where the UE is configured with a large BWP, which covers multiple 20 MHz bandwidths. This multiple of 20 MHz bandwidths with a guard band between two consecutive subbands was called a LBT bandwidth at the beginning of the NR−U SI [25−29]. The role of the guard band is to prevent interference between subbands caused by out-of-band power leakage. Here, the interference refers to the interference caused by one UE transmitting on a subband to the transmission of another UE on an adjacent subband, or even a transmission from equipment of other technologies (e.g., Wi-Fi). Such interference is called intersubband interference.

To reduce the impact of interference between subbands, some guard bands need to be reserved between subbands, so that the transmissions on adjacent subbands are farther apart in the frequency domain to minimize the intersubband interference to each other. Furthermore, these LBT subbands are all configured within the same carrier bandwidth and belong to the same cell subband, which is called an intracell guard band.

After many long discussions in RAN1 meetings, it was finally decided to adopt the design idea of the second option. The main reason is that this option offers better flexibility than the first option and it is aligned with the original intention and design principle of the NR.

After determining the design concept, RAN1 immediately devoted itself to discussing the details of some specific designs. The first design target was the design of the guard band and the LBT subband in the cell. For this purpose, the 3GPP first considered a default guard band value in a cell. Namely fixing the 20 MHz bandwidths within a frequency carrier together with the corresponding guard bands. The center frequencies of the 20 MHz bandwidths should be close to those used by other systems such as Wi-Fi. This is for the friendly coexistence with different systems in a shared spectrum. Then, according to the different subcarrier intervals, a guard band size is determined. These guard bands are derived based on an integer number of resource blocks [30], Table 5.3 1−1 (Fig. 18.19).

Based on the default guard band in a cell, the system can basically keep the intersubband interference under control, but it may not be suitable for other deployment scenarios. Following the NR's usual design principle of being flexible, it was proposed to support the configuration of the interval guard band in addition to the default interval guard band. The advantage is that the network can flexibly configure a larger interval guard band when the network encounters severe interference, but at the expense of spectrum efficiency. On the contrary, when the intersubband interference in the system does not affect the communication performance, the network can

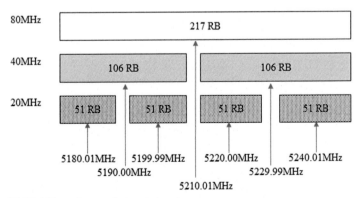

Figure 18.19 NR−unlicensed channel raster.

choose to configure a smaller interval guard band to obtain greater spectrum efficiency. It should be noted that the RAN4 at the 3GPP defines both RF requirement for the UE and the default guard band value for the cell. This includes limiting the out-of-band energy leakage requirement the UE must fulfill. In a case where the gNB configures an interval guard band that is smaller than the default value, the UE is then required to be capable of suppressing the out-of-band energy leakage to meet the requirement. In the RAN1#101-e meeting, in order to avoid introducing two different UE capabilities, the final agreement mandates that the gNB is only allowed to configure an interval guard band larger than the default value, and a non-zero interval guard band smaller than the default value can be configured as such. Furthermore, the LBT subband is also commonly referred to as a RB set. The configuration of RB sets and interval guard bands is as shown in Fig. 18.20. The gNB first configures a carrier bandwidth on a common RB grid (CRB) and configures one or more intracell guard bands in the carrier bandwidth. The configuration of the guard bands in the cell includes the CRB position of the starting CRB and the guard band length. When the configuration is completed, the entire carrier bandwidth is divided into multiple RB sets. Finally, the network configures the BWP and maps the RB set to the BWP. It is worth noting that 3GPP specification requires that network configuration of a BWP must include an integer number of RB sets. Similar to the BWP configuration method, the RB sets in the uplink carrier and downlink carrier are configured independently.

After the configuration of the RB sets, during uplink data scheduling process, the gNB needs to instruct the UE to transmit in one or more RB sets. This method of scheduling the UE according to the RB set is called the frequency-domain resource allocation (FDRA) type 2. In the following, the discussion is focused on the determination of the allocated RB set.

Uplink scheduling can be divided into multiple scenarios, including the scheduling for RRC-connected UEs and RRC idle UEs. Furthermore, the scheduling for the connected UE is divided into scheduling using a common search space set (CSS set) and a UE-specific search space set. In these

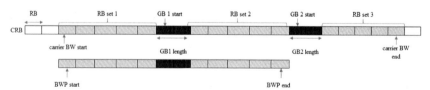

Figure 18.20 Intracell guard band.

different scenarios, the resource allocation determination methods are not exactly the same. When the gNB schedules the UE in the USS, the gNB DCI formats 0_0 and 0_1 can be used for scheduling. It is worth noting that the 3GPP reserves the possibility of using DCI format 0_2 to schedule NR−U UEs, but it is restricted to certain functions. The main reason is that the support for these functions requires further specification revisions. Considering that DCI format 0_2 in Release 16 incorporated some enhanced features for Ultra Reliable Low Latency Communications (URLLC), which is not a main focus of Release 16 NR−U to support URLLC services in the unlicensed spectrum, the support of these URLLC enhancements for NR−U is postponed to Release 17 for discussion. The FDRA information fields in both DCI formats have Y bits, which are used to indicate one or more scheduled RB sets and the value of Y is determined by the total number of RB sets on the active uplink BWP. The advantage of introducing Y-bit indication is that the gNB is able to flexibly indicate uplink transmission in any one or more RB sets without increasing the overhead of DCI.

When gNB uses DCI format 0_0 to schedule RRC-connected UEs in a CSS set, schedules RRC idle UEs in a type-1 PDCCH CSS set, or uses RAR UL grant to schedule Msg3-PUSCH, the DCI format 0_0 and the RAR UL grant will not include an information field that explicitly indicates the scheduled RB sets. Three candidate methods for the UE to determine the scheduled RB set were discussed and considered in RAN1 [31−35]. The first method is that the UE will always determine the first scheduled RB set is in the uplink active BWP (for RRC idle UEs, the initial uplink BWP is the uplink active BWP). The second option is that the UE will treat all the RB sets as the scheduled RB sets. The third option is that the UE will determine that the scheduled RB sets in the uplink active BWP are the ones that overlap with the RB sets in the downlink active BWP in which the UE receives the DCI format or RAR UL grant. Among the three options, the first option is rather simple to implement, but the scheduling restrictions are relatively large. Especially when the interference environment is different for each RB set. If the first RB set is in a strong interference state for a long time, the LBT at the UE will keep on failing, resulting in no uplink transmission at all.

The second option requires the UE to transmit with a large bandwidth every time and for which the LBT on each RB set is required to succeed as well. As such, this increases the failure probability of uplink transmission. On the other hand, the third option enhances the probability of

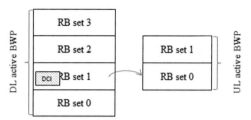

Figure 18.21 RB set allocation example.

successful uplink transmission for the UE. As shown in Fig. 18.21, when the downlink BWP and the uplink BWP has four and two RB sets, respectively, the gNB has a choice to send the scheduling DCI on downlink RB set 1 or RB set 2. Then the corresponding uplink RB set 0 or RB set 1 is the scheduled uplink RB set. If the gNB successfully completes the LBT and sends the uplink scheduling DCI in the downlink RB set 1, then the UE will have a high probability to also successfully pass the LBT and perform uplink transmission. Due to this enhanced advantage, the third option was adopted in the RAN1#100b-e meeting. However, there is a drawback in this option; that is, when the downlink RB set does not have a corresponding uplink RB set, the UE cannot determine the uplink scheduled resource (e.g., an example is given in Fig. 18.21 for the case where the DCI is sent in the downlink RB set 0 or RB set 3). After discussion, RAN1#100b-e finally decided that in this case, the UE should transmit in the uplink in RB set 0 of the uplink active BWP.

For the scheduling scenarios of RAR and type 1 PDCCH CSS set, RAN1 discussed the initial plan to reuse the previously introduced scheduling scheme in the CSS set to achieve a unified design. However, a problem was identified during the discussion that when a connected UE and an idle UE overlap in a RB set, the gNB cannot distinguish whether the scheduled uplink transmission is from the connected UE or idle UE by relying on the received PRACH only as shown in Fig. 18.22. In this case, the gNB needs to blindly monitor uplink transmission in RB set 0 from the connected UE and uplink transmission from the idle UE in the initial uplink BWP at the same time, and also to reserve two resources for these two transmissions, which will result in serious resource wastage. For this reason, it was eventually decided in the RAN1#101-e meeting that the uplink scheduling using the CSS set of RAR and type 1 PDCCH, and UE uplink transmission of PRACH are to be in the same RB set. The reason for adopting this scheme is that when a UE sends a PRACH

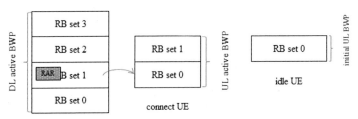

Figure 18.22 RAR (RAR UL grant) scheduling for connect and idle UE.

in a certain RB set and it is received by the gNB, then there is no ambiguity of which RB set has been used between the UE and the gNB. And since the UE is able to transmit PRACH, it implies that the LBT was successful in this RB set, the UE will face a small probability of continuous LBT failure in the same RB set.

18.4.2 PDCCH monitoring enhancement

In the downlink control channel design, there is no difference fundamentally between the NR and NR−U. As described in Section 6.2.3 and 6.2.4, UEs are configured with some periodic PDCCH search spaces for receiving downlink control channels. The concepts of NR PDCCH search space and PDCCH CORESET are fully reused in the NR−U system. However, due to the specific problem in the unlicensed spectrum, gNB cannot guarantee that the downlink control channel will be successfully transmitted in resources without scheduling grant every time as it depends on the success of LBT at the gNB. Therefore the related design for the control channel in NR−U is mainly dedicated to resolving the problems caused by the LBT failure.

18.4.2.1 CORESET and search space set configuration

In Section 18.4.1 it was discussed that the NR−U supports multiple RB sets in a BWP and each RB set can be regarded as an LBT subband. If a gNB intends to transmit in a given RB set/LBT subband, it must first ensure a successful LBT in this RB set. In practice, the success of LBT cannot be predicted in advance in a real communication, and as such a new question arises: how can a gNB successfully or ensure a higher chance in transmitting the downlink control signal in a shared spectrum: In the design of NR, the location of CORESET can be flexibly configured in BWP (i.e., a CORESET resource can span multiple RB sets). At the same time, when the LBT on a certain RB set is unsuccessful, then the gNB cannot send the downlink control channel in this RB set. This results in that some resources are available and some are not in the same CORESET. As shown in Fig. 18.23, when a UE's downlink active BWP contains three RB sets and the configured CORESET spans RB set 0 and RB set 1, if the LBT on the gNB side fails for RB set 1, then the resources of the CORESET on RB set 1 become unavailable [25].

In order to solve this problem, the 3GPP has considered different schemes. The first solution is up to network implementation for the gNB to avoid transmitting the downlink control channel using unavailable resources. The advantage of this scheme is that there is no specification impact, which saves time for standardization progress. But its shortcomings are also obvious, in that when downlink resources are unavailable due to LBT failure, the gNB has fewer resources to schedule the UE. And when the interleaving function in CORESET is enabled, the unavailability of some resources will cause the interleaving PDCCH candidate resources to be punctured, which can significantly reduce reliability in downlink control channel reception at the UE.

As these problems are identified in RAN1, a number of solutions are also proposed and one solution is shown in Fig. 18.24. In this case, the gNB configures a CORESET with large bandwidth and distributes each candidate PDCCH (PDCCH candidate) resource to each RB set as much as possible using interleaving. As such, the UE may still be able to decode the DCI even if puncturing occurred. Although this solution is intuitively simple, the reliability of PDCCH transmission will sorely rely on the implementation of the gNB. Therefore RAN1 decided to seek other enhanced solutions.

Figure 18.23 CORESET configuration example.

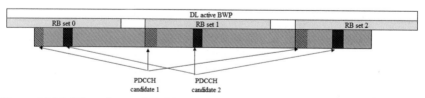

Figure 18.24 NR—unlicensed PDCCH monitoring alternative 1.

Another solution is to configure multiple CORESETs, and each CORESET is only included in one LBT subband/one RB set. In this way, the problem of interleaving is naturally solved, but on the other hand, this design solution requires the gNB to configure the same number of CORESETs as the RB sets. In the NR, the UE is required to support a BWP of up to a maximum bandwidth of 100 MHz in FR1. This corresponds to five RB sets and consequently five different CORESETs need to be configured in a BWP, which is beyond the capacity of the NR of supporting up to three CORESETs in a BWP. Similarly, the search space set needs to be associated with CORESET, and the same search space set index prohibits the association with different CORESETs in the NR. Therefore when the specification simply increases the number of configurable CORESETs, it will also need to support an increased number of associated search space sets, as shown in Fig. 18.25. After extensive discussions, RAN1 decided to carry out a revised design following this principle.

The solution finally adopted in the RAN1#98b meeting solved the problem of requiring an increased number of CORESETs and search space sets. The adopted solution is that a CORESET should be configured within a RB set, so that the gNB can configure a special search space set to be associated with the CORESET. Once it is associated with the special search space, a mirror CORESET is then generated and the resources of the mirror CORESET will be duplicated to other RB sets, as shown in Fig. 18.26. After the CORESET resource is mapped to the RB

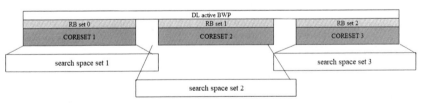

Figure 18.25 NR—unlicensed PDCCH monitoring alternative 2.

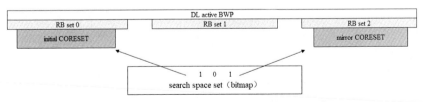

Figure 18.26 NR—unlicensed CORESET and search space set configuration.

set, the UE monitors the downlink control channel in other RB sets according to the mapped CORESET time-frequency-domain resources and the associated search space set. This special search space set configuration requires a bitmap parameter with each bit corresponding to a RB set. When a bit is 1, CORESET will be mapped to the corresponding RB set. When a bit is 0, CORESET does not need to be mapped to the corresponding RB set. With this solution, the number of CORESETs that need to be configured and the number of corresponding search space sets do not need to be increased.

The condition in which this solution can be implemented is that the frequency resources of the initial CORESET need to be confined with a set of resource blocks. Specifically, the resources of the initial CORESET must not exceed a range of a set of resource blocks. This will raise a new problem that the resource allocation of CORESET in the NR is indicated by the granularity of six resource blocks (refer to Section 5.4 for details on the CORESET design of the NR). Due to the large granularity of this indication, in order to ensure that the initial CORESET resource after configuration does not exceed the RB set, some PRB resources that are not integer multiples of six may not be used for CORESET configuration in the actual system.

The solution to this problem is to introduce a new CORESET resource configuration method as shown in Fig. 18.27 [30]. In this method, the initial CORESET-starting RB is determined by a RB offset

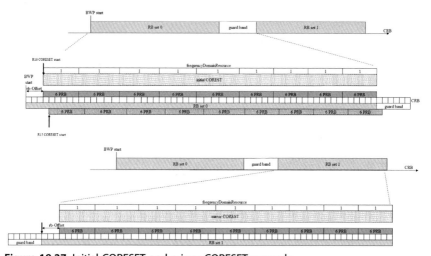

Figure 18.27 Initial CORESET and mirror CORESET example.

parameter (rb-Offset) from the first RB in the BWP, so that the resources of the initial CORESET can be adjusted within a RB set. Since the offset value ranges from 0 to 5, the offset allows at most by five resource blocks, which perfectly avoids the problem where some resource blocks exceed the set of resource blocks in a granularity [29,30] of six RBs. As shown in Fig. 18.27, the mapping of resource in the new RB for the mirroring COREST is determined by shifting the initial CORESET resource in the frequency domain, and the starting position from the boundary of the new RB set is determined by the rb-Offset offset. When determining the size of the initial CORESET resource, the gNB needs to consider that the mapping of the mirrored CORESET resource is also prohibited from exceeding the RB set.

Due to the introduction of the new CORESET configuration and new search space configuration scheme, the issue of maintaining backward compatibility was also discussed in RAN1 meetings. The question mainly occurs when a search space with Release 15 configuration is associated with a Release 16 CORESET configuration, and vice versa, and how should the UE interpret these configurations. The final rules addressing the issue are summarized in Table 18.8.

18.4.2.2 Search space set group switching

In this section, another new feature is introduced for the NR−U system: search space group switching. Before giving a detailed introduction of this feature, some of the pain points of wireless communication in the unlicensed spectrum described in Section 18.2 are reviewed. Firstly, since the gNB needs to pass the LBT before accessing the channel, and in most

Table 18.8 R16 CORESET and search space set configuration.

COREST configuration	Associated search space set version	CORESET resource determination
rb-offset not configured	Release 15	Follow Release 15 mechanism
rb-offset not configured	Release 16	Follow Release 16 mechanism assuming rb-offset = 0
rb-offset configured	Release 15	Follow Release 16 mechanism
rb-offset configured	Release 16	Follow Release 16 mechanism, expecting CORESET is restricted within 1 RB set

cases the gNB will use the LBT Type 1 method with random channel access time, then the issue is how to ensure the data is transmitted by the gNB at the end of a successful LBT without a long delay, which increases the risk of losing the channel access. This problem, however, does not exist in Wi-Fi, since it is fundamentally an asynchronous system, where Wi-Fi devices can initiate transmission at any time/immediately after the end of a successful LBT. In the NR−U system, on the other hand, it is a synchronous system where the operation is fully based on a frame structure with clear subframe and slots boundaries, over which the control channel, data channel and reference signals are transmitted. The locations (or resources) allocated for the transmission of these channels and signals are often predefined. Therefore the randomly generated LBT duration resulting in a random LBT ending time position does not fit well in a synchronous NR−U system. To solve this problem, ideally, the system is supposed to reserve as many locations as possible for the gNB to schedule UEs so that the gNB is able to send downlink control signals at any position and the LBT finishes. But this will create a new problem. In order to cooperate with the gNB to send downlink control signals flexibly, the UE needs to be in a state of frequently monitoring the downlink control signals, which is not energy consumption friendly on the terminal side [36−39]. As a result, in the RAN1#98b meeting it was decided to compare the channel-access success rate of the gNB against the UE's energy consumption, and the search space group switching scheme, as shown in Fig. 18.28, was adopted. In this scheme the UE can be configured with two different search space set groups. One group is activated for UE monitoring outside the gNB COT; while the other group is utilized for UE monitoring inside the gNB COT. In the NR−U, the search space set group outside the gNB COT is relatively denser. For instance, a mini-slot-based monitoring is used outside the gNB COT, for which the UE energy consumption is higher. Nevertheless, the UE will switch to a

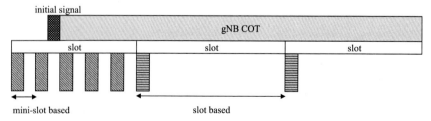

Figure 18.28 Search space set group switching concept.

more sparse search space set group (e.g., slot-based monitoring inside the gNB COT).

For the NR−U, as discussed a UE can be indicated by COT information of the gNB with the remaining COT duration that can still be used by the gNB. For the gNB COT information, as was described in the previous section, a gNB can establish a COT when the LBT is successful. During the COT, the gNB can perform downlink transmissions including control channels, data channels, or reference signals, and the gNB can also share its COT with a UE for uplink transmission. Therefore technically the UE receives useful messages from the gNB only within the gNB COT and any processing conducted outside the gNB COT is deemed as a waste of power. Along with this logic, the UE should only spend its processing power within the gNB COT. This issue received widespread attention during RAN1 meetings. After summarizing the views from multiple companies, the original intention was for the gNB to indicate its COT starting timing point to the UE, so that the UE is aware and can behave accordingly. To enable this, a straightforward solution was to design an initial signal or so-called preamble at the beginning of the gNB COT. The UE could then determine the start of the gNB COT as long as it detects the preamble signal. Moreover, RAN1 also started to look at the possibility of defining the same initial signal used in Wi-Fi. This proposal was led by the IEEE standards organization and it was proposed in the 3GPP meeting to design such an initial signal to be consistent with the current reference pilot used in Wi-Fi. The advantage is that the devices belonging to two different systems can mutually discover each other, leading to a more friendly coexistence environment. While the design idea can be greatly beneficial to Wi-Fi devices (with no extra cost), it imposes significant limitations for NR-based UEs. One difficulty is that it will require NR-based UE devices to start supporting the baseband processing mechanism of the Wi-Fi system, including channel coding and sampling frequency. Thus an original simple transplanting of the Wi-Fi preamble signal into NR-based UE devices will cause serious compatibility problems. On the other hand, if the Wi-Fi preamble design is significantly modified to be compatible with the existing NR receiver processing, it will take a long time for research and integration into the 3GPP system, eventually leading to a failure to complete the standardization work for the NR−U within the planned Release 16 timeframe. For this reason, the use of a new Wi-Fi-like initial signal design for the purpose of indicating the gNB COT starting time was abandoned.

The second alternative for a UE to determine whether the gNB has established a COT is to base it on one of the existing reference signal designs. During the discussion in 3GPP RAN1, it was determined that DMRS is a reference signal that can be considered for this purpose. The following are two key issues that need to be resolved: Firstly, how to design a common DMRS that can be recognized by different UEs; secondly, UE detection of the DMRS signal needs to be reliable. The first problem can be mainly understood as that in the design of the current NR system, except for the DMRS of the PDCCH of the system message and scheduling the system message, which can be identified by different UEs. Other existing DMRS in use for other control and data channels are all based on a UE-specific scrambling, and it cannot be detected or recognized by other UEs. On the other hand, system messages are all sent within a specified time window. If the gNB creates a new COT at a different time instance, the DMRS of the system message cannot be used to notify the UE of the gNB COT starting time. Therefore a new design of a common DMRS is required for the NR−U system in RAN1. Regarding the second question, which is also the most critical issue, the UE needs to constantly monitor the presence of the common DMRS. When the presence is detected, the UE thus is able to determine a starting time of the gNB COT. However, since the presence detection of the common DMRS is based only on a simple energy detection, it is thus prone to false alarms or missed detections at the UE, which has a great impact on the reliability of COT detection. The impact here can be understood as that once there is an ambiguity in understanding between the gNB and the UE, the UE and the gNB may receive and transmit control signals in different search space groups, which will make the entire system unreliable.

For these two reasons, the 3GPP finally determined in the RAN1#99 meeting that the design plan is to base it on PDCCH detection. The reason, as analyzed before, is that PDCCH detection needs to pass cyclic redundancy check (CRC). The current number of CRC bits is 16, so the false-alarm probability of the CRC check is 2^{-16}, which fully meets the system requirements. Based on this design idea, several more detailed schemes have been expanded under discussions [36−47]. The first scheme is based on the detection of the common PDCCH, namely DCI format 2_0. When the UE detects a DCI format 2_0, the DCI carries a 1-bit search space group switching trigger, and the UE determines whether to switch according to the indication. At this point, one could wonder why the gNB indicates no search space set group switching even inside gNB COT. The reason is that when the gNB COT duration is very short or DCI format 2_0 is transmitted close to the end of the COT, the gNB may indicate to the UE to maintain in the current search space set group, in order to prevent the UE from frequently switching between the two search space set groups.

On the other hand, as the monitoring of DCI format 2_0 is an optional configuration, the search space set group switching feature can also be utilized when DCI format 2_0 is not configured. In this case, the UE will follow an implicit search space group set switching mechanism (i.e., when the UE detects any DCI format in one group), the UE will switch to another group. It is worth noting here that the switching of the search space set group requires a certain delay. Specifically, after receiving the DCI format that triggers the switching, the UE needs to wait for the first slot boundary after P symbols to complete the switching. As shown in Fig. 18.29, when the UE receives the trigger DCI format in slot n, the actual switching occurs at the boundary of slot n + 2. The value of P here is configurable and is related to UE capabilities.

18.4.2.3 Enhancement on downlink control information format 2_0

DCI format 2_0 is a very special DCI format specified during NR Release 15, except that it is carried in a group-common PDCCH and the SFI information it carries can be used to cancel the higher-layer configured periodic uplink and downlink reception or transmission resources. In the NR−U, this DCI format 2_0 has been enhanced besides the SFI to provide more features as follows.

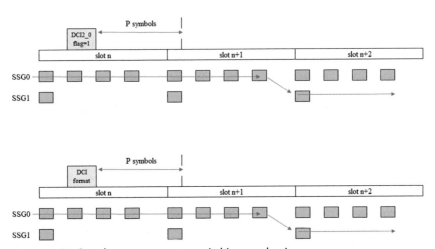

Figure 18.29 Search space set group switching mechanism.

COT information: As shown in Fig. 18.30, the gNB COT information is indicated in DCI format 2_0. The gNB first configures using RRC signaling a COT duration table, which includes up to 64 rows and requires up to 6 bits to indicate. Each row is configured with a COT duration represented by number of OFDM symbols. The maximum number of symbols is 560, which is equivalent to a COT duration of 20 ms in 30 KHz subcarrier spacing. Note that according to Table 18.1, the maximum value of a COT is 10 ms, but the COT duration that can be indicated here is up to 20 ms. The reason is that according to the regulations, it allows the transmitter to have a transmission pause/gap in the COT for communication in the unlicensed spectrum and the time of the gap is not included as part of the effective COT duration. Specifically, if the gNB initiates a COT with a duration of 10 ms and there is a 10 ms gap in the middle, then the 10 ms effective COT will take 20 ms to complete. For this reason, the maximum COT duration indicated in the design of NR−U can be up to 20 ms. The start timing for the COT is the start time of the slot in which the DCI format 2_0 is received. It is worth noting that the maximum absolute duration of the COT is 20 ms, and as such, the number of symbols that can be indicated is only 280 when the subcarrier spacing is 15 KHz. On the other hand, the COT information indicates that the configuration is not necessary when the DCI format 2_0 is configured. Therefore when the COT information is not configured, the UE determines the COT duration according to the SFI indication. That is, the last slot indicated by the SFI is the end of the COT. But why is the COT information so important in NR−U systems? For example, for the search space set group-switching feature in Section 18.4.2.2, the UE needs to know the starting time position and a duration of the COT so that the UE can switch back to the initial search in time when it reaches the end of the COT. In addition, the gNB COT-sharing feature introduced in Section 18.2.2 also requires the UE to determine whether the scheduled uplink resource is within the gNB's COT before it can share the gNB's COT.

Figure 18.30 NRU downlink control information format 2_0 contains channel occupancy time duration information.

Search space set group-switching trigger information: the gNB in the NR−U can configure trigger indication information in DCI format 2_0. If configured, the indication information is one bit. Since there are two search space set groups, namely group 0 or group 1, 1-bit indication information directly indicates the index number of the group. If the indication bit is 0, the UE is in the search space set group 0, otherwise in search space set group 1.

RB set availability indication: the gNB in the NR−U can configure a so-called RB set availability indication in DCI format 2_0. If configured, the indication field contains X bits, and the value of X is determined by the number of RB sets on the carrier. For example, if there are five RB sets on the carrier, the value of X is 5 and each bit is mapped to a dedicated RB set. When the mapped bit is 1, the corresponding RB set is available for reception, implying that the gNB has successfully passed LBT on the corresponding RB set. Otherwise, the corresponding RB set is not available for reception. At the time when this book was written, the 3GPP had only determined that the downlink signal here was a periodically configured CSI−RS reference signal. It is also necessary to discuss whether it is also applicable to the reception of SPS−PDSCH.

18.4.3 Enhancement on physical uplink control channel

This section introduces the enhancement designs for PUCCH in the NR−U system. The regulation for the use of the unlicensed spectrum mandates that radio transmissions must satisfy a minimum transmission bandwidth (occupancy channel bandwidth, OCB) requirement (i.e., each transmission needs to occupy at least 80% of the LBT subband (20 MHz) bandwidth). According to Section 5.5, the PUCCH format 0 and PUCCH format 1 designed and specified in NR Release 15 both occupy only one RB, which naturally cannot meet the OCB requirements. Thus an enhancement is needed to enable PUCCH format 0 and PUCCH format 1 in order to address the OCB requirement in the unlicensed spectrum. For this reason, RAN1 considered the use of an interlace structure design for PUCCH. The so-called interlace structure refers to two consecutive available resource blocks being separated by a fixed number of resource blocks. By using this interlaced structure, the number of PUCCH resources can be widened in the frequency domain to achieve the OCB requirements (Fig. 18.31).

Figure 18.31 Interlace with 30 KHz subcarrier spacing.

At the beginning of the interlace discussion, there were two different alternatives. The first alternative is an interlace structure with subcarrier-based granularity. Such an interlace structure is called a sub–PRB-based interlace, and it is beneficial for power boosting on the sub-PRB-based resources. This is due to the maximum power spectral density in the NR−U system being defined as a fixed value per 1 MHz. Then the less resources are used for transmission, the greater the power is allocated on the used resources such that the average received signal-to-noise ratio on the resources is also greater. Subsequently, this would result in an improved transmission quality and better performance can be achieved. However, during the discussion the drawback of sub-PRB-based interlace was also identified. Firstly, sub-PRB-based interlace has only limited use cases (e.g., mainly targeting small uplink payload transmission in a coverage limited case). For a common use case with a relatively larger payload, power boosting cannot be exploited with this interlace structure. Secondly, NR Release 15 uses RB-level scheduling. If in the NR−U system the scheduling resource is divided into a sub-PRB level that is smaller than a RB, this would require a redesign in the uplink scheduling mechanism, including resource allocation and reference signal design, thus potentially requiring a large amount of specification effort. The original intention of the NR−U design was to follow the design of NR Release 15 as much as possible, so that the NR−U system can be nicely integrated with the Release 15 NR. Other benefits of reusing the Release 15 design include the fact that not only can it simplify UE implementation by allowing the reuse of the Release 15 baseband module as much as possible, the operators may also easily integrate the NR−U into NR networks. With this consideration, the 3GPP finally abandoned the second alternative and adopted a RB-based interlace structure. In the following chapters we will introduce the design of PUCCH based on the RB-based interlace structure.

18.4.3.1 Interlace design

After it was agreed that the RB-based interlace structure was to be adopted for the uplink PUCCH transmission, the more concrete interlace structure design and configuration became the major work task in 3GPP

RAN1. First of all, since the NR system is able to support multiple sub-carrier spacings, during the early stages of the NR—U discussion it was determined that the subcarrier spacing of 15 and 30 KHz subcarrier spacings should be supported for FR1 (e.g., carrier frequency below 6 GHz) in the NR—U. Furthermore, since the number of subcarriers contained within a RB in the NR system is fixed to 12, the actual bandwidth of a RB will change according to the configured subcarrier spacing for the BWP. Secondly, except for special scenarios, since the regulations say the nominal channel bandwidth is 20 MHz for the unlicensed spectrum, then the requirements for OCB are also to be based on this nominal channel bandwidth. And therefore these two factors should be taken into account in the design principle. Note that the group of RBs of an interleace structure are indexed and called an interlace index. Moreover, the group of RBs for a given interlace index are uniformly distributed within the interlace structure and this group of RBs is called the interlaced RB (IRB). The consecutive IRBs of an interlace index are separated by MRBs in the frequency domain, where the value of M varies according to the subcarrier spacing. For the 15 KHz subcarrier spacing case, the value of M is 10. For the case of 30 KHz subcarrier spacing (SCS), since the bandwidth of each RB is doubled compared to the 15 KHz SCS case, the interval between the adjacent IRBs becomes five resource blocks, and the UE can allocate up to five interlaces. During the discussions in RAN1, some companies also proposed the need to support 60 KHz and adopt a design scheme with one RB interval, but it was not adopted due to lack of consensus (Fig. 18.32).

In the following, how the interlace is configured by the gNB is described. In Chapter 4, Bandwidth Part, it was explained that one or more BWPs can be configured by a gNB to a UE, and the UE's data reception and transmission will occur and be confined in the configured active BWP. As such, a more natural solution is to configure the interlace directly in the BWP. Specifically, the first IRB of the first interlace index starts from the first RB of the BWP. The advantage of a such configuration is that no additional signaling is required to achieve the effect of configuring the comb

Figure 18.32 Interlace design with different subcarrier spacing.

ruler. However, the potential drawback of this scheme is that it is not easy for the gNB to schedule multiple UEs simultaneously. It can be simply understood that when the gNB tries to schedule different UEs in a cell, their BWPs may not be completely aligned in the frequency domain. In NR Release 15, the gNB realizes that different UEs are separately allocated in the frequency domain (i.e., frequency-division multiplexing, FDM mode). However, if the interlace of different UE configurations are not aligned, due to there being multiple resource blocks in the interlace, they can still be regularly arranged. As a result, the network needs to increase its scheduling effort to ensure a FDM'ed UE scheduling. As shown in Fig. 18.33, when UE1 and UE2 are to be scheduled at the same time, the gNB needs to check constantly the frequency resources of respective interlace for both UEs to ensure that there is no frequency overlapping between the two interlaces; otherwise interference will occur.

However, if the interlace structure is fully aligned for different UEs and independent of the BWP configuration (Fig. 18.34), as long as the gNB uses different interlace indexces when scheduling the different UEs, frequency-domain collisions can be completely avoided. Therefore the scheduling complexity for the gNB can be greatly reduced. At the same time, it was decided that a cell-specific interlace configuration is adopted to align different UEs. The starting IRB of the interlace index 0 is at point A (for a more detailed introduction to Point A, refer to Chapter 4, Bandwidth part).

Figure 18.33 Interlace design within bandwidth part.

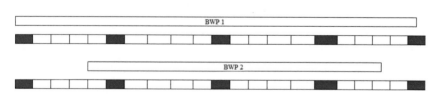

Figure 18.34 Interlace design independent of bandwidth part.

18.4.3.2 Physical uplink control channel design

For the design of the PUCCH format, the NR−U system adopts a similar idea as Release 15. In NR Release 15, the design of PUCCH includes format 0/1/2/3/4, among which the differences are in the payload size, the number of OFDM symbols (allocation in time), and the number of RBs (allocation in frequency). From a functional analysis, since PUCCH format 0 and 1 are used in the initial access phase, they are considered as the most essential formats to be adopted first in the NR−U. On the other hand, PUCCH format 2 and 3 are used for RRC-connected UEs and designed to carry a large amount of feedback, such as a large-size HARQ codebook or CSI feedback. The use case of PUCCH format 4 in Release 15 is for carrying small payload feedback in the case of limited coverage. Considering that the main/target deployment scenario of NR−U systems in the unlicensed spectrum is small cells, typically there is no coverage limitation problem under normal circumstances, and as such PUCCH format 4 is not really needed and hence is not supported in the NR−U. Moreover, it was also decided that a PUCCH transmission cannot cross multiple RB sets but is restricted to one RB set only.

- PUCCH format 0 and format 1

PUCCH format 0 and PUCCH format 1 have a relatively similar design as NR Release 15. They are both sequence-based signal transmission, so that the receiver gNB does not need to perform a cumbersome decoding process to extract the information, but only uses a cross-correlation function to determine the sequence (and hence the information) sent by the UE. According to the interlace structure, PUCCH format 0 needs to fit into an interlace and the resources of PUCCH are limited to one LBT subband bandwidth. Therefore the number of IRB of an interlace is 10 or 11. In NR Release 15, the frequency size of PUCCH format 0 overoccupies only contain 1 RB. Therefore the design goal in the NR−U is to extend one RB into 10 or 11 IRBs. During the discussions in RAN1, a simple extension scheme is to use a sequence design similar to Release 15 for the first IRB, and then duplicate them to the remaining IRBs. This solution is simple, but it has obvious drawbacks, such as due to repetition in the frequency domain, the PUCCH format 0 will have a larger peak-to-average power ratio (PAPR) in the time domain. This will makes the power amplifier less efficient for the UE, and could lead to nonlinear interference. For this reason, the 3GPP further considered a potential method to reduce the PAPR for the PUCCH format 0 and 1 (Table 18.9).

Table 18.9 R15 physical uplink control channel formats.

		PUCCH length			
		short (1 to 2 symbols)	long (4 to 14 symbols)		
payload	up to 2 bits	format 0		1	RB number
			format 1	1	
	more than 2 bits	format 2		1-16	
			format 3		
			format 4	1	

Two main alternative designs are introduced here: the first solution is to modulate a different phase offset on each IRB, and the second solution is to add a dedicated phase offset to each IRB. The difference between these two alternatives can be mathematically expressed as follows:

Alternative 1: for an initial base sequence $S(n)$, when it is duplicated on an IRB, it is modulated with a phase shift α, such as

$$S_1(n) = e^{j\alpha n} \cdot S(n), n = 0, \ldots, 11 \tag{18.3}$$

Alternative 2: as opposed to alternative 1, alternative 2 does not involve modulation but a simple phase shift multiplication; that is,

$$S_2(n) = e^{j\alpha} \cdot S(n), n = 0, \ldots, 11 \tag{18.4}$$

The advantage of Scheme 2 is that the receiver does not need to have a prior knowledge of α. And hence the phase information can be completely left for UE implementation. But the disadvantage is that the performance of PAPR cannot be unified across different implementation algorithms. As such, it is not ideal for the network to control PUCCH transmission power for different UEs since the gNB would not have any knowledge about the algorithm used in the UE implementation. Moreover the gNB's cross-correlation detection of the PUCCH is based on noncoherent detection, which will result in poor detection performance. For these reasons, alternative 1 was finally adopted. The design idea of PUCCH format 1 is basically the same as format 0, and hence will not be repeated here.

- PUCCH format 2

In NR Release 15, the frequency-domain resources of PUCCH format 2 can be configured with 1 to 16 resource blocks. In the NR−U system, as previously introduced, the PUCCH resource is to be restricted within a RB set. Moreover, the number of IRBs in a RB set is 10 or 11 depending on the interlace index. For example, the uplink control information (UCI) carried in PUCCH format 2 should reach a target transmission code rate to ensure it can be reliably received by the gNB. When the number of UCI bits is small, the number of required resources achieving the target code rate will also be small. On the contrary, when the payload gets larger, more resources in terms of IRBs are needed. To this end, it is also possible to configure two interlaces for PUCCH format 2 in the NR−U, resulting in up to 22 IRBs to achieve a similar resource capacity as designed in Release 15. On the other hand, the configuration for PUCCH format 2 should follow the same principle as used in Release 15 (i.e., semistatic configuration via RRC signaling). A UE selects between one interlace and two interlaces according to an expected payload size to be transmitted in the PUCCH. As a further enhancement, RAN1 also made further considerations on the resource occupation efficiency of PUCCH. The main problem here is that when the number of UCI bits is small, such as in extreme cases, only one RB transmission is needed to meet the reliability requirements. In this case, the use of interlace for transmission will cause a waste of resources. To balance between saving resource for spectrum efficiency and meeting the OCB requirement, it was decided in the RAN1#99 meeting that it will support multiplexing of multiple UEs within on a same interlace. Among them, multi−User multiplexing is based on orthogonal code (orthogonal cover code, OCC) and scrambling code (Fig. 18.35). The NR−U system supports two configurations of OCC length 2 and OCC length 4. The former can multiplex two users; while the latter can multiplex four users. The traditional HARDARMA code

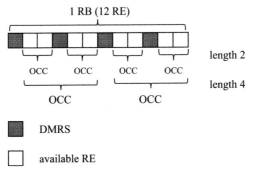

Figure 18.35 Physical uplink control channel format 2的with orthogonal cover code.

is chosen for the OCC code. However, a new problem was identified: the impact of the OCC code sequence on PAPR is uneven. For example, when UE1 is configured with OCC code [1, 1] by the gNB and UE2 is configured with OCC code [1, -1] by the gNB, then there will be a higher PAPR impact for UE1. To solve this problem, RAN1 adopted the idea from PUCCH format 0; that is, for different comb rule resource blocks in a comb ruler, the UE will use the method of cycling the OCC codes. Due to the same OCC code switching between the multiplexing users in the same interlace, OCC orthogonality can still be maintained.

- PUCCH format 3

The design principle for PUCCH format 3 is very similar to that for PUCCH format 2. Up to two interlaces can be configured and a UE can select between one index or two indexes for transmission according to the number of UCI bits. The main difference here is that PUCCH format 3 uses the DFT-s-OFDM waveform, so that the selection of the number of resource blocks is limited. That is, the principle of DFT length of $2 \times 3 \times 5$ still needs to be followed, such that the length needs to be dividable by 2, 3, and 5. When PUCCH format 3 is transmitted with an interlace, the UE must use the first 10 IRBs for transmission instead of 11. When PUCCH format 3 is transmitted with two interlaces, the UE must select 20 IRBs. PUCCH format 3 also supports OCC, but unlike PUCCH format 2, the OCC needs to be performed in the time domain for the impact of DFT (Fig. 18.36). The UCI after OCC scrambling is then mapped onto the interlace RB via DFT.

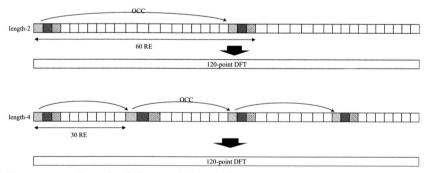

Figure 18.36 Physical uplink control channel format 3 with orthogonal cover code.

18.5 Hybrid automatic repeat request and scheduling

When the NR system is deployed in the unlicensed spectrum, it should support a network deployment in a SA manner. Specifically, the NR−U system should work independently in the unlicensed spectrum without relying on a licensed carrier to provide assisted access services. For this type of operating scenario, the initial access, mobility measurement and reporting, channel measurement and reporting, DCI, downlink data transmission, uplinking control information, and uplink data transmission from a UE all need to be performed on an unlicensed carrier. Since performing LBT is required before the start of any transmission on an unlicensed carrier, it is possible that a channel or signal cannot be transmitted due to LBT failure. In this case, enhancement to the normal HARQ reporting and scheduling to improve data transmission efficiency on unlicensed carriers is the main issue discussed in this section.

18.5.1 Hybrid automatic repeat request mechanism

In the Release 15 NR system, two types of HARQ feedback mechanisms are supported for the UE to report HARQ−ACK (Hybrid Automatic Repeat Request Acknowledgment) information of PDSCH decoding results to the network, namely the Type-1 codebook and Type-2 codebook. The Type-1 codebook and Type-2 codebook are also known as semistatic codebook and dynamic codebook, respectively. These HARQ feedback mechanisms from the Release 15 NR system are used as the basis for enhancement for the UE to report HARQ−ACK information in the unlicensed spectrum.

18.5.1.1 Hybrid automatic repeat request problems

During the discussion on HARQ feedback reporting in the unlicensed spectrum, some new problems were identified [48]:

- Problem 1: HARQ−ACK information cannot be fed back due to LBT failure.

    Since LBT is required before the start of any radio transmission on an unlicensed channel, the ability to be able to transmit a physical channel or signal depends all on the success or failure of the LBT carried out by the intended transmitter. That is, if the LBT process failed for an intended PUCCH transmission from a UE, the UE will not be able to transmit the PUCCH on the corresponding PUCCH resource. If the intended PUCCH transmission was meant to report HARQ−ACK information of decoding results for one or more received PDSCH(s), as shown in Fig. 18.37, then the gNB would not receive the corresponding HARQ—ACK information from the UE. Normally one PUCCH carries HARQ—ACK feedbacks for multiple PDSCHs/TBs. If UE fails to feed back HARQ—ACK information for multiple TBs, then the gNB may schedule retransmission for the same TBs regardless if one or more of them were actually successfully decoded by the UE. Consequently, such operation will significantly affect the transmission efficiency in the NR—U system. Therefore the inefficiency problem in PDSCH retransmission for the same HARQ process at the gNB side due to LBT failure at the UE and not feeding back the HARQ—ACK information, needs to be resolved.

- Problem 2: HARQ—ACK information cannot be fed back due to insufficient processing time.

Multiple channel-access procedures are supported in the unlicensed spectrum. If a UE is able share/make use of the COT acquired by the gNB, then the UE may use the Type-2 channel-access procedure, which offers a higher probability of accessing the channel (e.g., using Type-2A channel access, Type-2B channel access, or Type-2C channel access). These channel-access procedures are especially suitable for transmission of PUCCH with HARQ—ACK information, so that the UE has a higher probability to successfully feed back the HARQ—ACK information to the gNB. However, the use of Type-2A channel access, Type-2B channel access, or Type-2C channel access needs to fulfill certain constraints. For example, the gap between the starting time position of the intended PUCCH transmission and the ending time position of the previous downlink transmission should be less than or equal to 16 μs for Type-2C channel access. However, this gap duration is too short to provide

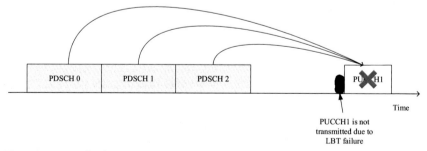

Figure 18.37 Hybrid automatic repeat request acknowledgment information cannot be transmitted due to listen before talk failure.

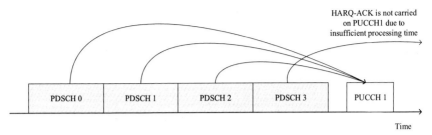

Figure 18.38 Hybrid automatic repeat request acknowledgment information cannot be transmitted due to insufficient processing time.

sufficient processing time for the UE to decode and prepare the HARQ—ACK report for the latest received PDSCH in the previous downlink transmission burst scheduled by the gNB, as shown in Fig. 18.38. Therefore the lack of processing time problem in preparing HARQ—ACK feedback for the latest received PDSCH at the end of a downlink transmission burst also needs to be resolved.

Therefore the HARQ feedback enhancements in NR—U mainly aim at solving the above two problems.

18.5.1.2 Hybrid automatic repeat request acknowledgment retransmission

For the above first problem, where a UE LBT failure leads to no reporting of HARQ—ACK feedback information to the gNB, there were two main solution approaches discussed in 3GPP RAN1: one of which is to provide more PUCCH resources in the time or frequency domain for HARQ—ACK feedback; and the other one is to introduce dynamic retransmission of HARQ—ACK information. Both of solution approaches can provide UE with more opportunities for HARQ—ACK transmission. In order to provide more flexibility for HARQ—ACK feedback in the unlicensed spectrum, the dynamic retransmission of HARQ—ACK information was mainly considered in the NR—U during the standardization phase. More specifically, two new feedback mechanisms for the HARQ—ACK codebook were introduced.

The first one is the enhanced dynamic HARQ—ACK codebook, commonly known as eType-2 HARQ—ACK codebook, which is enhanced based on Type-2 HARQ—ACK codebook. In the feedback mechanism of the eType-2 HARQ—ACK codebook, the gNB groups each schedule PDSCH and explicitly indicate the group-related feedback information to the UE, so that the UE can generate and transmit group-based HARQ—ACK codebook according to the received group-related feedback information for the scheduled PDSCH.

During the discussions in the standardization phase, as shown in Fig. 18.39, two methods of HARQ−ACK grouping for PDSCHs scheduled by the gNB were considered as follows:

- Method 1: when the scheduled PDSCHs are grouped by the gNB, the corresponding HARQ−ACKs for the grouped PDSCHs are also grouped. Once the HARQ−ACK group is constructed, the size of the HARQ−ACK codebook for this group remains unchanged for the initial transmission or retransmission. Specifically, for a group of scheduled PDSCHs with a valid PUCCH resource allocation for feeding back the corresponding group HARQ−ACK information, then no new PDSCHs should be included into the group. When the group-based HARQ−ACK feedback is triggered, more than one group of HARQ−ACK codebook can be triggered for the same PUCCH resource.

- Method 2: when the scheduled PDSCHs are grouped by the gNB, the corresponding HARQ−ACKs for the grouped PDSCHs are also grouped, and the size of the grouped HARQ−ACK codebook may be increased for the retransmissions compared to the initial transmission of this scheduled PDSCH group. That is, for a group of scheduled PDSCHs with a valid PUCCH resource allocation for feeding back the corresponding group HARQ−ACK information, new PDSCHs can still

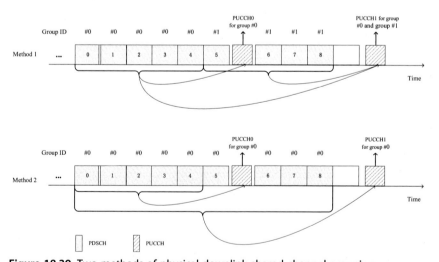

Figure 18.39 Two methods of physical downlink shared channel grouping.

be included into this group. When the group-based HARQ—ACK feedback is triggered, only one group of HARQ—ACK codebook should be triggered for an allocated PUCCH resource in this case.

During the standardization phase of the HARQ—ACK feedback mechanisms for the NR—U, the above method 1 and method 2 are merged into the same method. That is, the gNB groups the scheduled PDSCHs and explicitly indicates their group-related information, such that the UE can generate and transmit the corresponding HARQ—ACK feedback information based on the group-related information for the scheduled PDSCHs. If the UE fails to transmit the corresponding HARQ—ACK codebook group due to LBT failure, or the gNB fails to receive the expected HARQ—ACK codebook group on an allocated PUCCH resource from the UE, then the gNB triggers the UE via DCI to retransmit the HARQ—ACK codebook group by allocating another PUCCH resource. In the retransmission of the HARQ—ACK codebook group, new PDSCHs may or may not be included into the same group. Subsequently, the UE maintains the same HARQ—ACK codebook group size as the initial transmission if no new PDSCH is included into this group, or the UE increases the HARQ—ACK codebook group size for retransmission if new PDSCHs are included into this group. To construct the HARQ—ACK codebook group, the UE needs to know the starting time position of the HARQ—ACK codebook for the group, or the UE needs to know when to clear/reset the HARQ—ACK codebook for the group. Since the HARQ—ACK codebook group size for each transmission may be changed, the starting time position of the HARQ—ACK codebook group can be explicitly indicated by an indication. Specifically, it is indicated by a 1-bit signaling in the DCI [i.e., new feedback indicator (NFI) for which the details will be introduced later]. When the 1-bit NFI corresponding to group # 1 is toggled (i.e., the bit changes from 0 to 1 or from 1 to 0), the HARQ—ACK codebook for the group # 1 is reset, the existing HARQ—ACK codebook for the group # 1 is cleared, and a new HARQ—ACK codebook for the group # 1 should be regenerated.

In the NR—U system, one-shot HARQ—ACK feedback is introduced, also known as the Type-3 codebook. In the Type-3 HARQ—ACK codebook feedback mechanism, the HARQ—ACK codebook includes HARQ—ACK information of all the HARQ processes for all the configured cells in a PUCCH group. If a UE is configured with Type-3 HARQ—ACK codebook feedback by the gNB, a trigger bit is included in the DCI gNB triggering the UE to report a Type-3 HARQ–ACK codebook. The DCI that the gNB used to trigger a Type-3 HARQ–ACK codebook feedback from the UE may or may not be a DCI to schedule a PDSCH transmission.

18.5.1.3 Introduction and feedback of nonnumerical K1

For the above problem 2 where HARQ–ACK information cannot be fed back due to insufficient processing time, the solution is to introduce an invalid K1 indication (i.e., NNK1 indication in NR–U). In the NR–U, the HARQ timing indication in a DCI can be used to indicate a valid PUCCH slot for UE reporting of HARQ–ACK feedback information that corresponds to a received PDSCH scheduled by the DCI (HARQ timing indication indicates a numerical K1, NK1, in DCI). Alternatively, the HARQ timing indication in a DCI can be used to indicate an invalid PUCCH slot value so that the UE assumes the PUCCH slot for the HARQ feedback reporting corresponding to the scheduled PDSCH has not been determined yet and the corresponding HARQ–ACK feedback information should be withheld (HARQ timing indication indicates a NNK1 in DCI). This NNK1 indication is mainly supported by DCI format 1_1 for PDSCH scheduling. Specifically, the configured set of HARQ timing values for DCI format 1_1 may further include a NNK1 (in addition to NK1 values). When a UE receives a PDSCH scheduled by DCI format 1_1 and the HARQ timing indication indicates a NNK1 from the configured set of HARQ timing values, the UE assumes the PUCCH slot for the PDSCH scheduled by the DCI format 1_1 has not been determined temporarily and holds the corresponding HARQ–ACK feedback information.

The NNK1 indication can be applied to all Type-2 codebook, eType-2 codebook, and Type-3 codebook feedbacks, but it is not supported in Type-1 codebook feedback.

If the UE is configured with Type-2 HARQ–ACK codebook but not the eType-2 HARQ–ACK codebook, the PUCCH slot in which the HARQ–ACK feedback information should be reported for a received PDSCH with NNK1 indicated by the scheduling DCI format 1_1 is determined by the next DCI format indicating a NK1 value. For example, if a UE receives PDSCH#1 scheduled by DCI#1 with the HARQ timing indication in DCI#1 indicating a NNK1 from the configured set of HARQ timing values, then the UE reports the HARQ–ACK feedback information for the received PDSCH#1 on a PUCCH resource according to the HARQ timing indication in DCI#2, where the DCI#2 is the first DCI detected by the UE indicating a NK1 value after receiving DCI#1.

If the UE is configured with eType-2 HARQ−ACK codebook, the group-related indication will be included in a DCI format 1_1 with a HARQ timing indication parameter indicating the NNK1. The PUCCH slot in which the HARQ−ACK feedback information should be reported for a received PDSCH with NNK1 indicated by the scheduling DCI format 1_1 is determined by the next DCI format indicating a NK1 value and both DCI formats schedule PDSCHs for the same group. For example, if a UE receives PDSCH#1 scheduled by DCI#1 and it belongs to group#1, with the HARQ timing indication parameter in DCI#1 indicating the NNK1 from the configured set of HARQ timing values, then the UE transmits the HARQ−ACK feedback information for the received PDSCH #1 on a PUCCH resource according to the HARQ timing indication in DCI#2. The DCI#2 is the first DCI detected by the UE after receiving DCI#1 and it also schedules a PDSCH belongs to group#1, or triggers a HARQ−ACK codebook feedback for group#1 with the HARQ timing indication parameter indicating a NK1 value.

If the UE is configured with Type-3 HARQ−ACK codebook, when a DCI format schedules a PDSCH and provides a NNK1, the PUCCH slot, in which the HARQ−ACK feedback information for the PDSCH is reported, is determined by the next DCI format providing a NK1 and triggering a Type-3 HARQ−ACK feedback, if the processing timeline can be satisfied. For example, if the UE receives PDSCH#1 scheduled by DCI#1 with the HARQ timing indication parameter in DCI#1 indicating the NNK1 from the configured set of HARQ timing values, then the UE transmits the HARQ−ACK feedback information for the received PDSCH#1 on a PUCCH resource according to the HARQ timing indication in DCI#2 under the condition that there is sufficient processing time between PDSCH#1 and the PUCCH resource. Moreover, the DCI#2 triggers a Type-3 feedback and indicates a NK1 value by the HARQ timing indication parameter.

18.5.2 Hybrid automatic repeat request acknowledgment codebook

In the NR−U system, the HARQ feedback mechanisms for the enhanced Type-2 (eType-2) and Type-3 HARQ−ACK codebooks were introduced, in addition to Type-1 and Type-2 HARQ−ACK codebooks in Release 15. This section describes the generation of eType-2 and Type-3 HARQ−ACK codebooks.

18.5.2.1 eType-2 hybrid automatic repeat request acknowledgementcodebook

As mentioned earlier, if a UE is configured with the eType-2 HARQ−ACK codebook, the gNB groups all scheduled PDSCHs and explicitly indicates group-related feedback information for the PDSCHs. Then the UE generates and transmits the group-based HARQ−ACK codebook according to the received group-related feedback information for the PDSCHs. In the feedback mechanism of the eType-2 HARQ−ACK codebook, a UE can be configured with up to two PDSCH groups. This feature is mainly intended for PDSCH scheduled by DCI format 1_1. To support the feedback mechanism of the eType-2 HARQ−ACK codebook, the following group-related information parameters are included in DCI format 1_1.

- PUCCH resource indication: Used to indicate a PUCCH resource for HARQ−ACK transmission.
- HARQ timing indication: Used to dynamically indicate the slot in which the PUCCH resource is located. If the HARQ timing indication indicates the NNK1, it means that the PUCCH resource slot location is temporarily not determined.
- PDSCH group identity (ID) indication: Used to indicate the group ID of the PDSCH scheduled by this DCI format 1_1. Note that the PDSCH group for which this group ID is indicated is the scheduling group for the DCI, and the other PDSCH group not indicated by this PDSCH group ID is the nonscheduling group of the DCI.
- Downlink assignment index (DAI): In a single-carrier scenario, the DAI field represents C−DAI (Counter DAI) information. In a multiple-carrier scenario, the DAI field represents both C−DAI and T−DAI (Total DAI) information. The C−DAI information is used to indicate the location of this PDSCH scheduled by the DCI in the scheduling group of the DCI. The T−DAI information is used to indicate the number of PDSCHs that have been scheduled so far in the scheduling group.
- New feedback indicator (NFI): Used to indicate the starting time position of the HARQ−ACK codebook for the scheduling group of the DCI. If the NFI in the DCI is toggled, the HARQ−ACK codebook is reset.

- Number of requested PDSCH group(s): Used to indicate the number of PDSCH groups (i.e., one or two groups) the UE needs to feed back their corresponding HARQ−ACK codebooks. When this field is set to 0, the UE feeds back the HARQ−ACK codebook for the current scheduling group of the DCI. When this field is set to 1, the UE feeds back HARQ−ACK codebooks for the two groups (i.e., both the scheduling group and the nonscheduling group of the DCI).

When a UE is capable of feeding back HARQ−ACK information for up to two PDSCH groups, the gNB may configure the UE via higher-layer parameters with eType-2 codebook feedback and nonscheduling group-related information fields in DCI format 1_1 in order to help the UE generate more accurate HARQ−ACK codebook for the nonscheduling group.

- NFI for the nonscheduling group: Used jointly with the PDSCH group ID of the nonscheduling group to indicate the target HARQ−ACK codebook for the nonscheduling group.
- T−DAI for the nonscheduling group: Used to indicate the number of scheduled PDSCHs in the nonscheduling group for the target HARQ−ACK codebook.

Based on the group-related information in the DCI, the UE will be able to dynamically generate an eType-2 HARQ−ACK codebook for initial transmission and retransmission to the gNB. For the retransmission of the HARQ−ACK codebook for one PDSCH group, the UE maintains the same codebook size as in the initial transmission for the PDSCH group, or includes additional HARQ−ACKs for any newly scheduled PDSCHs based on gNB scheduling.

If a UE receives a PDSCH scheduled by DCI format 1_0, since there is no group-related information introduced for DCI format 1_0, the scheduled PDSCH is considered as belonging to PDSCH group#0 in certain conditions. Otherwise the PDSCH scheduled by DCI format 1_0 is considered as neither belonging to PDSCH group#0 nor PDSCH group#1 as well.

Generation of a eType-2 HARQ−ACK codebook for different cases is considered in the following.

- Case 1: UE is configured with nonscheduling group-related information fields and received PDSCHs scheduling by DCI format 1_1

Assume first that only one HARQ−ACK information bit corresponds to one HARQ process and a UE is scheduled for PDSCH reception on cell 1 and cell 2. Furthermore, DCI format 1_1 includes the following fields: PDSCH group ID indication (represented by G); DAI for the scheduling group (represented by C−DAI, T−DAI); NFI for the scheduling group (represented by NFI); HARQ timing indication (represented by K1 for a valid value or NNK1 for an invalid value); T−DAI for the nonscheduling group (represented by T−DAI2); NFI for the nonscheduling group (represented by NFI2); and the number of requested PDSCH group(s) (represented by Q).

As shown in Fig. 18.40, in slot n the UE receives a DCI scheduling PDSCH1 for cell 1 and in the DCI it is indicated that G = 0, NFI = 0, C−DAI = 1, T−DAI = 2, and the PUCCH resource for feeding back HARQ−ACK information for the scheduled PDSCH1 is in slot n + 3 by setting K1 = 3. Assuming Q is set to zero, meaning only the HARQ−ACK codebook for the scheduling group is requested by the gNB, the UE ignores T−DAI2 and NFI2 indications in the DCI for the nonscheduling group. For the PDSCH reception on cell 2, the UE receives a DCI scheduling PDSCH2 for cell 2 and in the DCI it is indicated that G = 0, NFI = 0, C−DAI = 2, T−DAI = 2, and the PUCCH resource for feeding back HARQ−ACK information for the scheduled PDSCH2 is in slot n + 3 by setting K1 = 3. Assuming Q is also set to zero in the scheduling DCI, the UE also ignores T−DAI2 and NFI2 indications in the DCI for the nonscheduling group.

In slot n + 1, the gNB transmits a DCI for scheduling PDSCH3 on cell 1 to the UE. For the purpose of demonstration, it is assumed the UE does not correctly detect this DCI.

In slot n + 2, the UE receives another DCI scheduling PDSCH4 on cell 2 and in the DCI it is indicated that G = 1, NFI = 0, C−DAI = 1, T−DAI = 1, and the HARQ timing indication in the DCI is set to NNK1, which means the PUCCH resource slot for reporting the HARQ−ACK information corresponding to PDSCH4 is not determined. Assuming Q is

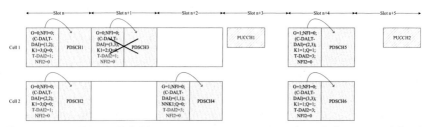

Figure 18.40 eType-2 Hybrid automatic repeat request acknowledgment codebook generation when nonscheduling group-related information fields are configured.

also set to zero in this scheduling DCI, then the UE also ignores T−DAI2 and NFI2 indication in the DCI for the nonscheduling group.

After UE decoding of PDSCH1 and PDSCH2, the UE determines the corresponding HARQ−ACK information for the group#0 HARQ−ACK codebook based on the same group ID $G = 0$ and the same $NFI = 0$ from their respective scheduling DCIs. Therefore the HARQ−ACK codebook generated by the UE for PUCCH 1 in slot $n + 3$ is shown in Fig. 18.41, where the first HARQ−ACK information bit corresponds to the decoding result of PDSCH1 and the second HARQ−ACK information bit corresponds to the decoding result of PDSCH2.

In slot $n + 4$, the UE receives a DCI scheduling PDSCH5 on cell 1 and in the DCI it is indicated that $G = 1$, $NFI = 0$, $C−DAI = 2$, $T−DAI = 3$, and the PUCCH resource for feeding back HARQ−ACK information for the scheduled PDSCH5 is in slot $n + 5$ by setting $K1 = 1$. Assuming the Q value is set to 1 in the scheduling DCI, meaning the HARQ−ACK codebook for both the scheduling group and the nonscheduling group is requested by the gNB, the UE further reads $T−DAI2 = 3$ and $NFI2 = 0$ contained in the DCI for the nonscheduling group. The UE further receives a DCI scheduling PDSCH6 on cell 2 and in the DCI it is indicated that $G = 1$, $NFI = 0$, $C−DAI = 3$, $T−DAI = 3$, and the PUCCH resource for feeding back HARQ−ACK information for the scheduled PDSCH6 is in slot $n + 5$ by setting $K1 = 1$. Assuming the Q value is also set to 1 in the scheduling DCI, the UE further reads $T−DAI2 = 3$ and $NFI2 = 0$ contained in the DCI for the nonscheduling group. After UE decoding of PDSCH 4, PDSCH5, and PDSCH6, the UE determines their corresponding HARQ−ACK information all belong to group#1 HARQ−ACK codebook based on the same group ID $G = 1$ and the same $NFI = 0$ from their respective scheduling DCIs. In the scheduling DCI for PDSCH6, the UE detects that the DCI also triggers feedback for another PDSCH group (i.e., group#0 according to $Q = 1$). Furthermore, the UE determines $NFI2 = 0$ for the group#0 and the total HARQ−ACK bits and the number of PDSCH reception for group#0 is 3 according to $T−DAI2 = 3$. Therefore the UE determines that both HARQ−ACK

PDSCH1   PDSCH2

Figure 18.41 Hybrid automatic repeat request acknowledgment codebook.

codebook of group#0 and group#1 should be transmitted using the PUCCH resource in slot n + 5, and the total number of HARQ−ACK bits for both group#0 and group#1 is 3. When the HARQ−ACK information for both groups need to be fed back on one PUCCH resource, the HARQ−ACK information bits for group #0 should be placed before the HARQ−ACK information bits for group #1.

Therefore the HARQ−ACK codebook generated by the UE for PUCCH 2 in slot n + 5 is as shown in Fig. 18.42, where the decoding result for PDSCH3 that was not received by the UE is set to NACK.

- Case 2: UE is not configured with nonscheduling group-related information fields and received PDSCHs scheduling by DCI format 1_1.

Similar to the above Case 1, it is assumed that only one HARQ−ACK information bit corresponds to one HARQ process and a UE is scheduled for PDSCH reception on cell 1 and cell 2. Furthermore, DCI format 1_1 includes the following fields: PDSCH group ID indication (represented by G); DAI for the scheduling group (represented by C−DAI, T−DAI), NFI for the scheduling group (represented by NFI); HARQ timing indication (represented by K1 for a valid value or NNK1 for an invalid value); and the number of requested PDSCH group(s) (represented by Q).

As shown in Fig. 18.43, in slot n the UE receives a DCI scheduling PDSCH1 for cell 1 and in the DCI it is indicated that G = 0, NFI = 0, C−DAI = 1, and T−DAI = 2, Q = 0, meaning the HARQ−ACK

1st bit	2nd bit	NACK	4th bit	5th bit	6th bit

PDSCH1   PDSCH2   PDSCH4   PDSCH5   PDSCH6

Figure 18.42 Hybrid automatic repeat request acknowledgment codebook.

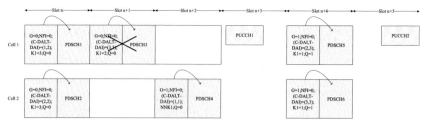

Figure 18.43 eType-2 Hybrid automatic repeat request acknowledgment codebook generation when nonscheduling group-related information fields are not configured.

codebook only for the scheduling group is requested, and the PUCCH resource for feeding back HARQ−ACK information for the scheduled PDSCH1 is in slot n + 3 by setting K1 = 3. Furthermore, the UE receives a DCI scheduling PDSCH2 for cell 2 and in the DCI it is indicated that G = 0, NFI = 0, C−DAI = 2, T−DAI = 2, and Q = 0, and the PUCCH resource for feeding back HARQ−ACK information for the scheduled PDSCH2 is in slot n + 3 by setting K1 = 3.

In slot n + 1, the gNB transmits a DCI for scheduling PDSCH3 on cell 1 to UE. For the purpose of demonstration, it is assumed the UE does not correctly detect this DCI.

In slot n + 2, the UE receives another DCI scheduling PDSCH4 on cell 2 and in the DCI it is indicated that G = 1, NFI = 0, C−DAI = 1, T−DAI = 1, Q = 0, and the HARQ timing indication in the DCI is set to NNK1, which means the PUCCH resource slot for reporting the HARQ−ACK information corresponding to PDSCH4 is not determined.

After UE decoding of PDSCH1 and PDSCH2, the UE determines the corresponding HARQ−ACK information for the group#0 HARQ−ACK codebook based on the same group ID G = 0 and the same NFI = 0 from their respective scheduling DCIs. Therefore the HARQ−ACK codebook generated by the UE for PUCCH 1 in slot n + 3 is as shown in Fig. 18.44, where the first HARQ−ACK information bit corresponds to the decoding result of PDSCH1 and the second HARQ−ACK information bit corresponds to the decoding result of PDSCH2.

In slot n + 4, the UE receives a DCI scheduling PDSCH5 on cell 1 and in the DCI it is indicated that G = 1, NFI = 0, C−DAI = 2, and T−DAI = 3, Q = 1, meaning the HARQ−ACK codebook for both the scheduling group and the nonscheduling group is requested, and the PUCCH resource for feeding back HARQ−ACK information for the scheduled PDSCH5 is in slot n + 5 by setting K1 = 1. The UE further receives a DCI scheduling PDSCH6 on cell 2 and in the DCI it is indicated that G = 1, NFI = 0, C−DAI = 3, T−DAI = 3, and Q = 1, and the

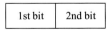

PDSCH1   PDSCH2

Figure 18.44 Hybrid automatic repeat request acknowledgment codebook.

PUCCH resource for feeding back HARQ−ACK information for the scheduled PDSCH6 is in slot n + 5 by setting K1 = 1.

After UE decoding of PDSCH 4, PDSCH5, and PDSCH6, the UE determines that their corresponding HARQ−ACK information all belong to the group#1 HARQ−ACK codebook based on the same group ID G = 1 and the same NFI = 0 from their respective scheduling DCIs. In the scheduling DCI for PDSCH6, UE detects that the DCI also triggers feedback for another PDSCH group (i.e., group#0 according to Q = 1). Furthermore, the UE determines NFI2 = 0 for the group#0 and the total number of HARQ−ACK bits for group#0 is 2 according to the previous PDSCH reception for group#0 (i.e., UE missed the reception of PDSCH3). Therefore the UE determines that both HARQ−ACK codebook of group#0 and group#1 should be transmitted using the PUCCH resource in slot n + 5, and the total number of HARQ−ACK bits for group #0 and group#1 is 2 and 3, respectively. Similarly, when the HARQ−ACK information for both groups need to be fed back on one PUCCH resource, the HARQ−ACK information bits for group#0 should be placed before the HARQ−ACK information bits for group#1.

Therefore the HARQ−ACK codebook generated by the UE for PUCCH 2 in slot n + 5 is as shown in Fig. 18.45, where the decoding result for PDSCH3 that was not received by the UE is not reflected in the codebook.

Since the nonscheduling group-related information fields are not configured in DCI format 1_1, in some cases there could be misalignment on the generated codebook size between the gNB and the UE, when the UE does not correctly receive the scheduling DCI transmitted by the gNB. In this example, the gNB expects the UE to feedback 6-bit HARQ−ACK information, but the UE feeds back only 5-bit of HARQ−ACK information.

- Case 3: UE receives PDSCHs scheduling by DCI format 1_0 and DCI format 1_1 with PDSCH group ID indicating group#0.

1st bit	2nd bit	3rd bit	4th bit	5th bit

PDSCH1 PDSCH2 PDSCH4 PDSCH5 PDSCH6

Figure 18.45 Hybrid automatic repeat request acknowledgment codebook.

Similar to the above two cases, it is assumed that only one HARQ—ACK information bit corresponds to one HARQ process and a UE is scheduled for PDSCH reception on cell 1 and cell 2. Furthermore, DCI format 1_1 includes the following fields: PDSCH group ID indication (represented by G); DAI for the scheduling group (represented by C—DAI, T—DAI); NFI for the scheduling group (represented by NFI); HARQ timing indication (represented by K1); and the number of requested PDSCH group(s) (represented by Q). In addition, assume the UE receives PDSCHs scheduling by both DCI format 1_0 and DCI format 1_1, and the indicated PDSCH group IDs from the two DCIs are both group#0 for two PUCCH resources.

As shown in Fig. 18.46, in slot $n + 1$ the UE receives DCI format 1_0 scheduling PDSCH1 on cell 1 and in the DCI format 1_0 it is indicated that C—DAI = 1 and the PUCCH resource for feeding back HARQ—ACK information for the scheduled PDSCH1 is in slot $n + 5$ by setting K1 = 4. In addition, the UE also receives DCI format 1_1 scheduling PDSCH2 on cell 2 and in the DCI format 1_1 it is indicated that G = 0, NFI = 0, C—DAI = 2, T—DAI = 2, and Q = 0, meaning the HARQ—ACK codebook only for the scheduling group is requested, and the PUCCH resource for feeding back HARQ—ACK information for the scheduled PDSCH2 is in slot $n + 5$ by setting K1 = 4.

On slot $n + 2$, the UE receives a DCI format 1_0 scheduling PDSCH3 on cell 1 and in the DCI format 1_0 it is indicated that C—DAI = 3 and the PUCCH resource for feeding back HARQ—ACK information for the scheduled PDSCH3 is in slot n + 5 by setting K1 = 3.

On slot $n + 3$, the UE receives another DCI format 1_1 scheduling PDSCH4 on cell 2 and in the DCI format 1_1 it is indicated that G = 0, NFI = 0, C—DAI = 4, T—DAI = 4, and Q = 0, and the PUCCH

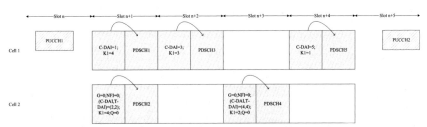

Figure 18.46 UE receives scheduling of downlink control information (DCI) format 1_0 and DCI format 1_1 for group #0.

resource for feeding back HARQ−ACK information for the scheduled PDSCH4 is in slot $n + 5$ by setting K1 = 2.

In slot n + 4, the UE receives DCI format 1_0 scheduling PDSCH5 on cell 1 and in the DCI format 1_0 it is indicated that C−DAI = 5 and the PUCCH resource for feeding back HARQ−ACK information for the scheduled PDSCH5 is in slot $n + 5$ by setting K1 = 1.

Therefore the HARQ−ACK codebook generated by the UE for PUCCH 2 in slot $n + 5$ is as shown in Fig. 18.47, where the PDSCHs scheduled by DCI format 1_0 is considered belonging to PDSCH group#0.

- Case 4: UE receives PDSCHs scheduling by DCI format 1_0 and not DCI format 1_1 with PDSCH group ID indicating group #0.

Again, similar to the previous cases, it is firstly assumed that only one HARQ−ACK information bit corresponds to one HARQ process and a UE is scheduled for PDSCH reception on cell 1 and cell 2. Furthermore, DCI format 1_1 includes the following fields: PDSCH group ID indication (represented by G); DAI for the scheduling group (represented by C−DAI, T−DAI); NFI for the scheduling group (represented by NFI); HARQ timing indication (represented by K1); and the number of requested PDSCH group(s) (represented by Q). In addition, assume the UE receives PDSCHs scheduling by DCI format 1_0 with PDSCH group ID indicating group#0 and not by DCI format 1_1.

As shown in Fig. 18.48, in slot $n + 1$ the gNB transmits a DCI format 1_1 scheduling PDSCH1 to the UE and in the DCI it is indicted that G = 0, NFI = 0, C−DAI = 1, T−DAI = 1, and Q = 0, and the PUCCH resource for feeding back HARQ−ACK information for the scheduled

1st bit	2nd bit	3rd bit	4th bit	5th bit

PDSCH1   PDSCH2   PDSCH3   PDSCH4   PDSCH5

Figure 18.47 Hybrid automatic repeat request acknowledgment codebook.

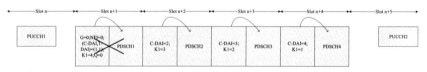

Figure 18.48 UE receives scheduling in downlink control information format 1_0.

PDSCH1 is in slot $n + 5$ by setting K1 = 4. For this illustrated case, it is assumed the UE did not correctly detect the scheduling DCI format 1_1 from the gNB.

In slot n + 2, the UE receives a DCI format 1_0 scheduling PDSCH2, and in the received DCI it is indicated that C−DAI = 2 and the PUCCH resource for feeding back HARQ−ACK information for the scheduled PDSCH2 is in slot $n + 5$ by setting K1 = 3.

In slot n + 3, the UE receives a DCI format 1_0 scheduling PDSCH3, and in the received DCI it is indicated that C−DAI = 3 and the PUCCH resource for feeding back HARQ−ACK information for the scheduled PDSCH3 is in slot $n + 5$ by setting K1 = 2.

In slot n + 4, the UE receives a DCI format 1_0 scheduling PDSCH4, and in the received DCI it is indicated that C−DAI = 4 and the PUCCH resource for feeding back HARQ−ACK information for the scheduled PDSCH4 is in slot $n + 5$ by setting K1 = 1.

Since the UE did not correctly detect the DCI format 1_1 scheduling PDSCH1 in slot $n + 1$ for group#0, the HARQ−ACK codebook generated by the UE for PUCCH 2 in slot $n + 5$ falls back to Type-2 HARQ−ACK codebook generation. As shown in Fig. 18.49, the PDSCHs scheduled by DCI format 1_0 are considered to not belong to any PDSCH groups, and therefore the retransmission of this HARQ−ACK codebook is not supported.

On the other hand, if the UE had detected the DCI format 1_1 scheduling PDSCH1 with group#0, the HARQ−ACK codebook generated by the UE for PUCCH 2 in slot $n + 5$ for DCI format 1_0 should be performed according to eType-2 HARQ−ACK codebook generation. As shown in Fig. 18.50, where the PDSCHs scheduled by DCI format

NACK	2nd bit	3rd bit	4th bit

PDSCH2   PDSCH3   PDSCH4

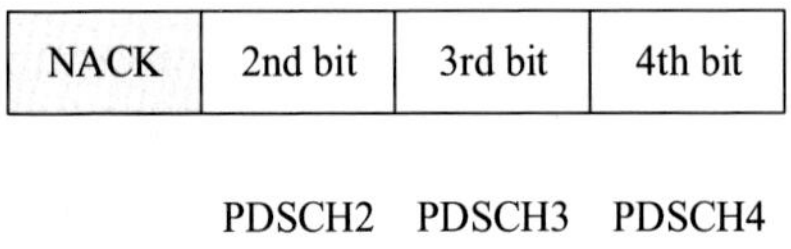

Figure 18.49 Hybrid automatic repeat request acknowledgment codebook.

1st bit	2nd bit	3rd bit	4th bit

PDSCH1   PDSCH2   PDSCH3   PDSCH4

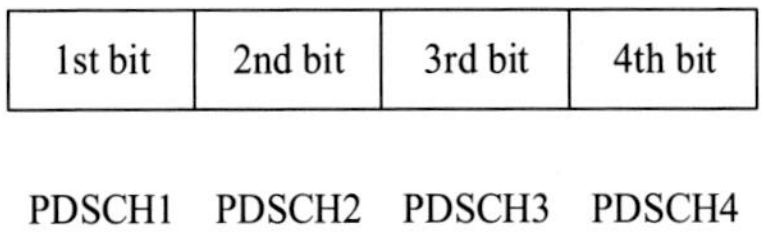

Figure 18.50 Hybrid automatic repeat request acknowledgment codebook.

1_0 are considered belonging to PDSCH group#0, the retransmission of this HARQ—ACK codebook is supported.

18.5.2.2 Type-3 hybrid automatic repeat request acknowledgment codebook

As mentioned earlier, when a UE is configured with the Type-3 HARQ—ACK codebook, a 1-bit trigger requesting feedback of Type-3 HARQ—ACK codebook (one-shot HARQ—ACK feedback request) is included in DCI format 1_1 and the gNB can trigger the request for one-shot HARQ—ACK feedback from the UE by setting the trigger bit to 1 in the DCI. The DCI format 1_1 the gNB triggers for Type-3 HARQ—ACK codebook feedback from the UE could be a DCI that schedules a PDSCH reception or does not schedule a PDSCH reception for the UE. When the UE receives a trigger from the gNB requesting Type-3 HARQ—ACK codebook feedback in DCI format 1_1, the UE must generate and report a Type-3 HARQ—ACK codebook. The Type-3 HARQ—ACK codebook includes HARQ—ACK feedback information for all of the HARQ processes and for all of the configured cells in a PUCCH group.

For Type-3 HARQ—ACK codebook, there can be two feedback scenarios: one is feedback of a Type-3 HARQ—ACK codebook with a new data indicator (NDI) information and the other one is without NDI information. The gNB can configure using higher-layer parameters whether or not the UE should include the NDI information when reporting the Type-3 HARQ—ACK codebook. The generation of Type-3 HARQ—ACK codebook for the two scenarios are described below.

- Scenario 1: Type-3 HARQ—ACK codebook feedback with NDI information

In NR communication, every transmission of a data Transport Block (TB) is associated with an NDI value. In the case of Type-3 HARQ—ACK codebook feedback with NDI information, the UE must report the NDI and HARQ—ACK information for the latest received TB (s) in each HARQ process. If the UE does not receive a TB for a HARQ process, then the UE sets the NDI to 0 and the HARQ—ACK information to NACK for this HARQ process. The order of information encoded in the Type-3 HARQ—ACK codebook is arranged in the following sequence: first in ascending order of code block group (CBG) or TB index, second in ascending order of HARQ process number, and finally in ascending order of cell index. For each data TB, HARQ−ACK information is placed before the NDI information.

Assuming that only one HARQ−ACK information bit corresponds to one HARQ process, a UE is configured with cell 1 and cell 2 in a PUCCH group, 16 HARQ processes are configured for each cell, and the UE may be scheduled for PDSCH reception in cell 1 and cell 2. Furthermore, DCI format 1_1 includes a trigger field for requesting feedback of Type-3 HARQ−ACK codebook (represented by T).

As shown in Fig. 18.51, in slot n the UE receives a DCI scheduling PDSCH1 in cell 1 and in the DCI it is indicated that HARQ = 4, NDI = 1, and the PUCCH resource for feeding back HARQ−ACK information for the scheduled PDSCH1 is in slot $n + 3$ by setting K1 = 3. In the scheduling DCI, T is set to 0, meaning the report of Type-3 HARQ−ACK codebook is not requested. In addition, the UE also receives a DCI scheduling PDSCH2 in cell 2 and in the DCI it is indicated that HARQ = 5, NDI = 0, and the PUCCH resource for feeding back HARQ−ACK information for the scheduled PDSCH2 is in slot $n + 3$ by setting K1 = 3. In the scheduling DCI this time, T is set to 0 again, meaning Type-3 HARQ−ACK codebook feedback is not requested.

In slot n + 1, the UE receives a DCI scheduling PDSCH3 in cell 1 and in the DCI it is indicated that HARQ = 8, NDI = 0, and the PUCCH resource for feeding back HARQ−ACK information for the scheduled PDSCH3 is in slot $n + 3$ by setting K1 = 2. In the scheduling DCI, T is set to 0, meaning the report of Type-3 HARQ−ACK codebook is not requested.

In slot n + 2, the UE receives a DCI scheduling PDSCH4 in cell 2 and in the DCI it is indicated that HARQ = 9, NDI = 1, and the

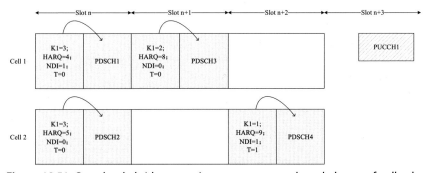

Figure 18.51 One-shot hybrid automatic repeat request acknowledgment feedback.

PUCCH resource for feeding back HARQ−ACK information for the scheduled PDSCH4 is in slot $n + 3$ by setting K1 = 1. In the scheduling DCI, T is set to 1, meaning the report of Type-3 HARQ−ACK codebook is requested this time.

The Type-3 HARQ−ACK codebook generated by the UE for PUCCH 1 in slot n + 3 includes NDI information for each TB. More specifically, the 9th and 10th bits correspond to the decoding result and NDI information for the scheduled PDSCH1. The 17th and 18th bits correspond to the decoding result and NDI information for the scheduled PDSCH3. The 43rd and 44th bits correspond to the decoding result and NDI information for the scheduled PDSCH2. The 51st and 52nd bits correspond to the decoding result and NDI information for the scheduled PDSCH4. The details of the Type-3 HARQ−ACK codebook for this example are shown in Fig. 18.52.

- Scenario 2: Type-3 HARQ−ACK codebook feedback without NDI information

For the feedback scenario of Type-3 HARQ−ACK codebook feedback without NDI information, the UE must report HARQ−ACK information for each HARQ process. When a UE does not receive a data TB for a particular HARQ process, then the UE sets the HARQ−ACK feedback information to NACK for this HARQ process. In addition, the UE must also reset the HARQ−ACK feedback information to NACK for a TB of a HARQ process if the UE has already reported ACK for this TB previously. The order of information encoded in Type-3 HARQ−ACK codebook is arranged in the following sequence: first in ascending order of CBG or TB index, second in ascending order of HARQ process number, and finally in ascending order of cell index.

NACK	0	NACK	0	NACK	0	NACK	0	9th bit	10th bit	NACK	0	NACK	0	NACK	0
HARQ0	NDI0	HARQ1	NDI1	HARQ2	NDI2	HARQ3	NDI3	HARQ4	NDI4=1	HARQ5	NDI5	HARQ6	NDI6	HARQ7	NDI7

17th bit	18th bit	NACK	0	NACK	0	NACK	0	NACK	0	NACK	0	NACK	0	NACK	0
HARQ8	NDI8=0	HARQ9	NDI9	HARQ10	NDI10	HARQ11	NDI11	HARQ12	NDI12	HARQ13	NDI13	HARQ14	NDI14	HARQ15	NDI15

NACK	0	NACK	0	NACK	0	NACK	0	NACK	0	43rd bit	44th bit	NACK	0	NACK	0
HARQ0	NDI0	HARQ1	NDI1	HARQ2	NDI2	HARQ3	NDI3	HARQ4	NDI4	HARQ5	NDI5=0	HARQ6	NDI6	HARQ7	NDI7

NACK	0	51st bit	52nd bit	NACK	0	NACK	0	NACK	0	NACK	0	NACK	0	NACK	0
HARQ8	NDI8	HARQ9	NDI9=1	HARQ10	NDI10	HARQ11	NDI11	HARQ12	NDI12	HARQ13	NDI13	HARQ14	NDI14	HARQ15	NDI15

Figure 18.52 Hybrid automatic repeat request acknowledgment codebook.

Considering the same example in Fig. 18.51, the Type-3 HARQ—ACK codebook generated by the UE for PUCCH 1 in slot $n+3$ does not include NDI information of each TB for this reporting scenario 2. More specifically, the 5th bit corresponds to the decoding result of PDSCH1, the 9th bit corresponds to the decoding result of PDSCH3, the 22nd bit corresponds to the decoding result of PDSCH2, and the 26th bit corresponds to the decoding result for PDSCH4. The details of Type-3 HARQ—ACK codebook for this example are shown in Fig. 18.53.

18.5.3 Multiple physical uplink shared channel scheduling

Since the processing of LBT is required before any transmission on an unlicensed carrier, when the gNB plans to schedule a PUSCH transmission for one UE, the PUSCH would not be transmitted if the LBT fails at the gNB side when the gNB is trying to transmit an UL grant to schedule the PUSCH transmission for the UE, or if the LBT fails at the UE side for the scheduled PUSCH resource after the UE received the corresponding UL grant. In order to introduce more opportunities for PUSCH transmission and reduce the impact of LBT failure on an unlicensed carrier, transmissions of multiple contiguous PUSCH are introduced for the NR—U. More specifically, multiple contiguous PUSCHs can be scheduled with a single UL grant from a gNB.

Scheduling of multiple contiguous PUSCHs can be supported by the nonfallback UL grant DCI format 0_1. The gNB can configure a set of time-domain resource assignments (TDRA) for a UE using higher-layer parameters. In the configured TDRA set, at least one row of TDRA parameters should be included and each row of the TDRA parameters includes a TDRA allocation for m contiguous PUSCHs, where the value range form is from 1 to 8.

If the value of m is 1, then the DCI format 0_1 is used to schedule one PUSCH transmission. In this case, the DCI format 0_1 may include parameter fields for the uplink shared channel (UL—SCH) indicator and

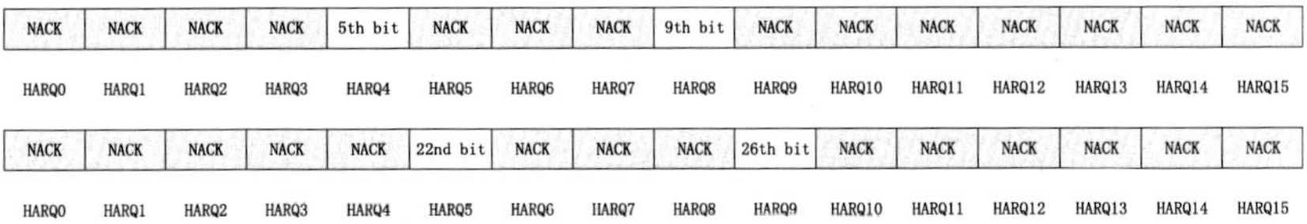

NACK	NACK	NACK	NACK	5th bit	NACK	NACK	NACK	9th bit	NACK	NACK	NACK	NACK	NACK	NACK	NACK
HARQ0	HARQ1	HARQ2	HARQ3	HARQ4	HARQ5	HARQ6	HARQ7	HARQ8	HARQ9	HARQ10	HARQ11	HARQ12	HARQ13	HARQ14	HARQ15

NACK	NACK	NACK	NACK	NACK	22nd bit	NACK	NACK	NACK	26th bit	NACK	NACK	NACK	NACK	NACK	NACK
HARQ0	HARQ1	HARQ2	HARQ3	HARQ4	HARQ5	HARQ6	HARQ7	HARQ8	HARQ9	HARQ10	HARQ11	HARQ12	HARQ13	HARQ14	HARQ15

Figure 18.53 Hybrid automatic repeat request acknowledgment codebook.

CBG transmission information (CBGTI), and a parameter field for the redundancy version (RV), which includes a 2-bit RV indication for this PUSCH.

If the value of m is larger than 1, then the DCI format 0_1 is used to schedule more than one contiguous PUSCH. In this case, the parameter fields for the UL−SCH indicator and CBGTI are not included in the DCI format 0_1, and an m-bit bitmapping between RV indication and PUSCH and m-bit bitmapping between NDI indication and PUSCH are respectively included in the DCI format 0_1.

If channel state information (CSI) feedback is triggered in a DCI format 0_1 for which also schedules m PUSCH transmissions, the selection of PUSCH in which the CSI feedback should be mapped based on the following principles:

- If the value of m is smaller than or equal to 2, the CSI feedback is mapped on to the last PUSCH of the m PUSCHs.
- If the value of m is larger than 2, the CSI feedback is mapped on to the second last PUSCH of the m PUSCHs.

18.6 NR−unlicensed with configured grant physical uplink shared channel

In this section, the extension of the NR−U for uplink transmission with CG (CG UL) is introduced. In Section 15.5, the design details of CG UL transmission in Release 15 and Release 16 for operating in the licensed spectrum were described. In this section, we focus on the special scenarios and necessary enhancements for the CG UL transmission in NR−U, and they are divided and explained according to the following topics: time-frequency-domain resource configuration, CG−UCI and repeated transmission, CG downlink feedback information (CG−DFI), and retransmission timer.

18.6.1 Configured grant resource configuration

In the unlicensed spectrum, the UE determines using channel-access detection whether it can send any data in the uplink to gNB. Due to this special operation and restriction, some considerations are given to the CG UL transmission in the NR−U system. The main enhancement introduced is that after the UE gains the right to access the channel via the LBT process, the UE acquires a channel occupation time, during which the UE should be allowed to continuously transmit multiple CG UL as much as possible. By doing so, the UE does not need to perform additional LBT, thereby greatly improving the efficiency in the UL transmission in the unlicensed spectrum. In Release 15, uplink transmission using CG resources cannot be configured with multiple consecutive CG−PUSCH resources in the time domain. Therefore the main goal of the time-domain configuration design in the NR−U is intentionally making multiple CG−PUSCH resources continuous in the time domain. During the discussions in RAN1, there were several candidate designs proposed [49−59]. The idea of one candidate scheme is for the gNB to configure a first CG−PUCH in the first slot and indicate the number of the CG−PUSCHs. From the second CG−PUSCH, the full slot is used as the CG−PUSCH resource. The advantage of this design is that the configuration information only needs to include the number of slots and the starting symbol position of the CG−PUSCH in the first slot. Then all of the CG−PUSCH resources are placed back to back, as shown in Fig. 18.54.

However, the limitation of this candidate scheme is that all CG−PUSCH transmissions starting from the second slot are full-slot transmission. The first disadvantage of this design is that the UE has a nonfull-slot transmission only for the first CG−PUSCH, which requires a transmission processing that is different from that of the subsequent full-slot transmissions. As such, a processing conversion is required. The second disadvantage is inflexibility in the resource configuration, which limits the possibility of future enhancement. For example, low-latency services need to be supported by the NR−U system in the future, but the limitation of full-slot transmission in this candidate scheme cannot would not be able to fulfill the low-latency requirements. So RAN1 decided to design a more flexible solution while maintaining its backward compatibility. As an alternative scheme, CG resources in subsequent slots starting from the second slot are configured according to the CG resources of the first slot to avoid that the CG resources in the subsequent slots are fixed to full-slot transmission and to improve the flexibility. However, as shown

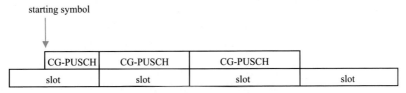

Figure 18.54 NR−unlicensed CG time-domain resource allocation option 1.

in Fig. 18.55, one problem of this alternative scheme is that the CG resources in consecutive slots may be discontinuous, such that additional LBT processes may be needed during continuous CG transmission across slots in the NR−U system. In order to solve this problem, as shown in Fig. 18.56, a simple modification can be done to the CG resource configuration by extending the first CG resource of the next slot forward in time until it is connected with the last CG resource of the previous slot continuously.

In addition, another issue is related to CG resource configuration in the first slot. When the gNB configures the starting time position and length of the first CG resource in the first slot, the first CG resource can be configured at any position in time within the slot such that the UE derives the time position for the remaining CG resources based on the configuration of the first CG resource. For example, as shown in Fig. 18.57, when the starting symbol of the first CG resource is symbol 2 and the length is three symbols, the UE can determine the following three CG resources. But if the starting symbol of the first CG resource is 2, and the length is 5 symbols as shown in Fig. 18.58, it is obvious that the UE cannot map an integer number of CG resources that occupy all the following symbols in the slot. In this case, even if the length of the first CG resource in the next slot is extended, the requirement for continuous CG resources cannot be met due to CG resources cannot be

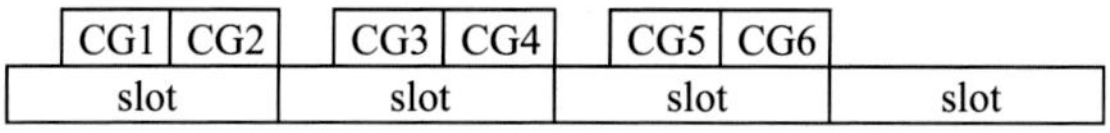

Figure 18.55 NR−unlicensed configured grant time-domain resource allocation option 2.

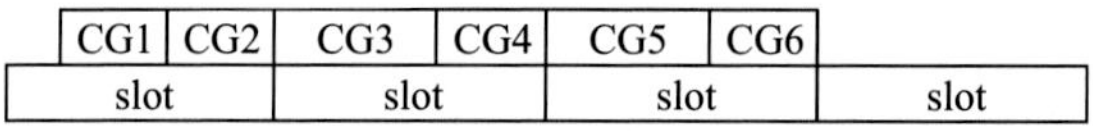

Figure 18.56 NR−unlicensed configured grant time-domain resource allocation option 3.

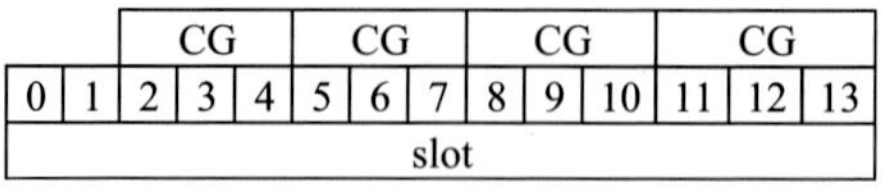

Figure 18.57 UE derives the rest of the configured grant (CG) resources based on the first CG resource within a slot.

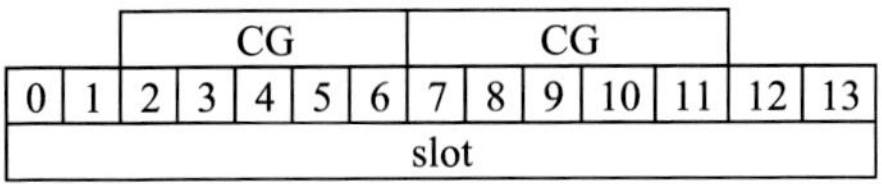

Figure 18.58 Example of configured grant resources do not fill up the remaining symbols of a slot.

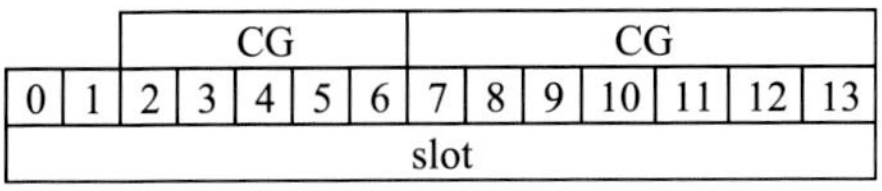

Figure 18.59 The last configured grant resource is extended to fill up the remaining symbols of a slot.

configured across slots. Therefore the most straightforward solution is to extend the last CG resource in the first slot to the last symbol as shown in Fig. 18.59.

This solution effectively solves the problem of continuous CG resource allocation. However, during the discussion in 3GPP RAN1, some opponents believed that if the gNB needs to configure continuous CG resources, then the gNB should ensure matching the size of the CG−PUSCH resource with its position within the slot without needing to do additional resource extension processing. On the contrary, if the gNB does not wish to configure continuous resources, it may deliberately do so to perform downlink retransmission on noncontiguous symbols. Finally, it was decided in the RAN1#99 meeting that the gNB only needs to configure the number of slots (N), number of CG−PUSCHs in the first slot (M), starting symbol position (S), and length (L) of the first CG−PUSCH in NR−U. For example, the gNB shown in Fig. 18.60 can be configured with four parameters: $S = 2$, $L = 2$, $M = 6$, and $N = 2$. That is, CG resources occupy two slots and each slot contains six CG−PUSCH resources. Each CG−PUSCH resource occupies two OFDM symbols, and the starting symbol of the first CG−PUSCH resource is 2. By doing so, the CG resources shown in Fig. 18.58 can be configured.

On the other hand, in a parallel discussion to the NR−U, the NR Release 16 URLLC/IIoT project was also discussing potential further enhancements to the CG UL transmission. According to the progress in that project, it is supported for the gNB to provide multiple sets of CG configurations in their latest design (see Section 15.3.3 for details). Therefore the same mechanism is also supported for the gNB in the

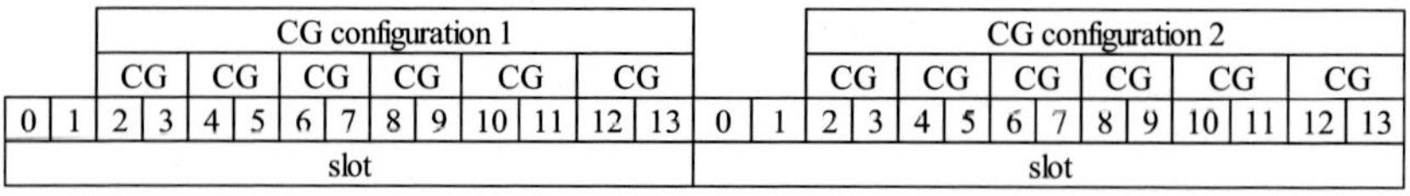

CG configuration 1														CG configuration 2												
	CG		CG		CG		CG		CG		CG				CG		CG		CG		CG		CG		CG		
0	1	2	3	4	5	6	7	8	9	10	11	12	13	0	1	2	3	4	5	6	7	8	9	10	11	12	13
						slot															slot						

Figure 18.60 Configured grant (CG) resource configuration is based on four parameters: start symbol, CG—physical uplink shared channel (PUSCH) length, slot number, and CG—PUSCH number within a slot.

NR−U system. In practice, the gNB may configure multiple sets of CG resource configuration to the UE and each set of configuration corresponds to a set of resources. That is, when multiple sets of CG resource are configured, they correspond to multiple groups of CG resources. If the gNB configures multiple sets of CG configuration, the gNB can choose to activate multiple sets or only some of the them. The activation method directly reuses the URLLC design from Release 16. For details, please refer to Chapter 15, 5G Ultra-Reliable and Low-Latency Communication: PHY Layer.

18.6.2 Configured grant−uplink control information and configured grant repetition

Similar to NR Release 15, CG transmission in the NR−U system supports transmission repetition for the same TB. However, since multiple sets of CG configuration are supported in NR−U, this enhanced feature makes it slightly different from the Release 15 operation. The main difference is that if the UE wants to perform multiple repeated transmissions of a TB (such as K repetitions), the NR−U UE is free to choose any CG resource for starting its continuous K transmissions of CG−PUSCH, which is different from the Release 15 operation where the continuous K transmissions can only start from a specified CG−PUSCH position. It should be noted that the K CG−PUSCH must be transmitted on consecutive CG resources, which must belong to the same CG resource configuration. For the example illustrated in Fig. 18.61, the gNB provides two sets of CG resource configurations (CG configuration 1 and CG configuration 2). If the UE is configured to transmit four consecutive repetitions, the UE can choose to transmit four repetitions in resources associated with either CG configuration 1 or CG configuration 2. However, the UE is not allowed to select CG resources for transmitting the repetitions cross different CG configurations. The number of consecutive transmissions is configured by the gNB via a RRC parameter (repK), and its values include (1, 2, 4, 8). When repK = 1, repeated CG transmission is

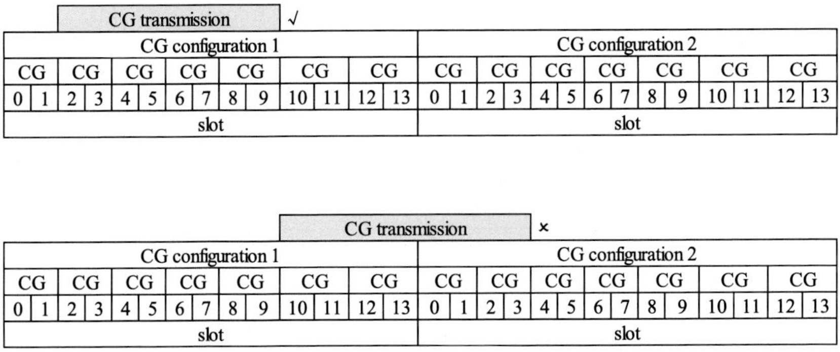

Figure 18.61 Configured grant (CG) repetition cannot cross CG resources belong to different CG configurations.

disabled. In addition, another enhanced feature of the CG transmission in the NR–U system is that the CG–PUSCH includes the transmission of UCI (CG–UCI) [51–57]. The CG–UCI carries some control signaling necessary for CG–PUSCH reception and the COT-sharing information. With this introduction of CG–UCI, the flexibility of CG transmission in the NR–U system has been further improved. In the following, some specific uses of CG–UCI are explained. Firstly, in Release15 NR, CG–PUSCH transmission corresponds to a predefined HARQ process and the process number is mapped to CG resources one by one. By doing so, if a process conflict occurs the UE will not be able to flexibly avoid it. In order to solve this problem, the UE in the NR–U system adds the HARQ process ID to the CG–UCI, so that the UE can schedule different HARQ process IDs flexibly. In the previous section, it is described that the UE can arbitrarily select the CG resource as the starting resource for the K repeated transmissions.

On the other hand, the choice of RV value during CG–PUSCH transmission is strictly controlled in Release 15. That is, the gNB mandates that the UE adopt a specific set of RV values and their selection order during the CG repetitions. In the NR–U, however, the UE is allowed to select the RV value by itself and use CG–UCI to notify the gNB of the selected RV value. In addition, the CG–UCI also includes the NDI indication of the PUSCH. The gNB can determine whether the data in the CG–PUSCH is newly transmitted data or retransmitted data according to the received HARQ process number and NDI indication. From the above aspects, it can seen that the introduction of CG–UCI has enhanced the flexibility of CG in the NR–U system. Table 18.10

Table 18.10 Configured grant—uplink control information information fields.

Information field	Bitwidth
HARQ	4
RV	2
NDI	1
COT-sharing information	variable

shows the parameter fields included in the CG—UCI and the corresponding number of bits. The COT-sharing information is used to indicate whether the COT of the UE can be shared with the gNB for downlink transmission (see Section 18.2.2).

Technically speaking, the CG transmission mechanism in the NR—U can be further optimized. For example, since HARQ, RV, and NDI are all included in the CG—UCI, the specification does not need to further mandate that the UE must transmit continuously for the K repetitions. The gNB can determine which CG—PUSCH carries this according to the indication in CG—UCI.

In the following, the UE's transmission design for CG—UCI is introduced. The specification mandates that the CG—UCI needs to be transmitted with the data part of each CG—PUSCH, but the two are to be independently encoded. Similar to the CG transmission in Release 15, when PUCCH and CG—PUSCH resources collide in the time domain, the UCI from the PUCCH and CG—PUSCH are multiplexed. The processing method can be simply understood as that part of the CG—PUSCH resources that are reserved for UCI transmission, and UCI and CG—PUSCH are encoded independently. The same method is reused for the NR—U system, but the difference is that when CG—PUSCH and PUCCH are collided within a slot, the UE needs to multiplex CG—UCI, UCI, and CG—PUSCH in the same channel and transmit them at the same time. This leads to a priority problem of how to determine the transmission priority of the abovementioned elements if the CG—PUSCH resources are insufficient to carry all of the information.

In Release 15 NR, the HARQ—ACK information is defined as the highest priority, followed by CSI type 1 information, and then CSI type 2 information. More specifically, the UE needs to drop low priority information and retain information with higher priority. In NR—U, due to the addition of CG—UCI, there has been continuing discussion on whose priority is higher than HARQ—ACK. Finally, it was decided in the

RAN1#98b meeting that CG–UCI and HARQ–ACK have the same priority, and it is required that CG–UCI and HARQ–ACK should be encoded jointly when such collision occurs. As shown in Fig. 18.62, when the PUCCH collides with the second CG resource transmission in a slot and if the UCI in the PUCCH carries the HARQ–ACK information, the UCI is multiplexed in CG–PUSCH and a joint encoding is performed for UCI and CG–UCI. On the other hand, when the control information carried in UCI is CSI, the multiplexing process should refer to the Release 15 rule. At this time, CG–UCI is regarded as HARQ–ACK.

In Section 18.5, the design of HARQ–ACK codebook was introduced. In practice, the gNB and the UE sometimes have ambiguities/misalignment on the total number of bits for the reported HARQ–ACK codebook. The main reason is due to the potential loss of one or more scheduling information (DTX). In the case of ambiguity/misalignment, the gNB would not be able to decode CG–UCI and HARQ–ACK feedback information. Therefore a set of fallback schemes is defined in 3GPP specification. In particular, the gNB can choose to directly cancel the function of UCI multiplexing in CG–PUSCH using the RRC configuration. If this function is disabled, the UE will choose to transmit UCI in PUCCH when it encounters PUCCH and CG–PUSCH transmission collision (Fig. 18.63).

In this section, it is introduced that a UE can notify its gNB that the CG–PUSCH transmitted to the gNB is a new transmission or a retransmission using the control information in the CG–UCI. In the next

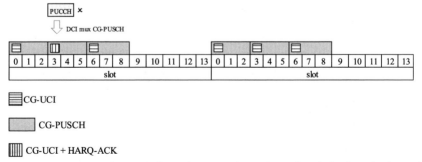

Figure 18.62 Case when Configured grant (CG)–physical uplink shared channel (PUSCH) and physical uplink control channel collides, uplink control information is multiplexed with CG–PUSCH.

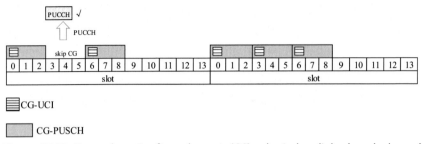

Figure 18.63 Case when Configured grant (CG)—physical uplink shared channel (PUSCH) and physical uplink control channel collides, skip CG—PUSCH and transmit only PUCCH.

section, the method by which the UE decides the timing to send its retransmission data is introduced.

18.6.3 Configured grant—downlink feedback information

In NR Release 15, the CG transmission does not support the retransmission mechanism. When the initial transmission of the CG is completed, the UE will start the CG timer (configuredGrantTimer). After the configuredGrantTimer expires, if the UE has not received a dynamic scheduling from the gNB for a retransmission of a given TB, the UE will consider that the initial transmission was successfully received by the gNB. As such, the UE will clear its data in the buffer. In the NR—U system, in order for the UE to obtain HARQ—ACK feedback information for the sent data, RAN1 introduced a new CG—DFI [49—59]. The CG—DFI is automatically activated after the CG transmission function is configured. The introduction of CG—DFI is designed to achieve two goals: (1) The gNB provides the UE with the ACK/NACK information of the CG—PUSCH in order for the UE to adjust the CW size during the next transmission (for CW adjustment-related content, refer to Section 18.2.2); (2) The gNB provides the ACK/NACK information of the UE CG—PUSCH and the UE will then be able to determine whether to retransmit or terminate the CG—PUSCH transmission in advance according to the ACK/NACK information.

The DCI format of CG—DFI reuses the DCI format 0_1 design and it is scrambled with CS—RNTI. When a UE is configured to detect DCI format 0_1 in the unlicensed spectrum and the CG transmission function is activated, the UE will also detect CG—DFI. The main information field in CG—DFI is the carrier indication information. This information field is used to indicate that the HARQ−ACK information in DFI is for the specifically indicated uplink carrier. In the early stages of 3GPP discussion, one solution suggested reusing the Type-3 HARQ−ACK codebook directly (for details of the Type-3 HARQ−ACK codebook, refer to Section 18.5.2). However, since the Type-3 HARQ−ACK codebook is based on a PUCCH Group (PUCCH group) structure, if there are multiple uplink carriers in a group and each uplink carrier can be configured with up to 16 processes, then the number of feedback bits would be large in this case. Due to limited capacity in the downlink DCI, it would not be able to carry such many information bits. Considering the overhead and transmission reliability of DCI, in the RAN1#99 meeting it was finally decided the solution is to indicate a selected uplink carrier in the DCI. One detail to note here is that the HARQ−ACK information in the CG−DFI is the result from the CRC check of the TB corresponding to the HARQ process. If the check passes, the HARQ−ACK feedback will be an ACK (1 bit indicates 1); otherwise it will be an NACK. Therefore even if the UE is configured for CBG transmission on an uplink carrier, the HARQ−ACK feedback of the CG−DFI for this carrier is also based on the TB feedback (Table 18.11).

When the UE receives the CG−DFI, it acquires the HARQ−ACK information corresponding to each HARQ process. However, in practice the UE may have just sent data in the uplink for some HARQ processes and the gNB has not have enough time to process the received uplink data. Thus the UE should check if the indicated HARQ−ACK information is valid when it receives CG−DFI from the gNB. The main design idea here is that if the UE knows that the gNB does not have enough time to process the uplink data, the UE will ignore the indicated HARQ−ACK information. The specific rule is: the gNB is to also used configure a minimum processing time cg−minDFI-Delay-r16 (D) during the configuration of CG transmission. Each time the UE receives the CG−DFI, it will determine whether the indicated HARQ−ACK information is valid under the condition whether the time between the last

Table 18.11 Configured grant−downlink feedback information content.

Indication field	Bitwidth
Carrier indicator	0 or 3
DFI indicator	1
HARQ−ACK information	16

symbol of the CG—PUSCH sent to the first symbol of the PDCCH carrying CG—DFI is greater than D. If this time length is greater than D, the HARQ—ACK information of the corresponding HARQ process indicated in the DFI is valid; otherwise it is invalid. For the illustrated example in Fig. 18.64, the DFI is valid for HARQ0 and HARQ1 in the HARQ—ACK information, but the DFI is invalid for HARQ2.

Similarly, when the CG transmission is a repetition and the HARQ—ACK information corresponding to a HARQ process is indicated as ACK in the DFI, at least one repetition of CG—PUSCH needs to meet the processing time D. Then the indicated HARQ—ACK information can be considered valid. Conversely, if the indicated HARQ—ACK information is an NACK, then all of the CG—PUSCHs need to meet the processing time D in order to for the HARQ—ACK information to be valid; otherwise, it is determined to be invalid.

18.6.4 Configured grant retransmission timer

In the radio communication over the unlicensed spectrum, the interference at the receiver side is generally more serious than that of the licensed spectrum. In some cases, the gNB is not able to timely schedule the UE for a retransmission before the configuredGrantTimer expires (e.g., due to LBT failure). In response to this situation, in addition to the CG—DFI introduced in Section 18.6.3, the 3GPP recommended the introduction of a new retransmission timer (cg—RetransmissionTimer) in the RAN2#105b meeting [60]. This cg—RetransmissionTimer is automatically reset every time the CG—PUSCH is successfully transmitted. The reason for successful transmission is that when the UE cannot transmit the CG—PUSCH due to LBT failure, the cg—RetransmissionTimer does not need to be started. After the expiration of the cg—RetransmissionTimer, if the CG—DFI from the gNB is not received and the previous CG transmission is indicated as an ACK, the UE considers the previous CG

Figure 18.64 UE validates the hybrid automatic repeat request acknowledgment information carried in configured grant—downlink feedback information based on a configured processing time D.

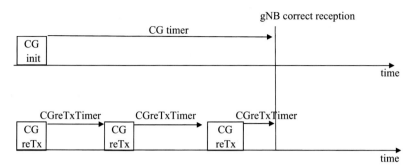

Figure 18.65 Configured grant (CG) timer and CG retransmission timer.

transmission to be unsuccessful and automatically initiates a retransmission. Furthermore, it was determined in the RAN2#107b meeting that when the UE needs to retransmit on its own, it needs to initiate a retransmission in the earliest available CG resources [61]. Then the UE can select the RV value by itself during CG retransmission, and the gNB determines whether it is a new TB transmission or retransmission according to the NDI value and HARQ process value. It should be noted that, as shown in Fig. 18.65, configuredGrantTimer is only started when the CG is first transmitted, and the timer is not reset when the CG is retransmitted. But once the configuredGrantTimer expires, no matter whether the cg−RetransmissionTimer is running or not, the UE will stop the cg−RetransmissionTimer and deem that the gNB has correctly received the CG transmission.

18.7 Summary

This chapter mainly introduced the necessary enhancements for operating NR communication in the unlicensed spectrum. Through six subsections, the channel monitoring process, initial access process, RB collection and control channel design, HARQ and scheduling, and CG UL-related aspects were discussed and explained.

References

[1] R1−1807389 Channel access procedures for NR unlicensed Qualcomm Incorporated, 3GPP RAN1#93 Busan, Korea, May 21−25, 2018.
[2] R1−1805919 Coexistence and channel access for NR unlicensed band operations Huawei, HiSilicon, 3GPP RAN1#93, Busan, Korea, May 21−25, 2018.

[3] R2−1901094 Detecting and handling systematic LBT failures in MAC MediaTek Inc. discussion Rel-16 NR_unliC−Core.

[4] R2−1910889 Report of Email Discussion [106#49][NR−U] Consistent LBT failures.

[5] R2−1904114 Report of the email discussion [105#49] LBT modeling for MAC Huawei, HiSilicon discussion Rel-16 NR_unliC−Core.

[6] R2−1912304 Details of the Uplink LBT failure mechanism Qualcomm Incorporated.

[7] R1−1901332 Feature lead summery on initial access signals and channels for NR−U, Qualcomm, RAN1 #AH-1901.

[8] R1−1906672 Physical layer design of initial access signals and channels for NR−U LG Electronics, RAN1#97.

[9] R1−1907258 Initial access signals and channels for NR−U Qualcomm Incorporated, RAN1#97.

[10] R1−1907258 Initial access signals and channels for NR−U Qualcomm Incorporated, RAN1#97.

[11] R1−1907451 Initial access signals and channels Ericsson, RAN1#97.

[12] R1−1906782 Initial access signals/channels for NR−U Intel Corporation, RAN1#97.

[13] R1−1903404 Feature lead summery on initial access signals and channels for NR−U, Qualcomm, RAN1 #96.

[14] R1−1912710 Enhancements to initial access procedure Ericsson, RAN1 #99.

[15] R1−1912710 Enhancements to initial access procedure Ericsson, RAN1#99.

[16] R1−2002028 Initial access signals and channels, Ericsson, RAN1# #100bis-e.

[17] R1−1908202 Considerations on initial access signals and channels for NR−U ZTE, Sanechips, RAN1#98.

[18] R1−1908137 Discussion on initial access signals and channles vivo, RAN1#98.

[19] R1−2001256 Email discussion for [100e-NR−UnliC−NRU-InitSignalsChannels-01] Qualcomm Incorporated.

[20] R1−1912939 Initial access and mobility procedures for NR−U Qualcomm Incorporated, RAN1#99.

[21] R1−1912198 Enhancements to initial access and mobility for NR−Unlicensed Intel Corporation, RAN1#99.

[22] R1−1912286 On Enhancements to initial access procedures for NR−U Nokia, Nokia Shanghai Bell, RAN1#99.

[23] R1−1912710 Enhancements to initial access procedure Ericsson, RAN1#99.

[24] R1−1912765 Initial access procedure for NR−U, Sharp, RAN1#99.

[25] R1−1908113 NRU wideband BWP operation Huawei, HiSilicon.

[26] R1−1908421 Wideband operation for NR−U OPPO.

[27] R1−1908688 On wideband operation in NR−U Nokia, Nokia Shanghai Bell.

[28] R1−1909249 Wideband operation for NR−U operation Qualcomm Incorporated.

[29] R1−1909302 Wideband operation for NR−U Ericsson.

[30] TS 38.101−1 NR; User Equipment (UE) radio transmission and reception; Part 1: Range 1 Standalone.

[31] R1−2001758 Discussion on the remaining issues of UL signals and channels OPPO.

[32] R1−2001651 Remaining issues on physical UL channel design in unlicensed spectrum vivo.

[33] R1−2001533 Maintainance on uplink signals and channels Huawei, HiSilicon.

[34] R1−2001934 Remaining issues of UL signals and channels for NR−U LG Electronics.

[35] R1−2002030 UL signals and channels Ericsson.

[36] R1−1906042 DL channels and signals in NR unlicensed band Huawei, HiSilicon.

[37] R1−1906195 DL signals and channels for NR−U NTT DOCOMO, INC.

[38] R1−1906484 DL signals and channels for NR−U OPPO.
[39] R1−1906656 On DL signals and channels Nokia, Nokia Shanghai Bell.
[40] R1−1906673 Physical layer design of DL signals and channels for NR−U LG Electronics.
[41] R1−1906783 DL signals and channels for NR−Unlicensed Intel Corporation.
[42] R1−1906918 DL signals and channels for NR−U Samsung.
[43] R1−1907085 DL Frame Structure and COT Aspects for NR−U Motorola Mobility, Lenovo.
[44] R1−1907159 Design of DL signals and channels for NR-based access to unlicensed spectrum AT&T.
[45] R1−1907259 DL signals and channels for NR−U Qualcomm Incorporated.
[46] R1−1907334 On COT detection and structure indication for NR−U Apple Inc.
[47] R1−1907452 DL signals and channels for NR−U Ericsson.
[48] R1−1807391 Enhancements to scheduling and HARQ operation for NR−U Qualcomm Incorporated, 3GPP RAN1#93, Busan, Korea, May 21−25, 2018.
[49] R1−1909977 Discussion on configured grant for NR−U ZTE, Sanechips.
[50] R1−1910048 Transmission with configured grant in NR unlicensed band Huawei, HiSilicon.
[51] R1−1910462 Configured grant enhancement for NR−U Samsung.
[52] R1−1910595 On support of UL transmission with configured grants in NR−U Nokia Nokia Shanghai Bell.
[53] R1−1910643 Enhancements to configured grants for NR−Unlicensed Intel Corporation.
[54] R1−1910793 On configured grant for NR−U OPPO.
[55] R1−1910822 Discussion on configured grant for NR−U LG Electronics.
[56] R1−1910950 Configured grant enhancement Ericsson.
[57] R1−1911055 Discussion on NR−U configured grant MediaTek Inc.
[58] R1−1911100 Enhancements to configured grants for NR−U Qualcomm Incorporated.
[59] R1−1911163 Configured grant enhancement for NR−U NTT DOCOMO, INC.
[60] R2−1903713 Configured grant timer(s) for NR−U Nokia, Nokia Shanghai Bell.
[61] R2−1912301 Remaining aspects of configured grant transmission for NR−U Qualcomm Incorporated.

CHAPTER 19

5G terminal power-saving

Zhisong Zuo, Weijie Xu, Yi Hu and Kevin Lin

19.1 Requirements and evaluation of power-saving techniques for 5G

The first basic version of the 5G-new radio (NR) radio system was developed in 3GPP R15, but it was not optimized for terminal power consumption. In the process of enhancing the 5G-NR system in R16, a comprehensive evaluation method was adopted to analyze various power-saving techniques that could be potentially employed by 5G terminals to reduce battery consumption. After careful evaluation and extensive simulation of these candidate techniques, the R16 version of the 5G-NR system was enhanced by techniques with higher power-saving gain.

19.1.1 Power-saving requirements for 5G terminals

The 5G-NR radio system standards developed by the 3GPP have been shown to provide extremely high system data rate and guarantee to meet the requirements of ITU's IMT-2020 5G data throughput minimum performance [1]. At the same time, the energy consumption on the terminal side is also an important performance requirement of IMT-2020. Starting from R16, the 5G-NR standard development at the 3GPP began to set up a work item to optimize energy efficiency for the device terminals.

The demand on terminal power-saving from IMT-2020 is expressed in two aspects. Firstly, the data should be transmitted efficiently when there is data to transmit. Secondly, it is required to switch to a very low-power consumption state quickly when there is no data to transmit. Power-saving for the terminal can be achieved by switching between states with different power consumption. The switching can be accomplished by network indication. When there is a data service, the terminal needs to be quickly "awakened" by the network and configured with resources for efficient transmission. When there is no data service, the terminal will enter into a low-power state in time.

Generally, the energy consumption for receiving and transmitting data by a terminal can be affected by several factors: the processing bandwidth

5G NR and Enhancements
DOI: https://doi.org/10.1016/B978-0-323-91060-6.00019-2

of the terminal, the number of carriers for reception and transmission by the terminal, the activated radio frequency (RF) link of the terminal, the time duration for reception and transmission by the terminal, and so on. According to the field trial data for long-term evolution (LTE), the terminal power consumption in RRC_CONNECTED mode accounts for most of the terminal power consumption. For data transmission in RRC_CONNECTED mode, the abovementioned energy consumption factors of terminal transceiver processing should match the data traffic. Hardware and processing resources used by the terminal in both time and frequency should match what is required by the received physical downlink control channel (PDCCH)/ physical downlink shared channel (PDSCH) and the transmitted physical uplink control channel (PUCCH)/ physical uplink shared channel (PUSCH). The matching process can be dynamic (e.g., at slot level), which can better achieve the effect of power-saving. Additionally, radio resource management (RRM) measurement that needs to be performed by the terminal also consumes energy. RRM measurement is required in all RRC_CONNECTED/IDLE/INACTIVE states when the terminal is powered on. Reducing unnecessary measurements will also play an important role in saving energy. When the terminal operates in a high-efficiency transmission mode, timely premeasurement can improve the transmission efficiency and reduce the amount of time for switching between the power-saving states for sending and receiving data.

While improving the energy efficiency in radio transmissions, terminals should still achieve low-latency and high-throughput performance. At the same time, it is also required that the adopted power-saving technology should not significantly reduce the performance of the network.

19.1.2 Candidate power-saving techniques

Candidate power-saving techniques for 5G terminals can be categorized into several groups of power-saving mechanisms. The first version of the 5G-NR radio system in 3GPP R15 already provided some basic features that can be helpful for reducing terminal power consumption, such as bandwidth part (BWP), carrier aggregation, and DRX mechanisms. Further power-saving enhancements for 5G terminals should expand on these basic power-saving features. In R16, candidate power-saving techniques considered for 5G-NR are divided into the following categories [2–4], some of which were then selected and standardized.

19.1.2.1 Terminal frequency bandwidth adaptation

In this power-saving technique, a terminal performs frequency adaption by adjusting the BWP within a carrier. The adjustment can be made according to its data traffic volume. As described in the BWP function in Chapter 4, Bandwidth Part, a narrower BWP in 5G-NR means RF processing with a smaller bandwidth can be performed by the terminal to achieve lower power consumption. Narrower BWP also reduces the processing energy consumption for the baseband. When a terminal is operating with multiple carriers, the support for rapid secondary carrier (SCell) activation and deactivation will also reduce energy consumption under the operation of CA/DC.

During the process of frequency adaptation, the terminal also needs to adjust its measurement RS accordingly. Under normal operation, a 5G terminal can only handle one BWP at a time. Assistance RS can help the terminal switching to a different BWP quickly. As shown in Fig. 19.1, if the measurement is performed within bandwidth of a larger target BWP before the BWP switching, the base station can provide more effective scheduling for the terminal, select a better MCS, and allocate a more suitable frequency position. Meanwhile, the terminal does not need to perform RF tuning to a BWP with larger-frequency bandwidth before the measurement is stable.

Physical-layer signaling should support fast switching between different BWPs. In a further optimization of signaling, BWP configuration is associated with DRX configuration [5,6].

These measurement techniques are not fully supported as part of the basic BWP mechanism in R15. For power-saving enhancement in R16,

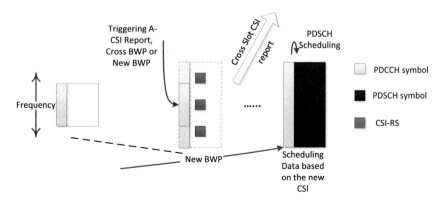

Figure 19.1 Power-saving actions through BWP adjustment. *BWP*, Bandwidth part.

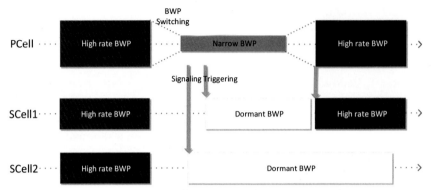

Figure 19.2 Carriers switching to different types of BWP. *BWP*, Bandwidth part.

necessary functions associated with the BWP switching mechanism should be considered.

In a multicarrier operating environment, some candidate techniques for fast switching are considered for power-saving optimization with a carrier as the switching granularity. In 5G-NR, up to 15 carriers can be configured as secondary carriers (SCells) in a CA. To achieve power-saving, a 5G terminal needs to turn off the reception for control channel, data channel, and measurement signals for a large number of secondary carriers when there is no data to receive. As shown in Fig. 19.2, when there is no active data, the terminal receives only downlink on the primary carrier, while other carriers enter into a sleep state (such as switching to a so-called sleep BWP). Moreover, the terminal can also be switched to a BWP with narrow bandwidth on the downlink of the primary carrier according to the data service in real-time.

In 3GPP R15, the basic version of 5G-NR already supports dynamic BWP switching. However, the basic version of carrier aggregation does not have a corresponding mechanism for fast switching of SCells (carriers) sleeping.

19.1.2.2 Time adaptive cross-slot scheduling

The power consumption for a terminal can be reduced by adjusting the order of control and data channel processing in time. Through sequential control and data processing, unnecessary consumption of RF and base-band processing power can be saved. The adaptive process of RF power consumption is shown in Fig. 19.3.

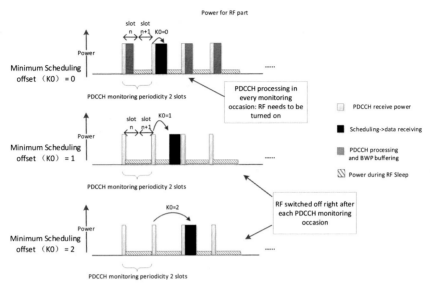

Figure 19.3 Schematic diagram of power change in cross-slot scheduling.

Why is there such a difference? In both 4G-LTE and 5G-NR systems, same-slot scheduling is always supported as the baseline. In 5G-NR, if K0 = 0 or K2 = 0 is included in the time-domain scheduling table configured for PDSCH/PUSCH, the terminal must prepare uplink/downlink data for transmission after the end of each PDCCH monitoring occasion. Since it takes time for the terminal to blindly detect scheduling information in PDCCH, the terminal has to cache/store downlink signals/channels for the full bandwidth of the active BWP in duration of several orthogonal frequency division multiplexing (OFDM) symbols after the PDCCH. Therefore it will consume RF power of the terminal for some time after each PDCCH monitoring occasion even when the terminal determines in the end that there was no data scheduled. In addition to the RF energy consumption shown in Fig. 19.3, caching/storing BWP data also consumes energy.

According to the data statistics collected from real networks, in most cases, a terminal only has control channel scheduling and data transmission in a small portion of the slots. For this small portion, the terminal still unnecessarily consumes power for detecting control channel information and buffering downlink signals and channels in slots where there was no data scheduled for the terminal (Fig. 19.4).

The advantage of cross-slot scheduling is that it allows the terminal to turn off its RF part after the PDCCH symbol and enter into a dormant state with relatively low-power consumption. Due to a transition time of hardware switching such as RF, the reduction of power consumption is often related to the time duration operating in low-power consumption. However, even for a time period of just one slot, sufficient power-saving can still be achieved. Meanwhile, for some specific terminal implementation, the effect of power-saving can also be observed for the same-slot scheduling, which only guarantees several symbols between PDCCH and PDSCH. If the time offset between the scheduling PDCCH and scheduled PDSCH is increased to more than one slot length, the terminal can even reduce the baseband hardware processing clock, voltage, and other parameters. In such a way, the terminal enters into the lower-power consumption state and adapt to the scenario with lower-data arrival rate. Adjusting baseband hardware parameters to a different level, the offset between PDCCH and PUSCH have to be configured to larger number of slots accordingly.

As the control information in PDCCH not only triggers the data reception in a terminal, but also triggers downlink measurement of RS and transmission of SRS. Therefore the corresponding time slots where RS and SRS are measured in the downlink also need to have corresponding offsets to ensure that the terminal enters a lower power consumption state.

19.1.2.3 Adaptive antenna number

The reception and transmission on multiple antennas will also affect the power consumption for a terminal. For a terminal equipped with multiple antennas, the number of receiving antennas can only be semistatically determined. The number of transmitting antennas can be dynamically determined according to an indication information from the serving base station. Therefore the mechanism to achieve power-saving from adapting the number of antennas for the terminal should be enhanced on the receiving side.

The adaptive process of terminal receiving antennas is mainly considered in the adjustment of control reception and data reception. For the control reception, there is a certain correlation between the number of receiving antennas of the terminal and aggregation level of PDCCH. The aggregation level is determined by scheduling adaptation at the base station side. The basic version of the 5G-NR system has an assumption for the number of receiving antennas for the terminal. Above the 2.5 GHz

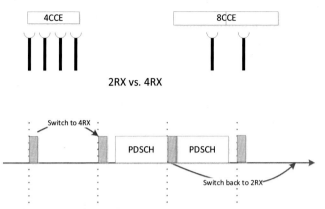

Figure 19.4 Adaptive switching of antenna.

frequency band, terminals are mandated with 4 receiving antennas. If the terminal is allowed to vary its receiving antenna number, the base station have to change PDCCH resources accordingly. According to the analysis of PDCCH simulation evaluation, the aggregation level is found to be correlated to the number of receiving antennas. When the terminal is instructed to switch from 4-antenna to 2-antenna reception, the number of PDCCH CCE resources is required to be doubled. Therefore the number of RX (receiving antennas) for the control channel can be reduced by allocating more resources in the downlink.

When the terminal receives data in PDSCH, the number of RX required is related to the RANK of the current wireless channel. If the RANK is very low, the performance of using a single RX antenna or two RX antennas would not be much different from that of four RX antennas for receiving single-layer transmission. But the reduced number of antennas can provide some power-saving benefit.

The maximum number of downlink multiple-input multiple-output (MIMO) layers in the basic version of the 5G-NR system is RRC configured in each cell, and the number of MIMO layers cannot be determined dynamically. Fast determination of the number of MIMO layers can help the terminal quickly switch the number of receiving antennas for the data reception.

19.1.2.4 Adaptive DRX

In both 4G-LTE and 5G-NR systems, a mechanism that allows terminal to turn off its receiver chain/unit from processing any signals and channels transmitted from the serving base station in the downlink, namely DRX,

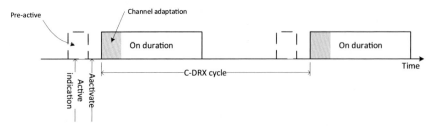

Figure 19.5 DRX adaptive switching DRX.

is supported as a power-saving feature. A DRX configured in a RRC-connected state is called C-DRX. Unless otherwise specified later, we all take C-DRX as DRX. A common DRX configuration is based on a timer control switch. The use of DRX is very simple, but it lacks a mechanism that is able to match with real-time data arrival. Therefore it is necessary to consider sending an indication to inform the terminal that there is downlink data arriving before the next DRX cycle starts. Upon receiving this indication/instruction to wake up, the terminal enters into a DRX ON period/state and starts to detect PDCCH. The manner in which the indication can be delivered to the terminal is using a dedicated wake-up signal or channel. During the DRX ON period, the terminal can also end/exit the DRX ON state in advance using power-saving triggering signal. It should be noted that the basic version of the 5G-NR system in R15 already supports a sleep indication based on MAC CE.

After the terminal receives the wake-up signal or the signal to wake up the channel, there is a certain preparation time before the start of DRX ON period. Initial measurements are desirably performed during this preparation time, as the traditional DRX mechanism often has a problem with untimely update of the wireless channel state. Due to terminal receiver being inactive during the DRX OFF period, past measurements of the wireless channel become outdated, and hence, unreliable. For the 5G-NR system, pretracking of beams transmitted from the base station is considered also part of the channel measurement. Some measurement signals can be configured during the preparation time. This definition of preparation time can also be extended to a few slots at the beginning of DRX ON.

19.1.2.5 Adaptive reduction of PDCCH monitoring

Out of all the power-saving schemes for terminals in 5G-NR, most of them are related to reducing the energy consumption corresponding to unnecessary PDCCH monitoring. A more direct power-saving technique would be to reduce the PDCCH monitoring itself. The monitoring of PDCCH was discussed as early as the initial study of 5G-NR. According to a measurement statistic collected during 4G-LTE deployments, PDCCH monitoring in terminal standby mode accounts for more than half of the daily communication power consumption. Because the terminal will conduct PDCCH blind detection and cache PDSCH symbols in every subframe, most of the slots actually have very little data or no data at all intended for the terminal.

In addition to DRX adaptation, PDCCH monitoring can be reduced by:

- Terminal dynamically interrupts PDCCH monitoring and skips PDCCH decoding for a period of time.
- Terminal is configured to perform fast switching of multi-CORESET/search space configuration.
- Physical-layer signaling indicates the number of blind detections for PDCCH.

19.1.2.6 User-assisted information reporting

Among the different power-saving techniques, factors and parameters that have influence on terminal power-saving would also depend partially on the implementation for the device terminal. For a device terminal implementation with parameters that are aligned with more power-saving techniques, it is expected this terminal device will achieve better power-saving. In order to fulfill this objective of achieving better power-saving, user assistance information can be used as terminal's recommendation to the base station. The base station can be configured based on the recommended parameters. The recommended user assistance information can include the processing time for K0 and K2, BWP configuration information, number of MIMO layers, DRX configuration, control channel parameters, etc.

19.1.2.7 Power-saving wake-up signal/channel

Solutions that are based on energy-efficient wake-up signal/channel are mostly built on the basic 5G-NR system design from R15. It includes a channel structure from the existing PDCCH design, extending the current TRS design, CSI-RS styled RS, SSS, and DMRS-based signal, and data-channel-based design. There are also newly introduced schemes such as sequence indication.

All of these signals or channels would be able to trigger terminal power-saving. However, further considerations that should be taken into account are signal resource efficiency, multiplexing capacity, terminal detection complexity, compatibility and coexistence with other signals or channels, and detection performance.

For the detection performance, there are multiple dimensions to consider. For the wake-up signals, a miss–detection rate of 0.1% and a false–alarm rate of 1% are generally required. The wake–up signal is related to the start of the subsequent DRX cycle, and the terminal will not try to detect control channel during the following DRX ON period when a miss–detection occurs. In this case, continuous loss reception of data should be avoided. On the other hand, false alarms will cause only a small increase in power consumption for the terminal, and as such, the related performance requirements are relatively low.

In addition to the consideration of detection performance, detection processing behavior for the terminal also requires some attention. If power-saving wake-up signal is not detected on the resources configured for the terminal, the terminal should be allowed to wake up at the start of the following DRX to reduce the impact of the miss–detection.

19.1.2.8 Power-saving assistance RS

The auxiliary RS is mainly for better synchronization, channel and beam tracking, channel state measurement, and measurement for wireless resource management. Compared with the existing measurement RS, power-saving auxiliary RS can also be used to help the terminal perform the power-saving process more efficiently and quickly. That is to say, the power-saving auxiliary RS should be sent before the power-saving process is performed by the terminal.

19.1.2.9 Physical-layer power-saving process

The aforementioned terminal power-saving techniques need to be combined with a corresponding terminal-side power-saving process. A variety of terminal power-saving techniques can be integrated in a power-saving process. Typically, different DRX adaptations are triggered by energy-efficient wake-up signals. Before starting a DRX, the auxiliary RS associated with power-saving wake-up signal allows the terminal to perform a quick measurement of the channel and beam tracking for some preprocessing work. The power-saving wake-up signal can also trigger BWP switching, MIMO layer adaptation, frequency premeasurement, and low-power processing mode in a terminal.

19.1.2.10 Higher-layer power-saving process

Enhancements that are to be bulit on top of the existing higher-layer process in 5G-NR are mainly considered. In a communication protocol, higher-layer power-saving processes and physical-layer power-saving techniques are always related to each other.

The baseline DRX cycle in R15 5G-NR does not support configuration of 10.24s. In R16, it is extended to 10.24s. The basic paging mechanism in 5G-NR, which supports paging of multiple terminals on a single resource, has a certain false-alarm rate that also needs to be optimized.

Effective switching between different RRC states (RRC_CONNECTED/RRC_IDLE/RRC_INACTIVE) can also contribute to terminal power-saving. The DRX mechanism is mainly defined in the MAC layer for the basic version of 5G-NR, and it needs to be combined with the power-saving wake-up signal of the physical layer. The purpose of the power-saving wake-up signal is to trigger MAC layer DRX ON duration timer. In addition, a time offset to the subsequent DRX cycle should be defined for the power-saving wake-up signal to allow a terminal processing time.

In addtion, a corresponding higher-layer process should also be defined for adapation of number of MIMO layers/number of antennas, reduced PDCCH detection, CA/DC power-saving, and assistance information reporting.

19.1.2.11 Power-saving in RRM measurement

Characteristics of RRM measurement in different RRC states may be different. Reducing unnecessary measurements can also help save energy for the terminal. Typically, the wireless channel changes slowly when the terminal is stationary or moving at a low speed. Therefore increasing the RRM measurment periodicity will have little impact on performance and accuracy.

Serving base station determines and sets corresponding RRM operations to achieve the purpose of power-saving for the terminal. The base station will require some information beforehand in order to determine the corresponding setting. Firstly, some information can be obtained directly by the base station, such as estimation of Doppler frequency shift and the cell coverage (e.g., macro or small cell). Secondly, the base station also uses assistance information reporting from the terminal, such as mobility management information, RS measurable by base station, and terminal measurement reporting. Then based on consideration of these information and thresholds, the base station can configure necessary measurements to effectively control energy consumption for the terminal.

19.1.3 Evaluation methodology for power-saving

Evaluation methodology for terminal power-saving in 5G-NR is based on a set of energy consumption models in Ref. [7].

The models of terminal energy consumption take into account all communication processing energy consumption factors. In order to facilitate comparison, the terminal energy consumption models introduce certain parameter assumptions. Corresponding to FR1, the subcarrier spacing (SCS) set for the evaluation is 30 kHz, a single carrier is used, carrier bandwidth is 100 MHz, duplex mode is TDD, and the uplink maximum transmission power is 23 dBm. Corresponding to FR2, the SCS for the evaluation work is 120 kHz, a single carrier, carrier bandwidth is 100 MHz, and the duplex mode is also TDD.

The power consumption model generally considers the average power consumption per slot. The power consumption value of a given terminal is shown in Table 19.1 for terminals in different states or processing different signals.

For the three sleep modes, there is a certain transition time for power consumption, and corresponding power consumption is defined according to Table 19.2.

In order to evaluate the changes in terminal power consumption under different configurations compared with a reference NR configuration, a power scaling model based on the reference NR configuration is also defined as follows:

- Receiving BWP with X MHz bandwidth, power scaling value $= 0.4 + 0.6 \, (X - 20)/80$, where $X = 10, 20, 40, 80$, and 100.
- Power scaling value for DL CA: $2CC = 1.7 \times 1CC$, $4CC = 3.4 \times 1CC$.
- Power scaling value for UL CA: $2CC = 1.2 \times 1CC$ when transmission power is 23 dBm.
- (RX) receive antenna: power scaling value for $2RX = 0.7 \times 4RX$ (applicable in FR1). Scaling value for $1RX = 0.7 \times 2RX$ (applicable in FR2).

Table 19.1 Terminal processing state and power consumption model.

Power state	Characteristic	Relative power FR1	FR2 (different to FR1)
Deep sleep	Time interval for sleep should be larger than the total transition time entering and leaving this state. Accurate timing may not be maintained.	1 (Optional: 0.5)	
Light sleep	Time interval for sleep should be larger than the total transition time entering and leaving this state.	20	
Micro-sleep	Immediate transition is assumed for power-saving study purpose from or to a nonsleep state	45	
PDCCH-only	No PDSCH and same-slot scheduling; this includes time for PDCCH decoding and any micro-sleep within the slot.	100	175
SSB or CSI-RS proc.	SSB can be used for fine time–frequency sync. and RSRP measurement of the serving/camping cell. TRS is the considered CSI-RS for sync.	100	175
PDCCH + PDSCH	PDCCH + PDSCH. ACK/NACK in long PUCCH is modeled by UL power state.	300	350
UL	Long PUCCH or PUSCH.	250 (0 dBm) 700 (23 dBm)	350

PDCCH, Physical downlink control channel; *PDSCH*, physical downlink shared channel; *PUCCH*, physical uplink control channel; *PUSCH*, physical uplink shared channel.

Table 19.2 Power state conversion time.

Sleep type	Additional transition energy: (relative power × ms)	Sleep type (ms)
Deep sleep	450	20
Light sleep	100	6
Micro–sleep	0	0

Table 19.3 Data traffic model.

	FTP traffic	Instant messaging	VoIP
Model	FTP model 3	FTP model 3	LTE VoIP.
Packet size	0.5 Mbytes	0.1 Mbytes	AMR (Adaptive
Mean interarrival time	200 ms	2 s	Multi–Rate) 12.2 kbps
DRX setting	Period = 160 ms Inactivity timer = 100 ms	Period = 320 ms Inactivity timer = 80 ms	Period = 40 ms Inactivity timer = 10 ms

LTE, Long-term evolution; *VoIP*, voice over Internet protocol.

- TX (transmit) antenna (FR1 only): $2TX = 1.4 \times 1TX$ (0 dBm) or $1.2 \times 1TX$ (23 dBm).
- PDCCH-only cross-slot scheduling $= 0.7 \times$ same-slot scheduling.
- When SSB is received, one SSB is 0.75 times the power of two SSBs.
- PDSCH only: 280 in FR1 and 325 in FR2.
- Power scaling value for short PUCCH $= 0.3 \times$ UL transmit power.
- Power scaling value for SRS $= 0.3 \times$ UL transmit power.

The evaluation of terminal power-saving techniques is done according to the power consumption model. The power-saving effect of different candidate techniques is analyzed and calculated using power consumption modeling. Link-level simulations and system-level simulations are also used as performance indicators for evaluation. A data traffic arrival model for system-level simulations also generates necessary power distribution for the power-saving analysis. The data traffic arrival model adopted can be categorized into three types as shown in Table 19.3.

Different types of slots can be calculated based on the data traffic arrival model and the power consumption of the corresponding slots can be obtained. Finally, the power consumption for the terminal can be calculated.

19.1.4 Evaluation results and selected terminal power-saving techniques

According to the above model, simulation platforms used by different companies were first calibrated, then various candidate techniques were evaluated. The amount of power-saving gain is the main evaluation objective for each technique. When comparing with the benchmark NR technology, terminal power-saving technology is expected to bring some performance loss, and the main loss is reflected in the User Perceived Throughput (UPT). In addition, there could also be a loss in end-to-end data latency. The purpose of the evaluation is to confirm that the terminal does not significantly cause performance losses when energy is saved.

Various power-saving techniques are evaluated under the aforementioned methodology [7].

19.1.4.1 Power-saving technology evaluation results

For evaluation of various power-saving techniques in terms of frequency-domain adaptation, a power-saving gain of 16%−45% for the terminal is observed from the evaluation of BWP switching. A gain of 12%−57.5% is observed from SCell adaptation operation, but the data delay is increased by 0.1%−2.6%.

In terms of time-domain adaptation, a power-saving gain of up to 2%−28% has been observed in the evaluation for cross-slot scheduling. However, the UPT is declined by 0.3%−25%. UPT is often related to the cross-slot scheduling offset, and the larger the offset, the smaller the perceived throughput. For same-slot scheduling, a power-saving gain of 15% can also be achieved by increasing the time-gap interval between the control and data symbols. However, the problem of resource fragmentation caused by same-slot scheduling with a time gap will bring up to 93% of resource overhead. Multislot scheduling is observed to bring less than 2% of power-saving gain.

In terms of spatial domain, the dynamic adaptation of number of MIMO layers or number of antennas can bring up to 3%−30% power-saving gain. At the same time, an increase of 4% in latency is observed in that evaluation. The semistatic antenna adaptation can save energy by 6%−30%, but it has an obvious delay and throughput loss. In addition, in order to compensate for a small number of transmitting antennas employed at the terminal, from the base station side it needs to configure more control and data resources for transmitting the same information.

Adaptation in the DRX domain showed a power-saving gain of 8%—50%. These gains are based on a reference DRX configuration in the evaluation assumption. A delay in data delivery is increased by 2%—13%. However, due to insufficient optimization in the design for DRX configuration in the basic version of the 5G-NR system, it increased the energy consumption by 37%—47% under the traffic model given for the simulation evaluation.

Dynamic PDCCH detection adaptation achieved a power-saving gain by 5%—85%. The loss in data delay and throughput was 0%—115% and 5%—43%, respectively.

In the evaluation, the power-saving wake-up signal is mainly used to trigger the DRX adaptation in a terminal. For some evaluations, the power-saving wake-up signal is also used to trigger the switch of BWP and PDCCH monitoring.

Due to some differences in the actual network configuration and terminal implementation, the reporting of user assistance information for power-saving is not directly reflected in the evaluation. However, based on the analysis results of the higher-layer process, it is still considered provide great benefit to terminal energy conservation.

Power-saving gain in the RRC_CONNECTED state can be up to 7.4%—26.6% using the time-domain adaptation and relaxation of RRM measurement. The observed power-saving gain in the RRC_IDLE and INACTIVE states can reach 0.89%—19.7%.

Power-saving gain in the RRC_CONNECTED state can reach up to 1.8%—21.3% when the RRM measurement is made adaptive and the requirement is relaxed in the frequency domain. The observed power-saving gain in the RRC_IDLE and INACTIVE states can reach 4.7%—7.1%.

The evaluation for additional allocation of RRM measurement resources also showed that up to 19%—38% of power-saving gain can be obtained.

In the analysis for power-saving using higher-layer paging process with a support of up to 10.24 seconds for a DRX cycle showed it can also provide power-saving effect for the terminal. Some enhanced rules for the RRC_IDLE and INACTIVE states can also save energy.

19.1.4.2 Power-saving techniques introduced in 5G-NR

According to the results of evaluation and analysis, 3GPP has selected several terminal power-saving techniques to enhance the basic version of 5G-NR [8].

The basis of the frequency-domain adaptation power-saving technique is the BWP framework adopted for the 5G-NR radio interface. The configuration of BWP(s) itself has already taken into account energy conservation in the design. Some optimized BWP configurations are adopted and measurement RSs are introduced to improve efficiency in BWP switching. However, no evidence was shown during the evaluation that these measurement RSs are needed for the BWP to obtain power-saving gains. To achieve optimization for measurement, corresponding measurements can be additionally configured, not necessarily bundled with the BWP switching mechanism. The enhancement of the BWP framework itself, such as the reduction of BWP switching delay based on downlink control information (DCI), has not been proven to be necessary.

Furthermore, for the frequency-domain adaptation, secondary carrier (SCell) adaptation has shown to provide a certain gain, and the cost to the performance is marginal. The adaptation of the auxiliary carrier for power-saving is undertaken as part of the carrier aggregation enhancement work for 5G-NR.

In terms of time-domain adaptation, the same-slot scheduling used in the basic version of 5G-NR does not need to be modified, since the enhancement for the same-slot scheduling evaluated will only reduce resource utilization for the system. For the cross-slot scheduling, since multiple terminals can be scheduled in different slots, it will not cause any loss in resource utilization for the base station.

In terms of spatial-domain adaptation, RX antenna adaptation has shown to provide power-saving gains. However, the number of antennas is not directly defined in the 5G-NR protocol. Instead, the number of MIMO layers is defined. Therefore only the enhancement for the number of MIMO layers is considered.

Considering changes in RRM measurement will cause impact on network performance, enhancements introduced to the basic version of 5G-NR only include extending the measurement interval and the corresponding trigger conditions.

Finally, terminal power-saving for 5G-NR system introduces a power-saving wake-up signal triggered DRX adaptation, cross-slot scheduling, MIMO layer configuration based on BWP, secondary carrier (SCell) dormancy, terminal assistance information reporting, and RRM measurement relaxation.

The power-saving wake-up signal for DRX adaption is defined as a special channel, which is mainly to be used for triggering PDCCH detection in the following DRX ON duration. The definition for the power-saving wake-up signal is based on PDCCH. The concept of DCI format, search space, and control resource aggregation from R15 are reused. Reception of the power-saving wake-up signal needs to be demodulated and polar decoded. Since the timing granularity for switching of cross-slot scheduling is finer than that for DRX, the trigger for cross-slot scheduling is not transmitted in the power-saving wake-up signal. The DRX triggering by the power-saving wake-up signal is adaptively performed by a corresponding higher-layer process, including the MAC entity.

In the R15 basic version of 5G-NR system, MIMO layer configuration is only supported at the cell level. A new MIMO layer configuration at the BWP level is introduced in R16 to allow a terminal to assume data channels with fewer layers when switching to a different BWP. For the implementation, this means that the terminal is also able to operate with a smaller number of receiving antennas in the BWP with fewer layers.

In R16, a higher-layer mechanism is introduced in the 3GPP standard for terminal reporting of expected parameters that are optimized for operating in the RRC_CONNECTED state. Additional higher-layer mechanisms are also introduced allowing the terminal to report the expected parameters for DRX, BWP, and SCell configurations. For these configurations, different configuration values are related to the terminal's RRC state and the terminal hardware implementation. Then the reporting from different terminals allows the base station to configure suitable power-saving parameters.

RRM measurement relaxation in higher layers is mainly limited to the RRC_IDLE and INACTIVE states. Parameters include longer measurement periods, reduced number of measurement cells, and reduced carriers for measurement. Relaxed triggering conditions for measurement is based on the terminal not being at the cell edge, but in a fixed position or in a low-mobility state. The higher layer also defines the corresponding conditions and thresholds in order to determine the state of these terminals.

A SCell enhancement scheme is introduced as part of the carrier aggregation enhancement work, for which the main purpose is to trigger SCell dormancy. SCell dormancy can be triggered by the power-saving wake-up signal before a DRX cycle (i.e., nonactivation time). Secondary cell (carrier) dormancy can also be triggered by a PDCCH during the activation time.

The following sections of this chapter describe the specific power-saving techniques adopted in R16 of 5G-NR.

19.2 Power-saving signal design and its impact on DRX

19.2.1 The technical principle of power-saving signal

Based on the previous analysis, since the power consumption in an RRC_CONNECTED state accounts for an important and major part of a 5G terminal's power consumption, a power-saving signal is introduced in R16 for the purpose of power-saving when the terminal is in the RRC_CONNECTED state.

The traditional terminal power-saving mechanism is mainly based on the configuration and use of DRX. When DRX is configured, the terminal monitors PDCCH during the DRX ON duration. If a data scheduling DCI is detected during the DRX ON duration, the terminal continuously monitors PDCCH controlled by a DRX timer until the data transmission has completed. Otherwise, if the terminal does not detect a DCI during the DRX ON duration, the terminal enters the DRX OFF period (discontinuous reception) for power-saving. As can be seen, the DRX is a power-saving mechanism with DRX cycle as the timing granularity, as such, the optimal control for the terminal power consumption is not realized. For example, even if the terminal has no data scheduled, the terminal still needs to periodically perform blind detection of the PDCCH when the DRX ON duration timer starts, so there is still power wastage (Fig. 19.5).

In order to realize further power-saving for the terminal, a power-saving signal is introduced. The power-saving signal is used in conjunction with the DRX power-saving mechanism. The technical principle is that the terminal receives a power-saving signal before the configured DRX ON duration, as shown in Fig. 19.6, when the terminal has data transmission in a DRX cycle, the power-saving signal "wakes up" the terminal to monitor PDCCH during the DRX ON duration. Otherwise, when the

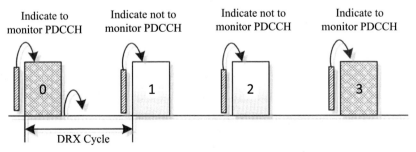

Figure 19.6 Wake-up mechanism.

terminal has no data transmission in a DRX cycle, the power-saving signal does not "wake up" the terminal, and the terminal does not need to blindly detect PDCCH during the DRX ON duration. Compared with existing DRX mechanism, when the terminal has no data transmission, the terminal can omit PDCCH monitoring during the DRX ON duration, and therefore achieving power-saving for the terminal. The time duration before the DRX ON duration is called the inactive time. The time duration in the DRX ON duration is called the active time.

In addition, when the terminal is operating in a carrier aggregation mode or dual-connectivity mode, since traffic load of the terminal may fluctuate over time, the number of carriers needed for data transmission may also change with time. However, in the current design the number of carriers for data transmission can only be changed by RRC configuration/reconfiguration, or carrier activation/deactivation with MAC CE. It usually takes a long time for the terminal and it is not able to match changes in data traffic. Consequently, when there is an insufficient number of active carriers for the terminal to transmit its data traffic, more carriers will need to be activated, which will increase the data transmission delay. On the other hand, if the number of activated carriers is more than the required and excess carriers cannot be deactivated timely when the terminal has no or less data to transmit, it will result in power consumption wastage.

In order to facilitate a fast adjustment in the frequency-domain adaptation for power-saving, the concept of SCell dormancy is introduced in 3GPP Released 16. With this mechanism, when the terminal has no data transmission, some of its configured SCells remain in the "active" state but a dormant BWP could be used on these carriers without the terminal monitoring PDCCH and receiving PDSCH. SCell dormancy and dormant BWP are described in Section 19.5.

The power-saving signal may also be used to indicate the SCell dormancy operation. Furthermore, the power-saving signal can be dynamically transmitted to the terminal, and thus the timing delay for the indication is reduced compared with RRC configuration/reconfiguration and MAC CE. As such, it can provide timely and accurate control of power consumption for the terminal.

19.2.2 Power-saving signal in R16

As described in the previous section, the purposes of the power-saving signal are to indicate whether to wake up to monitor PDCCH and SCell dormancy operation for the terminal. The indication to wake up the terminal requires only 1 bit. If the value of the indication bit is "1", the terminal needs to wake up to monitor PDCCH. Otherwise, if the value of the indication bit is "0," the terminal does not need to wake up to monitor PDCCH. In 5G-NR, a terminal can be configured with up to 15 secondary carriers/SCells. If one indication bit is used for each SCell, the maximum number of indication bits for SCell dormancy is 15, for which the overhead is considered to be high. As such, the concept of SCell grouping is adopted. In SCell grouping, the SCells are divided into at most five groups, and each group corresponds to an indicator bit. If the value of the indication bit is "1", the terminal should operate in a nondormant BWP on all the corresponding SCells. That is, if the terminal is operating in nondormant BWP before receiving the power-saving signal, the terminal remains in the nondormant BWP. If the terminal is operating in a dormant BWP on a SCell, the terminal needs to switch to a nondormant BWP. Similarly, if the value of the indication bit "0", the terminal should operate in a dormant BWP on all the corresponding SCells. It should be noted that, as described in Section 19.5, there are other DCI formats that can also trigger SCell dormancy operation during the DRX active time. The SCell group for power-saving signal and SCell group within the DRX active time are configured independently. Therefore in the power-saving signal design, the number of bits required for a single user is at most 6, including 1 wake-up indication bit and up to 5 SCell dormancy indication bits.

In the following, we need to solve the problem of how to deliver the terminal's power-saving indicator bits in a DCI. Obviously, if a DCI is allowed to carry only the power-saving indication bit for a single terminal, the transmission efficiency would be low. Firstly, it needs 24 bits of CRC parity bits for the DCI itself. Secondly, when the total number of bits for the DCI format is less than 12, the efficiency of polar coding would be low. Therefore it should be allowed to carry indication bits for multiple terminals by a power-saving signal to improve resource utilization efficiency. As shown in Fig. 19.7, the network only needs to inform each user its starting position for the power-saving indication bit in the DCI. The number of SCell dormancy indication bits can be implicitly obtained from the configured number of SCell groups for each terminal (note that wake-up indication bit will always appear, and the number of SCell dormancy indication bit can be 0). Furthermore, the network will also inform the terminal the DCI size and the PS-RNTI for the

Wake-up Scell dormancy

| UE 1 | UE2 | UE n |

Figure 19.7 DCI format of DCI 2—6. *DCI*, Downlink control information.

PDCCH. The DCI format specified for the power-saving signal in R16 is DCI format 2_6.

19.2.2.1 PDCCH monitoring occasion for power-saving signal

Similar to other PDCCHs, the PDCCH for the power-saving signal is also to be detected in a configured PDCCH search space. To support multibeam transmission of power-saving signals, up to three PDCCH CORESETs can be supported and the same MAC CE update mechanism as for R15 PDCCH should be followed. To reduce power consumption for the terminal, the aggregation levels and the number of PDCCH candidates for each aggregation level are configurable.

PDCCH monitoring occasions for the power-saving signal are discussed in detail by the 3GPP. The first problem is the starting position for the PDCCH monitoring. Since the power-saving signal is located before the DRX ON duration, the starting position for the power-saving signal detection can be obtained by a time offset relative to the starting position of the DRX ON duration. During the standards discussion at the 3GPP, two methods to obtain the PS-offset were considered:

- Option 1: the PS-offset is configured by explicit signaling.
- Option 2: the PS-offset is obtained by the PDCCH search space.

In the first option the network directly configures a PS-offset. In the second option, it does not require an explicit configuration but the terminal implicitly obtains the PS-offset from the PDCCH search space configuration. For example, after configuring the PDCCH search space for a terminal, the latest PDCCH monitoring occasion prior to and closest to the DRX ON duration can be taken as the PDCCH monitoring occasion. Alternatively, the period of the PDCCH search space can be configured to be the same as that of the DRX cycle. And an appropriate time offset should be set so that the PDCCH monitoring occasion is located before the DRX ON duration. Both options will result in reasonable PDCCH monitoring occasions for the power-saving signal. However, for the second option, because the PDCCH period supported by the existing

Table 19.4 Minimum time interval.

Subcarrier spacing (kHz)	Minimum time interval (slots)	
	Value 1	Value 2
15	1	3
30	1	6
60	1	12
120	2	24

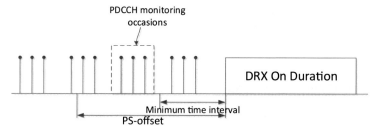

Figure 19.8 PDCCH monitoring occasion for power-saving signal. *PDCCH*, Physical downlink control channel.

protocol does not match the numerical range of the DRX cycle, from the perspective of easier standardization, the first option is chosen. That is, explicit signaling is used to configure the PS-offset.

After determining the starting time location for the PDCCH monitoring occasion, it is also necessary to further determine the end point in time for the PDCCH monitoring occasion, which is to be determined by the capability of the terminal. The terminal needs to perform operations such as device wake-up and initialization after wake-up within a minimum time gap before the DRX ON duration, so the terminal does not need to detect power-saving signals within a minimum time interval gap before the DRX ON. Terminals with faster processing speed can use a shorter minimum time interval (see value 1 in Table 19.4), while terminals with slower processing speed need to use a larger minimum time interval (see value 2 in Table 19.4).

Therefore the time position indicated by the PS-offset time is the earliest starting point for detecting the power-saving signal, and the power-saving signal should be monitored within the full duration of a PDCCH search space (defined by the parameter "duration" of the PDCCH search space), but it should not be detected during the minimum time interval.

As shown in Fig. 19.8, the power-saving signal is monitored within the dashed box.

19.2.2.2 Power-saving signal for short DRX cycle

An important aspect of power-saving signal design during the standardization discussion was about whether the power-saving signal could be applied to both long DRX and short DRX, or only one of them. As the name suggests, the long DRX has a longer DRX cycle and subsequently a longer DRX ON duration/period is configured, so that more terminal power-saving gain can be obtained when the power-saving signal is applied to a long DRX. In addition, long DRX generally has a regular periodicity, which is beneficial to multiplex power-saving signals for multiple users in the same DCI. And thus the overhead of power-saving signals is effectively reduced. So therefore the power-saving signal should be applied to long DRX.

During the discussion at 3GPP, however, there were different opinions on whether the power-saving signals should be used for short DRX. On the one hand, companies that support short DRX believe that short DRX can also be configured with larger DRX cycles, so power-saving signals can bring in additional power-saving gains in these cases. On the other hand, companies opposed to supporting short DRX hold the view that short DRX generally has a short cycle and can already achieve better power-saving, and the incremental power-saving effect brought by further use of power-saving signals is not obvious. In general, short DRX is triggered by data scheduling based on random arrival, and as such, short DRX does not always appear periodically in time, which makes it difficult to multiplex power-saving indication for multiple users within one power-saving signal.

After repeated discussion, it was in the end decided by the 3GPP that power-saving signals are only supported for long DRX in R16.

19.2.2.3 Whether to monitor power-saving signal during DRX active time

The power-saving signal is periodically configured before the DRX ON duration, so the terminal usually monitors the power-saving signal outside the DRX active time. However, in some cases there could be continuous data scheduling during a DRX cycle. Therefore even if the DRX ON duration timer expires, the terminal may have already started the DRX-inactivity timer due to data scheduling. With continuous scheduling of data from the base station, the DRX-inactivity timer may still be running at the time when the terminal is due to monitor a power-saving signal before the next DRX ON duration. That is, the terminal is still in DRX active time. Therefore it is necessary to specify whether the terminal needs to detect the power-saving signal during DRX active time.

Considering that the terminal may still have data transmission during the DRX active time, the DRX-inactivity timer can effectively control whether the terminal needs to continue to monitor PDCCH. Therefore it is not necessary to further control PDCCH monitoring using the power-saving signals. Therefore it is finally determined in 3GPP that the terminal does not need to monitor any power-saving signal during DRX active time.

19.2.2.4 Terminal behavior after detection of power-saving signal

The terminal is required to receive and detect power-saving signals based on a configuration from the base station. When the terminal detects a power-saving signal, it determines whether to wake up and monitor PDCCH based on the indication from the corresponding bit carried in the power-saving signal. For example, if the value of the wake-up indication bit is "1", the physical layer of the terminal sends an indication to an upper layer (the MAC layer) to start a DRX ON duration timer and monitors PDCCH while the timer is running. If the value of the wake-up indication bit is "0", the physical layer of the terminal sends an indication to the MAC layer to not start the DRX ON duration timer and it will not monitor PDCCH.

There are also some cases where the terminal does not need to detect the power-saving signal. These are defined in the next section. For these cases, the physical layer of the terminal should report to the MAC layer to fall back to a traditional DRX mode (i.e., to start the DRX ON duration timer as normal for PDCCH monitoring).

There can be also some abnormal conditions that will cause the terminal to fail to detect the power-saving signal. For example, due to sudden deterioration of the wireless channel or for other reasons, the power-saving signal may be not detected by the terminal. It may be because the instantaneous load of the network is too large and there is not enough PDCCH resources for sending the power-saving signal. In these cases, whether the terminal starts the DRX ON duration timer for PDCCH detection is configured by higher-layer signaling.

19.2.3 Impact of power-saving signal on DRX

One of the main functions of the power-saving signal is indicating to a terminal whether to wake up and start the drx-onDurationTimer in the subsequent DRX cycle, and also for the terminal to start monitoring PDCCH in the downlink [9]. In other words, the power-saving signal mainly affects the starting of the drx-onDurationTimer. Besides the drx-onDurationTimer, the power-saving signal has no impact on the operation of other timers. Therefore the power-saving signal is only used in conjunction with DRX. Only the terminal configured with DRX function can be configured with power-saving signal, as such.

During the study phase of the terminal power-saving work in 3GPP, two candidate options were considered for waking up the terminal based on the power-saving signal [10]:

- Option 1: The terminal decides whether to wake up based on reception of the power-saving signal. That is, if the terminal receives the power-saving signal before the next occurrence of drx-onDurationTimer, the terminal starts drx-onDurationTimer at the following occurrence of drx-onDurationTimer. Otherwise, the terminal does not start drx-onDurationTimer at the following occurrence of drx-onDurationTimer.

- Option 2: The terminal decides whether to wake up according to an explicit indication in the power-saving signal. That is, if the terminal receives a power-saving signal before the next occurrence of drx-onDurationTimer and the power-saving signal indicates the terminal to wake up, the terminal then starts the drx-onDurationTimer at the following occurrence of drx-onDurationTimer. If the terminal receives the power-saving signal before the next occurrence of drx-onDurationTimer and the power-saving signal indicates not to wake up, the terminal does not start the drx-onDurationTimer at the following occurrence of drx-onDurationTimer.

Since the power-saving signal is designed based on PDCCH, there is a certain probability of PDCCH misdetection. In the case that the terminal mis-detects the power-saving signal, according to the above Option 1 it may cause the terminal to further mis-detect scheduling signals sent from the network. That is, if the network sends a power-saving signal to the terminal but the terminal does not detect the power-saving signal, the behavior of the terminal in this case is not to start the drx-onDurationTimer at the following occurrence of drx-onDurationTimer. And therefore the terminal would not monitor any downlink PDCCH from the network during the DRX ON period. Then, if the network schedules the terminal during the DRX ON period, the terminal will not receive the scheduling PDCCH sent from the network and the scheduling performance would be affected as such. In addition, from the perspective of saving PDCCH resource overhead, the power-saving signal may be terminal specific or terminal group—specific. If a power-saving signal is shared by a group of terminals, the wake-up indication for different terminals in the same group may be different. Then it is obvious that the above Option 1 cannot realize the function of waking up some terminals while not waking up other terminals in a group. Considering the above two reasons, Option 2 was finally adopted.

In a different case, when the terminal monitors but does not detect the power-saving signal, terminal behavior can be configured by the upper layer. More specifically, if the terminal does not detect the power-saving signal before the next occurrence of drx-onDurationTimer, whether the terminal starts the drx-onDurationTimer at the following occurrence of drx-onDurationTimer can be configured by the network. If the network does not configure the behavior for the terminal, the default behavior is not to start the drx-onDurationTimer at the following occurrence of drx-onDurationTimer.

As described in Section 19.2.2, in the time domain, the terminal monitors power-saving signals in PDCCH monitoring occasion(s) located over a period of time prior to the occurrence of the drx-onDurationTimer. A maximum time offset of the power-saving signal relative to the next occurrence of the drx-onDurationTimer is configured by network. In addition, the terminal reports to the network a minimum time offset based on its processing capability. Thus the terminal monitors power-saving signals within the time between the maximum and minimum time offsets before the next occurrence of the drx-onDurationTimer, as shown in Fig. 19.9. In the frequency domain, the power-saving signal is configured per MAC entity and only configured on the PCell and PSCell.

The terminal does not monitor power-saving signals in the following cases:
- The PDCCH monitoring occasion of power-saving signal is overlapped with the DRX Active Time.
- The PDCCH monitoring occasion of power-saving signal is located during BWP switching.
- The PDCCH monitoring occasion of power-saving signal is located during measurement gap.

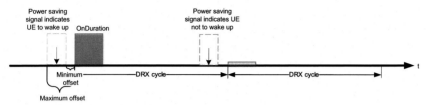

Figure 19.9 Power-saving signal impact on DRX.

When the terminal does not monitor power-saving signal in the above cases, the terminal will start the drx-onDurationTimer at the following occurrence of drx-onDurationTimer.

In R15 version of 5G-NR, for a terminal configured with DRX function, the terminal only sends periodic/semicontinuous SRS and periodic/semicontinuous CSI reports during DRX Active Time for the purpose of power-saving. The terminal sends SRS to facilitate channel estimation and frequency selective scheduling in the uplink by the network. The terminal also reports CSI to the network to facilitate downlink frequency selective scheduling as well as beam management for the terminal. In R15, for terminals configured with DRX function, since the terminals will periodically start a drx-onDurationTimer to enter into the DRX Active Time, the terminals can be configured with an appropriate CSI reporting period so that it can report the CSI to the network in every DRX cycle. After introducing the power-saving signal, if the terminal does not have uplink and downlink services for a long time, the network may not wake up the terminal, and will remain in the DRX Inactive Time for a long time. If the terminal does not report CSI during this long time, the network cannot adequately monitor the beam quality for the terminal, and in a more serious case, beam failure may occur. Considering the tradeoff between power-saving for the terminal and beam management for the network, whether the terminal should perform periodic CSI reporting during the period when the terminal should have started but did not start the drx-onDurationTimer due to power-saving signal is configurable by the network.

19.3 Cross-slot scheduling

19.3.1 Technical principles of cross-slot scheduling

Cross-slot scheduling is similar to time-domain adaptation. It belongs to the same category as DRX adaptation in principle. The difference is that cross-slot scheduling ensures the scheduling PDCCH and the scheduled

PDSCH/PUSCH are separated in the time domain to avoid processing time overlapping at the terminal. In the basic version of 5G-NR in R15, a timing offset between the scheduling PDCCH and the scheduled PDSCH/PUSCH in unit of slots is controlled by parameters K0/K2, respectively.

Same-slot scheduling refers to the case when the scheduling PDCCH and the scheduled PDSCH/PUSCH are colocated in the same slot. Same-slot scheduling of the control channel and data channel will result in time overlapping of decoding and receiving at the terminal side. As shown in Fig. 19.10, the receiver does not know in advance whether there is any data scheduled for the terminal before the control channel is decodded. Therefore the terminal will need to store signals or sample points for several OFDM symbols after the control channel covering the entire active BWP bandwidth. Only when the PDSCH scheduling information is decoded in PDCCH can the terminal then proceed with decoding of PDSCH RB occupied in the BWP.

When the scheduling PDCCH and the scheduled PDSCH/PUSCH are in different time slots, for which is commonly known as cross-slot scheduling, as mentioned earlier the time overlap for demodulating control channel and data channel can be avoided. As illustrated in Fig. 19.11, since there is a long offset between the control and the corresponding data channels, it is not required for the terminal to receive and buffer the entire BWP of signals and samples of data channel before the control channel is decoded. During the time offset period/interval between the control and data channels, receiver processing can be greatly simplified for

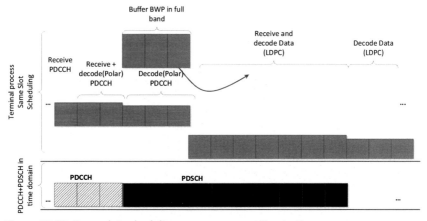

Figure 19.10 Same-slot scheduling power consumption in time.

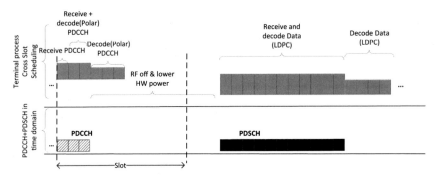

Figure 19.11 Cross-slot (K0 = 1) scheduling power consumption in time.

the terminal and the RF receiving module can be turned off to reach the energy consumption state of micro-sleep or light sleep.

In an extreme configuration of PDCCH and PDSCH in the frequency domain, the terminal will need to buffer the entire active BWP from the first symbol in a slot. The unnecessary power consumption in this case will be even higher.

For a typical data service, the data is usually scheduled using only a small portion of slots. In a real network, only about 20% of the total subframes are scheduled by PDCCH. However, for the case of same-slot scheduling, even when no data is scheduled in a slot for a terminal, the terminal will still perform some postprocessing for BWP caching after the control channel. Therefore statistically speaking, same-slot scheduling has a significant impact on the average power consumption for the terminal. However, when comparing the same-slot scheduling in Fig. 19.10, in the case of cross-slot scheduling in Fig. 19.11, the unnecessary postprocessing power consumption is minimized.

In various hardware implementations, the postprocessing time may be different from a terminal to another. However, terminal and chip vendors generally believe that one slot length is sufficient for the control channel decoding. As shown in Fig. 19.11, the same amount of data processing in cross-slot scheduling allows the RF chain to shut down and avoids data buffering, but at a cost of extra one-slot delay for the data service.

When the slot offset is greater than 1, the receiver terminal can further optimize its energy consumption. At the time of RF shutdown, the clock frequency and the level of hardware processing can be also reduced, leading to lower energy consumption. In Fig. 19.12, an optimization in power consumption is illustrated for the case of K0 = 2. When the

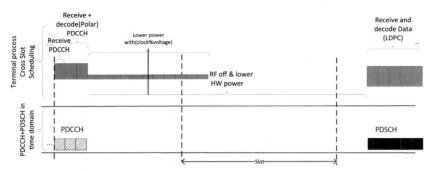

Figure 19.12 cross-slot (K0 = 2) scheduling power consumption in time.

processing timing requirement is relaxed, the encoding and decoding speed can be obviously slowed down for the terminal. The power consumption of a typical baseband digital processing circuit is roughly proportional to the voltage square. But the processing time is inversely proportional to the voltage. Therefore the cumulative power consumption for the same amount of coded data will be reduced. The design and the power consumption reduction range are also related to a specific implementation of hardware products. But achieving lower power consumption still generally depends on longer processing time among different terminal implementations.

The above diagram mainly illustrates a scheme of power-saving enhancement in downlink data scheduling. For uplink data transmission, there are two main considerations for power-saving that are different for cross-slot scheduling. Similar to LTE, uplink scheduling DCI and downlink scheduling DCI can be independently transmitted within the control channel search space in 5G-NR. The terminal does not have any prior knowledge about the scheduling DCI and the type of scheduling until the control channel intended for the terminal is decoded. The terminal also does not differentiate uplink or downlink scheduling in the power enhancement processing for DCI detection. Since as mentioned in Section 19.1.2 that DCI decoding speed can be reduced by adjusting parameters, K2 needs to be increased accordingly. Moreover, for uplink cross-slot scheduling, the preparation process for PUSCH transmission after detecting the control channel can be separated from the power-saving process. Reducing the uplink data preparation speed can also enhance the power consumption for the uplink data processing module. For these reasons, K2 needs to be configured to be greater than 1 and configured separately from K0. Terminal preferred Cross-slot parameters

K0 and K2 that are preferred by the terminal can also help the base station to set appropriate slot offsets for the terminal.

19.3.2 Flexible scheduling mechanism for cross-slot scheduling

As mentioned earlier, in the basic version of 5G-NR in R15, a timing offset K0/K2 between the scheduling PDCCH and the scheduled PDSCH/PUSCH can be respectively configured by the base station. A set of time offsets for the data channel is captured by a time-domain resource allocation (TDRA) table [11]. For both PDSCH and PUSCH scheduling, separate tables are configured for time-domain resources. For the flexible use of resources, each TDRA is associated with a data mapping type, slot offset K0/K2, start symbol S, and symbol number L. The K0 and K2 parameters determine whether or not cross-slot scheduling is applied. The TDRA table can configure up to 16 values, all of which are independent of each other. In the scheduling DCI, the table index indicates time-domain resource scheduling parameters that will be used for the scheduling. Specifically, the number of slots between the data is scheduled relative to the slot where the scheduling DCI is received, and also the starting and ending symbols in the data slot. The goal of terminal power-saving is achieved by configuring TDRA parameters with offset values greater than one slot (Table 19.5).

In the basic version of 5G-NR in R15, multiple TDRA parameters can be flexibly configured for the terminal, including both same-slot scheduling and cross-slot scheduling. During terminal blind detection of scheduling DCI in PDCCH, since the terminal does not have prior knowledge of any scheduling information and whether a cross-slot or same-slot scheduling is used until the DCI is decoded, the terminal must proceed as if the same-slot scheduling is used at each detection DCI opportunity and make necessary preparations accordingly. Therefore in the NR the cross-slot enhancement scheme introduces a dynamic DCI to instruct users to disable or enable all TDRA parameters with slot offset

Table 19.5 TDRA parameter.

Index	PDSCH mapping type	K_0	S	L
Index	PUSCH mapping type	K_2	S	L

PDSCH, Physical downlink shared channel; *PUSCH*, physical uplink shared channel; *TDRA*, time-domain resource allocation.

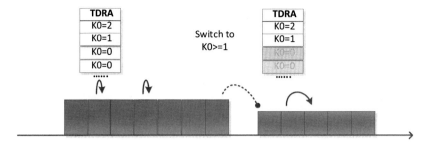

Figure 19.13 Cross-slot switching.

that is less than a certain limit. In Fig. 19.13, the minimum threshold is enabled and the minimum K0 value is 1 in this example.

For the basic version of 5G-NR system, K0 > 1 can also be configured semistatically for all of the PDSCH time-domain resource allocation entries in the table to ensure the preparation and processing of downlink data can be always performed in a cross-slot manner. By doing so, the terminal can operate semistatically in a longer delay mode for data, but this operation would not be able to catch up to rapid changing data services. Since there is no special enhancement for terminal power-saving in the basic version of 5G-NR, this power-saving mode is actually implementable under the basic version of the 5G-NR standard, but it is up to each terminal design. That is to say, even if all K values of the TDRA table configured by base station are greater than 1, the benefits of power-saving still cannot be guaranteed for all terminals.

19.3.3 Processing of dynamic indicating cross-slot scheduling

A series of processing mechanisms that enable dynamic indication cross-slot are introduced as the main enhancements for cross-slot scheduling in 5G-NR. The configuration of TDRA parameters is still largely based on the mechanism used in the basic version of 5G-NR. The parameter entries for K0/K2 in the TDRA table that are less than a preconfigured value can be dynamically enabled or disabled via the indication. This dynamic indication method is compatible with the data resource allocation framework in the time domain of the 5G-NR basic version. In uplink scheduling DCI format 0_1 and downlink scheduling DCI format 1_1, a bit field is added to indicate the minimum K0/K2 value that is applicable for the cross-slot scheduling. The minimum K0/K2 value is different from the K0/K2 indicated in TDRA. In the reminder of this chapter, the expressions for minimum K0/K2 are denoted as $minK_0/minK_2$.

Based on network configuration, combinations of minimum K0 and K2 values can be jointly indicated by DCI accordingly. The minimum K2 value for the PUSCH data preparation time is different from the minimum K0 value for PDSCH data decoding, so they need not be the same. The minimum K0/K2 values currently specified are not only applicable to the DCI format 0_1 and 1_1, but also applicable to other DCI formats in the special search space for the control channel. For DCI formats in common search space type 0/0A/1/2 and configured with the default TDRA table and beam recovery search space, the minimum K0/K2 values are not applicable. In this case, the control channel is decoded according to the requirement of same-slot scheduling.

In the dynamic indication mode, changing of the minimum K0/K2 value can be indicated in each slot. When the data arrival rate is high, the base station can immediately indicate a smaller minimum K0/K2 value for the terminal, or use the same-slot scheduling without a restriction. When the base station finds that the terminal data is inactive, the terminal is switched/indicated to use a larger minimum K0/K2 value.

The process of dynamic switching to cross-slot scheduling for a terminal is shown in Fig. 19.14. Taking scheduling for PDSCH as an example, in the figure the terminal is configured as follows: when the indicated state for cross-slot scheduling is 1, the minimum K0 value is assigned to 1;

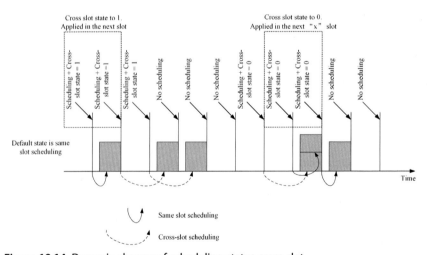

Figure 19.14 Dynamic change of scheduling status cross-slot.

when the indicated state for cross-slot scheduling is 0, the minimum K0 value is assigned to 0. It can be seen that the changes in cross–slot scheduling can be indicated in each slot. However, the cross–slot scheduling state indicated in the DCI does not take any immediate effect, but it goes through a delay of X. In Section 19.3.4, the determination for X will be explained in detail.

19.3.4 Application timing in cross-slot scheduling

When dynamic DCI is used, there is a small delay from the receiving of DCI to the applying of newly indicated cross-slot scheduling. Especially when the original minimum K value in effective for the terminal is larger, its processing of the control channel may take more time. Thus the terminal will take some time to update based on the new DCI indication to apply the new minimum K value. The newly indicated minimum K in cross-slot scheduling will generally take no less than one slot to be applies. Another factor is the minimum K value that was originally assumed by the terminal, which can be looked as the terminal's upper bound for the control channel decoding. RF hardware circuits in the terminal also need this time to lower or increase its processing capability. The application timing for cross-slot scheduling is in principle the maximum of a constant value and the previous value of the minimum K value.

In general, the control channel processing time of a terminal does not vary due to changes in SCS. Since the absolute time of a slot length reduces with an increase in SCS, the timing (number of slots) in which cross-carrier scheduling can be applied for a terminal will depend on the SCS in a carrier.

The minimum K0/K2 value for PDSCH/PUSCH processing and the TDRA table are configured with respect to the scheduled carrier. The PDSCH/PUSCH parameter is also configured per BWP. When cross-slot scheduling is used for a different SCS in carrier aggregation, the K value of the scheduled carrier active BWP needs to be converted to a K value for the scheduling carrier active BWP. The conversion of the new indication update time needs to be carried out according to the SCS coefficient μ of the scheduling and scheduled carriers (Table 19.6).

The conversion can be calculated as follows [11]:

$$\text{Max}(\text{ceiling}(\text{minK}_{0,\text{scheduled}} \times 2^{\mu_{\text{scheduling}}}/2^{\mu_{\text{scheduled}}}), Z) \qquad (19.1)$$

Table 19.6 Minimal constant Z.

μ	Z
0	1
1	1
2	2
3	2

where the $minK_{0,scheduled}$ is the original minimum K0 of the active BWP in the scheduled carrier, $\mu_{scheduling}$ is the SCS of the scheduling PDCCH, and $\mu_{scheduled}$ is the SCS of the active BWP for the scheduled PDSCH/PUSCH. If the scheduling PDCCH is after the first three symbols in the slot, one should be added to the Z value in the table.

Based on a delay confirmation mechanism, the minimum K0/K2 value is synchronized between the base station and terminal. Since the delay in application timing is reflected in PDCCH, it is only calculated based on K0.

19.3.5 Error handling in cross-slot scheduling

Although there is a delay confirmation mechanism to update the application timing in cross–slot scheduling, a DCI may be missed due to interference. And then, it is unavoidable for the terminal receiving an incorrect scheduling timing instruction in another DCI. Typically for a network, the block error rate for the control channel is maintained at the level of 1%. Loss of dynamic signaling can lead to incorrect indication. A typical case is that the first feedback reporting will indicate that the terminal did not receive correctly when the base station tries to update the cross–slot delay indication. The base station will then proceed to send DCI in a later slot, and the cross–slot indication field in the DCI will maintain the indication. As such, the timing in which the terminal updates the cross–slot operation will be later than the base station. Due to such inconsistent understanding of among the base station and terminals in a cell, it is necessary to introduce an error handling mechanism. Two main mechanisms were included in 3GPP R16.

Firstly, when the received K0/K2 parameter value for PDSCH/USCH in DCI is less than the minimum K0/K2 value applicable to a current scheduling, the terminal is allowed to not process the data. At this point in time, the terminal may be in a RF shutdown state and cannot respond at the specified time. Due to lack of response from the terminal,

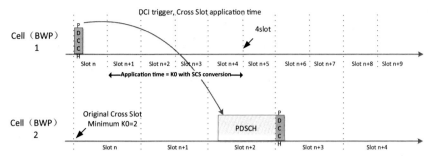

Figure 19.15 Application delay of cross-carrier scheduling and cross-slot indication, minK0 = 2.

it will also help the base station to determine whether an inconsistency has occurred based on the response from the terminal [12].

In the previous typical example, if multiple control signaling are lost continuously or the scheduling slot time gap is too long, the terminal may receive an opposite cross–slot indication within a certain period. In the second error handling mechanism, it is therefore to specify that the terminal should not respond to another update within the application delay of cross-slot indication [12].

19.3.6 Impact of cross-slot scheduling on uplink/downlink measurement

Terminal power-saving from cross-slot scheduling is mainly based on the time gap interval between reception and transmission of different signals/channels. In addition to the control and data channels, some downlink measurements based on reception of other signals also need to be considered in the terminal power-saving work. One of these cases is triggering of aperiodic downlink measurement RS and uplink SRS using downlink control DCI. The terminal also needs to ensure that the measurement of a downlink RS, which may be triggered aperiodically, is also after the offset. In addition, the triggered SRS cannot be sent earlier than PUSCH.

In the basic version of 5G-NR, some restrictions on downlink measurements are already incorporated. Specifically, the time gap interval allocated for CSI-RS resources is determined by the aperiodic trigger offset. However, when QCL-typeD is not included as a measuring condition in the triggering DCI, the CSI trigger offset is fixed at 0. This subsequently incur no power-saving for the terminal. Enhancements in cross-slot scheduling also place limits on the offset for CSI-RS. If a minimum

K0/K2 is configured, an offset for the measurement should be triggered and used according to the configuration, rather than performing measurement in the same slot.

For aperiodic SRS, there is no such restriction for QCL-typeD as in the above. Therefore terminal power-saving from cross-slots does not place further limits on the trigger offset value. As such, the 5G-NR standards do not further enhance the SRS, since the offset can be reasonably configured by the base station. When the base station configures a small offset value for the aperiodic SRS while the configured K2 is large, power-saving in the uplink processing will be limited by the SRS processing.

19.3.7 BWP switching in cross-slot scheduling

Since the configuration of minimum K0/K2 values and TDRA tables applies to the scheduled cross-slot BWP in cross-slot scheduling, the use of minimum K0/K2 also needs to change taking into account when BWP switching occurs according to the configuration.

When the BWP switching is triggered in either RRC semistatic configuration or dynamic triggered switch mode, the activation start of the new BWP will lead to updates for the minimum K0/K2 values and TDRA table. When a BWP is first configured, a set of values should be determined before the initial minimum K0/K2 indication. Currently, the approach is to take a set of the smallest sequences configured.

After a BWP is activated and if the terminal receives both a BWP switch indicator and a minimum K0/K2 indicator, two factors (1) and (2) should be taken into account by the terminal. Firstly for (1), how to maintain the condition of minimum K0/K2 value on the original BWP and at the same time without a corresponding conversion for the SCS on the new BWP. As such, a certain delay should be taken into account for this condition and the SCS conversion of BWP should be considered. Secondly for (2), combinations for the minimum K0/K2 configuration for the new BWP would not be dependent on the original BWP. The minimum K0/K2 value in the DCI should be indicated according to the configuration on the new BWP (Fig. 19.15).

For these two types of factors, an appropriate conversion is required. For the first factor (1), 5G-NR base station schedules in cross-slots by utilizing the transition time of DL (UL) BWP switching should not have another new DL (UL) receiving (transmitting). Section 4.3 of BWP switching triggered by DCI showed that this is to avoid terminal processing during the uncertain period of obtaining BWP parameters. Cross-slot scheduling in 5G-NR that defines the DCI scheduling that triggers BWP switching should fulfill (Fig. 19.16):

$K0 \geq \mathrm{ceiling}(\min K_{0,\mathrm{scheduling}} \times 2^{\mu_{\mathrm{scheduled}}}/2^{\mu_{\mathrm{scheduling}}})$ for time-domain resource allocation in downlink scheduling, where K0 is the indicated in the DCI, $\min K_{0,\mathrm{scheduling}}$ is the original minimum K0 specified by scheduling BWP, $\mu_{\mathrm{scheduling}}$ is the SCS of PDCCH, and $\mu_{\mathrm{scheduled}}$ is the SCS of the scheduled PDSCH.

$K2 \geq \mathrm{ceiling}(\min K_{2,\mathrm{scheduling}} \times 2^{\mu_{\mathrm{scheduled}}}/2^{\mu_{\mathrm{scheduling}}})$ for time-domain resource allocation in uplink scheduling, where K2 is the indicated in the DCI, $\min K_{2,\mathrm{scheduling}}$ is the original minimum K2 specified for the scheduling BWP, $\mu_{\mathrm{scheduling}}$ is the SCS of PDCCH, and $\mu_{\mathrm{scheduled}}$ is the SCS of the scheduled PUSCH.

For the second factor (2), a new K0 (K2) value for the scheduled PDSCH (PUSCH) is specified for cross-slot scheduling in 5G-NR due to configuration of new BWP and SCS. This also makes use of the fact that the terminal now does not need to perform any transmission and reception during the BWP switching time (specific reference to Chapter 4: Bandwidth part). As shown in boldface in Fig. 19.17, the 5G-NR standard introduces constraints to solve factors (1) and (2).

In a relatively infrequent/corner case, when one DCI indicates cross-slot scheduling and another DCI indicates BWP switching on the scheduling carrier with a change in SCS for the scheduled BWP, it is insufficient for the terminal to determine the time delay to apply the cross-slot scheduling based on the K value. As such, for 5G-NR cross-slot

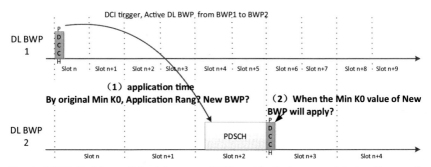

Figure 19.16 Problems, Schedule across BWP and indicate cross-slot simultaneously. *BWP*, Bandwidth part.

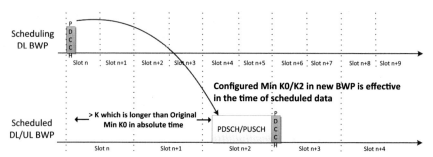

Figure 19.17 Timing requirements, Schedule across BWP and indicate cross-slot simultaneously. *BWP*, Bandwidth part.

scheduling, the delay definition in application timing (Section 19.3.4) is made that is made as equivalent to absolute time to resolve this situation.

19.4 MIMO layer restriction

19.4.1 Impacts of RX and TX antennas on energy consumption

Energy consumption of a device can be reduced by using a smaller number of antennas for both transmission and reception. As shown in Fig. 19.18, the RF for the transmission side has a corresponding relationship with the antenna panel, while reducing the group of antennas will reduce the power consumption of the corresponding RF. In the transmitter, the RF contains a power amplifier that consumes a relatively large proportion of power for the device. In the receiver, shutting down the RF chain will also produce a power-saving effect. Turning off all the RF when the device hardware is still running does not mean that there is no energy consumption energy at all for the device. The RF and corresponding circuits still need to enter a standby stage, where these inactive links remained readily available for operation at any time. These factors have already been taken into consideration in the power-saving evaluation methodology. However, according to the previous analysis, this will come at a cost of some loss in performance and wireless resource utilization. During scheduling, the base station will fully consider the impact of the performance loss and allow the terminal to reduce the number of antennas whenever feasible.

Based on these considerations, an NR power-saving technique at the transmitter side is based on limiting the number of MIMO layers used at the base station to control the power consumption for the terminal. From

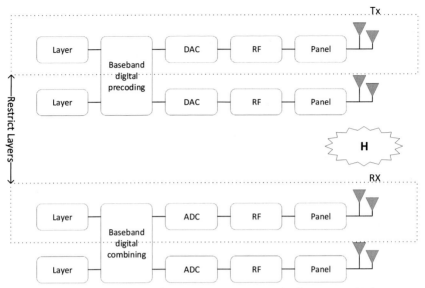

Figure 19.18 Limited number of MIMO layers. *MIMO*, multiple-input multiple-output.

the terminal side, the terminal is also able to fully control the energy consumption for control and data channel transmission and reception by limiting the number of MIMO layers. As shown in Fig. 19.18, reducing the number of transceivers makes it unnecessary to map data onto some groups of antennas. Although the number of MIMO layers does not necessarily equal the number of antennas (the framework of NR technology allows one layer to map to multiple antennas), due to the relatively fixed mapping relationship, there is no difference in terms of power-saving effect for the terminal between changing the number of MIMO layers and directly changing the number of antennas.

In 5G-NR, the number of MIMO layers configured is only for the data channel transmission and reception, and there is no specific/explicit configuration of number of antennas or MIMO layers for the control channel. For terminal implementation, however, the number of receiving layers in the control channel can be adjusted according to the number of associated data layers [13].

In the basic version of 5G-NR in R15, the specification supports mainly semistatic configuration of MIMO layers. The maximum number of downlink MIMO layers is configured in a cell via the parameter *PDSCH-ServingCellConFigure*. Additionally, changes in the maximum number of MIMO layers for the downlink can only be made using higher-layer reconfiguration. Due to a high-cost associate with long reconfiguration cycle, reconfiguring the number of MIMO layers would not make the process adaptive and subsequently provide energy saving for the terminal. Power-saving enhancement in 5G-NR reuses the framework of BWP in order to support the maximum number of MIMO layers in each configured BWP. Subsequently, terminal power-saving from adapting the number of MIMO layers is achieved from dynamic switching of BWP (Fig. 19.19).

19.4.2 DL MIMO layer restriction

As mentioned previously, the limitation on downlink MIMO layers should be incorporated as part of the BWP configuration framework. The mechanism is to configure the maxMIMO-Layers-r16 parameter in PDSCH-Config for every BWP. If each BWP is configured with a different maximum number of MIMO layers, an energy-efficient BWP would have fewer maximum number of layers configured and a BWP for better performance would be configured with a larger maximum number of layers. Therefore dynamic change of number of MIMO layers can be achieved by switching the BWP triggered by DCI.

However, the above limitation on the maximum number of layers per BWP can lead to compatibility problems with the maximum number of layers per cell (carrier). On one hand, when the maximum number of MIMO layers per BWP is configured, it cannot exceed the maximum number of layers for the cell. When the maximum number of layers per BWP is not configured, the maximum number of layers for the cell should be applied. On the other hand, the maximum number of layers in the cell configuration will affect the calculation for the coefficient of TBS finite soft buffer size (dl-sch TBSLBRM) under multiple carriers. In the basic version of 5G-NR, the coefficients for each carrier are determined based on the cell-configured maximum number of layers (which is four). In order to maintain the compatibility for the finite soft-bit buffer size calculation, it is mandated that the maximum number of layers for one BWP should be equal to the maximum layers for the cell when all BWPs are configured with the maximum number of layers.

19.4.3 UL MIMO layer restriction

For uplink MIMO layer limitation, the basic version of 5G-NR supports limiting the maximum RANK per BWP for codebook-based uplink transmission, which is equivalent to limiting the number of MIMO layers. Based on this RANK limitation, the effect of power-saving for the terminal is achieved. For noncodebook-based uplink transmission, the base station can still indirectly limit the number of layers of uplink transmission via configuration of SRS resources and ports for the BWP. Therefore uplink MIMO does not need to be further enhanced from the basic version of 5G-NR.

19.5 SCell dormancy

19.5.1 Multicarrier power-saving based on carrier aggregation

As previously described in the candidate power-saving techniques, the dynamic activation and deactivation of SCell is a frequency-domain adaptation method for a terminal operating in carrier aggregation. 5G-NR supports up to 16 aggregated carriers (Carrier Aggregation). The energy consumption of sending and receiving signals and channels on multiple carriers simultaneously would be significant for a terminal, but the terminal may not be required to transmit and receive data on multiple carriers simultaneously in every slot. On an activated carrier, PDCCH detection should be carried out according to the periodic search space configuration. As in the case of a single carrier, PDCCH detection alone takes up more energy consumption. During the feasibility study stage of the terminal power-saving item in 3GPP R16, evaluation showed dynamic dormancy of SCell can reduce terminal energy consumption by at least 12%.

In principle, only one carrier is required to be active in carrier aggregation mode. This carrier acts as an anchor carrier to trigger sleep or non-sleep mode for other carriers. The primary carrier in carrier aggregation is well suited as such an anchor carrier. Instant sleep mode cannot be supported on the primary carrier since the terminal is required to receive and decode system broadcast messages. Therefore terminal power-saving techniques for multicarrier "sleep" operation can be supported by introducing SCell dormancy on the framework of carrier aggregation in 5G-NR [14,15].

Due to the dynamic nature of data arrival, activation and deactivation of SCells need to be indicated instantly to the terminal. The dynamic signaling is introduced under carrier aggregation enhancement work for 5G-NR. The downlink control DCI transmitted on the primary carrier can trigger or terminate dormancy of multiple SCells (carriers) at one time.

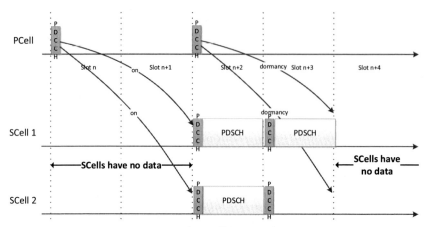

Figure 19.19 Dynamic trigger secondary cell (carrier) dormancy and exit dormancy.

During the evaluation process, it was found that dynamic triggering provided higher power-saving gain than that triggered by using a timer. Moreover, the dynamic triggering has less impact on data delivery delay.

19.5.2 Power-saving mechanism of SCell (secondary carrier)

In 4G-LTE, terminal power-saving is realized by introducing a secondary carrier dormant state. The dormant state is defined as an RRC state it takes a long time for a terminal to enter into and leave the dormant state, and the state transition is complicated. For 5G-NR, the support of SCell dormancy is doen using dynamically switched carriers. Two subschemes were discussed by the 3GPP.

The first subscheme is based on a dynamic and direct indication using PDCCH. In this subscheme, PDCCH detection and uplink/downlink data transmission/reception are disabled for one or a group of SCells based on PDCCH instruction. A corresponding delay mechanism for disabling the PDCCH detection on a secondary carrier is also required to be defined. This mechanism is also introduced in cross–slot scheduling.

The second subscheme is to utilize the BWP switching mechanism to achieve the SCell dormancy for the terminal. In the basic version of 5G-NR in R15, up to four BWPs per carrier can be configured for a terminal subject to its capability. When a BWP is not configured for PDCCH detection, the BWP is considered as a dormant BWP. When a SCell switches to a carrier with dormant BWP, the whole carrier enters into a dormant mode. The framework and features of a BWP are then applied to the SCell dormancy indication.

Of the two schemes, the latter is less complex and does not require much of specification work in 3GPP. Therefore terminal power-saving enhancement adopts the BWP switching mechanism for SCell dormancy. Energy-efficient dormancy is defined only in a downlink BWP. When there is no PDCCH detection on the downlink BWP, and the corresponding uplink PUSCH transmission will also cease.

The sleep indication can be sent in a power-saving wake-up signal or in a normal control channel during the DRX active time. The combination with the wake-up signal is to confirm and continue with the dormant state for the indicated SCell before the next DRX active period, and to avoid unnecessary activation of the SCell at the beginning of each DRX active period. During the active period of DRX, a normal DCI in the primary carrier can indicate a SCell for a terminal at any time. Furthermore, the switching time for a secondary cell dormancy triggered by a normal DCI uses the same time delay as BWP switching across carriers.

For a dormant BWP, a periodic CSI measurement resource can still be configured for a terminal. Interruptions in control and data reception do not impact the measurement. When the SCell exits the dormancy state, the measurement for MCS selection on the new BWP can be performed according to the dormancy BWP (Fig. 19.20).

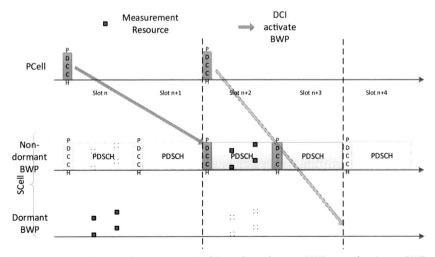

Figure 19.20 Carrier dormancy switching based on BWP mechanism. *BWP,* Bandwidth part.

Due to the reuse of the BWP switching mechanism, after the bwp-InactivityTimer has run out, the dormant BWP of the secondary cell will return to the default BWP. It is not precluded for the default BWP to be configured as a dormancy BWP. If the default BWP is configured as a dormancy BWP, secondary cell dormancy is triggered when the bwp-InactivityTimer has run out.

19.5.3 Secondary cell (carrier) dormancy trigger outside DRX active time

The secondary cell dormancy triggering outside the DRX activate time is indicated by a dedicated bit in DCI format 2_6 of the power-saving wake-up signal. Due to the ability of the power-saving wake-up signal to provide indication for multiple terminals at the same time, the dormancy indication for each terminal is limited to 5 bits or less. Each bit is used to indicate dormancy or nondormancy for a group of SCells. In terms of the energy consumption, it is associated with terminal RF design for carrier frequency switch in the same frequency band. There is a little difference for power consumptions between simultaneous transmission for multiple carriers and individual carriers within a frequency band. Grouping carriers are determined by base station configuration and could be based on frequency band.

The terminal can only be configured with one dormant BWP on a secondary cell. When the terminal is instructed to switch out of the dormant BWP, it is necessary to determine a BWP which is nondormant and to be configured as the initial nondormant BWP. Similar to SCell grouping, this parameter can be configured outside or within the DRX activate period to provide some flexibility. In order to reduce unnecessary BWP switching for both the terminal and the base station, it is mandated in 5G-NR that the terminal should switch to the initial nondormant BWP only when the current active BPW is a dormant BWP and has instructed to switch out of the dormant BWP. When the current active BWP for the terminal is already a nondormant BWP, then the current active BWP should be kept if an instruction is received to switch out of dormancy.

The instruction to dormancy can only be triggered once outside the DRX active time. In the case when the power-saving wake-up signal is lost but the configuration behavior of the terminal is still alive, the terminal simply keeps the last activated BWP.

19.5.4 SCell dormancy trigger of SCell in DRX active time

The secondary cell dormancy trigger during DRX active time is indicated by a normal PDCCH. The indications can be configured using two different approaches by the base station.

The first approach is to add special bits to the scheduling DCI commonly used for the primary carrier. Since the PDCCH scheduling DCI already has a large load, the dormancy indicator domain is limited to 5 bits or less. Each bit is used to indicate dormancy or nondormancy of a group of SCells (carriers). The base station can also utilize the characteristics that carriers in the same frequency band carrier share RF components. Dedicated parameter fields can be added in DCI format 0_1 and DCI format 1_1. The advantage of this approach is that the instructions to dormancy can be indicated at any time by the scheduling DCI of the primary carrier. However, the scheduling DCI of the primary carrier still needs to transmit this parameter bit field even without changing the dormancy state of the SCell, and hence, this may lead to unnecessary costs to the terminal.

The second approach is to redefine the parameter field in the scheduling DCI specifically for secondary cell dormancy indication. When the frequency resource indication field in DCI format 1_1 is all zeros or all ones, the following parameter filed in the DCI are used to indicate dormancy for 15 SCells (carriers):

- Modulation order and coding rate for transport block 1
- New data indicator for transport block 1
- Redundant version of transport block 1
- HARQ process number
- Antenna port(s)
- DMRS initialization value

The second approach does not involve data scheduling, but still uses the entire DCI format. When the number of configured carriers is small, the overhead will be comparatively large. HARQ-ACK feedback in 5G-NR is based on the decoding outcome of the scheduled data. The data scheduling in the first approach naturally enables the communication handshake between the base station and the terminal. The second approach of not scheduling data requires another method to support the communication handshake. Therefore a HARQ-ACK feedback timing that is based on a fixed timing relationship to the scheduling PDCCH is adopted in 5G-NR.

Similar to the outside of the DRX active time discussed in Section 19.5.3, the base station first configures the terminal with an initial nondormant BWP. Regardless of the bwp-InactivityTimer timeout, the terminal switches to the first nondormant BWP during the active time only if the current active BWP is a dormant BWP and the instruction to switch out of the dormancy is received. When the current active BWP for the terminal is already a nondormant BWP, then the current active BWP should be kept even if an instruction is received to switch out of dormancy.

During the active time, the first and second approaches can be flexibly configured by the base station according to the requirements of the system setting. The method triggering a BWP to dormancy is highly overlapping to the method to trigger BWP switching by DCI in the basic version of 5G-NR. This comes with a compatibility issue where the BWP ID in the triggering DCI is the same as the BWP ID for the dormant BWP. In 5G-NR, this kind of configuration and instructions are not precluded. If the BWP ID of the switching triggered by DCI is the dormant BWP, then the switching becomes unreliable since there is no downlink data and HARQ-ACK feedback.

19.6 RRM measurement relaxation

19.6.1 Power-saving requirement in RRC_IDLE or RRC_INACTIVE mode

Terminals in the RRC_IDLE and RRC_INACTIVE modes based on network configuration should perform RRM measurement on the camping/serving cell and neighboring cells to support mobility management, such as cell reselection. In R15 of 5G-NR, for the purpose of power-saving, if a terminal's channel quality in a camping/serving cell is good, the terminal is not required to perform intrafrequency measurements, measurement on NR interfrequencies, and inter-RAT frequencies with an equal or lower reselection priority than the reselection priority of the current NR frequency [16]. For NR interfrequencies and inter-RAT frequencies with a higher reselection priority, the terminal is allowed to perform a relaxed measurement with a larger measurement interval. More specifically:

- If the serving cell fulfills $Srxlev > S_{IntraSearchP}$ and $Squal > S_{IntraSearchQ}$, the terminal may choose not to perform intrafrequency measurements.

- If the serving cell fulfills Srxlev $>$ $S_{nonIntraSearchP}$ and Squal $>$ $S_{nonIntraSearchQ}$, the terminal may choose not to perform measurements of NR interfrequencies or inter-RAT frequency cells of equal or lower priority. At the same time, for NR interfrequencies or inter-RAT frequency cells of higher priority, the terminal can perform a more relaxed measurement with a measurement interval Thigher_priority_search = (60 * Nlayers) seconds, where Nlayers is the number of NR interfrequencies or inter-RAT frequencies of higher priority.

For terminals that need to perform measurement on neighboring cells, it is necessary to introduce a measurement relaxation enhancement mechanism to further save power.

19.6.2 Relaxed measurement criterion

Two criteria are introduced and need to be fulfilled in order for a terminal to perform relaxed measurements (i.e., terminal with low-mobility criterion and terminal not at cell-edge criterion). Both criteria are based on "cell level" measurement results on the serving cell.

19.6.2.1 The terminal not at cell-edge criterion

A RSRP threshold $S_{SearchThresholdP}$ and/or a RSRQ threshold $S_{SearchThresholdQ}$ are configured by the network, where $S_{SearchThresholdP}$ is mandatory and $S_{SearchThresholdQ}$ is optional. If both $S_{SearchThresholdP}$ and $S_{SearchThresholdQ}$ are configured, a terminal can perform relaxed measurement on neighboring cells only when both RSRP and RSRQ based criteria are met.

Network configuration should ensure that $S_{SearchThresholdP}$ $\leq$ $S_{IntraSearchP}$, and $S_{SearchThresholdQ}$ $\leq$ $S_{IntraSearchQ}$ as well as $S_{SearchThresholdP}$ $\leq$ $S_{nonIntraSearchP}$, and $S_{SearchThresholdQ}$ $\leq$ $S_{nonIntraSearchQ}$.

19.6.2.2 The terminal with low-mobility criterion

A RSRP relative threshold $S_{SearchDeltaP}$ and a time period threshold for evaluating the RSRP change $T_{SearchDeltaP}$ are configured by the network. If RSRP measurements on the serving cell do not change more than $S_{SearchDeltaP}$ during a time period of $T_{SearchDeltaP}$, the terminal can perform relaxed measurements on neighboring cells.

The terminal will perform intrafrequency and interfrequency neighbor cell measurements during $T_{SearchDeltaP}$ after cell selection/reselection (Fig. 19.21).

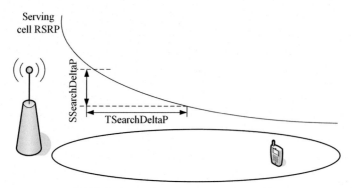

Figure 19.21 Terminal with low-mobility criterion.

The network broadcasts these corresponding parameters of triggering criteria to enable the RRM measurement relaxation feature in the cell. The configured triggering criterion/criteria can be either for terminals not at cell-edge only, terminals with low-mobility, or both. When the network configures the parameters with triggering criteria of both low-mobility and noncell edge, a terminal performs measurement relaxation according to one of the following options, which is also indicated by the network:

- Option 1: The terminal complies with both low-mobility and noncell-edge criteria
- Option 2: The terminal complies with either the low-mobility or the noncell-edge criterion.

In 5G-NR R15, terminals always perform RRM measurements on NR interfrequencies and inter-RAT frequencies with a higher priority for the purpose of load balancing but not for mobility management. In R16, with the introduction of RRM measurement relaxation enhancement, whether RRM measurements for NR interfrequencies and inter-RAT frequencies with a higher priority can be further relaxed compared to the maximum measurement interval supported by R15 is configurable by network.

The network can configure logged MDT for terminals in RRC_IDLE and RRC_INACTIVE modes to collect measurement results for network performance optimization. A terminal in RRC_CONNECTED mode starts timer T330 after receiving the recording measurement configuration message. When the terminal enters RRC_IDLE mode or RRC_INACTIVE mode, RRM measurement results will be recorded during the running of T330. After the introduction of RRM measurement relaxation for RRC_IDLE and RRC_INACTIVE mode, whether the terminal can perform relaxed RRM measurement during the running of T330 is an issue to be considered. Considering that the network usually optimizes the network according to the collected measurement records from multiple terminals, and the terminal only performs relaxed RRM measurement when the RRM relaxation criteria are met, the relaxed RRM measurement usually does not have a significant impact on the measurement performance. Therefore it was determined that the terminal is allowed to perform the relaxed RRM measurements during the running of T330.

19.6.3 Relaxed measurement method

For intrafrequency and NR interfrequencies or inter-RAT frequencies with an equal or lower reselection priority than the reselection priority of the current NR frequency, RRM measurement relaxation methods are defined individually for different measurement relaxation criteria. More specifically,

- For the low-mobility scenario, RRM measurement relaxation with longer measurement intervals is applied. The scaling factor of measurement interval is fixed.
- For the noncell-edge scenario, RRM measurement relaxation with longer measurement intervals is applied. The scaling factor of measurement interval is fixed.
- When both the low-mobility and noncell-edge criteria are fulfilled, the terminal is not required to perform intrafrequency, interfrequency, and inter-RAT neighboring cell measurements if the last/most recent cell reselection measurement was performed less than 1 hour ago.

For NR interfrequencies and inter-RAT frequencies with a higher reselection priority than the reselection priority of the current NR frequency, the RRM measurement relaxation method is performed as follows.

- Case 1: When $S_{rxlev} > S_{nonIntraSearchP}$ and $S_{qual} > S_{nonIntraSearchQ}$

  No relaxation on the measurements for NR interfrequencies and inter-RAT frequencies with a higher priority than $T_{higher\_priority\_search}$ is expected, when the low-mobility criterion is not configured or not fulfilled.

  A terminal can stop both the equal/low-priority and the high-priority interfreq/inter-RAT measurements, when the low-mobility criterion is configured and fulfilled.

- Case 2: When $S_{rxlev} \leq S_{nonIntraSearchP}$ or $S_{qual} \leq S_{nonIntraSearchQ}$

The relaxed requirement for NR interfrequencies and inter-RAT frequencies with a higher priority uses the same relaxed measurement requirement as those for the frequency with equal/lower priority.

19.7 Terminal assistance information for power-saving

19.7.1 Terminal assistance information procedure

In order to help the network configure appropriate parameters to achieve the purpose of power-saving for the terminal, the network can configure/request the terminal to report terminal assistance information for power-saving. The following terminal assistance information for power-saving was introduced for 5G-NR in R16 (Fig. 19.22).

- Terminal's preference on DRX parameters for power-saving
- Terminal's preference on the maximum aggregated bandwidth for power-saving
- Terminal's preference on the maximum number of secondary component carriers for power-saving
- Terminal's preference on the maximum number of MIMO layers for power-saving
- Terminal's preference on the minimum scheduling offset for cross–slot scheduling for power-saving
- Assistance information to transition out of RRC_CONNECTED state when the terminal does not expect to send or receive data in the near future

For each terminal assistance information type, the terminal should first inform the network that it has the capability to report a certain type of

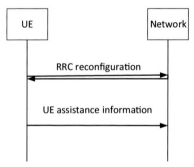

Figure 19.22 Terminal assistance information.

terminal assistance information through the terminal capability reporting mechanism, and then the network configures the terminal with the terminal assistance information reporting function for the certain type through the RRC reconfiguration message. Different types of terminal assistance information reporting functions can be individually configured by the network. For each type of terminal assistance information, the terminal can only report terminal to the network only when the network configures the terminal with a reporting function for a certain type of terminal assistance information.

In order to prevent terminals from reporting terminal assistance information frequently, the network can configure a prohibit timer for a certain type of terminal assistance information. The terminal will start this prohibit timer every time it reports this certain type of terminal assistance information. During the running of the prohibit timer, the terminal cannot report this certain type of terminal assistance information. Only when the prohibit timer is not running and the reporting trigger condition for this certain type of terminal assistance information is met, can the terminal report this certain type of terminal assistance information. For each type of terminal assistance information, the maximum prohibit timer length can support is 30s.

In order to save signaling overhead for the reporting of terminal assistance information, delta reporting is supported for each type of terminal assistance information. That is, for a certain type of terminal assistance information, if the terminal's current preference for the feature does not change from the last preferred configuration reported by the terminal for the feature, the terminal may choose not to report the terminal assistance information for the type this time. From the perspective of the network side, when the network receives the terminal assistance information from the terminal for a certain feature, the network will maintain the reported information until it receives the next report of terminal assistance information from the terminal for the same feature.

When a terminal reports terminal assistance information for a certain feature to the network, the terminal will include the parameters for the feature for which the terminal has a preference in the terminal, while the parameters the terminal has no preference for are not included in the reported terminal assistance information. If the terminal has no preference for any of the parameters for a certain feature, the terminal can inform the network that it has a preference for none of the parameters by reporting an empty terminal assistance information for the feature (i.e., the terminal assistance information does not contain any parameters).

In Mr-DC, cell group-specific terminal assistance information for power-saving can be configured by the network. For both MCG and SCG, the terminal reports the terminal assistance information for power-saving only when the network configures the terminal to report the terminal assistance information for this CG. The network can configure the terminal with terminal assistance information for NR SCG through Signalling Radio Bearer (SRB1) on MCG or SRB3 on SCG. The terminal can report the terminal assistance information for power-saving for NR SCG to the network through SRB1 on MCG or SRB3 on SCG.

19.7.2 Terminal assistance information content

As mentioned in the previous section, six types of terminal assistance information for power-saving are introduced in 5G-NR. In this section, each type of terminal assistance information is described.

19.7.2.1 Terminal's preference on DRX parameters for power-saving

Terminal's preference on DRX parameters for power-saving include terminal's preference on drx-InactivityTimer, DRX long cycle, DRX short cycle, and drx-ShortCycleTimer. All four parameters are optional for the terminal to report. As mentioned before, if the terminal has no preference for all of the parameters for DRX, the terminal can inform the network that it has preference for none of the parameters by reporting an empty terminal assistance information for DRX.

For a terminal configured with a secondary DRX group, the terminal could report its preference on the drx-InactivityTimer for the default DRX group and that for the secondary DRX group, respectively.

For each parameter, the terminal can report any value within the supported value range of the parameter. When the network configures a terminal with the DRX function, if the network configures a DRX short cycle for the terminal, the configured value of DRX long cycle and DRX short cycle should satisfy that the value of the DRX long cycle is an integer multiple of the DRX short cycle. For the same reason, when the terminal reports its preference on DRX parameters to the network, if the terminal reports both the preferred DRX long cycle and the preferred DRX short cycle, the preferred DRX long cycle reported by the terminal should be an integer multiple of the preferred DRX short cycle.

19.7.2.2 Terminal's preference on the maximum aggregated bandwidth for power-saving

The terminal can report its preference on the maximum aggregated bandwidth for uplink and downlink and for FR1 and FR2 individually to the network. For a certain FR, the terminal can report the preferred maximum aggregated bandwidth for the FR only when the network configures at least a service cell for the terminal in that FR.

19.7.2.3 Terminal's preference on the maximum number of secondary component carriers for power-saving

The terminal can report its preference on the maximum number of secondary component carriers for uplink and downlink separately. In Mr-DC, the terminal can implicitly indicate a preference for NR SCG release by reporting the maximum aggregated bandwidth preference (if configured) and maximum number of secondary component carriers (if configured) as zero.

19.7.2.4 Terminal's preference on the maximum number of MIMO layers for power-saving

The terminal can report its preference for the number of MIMO layers for uplink and downlink and for FR1 and FR2 individually. The terminal can report its preferred MIMO layer to be as low as 1.

19.7.2.5 Terminal's preference on the minimum scheduling offset for cross-slot scheduling for power-saving

The minimum scheduling offset value expected by the terminal for the cross-slot scheduling includes a minimum K0 value and a minimum K2 value. The minimum K0 and K2 values expected by the terminal can be reported separately for different SCS. These two parameters are optional for terminal reporting. As mentioned above, when the terminal does not have any preference for a certain parameter, the terminal will not include the parameter in the terminal assistance information for this feature.

19.7.2.6 RRC state transition

If the terminal does not expect downlink data reception and uplink data transmission in the near future, the terminal can report to the network that it expects to leave the RRC_CONNECTED state. At the same time, the terminal may further indicate to the network whether it is desired to transition to the RRC_IDLE or RRC_INACTIVE state.

If the terminal wishes to cancel the expectation of leaving the RRC_CONNECTED state that was previously reported to the network, for example, if the terminal has new data arrival for uplink transmission, the terminal should report to the network the expectation of staying in the RRC_CONNECTED state. In this case, whether the terminal can report to the network that it expects to remain in the RRC_CONNECTED state also depends on the network configuration.

19.8 Summary

5G-NR standards for the terminal power-saving technology include both the basic version of NR power-saving in R15 and power-saving enhancements introduced in R16. Th 5G-NR basic version of BWP design, flexible scheduling, DRX configuration, and other functions provide a certain level of terminal power-saving. On top of these, the power-saving enhancements for 5G-NR further allow the terminal to better adapt its operation in all areas including both the frequency and time domains, number of TX and RX antennas, number of MIMO layers, DRX wake-up/sleep, terminal's preferred configuration parameter reporting, and relaxation on RRM measurements.

References

[1] M.2410-0 (11/2017), Minimum requirements related to technical performance for IMT−2020 radio interface(s), ITU.

[2] R1-1901572, Power-saving schemes, Huawei, HiSilicon, 3GPP RAN1#96, Athens, Greece, February 25th−March 1st, 2019.

[3] R1-1903016, Potential techniques for UE power-saving, Qualcomm Incorporated, 3GPP RAN1#96, Athens, Greece, February 25th−March 1st, 2019.

[4] R1-1903411, Summary of UE power-saving schemes, CATT, 3GPP RAN1#96, Athens, Greece, February 25th−March 1st, 2019.

[5] R1-1813447, UE adaptation to the traffic and UE power consumption characteristics, Qualcomm Incorporated, 3GPP RAN1#95, Spokane, Washington, USA, November 12th−16th, 2018.

[6] R1-1900911, UE adaptation to the traffic and UE power consumption characteristics, Qualcomm Incorporated, 3GPP RAN1 Ad-Hoc Meeting 1901, Taipei, Taiwan, 21st−25th January, 2019.

[7] TR38.840 v16.0.0, Study on user equipment (UE) power-saving in NR.

[8] RP-191607, New WID: UE power-saving in NR, CATT, 3GPP RAN#84, Newport Beach, USA, June 3rd−6th, 2019.

[9] TS38.321 v16.0.0, Medium access control (MAC) protocol specification.

[10] R2-1905603, Impacts of PDCCH-based wake-up signaling, OPPO, 3GPP RAN2#106, Reno, USA, 13rd−17th May, 2019.

[11] 3GPP TS 380.214 v16.1.0, Physical layer procedures for data.

[12] 3GPP TS 38.213 v16.1.0, Physical layer procedures for control.

[13] R1-1908507, UE adaptation to maximum number of MIMO layers, Samsung, 3GPP RAN1#98, Prague, CZ, August 26th—30th, 2019.

[14] R1-1912786, Reduced latency Scell management for NR CA, Ericsson, 3GPP RAN1#99, Reno, USA, November 18—22, 2019.

[15] R1-1912980, SCell dormancy and fast SCell activation, Qualcomm Incorporated, 3GPP RAN1#99, Reno, USA, November 18—22, 2019.

[16] 3GPP TS 38.304 v15.6.0, User equipment (UE) procedures in Idle mode and RRC Inactive state.

Further reading

R1-2002763, Summary#2 for procedure of cross-slot scheduling power-saving techniques, MediaTek, 3GPP RAN1#100bis, e-Meeting, April 20th—30th, 2020.

R2-2004943, CR for 38.331 for power-savings, MediaTek Inc., 3GPP RAN2#110e, June 1st—12th, 2020.

R4-2005331, Reply LS on RRM relaxation in power-saving, 3GPP RAN4#94ebis, April 20—30, 2020.

3GPP TS 38.212 v16.1.0, Multiplexing and channel coding.

3GPP TS 38.133 v15.9.0, Requirements for support of radio resource management.

CHAPTER 20

Prospect of R17 and B5G/6G

Zhongda Du, Jia Shen, Han Xiao and Li Guo

20.1 Introduction to Release 17

The first half of 2019 is the warm-up period for the R17 project, during which the proposals of the R17 project were submitted by the 3GPP members one after another at the 3GPP plenary session. In June 2019, at the RAN#84 meeting, the 3GPP decided to discuss the details of each topic of R17 through email discussions. At the RAN#86 meeting, the 3GPP finally passed the R17 project package, and also decided that the entire work cycle of R17 would be 15 months. Unfortunately, the outbreak of COVID-19 in early 2020 directly led to the cancellation of the 3GPP meetings in the first half of 2020. At the RAN#e87 meeting, the 3GPP decided to shift the overall work plan of R17 by one-quarter. The work plan for R17 after such adjustments is shown in Fig. 20.1.

Overall, the project of R17 continues to improve the technologies for eMBB business, and further enhances the vertical industry-related technologies that have been introduced to the NR in R16. Both eMBB services and vertical industries are related to the improvement of network coverage, including the standardization of nonterrestrial communications, which extends the coverage of cellular networks to three-dimensional coverage in the sea, land, and air. R17 also includes new research and discussion on 5G NR in terms of spectrum, application, broadcast communication mechanisms, and network maintenance. Fig. 20.2 shows a schematic diagram of the topic classification of the entire R17 project.

All the technical enchantments for eMBB in R17, except multi-SIM card coordination and RAN slicing, are basically the continuation or expansion of the technical functions of R16. For example, in the work item of MIMO [1], the functions of signaling overhead and latency reduction for the FR2 system is extended from the scenarios of medium and low speed to the scenarios of high speed, and it is assumed that UE has multiple antenna panels; the feature of beam management and channel robustness and reliability improvement under multipoint transmission and multipanel reception is extended from PDSCH channel to PDCCH, PUSCH, and PUCCH channel; more antenna ports of SRS are added to increase the coverage and capacity of SRS; in the

5G NR and Enhancements
DOI: https://doi.org/10.1016/B978-0-323-91060-6.00020-9
981

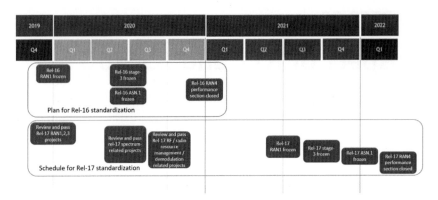

Figure 20.1 R17 work plan [34].

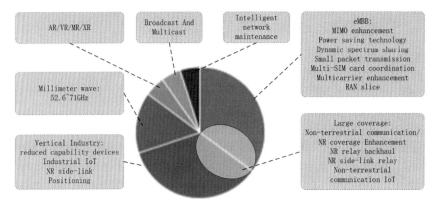

Figure 20.2 Schematic diagram of R17 project.

FDD band, further enhancement is considered to explore the use of channel reciprocity to enhance the measurement and reporting of CSI, and so on. The DCI-based solution for power-saving technology in R16 can mainly avoid unnecessary waking up of the UE, while in R17, the needs of higher-speed services are taken into account and the objective is to reduce the PDCCH monitoring when the UE is not in sleep mode. The system overhead of monitoring paging messages and reference signals in the states of RRC_IDLE and RRC_INACTIVE is going to be reduced [2]. The motivation of supporting small packet transmission in the RRC_INACTIVE state is mainly to avoid resuming RRC connection or reduce the overhead of signaling and power consumption caused by the UE entering the RRC_CONNECTED state, which is especially suitable for wearable devices [3].

A single UE with multiple SIM cards has been popular since the 4G era. The conflict between the two systems corresponding to two SIM cards is often caused by the limited RF resources of UE, such as limited RF links and antennas. In 4G LTE, such a problem is resolved by the implementation method of a UE without standardized solution. The result is not desirable. The RF resources of the 5G NR UE have increased and the typical configuration is 2T4R. However, the mid-range/low-end smartphones or wearable devices in the market will also use relatively low configurations for RF resources such as 1T2R. So similar problems still exist. A standardized solution adopted in R17 can reduce the paging conflicts between the two systems as much as possible. When a UE decides to leave the current serving system (usually 5G NR) and respond to the paging from another system (usually 4G LTE), the solution can reduce the impact on the ongoing services in the current system and improve network awareness [4]. In such a scenario, supporting a response to the paging of voice service is the focus.

The main objective of the project of RAN slicing is to help a UE to quickly access a cell that supports the service slices that the UE wants to initiate. The targets include: Matching the slice the cell can support and the slice the UE wants to initiate during cell selection/reselection procedure, and initiating the random access procedure based on the slice information. When a mobility event, such as handover, happens, the solution can ensure the continuity of traffic when the source cell and the target cell are not compatible. Further enhancing the concept of slicing on the RAN side is also a trend of 5G technology [5].

Enhancements for vertical industry-related technology are a notable feature of R16. And new breakthroughs have been made in R17 including the introduction of compact NR UE and high-precision indoor positioning. The three major application scenarios of NR 5G have their own focus: eMBB wants high speed, mMTC requires wide coverage and multiconnection, while URLLC pursues the ultimate performance in terms of reliability and delay. Generally, one terminal can only support at most two out of three types of NR applications because in the same communication system the requirements of these three application scenarios often contradict each other. However, some terminals might be requested to take into account all of those requirements, but with a greatly reduced degree of support for those requirements. Logically, the requirements for the NR UE with reduced capability can be found Fig. 20.3.

At the beginning of the discussion of this topic at the 3GPP, such a UE was called "NR-Lite," which means a "light" terminal. "Light" is not

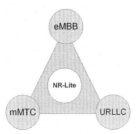

Figure 20.3 NR-Lite.

only reflected in the size and weight of the UE, but also in the configuration and processing capacity of the hardware. Such a UE has fewer antenna ports and narrower working bandwidth than smart terminals. One typical application is the industrial sensors. Furthermore, it requires a long standby time, more connections, and also needs to maintain similar coverage. When 3GPP finally launched the project, the name was changed to "reduced capability NR devices." During the discussion of 3GPP such a terminal type also included smart wearable devices such as smartwatches, video surveillance, and tracking devices for industrial or smart cities. However, the NB-IoT and mMTC terminals are not included, which have been developed based on LTE systems, in order to avoid the impact on mature products on the market and also to reduce the workload of 3GPP standardization [6].

The positioning technology based on NR positioning reference signal in R16 is actually mature and can provide the positioning accuracy of 3 m (with 80% probability) in indoor scenarios and 10 m (80% probability) in outdoor scenarios. At the same time, the UE-based positioning solution is also introduced where the UE calculates the final location information according to the assist information of the network. However, the solutions of R16 cannot meet the performance requirements of indoor industrial positioning in terms of accuracy (≤ 0.2 m) and latency (≤ 100 ms). At the same time, commercial applications related to positioning also require a positioning accuracy of less than 1 m. In addition in the positioning project of R17 the reliability and integrity requirements of positioning are proposed for the first time, which means that the positioning system must be available all the time. If there is a problem with the positioning system it is necessary to notify the UE using a positioning service in time to avoid accidents caused by mistrust. This requirement is especially important for the positioning applications used in traffic, electricity, or emergency calls.

The technical discussion and design will focus on improving the positioning reference signals, positioning methods, and related procedures specified in R16 [7].

The vertical industry-related technologies in R17 are enhanced mainly on the basis of R16. The basic idea of R16 URLLC in improving reliability is to increase the repeated branches of PDCP PDU (up to 4) and also to use the method of "cutting in line," which means sacrificing the transmission or feedback of logical channels with low priority to improve the reliability of high-priority logical channels. This approach is more or less "simple and rude." R17 intends to refine the solutions of R16 to minimize the impact on logical channels with relatively low priority. For example, certain channel multiplexing methods can be used when processing transmissions with different priorities within one UE or between UEs so that the logical channels with relatively low priority will not be completely discarded when conflict occurs. In addition, the use of unlicensed spectrum in NR is also a big breakthrough. Generally, the unlicensed spectrum is not comparable to the licensed spectrum in terms of communication reliability and delay performance because of the spectrum sharing. But in an industrial scenario with an enclosed environment, we can basically achieve "exclusive use" of the spectrum. Under this premise, the reliability and delay performance can be improved to a certain extent through technical solutions, such as using FBE (frame-based equipment), channel access mode, and so on [8].

The most important application scenario of sidelink communication technology in R16 is V2X (i.e., vehicle networking where the form of UE is often a vehicle-mounted terminal). In R17, this sidelink communication technology based on PC5 interface will be extended to public safety applications and commercial applications. In these new application scenarios UEs are often handheld terminals. Therefore the enhancements for sidelink technology in R17 focus on the solutions related to power-saving such as DRX mechanism, coordination mechanism between UEs, and partial sensing in resource allocation to reduce UE monitoring on control channel [9]. Another topic directly related to the sidelink communication technology is the relay based on the PC5 interface, including the relay from UE to the network and the relay from UE to UE. Both relay methods are important for public safety applications, while some commercial applications, such as smartwatches or bracelets connected to the network through smartphones, are more interested in terminal-to-network relay. For smartwatches or bracelets, PC5-based short-distance communications with smartphones consume almost zero power, which can greatly increase the standby time. In indoor scenarios, with the assist of the smartphones with higher configuration, wearable devices can communicate with the network smoothly, which can be regarded as a technical solution to expand network coverage [10].

Regarding the coverage issue, R17 provides corresponding solutions for both eMBB business and vertical industries. The main purpose of NR coverage enhancement project is to solve the coverage problem of uplink traffic channel. The most typical global spectrum in 5G NR is 3.5 GHz (band N78, N79). Compared with the spectrum used in LTE, 3.5 GHz does not have much advantage in indoor coverage in urban areas even with the help of beamforming technology because of fast propagation road loss and large building penetration loss. In the central area of the building, the coverage of the 5G NR is even worse than that of the LTE. At present, the global deployment of SA network is the trend of 5G NR, and it will be deployed on a large scale in China from the very beginning. In this case, coverage becomes bottleneck for 3.5 GHz SA networks. In addition to designing enhancements directly for 3.5 GHz, the uplink coverage problem can also be effectively solved through the use of low-frequency spectrum, such as 1.8 or 2.1 GHz FDD spectrum. In this project, the research mainly focus on voice services and low-to-medium-rate data services at the FR1 spectrum. Although the coverage of FR2 will also be studied in this topic, but the coverage of FR2 already has a better solution, that is, the use of relay backhaul technology [11].

The abovementioned coverage issues, whether it is FR1 or FR2 spectrum, are aimed at traditional cellular-based terrestrial networks. The issue in nonterrestrial communication is wide area coverage. In R16, the research objectives of nonterrestrial communication are high-orbit satellites (i.e., synchronous satellites) and low-orbit satellites. The main challenges faced by the 5G NR communication system are ultralong signal transmission delay and large frequency offset. It should be noted that, besides requiring Global Navigation Satellite System (GNSS) positioning capability for the UE, nonterrestrial communications do not request any additional requirements on the transmission power. The third type of power level, which is 23dbm, is still used. The solutions adopted in R17 standardization of nonterrestrial communications can also be extended to high-altitude balloons and air-to-ground communications. In air-to-ground communication the aircraft moves quickly, which is similar to a special UE; while in satellite communication, the deployment in the air is more like the radio frequency part of the base station or DU [12]. Another extension of nonterrestrial communications in R17 is transferring the similar solutions to the IoT functions of LTE systems, which then can be used in scenarios such as tracking the location of cargo on ocean-going tankers and providing simple communications [13]. One important reason nonterrestrial communication can gradually enter the market and begin the industrialization is the mature and reliable space launching technologies, which makes the launching of small satellites more economical.

R17 has also made bold explorations in millimeter wave (mmWave), XR, broadcast multicast, and network intelligence. At present, the frequency spectrum involved in 5G NR is divided into two ranges: FR1 (400 MHz−7.125 GHz) and FR2 (24.25−52.6 GHz). At the beginning of R17 some companies and operators in the United States proposed to systematically study the 52.6−114.25 GHz spectrum starting with the waveform. The spectrum management regulations in various regions of the world were first studied. It is gradually discovered that the industry is most interested in the spectrum close to 60 GHz which is 52.6−71 GHz. Especially in many countries, the unlicensed spectrum near this section of spectrum (see Table 20.1) may be industrialized. This spectrum was also officially designated as a cellular radio spectrum by ITU−R at the WARC19 conference in Egypt in 2019.

For this reason, the 3GPP finally decided to study and standardize only this section of the spectrum. In order to reduce the workload of standardization, the major principle is to directly extend several physical layer parameters, such as subcarrier spacing (SCS) and channel bandwidth, on the basis of the original 5G NR waveform. In this way, the existing physical layer protocol framework can basically be directly inherited. In addition, the transmission and reception of the mmWave beam in this frequency band are more directional with the so-called "pencil beam."

Table 20.1 Unlicensed spectrum allocation.

Region	Spectrum (GHz)	Max TX power (mW)
United States	57−64	500
Canada	57−64	500
Japan	59−66	10
European Union	57−66	20
Australia	59.4−62.9	10
South Korea	57−64	10

Under this premise, the channel access solution will be different from the existing channel access methods of LBE and FBE, and there is a higher possibility of coexistence in the spatial dimension [14,15].

XR is a general term that can stand for augmented reality, virtual reality, hybrid reality, and so on. In order to achieve the visual effect, XR-related applications require communication systems to provide high reliability and low delay while maintaining high traffic. Another similar application is cloud gaming. XR and cloud gaming are both challenges and also opportunities for 5G NR systems. First of all, in the network architecture, the 5G network must increase the edge computing nodes, which means to shorten the distance between the cloud computing nodes and terminals. Otherwise the delay requirements are often not met. Secondly, both helmets and glasses have high requirements for power consumption. One reason is that users will continuously use the XR devices for a long time. Another reason is that the computing load of the XR devices is relatively high. How to reduce power consumption, increase standby time, and improve user comfort are all areas worth exploring. Finally, the traffic requirements of XR and cloud games seem to be a posture of "eat one's cake and have it too." Even the communication system such as 5G NR can hardly provide multiple users with services of "high traffic, low delay and high reliability" for a long time. In order to achieve the effect of "good steel is used in the blade," it is necessary to carefully study the traffic model of XR and cloud games, especially the distribution range of packet size, the arrival rule in the time domain and key performance such as delay and packet loss rate so that the communication system can be adapted properly. The main purpose of the XR project of R17 is to find a suitable evaluation method to conduct the abovementioned research [16].

Broadcast multicast technology has a long history at the 3GPP. Both UMTS and LTE systems standardize the MBMS scheme based on SFN. The LTE system also supports the PDSCH-based, single-cell broadcast technology (SC−PTM). However, wide application of broadcast multicast technology in the market is rare. Broadcast multicast technology can be a useful supplement to certain technologies. For example, in IoT and V2X, broadcasting can improve spectrum efficiency. In applications such as video-on-demand, multiple users in a cell may order the same video content on demand. Under this premise, combining multiple unicast threads into one broadcast multicast thread can save wireless resources. In addition, broadcasting related to public safety is obviously more efficient than unicast. For this reason, the 3GPP finally decided to study a simplified broadcast and multicast solution based on 5G NR in R17. The scheme does not request using traditional SFN to increase cell-edge coverage. It can adopt the mixed networking with unicast, and the switching mechanism between broadcast multicast and unicast will be introduced. When a UE is in RRC_CONNECTED state, unicast can provide uplink feedback for broadcast multicast to increase the reliability [17].

The cells with same color in the examples shown in Fig. 20.4 constitute the broadcast multicast synchronize coverage area.

Finally, let us talk about the intelligence of network maintenance. The 3GPP has introduced SON and MDT mechanisms into the system since LTE and uses the measurement and statistical information provided by the UE to model the operating status of the network. With the gradual maturity and popularization of computing power (hardware) and deep learning algorithms (software) required by artificial intelligence (AI), big data collection and intelligent processing are being introduced into network maintenance. On this basis, AI will be gradually introduced into other modules of the 3GPP communication system [18].

20.1.1 Prospect of B5G/6G

Mobile communication technology evolves every ten years, and the preresearch and planning of the next generation is started after completing the standardization of the first generation. This is because a globally unified standard is necessary for the mobile communications industry. And the standardization work can be initiated only when the international industry has achieved consensus on the evolution direction and the core technologies. The formal standardization work usually takes 3−4 years, thus only 6−7 years are left for the preresearch stage. The standardization

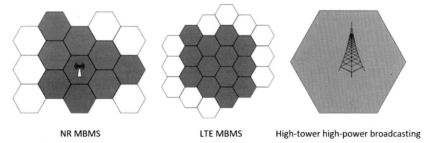

NR MBMS   LTE MBMS   High-tower high-power broadcasting

Figure 20.4 Comparison of NR MBMS, LTE MBMS, and high-tower high-power broadcasting.

of the 5G NR R15 was initiated in 2015 and completed at the end of 2018. As an indispensable part of 5G, the R16 standard was frozen in 2020. Assuming that the standardization work of the new generation (the official name is undecided yet, let us call it 6G for the time being) will start in 2025, there are about 4 years for preparation from today. In fact, the envisioning, anticipating, and preresearch of key technologies for 6G started in 2019, which can be referred to as B5G (Beyond 5G) research. Why is it called B5G instead of 6G? Is there any difference between the two names? There is usually a series of incremental enhancements and optimizations between two generations of mobile communications standards. Between 5G and 6G, there might be some intermediate release such as "5.5G." In the early stage of research, it is difficult to judge whether a technology will be introduced in an intermediate standard release (i.e., 5.5G) or will be standardized until 6G. So we can call the study beyond 5G as B5G. The preresearch of B5G technology can serve both 6G and the intermediate releases such as 5.5G, depending on when the actual market demand arrives and when the key technology becomes mature. If a new technology originally reserved for 6G is backward compatible with the 5G standard, the market demand appears before 2025, and if the technology is mature soon enough, it can be standardized in 5.5G. Therefore at this stage it is impossible to draw a clear line between B5G research and 6G research, and there is no need to make a hard partition.

20.1.1.1 Vision and requirements of B5G/6G

Before studying the new generation mobile communication technology and formulating the new generation mobile communication standard, we should first think clearly on a few questions. What are new goals of the new generation technology? What are the new requirements of people and the human society? What applications that are not supported by the legacy systems will become important in the future? To answer these questions, we will start with the study of B5G/6G vision and requirements.

In the history of mobile communications development, each generation of technology has carried the task of "improving service performance and expanding the application scope," as shown in Fig. 20.5. Generally, it will take two generations of technology for every new mobile service to grow from initial introduction to popularization. The 1G system had been able to provide mobile voice service, but mobile voice was not really

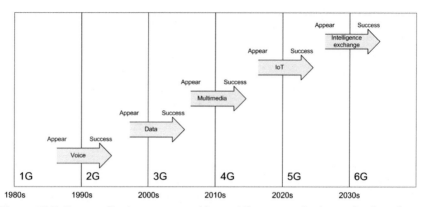

Figure 20.5 New applications supported by mobile communication technology from 1G to 5G.

mature until the 2G era. The 2G system began to support mobile data, but 3G system (HSPA) provided the ability to transmit data at high speed. Mobile multimedia applications represented by mobile video appeared in the 3G era, but did not actually become popular until the 4G era. The Internet of Things (IoT) was introduced at the later stage of 4G, but it turned into a real core application in the 5G system. After voice, data, multimedia, and IoT services have all been well supported, what new value and changes can B5G/6G bring to people's work and life? We believe that this new addition would be the exchange of intelligence. It is expected that this new type of service will be introduced from the late 5G stage and become a major feature of 6G.

Mobile communication technology was invented to realize information exchange over the world:

- In the 1G to 4G era, the focus was mainly on information exchange between people (i.e., realizing the interaction of information and emotion, and providing sensory enjoyment). Parts or all of traditional lifestyles such as letter conversation, reading books and newspapers, art appreciation, shopping and payment, sightseeing, sports, and games have been transferred onto mobile phones. To a considerable extent, the "mobility of life and entertainment" has been realized. So we can call 4G the "mobile Internet."
- In addition to continuing to improve the experience of mobile life, 5G will shift the focus to the "mobility of work and production." Based on 5G, the technologies of mobile IoTs, V2X, and Industrial IoT are intending to replace the traditional ways of production and work.

• However, when we studied the service requirements of 5G a decade ago, what we did not expect was the rapid popularization of AI technology. Therefore a deficiency 6G needs to fix is "the mobility of thinking and learning," which we can call "Mobile Internet of Intelligence"(Fig. 20.6).

The way information flows was gradually formed throughout the history of the world and mankind, which has its own rationality and scientificity. Reviewing the history of development of mobile communication and information technology, the successful services are always those that use the natural way of information flow in traditional life and work to the mobile network. In traditional life and work, data exchange and sensory interaction cannot replace the interaction of intelligence. For example, if we want a person to finish a job, we would not always stand beside him and direct his every move like a "puppeteer." Instead, we will teach him the knowledge, methods, and skills that are needed for completing the job, and then let him use the "intelligence" he has learned to finish the work on his own. At present, what we have achieved in 4G and 5G systems is still the "puppeteer" IoTs, where every sensing data of the terminal is collected to the cloud, and every move of the terminal is remotely controlled by the cloud. Only the cloud has the intelligence of inference and decision-making, and the terminal only mechanically performs "reporting and execution," which is contrary to the natural way of work

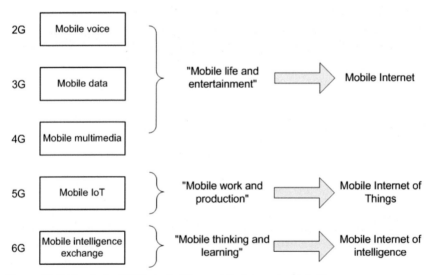

Figure 20.6 The task of 6G is to build a mobile internet of intelligence.

in the real world. Although 5G has made a lot of breakthrough innovations in the aspects of low latency, high reliability, and massive connectivity, excessive system resources are still consumed to meet these KPIs. It might not be a "sustainable" development model to only rely on limited radio resources to meet the challenging KPIs for the growing number of IoT terminals.

With the rapid development of AI technology, it is possible to realize the machine-to-machine information exchange in a more reasonable way. More and more mobile terminals have the computation capability and architecture for AI inference, which can support the mode of "learn before work." However, the existing mobile communication network still cannot well support the interaction of "intelligence," which is a new type of traffic flow. The interaction of data information (including human data and machine data) and sensory information (all kinds of audio and video information) has been efficiently supported in 4G and 5G systems. However, the interaction of intelligent information (knowledge, meanings, methods, strategies, etc.) has not been fully considered. The intelligent interaction between humans (learning, teaching, referencing, etc.) can naturally be accomplished through human language and sensory interaction. However, the intelligence interaction between other types of intelligent agents (e.g., robots, smart cars, intelligent machines) needs to be realized by more efficient and direct communication, which is believed to be one of the core motivations of B5G/6G technology. Therefore intelligence streams may become a new service flow in mobile communication system in addition to the data stream and media stream, and interaction of intelligence will be a new service type in 6G system (Fig. 20.7).

With the rapid popularization of AI technology, it is expected that in the near future, the number of other types of intelligent agents in the world (such as smartphones, smart machines, smart cars, drones, robots, etc.) will far exceed the human population. 6G and other next generation communication systems should serve all kinds of intelligent agents, not just human and nonintelligent machines. Therefore we should also design the next generation of mobile communication system that can serve as a tool for "mutual learning, mutual teaching and intelligent cooperation" among all intelligent agents (especially nonhuman agents) (Fig. 20.8).

In the current development stage of AI, the main format of intelligence interaction between nonhuman intelligent agents is the interaction of intermediate data during the AI inference process. At the end of 2019, the 3GPP SA1 (system architecture working group 1, responsible for

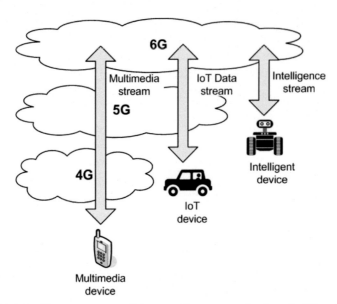

Figure 20.7 Intelligence stream will become the new service type in 6G system.

Figure 20.8 6G Vision: a mobile communication system for all intelligent agents.

service requirements) launched the study item "traffic characteristics and performance requirements for AI/ML model transfer in 5GS" to study the new service requirements and KPIs for the intelligence traffic flows over 5G networks [19]. The study item focuses on three application scenarios:

- Split AI/ML operation
- Distribution and sharing of AI/ML model and data
- Federated learning (FL) and distributed learning

This book is not about AI technology and does not intend to introduce the fundamental concept of AI and Machine Learning (ML) technologies. It will only discuss the potential requirements for transferring AI/ML models and data in B5G and 6G systems. At present, the most popular AI/ML model is the Deep Neural Network (DNN), which is widely used in speech recognition, computer vision, and other fields. Taking the Convolutional Neural Network (CNN) used for image recognition for example, if the split model inference, model downloading, and training are carried out cooperatively between the mobile terminals and the network, the mobile communication system will need to support much higher performance KPIs than the existing 5G system.

Split inference technology can jointly utilize the AI computation resource of mobile devices and network by splitting the AI inference task to both sides. Compared with the device-only or network-only inference, inference with device-network synergy can effectively alleviate pressure of computation, memory footprint, storage, power, and required data rate on devices, reduce the inference delay, and improve the inference accuracy and efficiency [20−24]. Taking the CNN AlexNet [25] analyzed in [23] as an example (shown in Fig. 20.9), the candidate split points can be set after certain pooling layers in the middle of the network. The device executes the inference up to a specific CNN layer before the split point, and then the generated intermediate data is transmitted to the network server through the B5G/6G network. The network server completes the inference operation of the remaining layers after the split point. Different split points lead to different requirements on device computation resources. And with a certain AI inference delay requirement, different split points result in different intermediate data amounts that require different UL data rates. An example of the transmission data rate of the mobile network required for split-image recognition of some applications

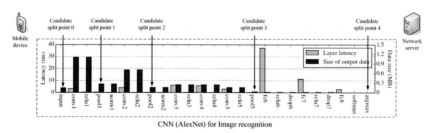

Figure 20.9 Example of DNN-based split AI inference.

is shown in the second column of Table 20.2. We can see that for some services with high requirements for real-time operation, image recognition for a frame, need to be finished within dozens of Ms or even several Ms. The maximum single-user uplink data rate requirement for some split points can exceed 20 Gbps.

Due to the limited computation and memory resource of a mobile device, it is difficult to use the large-scale AI model with strong generalization capability. The AI model using a small amount of computation, memory, and storage is often only suitable for a specific AI task and environment. When the task or environment changes, the model also has to be changed. If the device does not store the required model due to the limited storage space, it has to download the model from the network

Table 20.2 Example of data rate requirements for AI/ML services.

User application	Data rate requirement for intermediate data uploading	Data rate requirement for AI model downloading	Data rate requirements for two-way transmission of federated learning data
One-shot object recognition at smartphone	1.6−240 Mbps	2.5−5 Gbps	6.5 Gbps
Person identification in security surveillance system	1.6−240 Mbps	2.5−5 Gbps	11.1 Gbps
Photo enhancements at smartphone	1.6−240 Mbps	2.5−5 Gbps	16.2 Gbps
Video recognition	16 Mbps−2.4 Gbps	8.3−16.7 Gbps	19.2 Gbps
Augmented reality display/gaming	160 Mbps−24 Gbps	250−500 Gbps	20.3 Gbps
Remote driving	>16 Mbps−2.4 Gbps	>250−500 Gbps	6.5 Gbps
Remote-controlled robotics	16 Mbps−2.4 Gbps	250−500 Gbps	11.1 Gbps

server. The size of common CNN models for image recognition can reach tens to hundreds of MByte [25−30]. An example of the DL data rate required for downloading the model is shown in the third column of Table 20.2. Since the working environment of the mobile device is uncertain and can abruptly change, it may be necessary to download the model within 1 s or even 1 Ms. For some applications with high requirements for real-time operation, the required single-user downlink data rate can reach up to 500 Gbps.

Because of the diversity and completeness of the training data needed for AI model training, the training set collected by the mobile devices may be considerably valuable. In order to protect user privacy, FL based on mobile devices has become an attractive AI training technology [25,31−33]. The mobile FL server iteratively distributes the model to be trained and the federated devices iteratively report the gradients to the server. The typical amount of data exchanged between network and devices per iteration can reach hundreds of megabytes. The training time should be minimized since mobile devices may stay in an environment for only a short time period. Furthermore, considering the limited storage space at a device, it may not be practical to require the training device to store a large amount of training data in the memory for a training after it moves outside the environment. The mobile communication transmission will be able to catch up with the AI training capability of the mobile device to avoid becoming the bottleneck. For that, one training iteration must be completed within tens of Ms, which would require a single-user uplink and downlink transmission rate to reach more than 20 Gbps.

Thus it can be concluded that transmission of AI model and data requires tens or even hundreds of Gbps experienced data rate per user, which cannot be achieved with the current 5G system. Hence, the AI services may bring an essential service type for defining key requirements of the B5G/6G system. On the other hand, people's higher-level sensory requirements may also continue to promote the evolvement of multimedia services. For example, holographic video may also require mobile communication systems to provide tens to hundreds of Gbps data rate, due to the requirement of transmitting dozens of high-definition video streams at one time. Of course, we need further study on how strong the actual demand of such multimedia services that consumes excessive hardware and wireless resources is.

The pursuit of the new generation of mobile communication technology in terms of higher data rate is always the same. The new 6G services represented by mobile AI and mobile holographic video may require a peak rate of 100 Gbps—1 Tbps level and an experienced data rate of tens of Gbps, which needs nearly a two orders of magnitude increase over 5G.

In addition, the efforts of "revolutionize the vertical industries" in the 4G and 5G phases will not stop, and the potential performance improvements being discussed include:

- Ubiquitous coverage: the 5G system, which mainly focuses on terrestrial mobile communications, will be expanded to cover all the users (e.g., including those in desert, on airplanes and ocean-going vessels) on the planet in the 6G era.
- Support much lower latency (0.1 Ms or even zero delay) and much higher reliability (e.g., 99.9999999%).
- Support the moving speed of up to 1000 km/h.
- Support more accurate positioning (e.g., on cm level).
- Support millions of UE connections per km^2.
- Support IoT network with extremely low power consumption (or even "zero power consumption").
- Support higher network security (Fig. 20.10).

Many of the new requirements about vertical industries are still in the preliminary stage of imagination. There is still a lack of solid and pragmatic research and analysis on the necessity of these KPI, such as:

- "Ubiquitous 6G coverage" can indeed cover the areas and scenarios that cannot be reached by the current 5G system. However, the

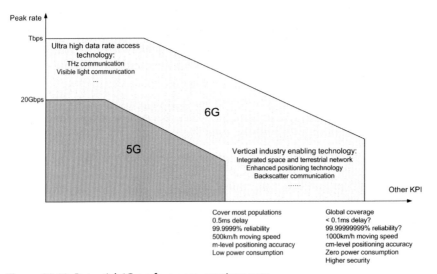

Figure 20.10 Potential 6G performance requirements.

question is how many users exist and how much capacity is actually required in these remote areas and extreme scenarios. Is there a sufficient market scale to support commercial deployment of the 6G network?

• Zero delay and 99.9999999% reliability are not technically impossible. But they will inevitably bring huge communication redundancy and consume excessive system resources in exchange for that. After all, the 6G system will be a commercial communication network that needs to take into account the input–output ratio. Does there exist any vertical application that is worthwhile for such a high cost? In the oncoming AI era, is it necessary to pursue such stringent performance for the intelligent machines that may be capable of fault tolerance and delay tolerance?

Therefore to finally identify these vertical industry requirements, we need to understand real demands from the target vertical industry. The requirements of delay, reliability, coverage, number of users, positioning accuracy, power consumption, and security can be introduced at any time when the demands from vertical industries appear, and then can be added to the existing 5G requirements. If there is a real requirement, we do not have to wait for 6G to support it. It can be covered in an enhanced 5G release (such as 5.5G). Only the data rate with "order of magnitude increase" would be an "iconic" indicator with which 6G can significantly differentiate from 5G.

In addition, from the above introduction of "AI services over B5G/6G systems," we can see that the "intelligence-domain" resources (e.g., AI computation resource) can be exchanged with the resource in traditional domains (e.g., time domain and frequency domain). Therefore this new domain may be added to the B5G/6G performance KPI system. The combination of AI and B5G/6G will start from "serving AI services" (for the AI) to "enhancing B5G/6G by AI" (by the AI), and finally reach the full integration of application-layer AI and access-layer AI. Then a unified 6G AI system (of the AI) is formed. Therefore the design target of 6G system should be to maximize the efficiency of cross-layer AI operation, not just to optimize the performance of communication links. For example, AI inference accuracy and AI training loss could be used instead of communication bit error rate. AI inference latency and AI training latency could be used instead of communication latency. AI operation efficiency could be introduced to replace the data transmission efficiency. Compared with the 5G KPI system (the left side of Fig. 20.11), the 6G KPI system may

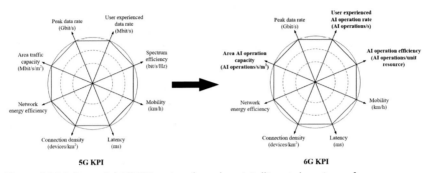

Figure 20.11 Potential 6G KPI system based on intelligent-domain performance.

introduce these intelligent-domain KPIs, as show in the example on the right side of Fig. 20.11.

- A more reasonable KPI than the user-experienced data rate should be the user-experienced AI operation rate.
- A more reasonable KPI than spectrum efficiency should be AI operation efficiency.
- A more reasonable KPI than area traffic capacity should be area AI operation capacity.

The AI operation rate can be defined according to Eq. (20.1). For this KPI, the design target of the 6G system should be to maximize the number of AI operations (inference, training) the user can complete per second, rather than simply the number of bits per second transmitted over the air interface. In order to improve the AI operation rate, AI parameters, and 6G parameters need to be jointly optimized. For example, under a certain achievable data rate, a proper DNN split point fitting can achieve the highest AI operation rate. If a certain data rate is used to download an appropriate AI model, a higher AI operation rate may be achieved than when only using it to transmit traditional data.

$$\text{AI operation rate} = \frac{\text{Number of AI operations}}{\text{Time}} \qquad (20.1)$$

AI operational efficiency can be defined by Eq. (20.2). For the KPI, the system design target should be to maximize the number of AI operations per unit of resources (including time-domain resource, frequency-domain resource, computation resource, storage resource, power consumption, etc.). The resources in the denominator can be flexibly exchanged to avoid the "bottleneck" domain, and achieve the optimal multidomain resource utilization. For example, using the optimal AI model and AI split points for

AI inference can actually save air interface resources. Applying FL only on the UEs with good network coverage can actually save storage resources and power consumption of the federated devices, thereby improving the efficiency of AI operation.

$$\text{AI operation efficiency} = \frac{\text{Number of AI operations}}{\text{Time} \times \text{Frequency} \times \text{Computing power} \times \text{Storage} \times \text{Power consumption}}$$

(20.2)

20.1.1.2 Candidate technologies for B5G/6G

In order to achieve performance KPIs that are considerably higher than that of the 5G, it is necessary to find the corresponding enabling technologies. This book does not intend to introduce B5G/6G candidate technologies in detail. Here is just a brief introduction to B5G/6G candidate technologies that have been widely studied in the industry. Corresponding to the various potential 6G requirements, they can be roughly divided into the following categories.

20.2 Technologies targeting high data rate

The pursuit of higher data rate has always been the objective of mobile communication technology. Before 4G, the data rate was increased mainly by expanding bandwidth and improving spectrum efficiency. 3G introduced CDMA (code division multiple access) and link adaptation, and expanded the system bandwidth to 5 MHz. 4G introduced OFDMA and MIMO and expanded the system bandwidth to 20 MHz. However, the 5G seems lack a technology that can greatly improve the spectrum efficiency. For example, NOMA (nonorthogonal multiple access), which can improve the spectrum efficiency slightly, has not been adopted. The improvement of the data rate mainly depends on the bandwidth expansion. And the larger bandwidth can mainly be supported in the higher frequency band (such as mmWave band). It is expected that 6G will go further in this direction (i.e., "looking for a larger bandwidth in a higher spectrum, and then turn it into a high data rate"). Candidate technologies for higher spectrum include THz transmission and visible light communications (VLC).

The use of THz technology in 6G inherits the idea of "gradually increasing the spectrum band." The mmWave technology for frequency band below 100 GHz has been standardized in the 5G stage. Going upward beyond 100 GHz, the spectrum would enter the THz band in the general sense (as shown in Fig. 20.12). There exist different opinions about the frequency range of "THz band." A common understanding is that THz band is the range of 0.1THz−10THz, which is expected to provide tens or hundreds of GHz spectrum available for wireless communications and to support 100 Gbps or even 1 Tbps peak data rate. A THz system will inherit the massive antenna technology in the 5G MIMO system, while the number of antenna elements will be increased to several hundreds or even several thousands. This may be the reason the wireless communication industry pays the most attention to this technology. However, the measurement and test of propagation characteristics of THz channels is still in its early stage. THz signal has been used in the field of security testing, which indicates that it can penetrate some materials such as clothing and plastic at a very close distance. But this does not mean that broadband communication technology of THz can work in a nonline-of-sight environment like the wireless communication technology below 6 GHz. Whether the high-speed data can be supported by the THz signal experiencing penetration has not been verified by tests and trials. In addition, the THz signal strength drops very fast with distance. Hence, its coverage range will be much smaller than mmWave, and will be mainly used for very short-range wireless access or backhauling in specific scenarios. Finally, the research and development of THz transceivers and related RF modules are still in their early stage and it will be difficult to productionalize and commercialize them in the near future. It will be a great challenge to provide low-cost THz equipment and devices for large-scale commercial deployment before 2030.

VLC is a wireless communication technology that uses visible light spectrum. Optical fiber communication has become the most successful

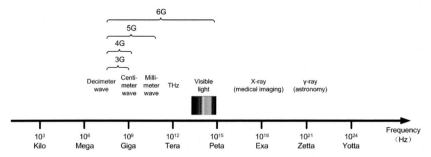

Figure 20.12 THz and visible spectrum.

wired broadband communication technology, and has been gradually used in wireless communication scenarios in recent years. Wireless laser communication has been used in satellite-to-satellite and satellite-to-ground communication. The indoor broadband wireless access technology based on LED lighting (called LiFi) has been developed for over 10 years. Like THz, the biggest advantage of VLC is the rich spectrum resource. The bandwidth of LED communication is mainly limited by the working bandwidth of the transceiver, which is only a few hundred MHz, resulting in the highest data rate of about 1 Gbps. In the deployment scenarios with a widespread angle of coverage, VLC hardly achieves a higher data rate than high-speed WiFi and 5G. This might be the main reason VLC technology has not been widely used yet. A laser diode (LD) transmitter can achieve a transmission bandwidth of several GHz, and then achieve a transmission rate of tens of Gbps through multichannel parallel transmission (i.e., optical version of MIMO transmission). Compared with THz communication, VLC has the advantage that VLC communication equipment can be developed on the basis of mature optical fiber communication devices, which builds a much better foundation for R&D and commercialization. However, VLC also has some disadvantages. For example, the VLC is easily disturbed by natural light and blocked by clouds, and requires separate hardware modules for transmitter and receiver. In particular, VLC is designed based on optical systems (such as lenses, mirrors, gratings, etc.), which is completely different from the technical route of traditional wireless communication based on RF antenna array. Compared with THz communication (which still uses large-scale antennas), VLS would require completely new research and system designs, with which 3GPP is not familiar.

For an ultrahigh-speed access technique (either THz communication or VLC), implementing the transceiver and receiver only solves the first problem. Another thorny problem is how to align the beam (THz beam or optical beam). From the perspective of energy conservation, it can be concluded that the sharp increase of transmission bandwidth will inevitably lead to a sharp decrease of power spectral density, assuming no significant increase in transmission power. To maintain a usable coverage range the only way is to concentrate the energy in a narrow spatial direction. The reason 5G mmWave technology uses beamforming based on massive MIMO antennas is to concentrate energy in a very narrow beam (vividly called pencil beam) in exchange for effective coverage. If the transmission bandwidth of THz communication or VLC is wider than that of the mmWave, it is necessary to further narrow the beam width. The LD-based VLC transmitter can naturally generate a very narrow beam and concentrate the energy to a small spot at the receiver side. However, even the imperceptible move of the device could break the communications link. Therefore no matter which ultrahigh-speed access technique is adopted in 6G system, we will have to resolve the essential issue of fast directing, maintenance and recovery of the extremely narrow beam. In fact, the functions of beam management and beam failure recovery used in 5G mmWave can be regarded as the first version of beam directing and recovery technology. Even for the pencil beam of 5G mmWave, such a complicated beam management mechanism is needed. It is definitely a difficult task to direct the narrower beams in THz communication or VLC.

Finally, from the perspective of network topology, sidelink (UE-to-UE communication) may play a more important role in 6G high data rate systems. As the coverage range of ultrahigh data rate (100 Gbps—1 Tbps) connections may be further shortened (for example, within 10 m), it is harder to deploy the system with cellular topology. In contrast, sidelink would be more feasible for the ultrahigh data rate. Therefore unlike previous generations of mobile communication technologies, 6G system will achieve peak data rate in sidelink, rather than in downlink.

20.3 Coverage extension technology

The coverage extension technology of 6G can be roughly divided into two categories: line-of-sight (LOS) channel expansion for high-speed access and coverage extension for special deployment scenarios.

Starting from the 5G mmWave, the ultrahigh-speed wireless communication technologies can only be used in LOS areas. In a deployment environment with many buildings (such as urban areas), the channel between a UE and the base station is mostly NLOS and thus the mmWave system cannot provide a satisfactory coverage rate. Therefore the topic of how to obtain as many LOS channels as possible and convert some NLOS channels into LOS channels becomes is being studied in academica and industry. One possible technology is the intelligent reflective surface (IRS). By using a configurable array as the reflective surface, the IRS can reflect the incoming beam to a specific direction (as shown in the left of Fig. 20.13), and bypass the obstacles between the transmitter and receiver. This can be considered as a "channel transform" scheme that converts the NLOS channel into a LOS channel. Different from the

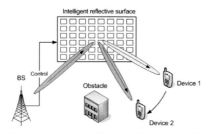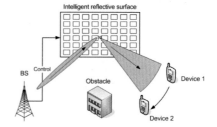

Figure 20.13 Intelligent reflective surface.

traditional reflective surface, the IRS does not need to obey the optical principle that the incident angle is equal to the exit angle. Through the configuration of the reflection unit in the array, the angle of the reflection beam can be changed to cover different UEs behind the obstacles or track the moving UE. In addition, the IRS can also expand a narrow beam into a relatively wide beam, covering multiple UEs at the same time (see left side of Fig. 20.12). Compared with a relay station, the IRS will become an attractive LOS channel expansion technology if it can substantially lower the deployment cost and simplify deployment requirements.

Another LOS channel expansion technology is mobile base station technology. In this scheme, base stations deployed in vehicle or UAV can actively move according to the position of the served UEs, in order to bypass obstacles and expand LOS channels for better supporting mmWave, THz, or VLC communications.

The goal of coverage extension for special deployment scenarios is to cover remote areas such as space, ocean, and desert. Although there is not a large number of users in these deployment scenarios, the system can still be used for special use cases such as emergency disaster relief, ocean transportation, and field exploration. In Section 20.1, it was mentioned that the standardization of nonterrestrial networks (NTN, nonterrestrial communication network) will be carried out in R17 NR [12]. The above-mentioned use cases can be supported with the seamless coverage and survivability of satellite communication network. However, satellite communication networks also have some problems such as limited system capacity, lack of indoor coverage, relatively large time delay, and the need for special terminals. Therefore an important area of 6G technology is to build an integrated space—terrestrial network. Terrestrial mobile communication networks, air-based communication networks (aircraft, drones), and space-based communication networks (satellites) are integrated to achieve the intersystem interconnection (shown in Fig. 20.14).

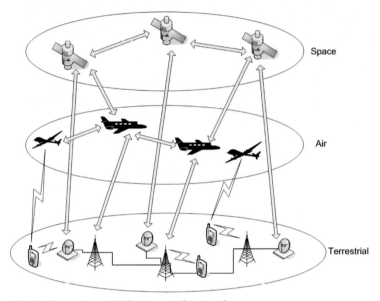

Figure 20.14 Space—terrestrial integrated network.

The air—space—terrestrial network integration is a networking technology that mainly achieves the joint networking and flexible resource allocation between the air-based, space-based, and terrestrial networks.

20.4 Vertical application enabling technology

Like 5G, B5G/6G technology will continue to be optimized for various vertical industry applications. The NB-IoT system is designed to support low-cost and low-power machine-to-machine communication for use cases such as remote meter reading and remote control. However, even if the power consumption is reduced to a very low level and there is no need to replace the battery for several years, the battery life still cannot meet the needs of some use cases. For example, for many IoT devices embedded in closed environments, buildings, or hazardous areas, it is desired that the battery does not have to be replaced during the entire life cycle of the device (e.g., over 10 years). To realize that, relying on the electricity in the battery is not sufficient, and the ambient energy must be utilized. Such a solution can be referred to as "zero power communication" technology. There are several possible methods to achieve the "zero power communication," such as energy harvesting, wireless power transfer, backscatter communications, and so on. Power harvesting converts the energy in the surrounding environment (such as solar energy, wind energy, mechanical energy, and even human energy) into electrical energy, which can be used by the IoT module to transmit and receive signals. Wireless power transfer is similar to wireless charging technology, which can transmit power to IoT devices over the air. Backscatter communication has been widely used in short-distance communication systems such as RFID, and it transmits data through passive reflection and modulation of the received ambient energy. In the traditional backscatter communication, the reader transmits a continuous-wave signal directly to the electronic tag, which is reflected by the circuit of the tag. The reflected signal is modulated with data in the reflection, and received by the reader (see the left side of Fig. 20.15). However, due to the limited transmission power of the reader, the traditional method can only work within a very short distance. One improved method is to set up some high-power beacons dedicated to emitting backscatter signals in the working area. And the tag reflects the backscatter signals from power beacon (the right side of Fig. 20.15). This method may be able to support backscatter communication over relatively long distances. The core technology of "zero power communication" is to develop effective, reliable, and low-cost "zero power" RF modules. However, due to the discontinuity and instability of the ambient energy acquisition, the specific communication system design also needs to be considered.

Another important vertical application enabling technology is high-precision wireless positioning. The positioning technologies based on GNSS are widely used in mobile internet services. However, one problem of GNSS-based positioning is that it cannot cover indoor environments. As mentioned in Section 20.1, the R16 NR standard already supports

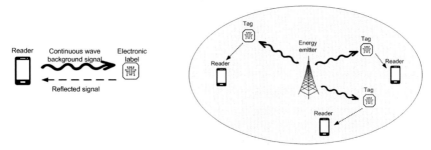

Figure 20.15 Backscatter communication.

positioning technology based on the 5G network with an accuracy of less than 3 m, which can be used in indoor scenarios without GNSS coverage. In R17 NR, the 3GPP plans to further improve the positioning accuracy to the submeter level. However, various mobile services actually even require a higher positioning accuracy. The goal of 6G system may be to achieve centimeter-level positioning accuracy. At present, it is not clear what new technology can achieve such a high positioning accuracy. At the same time, positioning based on wireless signals is not the only technical route for high-precision positioning. In fact, positioning based on computer vision and AI technology is more widely used in indoor environments. High-precision positioning technologies based on 5G and 6G also have to compete with these positioning technologies to decide which technology can eventually win in the market.

In addition to the above two examples, some other potential technologies enabling vertical applications may also be adopted in B5G/6G systems. Of course, as mentioned above, since these technologies do not rely on the evolvement of 6G core technology and do not use the 6G new spectrum, their R&D and standardization can start at any time according to the needs of the market. Then they may become a part of enhanced 5G technology, and enter the international standard before the 6G era starts.

In short, research on the key technologies of 6G is still in the early stage. The industry is extensively investigating and evaluating various candidate technologies. It is too early to talk about the selection and integration of the technologies. A technology we are optimistic about today may not necessarily become a part of 6G in the end. On the contrary, there may still be competitive 6G technologies that have not yet entered our vision. At present, the high-priority work is to clarify the service and technical requirements of 6G as soon as possible, and to figure out what kind of 6G system we should design and what new values and experiences we should bring to users and the industry.

Finally, another question is: Will the legacy spectrum already utilized by the 5G system be refarmed by the 6G system? From 2G, 3G, to 4G, mobile communication systems are designed to completely replace the previous generation. During the early stage of 5G technology study, two different ideas were considered: the first one was to only design the 5G NR system for the newly allocated high-frequency spectrum (i.e., the mmWave spectrum), and just reuse 4G LTE design in the legacy spectrum (i.e., below 6 GHz). This method means 5G and 4G systems coexist and complement each other for a long time. The second way was to redesign the system for the spectrum below 6 GHz as a part of 5G NR, so that it gradually replaced the 4G system. Obviously, the 5G system was finally designed by following the second way. This is because when designing the 4G system in 2004–2008, considering the realizable capability of equipment and chipset at that time, the LTE system only adopted a simplified version of the OFDMA design, which did not fully release the potential of OFDMA system in terms of flexibility and efficiency. Therefore redesigning a complete version of the OFDMA system for 5G NR and refarming the 4G spectrum using 5G system was reasonable. However, what is the relationship between 6G and 5G? Will the system below 100 GHz be redesigned in 6G standardization, which is aimed to eventually replace 5G? Or should we only design the 6G system for the new spectrum above 100 GHz and just reuse the 5G system for the spectrum below 100 GHz, so that 6G and 5G can coexist and complement each other in the 6G era? The answer depends on whether there is enough room for enhancements and optimizations under the 100 GHz, and whether it is worth redesigning the system. It also depends on the consensus of the industry in the next few years.

20.5 Summary

From 2020 to 2025, the 5G standard will continue evolving to meet the needs of more vertical industry applications, while optimizing and enhancing the 5G core designs. According to past experience, 6G standardization may start from 2025. The preresearch on the service and technical requirements, key technologies, and system architecture of 6G is still in the early stages. However, it can be expected that many 5G technologies and designs will be reused and enhanced in the 6G era. Therefore an in-depth understanding of the 5G standard will still be necessary for the system design and standardization of 6G in future.

References

[1] RP-193133. WID proposal for Rel.17 enhancements on MIMO for NR, Samsung, 3GPP TSG RAN meeting #86, Sitges, Spain, Dec. 9 − 12, 2019.
[2] RP-193239. New WID: UE power-saving enhancements, MediaTek Inc, 3GPP TSG RAN meeting #86, Sitges, Spain, Dec. 9 − 12, 2019.
[3] RP-193252. New WID on NR small data transmissions in INACTIVE state, ZTE Corporation, 3GPP TSG RAN meeting #86, Sitges, Spain, Dec. 9 − 12, 2019.

[4] RP-193263. New WID support for multi-SIM devices in Rel-17, Vivo, 3GPP TSG RAN meeting #86, Sitges, Spain, Dec. 9 − 12, 2019.

[5] Study on enhancement of RAN slicing, CMCC Verizon, 3GPP TSG RAN meeting #86 9 − 12:2019.

[6] New SID on support of reduced capability NR, Ericsson, 3GPP TSG RAN meeting #86 9 − 12:2019.

[7] RP-193237. New SID on NR positioning enhancements, Qualcomm, 3GPP TSG RAN meeting #86, Sitges, Spain, Dec. 9 − 12, 2019.

[8] RP-193233. New WID on enhanced industrial Internet of Things (IoT) and URLLC support, Nokia, 3GPP TSG RAN meeting #86, Sitges, Spain, Dec. 9 − 12, 2019.

[9] RP-193257. New WID on NR sidelink enhancement, LG Electronics, 3GPP TSG RAN meeting #86, Sitges, Spain, Dec. 9 − 12, 2019.

[10] RP-193253. New SID: study on NR sidelink relay, OPPO, 3GPP TSG RAN meeting #86, Sitges, Spain, Dec. 9 − 12, 2019.

[11] New SID on NR coverage enhancement, China Telecom, 3GPP TSG RAN meeting #86 9 − 12:2019.

[12] RP-193234. New WID: solutions for NR to support non-terrestrial networks (NTN), THALES, 3GPP TSG RAN meeting #86, Sitges, Spain, Dec. 9 − 12, 2019.

[13] RP-193235. New study WID on NB-IoT/eTMC support for NTN, MediaTek Inc, 3GPP TSG RAN meeting #86, Sitges, Spain, Dec. 9 − 12, 2019.

[14] RP-193259. New SID study on supporting NR above 52_6 GHz, Intel Corporation, 3GPP TSG RAN meeting #86, Sitges, Spain, Dec. 9 − 12, 2019.

[15] RP-193229. New WID proposal for extending NR operation up to 71 GHz, Qualcomm, 3GPP TSG RAN meeting #86, Sitges, Spain, Dec. 9 − 12, 2019.

[16] New SID on XR evaluations for NR, Qualcomm Korea, 3GPP TSG RAN meeting #86 9 − 12:2019.

[17] RP-193248. New WID proposal: NR multicast and broadcast services, HUAWEI, 3GPP TSG RAN meeting #86, Sitges, Spain, Dec. 9 − 12, 2019.

[18] New WID SON/MDT for NR, CMCC, 3GPP TSG RAN meeting #86. 9−12:2019.

[19] S1-193606. New WID on study on traffic characteristics and performance requirements for AI/ML model transfer in 5GS, OPPO, CMCC, China Telecom, China Unicom, Qualcomm, 3GPP TSG-SA WG1 Meeting No. 88, Reno, Nevada, USA, November 18−22, 2019.

[20] Zhou Z, Chen X, Li E, Zeng L, Luo K, Zhang J. Edge intelligence: paving the last mile of artificial intelligence with edge computing. Proc IEEE 2019;107(8).

[21] Chen J, Ran X. Deep learning with edge computing: a review. Proc IEEE 2019;107(8).

[22] Stoica I, et al. A Berkeley view of systems challenges for AI, arXiv:1712.05855 [Online], https://arxiv.org/abs/1712.05855; 2017.

[23] Kang Y, et al. Neurosurgeon: collaborative intelligence between the cloud and mobile edge. ACM SIGPLAN Notices 2017;52(4):615−29.

[24] Li E, Zhou Z, Chen X. Edge intelligence: on-demand deep learning model co-inference with device-edge synergy. In: Proceedings of the workshop mobile edge communication (MECOMM); 2018. p. 31−6.

[25] Krizhevsky A, Sutskever I, Hinton GE, ImageNet classification with deep convolutional neural networks. In: Proc. NIPS; 2012. p. 1097−105.

[26] Simonyan K, Zisserman A, Very deep convolutional networks for large-scale image recognition, arXiv:1409.1556, https://arxiv.org/abs/1409.1556; 2014.

[27] He K, Zhang X, Ren S, Sun J, Deep residual learning for image recognition. In: Proc. IEEE CVPR, June 2016. p. 770−78.

[28] Howard AG et al., MobileNets: efficient convolutional neural networks for mobile vision applications, arXiv:1704.04861, https://arxiv.org/abs/1704.04861; 2017.

[29] Szegedy C, et al., Going deeper with convolutions. In: Proc. CVPR, 2015, pp. 1−9.

[30] Ioffe S, Szegedy C. Batch normalization: accelerating deep network training by reducing internal covariate shift. In: ICML; 2015.

[31] Nishio T, Yonetani R, Client selection for federated learning with heterogeneous resources in mobile edge, arXiv:1804.08333, https://arxiv.org/abs/1804.08333; 2018.

[32] Federated Learning, https://justmachinelearning.com/2019/03/10/federated-learning/.

[33] Tran NH, Bao W, Zomaya A, Nguyen MNH, Hong CS, Federated learning over wireless networks: optimization model design and analysis. In: IEEE INFOCOM 2019-IEEE conference on computer communications.

[34] RP-200493, 3GPP release timelines, RAN Chair, RAN1/RAN2/RAN3/RAN4/RAN5 Chairman, 3GPP TSG RAN meeting #87e, Electronic Meeting, March 16 − 19, 2020.

Index

Note: Page numbers followed by "*f*" and "*t*" refer to figures and tables, respectively.